Applied Analog Electronics

A First Course in Electronics

Applied Analog Electronics

A First Course in Electronics

Kevin Karplus

University of California, Santa Cruz, USA

NEW JERSEY • LONDON • SINGAPORE • BEIJING • SHANGHAI • HONG KONG • TAIPEI • CHENNAI • TOKYO

Published by

World Scientific Publishing Co. Pte. Ltd.
5 Toh Tuck Link, Singapore 596224
USA office: 27 Warren Street, Suite 401-402, Hackensack, NJ 07601
UK office: 57 Shelton Street, Covent Garden, London WC2H 9HE

Library of Congress Cataloging-in-Publication Data
Names: Karplus, Kevin, author.
Title: Applied analog electronics : a first course in electronics /
Kevin Karplus, University of California, Santa Cruz, USA.
Description: New Jersey : World Scientific, [2023] | Includes bibliographical references and index.
Identifiers: LCCN 2022008146 | ISBN 9789811254413 (hardcover) | ISBN 9789811254963 (paperback) |
ISBN 9789811254420 (ebook for institutions) | ISBN 9789811254437 (ebook for individuals)
Subjects: LCSH: Analog electronic systems--Textbooks. | Electronic circuits--Textbooks. | Electronics--Textbooks.
Classification: LCC TK7867 .K267 2023 | DDC 621.3815--dc23/eng/20220523
LC record available at https://lccn.loc.gov/2022008146

British Library Cataloguing-in-Publication Data
A catalogue record for this book is available from the British Library.

For any available supplementary material, please visit
https://www.worldscientific.com/doi/10.1142/12781#t=suppl

Desk Editors: Nandha Kumar Krishnan/Steven Patt

Typeset by Stallion Press
Email: enquiries@stallionpress.com

Printed in Singapore

Acknowledgments

I thank the University of California, Santa Cruz bioengineering students, who were the guinea pigs for the development of the applied electronics for bioengineers course. The questions that students ask in class, on class forums, and in office hours have been invaluable in developing the book and finding places where the presentation is more confusing than it needs to be. There is no substitute for feedback from the genuine audience for the book.

Some students have provided extensive feedback on the book drafts. I particularly want to thank Lon Blauvelt, who has found many copy-editing errors in an early version and who pointed out many places where the text was more confusing than it needed to be; Nicholas Hammond and Mykhaylo Dudkin, who between them pointed out more errors than all other readers combined; and Charles Cai.

I want to thank the group tutors who helped me run the instructional labs, particularly Henry Hinton and Ali Fallahi, allowing me to have up to 50 students at a time in a lab section. Some of them also provided useful feedback on ways to improve the labs.

I thank my son, Abraham Karplus, who developed the PteroDAQ data-acquisition software used throughout the course. Without PteroDAQ, the course would be a very different (and inferior) course.

I thank my wife, who tolerated my turning our bedroom into an electronics lab and my spending many long hours playing with the labs for the course and writing this book.

Chapter 7 started as a very short handout prepared by Steve Petersen for EE 157 at UCSC, but I have extensively modified and extended it. I thank Steve for giving me the seed to work from, but I take full blame for any errors that remain.

Preface

This preface is intended primarily for people who plan to teach from this book, including autodidacts who plan to teach themselves. The introduction for students is in Chapter 1.

Why I wrote this book

I started creating the course that this book is based on in June 2012, teaching it for the first time in January 2013. There was a pressing need to have a more accessible and useful electronics course for bioengineering majors—the existing EE circuits course that they were required to take was really an applied math course, with almost no engineering design. The bioengineers (except the few who went into bioelectronics) had no idea why they were taking circuits, saw no point to all the math, and generally did poorly in the course.

Thinking about the applied circuits course started from what design labs I wanted the students to do, and only afterwards filled in what concepts were needed to allow them to do the designs. I had lots of ideas for labs, and I tried out several of them at home. Many of the labs that I initially thought would be great got rejected, because they were too hard, too easy, didn't teach anything useful, or would require working in a wet lab, rather than an electronics lab. Mixing wet-lab and electronics equipment is a bit risky, particularly for first-time lab students.

Once I'd chosen several of the labs, I started looking for a textbook. I was not able to find one that came close to teaching what I wanted in the order I wanted. Almost all wanted to do 5–10 weeks of preparation before students did any design (if they ever got to design at all), but I wanted students doing design from the first or second week.

I tried putting together an online text out of Wikipedia pages, but that turned out to be difficult—the pages either had so little information that students learned nothing from them, or so much esoterica that students couldn't extract the key ideas from them.

I ended up writing detailed lab handouts that provided background information students needed, as well as the design goals and requirements of the labs.

Unfortunately, bioengineering students have been conditioned to look at lab handouts as protocol sheets to glance at ahead of time, but only really read once they are in the lab. As a result students were coming to lab unprepared, with none of the design work done, and large amounts of lab time was being wasted on students reading and doing design work that should have been done at home.

This book is an attempt to rewrite those lab handouts, separating out the background material from the design assignment, and supplementing the background with material that had been covered in lectures, but had not worked its way into the lab handouts. It is my hope that the extra gravitas of the *textbook* format will make students more willing to do the reading and the homework ahead of time, so that lab time can be used more productively.

This book is designed for a specific course and includes some material that is normally found in the course syllabus, rather than in a textbook. As the book has matured, I've removed much of the most course-specific material, but I've chosen to retain some in the book to try to convey some of my teaching style and to make it easier for an instructor to teach the course the way I would teach it. Anyone using the book can, of course, override these sections with policies presented in their own syllabus.

Detailed notes on the development and presentation of the course can be found on my blog, where I have over 500 posts specific to the course development: https://gasstationwithoutpumps.wordpress.com/circuits-course-table-of-contents/ It is not possible, nor desirable, to include all that information in this book, but anyone thinking of duplicating or adapting this course is likely to find some food for thought on the blog.

If instructors at other institutions or in other fields find this course design to be useful, I'll be pleased. I firmly expect that any instructor who undertakes such a course will find much that they want to do differently (the course has come out different with somewhat different labs and different order of instruction each time I've taught it), so this book should be treated more as a jumping-off place for exploring the world of applied circuits classes than as a finished product.

Who the book is for

This book and the course it is associated with are intended for anyone who wants a practical introduction to electronic circuits. We work at the op-amp level, not the transistor level, so that simple models are sufficient for most of the design work.

Although examples were chosen for bioengineers, the applications are mostly ones of interest to anyone: blood pressure, pulse, EKG, temperature, and sound. Only the electrode characterization is rather bio-specific.

The course expects students to have seen some circuits before—at the level covered in a high-school or college freshman physics class, for example. Students should have heard of Ohm's law ($V = IR$) and of capacitors ($Q = CV$), and know that current is the movement of charged particles (generally electrons in metal wires, though we do look at ionic currents in the electrode labs, and talk about holes as charge carriers in p-type semiconductors). I try to review this material in Chapter 2, so that students with a few holes in their education can fill in the gaps. Over the years of teaching the course, I've made fewer and fewer assumptions about what students retain from prior courses, so it may now be possible for students to take the course with no prior notion of circuits.

Students are expected to have been exposed to complex numbers, so that Euler's formula $e^{j\theta} = \cos(\theta) + j\sin(\theta)$ does not come as a shock to them. We also do some differentiation ($I = \frac{dQ}{dt}$ and derivatives of sinusoids when looking at complex impedance). They are also expected to understand and be able to manipulate logarithms and fractions. Even college engineering students seem to have difficulty understanding logarithms or adding and multiplying fractions—skills that used to be considered high-school and middle-school math.

Because of the math and physics requirements, the course is aimed primarily at sophomores in college, but it should be accessible to interested high-school students who have had calculus. Generally, students who have had single-variable differential calculus (AP Calculus AB) should have no trouble with any of the math in this book, and students might be able to get by with a good deal less math. Facility with simple algebra and fractions is essential, though.

Autodidacticism

As is the case for many of the courses I've created, I taught myself circuits. I had a little help (my Dad taught me a little when I was in high school, and I had a digital logic course and a VLSI design course as a grad student), but mostly I learned on my own from books and from experimentation.

Because so much of what I've taught is material that I've had to teach myself, I tend to take a different approach to teaching than many other faculty. I see my role as trying to provide guidance for students to learn the material faster than I did, with less time chasing down blind alleys, not to just dump some predigested knowledge into their heads for them to memorize and regurgitate. I don't teach them as I've been taught, but as how I wish I had been taught. I tend to pose them problems to guide their learning, rather than giving them information that they are expected to repeat back to me.

I've tried to write this book so that other autodidacts can teach themselves the rudiments of electronics from this book, providing pointers to other sources (mainly Wikipedia) where there is no room in the book for all the material that one should know. Citations in the book are heavy on web pages and light on traditional (book and journal sources), to make it easier for people without access to academic libraries to find the material.

Because the book is intended as a jumping-off place for exploring electronics, in places it contains more than the bare minimum needed to do the labs. In some places, students will complain that the book provides insufficient detail. My goal is *not* to provide a lab handbook that tells students precisely what they have to do, but to give them enough resources to figure things out for themselves—they will certainly have to supplement the book with data sheets, and occasionally with other outside sources.

To enable people to work through the material themselves, I've tried to design the labs in three levels:

1. using the professional-level test equipment in some of our university teaching lab (setups that cost about $10,000 a station),
2. using USB-based test equipment (requiring about $450 a station, plus a computer),
3. using home-brew equipment (requiring about $100 a station, plus a computer).

The labs with professional-level equipment have been used repeatedly in the classroom, but we switched to the USB-based approach in the 2018 offering of the course. The USB-based test equipment allowed larger lab section sizes and better data collection, because of the integration of the test equipment with the student laptops that they prepared their reports on. Unless the course is being taught in an already-equipped lab with training on that lab equipment as a major goal, I recommend using USB-based test equipment.

Don't be too helpful

This section has previously appeared in a slightly different form on my blog as Showing is better than telling, but not by much [52].

Robert Talbert, in a blog post *Examples and the light bulb* [96] wrote

> *I have a confession to make: At this point in the semester (week 11), there's a question I get that nearly drives me to despair. That question is:*
>
> **Can we see more examples in class?**
>
> *Why does this question bug me so much? It's not because examples are bad. On the contrary, the research shows (and this is surely backed up by experience) that studying worked examples can be a highly effective strategy for learning a concept. So I ought to be happy to hear it, right?*

The difficulty, of course, is that the students are asking to see examples, rather than working on the examples themselves—they are asking to be spoon-fed mush rather than chewing for themselves.

I have found in my own learning that I can get a certain amount by reading, but that really understanding material requires me to work out problems for myself. Sometimes this just means doing exercises from the textbook (a boring task which I have trouble forcing myself to do without the structure of a course), and sometimes it means struggling with making something work to solve a real problem. Real problems are both motivating and frustrating—just doing carefully drafted exercises that are designed to work out easily doesn't always help much in applying ideas to the real world.

Talbert gets the point across well:

> *Of course at the beginning of a semester, students aren't experts, and showing them examples is important. But what I also have to do is (1) teach students how to study examples and (2) set and adhere to an exit strategy for giving examples. My job is not to give more and more examples. Instead it's to say: Rather than give you more examples, let me instead give you the tools to create and verify your own examples. And then, at some point in the semester, formally withdraw from the role of chief example-giver and turn that responsibility over to the students.*

This is the same idea as in my post *Descaffolding* [48], which was prompted by a post by Grant Wiggins, *Autonomy and the need to back off by design as teachers* [104]. It also fits in with Dan Meyer's theme to "be less helpful" [70]

Given how frequently teachers and teacher leaders discuss it, I think that over-scaffolding is a common problem for many teachers. We all want to help the struggling student succeed, but too often we make them incapable of succeeding without us. If they always outsource their thinking, they'll never develop their own skills.

To use analogies from other fields: over-scaffolding is like showing the students only great literature and telling them about the writing process, but never having them struggle through 5 to 10 drafts of a piece of writing, or teaching art by showing only cast bronzes and mosaics, but never having them do a sketch or sculpt in clay. Showing or telling students how to do something is often necessary (students can't be expected to guess non-obvious methods), but it needs to be followed by students doing things for themselves.

A lot of us put a lot of time into polishing our presentations so that the students see the cleanest, most elegant way of doing a proof or solving a problem, but never see the debugging and refinement process that creates such elegant results. I've never been guilty of the over-polished lecture: I give my lectures as extemporaneous performances that are never the same twice. For one course (not the electronics course that this book is for), I did not even prepare any lectures, but had the students give me problems from the homework that they wanted to see how to do, a process I called *live-action math.* That approach required a thorough understanding of the material and a confidence that I could do any of the problems in front of an audience without prior prep.

Not all my classes are so extreme, but when I give examples I always try to make them examples of problem solving (as opposed to examples of solved problems). In the first, prototype run of the electronics course I probably did about the right number of examples in class and got the students involved in solving them, but I did not give the students enough simple problems to practice on. I was withdrawing the supports too quickly and trying to have them jump from the material in the reading (which they were not doing) directly to design problems. In subsequent offerings I have gradually increased the number of routine exercises (though I've always hated the drill work) to help them build their skills, but more are probably still needed.

So too many examples is not a big problem in my teaching style. The bigger teaching difficulty I have is keeping myself from doing debugging for the students. In labs and programming courses I can find student errors much more quickly than they can, and I have to restrain myself from just pointing out the (to me) obvious problem. I can think of several times in the electronics lab when I glanced at a breadboard that students had asked for help with and just asked them "where's the connection to ground for this component?" or "why are all these nodes shorted together?" That was not quite the right approach—it got them unstuck and left them some of the debugging still to do (that is, it was better than just moving the wires around for them), but did not help them develop the skills needed to see the problem at a glance themselves.

Some other approaches, like "Show me your schematic—I can't debug without a clear schematic of what you are trying to build," were probably more effective—walking away from the students and telling them to call me back when they had a schematic to debug from was very useful in the second run of the class. By the third week, everyone had a schematic drawn before asking for help. (The students requested that 2017 T-shirts have "Show me your schematic!" on them, because I said it so much in lab.)

It might be better for me to go through a checklist with the students—for example, having them check that each component has the right number of connections and check the breadboard against the schematic to see if the wiring is the same. I tried to do this in the second run of the class, particularly in consistently checking that the breadboard had the same circuit as the schematic, one wire at a time. A few of the students picked up that habit, but many still called me over for help before they had done consistency checks themselves. I've now incorporated many of the suggestions I give students into Chapter 13.

Occasionally I still have to step in to correct a misunderstanding (particularly at the beginning when some students don't understand how the holes of the breadboard are connected together underneath and put components in sideways), but by stepping them through a process I think I could eventually get more of them debugging on their own.

After all, the point of programming assignments and labs is to teach students how to debug, not just to get them to produce working programs or circuits. It is much harder to teach a student how to debug than to demonstrate debugging—I'm still working on better ways to do that. I think that what I've done in the applied electronics course worked for some students (they were debugging pretty independently by the end of the course), but others were still relying too much on help even at the end.

A big chunk of learning how to teach is figuring out how to withdraw the initial support without students failing. Suddenly yanking it out from under them will make many collapse, but being too slow to remove support will leave them still leaning on the crutch when they should be running on their own.

Setting up a course based on the book

To clone this course, an instructor should first go through and do all the labs him- or herself. Even people who know the theory very well and can do the designs in their sleep will learn a lot about the little problems that will plague the students. Doing the designs and making the measurements will give a better understanding of where the students might stumble, where they will need to be very efficient to finish on time, and what results to expect.

Video lectures for much (but not all) of the material in the book is available on YouTube on two playlists that correspond to the two halves of the course as it was taught in 2020–2021. The playlists are at https://tinyurl.com/electronics-A (about 27 hours) and https://tinyurl.com/electronics-B (about 12 hours).

This course was originally intended to take about 220±10 hours over one 10-week quarter, but has been redesigned to take 300±15 hours over two 10-week quarters. Squeezing it into less than 200 hours is not advisable—the labs and the write-ups take time, especially if the students are expected to make mistakes and correct them.

The course is centered around the lab, not the lectures. The one-quarter version of the course scheduled 6 hours of lab a week, 3.5 hours of lecture, and 11.5 hours a week for reading, doing pre-lab exercises, and writing up design reports for 10 weeks. The revised two-quarter version schedule 190 minutes (3:10) of lab a week, 195 minutes (3:15) of lecture, and 515 minutes (8:35) of reading, homework, and write-ups for two 10-week quarters.

The primary instruction occurs in the lab, with the lectures as support sessions to learn tools and concepts needed for doing the labs. Trying to do this course as a huge lecture with lots of barely trained TAs running the labs is almost a guarantee of an unsuccessful course.

The lab equipment needed for the course is described in the next section and the parts needed in Section 3.1.

In addition to standard, off-the-shelf parts, I custom made several parts:

stainless-steel-electrode pairs: I cut up some 1/8″ 316L stainless-steel welding rods (the type for inert-gas welding, not coated for arc welding) into 13 cm pieces and ground the ends a bit to eliminate any sharp burrs. I also cut up an old plastic cutting board into approximately 35 mm squares, and drilled two 1/8″ holes in them 2 cm apart. I hammered the welding rod into the holes with about 3 cm sticking out on one side. The short end can be immersed in salt water, and the long end used to attach a wire with an alligator clip (see Figure 39.2).

holders for silver electrodes: I used a laser cutter to cut several holders for 24-gauge silver wire out of acrylic (see Figure 39.1(a)). The SVG and DXF files for the holder design can be found at https://users.soe.ucsc.edu/~karplus/bme51/pc-boards/electrode-holder/ (the DXF files are more likely to be useful).

pressure sensors: Because the pressure sensors were too expensive at about $12 each to require each student to buy one, I bought one per station for the lab and soldered them to custom breakout boards for easier handling.The design can be found at https://users.soe.ucsc.edu/~karplus/bme51/pc-boards/MPX2050DP-breakout-v1_3/

hysteresis-oscillator boards: To give students practice with soldering on a board with few parts, I designed a custom PC board (2.5 cm by 5 cm) that they could use as soldering practice. The design can be found at https://users.soe.ucsc.edu/~karplus/bme51/pc-boards/hysteresis-oscillator-rev0.7/.Because the Teensy boards come without headers, the hysteresis-oscillator board is now their second soldering project, but it is still good practice at through-hole soldering.

capacitance touch sensors: Students make their own capacitance sensors out of aluminum foil and packing tape in Lab 4.

op-amp prototyping boards: I designed 5 cm-by-5 cm prototyping boards with places for an INA126P instrumentation amp and an MCP6004 op-amp chip, as well as several resistors, 2 4-pin screw terminals, and a potentiometer. I replaced these with a revised design that eliminated the INA126P chip and provided more room for resistors and a less confusing wiring grid (see Figures 31.2–31.4.) The new design can be found at https://users.soe.ucsc.edu/~karplus/bme51/pc-boards/op-amp-proto-rev0/.

unknown impedance boards: My son, Abraham Karplus, designed an "impedance token" board for me to make simple RC circuits that students can use for practice at impedance spectroscopy and fitting impedance models. The board designs are available at https://users.soe.ucsc.edu/~karplus/bme51/pc-boards/ImpedanceToken/. The directory there also has a Python program for selecting resistor and capacitor values from a set of values, to make each impedance token different.

I've considered ordering a large number of the PC boards used in the course and reselling them, to get lower costs per board through volume production, but even buying the boards in 10s of units, the cost is only about $2.50 per student. Eagle design files and Gerber files for all the PC boards can be found at https://users.soe.ucsc.edu/~karplus/bme51/pc-boards/

This course is also a writing course, with 10 substantial design reports required, each going through two drafts. Detailed feedback on the design reports is essential for students to learn engineering writing. Insistence that serious errors be corrected before work is accepted is also important, even though this usually means more time spent on reading the student work that is most painful to read. The instructor should budget at least 30 minutes per student per week for providing feedback to the students on their writing.

I generally have design reports due shortly after labs are over, with a grading schedule that returns them well before the next labs are due. The short writing time between the end of lab and the due date for the report encourages students to write stuff up before lab, rather than leaving the documentation to the end. Writing as you go is an extremely important habit to develop in engineers, but difficult to instill. Furthermore, this course is a design course, so most of the thinking and writing should be happening *before* students enter the lab, not after the lab is over.

With 10 design reports in the course, the grading schedule for a one-quarter course was insane, with reports due the Friday morning after the Thursday lab session and graded reports returned on Monday. The same 10 reports spread over two quarters is a much more reasonable schedule, with a report due about every two weeks and a week to do the grading. Students can be given a weekend to finish a report, with still a weekend for grading before the next report is due. Even this schedule became insane with 80 students in the course, as 40 reports took over 40 hours to grade.

What bench equipment is needed for the course

The book, as currently written, assumes that every student has a microcontroller and computer with the PteroDAQ software installed.

The instructions for the labs assume that students will be using an Analog Discovery 2, which combines an oscilloscope, function generator, and power supply in one compact package. Using separate function generators, bench power supplies, and oscilloscopes is certainly possible, but would increase the time needed for data collection and recording, especially in Labs 7, 8, 10, and 12.

Table P.1 lists the equipment needed for each lab for the versions of the labs using professional equipment. The rest of this section will describe both what equipment is available in the lab that we use for the course, and what minimal features are needed for setting up a lab elsewhere, as most of the equipment in the lab is overkill for what we need.

Soldering equipment: We need a soldering station, preferably one with a small tip, like the Weller ETU 0.015″ or ETV 0.024″ sloped tip and a solder vacuum (also called a *solder sucker*) for unsoldering. Some engineers prefer to use solder wick to a solder sucker, but I've had less success with solder wick. Useful, but not essential, is a board holder for holding PC boards while soldering—I like the Pana-vise Jr. PV-201 holders, but much cheaper ones are usable. Soldering equipment is needed for all levels of the labs, but can be obtained for about $25–$30 a station, though a temperature-controlled station like the Weller WESD51 Digital Soldering Station is a good investment at about $130. We've had pretty good luck with the Aoyue 9378 soldering stations at $90 a station. For at-home labs, we have used extremely cheap soldering irons ($5 from Harbor Freight), as shipping the soldering stations was too expensive.

Equipment	Which labs												
Laptop	1	2	3	4	5	6	7	8	9	10	11	12	13
Soldering station	1			4		6			9	10			13
Fume extractor	1			4		6			9	10			13
Board holder(optional)				4					9	10			13
Microcontroller (for PteroDAQ)	1	2	3	4	5	6	7						13
Function generator			3			6	7	8		10	11	12	
Ohmmeter		2	3	4	5	6	7	8	9		11	12	13
Voltmeter		2					7	8				12	
Oscilloscope			3	4	5	6	7		9	10	11		13
Bench power supply							7			10	11	12	
Thermometer		2										12	
Hot water & ice		2											
Beakers		2										12	
Secondary containment		2										12	
Micrometer				4								12	
Calipers												12	
Drill press (or drill)					5	6							
Pressure sensor					5								
Blood pressure cuff					5								
Electrodes/holders												12	
NaCl solutions												12	
EKG electrodes												12	13

Table P.1: *The equipment needed is listed here, along with which labs it is needed for. The numbers are the numbers of the labs that use that piece of equipment. The voltmeter, oscilloscope, and power supply can all be replaced by an Analog Discovery 2, but an ohmmeter is needed for Lab 2. For other labs, an ohmmeter is useful for checking that resistor values are correct, as students often misread the color codes.*

Host computer for PteroDAQ and Analog Discovery 2: Students should have their own laptops in lab—PteroDAQ works with Windows, linux, and Mac OS X operating systems. Getting PteroDAQ to work on Windows 8 or 10 systems is possible but difficult, because the serial drivers are unsigned, and so require rather awkward workarounds to circumvent Windows 8's insistence on signed drivers. The installation instructions on the PteroDAQ site provide the necessary instructions (see also Section 3.5).

Function generator: The circuits lab that we used earlier had Agilent 33120a arbitrary waveform function generators, which is definitely overkill for the needs of the course. A low-frequency sine-wave generator is sufficient for Lab 3 and a medium-frequency (around 50 kHz–100 kHz) triangle wave generator suffices for Lab 11. The impedance measurements lab benefit from having an oscillator that goes up to 1 MHz, but that isn't strictly necessary.

The function generators in the Analog Discovery 2 are better for this course than a standalone function generator, as they can be coupled to the oscilloscope using the network-analyzer and impedance-meter functions to automate measurements in Labs 8, 11, and 12.

The function generator built into the Analog Discovery 2 USB oscilloscope provides much cleaner signals than most of the low-cost function generators I've used, it has good offset voltage control, and it has powerful modulation and sweep capabilities. The Analog Discovery 2 is certainly more function generator than needed for the course, and the academic price makes it very cheap.

A low-cost function generator kit (there are several on the market) probably suffices for this course. I've tested JYE Tech's FG085 function generator, and found it usable, but the waveforms are not high quality and the buttons on the device are unreliable.

I've also tried using the function generator built into the BitScope BS10 USB oscilloscope, but it does not produce outputs centered at 0 V—its outputs are between 0 V and 3.3 V, which is problematic for measuring the impedance of electrodes in Lab 12. Adding external circuitry to recenter the voltage is doable, but a nuisance for most uses of a function generator.

Oscilloscope: In the instructional lab used for the applied electronics course at UCSC, we had Tektronix TDS3054 4-channel and TDS3052 2-channel digital scopes, and Kikusui C0S5041 analog scopes. All of these have far more bandwidth and functionality than are needed for this course. Initially, I favored teaching with the analog scopes, which are easier to set up, but I now prefer to have students use digital scopes, because many of the waveforms we look at are based on heartbeats, and so are rather too low frequency for an analog scope. The digital scopes are also capable of measuring (rather crudely) a number of properties of the waveforms. The university has since replaced the instructional scopes with Keysight EDUX1002A, which are much cheaper and easier to use than the old Tektronix and Kikusui scopes.

Standalone oscilloscopes are a very traditional engineering tool, and students probably should have some exposure to them, but they are being replaced more and more with USB-connected and wireless instruments that use a laptop or phone for the user interface. We have switched to using the Analog Discovery 2 USB oscilloscope, which combines a dual power supply, a voltmeter, a pair of function generators, and a 2-channel digital oscilloscope in a single low-cost unit. This allows us to equip a lab for about $300 a bench ($450 a bench including a soldering station), rather than over $10,000 a bench. Overall, the Analog Discovery 2 looks like the best investment that a student could make in a "bench" instrument.

For a hobbyist, a USB oscilloscope with a 10–20 Msample/s sampling rate may be enough. I have a BitScope BS10, because it was one of the first USB oscilloscopes to provide software that runs on Mac OS X, Linux, and Windows. It is a more featureful instrument than their newer BitScope Micro, which should still be adequate for most of the labs in the book.

Digilent also makes a low-cost open-source instrument (the OpenScope MZ), which is not as good as the Analog Discovery 2, but which should be adequate if budgets are very tight. It looks like a better deal than the BitScope instruments, but not nearly as useful as the Analog Discovery 2.

For a student who cannot afford a $100–$400 USB oscilloscope, the PteroDAQ data-acquisition system with a $12 Teensy LC board should be adequate for most of the labs, though a function generator will also be needed. The sampling rate is limited to about 20,000 samples per second, depending on the speed of the host computer, which is adequate for looking at EKGs and other heart-rate-based signals, but it is a bit slow for looking at audio signals for the audio-amplifier labs. The analog-to-digital converters on the microcontroller boards also have a rather limited voltage range (0 V–3.3 V on the Teensy boards), which can make debugging signals outside that range difficult. The rather limited sampling rate limits the frequency at which impedance measurements can be made—this limitation may make Labs 8 and 12 much less doable.

Multimeter: In the instructional lab, each bench used to have two Agilent 34401A multimeters, which was very convenient, but we could get by with just one. The multimeters got used for two functions: measuring resistor values and measuring AC voltage for the impedance characterization in Labs 8 and 12. For the impedance characterization, it was useful to have a multimeter like this one that can measure RMS voltage from 3 Hz to 300 kHz (and we can push the 34401A meters to 1 MHz if we sacrifice a little accuracy), but the network analyzer function of the Analog Discovery 2 is better for this purpose, and the impedance analyzer function better still.

When measuring currents in this course, we always provide our own current-sensing resistor, and measure the voltage across it, rather than using an ammeter. That's a safer way to measure current anyway, as under-supervised students in EE courses are always blowing the fuses in the ammeters, rendering them useless for other lab courses. (The joys of shared teaching labs!) The voltmeter function of the Analog Discovery 2 can replace the bench multimeters for both voltage and current measurement.

The only function we used of the bench multimeters that the Analog Discovery 2 lacks is an ohmmeter for measuring resistance. For this purpose, a handheld digital multimeter suffices, though students put too much faith in the numbers reported by the cheap multimeters, which are often highly inaccurate.

I had an old hand-held Fluke 8060A meter that worked from 20 Hz to 100 kHz for AC voltage measurement, which is sufficient, especially as it can be pushed to 5 Hz–300 kHz with some loss of accuracy. More recent Fluke meters seem to have sacrificed bandwidth for price—for the Fluke 175, 177, and 179 meters, the AC response is only 45 Hz to 1 kHz, which is too narrow to be useful for the impedance labs. The Fluke 287 and 289 meters still have 100 kHz bandwidth, but these cost $440 or more, and the Agilent (now Keysight) U1252B has 100 kHz bandwidth, but costs over $500 (the slightly cheaper U1251B has only 30 kHz bandwidth for AC voltage measurement).

Cheap hand-held multimeters like those sold at hardware stores often don't measure reliably above 2 kHz, which is too low for the impedance labs (some only go to 400 Hz!). The cheap meters often do not have a specification for the AC bandwidth, presuming that the user will only be measuring line voltage at 60 Hz (or 50 Hz, depending on what part of the world you live in). They are generally fine for measuring DC voltage and DC resistance, though, at least with the low precision needed for this course.

The $10 aiyun DT-9205A multimeter can be used, as it can measure voltages and get the ratios right up to about 40 kHz (though for accurate single-voltage measurements, the range is limited to about 1 kHz). The voltages we are interested in measuring are often fairly small, so having a low-voltage range on the meter is essential. A cheap digital multimeter like this one, combined with a good USB oscilloscope like Digilent's Analog Discovery 2, is probably the best investment for a hobbyist.

A cheap multimeter and a good USB oscilloscope is probably a better investment than a good multimeter and low-quality oscilloscope.

In a pinch, the PteroDAQ data-acquisition system can be used for measuring small DC voltages fairly reliably (between 0 V and 3.3 V), and the range can be extended by adding a voltage divider, so a DC multimeter is a low-cost convenience rather than an absolute necessity for the lab.

Thermometer: We used to use a $2 28-cm-long glass thermometer with markings every 1°C. These worked fairly well, but were easily broken when students carried them in their backpacks. We switched from including them in the parts kit to having them as lab equipment, then in 2014 we were loaned some digital thermometers used in another lab class. These thermometers turned out to be a very poor choice, despite the easy-to-read 0.1°C resolution—several of the digital thermometers read 1°C to 2°C off, making the students' calibrations of the thermistors that far off from the specs. Accuracy is more important than precision for Lab 2, and that is the only lab we use the thermometers for.

There are low-cost digital thermometers available that can be easily calibrated with ice water. We are now using "CDN DTQ450X Digital ProAccurate Instant-Read Thermometers", which seem to be accurate enough for this course and cost only about $13 each.

Beakers, secondary containment tub, hot water, and ice bucket:

For beakers, I bought ceramic coffee cups for about $0.50 each at the thrift store (as opposed to 150 mL beakers at around $2.50 each). One could use disposable coffee cups for the thermistor lab, Lab 2, but I prefer having as little waste produced as possible. The ceramic cups also have a fairly high thermal mass, which provided opportunity for reminding students of that concept, as they could not get the highest and lowest temperatures unless they preheated or prechilled the cups before adding the hot or cold water.

Clear plastic cups are more useful for the electrode lab, Lab 12, so that the electrolyte solution can be added to a calibrated depth on the electrodes.

Secondary containment tubs to prevent spills from spreading to surrounding electronics equipment were provided by the University, but dish-washing tubs would work as well.

For Lab 2, hot water was provided by a coffee urn, which only goes up to about 70°C–80°C and by a tea kettle that provided 100°C water. Ice water was provided from a large Thermos (another thrift store purchase), but the ice needed to be replenished from an ice machine every couple of hours.

The hobbyist or student at home probably has better access to hot and cold water sources than the electronics lab, so the thermistor lab may be easier to do at home than in the professionally equipped labs.

Bench power supply: Each station in our lab had an Agilent E3631A bench power supply. These power supplies provide three independently adjustable power sources (one up to 6 V and 5 A, one up to 25 V and 1 A, and one up to −25 V and 1 A).

The settable current limitation on the power supplies is a very useful feature, as it prevents blowing fuses on ammeters and can reduce the chance of damaging chips if things are miswired, but settable voltage is sufficient.

Formerly, the power-amp lab (Lab 11) used all three supplies, but that lab has been redesigned to work with a single power supply. If the power amp is powered directly from a USB cable, the current is usually limited to about 500 mA, but a 5 V or 6 V AC/DC adapter with a barrel plug could be used for the power-amp lab to get more power. The Meanwell brand sold by Jameco and Mouser seem pretty reliable.

The power-amp lab now recommends using the power supply of the Analog Discovery 2, and discusses the limitations of that power supply.

Electrodes and electrode holders: There are two sets of electrodes to be tested in Lab 12: a pair made from stainless steel (polarizable) and a pair made from Ag/AgCl using fine silver wire (nonpolarizable).

The stainless-steel ones are made from 316L stainless-steel 1/8″ welding rods (for inert gas welding, so no coating). I cut two pieces about 5″ (12.5 cm) long with a pair of big bolt cutters, then ground the ends to round them. I then drilled two 1/8″ holes about 2 cm apart in a scrap of plastic from an old cutting board (probably made of HDPE), and hammered the rods through until they stuck out about 1″ (2.5 cm) on one side. The short end gets immersed in the solution being measured, and the long end provides a place to clip alligator clips on. (See Figure 39.2.)

The silver/silver-chloride electrodes are made by wrapping 24-gauge fine silver wire around an electrode holder that provides markings for immersing the electrodes to a measured depth (Figure 39.1(a)). The holders I made were cut from clear acrylic with a laser cutter—the design files are at https://users.soe.ucsc.edu/~karplus/bme51/pc-boards/electrode-holder/ as files holder5.svg and holder5.dxf.

Bottles of salt water: I cleaned out some plastic mineral water bottles and had a colleague make up stock solutions of 1 M, 0.1 M, 20 mM, and 5 mM NaCl. I needed about 100 mL per student of each solution, with the students working in pairs. Individual measurements take much less solution, but the students needed to measure two different sets of electrodes, and many needed to remeasure at least one of the sets.

For Lab 12, the bottles are kept in a secondary containment tub well away from electronics equipment and students bring their cups (in secondary containment tubs) to the bottles to pour what they need.

If you have to make the stock solutions yourself, a centigram scale for measuring the NaCl and a graduated cylinder for measuring distilled or deionized water would be needed.

Drill press: I brought in a drill press from home, so that students could drill 2 mm holes in PVC elbows for the breath-pressure apparatus (Lab 5) and 3 mm holes in LEGO® bricks for holding optoelectronics (Lab 6).

Micrometer: We bought one micrometer for every 16 students in the lab, for measuring the thickness of the insulator in Lab 4 and the wire diameter in Lab 12.

Calipers: We bought one set of stainless-steel vernier calipers for every 16 students in the lab, for measuring the dimensions of the electrodes.

Time expectations for the course

Hours needed

This book is now designed around two 5-unit courses—that is, the total time expected of students is about 300–330 hours. For two 10–11-week classes, that's 15 hours a week: about 3.2 in lecture, 3.2 in lab, 2 reading in preparation for class, 2 doing homework exercises, and 4.6 doing design work and write-ups (either pre-lab or post-lab).

If time constraints are tight for a course, it is better to reduce the lecture/discussion hours than the lab hours. Most of the learning happens in design work, lab time, and report writing. The reading time and class time are to enable the learning in the remaining time.

The time spent reading will not be entirely for this text book, but also reading supplemental material (particularly from Wikipedia).

The design reports are due shortly after the lab is completed, so the writing of the report should not be left until the lab is done, but should be written as much as possible while doing the design and analysis work. I am now requiring a complete draft of the report (excluding measurements and conclusions) *before* labs. At the very least, schematics should be drawn neatly before coming to lab. That way the short time after the lab is used only for making corrections and incorporating final results, not for trying to describe the entire process. The time spent each week for writing should be spread out, not concentrated into a single evening.

Some students, pressed for time, try to skimp on the reading and the pre-lab design work. This usually results in their wasting much of the lab time trying to do pencil-and-paper or calculator work, rather than building, testing, and debugging their designs. The overall result is that they run out of lab time, rush through the lab work keeping poor lab notes, and write up poor design reports. They often have to redo the design reports, costing them more time than if they had invested up-front in reading the material and doing the designs before lab.

This is a *design* class, not a science demo-lab course. The lab is where the designs are built and tested, not where the design is first thought about. Most of the thinking and writing in this class should happen *before* each lab, not after.

Lab	Sessions	Time
1	2*	3:10
2	4	6:20
3	2	3:10
4	4	6:20
5	4	6:20
6	4*	6:20
7	2	3:10
8	2	3:10
9	5*	7:55
10	2	3:10
11	3*	4:45
12	3	4:45
13	3*	4:45
total	40	63:20

Table P.2: *Time allocation used in 2019. There was no lab report for Lab 1, and there were combined reports for Lab 7 and Lab 8 and for Lab 10 and Lab 11. The labs with asterisks required an extra weekend lab session for some students to complete the work. It is not clear that scheduling extra lab sessions for the labs would reduce this need—some students don't ask for help until too late to finish the work, no matter how much time they have.*

Possible time allocation for labs

The course that this book was written for was originally designed around a 10-week schedule, with 3 days a week of lecture/discussion (3.5 hours/week) and 2 days a week of lab (6 hours/week). That schedule turned out to be rather intense, both for the students and for the instructor, so the course was redesigned to take two 10-week quarters, with 3:15 hours a week of lecture and 3:10 hours a week of lab. For ease of scheduling, the lectures and labs alternate, with MWF lectures and TTh labs. Longer lab times may be more effective in general, but it is useful in several of the labs to have a lecture between parts of the lab, to help students learn how to analyze the data they have collected while they still have time to collect more data. Live demos/tutorials on using gnuplot in the lectures between lab sessions have been particularly useful.

In any schedule, the time spent on reading, homework, and design exercises is heavily loaded toward the beginning of the course. I have experimented with having students do the homework before lectures on the material (to encourage learning from reading and to encourage asking questions rather than passive absorption), but many students have had great difficulty with this approach.

Every time I teach the course, I end up with a slightly different schedule for the labs and lectures, based on what the class gets quickly and what they need more time on. Table P.2 gives a representative assignment of lab times, assuming that there are 20 lab sessions of 95 minutes each in each quarter. Schedules of labs and lectures for each run of the course can be found at https://users.soe.ucsc.edu/~karplus/bme51/.

I have not attempted to make a 15-week semester schedule for the course. The lab time per week would need to be increased to at least four hours to fit in 60 lab hours. Alternating days of lecture and lab would still be valuable.

Contents

Acknowledgments v

Preface vii

List of Figures xxxi

List of Tables xxxix

1. Why an Electronics Class? 1
   - 1.1 First (and sometimes last) course on electronics 1
   - 1.2 Why teach electronics to non-EE majors? 1
   - 1.3 Teaching design 2
   - 1.4 Working in pairs 4
   - 1.5 Learning outcomes 4
   - 1.6 Videos for the course 5

2. Background Material 7
   - 2.1 Metric units 7
   - 2.2 Dimensional analysis 8
   - 2.3 Logarithms 10
     - 2.3.1 Definition of logarithms 10
     - 2.3.2 Expressing ratios as logarithms 10
     - 2.3.3 Logarithmic graphs 12
   - 2.4 Complex numbers 14
   - 2.5 Derivatives 17
   - 2.6 Optimization 18
   - 2.7 Inequalities 19

3. Lab 1: Setting Up 23
   - 3.1 What parts are needed for the course 23
   - 3.2 Sorting parts 24

3.3 Soldering 24
3.3.1 General soldering advice 24
3.3.2 Soldering Teensy headers 27
3.4 Installing Python 30
3.5 Installing data-acquisition system: PteroDAQ 30
3.6 Installing plotting software (gnuplot) 31
3.7 Using voltmeter 32
3.8 No design report 32

4. Voltage, Current, and Resistance 33
4.1 Voltage 33
4.2 Current 34
4.3 Resistance and Ohm's law 35
4.4 Resistors 36
4.5 Series and parallel resistors 39
4.6 Power 42
4.7 Hydraulic analogy 42

5. Voltage Dividers and Resistance-based Sensors 45
5.1 Voltage dividers 45
5.1.1 Voltage divider—worked examples 47
5.1.2 Thévenin equivalent of voltage divider 49
5.1.3 Potentiometers 51
5.1.4 Summary of voltage dividers 54
5.2 Thermistors 54
5.3 Other temperature sensors 57
5.4 Other resistance sensors 59
5.5 Example: Alcohol sensor 59
5.6 Block diagram 60

6. Signals 63
6.1 Signals 63
6.2 Measuring voltage 63
6.3 Time-Varying Voltage 65
6.4 Function generators 68
6.5 Data-acquisition systems 68

7. Design Report Guidelines 71
7.1 How to write up a lab or design 71
7.2 Audience 71
7.3 Length 72
7.4 Structure 73
7.5 Paragraphs 75
7.6 Flow 75

7.7 Tense, voice, and mood . . . 75
7.8 Formatting with LaTeX . . . 77
7.9 Math . . . 78
7.9.1 Number format . . . 78
7.9.2 Math formulas . . . 79
7.10 Graphical elements . . . 82
7.10.1 Vector and raster graphics . . . 84
7.10.2 Block diagrams . . . 85
7.10.3 Schematics . . . 88
7.10.4 Graphs . . . 91
7.10.5 Color in graphs . . . 97
7.10.6 Listing programs and scripts . . . 99
7.11 Word usage . . . 99
7.12 Punctuation . . . 110
7.12.1 Commas . . . 110
7.12.2 Colons . . . 111
7.12.3 Periods . . . 111
7.12.4 Apostrophes . . . 111
7.12.5 Capitalization . . . 112
7.12.6 Spaces . . . 113
7.12.7 Dashes and hyphens . . . 114
7.12.8 Fonts . . . 116
7.13 Citation . . . 117

8. Lab 2: Measuring Temperature 121

8.1 Design goal . . . 121
8.2 Pre-lab assignment . . . 121
8.3 Setting up the thermistor . . . 125
8.4 Measuring resistance . . . 127
8.5 Fitting parameters with gnuplot . . . 128
8.6 Using a breadboard . . . 129
8.7 Measuring voltage . . . 134
8.8 Recording voltage measurements . . . 135
8.9 Demo and write-up . . . 136

9. Sampling and Aliasing 139

9.1 Sampling . . . 139
9.2 Aliasing . . . 141

10. Impedance: Capacitors 145

10.1 Capacitors . . . 145
10.1.1 Ceramic capacitors . . . 147
10.1.2 Electrolytic capacitors . . . 149
10.2 Complex impedance . . . 152
10.2.1 Impedances in series and parallel . . . 153
10.2.2 Impedance of capacitor . . . 154

11. Passive RC Filters 159

11.1 RC filters . . . 159
11.2 RC voltage divider . . . 160
11.3 Simple filters—worked examples . . . 162
11.4 RC time constant . . . 166
11.5 Input and output impedance of RC filter . . . 169
11.6 Recentering a signal . . . 171
11.7 Band-pass filters . . . 173
11.7.1 Special cases . . . 175
11.7.2 Examples and exercises . . . 176
11.7.3 Cascaded high-pass and low-pass filter . . . 179
11.8 Band-stop filters . . . 180
11.9 Component tolerance . . . 183
11.10 Bypass capacitors . . . 184

12. Function Generator 187

12.1 Agilent 33120A function generators . . . 188
12.2 Analog Discovery 2 function generator . . . 189

13. Debugging 191

13.1 Expectation vs. observation . . . 191
13.2 Show me your schematic! . . . 193
13.3 Color code for wires . . . 194
13.4 Good breadboard practice . . . 196
13.5 Limitations of test equipment . . . 196

14. Lab 3: Sampling and Aliasing 199

14.1 Design goal . . . 199
14.2 Pre-lab assignment . . . 199
14.3 Using function generator with offset . . . 202
14.4 Wiring high-pass filter . . . 204
14.5 Using gnuplot . . . 205
14.6 Demo and write-up . . . 206

15. Oscilloscopes 209

15.1 Analog oscilloscopes . . . 209
15.2 Digital oscilloscopes . . . 209
15.3 Differential channels . . . 211
15.4 DC and AC coupling . . . 211
15.5 Triggering an oscilloscope . . . 212
15.6 Autoset . . . 213
15.7 Oscilloscope input impedance and probes . . . 214

16. Hysteresis 217

16.1 What is hysteresis, and why do we need it? . . . 217
16.2 How a hysteresis oscillator works . . . 221

16.3 Choosing RC to select frequency 224
16.3.1 Improved model of 74HC14N 225
16.3.2 Minimum value for R 226
16.3.3 Maximum value for C 226
16.3.4 Minimum value for C 227
16.3.5 Maximum value for R 227
16.4 Feedback capacitance 228
16.5 Capacitance touch sensor 230
16.6 Multi-dielectric capacitors 232

17. Lab 4: Hysteresis 235

17.1 Design goal 235
17.2 Design hints 235
17.3 Pre-lab assignment 236
17.4 Procedures 238
17.4.1 Characterizing the 74HC14N 238
17.4.2 Breadboarding the hysteresis oscillator 240
17.4.3 Using hysteresis to clean up a noisy analog signal 243
17.4.4 Soldering the hysteresis oscillator 246
17.5 Demo and write-up 246

18. Amplifiers 247

18.1 Why amplifiers? 247
18.2 Amplifier parameters 248
18.2.1 Gain 248
18.2.2 Gain-bandwidth product 249
18.2.3 Distortion and clipping 250
18.2.4 Input offset 252
18.2.5 Input bias 252
18.2.6 Common-mode and power-supply rejection 253
18.2.7 Other amplifier parameters 253
18.3 Multi-stage amplifiers 253
18.4 Examples of amplifiers at block-diagram level 256
18.4.1 Example: Temperature sensor 256
18.4.2 Example: pH meter 257
18.4.3 Example: Ultrasound imaging 258
18.5 Instrumentation amplifiers 259

19. Operational Amplifiers 263

19.1 What is an op amp? 263
19.2 Negative-feedback amplifier 265
19.3 Unity-gain buffer 270
19.4 Adjustable gain 272
19.5 Gain-bandwidth product in negative feedback 274

20. Pressure Sensors 279
20.1 Breath pressure . . . 279
20.2 Blood pressure . . . 281
20.3 Pressure sensors and strain gauges . . . 287

21. Lab 5: Strain-Gauge Pressure Sensor 291
21.1 Design goal . . . 291
21.2 Pre-lab assignment . . . 291
21.2.1 Sensor values . . . 291
21.2.2 Block design . . . 293
21.2.3 Schematics . . . 293
21.3 Procedures . . . 295
21.4 Breath pressure . . . 297
21.5 Blood pressure . . . 297
21.6 Demo and write-up . . . 299
21.7 Bonus activities . . . 299

22. Optoelectronics 301
22.1 Semiconductor diode . . . 301
22.2 Light-emitting diodes (LEDs) . . . 302
22.3 Photodiode . . . 304
22.4 Phototransistor . . . 305
22.5 Optical properties of blood . . . 309

23. Transimpedance Amplifier 313
23.1 Transimpedance amplifier with complex gain . . . 314
23.2 Log-transimpedance amplifier . . . 315
23.3 Multistage transimpedance amplifier . . . 319
23.4 Compensating transimpedance amplifiers . . . 321

24. Active Filters 327
24.1 Active vs. passive filters . . . 327
24.2 Active low-pass filter . . . 328
24.3 Active high-pass filter . . . 330
24.4 Active band-pass filter . . . 334
24.5 Voltage offset for high-pass and band-pass filters . . . 341
24.6 Considering gain-bandwidth product . . . 342
24.7 Multiple-feedback band-pass filter . . . 343

25. Lab 6: Optical Pulse Monitor 347
25.1 Design goal . . . 347
25.2 Design choices . . . 347
25.3 Procedures . . . 348
25.3.1 Try it and see: LEDs . . . 348
25.3.2 Set up log amplifier . . . 349

25.3.3 Extending leads . . . 349
25.3.4 Assembling the finger sensor . . . 350
25.3.5 Try it and see: Low-gain pulse signal . . . 351
25.3.6 Procedures for second stage . . . 354
25.4 Demo and write-up . . . 356

26. Microphones 359
26.1 Electret microphones . . . 360
26.2 Junction field-effect transistors (JFETs) . . . 361
26.3 Loudness . . . 362
26.4 Microphone sensitivity . . . 364
26.4.1 Microphone DC analysis . . . 366
26.4.2 Power-supply noise . . . 367
26.4.3 Microphone AC analysis . . . 368
26.4.4 Sound pressure level . . . 372

27. Lab 7: Electret Microphone 375
27.1 Design goal . . . 375
27.2 Characterizing the DC behavior . . . 375
27.2.1 DC characterization with Analog Discovery 2 . . . 377
27.2.2 DC characterization with PteroDAQ . . . 378
27.2.3 DC characterization with a voltmeter . . . 380
27.2.4 Plotting results . . . 380
27.2.5 Optional design challenge . . . 381
27.3 Analysis . . . 381
27.4 Microphone to oscilloscope . . . 382
27.5 Demo and write-up . . . 385

28. Impedance: Inductors 387
28.1 Inductors . . . 387
28.2 Computing inductance from shape . . . 388
28.3 Impedance of inductors . . . 390
28.4 LC resonators . . . 392

29. Loudspeakers 395
29.1 How loudspeakers work . . . 395
29.2 Models of loudspeakers . . . 396
29.2.1 Models as electronic circuits . . . 396
29.2.1.1 R and RL models for loudspeaker . . . 396
29.2.1.2 Loudspeaker model with RLC for mechanical resonance . . . 398
29.2.1.3 Loudspeaker model with nonstandard impedance . . . 398
29.2.1.4 Resonance with nonstandard impedances . . . 400
29.2.2 Fitting loudspeaker models . . . 401
29.3 Loudspeaker power limitations . . . 405
29.4 Zobel network . . . 407

30. Lab 8: Loudspeaker Modeling 409

30.1 Design goal . . . 409
30.2 Design hints . . . 409
30.3 Methods for measuring impedance . . . 409
30.3.1 Using the impedance analyzer . . . 410
30.3.1.1 Setting up the impedance analyzer . . . 410
30.3.1.2 How compensation works for the impedance analyzer . . . 412
30.3.2 Using voltmeters . . . 415
30.4 Characterizing an unknown RC circuit . . . 416
30.5 Characterizing a loudspeaker . . . 417
30.6 Demo and write-up . . . 417

31. Lab 9: Low-Power Audio Amplifier 419

31.1 Design goal . . . 419
31.2 Power limits . . . 419
31.3 DC bias . . . 420
31.4 Pre-lab assignment . . . 421
31.5 Power supplies . . . 425
31.6 Procedures . . . 425
31.7 Soldering the amplifier . . . 426
31.8 Bonus . . . 431
31.9 Demo and write-up . . . 431

32. Field-effect Transistors 433

32.1 Single nFET switch . . . 436
32.2 cMOS output stage . . . 439
32.3 Switching inductive loads . . . 442
32.4 H-bridges . . . 444
32.5 Switching speeds of FETs . . . 446
32.6 Heat dissipation in FETs . . . 449

33. Comparators 453

33.1 Rail-to-rail comparators . . . 453
33.2 Open-collector comparators . . . 455
33.3 Making Schmitt triggers . . . 456
33.3.1 Inverting Schmitt trigger with rail-to-rail comparator . . . 458
33.3.2 Inverting Schmitt trigger with open-collector comparator . . . 460
33.3.3 Non-inverting Schmitt trigger with rail-to-rail comparator . . . 461

34. Lab 10: Measuring FETs 463

34.1 Goal: Determining drive for FETs as switches . . . 463
34.2 Soldering SOT-23 FETs . . . 463
34.3 FETs without load (shoot-through current) . . . 464
34.4 FET with load . . . 466

34.5 Write-up . . . 468
34.6 Bonus lab parts . . . 468

35. Class-D Power Amplifier 471
35.1 Real power . . . 471
35.2 Pulse-width modulation (PWM) . . . 473
35.3 Generating PWM signals from audio input . . . 475
35.4 Output filter overview . . . 477
35.5 Higher voltages for more power . . . 479
35.6 Feedback-driven class-D amplifier . . . 483

36. Triangle-Wave Oscillator 487
36.1 Integrator . . . 487
36.2 Fixed-frequency triangle-wave oscillator . . . 488
36.3 Voltage-controlled triangle-wave oscillator . . . 491
36.3.1 VCO: Frequency linear with voltage . . . 491
36.3.2 Sawtooth voltage-controlled oscillator . . . 494
36.3.3 VCO: Frequency exponential with voltage . . . 495

37. Lab 11: Class-D Power Amp 499
37.1 Design goal . . . 499
37.2 Pre-lab assignment . . . 499
37.2.1 Block diagram . . . 499
37.2.2 Setting the power supply . . . 500
37.3 Procedures . . . 502
37.4 Demo and write-up . . . 502
37.5 Bonus lab parts . . . 503

38. Electrodes 505
38.1 Electrolytes and conductivity . . . 505
38.2 Polarizable and nonpolarizable electrodes . . . 508
38.3 Stainless steel . . . 509
38.4 Silver/silver chloride . . . 509
38.5 Modeling electrodes . . . 510
38.6 Four-electrode resistivity measurements . . . 513

39. Lab 12: Electrodes 515
39.1 Design goal . . . 515
39.2 Design hint . . . 515
39.3 Stock salt solutions . . . 515
39.4 Pre-lab assignment . . . 516
39.5 Procedures . . . 518
39.5.1 Characterizing stainless-steel electrodes . . . 518
39.5.2 Interpreting results for stainless-steel electrodes . . . 520

39.5.3 Electroplating silver wire with AgCl 521
39.5.4 Characterizing Ag/AgCl electrodes 524
39.5.5 Characterizing EKG electrodes 524
39.6 Demo and write-up 524

40. Instrumentation Amps 527
40.1 Three-op-amp instrumentation amp 527
40.2 Two-op-amp instrumentation amp 529

41. Electrocardiograms (EKGs) 533
41.1 EKG basics 533
41.2 Safety 536
41.3 Action potentials 537

42. Lab 13: EKG 539
42.1 Design goal 539
42.2 Pre-lab assignment 539
42.3 Procedures 541
42.4 Demo and write-up 544

A: PteroDAQ Documentation 547

B: Study Sheet 553
B.1 Physics 553
B.2 Math 553
B.3 Op amps 553
B.4 Impedance 553

References 555

Index 567

List of Figures

2.1	Linear and logarithmic plots	13
2.2	Complex numbers in Cartesian and polar forms	15
3.1	Two types of male header pins	24
3.2	Cross section of a good through-hole solder joint	25
3.3	Soldering iron tip	26
3.4	Basic technique for soldering	27
3.5	Supporting Teensy LC to solder on 3-pin female header	27
3.6	Teensy LC with female header installed	28
3.7	Breaking breakaway male headers before soldering	28
3.8	Using a breadboard as a jig to hold the male headers	29
3.9	Completed Teensy LC board	29
3.10	Several bad solder joints	30
4.1	Series and parallel connections	34
4.2	Plot of resistance values to show uniform spacing	38
4.3	Resistors in serial and parallel arrangements	40
5.1	Voltage divider circuit	46
5.2	Visualization of voltage-divider formula	46
5.3	Voltage divider circuit ($10\,\mathrm{k\Omega}$, $22\,\mathrm{k\Omega}$)	47
5.4	Exercise: What is the output voltage for this voltage divider?	49
5.5	Thévenin equivalent of voltage divider	49
5.6	Schematic for Thévenin equivalence	51
5.7	Ladders of resistors for Exercise 5.7	52
5.8	Potentiometer schematic and photo	52
5.9	Symbols for potentiometer as variable resistor	53
5.10	Potentiometer with series resistors	53
5.11	Potentiometers used as voltage dividers (exercise)	54
5.12	Symbol for thermistor	55
5.13	Voltage response for thermistor circuit with different resistors	56
5.14	Thermocouple diagram	58
5.15	Temperature-sensor block diagram	61
5.16	Expanded temperature-sensor block diagram	62

6.1 Schematics for voltmeter connection 64
6.2 Plot of sine wave illustrating amplitude and peak-to-peak voltage 67
6.3 Block diagram of PteroDAQ data-acquisition system 69

7.1 Four Cs of technical writing 72
7.2 Generic block diagram 87
7.3 Symbols for schematic diagrams 89
7.4 Script definitions.gnuplot to set up Bode plots 93
7.5 Gnuplot script for plotting data with lines 94

8.1 Attaching wire to alligator clip with a screw 125
8.2 Thermistor taped to thermometer 126
8.3 Plot of temperature vs. resistance with fitted model 129
8.4 Bare breadboard 130
8.5 Using double-length male headers for power and ground 130
8.6 Components correctly placed in breadboard 131
8.7 Components incorrectly placed in breadboard 131
8.8 Resistors in breadboard 132
8.9 Resistor horizontal in breadboard 132
8.10 Three stripped wires: 5 mm, 3 mm, and 1 cm 133
8.11 Too much bare wire exposed 133

9.1 Changing sampling frequency 140
9.2 Sampling and reconstructing a 19 Hz sine wave from 40 Hz samples 140
9.3 Moiré pattern for parallel lines 141
9.4 Example of wheel turning to illustrate negative frequency 142
9.5 Aliasing of asymmetric waves 142
9.6 Apparent frequency as function of actual frequency 143
9.7 Sampling at 30 Hz to suppress 60 Hz interference 144
9.8 Too low a sampling frequency misses information 144

10.1 Ceramic capacitors, showing markings 147
10.2 Electrolytic capacitors 150
10.3 Plot of the impedance of a nominally $10\,\mu$F electrolytic capacitor 151
10.4 Parallel impedances 154
10.5 Phase shift for voltage with respect to current with a capacitor 156

11.1 Voltage dividers that act as high-pass and low-pass filters 159
11.2 Plot of the effect on amplitude of sinusoids of simple RC filters 161
11.3 First four harmonics of ramp-down signal with filtering 162
11.4 Gnuplot script to create Figure 11.2 163
11.5 RC charge/discharge circuit 167
11.6 RC charging curve 167
11.7 Two circuits that shift the DC offset of a signal to be centered at 1.65 V 171
11.8 Passive band-pass filter 173
11.9 Bode plots for passive band-pass filters 175
11.10 Bode plot for passive band-pass filter, show effect of swapping capacitors 176
11.11 Bode plot for high-gain passive band-pass filter 177
11.12 Band-pass filter from cascaded high-pass and low-pass 179

11.13 Bode plot for bad cascaded high-pass and low-pass filters 180
11.14 Bode plot for fixed cascaded high-pass and low-pass filters 180
11.15 Passive band-stop filter 181
11.16 Bode plots for passive band-stop filters with constant RC 182
11.17 Bode plots for passive band-stop filters of varying widths 183

12.1 Function generator model 188
12.2 WaveForms 3 screenshot showing choice of function type 190

13.1 Amplifier schematic for use as a color-coding exercise 195
13.2 Schematic showing noise injection at A-to-D input 197

14.1 DC-blocking block diagram 200
14.2 DC-blocking capacitor and voltage divider for sampling lab 201
14.3 Connecting function generator to Teensy LC 203
14.4 Gnuplot script for plotting data with lines 205
14.5 Example plot from sampling lab 206

15.1 Oscilloscope probe showing voltage divider 214

16.1 Representative transfer function for a digital input 218
16.2 Results of interpreting a noisy input with the simple function of Figure 16.1 218
16.3 Representative transfer function for a digital input with hysteresis 219
16.4 Results of interpreting a noisy input with the hysteresis of Figure 16.3 219
16.5 Approximate transfer function for the 74HC14N at 3.3 V 220
16.6 Schematic symbols for Schmitt trigger 220
16.7 Simple hysteresis oscillator 221
16.8 Hysteresis oscillator with capacitive sensor 221
16.9 Hysteresis oscillator as touch sensor 222
16.10 Oscillation of a Schmitt-trigger hysteresis oscillator 223
16.11 Charge/discharge curves for RC charging 223
16.12 Repeated charge/discharge cycles 224
16.13 Equivalent circuit for a 74HC14N Schmitt trigger 226
16.14 Relaxation oscillator with 60 Hz interference 227
16.15 Oscilloscope trace of oscillator with and without feedback capacitor 229
16.16 Hysteresis oscillator output showing jumps in capacitor voltage 230
16.17 Ringing from 74HC14N 231
16.18 Plot of the frequency of an oscillator with 60 Hz frequency modulation 232
16.19 Frequency of the oscillator in the capacitive touch sensor vs. time 233
16.20 Proximity sensor as multi-dielectric capacitor 233
16.21 Capacitance vs. distance for a proximity sensor 234

17.1 Numbering of pins on a dual-inline package 239
17.2 Schematic for the hysteresis oscillator PC board 240
17.3 Touch sensor made with aluminum foil and packing tape 241

18.1 Schematic symbols for a differential amplifier 248
18.2 Voltage transfer function illustrating clipping 251
18.3 Clipping a sine wave 252
18.4 Block diagram of multistage amplifier 254
18.5 Ultrasonic echo for Exercise 18.1 255

18.6 Block diagrams of multistage amplifier for eliminating DC offset . . . 256
18.7 Block diagram of a pH meter . . . 258
18.8 Instrumentation amplifier symbol . . . 260
18.9 Block diagram for 2-stage instrumentation amplifier . . . 261

19.1 Open-loop gain . . . 264
19.2 Negative-feedback amplifier using an op amp . . . 265
19.3 Example of negative-feedback amplifier with gain –22 . . . 269
19.4 Example of a negative-feedback amplifier with gain 34 . . . 270
19.5 Unity-gain buffer . . . 271
19.6 Unity-gain buffer to make virtual ground . . . 271
19.7 Potentiometer between stages of amplifier as gain control . . . 273
19.8 Adjustable gain in a single stage . . . 273
19.9 Multi-stage amplifier for Exercise 19.3 . . . 274
19.10 Measured gain of MCP6004 amplifiers . . . 275
19.11 Comparison of measurement of non-inverting amplifier and model with gain-bandwidth product limitations . . . 276

20.1 Apparatus for breath-pressure measurements . . . 280
20.2 Breath-pressure measurements . . . 281
20.3 Example of an amplitude envelope . . . 283
20.4 Cuff pressure measurements . . . 284
20.5 Filtered cuff pressure measurements . . . 285
20.6 Band-passed blood cuff pressure vs. low-passed blood cuff pressure . . . 286
20.7 Python script for filtering blood-pressure signals . . . 288
20.8 MPX2053DP sensor, mounted on a *breakout board* . . . 289
20.9 Equivalent circuit for NXP temperature-compensated pressure sensors . . . 290

21.1 Blood-pressure cuff . . . 298

22.1 Schematic symbols for diodes, light-emitting diodes (LEDs), and photodiodes . . . 302
22.2 Diode junctions, unbiased and forward biased . . . 302
22.3 LED with current-limiting resistor . . . 304
22.4 Photodiode dark current vs. temperature . . . 305
22.5 Schematic symbols for optoelectronic parts . . . 306
22.6 Cross section of an NPN phototransistor . . . 307
22.7 Phototransistor current vs. voltage plot, log current scale . . . 307
22.8 Phototransistor current vs. voltage plot, linear current scale . . . 308
22.9 Phototransistor current vs. voltage test fixture . . . 309
22.10 Hemoglobin absorption spectrum . . . 310

23.1 Transimpedance amplifiers with photodetectors as the current source . . . 314
23.2 Generic voltage-vs.-log-current plot . . . 316
23.3 Voltage vs. current for a diode . . . 318
23.4 Generic log transimpedance amplifier . . . 318
23.5 Two designs for logarithmic light detectors . . . 319
23.6 Open-loop gain of op amp . . . 322
23.7 Uncompensated transimpedance amplifier . . . 323
23.8 Unstable oscillation of transimpedance amplifier . . . 324

23.9 Fourier transform of unstable oscillation amplifier . . . 324
23.10 Bode plots for uncompensated and compensated transimpedance amplifiers . . . 325
23.11 Fourier transform of compensated transimpedance amplifier . . . 325

24.1 Active low-pass filter . . . 328
24.2 Active high-pass filter . . . 331
24.3 Stable and unstable high-pass filters . . . 333
24.4 Active band-pass filter . . . 335
24.5 Bode plot for active band-pass filter examples . . . 338
24.6 Bode plots for worked example of active band-pass filter . . . 339
24.7 Comparing models and measurement for an active high-pass filter . . . 342
24.8 Multiple-feedback band-pass filter (2 kHz) . . . 343
24.9 Frequency response of multiple-feedback active filter . . . 345

25.1 Lengthening leads on optoelectronic devices . . . 351
25.2 Using LEGO® bricks to hold optoelectronics . . . 352
25.3 Drill jig for drilling LEGO® bricks . . . 353
25.4 Pulse monitor signals with only first stage . . . 354
25.5 Unfiltered and band-pass-filtered pulse waveform . . . 355
25.6 Scatter plot of pulse period pairs . . . 357

26.1 Electret and JFET making up an electret microphone . . . 360
26.2 Cross section of junction field-effect transistor (JFET) . . . 361
26.3 dBA weighting . . . 363
26.4 Schematic of electret microphone with bias resistor . . . 365
26.5 Current-versus-voltage for a typical electret microphone with load line . . . 366
26.6 Noise on the Analog Discovery 2 power supply . . . 367
26.7 FFT of power-supply noise . . . 368
26.8 Electret microphone with noise-filter for power supply . . . 369
26.9 Thévenin equivalent of microphone with bias resistor . . . 369
26.10 Microphone with bias resistor with an active filter . . . 370

27.1 Electret microphone test fixture with function generator . . . 376
27.2 Back of the electret microphone . . . 377
27.3 Test setup for I-vs.-V measurements of an electret microphone . . . 377
27.4 Electret microphone test fixture with potentiometer . . . 379
27.5 Current-versus-voltage plots for electret mic . . . 382
27.6 Schematic for the microphone, bias resistor, and DC-blocking capacitor . . . 383

28.1 Illustration of right-hand rule . . . 388
28.2 Voltage and current convention for inductors and capacitors . . . 388
28.3 Modeling inductors as an inductance in series with a resistance . . . 391
28.4 Plot of the impedance of an AIUR-06-221 inductor, using Figure 28.3 model . . . 391
28.5 Gnuplot script used to generate Figure 28.4 . . . 391
28.6 Series and parallel arrangements of inductors and capacitors . . . 392

29.1 Cross-sectional view of a typical loudspeaker . . . 396
29.2 Cross section of the magnetic field around the voice coil . . . 396
29.3 Schematic symbol for loudspeaker . . . 396
29.4 Electrical models for loudspeakers . . . 397

29.5 Fits for 3 models of a "bass-shaker" loudspeaker 399
29.6 Resonance of two series nonstandard impedances 401
29.7 Gnuplot script for fitting loudspeaker models 403

30.1 Impedance networks for computing compensation in impedance analyzer 412
30.2 Impedances for short-circuit and open-circuit compensation 414

31.1 TRS audio plug 421
31.2 Drawing of op-amp protoboard 427
31.3 Photo of op-amp protoboard 427
31.4 Layout planner for op-amp protoboard 429
31.5 Photograph of soldered microphone preamplifier 431

32.1 Symbols for MOSFETs 433
32.2 Schematic for measuring R_{on} for 2N7000 nFET 435
32.3 R_{on} vs. V_{gs} for 2N7000 nFET 435
32.4 Cross section of a power nFET 436
32.5 Four different packages for FETs 437
32.6 Simple low-side switch with a single nFET 437
32.7 cMOS output stage 439
32.8 Melted breadboard 440
32.9 Test jig and measurements for I-vs.-V plot of inverter outputs 441
32.10 cMOS output stages with different gate drivers 442
32.11 Schematic of an inductor that can be switched on and off 443
32.12 cMOS driver for inductor 444
32.13 Single-nFET driver for inductor 444
32.14 H-bridge 445
32.15 Gate voltage, drain voltage, and load current for an nFET 447
32.16 Resistive load on nFET for computing power dissipation 450
32.17 Thermal plot for Exercise 32.4 451

33.1 Current vs. voltage for TLC3702CN comparator output 454
33.2 Test jig for LM2903 open-collector I-vs.-V data 456
33.3 Output of the LM2903 has a maximum current of around 15 mA–19 mA 456
33.4 Generic form of a Schmitt trigger, with positive feedback 457
33.5 Schmitt trigger from rail-to-rail comparator 458
33.6 Schmitt trigger from open-collector comparator 460
33.7 Non-inverting Schmitt trigger 462

34.1 Half-H-bridge breakout boards for SOT-23 FETs 464
34.2 Schematic for half-H-bridge breakout board 464
34.3 Soldering strategy for adding SOT-23 FETs to half-H-bridge board 465
34.4 Using half-H-bridge as low-side switch 467

35.1 PWM signal with 80% duty cycle 474
35.2 Output of a comparator with a triangle wave on the negative input 476
35.3 Two ways to drive the two sides of an H-bridge 477
35.4 Two LC low-pass filters combined with a loudspeaker model 478
35.5 Gnuplot script to choose capacitor and inductor values for low-pass filter 480
35.6 Result of running the script in Figure 35.5 481

35.7 Separate controls for the gates of the nFET and pFET 482
35.8 Power driver with logic-level inputs 483
35.9 Block diagram for class-D amplifier using feedback 484
35.10 Summing integrator 485
35.11 Frequency response of class-D amplifier using feedback 485
35.12 Block diagram for a class-D amplifier with feedback from after LC filter 486

36.1 Integrator using an op amp with a negative-feedback capacitor 488
36.2 Fixed-frequency triangle-wave oscillator 488
36.3 Waveforms from fixed-frequency triangle-wave oscillator 490
36.4 Triangle wave and square wave from oscillator with $22\,\mu\mathrm{s}$ RC time constant 490
36.5 Voltage-controlled triangle-wave oscillator (linear) 491
36.6 Output waveforms of VCO 493
36.7 Frequency vs. voltage for linear VCO 493
36.8 VCO waveforms with small and large resistor 494
36.9 Rising edge of sawtooth waveform 495
36.10 Voltage-controlled triangle-wave oscillator (exponential) 496
36.11 Frequency vs. voltage for VCO with diodes for exponential response 497

38.1 Graph of conductivity vs. concentration for NaCl solutions 507
38.2 Simplified equivalent circuit for a polarizable electrode 510
38.3 Measurement circuits for electrode impedance 512
38.4 Using four electrodes to measure resistivity 512

39.1 Electrode holder for Ag/AgCl electrodes 517
39.2 Stainless-steel electrodes made from 1/8″ 316L stainless-steel welding rod 519
39.3 Circuit for providing a controlled current to electrodes 522
39.4 Voltage across electrodes while electroplating 523

40.1 Three-op-amp instrumentation amplifier with external resistor to set gain 528
40.2 Block diagram of the three-op-amp instrumentation amplifier 528
40.3 Two-op-amp instrumentation amplifier with external resistor to set gain 530

41.1 Approximate electrode placement for the three electrodes 534
41.2 An EKG trace of Kevin Karplus's heartbeat 535
41.3 Gnuplot commands used to set up the grid and change the shape of the plots 535
41.4 Protection circuits using diode clamps 537

42.1 Voltage divider to make dummy inputs for EKG 542
42.2 Series chain folded for compact layout 544

B.1 Generic negative-feedback op-amp circuit 553

List of Tables

P.1 Equipment needed for labs xiv
P.2 Possible lab allocation xix

2.1 Table of metric prefixes 8
2.2 Table of common metric units 9

4.1 Resistor color code 36
4.2 Table of standard resistor values 37

10.1 Capacitor codes for capacitors in the lab kit 148
10.2 Tolerance values for capacitors 148
10.3 Capacitor temperature codes 149

28.1 Inductance from shape 389

29.1 Crest factors for common waveforms 406

38.1 Conductivity of NaCl 506

A.1 List of boards supported by PteroDAQ 547
A.2 Sample periods for PteroDAQ 551

Chapter 1

Why an Electronics Class?

1.1 First (and sometimes last) course on electronics

This course is intended to be a first course in electronics for students who have not had anything more than a high-school physics course that covered circuits. Students should have heard of resistors and capacitors and know something about how electrons flow in wires to make circuits. Important concepts (like voltage, current, and resistance) will be reviewed, but more time will be spent on concepts that are likely to be new to students.

> The course is structured around three big concepts that are used repeatedly in different ways: *voltage dividers*, introduced in Section 5.1; *complex impedance*, introduced in Section 10.2, and *negative-feedback amplifiers* using op amps, introduced in Section 19.2. We'll use these three ideas over and over again to design a variety of different circuits.

I'll try to keep the theory and the math to a minimum, introducing just enough to make the designs in the labs possible. Students wishing to go on to more advanced electronics courses will need to follow this course with a conventional electrical engineering (EE) circuits course, which provides the math and theory in abundance.

EE circuits courses are usually taught as applied math courses, preparatory to later using the math to do design. That can work well if you later take design courses that use it, but is pretty useless if you stop with just the math and never do design. If you only take one electronics course, it should be one that does a lot of design, not one that prepares you for something you then don't do.

> This course turns around the conventional EE pedagogy, emphasizing the design elements first. If you go on in electronics, the math in the EE circuits course will make a lot more sense after this course, as you'll know what the math is useful for. But even if you don't go further in electronics professionally, this course will teach you how to design and build some simple amplifier circuits that can be useful in a lab and allow you to explore electronics at a hobbyist level.

1.2 Why teach electronics to non-EE majors?

This book was originally created for a course for bioengineers, not electrical engineers. The justification for an electronics course for bioengineers can be summed up in one word: *sensors*.

A sensor is a device that converts some physical or chemical property of interest into a more easily measured or recorded property, generally an electrical parameter: a voltage, a current, a resistance, a capacitance, an inductance, and so forth. That electrical property can be amplified, filtered, and manipulated by electronic circuits, after which it is usually converted to a numeric value that can be recorded on a computer or in a lab notebook.

This class focuses on circuits needed to connect common sensors to computers, where the information can be processed, recorded, or acted on. Because this is a first course in electronics, not a computer engineering course, we will look only at sensors that produce one-dimensional analog outputs, not more sophisticated sensors like digital cameras nor digital interfaces like I^2C and SPI. That is not to say that such topics are unimportant, just that they are beyond the scope of this course.

This course emphasizes *analog* electronics that converts the signal from whatever the sensor produces to a voltage that can be read by a low-cost computer.

We'll cover several different sensors: thermistors for measuring temperature, microphones for sound, electrodes for converting ionic current to electronic current (including EKG measurements), pressure sensors for breath pressure or blood pressure, and phototransistors for light measurements (for optical pulse monitoring).

There are, of course, other applications of electronics, and we'll look at one of them: audio amplification of sound waves. In addition to sound being a useful signal type for bioengineering work, it is pedagogically convenient, as students are already familiar with sound and electronic devices for dealing with sound. Furthermore, the pulse-width modulation and power amplifier design (Lab 11) are similar to motor-control applications in powered prostheses, wheelchairs, and mobility devices, though we'll use a loudspeaker as our output device, rather than a motor, and only use a few watts of power.

At UCSC, many students in the biomolecular concentration work in the nanopore and nanopipette labs, where electronic sensing of ionic current through small holes is a primary lab technique.

1.3 Teaching design

One reason for teaching electronics is that it provides a medium for teaching *engineering design*, which is the process of converting a specification or goal into a detailed design that achieves that goal.

To do engineering design, we need to have mental models of the real-world phenomena that we wish to measure and of the electronic components we can use for these measurements.

Models in electronics are chosen to represent just enough of the real world to allow us to do design. If we choose too simple a model, the design decisions we make will be incorrect, due to missing some important effect. If we choose too complex a model, we'll have a hard time making any design decisions, because we won't be able to do optimization or predict the effect of changes easily.

The same physical device may be modeled with many different models for different design purposes. For example, we'll look at several different models of a loudspeaker in Lab 8, ranging from a simple $8\,\Omega$ resistor to a complicated model that captures both the mechanical resonance of the loudspeaker and its electrical behavior at high frequencies. None of these models is *correct* in any absolute sense—they are just more or less useful for various design tasks.

Of course, when we work with models, we never know for sure whether the model is really capturing all that it needs to—so we must check our thinking by querying the real world and not just the model. "Try it and see" is my standard answer to any question of the form "Is this right?" or "Will this work?" (unless there is a safety concern that needs to be addressed). I reply that way so often that it becomes a mantra for the lab—we've even put "Try it and see!" on the class T-shirts.

In any given design problem we may need to use several different models of the same phenomenon. These are often at different levels of abstraction—for example, in most circuits classes, a wire is treated as a *node*, that is as if all parts of the wire have the identical voltage at all times. For many purposes, this is an excellent level of abstraction, allowing us to ignore many irrelevant properties of the wire. But when we are passing a large current through the wire, we may need to worry about the resistance of the wire, which produces a voltage drop $V = IR$ and dissipates power in the wire that may heat it up. (I have some wires whose insulation melted together when I tried using them to supply current to a motor—the wires had a much higher resistance than I expected.)

Even the model of a wire as a resistor is not always enough—if we are dealing with high frequency signals, then the inductance of the wire may become important. In still other applications, we may need to model the wire as a distributed transmission line (which is beyond the scope of this course).

Picking the right level of abstraction—the right model to use—depends partly on experience, but also on checking whether the model chosen works well enough.

Besides checking that models are reasonable, I also want students to do what engineers refer to as *sanity checks*—checks that part values, amplifier gain, voltages, currents, or other numbers in their design "make sense".

Often these sanity checks are just arithmetic checks: for example, if I have an amplifier with a gain of 1000 and an input signal of 200 mV, I expect an output signal of 200 V, but my power supply can't deliver that, so something is wrong. Is my input signal only 200 μV not 200 mV? Is the amplifier gain only supposed to be 10? Is the circuit one that is supposed to saturate the amplifier at one of the power rails, like a comparator, so that the model of a linear amplification is the wrong model?

Sometimes the sanity check is just a polarity check: increasing current causes a larger voltage drop across a resistor—does that cause the voltage at one end of the resistor to go up or down relative to ground? (The answer depends on which way the current is flowing!)

Some sanity checks are just completeness checks: does every 2-port component have both ports hooked up in a schematic? Are both power and ground wired to every amplifier? Is every block of the block diagram expanded into a schematic?

Some are simple consistency checks—beginning students often draw wires that create short circuits to ground or power in their schematics. Is everything connected by wires in the schematic supposed to be at the same voltage?—that is what putting the wire in the schematic means.

There are lots of other sanity checks that can be done, and students need to get into the habit of looking for sanity checks that they can do on their designs—not relying on an instructor to check their work for them (or even tell them explicitly what sanity checks to do).

I want students to learn *skills*, not facts, in this course. (All that students need to memorize is summarized on the study sheet Appendix B.) The skills I want students to have by the end of the course are to be able to design and build simple amplifier circuits and to write design reports. I don't care much whether they can work textbook problems—what I want is that they acquire the mental attitudes of engineers: that they can design and build things, that data sheets are worth consulting, that precise and accurate recording of what was designed and measured is essential, that often you have to check things for yourself (not blindly trusting the data sheets or simple models), that consistency and sanity checks are an important part of any problem solving, that breaking a problem into subproblems is an essential element of design in any engineering field, that one can improve one's ability with practice, and so forth.

Of course, developing these attitudes takes more than one textbook or one course, but for many students this course will be their first exposure to engineering ways of thinking, so I hope to make the most of it.

1.4 Working in pairs

Because of space limitations in the lab and to improve learning in the lab, all labs will be done by pairs of students (unless there are an odd number of people in class, in which case we will have a singleton, not a triple). For each lab the pairing will be different, so that no one has an unfair advantage or disadvantage from consistently being paired with a more or less competent partner.

Rotating partners for labs has the further pedagogic advantage of learning to work with people who have different styles of work—and realizing what work behaviors are particularly annoying, so that you can try to avoid those behaviors yourself.

For each lab, the partners have to choose whether to turn in a joint report with both names on it as co-authors, or separate reports with one author each, but explicitly acknowledging in writing the work done by the other partner. Both partners should keep their own lab notebooks, as they may not have access to their partner's lab notebook later in the quarter.

On joint reports, both partners are fully responsible for everything in the report and get identical grades. It is very important that you check your partners work at least as carefully as you check your own. As the Russian proverb goes Доверяй, но проверяй *doveryai, no proverayai* (trust, but verify). President Ronald Reagan became very fond of this Russian proverb after he learned it.

1.5 Learning outcomes

UCSC requires faculty to come up with a list of learning outcomes for any course that they create. This section lists the outcomes I used for the 2018 offering of the course.

The student will be able to

- draw useful block diagrams for amplifier design.
- use simple hand tools (screwdriver, flush cutters, wire strippers, multimeter, micrometer, calipers, etc.).
- hand solder through-hole parts and SOT-23 surface-mount parts.
- use USB-controlled oscilloscope, function generator, and power supply.
- use Python, gnuplot, PteroDAQ data-acquisition system, and WaveForms 3 on own computer.
- do computations involving impedance using complex numbers.
- design single-stage high-pass, band-pass, and low-pass RC filters.
- measure impedance as function of frequency.
- design, build, and debug simple op-amp-based amplifiers.
- draw schematics using computer-aided design tools.
- write design reports using LaTeX and biblatex.
- plot data and theoretical models using gnuplot.
- fit models to data using gnuplot.

1.6 Videos for the course

Video lectures for much (but not all) of the material in the book is available on YouTube on two playlists that correspond to the two halves of the course as it was taught in 2020–2021. The playlists are at https://tinyurl.com/electronics-A (about 27 hours) and https://tinyurl.com/electronics-B (about 12 hours).

These videos are intended to supplement the textbook, not replace it. It is particularly important for students to read the instructions for each lab and not rely just on the partial demos in the videos.

Chapter 2

Background Material

Although this book was originally intended for students who have already successfully completed a calculus-based physics course on electricity and magnetism, students with less background have successfully completed courses based on the book, and the prerequisite for the course was reduced to just a first course in differential calculus.

In this chapter, I'll try to review briefly the material we'll use from algebra, calculus, and physics courses, so that students can fill in gaps in their preparation by looking up and learning the background material.

2.1 Metric units

This book will use, as much as possible, standard metric prefixes and units.

Table 2.1 lists the standard metric prefixes, but not all of them are in common use for electronics. Most often, we use the powers of 1000 (kilo-, mega-, giga- going up and milli-, micro-, nano-, pico- going down). Occasionally we need slightly larger or smaller prefixes (tera-, femto-, atto-). Know the prefixes and use them correctly, paying particular attention to whether they are upper-case or lower-case. I don't want to see anyone calling for 100 MV signals, when they mean 100 mV—an error that is a factor of 1,000,000,000 is not negligible.

Some people use "u" instead of "μ" when typing documents, because "μ" was difficult to type before Unicode was defined and became widely used. In some applications, such as computer files that are encoded in ASCII rather than Unicode, this usage is still acceptable, but for the design reports in this class "μ" should be used whenever a multiplier of 10^{-6} is needed.

> The most common usage of metric prefixes is to make all numbers be in the range $[1.0, 1000.)$—whenever a number would be 1000 or larger, the next larger metric prefix should be used, and whenever a number would less than 1, then the next smaller metric prefix should be used.
>
> If you *must* use a number smaller than 1, then the leading zero has to be included, as the decimal point is lost too easily. It is OK to report "0.5 V" instead of "500 mV", but it is never OK to write ".5 V".

For some unknown reason, many electrical engineers avoid the use of "nF", preferring to report a capacitor size of 0.047 μF, rather than 47 nF. Personally, I prefer to consistently keep the numbers in the range 1 to 999.999..., and I see no reason to avoid "nF".

Multiple	Prefix	Abbreviation
10^{30}	quetta-	Q
10^{27}	ronna-	R
10^{24}	yotta-	Y
10^{21}	zetta-	Z
10^{18}	exa-	E
10^{15}	peta-	P
10^{12}	tera-	T
10^{9}	giga-	G
10^{6}	mega-	M
10^{3}	kilo-	k
10^{2}	hecto-	h
10^{1}	deka-	da
10^{-1}	deci-	d
10^{-2}	centi-	c
10^{-3}	milli-	m
10^{-6}	micro-	μ
10^{-9}	nano-	n
10^{-12}	pico-	p
10^{-15}	femto-	f
10^{-18}	atto-	a
10^{-21}	zepto-	z
10^{-24}	yocto-	y
10^{-27}	ronto-	r
10^{-30}	quecto-	q

Table 2.1: *These are the standard prefixes in the metric system. Case is very important: M and m have a ratio of* 10^9 *and Y and y have a ratio of* 10^{48}. *Warning: k for kilo- is always lower-case—"K" is used for kelvin. In electronics we mostly use the prefixes that are powers of 1000, from* pico- *to* giga-. *The prefixes* hecto- *and* deka- *are not used, and* deci- *is used only for* decibel *(dB).*

Avoid writing units like mm^2 or cm^2, because it is not immediately clear whether you mean $\text{m}(\text{m}^2)$ or $(\text{mm})^2$, though most engineers will read it as $(\text{mm})^2$. Either include the parentheses explicitly, or use floating-point notation for the numbers and use the raw metric unit without prefixes: $3.3\,(\text{mm})^2 = 3.3\text{E-}6\,\text{m}^2$.

Table 2.2 has a table of the standard metric units that we'll be using repeatedly in this course. It isn't necessary to remember all the translations to fundamental units, but which unit is associated with which concept is essential knowledge.

If you need to use a unit as a noun, you must spell it out, not use the unit abbreviation. This restriction is particularly important for plurals: Ω already means "ohms", and Ωs means "ohm-seconds" (a unit that has no use that I'm aware of). Be careful about capitalization—we use both seconds (s) and siemens (S) in this course.

2.2 Dimensional analysis

One important method for making sure your computations are meaningful is to keep the units with the numbers for all intermediate results.

When you multiply two numbers together, the units multiply also, and you can only add or subtract numbers when they have identical units.

Property	Unit	Symbol	Fundamental units
Time	second	s	s
Distance	meter	m	m
Volume	liter	L	$1000\,\mathrm{L} = \mathrm{m}^3$
Mass	gram	g	g
Temperature	kelvin	K	K
Frequency	hertz	Hz	s^{-1}
Force	newton	N	$\mathrm{kg\,m\,s}^{-2}$
Pressure	pascal	Pa	$\mathrm{N/m}^2 = \mathrm{kg\,m}^{-1}\,\mathrm{s}^{-2}$
Energy	joule	J	$\mathrm{N\,m} = \mathrm{kg\,m}^2\,\mathrm{s}^{-2}$
Power	watt	W	$\mathrm{J/s} = \mathrm{kg\,m}^2\,\mathrm{s}^{-3}$
Current	ampere	A	A
Charge	coulomb	C	A s
Potential difference	volt	V	$\mathrm{W/A} = \mathrm{kg\,m}^2\,\mathrm{s}^{-3}\,\mathrm{A}^{-1}$
Resistance	ohm	Ω	$\mathrm{V/A} = \mathrm{kg\,m}^2\,\mathrm{s}^{-3}\,\mathrm{A}^{-2}$
Conductance	siemens	S	$1/\Omega = \mathrm{s}^3\,\mathrm{A}^2\,(\mathrm{kg})^{-1}\,\mathrm{m}^{-2}$
Capacitance	farad	F	$\mathrm{C/V} = (\mathrm{kg})^{-1}\,\mathrm{m}^{-2}\,\mathrm{A}^2\,\mathrm{s}^4$
Inductance	henry	H	$\mathrm{Vs/A} = \mathrm{kg\,m}^2\,\mathrm{A}^{-2}\,\mathrm{s}^{-2}$

Table 2.2: *Metric units used in this book. Be careful about capitalization—unit names are not capitalized, even when the unit is named after a person, but many of the symbols are capitalized.* ***Capitalization matters:*** *We'll use both picoamps (pA) and pascals (Pa) in this book, and it is not always easy to tell which units are meant from context. Although international usage allows either lower-case or upper-case L for liter, US standards call for upper-case only (to avoid confusion between "l" and "1").*

Table 2.2 is useful for translating units to the underlying fundamental units, when combining units that look like they are different.

Worked Example:
For example, one formula we often deal with is product of a resistance and a capacitance (*as in* Section 11.2).

If we have $470\,\Omega$ and $10\,\mu\mathrm{F}$, their product is $4.7\,\mathrm{mF}\Omega$, but we can use the definitions of ohms and farads to convert to standard units:

$$470\,\Omega\ 10\,\mu\mathrm{F} = 4.7\,\mathrm{mF}\Omega = 4.7\,\mathrm{mC/V}\ \mathrm{V/A} = 4.7\,\mathrm{mC/A} = 4.7\,\mathrm{ms}\ .$$

We use this particular product often enough that is worth remembering that ohms times farads is seconds ($\mathrm{F}\ \Omega = \mathrm{s}$).

We can often use dimensional analysis to help solve problems and remember formulas. For example, if we are trying to determine a capacitance (in F) and are given a voltage (in V) across the capacitor, then we can look for a charge (in C), which may come from integrating a current (in A = C/s).

Measurements in the US are often given in awkward, non-metric units (inches, feet, miles, pounds force, pounds mass, ounces, fluid ounces, pints, gallons, cubic feet, cubic yards, square inches, acres, etc.).

One of the first steps in using such measurements is to convert to standard units:

$$17\,\text{inch} \to 17\,\text{inch}\ 25.4\,\text{mm/inch} = 431.8\,\text{mm}$$
$$4\,\text{lb}_{\text{force}} \to 4\,\text{lb}_{\text{force}}\ 4.44822\,\text{N/lb} = 17.19\,\text{N}$$
$$20\,\text{psi} \to 20\,\text{psi}\ 6894.76\,\text{Pa/psi} = 137895\,\text{Pa}\ .$$

Exercise 2.1

If we have a current of 22 mA running for 200 ms, how much charge has been transferred? (Write out the dimensional analysis, not just the final value.)

Exercise 2.2

If we have a voltage drop of 2.7 V across a 47 Ω resistor, what is the current through the resistor? (Write out the dimensional analysis, not just the final value.)

2.3 Logarithms

2.3.1 *Definition of logarithms*

Logarithms are used extensively in electronics, in three forms: *base-10 logarithms* (sometimes called *common logarithms*), *base-2 logarithms*, and *natural logarithms*. All are inverses of exponentiation: $x = \log_{10}(y)$ means $10^x = y$, $x = \log_2(y)$ means $2^x = y$, and $x = \ln(y)$ means $e^x = y$.

The interesting properties of logarithms come from the interesting properties of exponentiation:

- $e^{(x+y)} = e^x e^y$, so $\ln(a) + \ln(b) = \ln(ab)$.
- $e^{(x-y)} = e^x / e^y$, so $\ln(a) - \ln(b) = \ln(a/b)$.
- $e^0 = 1$, so $\ln(1) = 0$.
- $e^{-y} = 1/e^y$, so $-\ln(b) = \ln(1/b)$.
- $\frac{de^x}{dx} = e^x$, so $\frac{d\ln(a)}{da} = 1/a$.
- For small ϵ, $e^\epsilon \approx 1 + \epsilon$, so $\ln(1+\epsilon) \approx \epsilon$.

2.3.2 *Expressing ratios as logarithms*

Electrical engineers often express ratios in logarithmic terms, using *decibels* (*dB*):

$$D = 20 \log_{10}\left(\frac{A}{A_{\text{ref}}}\right),$$

where D is in decibels, A is the amplitude of the signal, A_{ref} is the amplitude of the reference being compared to. You will often see A_{ref} given in the form "0 dB is A_{ref}".

For example, an amplifier whose output is 10,000 times its input can be described as having a gain of $20\log_{10}(10000) = 80\,\text{dB}$. Decibels are usually used only when the ratio is unitless—that is, both the numerator and denominator of the ratio have the same units, so that they cancel.

Sometimes we have a gain of less than 1 (from voltage dividers or passive filters, for example), which results in a negative value in decibels. For example, a gain of 0.1 could be expressed as $-20\,\text{dB}$. Because some people are uncomfortable with negative numbers, such gains are sometimes expressed as *attenuations*: a gain of $-20\,\text{dB}$ may be called an attenuation of 20 dB. Whichever terminology is used, the ratio of the output to the input is 0.1.

Because amplifiers and filters are usually set up in way that causes their gains to multiply, the decibel gains can be added. For example, if we have an amplifier that multiplies by 10, followed by a filter that cuts our signal in half, followed by another amplifier with a gain of 100, the overall gain of the system is $20\log_{10}(10\frac{1}{2}100) = 20 - 6.02 + 40 = 53.98\,\text{dB}$. It is common to approximate a factor of 2 as 6 dB.

The definition of decibels given here is for amplitude (of voltage, current, or other signals)—when we are talking about *power*, the definition changes to $10\log_{10}(P/P_{\text{ref}})$. That was the original definition of decibel (hence, the *deci-* prefix), and the more commonly used amplitude definition comes from the relationship between power and voltage or current with a resistor:

$$P = VI = V^2/R = I^2R\ .$$

Occasionally one will see base-10 logarithms expressed in *decades* rather than scaled to decibels. Frequency ranges are more often given in decades than in decibels, which are used more for voltage, current, or power ratios. Each decade is a factor of 10, so a frequency range from 1 Hz to 1 MHz might be expressed as six decades. Each decade is twenty decibels, so the range could also be expressed as 120 dB.

Base-2 logarithms are mainly used in talking about numbers represented in a computer, and for the analog-to-digital and digital-to-analog converters used to convert between voltages or currents and digital numeric representations. A 16-bit analog-to-digital converter can recognize $2^{16} = 65536$ different values. You will sometimes see fractional numbers of bits, when the number of distinguishable values is not a power of two. For example, a 16-bit analog-to-digital converter may be described as having 13.7 *effective bits*, if the noise level is ± 2.5 counts, so that only $65536/5 \approx 2^{13.7}$ levels are really meaningful.

Base-2 logarithms also come up when talking about frequencies, with the unit being the *octave*: a ratio of two in frequency. This usage comes originally from music applications, where frequency ratios of factors of two are particularly important.

Musical instruments are usually tuned so that the A above middle C is 440 Hz, a standard established in 1834 in Stuttgart, though still not universally used [117]. Three octaves lower would be a frequency of $2^{-3}440 = 55\,\text{Hz}$.

Natural logarithms occur ubiquitously in physics and electronics as the solutions of first-order differential equations. Perhaps the most powerful use of natural logarithms comes from their use with complex numbers (see Section 2.4). Sometimes authors use the unit *nat* for steps of 1 in natural logarithm, in analogy to *bit* for steps of 1 in logarithms base 2. We will not use nats in this book.

Exercise 2.3

Simplify $-\ln\left(\frac{a-b}{c-d}\right)$.

Exercise 2.4

How many decibels is a gain of 2000? How many decades?

2.3.3 *Logarithmic graphs*

Logarithms are often used to scale the axes of graphs. This rescaling is usually visible as a nonuniform set of tick marks on the axis, with ticks at 1, 2, 3, 4, 5, 6, 7, 8, 9, 10, 20, 30, ... In gnuplot, specifying a log scale on the y-axis is done with the commands

```
set logscale y      # specifies the logarithmic scaling
set mytics 10       # specifies the standard tick marks for log scale
```

Using a logarithmic scale does not change the label of the axis nor the labeling of the tick marks—if I have frequency on a log scale, I still call it "frequency [Hz]", not "log frequency", and the values on the major ticks are 1, 10, 100, ..., not 0, 1, 2, ...

The main virtue of a logarithmic scale is that distances correspond to *ratios* of the corresponding numbers, while on a linear scale distances correspond to *differences* of the corresponding numbers.

When a ratio is expressed in decibels, a linear scale is used rather than a logarithmic one, because the logarithmic conversion has already been done in the definition of decibels (see Section 2.3.2)—differences of decibels already correspond to ratios of the underlying quantities, so we want distances to correspond to differences of decibels, not ratios of decibels. (Ratios of decibels are almost never meaningful.)

Judicious use of log-scaling for axes can make functions much easier for people to understand visually. As a first approximation, people only understand straight-line graphs visually—they almost always project graphs to continue in a straight line past the ends of what is plotted, and interpolate between points with straight lines.

So what functions are easily understood? Refer to Figure 2.1, which shows the same functions on each of the four major plot types.

lin-lin Graphs with both axes expressed linearly have straight lines that follow formulas like $y = ax+b$, which are known as *affine functions*. Affine functions are more general than *linear functions*, which have the form $y = ax$ and must provide a 0 output for a 0 input. Affine functions include the linear functions, but they also include the *constant functions*, $y = b$, which are most definitely not linear.

The parameters of the straight line are the slope a, which is given in y-axis units per x-axis unit, and the offset (or intercept) b, which is given in y-axis units.

log-lin Graphs with a logarithmic y-axis and linear x-axis (known as log-lin or semilogy graphs) have straight lines that follow formulas of the form $\log(y) = ax + b$, $y = e^{ax+b}$, or $y = BC^x$ (where $B = e^b$ and $C = e^a$), called *exponential functions*.

Growth for bacteria is often well-modeled as an exponential function of time in the early stages, and so growth curves are often appropriately plotted on semilogy plots. One often sees exponential functions in electronics as a result of RC discharge curves (see Section 16.2, for example)—semilogy plots of voltage vs. time are appropriate for such curves, but only when the destination voltage is 0 V.

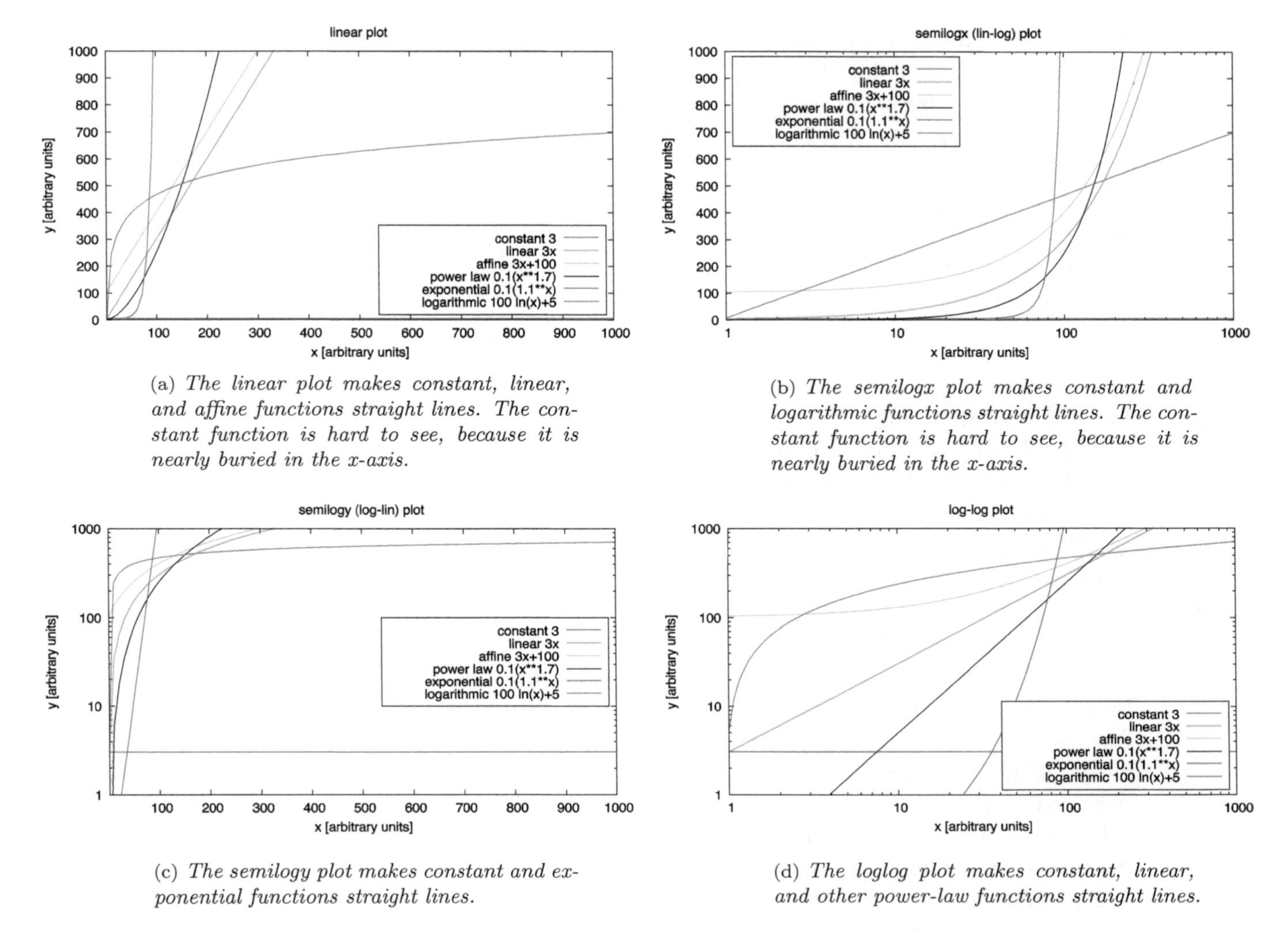

(a) *The linear plot makes constant, linear, and affine functions straight lines. The constant function is hard to see, because it is nearly buried in the x-axis.*

(b) *The semilogx plot makes constant and logarithmic functions straight lines. The constant function is hard to see, because it is nearly buried in the x-axis.*

(c) *The semilogy plot makes constant and exponential functions straight lines.*

(d) *The loglog plot makes constant, linear, and other power-law functions straight lines.*

Figure 2.1: *These four plot types are the main ones we'll use in this course. The constant function is a flat, straight line in all plots types, but other functions are straight lines in only one or two of the plot types. Choose your plot type to make the function you are interested in showing as near to a straight line as you can. Logarithmic scales are also useful for showing functions or data that have a very wide range, when the ratios of values are more interesting than the differences between values.*
Figure drawn with gnuplot [33].

The slope of the line on a semilogy plot corresponds to the growth rate of the function and can be expressed in units like dB/s or doublings per day for exponential functions of time. An exponential decay will have a negative slope, while exponential growth has a positive slope.

The offset of the line corresponds to the initial size of the function.

Semilogy plots are also useful for plotting probability density functions, when what we are interested in are the tails of the function. Probability distributions are often easier to identify and more robustly extrapolated on a log y-axis than on a linear y-axis. For example, a Gaussian distribution changes shape from a "bell-shaped" plot on linear y-axis to a simple parabola on a log y-axis. All the interesting variation in the tails disappears with the linear scaling.

lin-log Graphs with linear y-axis and logarithmic x-axis (know as lin-log or semilogx graphs) have straight lines that follow formulas of the form $y = a\log(x) + b$. Such logarithmic functions are occasionally found in electronics (as the inverses of exponential functions, swapping the roles of x and y). We will encounter them in the logarithmic transimpedance amplifier (see Section 23.2).

The proper units for slope are y-axis units per ratio unit: y-axis units/dB, y-axis units/octave, or y-axis units/decade. For example, if we want to convert a musical pitch as a frequency to the corresponding key number for a Musical Instrument Digital Interface (MIDI), the function is

$$k(f) = 69 + 12\log_2 \frac{f}{440\,\mathrm{Hz}} ,$$

and the corresponding slope is 12 keys per octave [137].

The offset b moves the curve vertically and corresponds to the function value with an input of 1. It is often easier, however, to give the function value for some other input, as 1 may not be in the domain where the function is a good model for the phenomenon. In the example of MIDI tuning, key 69 is the key for A440, the A above middle C that has a frequency of 440 Hz. Note how the expression for the function given above expresses this offset.

log-log Graphs in which both axes as logarithmic have as straight lines functions of the form $\log(y) = a\log(x) + b$ or $y = Bx^a$, where $b = \log(B)$. These are known as *power-law functions*, and include the linear functions $y = Bx^1$. This means that linear functions form straight lines on both lin-lin (as special cases of affine functions) and log-log plots (as special cases of power-law functions).

The slope of a line on a log-log graph corresponds to the exponent on the x term in the power law—the slope is technically unitless, but can be expressed in terms like dB/decade or dB/octave. A linear function ($b = 1$) would be 20 dB/decade or approximately 6 dB/octave. An inverse linear relationship ($b = -1$) would be -20 dB/decade or about -6 dB/octave.

The offset b sets the height of the curve, by changing the scaling factor B.

Log-log plots are used extensively in electronics, because many of the phenomena we are modeling are well approximated by power laws. A particularly important class of log-log plots are the Bode plots that plot gain or impedance vs. frequency (see, for example, Section 11.2).

2.4 Complex numbers

Electrical and electronics engineers use complex numbers extensively, as they provide the most convenient way to talk about and manipulate the amplitude and phase of sinusoidal signals.

The first bit of unfamiliar notation may be the definition of j:

$$j = \sqrt{-1} .$$

Mathematicians shudder at the use of the symbol j rather than i for the square-root of minus one, but electrical engineers reserve i for current, and so have adopted j for imaginary numbers. This tradition has gone on for over 100 years, and neither the mathematicians nor the electrical engineers are likely to change their notation. For this class, we'll use the electrical engineering notation.

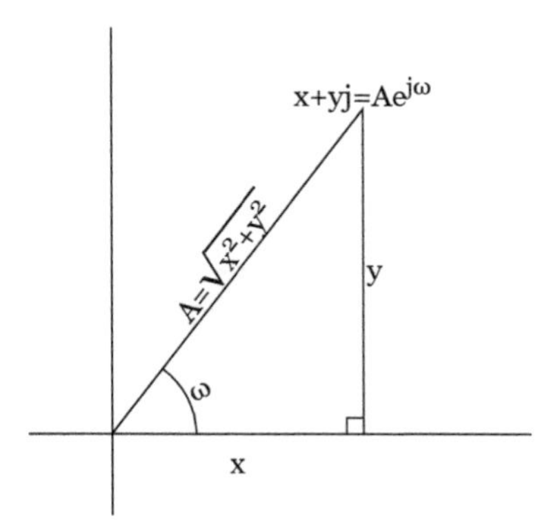

Figure 2.2: *Complex numbers may be viewed in Cartesian coordinates, $x + jy$ at (x, y) or polar coordinates (radius $A = \sqrt{x^2 + y^2}$, angle ω radians).*

Euler's Formula

$$e^{j\theta} = \cos(\theta) + j\sin(\theta)$$

has been called the most beautiful formula in mathematics, connecting exponentials and trigonometric functions in a profound way.

Many of the trigonometric identities so painfully learned in trigonometry classes (and forgotten by a year later) can be replaced by simple algebra on Euler's Formula.

For example, adding two angles:

$$\begin{aligned}\cos(\phi + \theta) + j\sin(\phi + \theta) &= e^{j(\phi+\theta)} \\ &= e^{j\phi}e^{j\theta} \\ &= (\cos(\phi) + j\sin(\phi))(\cos(\theta) + j\sin(\theta)) \\ &= (\cos(\phi)\cos(\theta) - \sin(\phi)\sin(\theta)) \\ &\quad + j(\sin(\phi)\cos(\theta) + \cos(\phi)\sin(\theta)).\end{aligned}$$

Two complex numbers are equal if, and only if, their real parts are equal and their imaginary parts are equal, so the derivation above gives the addition formulas for both cosines and sines, using only simple properties of exponentials and the distributive law.

It is often useful to view the complex numbers as being on a plane, with $z = x + jy$ being the point at coordinates (x, y) (Figure 2.2). In this representation, $z = Ae^{j\theta}$ has a simple geometric interpretation: the point z is distance A away from the origin, along a ray that has angle θ (in radians) counterclockwise from the x-axis.

In this view, Euler's formula expresses the relationship between Cartesian and polar coordinates.

A further useful application of Euler's formula (or of polar coordinates) is the interpretation of multiplication of complex numbers. If we have $z_1 = A_1e^{j\theta_1}$ and $z_2 = A_2e^{j\theta_2}$, then

$$z_1z_2 = A_1e^{j\theta_1}A_2e^{j\theta_2} = A_1A_2e^{j(\theta_1+\theta_2)}.$$

That formula means that multiplying by z_1 corresponds to scaling by its magnitude (A_1) and rotating about the origin by its phase (θ_1).

We'll be using complex numbers extensively for discussing impedance (Section 10.2), so thorough understanding is important.

Exercise 2.5

If $A = 1 + j$ and $B = 1 - j$, what are

- $A + B$,
- $A - B$,
- A^2,
- B^2,
- AB,
- Aj,
- Bj,
- A/B, and
- B/A,

expressed in the standard $x + yj$ form?

Exercise 2.6

Express the following in polar $re^{\theta j}$ form:

- 1,
- $1 + j$,
- $3 + 4j$,
- $5 + 12j$, and
- $-5 - 12j$.

Exercise 2.7

Express the following in standard $x + yj$ form:

- $3e^{\pi j}$,
- $e^{(\pi/4)j}$,
- $2e^{2\pi j}$, and
- $13e^{13j}$.

Exercise 2.8

Plot (by hand or with a program) all the complex numbers in Exercise 2.6 and Exercise 2.7.

2.5 Derivatives

The book assumes that students have had a semester of differential calculus, but we don't need all the material from such a course, as only a few simple rules for taking derivatives are needed.

There are three fundamental derivatives you should know:

$$\frac{\mathrm{d}}{\mathrm{d}t}\sum_n a_n t^n = \sum_n a_n n t^{n-1},$$

$$\frac{\mathrm{d}}{\mathrm{d}t} e^{\lambda t} = \lambda e^{\lambda t},$$

$$\frac{\mathrm{d}}{\mathrm{d}t}\ln(f(t)) = \frac{\frac{\mathrm{d}f(t)}{\mathrm{d}t}}{f(t)} \ .$$

To reduce writing, engineers often use the shorthand notation $f'(x)$ for $\frac{\mathrm{d}f(x)}{\mathrm{d}x}$ and $\dot{f}(t)$ for $\frac{\mathrm{d}f(t)}{\mathrm{d}t}$.

You should also know the following combining forms:

$$(u+v)' = u' + v',$$

$$(uv)' = u'v + uv',$$

$$(u/v)' = \frac{u'v - v'u}{v^2},$$

$$(f(g(x))' = f'(g(x))g'(x) \ .$$

Because this course is an electronics course, not a calculus course, it will be acceptable on homework and design reports to use computer tools such as Mathematica and WolframAlpha to take derivatives and solve equations. Electronics knowledge is needed to set up the appropriate equations—solving the equations does not have to be done by hand.

One common use for derivatives in this course is talking about *gain* or *sensitivity*, both of which are defined as the derivative of the output of a system with respect to its input. When the inputs and outputs are both electrical signals (as in an amplifier), we usually use the term *gain*, but when the input is a physical measurement (like temperature or pressure) and the output is electrical, we usually use the term *sensitivity*. Some authors refer to the sensitivity of a sensor as its *responsivity*, but that term has still not gotten widespread acceptance, so we will continue to use *sensitivity*.

The *gain* of a voltage amplifier is

$$G = \frac{\mathrm{d}V_{\mathrm{out}}}{\mathrm{d}V_{\mathrm{in}}} \ .$$

The *sensitivity* of a pressure sensor whose input is pressure P and whose output is voltage V is

$$S = \frac{\mathrm{d}V}{\mathrm{d}P} \ .$$

Some people mistakenly think of sensitivity or gain as "output over input", but this leads to many mistakes due to DC offsets in voltage. It is much better to think of them as "change in output over change in input".

Pay attention to units when taking derivatives. The units for $\frac{\mathrm{d}f(x)}{\mathrm{d}x}$ are the units for f divided by the units for x. For example, if both the input and the output of the function are in volts, then the gain is unitless (but is sometimes written as V/V to remind the reader that it is the ratio of voltages). If the input is pressure (in pascals) and the output is voltage (in mV), then the units for the sensitivity would be mV/Pa.

Exercise 2.9

If an amplifier circuit has the function $V_{\text{out}} = 27V_{\text{in}} + 20\,\text{V}$, what is the gain of the amplifier?

Exercise 2.10

If a temperature sensor with a resistance output has the function

$$R = 10\,\text{k}\Omega e^{3977\,\text{K}/T - 3977\,\text{K}/298.15\,\text{K}},$$

where T is the temperature in kelvins, and R is the resistance in ohms, then what is the sensitivity of the sensor at an arbitrary temperature T? What is it at 35°C?

2.6 Optimization

Another use for derivatives is in *optimization*, which is any method that finds the values of the inputs to a function that maximizes the output. We usually apply optimization to mathematical functions that describe the behavior of a physical or electronic system that we are interested in.

There are many optimization techniques, and which one to use depends on the nature of the function that we are optimizing, any constraints on the inputs to the functions, and whether we need exact or approximate solutions.

For this course, we will look only at one of the simplest optimization techniques, which is useful for continuous, differentiable functions of one variable with no domain limitations. For such functions, we can use the derivative to find the maxima. If the function $f(x)$ has a maximum at $x = x_m$, then the derivative of the function must be 0 there: $f'(x_m) = 0$. To find the maximum of f, we look at every value of x for which $f'(x) = 0$ and determine whether those points are maxima or not.

If we have simple domain constraints on a function (for example, that the input needs to be non-negative), we can add the boundaries of the domain to the set of points to check for maxima—this is easy when we have only a single variable to optimize, but more sophisticated approaches are needed when we have multiple variables to optimize. We will not need the more sophisticated optimization techniques in this book.

Exercise 2.11

For what value(s) of x is $f(x) = -14x^2 + 7x + 32$ maximized?

Exercise 2.12

If v is constrained to the interval $v \in [0,5]$, for what value(s) is $f(v) = v^3 - 9v^2 + 24v - 20$ maximized?

2.7 Inequalities

Many constraints and design goals are expressed as *inequalities*, not equations, and engineers are expected to be able to manipulate the inequalities to convert constraints on inputs or outputs to constraints on component values. Converting inequalities to equations is a bad idea, because doing so causes confusion about which side of the "solution" has the legal values, and it encourages engineers to design right at the limits of the constraints, rather than staying safely far from the limits.

When we get constraints on component values, voltages, or currents, we almost always want to stay far away from the constraints in real designs. Otherwise small variations in any of the values that go into the inequality can cause the constraint to be violated.

If we have an inequality of the form

$$A < B \ ,$$

we can do the following operations to change the inequality:

$$\begin{aligned} A + x &< B + x, && \text{for any real } x, \\ A - x &< B - x, && \text{for any real } x, \\ Ax &< Bx, && \text{for any positive } x, \\ Bx &< Ax, && \text{for any negative } x, \\ A/x &< B/x, && \text{for any positive } x, \\ B/x &< A/x, && \text{for any negative } x, \\ e^A &< e^B, && \\ A^k &< B^k, && \text{for any positive odd } k \ . \end{aligned}$$

If we have an inequality of the form

$$0 < A < B$$

(equivalent to the pair of inequalities $0 < A$ and $A < B$), we can do the following additional operations to change the inequality:

$$\begin{aligned} 0 < 1/B &< 1/A \\ \log(A) &< \log(B) \\ 0 < \sqrt{A} &< \sqrt{B} \\ 0 < A^x &< B^x, && \text{for any positive } x\,, \\ 0 < B^x &< A^x, && \text{for any negative } x\ . \end{aligned}$$

If we have two inequalities

$$A < B ,$$
$$C < D ,$$

we can combine them to get

$$A + C < B + D ,$$
$$A - D < B - C .$$

If all the numbers are positive,

$$0 < A < B ,$$
$$0 < C < D ,$$

then we can do more combining:

$$AC < BD ,$$
$$A/D < B/C .$$

Inequalities can only be applied to real numbers, not complex numbers, because complex numbers are not an ordered field.

Let's look at a few examples.

Worked Example:
If we have a 5 V power supply with a 2 A current limit, what resistances can we use as a load?

We have to combine an equation (Ohm's law—see Section 4.3)

$$V = IR$$

with an inequality

$$I < 2\,\mathrm{A} ,$$

to get

$$5\,V/R = I < 2\,\mathrm{A} .$$

We can multiply both sides by R and divide both sides by 2 A to get

$$2.5\,\Omega < R ,$$

which is the desired constraint on resistance. Note that this solution does *not* mean that $2.5\,\Omega$ is a desirable value for R—in fact, it would almost always be a bad choice, as slight variations in any of the parameters (voltage, current limit, or resistance) would cause the constraint to be violated.

Worked Example:
If we are making an RC filter with corner frequency $f_c < 2\,Hz$ using a capacitor with capacitance $C \leq 10\,\mu F$, what values can we use for the resistor R?

The concepts and formulas for RC filters are given in Chapter 11, and the formula for the corner frequency is

$$f_c = \frac{1}{2\pi RC} .$$

We can apply that formula and the first inequality in the question to get

$$\frac{1}{2\pi RC} < 2\,\mathrm{Hz} ,$$

which we can invert (because all values are positive) to get

$$0.5\,\mathrm{s} < 2\pi RC .$$

Because the constraint on C sets a maximum value for C, we can do division to get

$$\frac{0.5\,\mathrm{s}}{10\,\mu\mathrm{F}} < 2\pi R ,$$

which we can simplify to

$$7957.747\,\Omega < R .$$

We can pick any larger value of R—staying close to the minimum is rarely desirable. For example, we could pick $C = 100\,\mathrm{nF}$ and $R = 820\,\mathrm{k}\Omega$ to get a corner frequency of

$$f_c = \frac{1}{2\pi RC} = \frac{1}{2\pi 820\,\mathrm{k}\Omega\ 100\,\mathrm{nF}} \approx 0.97\,\mathrm{Hz} .$$

Worked Example:
If we have a differential amplifier with a gain of 10, and an output range of 1 V to 2 V with a 0 V input centered at 1.5 V, what is the legal range of the input signal?

Again we need to find an equation for the behavior of the amplifier, to combine with the inequalities of the constraint

$$1\,\mathrm{V} < V_{\mathrm{out}} < 2\,\mathrm{V} .$$

In this case, the equation is the gain equation of a differential amplifier, from Section 18.2.1,

$$G = \frac{V_{\mathrm{out}} - V_{\mathrm{ref}}}{V_{\mathrm{p}} - V_{\mathrm{m}}} ,$$

where G is the gain, $V_{\mathrm{p}} - V_{\mathrm{m}}$ is the differential input signal, and $V_{\mathrm{ref}} = 1.5\,\mathrm{V}$ is the reference voltage that the amplifier uses with a 0 V input.

We can rearrange the gain equation to get

$$V_{\mathrm{out}} = V_{\mathrm{ref}} + G(V_{\mathrm{p}} - V_{\mathrm{m}}) ,$$

and combine with the inequalities to get

$$1\,\text{V} < V_{\text{ref}} + G(V_{\text{p}} - V_{\text{m}}) < 2\,\text{V} .$$

Plugging in the given values for V_{ref} and G gives us

$$1\,\text{V} < 1.5\,\text{V} + 10(V_{\text{p}} - V_{\text{m}}) < 2\,\text{V} ,$$

which we can simplify in two steps:

$$-0.5\,\text{V} < 10(V_{\text{p}} - V_{\text{m}}) < +0.5\,\text{V}$$

$$-50\,\text{mV} < V_{\text{p}} - V_{\text{m}} < +50\,\text{mV} ,$$

which is the desired constraint on the input signal.

Exercise 2.13

We have an amplifier whose gain equation is $V_{\text{out}} - 1.65\,\text{V} = G(V_{\text{in}} - 1.65\,\text{V})$ and with the constraint on the output $0.05\,\text{V} < V_{\text{out}} < 3.25\,\text{V}$. If the input signal is $1.65\,\text{V} \pm 0.1\,\text{V}$, what are the constraints on gain G?

Exercise 2.14

If we want a corner frequency $f_c = 1/(2\pi RC)$ between 1 Hz and 2 Hz, and the capacitance C is 100 nF, what are the constraints on R?

Chapter 3

Lab 1: Setting Up

Bench equipment: Soldering station, fume extractor

Student parts: Teensy board, breadboard, male header pins, female headers

This is a short lab to familiarize students with the parts and prepare the microcontroller boards for later use with PteroDAQ. No report is expected.

The tasks for this lab are to

- get parts kit and identify parts,
- solder headers onto microcontroller board,
- install Python (preferably Anaconda distribution),
- install PteroDAQ data-acquisition system, and
- install gnuplot.

3.1 What parts are needed for the course

Because the parts are different from year to year (each year some part becomes unavailable and needs to be replaced by a different one), I've not put a parts list in the book directly. Instead, I maintain parts lists on my blog for each offering of the course. For example, the 2020 parts lists are at https://gasstationwithoutpumps.wordpress.com/parts-and-tools-list-for-bme-51a-winter-2020/ and https://gasstationwithoutpumps.wordpress.com/parts-list-bme-51b-spring-2020/, though the spring parts list had to be expanded when the course was moved to Fall 2020 and changed to at-home labs because of the COVID-19 pandemic. Some of the shared equipment had to be replaced by equipment in the kits (cheap soldering irons, solder suckers, and safety glasses), and some normally shared parts had to be sent to everyone (inductors, wire, solder).

If the book is ever sold with a kit of parts for self-study, the kit should come with photographs to help identify the parts. In a lab course setting, however, it suffices to have the instructor able to identify each part for the class and to have the students look through their kits to find the parts.

Several colors of wire will be provided in the lab, so that color coding can be used to indicate what signals are on each wire. For example, black is used only for the *ground* (0 V) wire, and red for the positive power supply (usually +3.3 V in this course). All the wires provided are 22-gauge solid wire, which is best for working with breadboards. The breadboards *will* work with finer 24-gauge wire, but there is a tendency for the wires to fall out or lose contact if jostled, which makes debugging very frustrating.

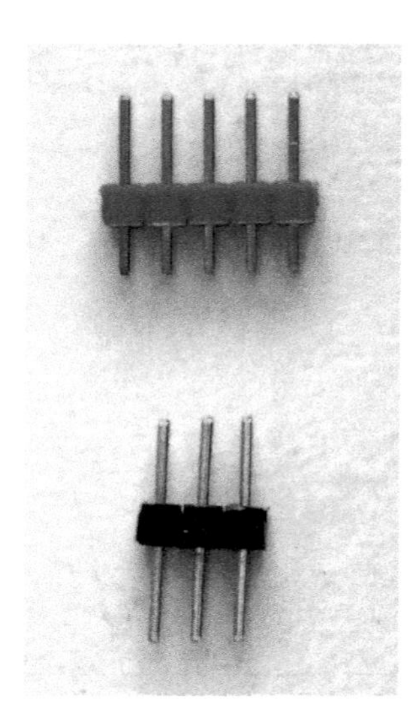

Figure 3.1: *Two types of male header pins for use with breadboards. The red ones in this picture have a short side for soldering to a printed-circuit board, while the long side is used for insertion into a breadboard (or connection to a female header). The black ones have two long sides—they are useful for connecting oscilloscopes and other bench equipment to breadboards. The headers in your lab kits may be any color—the color coding here is just for clarity in the picture.*

3.2 Sorting parts

For the first part of the lab, students will be learning to identify the parts in the parts kit, and doing some organizational work that will save time later on (like writing the value of the capacitors with a Sharpie on all the bags of capacitors). Generally, one partner will work on sorting his or her parts and installing software, while the other works on soldering headers onto the microcontroller board, as described in Section 3.3.2.

The lab kits come with two types of male header pins, as shown in Figure 3.1. One set have short and long sides—these are intended for soldering to the Teensy board, with the short side going through the holes in the board and being soldered, and the long side extending out to make contact with the breadboard. The other set have two long sides—these double-headed male headers are intended for making connection points on breadboards for oscilloscopes and other external equipment. See Section 8.6 for more information about using double-headed male headers.

3.3 Soldering

3.3.1 *General soldering advice*

Several of the boards that PteroDAQ can be used with come without *headers* (pins or sockets for connecting to wires or other boards) attached, both to save manufacturing cost and because different applications call for different header configurations.

We'll need to solder on the appropriate headers.

At UCSC, we are required to use lead-free solder, like most of industry. In Europe, the Restriction of Hazardous Substances Directive (RoHS) prohibits lead in consumer electronics (RoHS started in July 2006), so most manufacturers now make RoHS-compliant products using lead-free solder. Modern electronics uses lead-free solders because of the difficulty of disposing of products that contain lead, but the small quantities of lead in the solder used by hobbyists is not a serious hazard.

The lead-free solders are more brittle than tin/lead solders, which can cause problems in some applications, and they can end up growing "tin whiskers" leading to failure.

If you are working at home, you may prefer to use a classic lead-tin solder, 60%Sn/40%Pb or 63%Sn/37%Pb, which is easier to work with than lead-free solders. Commonly used solders in electronics have melting temperatures from 183°C (for 63% tin, 37% lead) to 227°C (for 99.3% tin, 0.7% copper), and we commonly use soldering irons set to 300°C–400°C (570°F–750°F), because we need to heat the metals being joined hot enough that the solder will wet them.

A bigger problem for safety than the use of lead is the smoke from heating the flux in the core of the solder, especially if the electronics lab is not as well ventilated as it should be.

> Whenever you are soldering, you should have a fume extractor (a small fan with a filter) right next to where you are working, to keep the fumes of the burning flux from being breathed in. The fan inlet should be within about 30 cm (a foot) of the part being soldered.

The basic principle of soldering is simple: two pieces of metal are heated up, melted solder is allowed to flow over the surfaces using capillary action, then cools to form a solid metal-to-metal connection. Figure 3.2 shows a cross section of what a good through-hole solder joint looks like.

Printed-circuit boards (PCBs) generally have one of two surface preparations: hot-air solder leveling (HASL) or electroless nickel with immersion gold (ENIG). Both methods coat all exposed copper on the printed-circuit board with a metal that is less prone to corrosion and more easily soldered to. The HASL coating is just solder (often lead-free these days) and is very cheap, but the coating is not very uniform, so soldering very fine-pitch parts is difficult. The ENIG coating consists of a thick (3 μm–6 μm) layer of nickel followed by a very thin layer (50 nm–200 nm) of gold [168]—it is a more expensive process, but allows for finer pitch of the soldering pads and better corrosion resistance for a longer shelf life for the boards before soldering. For hand-soldering of through-hole components, we have rather large spacing between solder pads (typically a pitch of 2.54 mm or 0.1 inch), so the cheaper HASL coating is all we need for the custom PC boards used in the course, but microcontroller boards have fine-pitch surface-mount components, so they use an ENIG surface.

The first step of soldering is to clean the metal surfaces that need to be soldered, to make them free of dirt, grease, and oxide coatings that can prevent the solder from wetting them. Generally, there are two cleaning steps: the first uses alcohol or some other mild organic solvent to remove dirt and grease, while the second uses "flux" to remove thin oxide layers. We will usually skip the first step, as it is not often necessary with new boards that have not been handled.

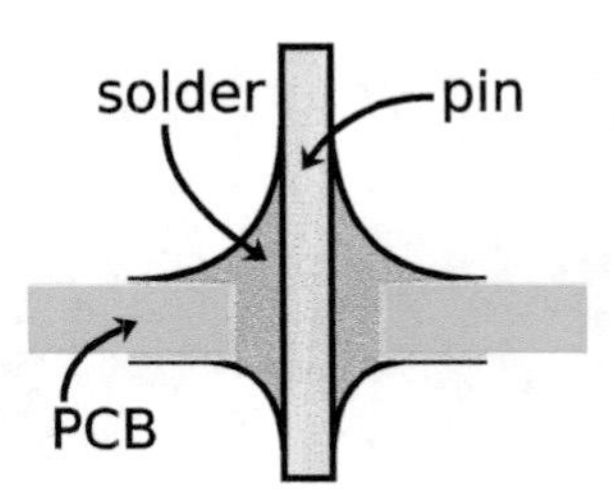

Figure 3.2: *A good through-hole solder joint has the solder forming a smooth meniscus making good connection to both the pad on the printed-circuit board (PCB) and the pin going through the hole. The solder should fill the hole but not flow too much out the other side. To get the proper shape, both the pad and the pin need to be hot enough for the solder to flow by capillary action, and the right amount of solder applied.*

The hole here is a little large for the size of the pin—the fit should be a little tighter, so that the solder layer is thinner.

There are many different types of flux in use, with different levels of corrosiveness and different methods for cleaning up flux residues. For hand soldering, we usually use a "flux-core" solder, where the flux is heat-activated and does not require cleaning the board after use.

> The corrosive flux in the solder will damage the tip of the soldering iron after a while. The soldering iron tip should be wiped clean of any flux residue frequently—certainly every time the iron is put back in the stand—and the tip should be kept coated with solder.

Warning: plumbers use *acid-core* solder for soldering copper pipes—the acid is a very aggressive flux that can damage electronic components, and acid-core solder should never be used for soldering electronics.

Damaged soldering-iron tips are the most frequent problem in student soldering, and the best preventative is keeping the tip tinned (coated with solder) and frequent cleaning off of flux residues. Hakko, a major manufacturer of soldering irons, has excellent advice on the proper use and care of irons—the advice is distributed by Adafruit (https://cdn-shop.adafruit.com/datasheets/hakkotips.pdf) and I highly recommend reading it before each soldering lab.

Figure 3.3 shows a good size of soldering-iron tip for soldering through-hole parts. Too small a tip results in inefficient heat conduction, increasing the risk of cold-soldered joints. Too large a tip results in heating up pads next to the intended one, risking bridges between adjacent pads.

> The basic technique of soldering is fairly simple, as shown in Figure 3.4. The two pieces of metal to be joined must be clean and hot enough for molten solder to wet them. We get them hot by touching both of them for a few seconds with a soldering iron that is well above the melting temperature of the solder (say, 370°C or 700°F for lead-free solder).

When the pin and the ring on the printed-circuit board are hot, flux-core solder is touched to the iron, pulled around to the other side of the joint, and then removed. The solder is melted by the iron, and the end of the solder is used to wick the molten metal to flow over the heated metals to be joined, as shown in Figure 3.4(b).

The *flux* in the core of the solder is an organic material that forms an acid when heated, to remove any thin layers of corrosion from the metal. This flux needs to be boiled away after it has done its job—if

Figure 3.3: *A good soldering iron tip for soldering through-hole parts is one that has a "screwdriver" shape and that is narrower than the 0.1″ pitch of the pads. The one I use is a Weller ETH 0.8 mm tip, shown here.*

(a) *To solder a pin to the board, touch* both *the pin and the copper ring around the hole with the soldering iron for about three seconds to heat both.*

(b) *Then touch the solder to the pin and the soldering iron until the solder melts and flows onto the pin. Remove the solder but leave the iron in contact with the pin for another second.*

Figure 3.4: *The basic technique for through-hole soldering is simple conceptually, but takes some practice to do well.*

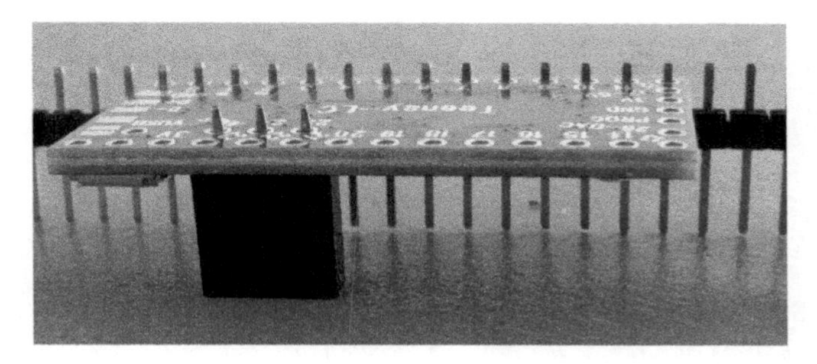

Figure 3.5: *The 3-pin female header is inserted from the component side of the board and the top of the header rests on the bench. To make a clean 90° angle, the other side of the board needs to be supported at the same height. Another female header (or, as here, some male header pins) can be temporarily inserted to provide that support.*
If you wish to add a 5-pin female header at the end of the board opposite the USB connector, that should be added after the 3-pin header has been soldered, to avoid having to manipulate four loose parts at once.

you see black specks in your solder joints, reheat the joint to remove the extra flux. It is also possible to get *flux pens* for adding flux before soldering, but these are not usually needed for through-hole soldering (they are commonly used when soldering surface-mount components by hand).

There are many learn-to-solder videos and tutorials on the web—if you've never soldered before, it might be worth looking at a few of them.

3.3.2 *Soldering Teensy headers*

To use a Teensy LC or Teensy 3.2 board for PteroDAQ, we'll need to have male header pins sticking out the bottom of the board to make contact with a breadboard and a 3-pin female header sticking out the top of the board, for access to the differential channel.

> The female header must stick out of the component side of the board. The male headers must stick out of the other side of the board. Fixing a mistake here is very difficult (buying a new board is about the best option).

When soldering on male or female headers, it is important to have the headers held at a right angle to the board so that connections can easily be made. This often requires some ingenuity in setting up

(a) *A close-up view of the three solder joints that hold the female header to the Teensy LC board.*

(b) *The top of the Teensy LC board showing the female header correctly soldered in place.*

Figure 3.6: *The top of the Teensy LC board showing the female header correctly soldered in place. Good solder joints should be shiny and round, completely covering and wetting both the ring on the PC board and the wire, as shown in Figure 3.6(a).*

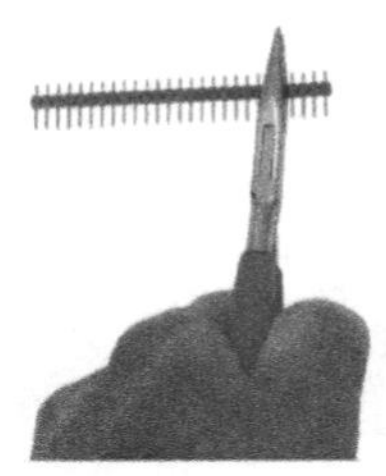

(a) *Hold the headers right next to the spot where you want to break them—here they are being held for breaking off four header pins. If you have only one pair of pliers, use your fingers for this first grip.*

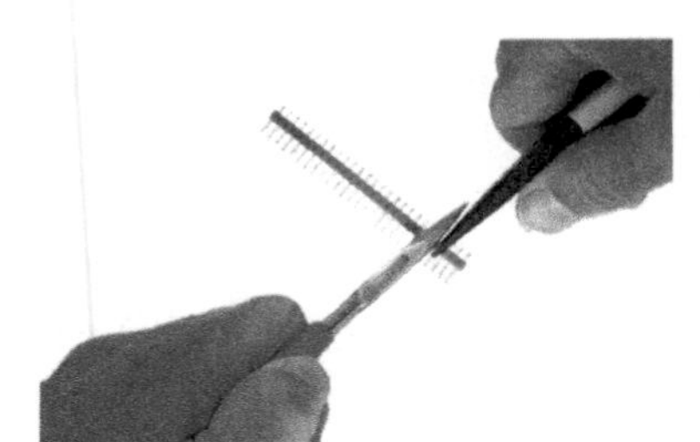

(b) *Grip the plastic of the adjacent pin with pliers and twist so that the pins are not parallel—the thinner plastic between the two grips will snap (usually cleanly).*

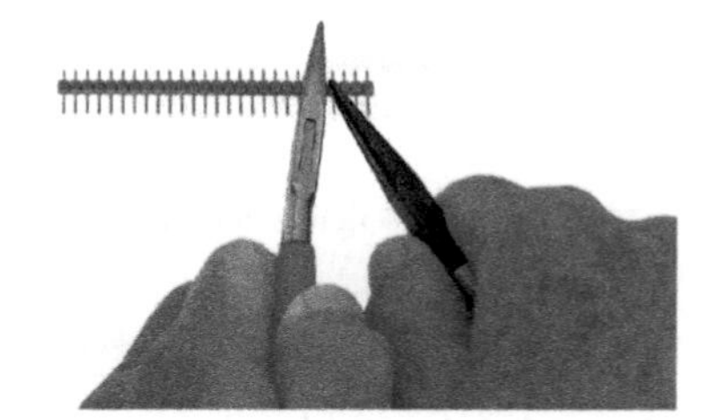

(c) *The second grip on the headers can come from the same side, instead of opposite sides as in Figure 3.7b. The twisting action is the same.*

Figure 3.7: *Count carefully the number of header pins you want to break off, and grip the plastic immediately next to where you want to break, twisting the headers on either side so that they are no longer parallel.*

the work. For example, Figure 3.5 shows one way to support a Teensy LC board upside-down, so that the three-pin female header can be soldered on.

The result of soldering the three pins of the female header onto a Teensy LC board is shown in Figure 3.6.

After every few solder points, it is a good idea to wipe off the tip of the iron on a wet sponge or a brass or copper metal tip cleaner, to keep the tip clean. Otherwise burnt residue of the flux builds up on the iron and makes good thermal contact difficult. Always wipe the tip before putting the iron back in its stand and just before turning the soldering iron off.

(a) *Top view of the 28 pins to be soldered to the Teensy LC. The long ends of the headers are inserted into the breadboard.*

(b) *Side view of the Teensy LC resting on the 28 pins and ready to be soldered.*

Figure 3.8: *Using a breadboard as a jig to hold the male headers works well to make sure that the headers are the right spacing and are kept at right angles to the board.*

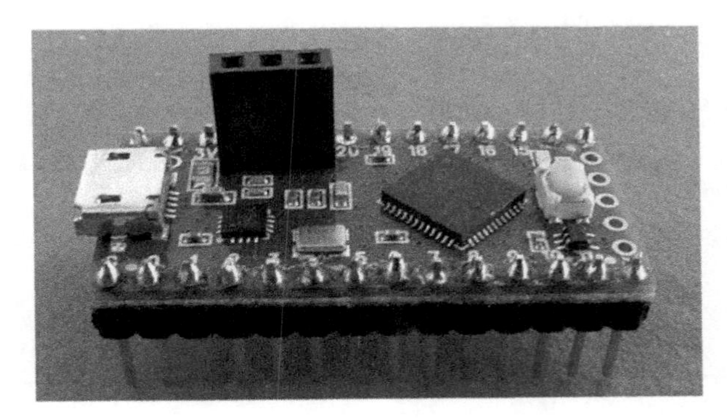

Figure 3.9: *The completed Teensy LC board with 3-pin female header on top and 28 male header pins on the bottom.*

After the female header is installed sticking up out of the component side of the board, we want to install male headers on the other side of the board, to make it easy to attach the Teensy LC board to a breadboard. The headers should be snapped into two 14-pin pieces, as explained in Figure 3.7.

Insert the long side of the headers into the breadboard at the appropriate spacing, as shown in Figure 3.8.

Figure 3.9 shows a completely soldered Teensy LC board.

Figure 3.10 shows several examples of poorly made solder joints.

(a) *This picture shows several bad solder joints. Pin 3 did not get heated enough and does not have enough solder—both the pin and the ring on the PC board still have areas that the solder doesn't cover. Pin 4 has way too much solder. Pins 5 and 6 have too little solder—the bare ring on the PC board is visible—one can also have the ring visible if there is enough solder, but the ring was not hot enough for the solder to wet it. Pins 7 and 8 have been bridged together. Pins 9 and 11 have not heated the pin enough—the solder wets the board, but not the pin—pin 12 may also have this problem. Pin 10 has a blobby appearance with black specks—the flux has probably not been boiled out of the joint.*

(b) *The solder ball on the middle pin of this figure is a more subtle bad solder joint. Notice how the solder curves inward at the bottom forming a ball shape—it may be sitting on a layer of corrosion or flux and not making proper contact with PC board underneath. Reheating the joint might help, as would using a solder sucker to remove the excess solder then resoldering.*

Figure 3.10: *Most of the problems with the bad solder joints can be fixed by reheating the solder, adding more solder, or using a solder vacuum (colloquially called a* solder sucker*), to remove excess solder. If solder is not behaving at a joint, it often helps to reheat the joint, remove the molten solder with a solder sucker, then resolder the joint.*

3.4 Installing Python

You will need either Python 2.7 or Python 3 to run PteroDAQ, and we will also be using the SciPy Python package later in the course. If you do not have SciPy already installed on your laptop, I recommend installing the Anaconda Python distribution from https://www.anaconda.com/products/distribution.

The instructions on the Anaconda site seem to work well.

3.5 Installing data-acquisition system: PteroDAQ

We have provided data-acquisition software that can be run on several different microcontroller boards: ARM-based boards (Teensy LC, Teensy 3.1, and Teensy 3.2) and several ATMega-based Arduino boards. (I recommend the Teensy LC as the best value for money, unless you already have one of the other boards.) The Arduino boards provide much lower resolution measurements and don't currently support frequency channels. The microcontroller boards are connected with a USB cable to a larger host computer—generally your own laptop.

> Make sure that your USB cable is a real data cable and not a "charging" cable that has only the power lines and not the data lines.

> You can get a copy of PteroDAQ to run on your own computer from the repository at https://github.com/karplus/PteroDAQ.

The latest version (which is sometimes a beta release) can be downloaded from https://codeload.github.com/karplus/PteroDAQ/zip/master This program was initially written by Abe Karplus when he was a high-school student, but it is now maintained by Kevin Karplus.

The software provides options for getting either the raw ADC readings (integer values from 0 to $2^{16}-1$) or converting to voltages, by scaling the 2^{16} value to be the reference voltage of the analog-to-digital converter. For many measurements, we are mainly interested in the ratio of the measured voltage to the reference voltage, rather than the voltage itself, for which the raw reading may be more useful, as it always has 2^{16} as the full-scale value, while the full-scale value on the voltage scale may vary (it is reported in the meta-data of the output file).

The software consists of two parts: a small program that runs on the microcontroller board and a Python program that runs on a laptop or desktop machine. The microcontroller code makes measurements and communicates them over the USB connection to the Python program on the host computer, which provides a graphical user interface, file I/O, and configuration information.

To use the data-acquisition software, you need to download the microcontroller program to the board. This only needs to be done once, as the program is stored in flash memory and will remain on the board until another program is downloaded to replace it. On the Teensy boards, accidentally pressing the "program" button while the board has power can result in wiping out the firmware, which then needs to be reloaded.

You must have the Arduino development software installed to download to Arduinos and Teensy boards, as the program is recompiled from source code every time you do a download. For Teensy boards, you must have Teensyduino installed in the Arduino environment. See the instructions on the PteroDAQ wiki site: https://github.com/karplus/PteroDAQ/blob/master/Installation.md

To download the program with the Arduino IDE, connect the board to the computer with a USB cable, and open the `firmware/firmware.ino` file in the Arduino environment. All the files in the firmware directory are needed, and the directory must be called "firmware" to match the "firmware.ino" name.

Set the board type and serial port in the *Tools* menu, and download firmware.ino to the board. Once the program is downloaded, the board will retain it in flash memory, even if power is removed. The Arduino environment can be closed after the download is done, as it is not needed again, unless the firmware on the board somehow gets corrupted.

3.6 Installing plotting software (gnuplot)

You will need to have plotting software that can produce scatter plots and line plots and that can fit arbitrary complex-valued functions to data. I've found that gnuplot is reasonably powerful, reasonably easy to use, stable, and free. Other powerful free plotting software includes R and Python (with the Matplotlib package [66]), but they are a little more difficult to work with—R's default settings for plots are terrible, and both Python and R have much harder to use parameter-fitting routines than gnuplot.

Install gnuplot on your own computer. Gnuplot can be found at http://www.gnuplot.info/, with downloadable executables at http://www.gnuplot.info/download.html. These downloads are generally available for Windows and Linux systems, but easily installed Mac OS executables are not usually available from the gnuplot developers. When doing installation on Windows, make sure that you check the box that requests that the command-line version be added to the path so that Windows can find the executable—otherwise it is hidden in some directory that you will have to type the full path-name for, a serious inconvenience.

Installing on a Mac OS X system is still a two-step process, as the gnuplot developers do not provide a MacOS installer. The best approach on Macs currently seems to be the following steps:

- Install brew from https://brew.sh/, following the instructions there. The command given on their homepage has to be run from bash, not tcsh, but most Macintosh users have bash as their default shell when running the Terminal application.
- Run the command `brew install gnuplot` to install gnuplot. (You used to be able to give options to the installation to change which terminals were installed, but that is no longer possible in brew.)

3.7 Using voltmeter

This section is optional, additional material, if there is sufficient lab time.

If you get all your soldering done and all the software installed, then you should try out the digital multimeter. Here are some things to try doing:

- Select two different resistors from your lab kit, one in the range 1 kΩ–10 kΩ, the other about 20 times larger. Measure each with the ohmmeter function of the multimeter. If the multimeter is not autoranging, be sure to pick the range for each that gives the most precision for the measurement.
- Put the two resistors in series on a breadboard (see Section 8.6). Measure the resistances again on the breadboard, to make sure you have not shorted them out. Measure the resistance of the two in series—does it match your expectation?
- Connect one end of the resistor chain to ground and the other to the 3.3 V or 5 V power supply of your microcontroller. (Use black wire for ground and red wire for the power supply.) Use the voltmeter to measure the voltage across the entire series chain and across each resistor separately. Use Ohm's law to determine the current through each resistor.

3.8 No design report

Because this lab has no design and no important measurements, there is no design or analysis report to write. You will have to have your solder joints inspected, to make sure that you will have a working microcontroller board for the next lab.

Chapter 4

Voltage, Current, and Resistance

4.1 Voltage

A *voltage* is defined as the difference in electrical potential between two points in space—this course will not go into electric fields and their properties (that is what physics classes are for), but we will rely very heavily on the notion that voltages are *differences*. Whenever we talk about a voltage, we are talking about a pair of points, not a single point.

We measure potential differences in *volts*, with the standard one-letter abbreviation V, which should be capitalized. Technically, two points differ by one volt if it takes one joule of energy to move one coulomb of charge from one point to the other—that is, $1\,\mathrm{V} = 1\,\mathrm{J}\,1\,\mathrm{C}^{-1}$.

Because wires conduct electricity well, the voltage difference between two points on the same wire is often small. When the difference is small enough to be negligible, we refer to the points on the wire as having the same voltage, and lump all such points together as a *node*.

> A *node* is a set of points that are treated as all having the same voltage. It is represented in a schematic by a line or connected lines.

A node is a modeling convention (points connected by conductors that we treat as having the same voltage), not a physical phenomenon. At different times we may model the same piece of wire as a node, a resistor, an inductor, part of a capacitor, a transmission line, or more complicated circuits, depending on what properties of the wire are important for the design being done. Our models generally take the form of a simplified mathematical idealization of some part of the behavior of the thing we are modeling.

The notion that different models are appropriate for different designs (or for different analyses of the same design) is an important one in all branches of engineering. We generally try to use the simplest model that adequately captures the phenomena we need to model. Throughout this book, we'll be looking at building models for various phenomena, then using the models to do design.

> When we are discussing voltages, we are always talking about the difference in potential between two nodes. To simplify the formulas we write down, we often designate one node in the circuit as *ground* and define all voltages as differences between nodes and ground. This makes ground always have a voltage of 0 V, by definition. When using this convention, we can talk about the voltage *of* a node, meaning (always) the difference in potential between that node and ground.

We use the symbol V to refer to a voltage, with subscripts to indicate which voltage we are talking about. It is quite common to give two subscripts, to indicate the two nodes whose difference is being taken. For example, when talking about field-effect transistors, we use the symbol V_{GS} to mean the

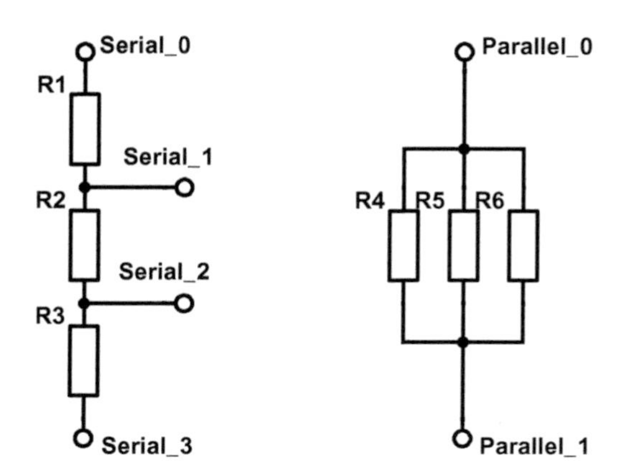

Figure 4.1: *Components are often connected in series, as shown on the left, or in parallel, as shown on the right. We often want to look at a cluster of components connected together in one of these ways as if it were a single, perhaps more complicated, component. Figure drawn with Digi-Key's Scheme-it [18].*

voltage from the gate to the source of the transistor, V_G to mean the voltage from the gate to ground, and V_S to mean the voltage from the source to ground, with the relationship $V_{GS} = V_G - V_S$.

> A *component* is a part of a circuit that does something—often something fairly simple. If the component connects exactly two nodes together, then we can talk about the voltage *across* the component (never other prepositions).

When describing the connections of a component, we usually talk of it *connecting* the two nodes, but some other phrases are used. For example, if the two nodes are the output of a power supply, function generator, or amplifier, we can talk about putting the component *across* the output.

Some consequences of the notion of voltages as differences between nodes include the following:

- If two components are wired in parallel (as shown in Figure 4.1), then they have the same voltage across them. They connect the same nodes, and the voltage is the difference in potential of the two nodes.
- If two components are wired in series, then the voltage across the pair is the sum of the voltages across the individual components. This is just algebra on the node potentials: $V_{\text{serial0}} - V_{\text{serial3}} = (V_{\text{serial0}} - V_{\text{serial1}}) + (V_{\text{serial1}} - V_{\text{serial2}}) + (V_{\text{serial2}} - V_{\text{serial3}})$.
- (Kirchhoff's voltage law) If you connect multiple components together, and add up the voltages across each of the components along a path, then any path that begins and ends at the same node will have a sum of zero.

4.2 Current

> Current is movement of charge and is measured in coulombs per second, more commonly called *amperes* or *amps*. The term *amp* is used as a short form both for *ampere* and for *amplifier*. Which meaning is intended will usually be clear from the context.
>
> Because current is movement of charge, it is correct to talk about current *through* a component, but not *across* one nor *at* a node.

We use the symbol I to refer to current, with a subscript to indicate which current in a circuit we are talking about. On a schematic, we use an arrow labeled with the current to indicate which direction of current flow is positive. If the current flows in the direction opposite the arrow, then the current is negative.

If we look at all the components connected to a node and sum the currents from the components into the node (some of which could be negative, indicating current flow from the node through the component), the results could be positive, zero, or negative. By convention, we do not allow charge to build up in nodes (you can build up charge in capacitors, but those are components, not nodes), so all the current flow into a node has to sum to zero (this is known as *Kirchhoff's current law*).

If you connect a number of components in series as in Figure 4.1, then Kirchhoff's current law implies that the current through each component is identical.

If you connect a number of components in parallel, so that they all share the same two nodes, then the currents through the components all add. In Figure 4.1, the current from Parallel_0 to Parallel_1 is just the sum of the currents through R_4, R_5, and R_6.

4.3 Resistance and Ohm's law

About all that many students remember about electricity from their physics classes is

> Ohm's law: $V = IR$,

which is a very handy relationship, but Ohm's law is often abused by students, who take unrelated voltages, currents, and resistances and plug them into the formula blindly.

The law is really the definition of the absolute resistance of a component. The resistance R of the component is the voltage V across the component divided by the current I through the component. All three variables refer to the *same* component—the only relevant voltage is the difference in voltage between the two ports of the component, and the only relevant current is the current through the component. Don't mix random other voltages or currents in the formula!

To characterize the behavior of components, we often measure and plot current through the component vs. the voltage across the component (I-vs.-V plots). The slope of the curve $\frac{dI}{dV}$ is the *dynamic conductance* of the component. If we swap the axes, plotting voltage vs. current, the slope of the curve, $\frac{dV}{dI}$, is the *dynamic resistance* of the component. The dynamic conductance or dynamic resistance can be different at different voltages or currents.

If the dynamic resistance is constant, that is, if the I-vs.-V or V-vs.-I curve is a straight line through the origin, then we can drop the modifier *dynamic* and just talk about the *resistance*, defined as the ratio of the voltage across a component to the current through the component $R = V/I$. The observation that conductors such as metals have this sort of linear behavior for voltage and current is referred to as *Ohm's law.*

> Resistance is measured in *ohms*, symbolized Ω. One ohm is the resistance that would require one volt to get a one-amp current $1\,\Omega = 1\,\text{V}/1\,\text{A}$. We have relevant resistances in this course from milliohms (mΩ) to gigaohms (GΩ).

Conductance, which is simply the inverse of resistance, is measured in siemens, with $1\,\text{S} = 1\,\text{A}/1\,\text{V}$. In some older texts, you will see the unit of conductance call a *mho* (from reversing *ohm*) with the symbol ℧, which I rather like, but this unit is now considered incorrect, and siemens should be used instead.

If you are unfamiliar with the notion of electrical resistance, you may wish to read the Wikipedia article on electrical resistance [125].

4.4 Resistors

Although wires have resistance, it is usually so small that we ignore it in our modeling of circuits.

> When we want resistance, we usually add components designed to have specific resistance, called *resistors.*

Although modern electronics usually uses tiny *surface-mount* resistors that are about 1 mm × 0.5 mm, such tiny parts are difficult for us to work with while prototyping, so we will use *through-hole* resistors that are cylinders about 6 mm long and 2 mm in diameter with a wire coming out each end.

Even the through-hole resistors are a bit small for printing text on, so the resistance of the resistor is marked on each resistor with a series of bands, called the *color code*, as explained in Table 4.1.

The color codes are not a very reliable way of distinguishing resistors, as the colors are sometimes hard to tell apart (red and orange on a resistor with blue body can be hard to tell apart, and white, grey, and silver are also often difficult to distinguish). Furthermore, it can be difficult to tell which end of the

Color	Value as digit	Value as multiplier	Value as tolerance
none			±20%
silver		0.01	±10%
gold		0.1	±5%
black	0	1	
brown	1	10	±1%
red	2	10^2	±2%
orange	3	10^3	
yellow	4	10^4	*
green	5	10^5	±0.5%
blue	6	10^6	±0.25%
violet	7	10^7	±0.1%
grey	8	10^8	±0.05%*
white	9	10^9	

Table 4.1: *The resistor color code consists of bands around the body of a cylindrical resistor. There may be 3, 4, or 5 bands. For three-band or four-band resistors, the first two bands are read as digits, and the third band as a multiplier. The fourth band (if present) is read as a tolerance. Thus, a resistor labeled red-red-orange-brown is read as 2-2-10^3-±1%, that is, as a 22 kΩ resistor whose tolerance is ±1%. A five-band resistor has three digits, a multiplier, and a tolerance. A resistor banded brown-black-black-yellow-brown would be read as 1-0-0-10^4-±1%, that is as 1 MΩ±1%. The resistor brown-red-black-black-brown would be read as 1-2-0-10^0-±1%, or 120 Ω, but if you flip the resistor around, it looks like brown-black-black-red-brown or 1-0-0-10^2-±1%, or 10 kΩ. I can't tell 120 Ω and 10 kΩ 1% resistors apart using the color code and blue-background resistors often make red and orange or orange and yellow difficult to distinguish, so I try to measure every resistor before I use it.*

Tight-tolerance resistors may have an additional band for the temperature coefficient. Just to confuse matters more, resistors made for military applications may have five bands that are digit-digit-multiplier-tolerance-failure rate.

(*) In high-voltage resistors, yellow and grey replace gold and silver for ±5% and ±10%, to avoid having metallic particles on the outside of the resistor that may cause arcing.

In this course, we'll be using 5-band 1% metal-film resistors with digit-digit-digit-multiplier-tolerance codes, because those are currently the cheapest ones available in assortments.

resistor to start from when reading the bands. Is a resistor brown-black-black-red-brown ($10\,\mathrm{k\Omega} \pm 1\%$) or brown-red-black-black-brown ($120\,\Omega \pm 1\%$)? Both are standard values.

> Before putting a resistor into a breadboard or printed-circuit board, it is a good idea to check its resistance with an ohmmeter. It is also a good idea to pick a standard orientation for all resistors in your prototype, so that you can read the color codes when debugging.

Although design calculations can end up with arbitrary desired values for resistors, getting custom resistors made in an unusual size is very expensive, therefore electronics engineers try to adjust their designs so that standard, widely-manufactured sizes are used (see Table 4.2). The available resistor sizes are powers of 10 times the sizes listed in the table, so 120 means that one can get $12\,\mathrm{m\Omega}$, $120\,\mathrm{m\Omega}$, $1.2\,\Omega$, $12\,\Omega$, $120\,\Omega$, $1.2\,\mathrm{k\Omega}$, $12\,\mathrm{k\Omega}$, $120\,\mathrm{k\Omega}$, and $1.2\,\mathrm{M\Omega}$, $12\,\mathrm{M\Omega}$, and $120\,\mathrm{M\Omega}$ resistors. At very small and very large sizes, only a few sizes might be available, not everything in the series.

There are six different series of resistor sizes, defined by the Electronics Industry Association: E6, E12, E24, E48, E96, and E192. The number of the series specifies how many logarithmic steps there are in each decade. The ratio between adjacent values in the series is approximately constant, for example, in the E96 series the ratio is about $10^{1/96} \approx 1.024$. The uniformity of the spacing can be seen in Figure 4.2.

The E6 series is contained in E12, which is contained in E24, and E48 is contained in E96, which is contained in E192, but the E24 and E48 series are different. Generally, loose tolerance resistors ($\pm 5\%$) are made in the E24 series, and tighter tolerance resistors ($\pm 1\%$) are made in the E96 series. But most manufacturers also make the 1% resistors in the E24 series, since those values are nice round numbers and have been very popular with designers. (Very tight tolerance resistors, down to $\pm 0.01\%$, are sold in the E192 series, but they can be very expensive.)

Many designs have a fair amount of freedom in choosing the resistor sizes, and so the most commonly used and stocked resistors are on the E6 scale, with 100, 330, and 470 being the most popular multipliers.

E6	E12	E24	E48	E6	E12	E24	E48	E6	E12	E24	E48
100	100	100	100	220	220	220	215	470	470	470	464
			105				226				487
		110	110			240	237			510	511
			115				249				536
	120	120	121		270	270	261		560	560	562
			127				274				590
		130	133				287			620	619
			140			300	301				649
150	150	150	147				316	680	680	680	681
			154	330	330	330	332				715
		160	162				348			750	750
			169			360	365				787
	180	180	178		390	390	383		820	820	825
			187				402				866
		200	196			430	422			910	909
			205				442				953

Table 4.2: *There are several different series of standard resistor values specified by the Electronic Industries Association (E96 and E192 are not shown here). The number of the series specifies how many logarithmic steps there are in each decade. The available resistor sizes are powers of 10 times the sizes listed (so 150 means $0.15\,\Omega$, $1.5\,\Omega$, $15\,\Omega$, $150\,\Omega$, $1.5\,\mathrm{k\Omega}$, $15\,\mathrm{k\Omega}$, $150\,\mathrm{k\Omega}$, $1.5\,\mathrm{M\Omega}$, ...).*

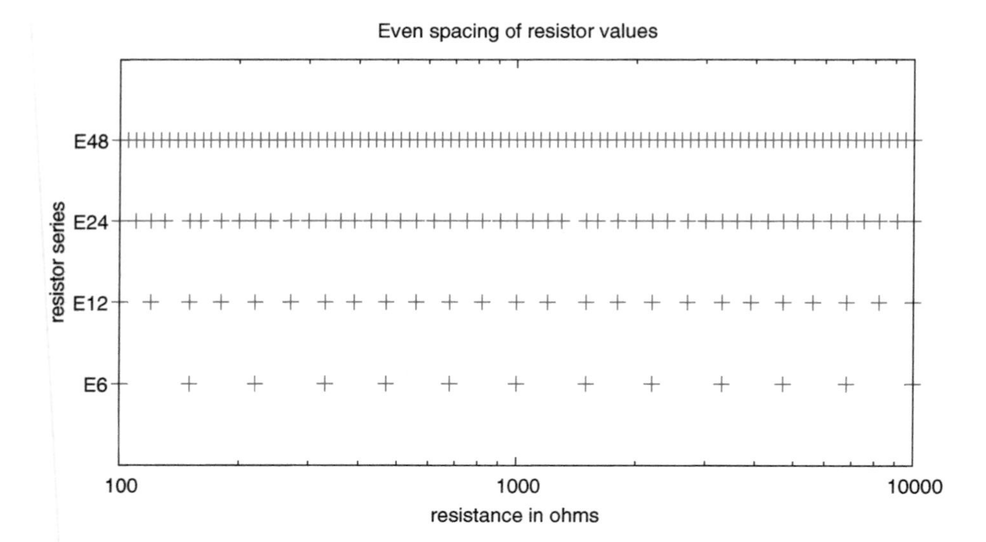

Figure 4.2: *The apparently arbitrary spacing of the resistance values in Table 4.2 makes sense when viewed on a log scale—the resistances have roughly uniform spacing there, reflecting the almost constant ratio between adjacent values.*
There is some unevenness in the spacing of the E24 series, which tries to keep numbers from the E12 series, even when they are not very close to the ideal spacing. The biggest gap is $150/130 = 1.154$ *and the smallest is* $160/150 = 1.067$, *but retaining the value of 150 was seen as more desirable than more uniform spacing.*
Figure drawn with gnuplot [33].

If you can adjust your designs to use a common size, you can reduce manufacturing and inventory problems. A lot of labs stock the E48 series (rather than the E96 series) in 1% tolerance resistors, to reduce the size of the inventory, though the values of the E48 series are about 5% apart, so there are "holes" in the series.

In this class, we will *always* use single resistors to get desired values, but some design styles allow two resistors in series to get more values from a given inventory of resistors.

The largest spacing between adjacent obtainable values is much closer if you allow pairs:

Series	R_{i}	$R_{\text{i}+1}$	Ratio
E6	1000+220=1220	1000+330=1330	1.09016
E12	1200+220=1420	1000+470=1470	1.03521
E24	2400+240=2640	2400+270=2670	1.01136
E48	2050+953=3003	3010	1.00233

Resistors serve many different purposes in a design, and they often are given different names to reflect these differences. The different functions overlap, and different engineers have their own favorite choices for which words to use when several apply.

sense resistor A sense resistor is used for sensing current and turning it into a voltage drop using Ohm's law. Sense resistors are designed into a circuit, and their size depends on how sensitive the circuit needs to be (how much the voltage must change for a change in current).

shunt resistor A shunt resistor is the same thing as a sense resistor, but is generally a very small resistor used to make a voltmeter into an ammeter—the small resistance replaces a wire in the circuit where the current is being measured. Shunt resistances are sometimes temporary insertions into a circuit for measuring and debugging, rather than permanent design features.

load resistor A load resistor is a resistor that we are delivering power to—we often care about power delivered to it: $P = IV = I^2R = V^2/R$. Sense resistors that also serve as the conduit for power to a circuit are sometimes called load resistors. When we are delivering voltage but no current from a circuit, its load resistance is effectively infinite, and we say that the circuit (often an amplifier or a voltage divider) has no load.

current-limiting resistor Many nonlinear components (such as light-emitting diodes) have an effective resistance that drops rapidly as the voltage across them increases. If you try to power them from too high a voltage source, the component overheats and burns out. By putting a resistor in series with the component, the current can be limited to values that remain within the specifications.

bias resistor A bias resistor is a resistor being used to provide a DC voltage or current to a transistor, sensor, or amplifier, to make sure that it has the correct DC value to behave as intended for AC signals.

pull-up resistor A pull-up resistor (often just abbreviated to *pull-up*) is a bias resistor that has one end connected to a positive power supply. This terminology is used mainly in digital design, where such resistors are used so that unconnected wires will be treated as 1s.

pull-down resistor A pull-down resistor is a bias resistor with one end connected to the lowest voltage power supply (negative or ground). This terminology is again used mainly in digital design, where pull-downs are used so that unconnected wires are interpreted as 0s.

Resistances often change with temperature or other physical properties, and there are many sensors based on this change in resistance. There are temperature sensors (both thermistors and resistance temperature detectors), chemical sensors (chemiresistors), force sensors (piezoresistors), and light sensors (photoresistors). In this course, we will use thermistors (Lab 2) and piezoresistors (Lab 5), but the electronics are very similar for any resistance-based sensor.

Exercise 4.1

If you have a $47\,\Omega$ resistor with $30\,\mathrm{mA}$ of current through it, what is the voltage across the resistor?

Exercise 4.2

If you have a $47\,\mathrm{k}\Omega$ resistor with $3.3\,\mathrm{V}$ across it, what is the current through the resistor?

Exercise 4.3

If you have a resistor with $2.5\,\mathrm{V}$ across it and $40\,\mu\mathrm{A}$ through it, what is its resistance?

4.5 Series and parallel resistors

Although there are many arrangements for connecting components, two are so common that they are given special names: *series* and *parallel*. Figure 4.3 shows three resistors connected in series and another three connected in parallel.

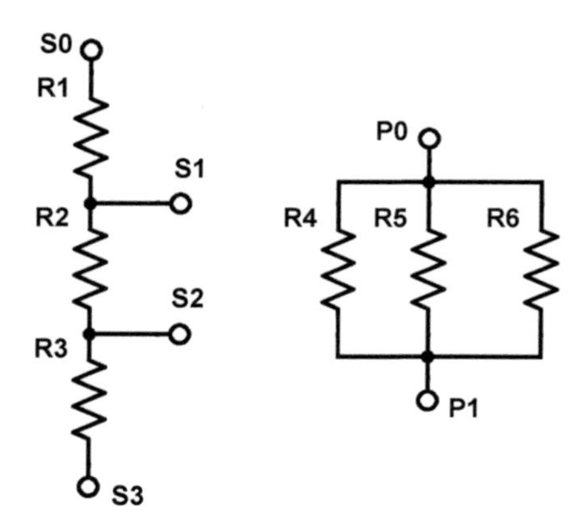

Figure 4.3: *The three resistors on the left (R_1, R_2, and R_3) are in* series. *The three resistors on the right (R_4, R_5, and R_6) are in parallel.*

The little circles are "ports", used for giving names to the nodes they are connected to.

Figure drawn with Digi-Key's Scheme-it [18].

We can use Ohm's law, together with simple rules about voltage and current (Kirchhoff's laws), to figure out how series-connected and parallel-connected resistors behave.

For the series connection, we have that the voltage across resistor R_1 is $V_{S_0,S_1} = V_{S_0} - V_{S_1}$, across resistor R_2 is $V_{S_1} - V_{S_2}$, and across R_3 is $V_{S_2} - V_{S_3}$. If no current is taken through ports S_1 and S_2, then the current through each resistor must be the same:

$$I = \frac{V_{S_0} - V_{S_1}}{R_1} = \frac{V_{S_1} - V_{S_2}}{R_2} = \frac{V_{S_2} - V_{S_3}}{R_3} \ .$$

Alternatively, we could express each of the voltages:

$$V_{S_0} - V_{S_1} = IR_1$$
$$V_{S_1} - V_{S_2} = IR_2$$
$$V_{S_2} - V_{S_3} = IR_3 \ .$$

We can add these equations together to get

$$V_{S_0} - V_{S_3} = I(R_1 + R_2 + R_3) \ ,$$

that is, the three resistors in series behave as if their resistances were added, *as long as no current is taken out of the intermediate nodes S_1 and S_2.*

For the parallel connection, the voltage across each of the resistors is the same: $V_{P_0} - V_{P_1}$. Each resistor has a current given by Ohm's law:

$$I_4 = (V_{P_0} - V_{P_1})/R_4$$
$$I_5 = (V_{P_0} - V_{P_1})/R_5$$
$$I_6 = (V_{P_0} - V_{P_1})/R_6 \ ,$$

and the total current from P_0 to P_1 is just the sum of these currents:

$$I = (V_{P_0} - V_{P_1})(1/R_4 + 1/R_5 + 1/R_6) \ .$$

That means that the equivalent resistance for the three resistors in parallel can be expressed as

$$R_4 \parallel R_5 \parallel R_6 = \frac{1}{\frac{1}{R_4} + \frac{1}{R_5} + \frac{1}{R_6}} \ .$$

For two resistors in parallel, a "simpler" form is often given:

$$R_4 \parallel R_5 = \frac{1}{\frac{1}{R_4} + \frac{1}{R_5}} = \frac{R_4 R_5}{R_4 + R_5} \ ,$$

but this form is not really simpler to use—on my pocket calculator, I can use the general form while entering each resistance value only once, but the "simpler" form calls for storing the values or entering them twice. Also, the general form works well for any number of resistors in parallel, but the "simpler" one does not generalize nicely.

Exercise 4.4

Draw a schematic diagram for three resistors in series, with values $1\,\text{k}\Omega$, $4.7\,\text{k}\Omega$, and $10\,\text{k}\Omega$. What is the resistance of the series chain?

Exercise 4.5

Draw a schematic diagram for three resistors in parallel, with values $1\,\text{k}\Omega$, $4.7\,\text{k}\Omega$, and $10\,\text{k}\Omega$. If you put $1\,\text{V}$ across the three resistors, what is the current through each resistor, and what is the total current? Use Ohm's law to determine the effective resistance of the whole circuit.

Exercise 4.6

Draw the schematic for 2 resistors in series, each with the value $22\,\text{k}\Omega$. What is the resistance of the series chain?

Exercise 4.7

If you wire n resistors with identical resistance R in series, what is the resistance of the series chain?

Exercise 4.8

If you wire a $22\,\text{k}\Omega$, a $10\,\text{k}\Omega$, and a $4.7\,\text{k}\Omega$ resistor in series, what is the resistance of the series chain? Draw the schematic diagram.

Exercise 4.9

If you wire two $22\,\text{k}\Omega$ resistors in parallel, what is the resulting resistance?

Exercise 4.10

If you wire n resistors with identical resistance R in parallel, what is the resistance?

Exercise 4.11

If you wire a $22\,\text{k}\Omega$, a $10\,\text{k}\Omega$, and a $4.7\,\text{k}\Omega$ resistor in parallel, what is the resulting resistance? Draw the schematic diagram.

4.6 Power

Power is the derivative of energy with respect to time and is measured in watts (W). For direct-current (DC) circuits, the power dissipated in a component can be computed as the voltage across the component times the current through the component:

$$P = VI\ .$$

We speak of the power as being *dissipated*, because most of the time the energy is converted into heat.

For a simple resistor, we can apply Ohm's law, $V = IR$, to get power dissipated in a resistor:

$$P = V^2/R = I^2R\ .$$

We will look at alternating-current (AC) power in Sections 29.3 and 35.1.

Exercise 4.12

How much power is dissipated in a $330\,\Omega$ resistor that is connected across a $5\,\text{V}$ power supply?

Exercise 4.13

Many resistors are limited to only $0.25\,\text{W}$ of power. What is the largest $1/4\,\text{W}$ resistor that you can put $100\,\text{mA}$ of current through? What is the largest resistor in the E24 series you could use?

4.7 Hydraulic analogy

Many students find the discussion of voltage, current, and resistance too abstract, and so they have a hard time reasoning about circuits. There is a more concrete analogy that works well for some people to understand what is going on: the *hydraulic analogy*.

In this analogy, we are thinking about the flow of water through pipes

- *charge* is represented by volume of water in cubic meters,
- *current* is represented by the flow rate of water in m^3/s,
- *voltage* is represented by pressure difference in N/m^2, and
- *resistance* is a property of a length of pipe—pressure difference between the ends of a length of pipe divided by the flow rate for that pressure difference. If we increase the cross-sectional area of the pipe, we can increase the flow for a given pressure (reducing the resistance). Similarly, making the pipe longer increases the pressure difference needed for the same flow rate.

If we multiply voltage and current, we should get power, and indeed pressure times flow rate is power in units of Nm/s, which is watts.

We will revisit the hydraulic analogy again in Section 10.1, when we look at capacitors, and in Section 28.1, when we look at inductors.

Chapter 5

Voltage Dividers and Resistance-based Sensors

5.1 Voltage dividers

In Section 1.1, I promised three big topics. This section introduces the first of them: the voltage divider.

A *voltage divider* consists of a series connection of two resistances, as shown in Figure 5.1. A voltage is applied across the pair of resistors and measured across one of the resistors. We will be using voltage dividers in one form or another for almost every lab this quarter—they are a fundamental design tool for electronics.

The theory of voltage dividers is easily derived from *Ohm's law*: $V = IR$ (Section 4.3) and *Kirchhoff's current law*: in equilibrium, all the current into a node sums to 0 A (Section 4.2).

If there is no current through the output wire ($I_{\text{out}} = 0\,A$), then the current through R_{d} must be the same as the current through R_{u}, $I_{\text{d}} = I_{\text{u}} = I$, and we can compute the voltages across the two resistors:

$$V_{\text{in}} - V_{\text{out}} = IR_{\text{u}} \ , \qquad (5.1)$$

$$V_{\text{out}} - V_{\text{gnd}} = IR_{\text{d}} \ . \qquad (5.2)$$

Adding the two equations, we can see that $V_{\text{in}} - V_{\text{gnd}} = I(R_{\text{d}} + R_{\text{u}})$, which gives us the rule for resistances in series: they add.

> The resistance $R_{\text{d}} + R_{\text{u}}$ is sometimes called the *input resistance* of the voltage divider, because it expresses the relationship between the current through the voltage divider and the input voltage.

We can solve for the current: $I = \frac{V_{\text{in}} - V_{\text{gnd}}}{R_{\text{d}} + R_{\text{u}}}$, and plug that formula into the equation for V_{out} (Equation 5.2) to get the formula for the measured voltage:

$$V_{\text{out}} - V_{\text{gnd}} = (V_{\text{in}} - V_{\text{gnd}}) \frac{R_{\text{d}}}{R_{\text{d}} + R_{\text{u}}} \ ,$$

where R_{d} is the resistance across which the voltage $V_{\text{out}} - V_{\text{gnd}}$ is measured, R_{u} is the other resistance, and $V_{\text{in}} - V_{\text{gnd}}$ is the voltage across the two resistors in series.

You can visualize voltage dividers with similar triangles, as shown in Figure 5.2.

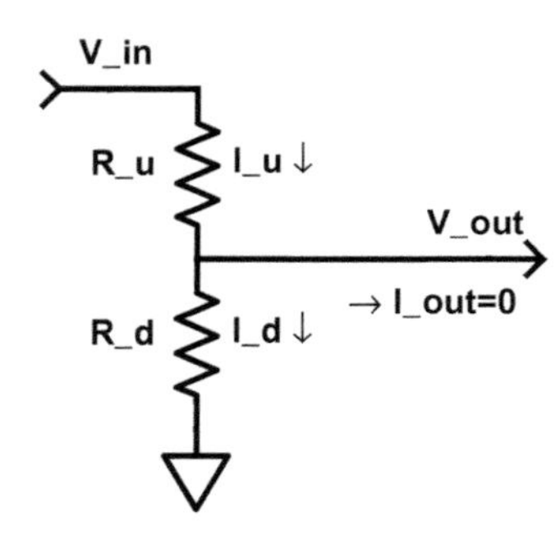

Figure 5.1: *Voltage divider circuit. If there is no current through the output wire, then the current through the two resistances is the same, and the output voltage can be easily calculated by using ratios. Reminder: the triangle pointing down is the symbol that we use for ground, the node that we measure other node voltages from. The triangle must always point down—triangles in other orientations do not mean ground.*
Figure drawn with Digi-Key's Scheme-it [18].

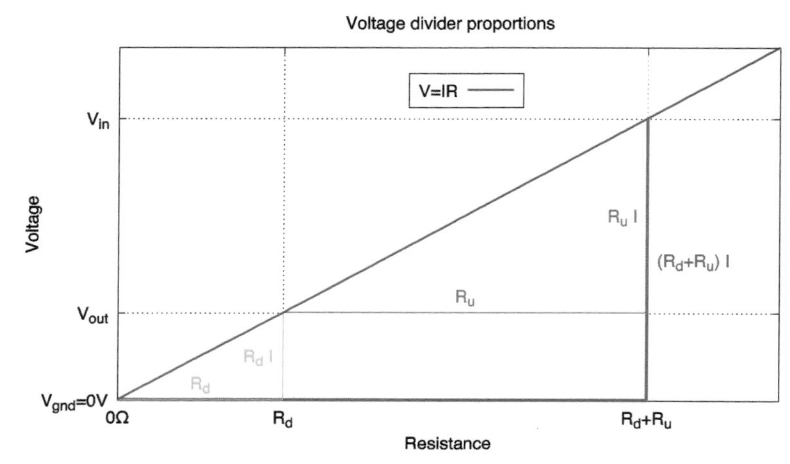

Figure 5.2: *The voltage-divider formula can be visualized using similar triangles. We have the same current I through both resistors, so we can draw the straight line with slope I for Ohm's law: $V = IR$. The voltage across each resistor is proportional to its resistance, and the voltage across the series chain is proportional to the sum of the resistances.*
Figure drawn with gnuplot [33].

The formula for voltage dividers is one of the few formulas worth memorizing in this course:

Voltage-divider criterion: no current out of the output port. When the criterion is met, the *gain* of a voltage divider is

$$\frac{V_{\text{out}} - V_{\text{gnd}}}{V_{\text{in}} - V_{\text{gnd}}} = \frac{R_{\text{d}}}{R_{\text{d}} + R_{\text{u}}} .$$

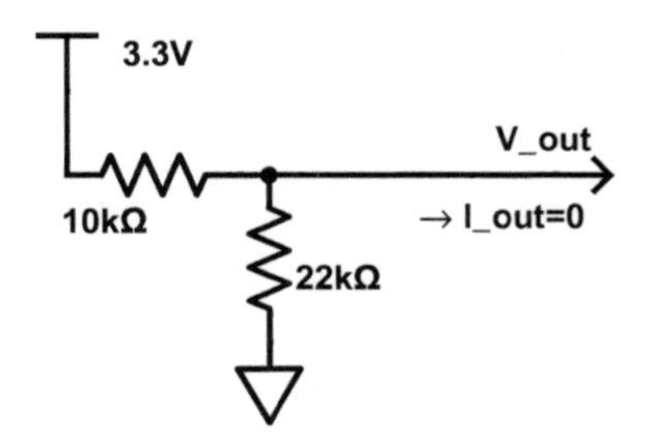

Figure 5.3: *Example of a voltage divider with specific resistances given.*
The orientation of resistors in a schematic does not matter—it is the connection between the resistors that matters, not whether they are drawn vertically or horizontally.
Figure drawn with Digi-Key's Scheme-it [18].

5.1.1 *Voltage divider—worked examples*

Let's work through a couple of examples of voltage dividers—one analyzing a divider, the other designing one.

Worked Example:
First, let's analyze the voltage divider in Figure 5.3. What is the voltage at V_{out}*? How much current flows through the resistors?*

Because no current flows through the output port, we can use the voltage-divider equation to compute

$$V_{\text{out}} = 3.3\,\text{V} \frac{22\,\text{k}\Omega}{22\,\text{k}\Omega + 10\,\text{k}\Omega} = 2.269\,\text{V} .$$

When writing down formulas that have numbers rather than variables, be sure to keep the units with numbers throughout the computation—one of the most common design errors is to mix different units together (like Ω and kΩ) to get numbers that are off by a factor of 1000 or more.

The current through the resistors is the same for both resistors and is most easily computed as

$$I_{\text{R}} = \frac{3.3\,\text{V}}{22\,\text{k}\Omega + 10\,\text{k}\Omega} = 103.1\,\mu\text{A} .$$

One could also compute it for just a single resistor,

$$\begin{aligned} I_{\text{R}} &= \frac{V_{\text{out}}}{22\,\text{k}\Omega} = \frac{2.269\,\text{V}}{22\,\text{k}\Omega} = 103.1\,\mu\text{A} \\ &= \frac{3.3\,\text{V} - V_{\text{out}}}{10\,\text{k}\Omega} = \frac{1.031\,\text{V}}{10\,\text{k}\Omega} = 103.1\,\mu\text{A} . \end{aligned}$$

Worked Example:
Now let's design a voltage divider. Let's say that we want to create a voltage reference of 1.8 V from a 3.3 V power supply, that we don't need to take any current from the 1.8 V reference, and that the circuit can't use more than 30 μA of current. How can we do that?

Because we don't need to take any current from the reference, we can use a voltage divider, similar to the one in Figure 5.3, but with different resistor values—let's call them R_u and R_d as in Figure 5.1. The resistors have to be large enough that the current through them is less than the specified $30\,\mu\mathrm{A}$, so

$$\begin{aligned} 30\,\mu\mathrm{A} &> \frac{3.3\,\mathrm{V}}{R_\mathrm{u}+R_\mathrm{d}} , \\ R_\mathrm{u}+R_\mathrm{d} &> \frac{3.3\,\mathrm{V}}{30\,\mu\mathrm{A}} = 110\,\mathrm{k}\Omega . \end{aligned}$$

The output voltage is defined by the voltage-divider equation:

$$\begin{aligned} 1.8\,\mathrm{V} &= 3.3\,\mathrm{V}\frac{R_\mathrm{d}}{R_\mathrm{u}+R_\mathrm{d}} , \\ R_\mathrm{u}+R_\mathrm{d} &= \frac{3.3\,\mathrm{V}}{1.8\,\mathrm{V}}R_\mathrm{d} = 1.833R_\mathrm{d} , \\ R_\mathrm{u} &= 0.833R_\mathrm{d} . \end{aligned}$$

If we limit ourselves to the standard resistor sizes in Table 4.2, we see that $820/1000 = 0.82$ is a ratio close to the 0.833 we want, but $680/820 = 0.8293$ is closer, and $150/180 = 0.8333$ is exactly the ratio we want.

If we choose $R_\mathrm{d} = 180\,\mathrm{k}\Omega$ and $R_\mathrm{u} = 150\,\mathrm{k}\Omega$, so that $R_\mathrm{d}+R_\mathrm{u} > 110\,\mathrm{k}\Omega$, then we have

$$\begin{aligned} V_\mathrm{out} &= 3.3\,\mathrm{V}\frac{180\,\mathrm{k}\Omega}{150\,\mathrm{k}\Omega+180\,\mathrm{k}\Omega} = 1.8\,\mathrm{V} , \\ I_\mathrm{R} &= \frac{3.3\,\mathrm{V}}{150\,\mathrm{k}\Omega+180\,\mathrm{k}\Omega} = 10\,\mu\mathrm{A} , \end{aligned}$$

which meets our voltage goal exactly and is well within our current limit.

Exercise 5.1

In Figure 5.1, if $V_\mathrm{in} = 5\,\mathrm{V}$, $R_\mathrm{d} = 1\,\mathrm{k}\Omega$, and $R_\mathrm{u} = 6.8\,\mathrm{k}\Omega$, what is V_out? Assume that there is no current through the V_out port. Remember that when a voltage is given for a single node, it means the difference in potential between that node and the specially marked "ground" node.

Exercise 5.2

Design a voltage divider (like the one in Figure 5.1) whose input is a 3.3 V power supply, and whose output is approximately 2.5 V. There will be no current taken from the output (so the voltage divider criterion is met). The current for powering the voltage divider (through R_u and R_d) should be no more than $100\,\mu\mathrm{A}$. Use standard resistance values from the E12 series. Report resistance values, the current, and the output voltage that is achieved. (As a general rule, whenever you round component values to standard values, you should recompute the effect of that choice and report the results—don't wait to be asked for them.)

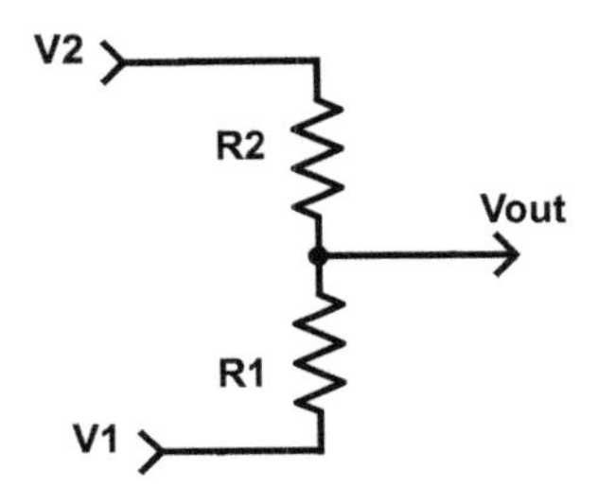

Figure 5.4: *For Exercise 5.3: what is the voltage V_{out}?*
Figure drawn with Digi-Key's Scheme-it [18].

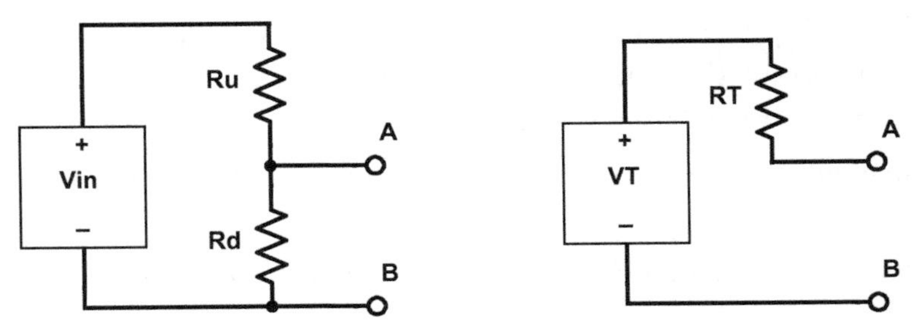

Figure 5.5: *The voltage divider on the left is equivalent to the simpler circuit on the right, with $V_T = V_{in}\frac{R_d}{R_u+R_d}$ and $R_T = \frac{R_u R_d}{R_u+R_d}$. The equivalence ignores the power lost as heat in the left-hand circuit: $V_{in}^2(R_u + R_d)$.*
Figure drawn with Digi-Key's Scheme-it [18].

Exercise 5.3

For the circuit in Figure 5.4, what is the voltage V_{out}? I don't want to know $V_{\text{out}} - V_1$ or $V_{\text{out}} - V_2$, but V_{out}, which is referenced to some ground node that is not shown in the schematic of Figure 5.4. For this exercise, assume that no current flows through the V_{out} connection.
Bonus question: Write a formula for V_{out} in terms of V_1, V_2, R_1, R_2, and I_{out}, where I_{out} is the current flowing out the output node to other (undrawn) circuitry. Do a "sanity check" that your formula simplifies to the formula for the first part of the exercise, if $I_{\text{out}} = 0\,\text{A}$.

5.1.2 *Thévenin equivalent of voltage divider*

There is a theorem in circuit theory, Thévenin's theorem [165], that says that any linear circuit with two terminals A and B, consisting of only voltage sources, current sources, and resistors, can be replaced by a single voltage source in series with a single resistor.

We won't prove that theorem in this class (that is routinely done in circuits classes), but we will show how to compute the Thévenin equivalent of a voltage divider, as shown in Figure 5.5.

If we take an external current, I_{AB}, from the terminal A to the terminal B, what is the voltage, V_{AB}, we would see between the terminals?

By Ohm's law, the current through the lower resistor is $I_d = V_{\text{AB}}/R_d$, and the current through the upper resistor is $I_u = (V_{\text{in}} - V_{\text{AB}})/R_u$. By Kirchhoff's current law, the current through the upper resistor is the sum of the currents through the lower resistor and the external current: $I_u = I_d + I_{\text{AB}}$.

Putting these three equations together gives us

$$(V_{\text{in}} - V_{\text{AB}})/R_{\text{u}} = V_{\text{AB}}/R_{\text{d}} + I_{\text{AB}} \ ,$$

which can be rearranged to

$$V_{\text{AB}}(1/R_{\text{d}} + 1/R_{\text{u}}) = V_{\text{in}}/R_{\text{u}} - I_{\text{AB}} \ ,$$

which can be further simplified to

$$V_{\text{AB}} = V_{\text{in}}\frac{R_{\text{d}}}{R_{\text{u}} + R_{\text{d}}} - I_{\text{AB}}\frac{R_{\text{u}}R_{\text{d}}}{R_{\text{u}} + R_{\text{d}}} \ .$$

What values do we need for the Thévenin-equivalent circuit on the right of Figure 5.5? If we set $V_{\text{T}} = V_{\text{in}}\frac{R_{\text{d}}}{R_{\text{u}}+R_{\text{d}}}$ and $R_{\text{T}} = \frac{R_{\text{u}}R_{\text{d}}}{R_{\text{u}}+R_{\text{d}}}$, we get exactly the same relationship between voltage and current for the voltage divider circuit as for the Thévenin-equivalent circuit.

The Thévenin-equivalent voltage, V_{T}, is just the voltage computed by the voltage-divider formula $V_{\text{T}} = V_{\text{in}}\frac{R_{\text{d}}}{R_{\text{u}}+R_{\text{d}}}$.

The Thévenin-equivalent resistance, R_{T}, is just the resistance of the two resistors, R_{u} and R_{d}, in parallel.

A common special case is when the two resistors are equal, $R_{\text{u}} = R_{\text{d}}$, then $V_{\text{T}} = V_{\text{in}}/2$ and $R_{\text{T}} = R_{\text{u}}/2$.

There is an easy way to determine what the voltage source and resistance are: the voltage source has whatever voltage you would see across the ports if they are not connected to any external circuitry (the *open-circuit voltage*), and the resistance is what you would get if all the voltage sources were replaced by short circuits and all the current sources by open circuits.

Another way to look at Thévenin equivalence for DC circuits is to consider the *open-circuit voltage* (the voltage with nothing connected to the two ports) and the *short-circuit current* (the current you would get through a wire connecting the two ports). The resistance is just the open-circuit voltage divided by the short-circuit current.

If we take no current externally from A to B, $I_{\text{AB}} = 0\,\text{A}$, then the voltage is V_{T}, which is called the *open-circuit voltage* for the Thévenin equivalent.

If we look at the current we get when V_{AB} is set to zero, we get $I_{\text{AB}} = V_{\text{T}}/R_{\text{T}}$, which is called the *short-circuit current*. The open-circuit voltage and short-circuit current are often easy to compute (or measure), and provide a quick way to determine the Thévenin-equivalent circuit.

For a voltage divider, the open-circuit voltage is the voltage we compute from the voltage-divider formula: $V_{\text{T}} = V_{\text{in}}\frac{R_{\text{d}}}{R_{\text{u}}+R_{\text{d}}}$, and the short-circuit current is $I_{\text{AB}} = V_{\text{in}}/R_{\text{u}}$.

Thévenin's Theorem also applies to AC voltage and current sources, using complex impedances rather than resistances.

Exercise 5.4

If you have a 5 V power supply, and make a voltage divider with a 3.3 kΩ resistor to 5 V and a 2.2 kΩ resistor to ground, what is the open-circuit output voltage? The short-circuit current? What is the Thévenin-equivalent resistance?

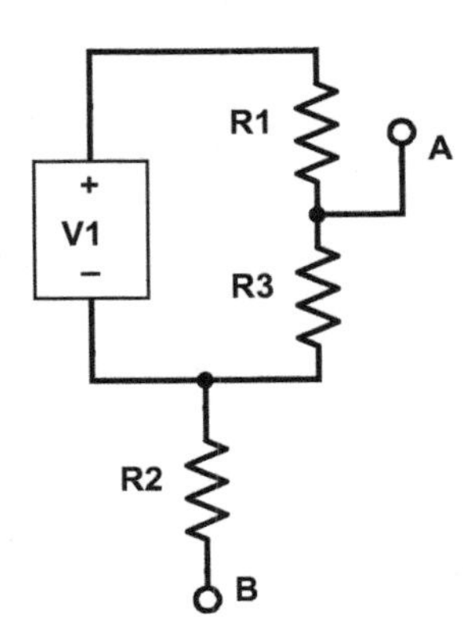

Figure 5.6: *Schematic for Exercise 5.5. The goal is to come up with an equivalent circuit between ports A and B, using only a voltage source and one resistor.*

Exercise 5.5

What is the open-circuit voltage V_{AB} in the schematic in Figure 5.6? If you add a wire from A to B, what is the short-circuit current through that wire? Come up with an equivalent circuit consisting of just one voltage source and one resistor (giving the voltage and resistance).

Exercise 5.6

Given a 3.3 V power supply, design a voltage divider that provides a Thévenin-equivalent voltage source with an open-circuit voltage of 2.1 V and an equivalent resistance of 1 kΩ. Round to the E12 series of resistors and report the resulting open-circuit voltage and equivalent resistance.

Exercise 5.7

For the resistor ladders in Figure 5.7, determine the Thévenin equivalent. Because there are no voltage or current sources, this will be just a single resistor between ports A and B. If we use R_n to designate the resistance of the circuit with n rungs, what is the value of R_{n+1}? Is there a limit as $n \to \infty$, and, if so, what is the limit?

5.1.3 *Potentiometers*

An adjustable voltage divider is called a *potentiometer*, sometimes abbreviated to *pot*. If the potentiometer is very small and is adjusted with a screwdriver, it is often called a *trimpot*. See Figure 5.8 for the schematic symbol for a potentiometer and photographs of a few types.

When a value is given for a potentiometer, it is the total resistance from one end to the other. There is a fixed resistor in the potentiometer between these points and a wiper arm that moves along the resistor, changing where the split between the "upper" and "lower" resistances is made. Turning the screw all the way one way moves the wiper arm to one end of the resistor, and turning the screw all the way the other way moves the wiper arm to the opposite end. Some trim pots have a little "CW" mark next to one of the end pins, indicating that the wiper arm is at that end when the screw is turned all the way clockwise.

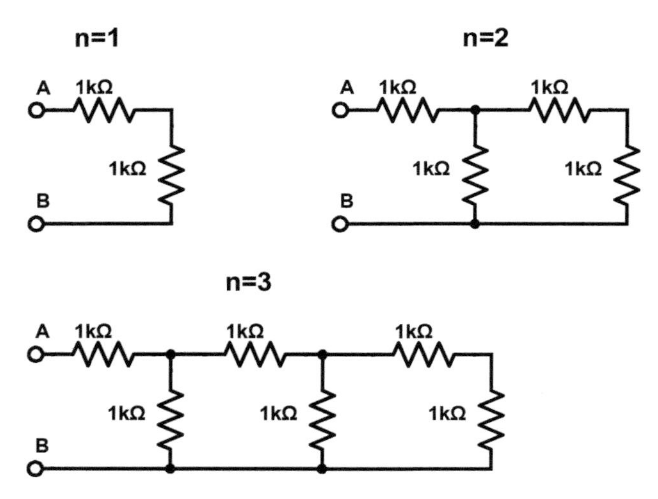

Figure 5.7: *These resistor ladders are for Exercise 5.7. The goal is to use Thévenin equivalence to figure out the resistance between ports A and B.*

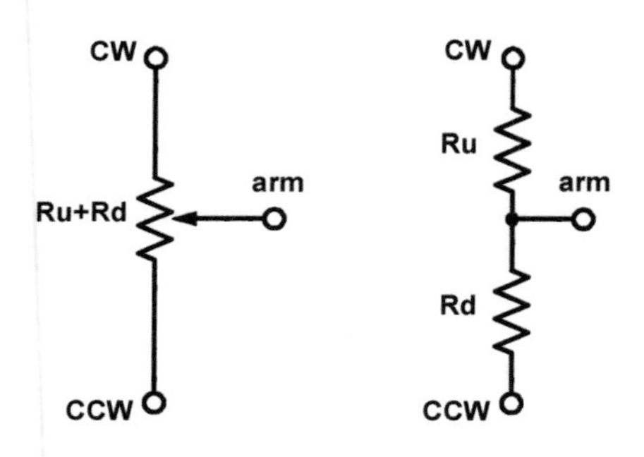

(a) *Schematic symbol for a potentiometer and equivalent circuit as two resistors.* R_u *is* $0\,\Omega$ *when the potentiometer shaft is fully clockwise.* R_d *is* $0\,\Omega$ *when the potentiometer shaft is fully counterclockwise. Figure drawn with Digi-Key's Scheme-it [18].*

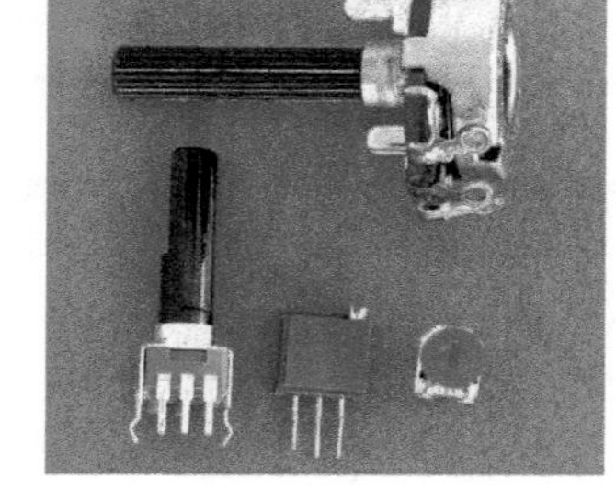

(b) *Potentiometers come in many shapes and sizes. Small ones that need a screwdriver for adjustment (like the two on the bottom row on the right) are referred to as* trimpots.

Figure 5.8: *Potentiometers are adjustable voltage dividers. There is a fixed total resistance from one end to the other,* $R_u + R_d$, *and an adjustable wiper arm or tap that determines how much of the resistance is on each side of the output. A potentiometer in a schematic diagram should always be labeled with the total resistance.*

A lot of trimpots have a worm gear inside converting a turn of the screw to a small change in the wiper arm position. These multi-turn trimpots allow very precise setting, but take a long time to sweep from one end to the other. For Lab 9, we'll use a trimpot to set the gain of an amplifier.

Sometimes we don't need a full voltage divider, but only a variable resistor. We can use a potentiometer for this purpose, by using the wiper arm and one of the end connections. The other end can either be left not connected, or connected to the wiper arm, as shown in Figure 5.9.

If we use a potentiometer as a voltage divider, then the range of voltages on the wiper arm goes all the way from the voltage at one terminal to the voltage at the other terminal. Quite often we want an adjustable voltage divider, but one with a smaller range. If we put fixed resistors in series with a potentiometer, then the output range of the voltage divider will be limited.

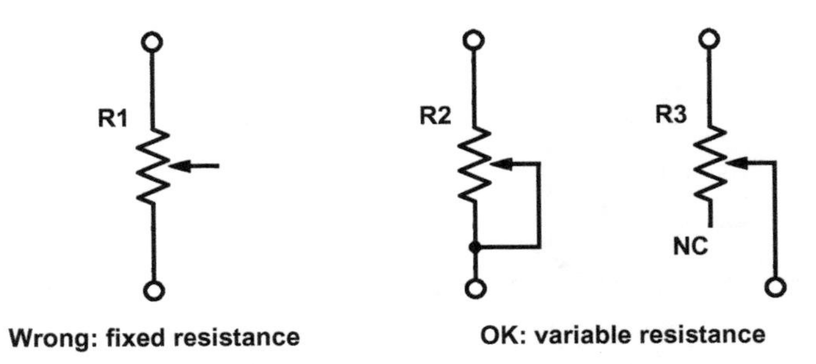

Figure 5.9: *A potentiometer can be used as a variable resistor, by connecting to the wiper arm and one of the end points. If the other end is left unconnected, it should be labeled "NC" on the schematic, to indicate that it is deliberately not connected, rather than the schematic being incomplete. The traditional symbol for a variable resistor (a resistor with a diagonal arrow across it) is not currently provided by Scheme-it.*
Figure drawn with Digi-Key's Scheme-it [18].

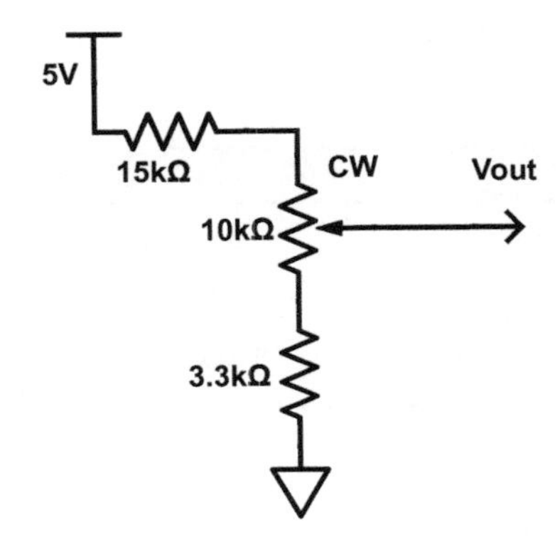

Figure 5.10: *A potentiometer with resistors in series can be used to limit the range of the gain of the voltage divider.*
Figure drawn with Digi-Key's Scheme-it [18].

Consider the example in Figure 5.10. The total resistance is $15\,\mathrm{k}\Omega + 10\,\mathrm{k}\Omega + 3.3\,\mathrm{k}\Omega = 28.3\,\mathrm{k}\Omega$, so the current through the resistors is

$$I_{\mathrm{R}} = \frac{5\,\mathrm{V}}{28.3\,\mathrm{k}\Omega} = 176.7\,\mu\mathrm{A}\ ,$$

assuming that no current flows through the output port V_{out}.

If the potentiometer is adjusted all the way clockwise, then the wiper arm is at the top of the potentiometer symbol, and the resistors of the voltage divider are $15\,\mathrm{k}\Omega$ and $13.3\,\mathrm{k}\Omega$, giving an output voltage of

$$V_{\mathrm{out}} = 5\,\mathrm{V}\frac{13.3\,\mathrm{k}\Omega}{28.3\,\mathrm{k}\Omega} = 2.35\,\mathrm{V}\ .$$

If the potentiometer is adjusted all the way counterclockwise, then the wiper arm is at the bottom of the potentiometer symbol, and the resistors of the voltage divider are $25\,\mathrm{k}\Omega$ and $3.3\,\mathrm{k}\Omega$, giving an output voltage of

$$V_{\mathrm{out}} = 5\,\mathrm{V}\frac{3.3\,\mathrm{k}\Omega}{28.3\,\mathrm{k}\Omega} = 583\,\mathrm{mV}\ .$$

Any output voltage between $583\,\mathrm{mV}$ and $2.35\,\mathrm{V}$ can be obtained by adjusting the position of the wiper arm appropriately.

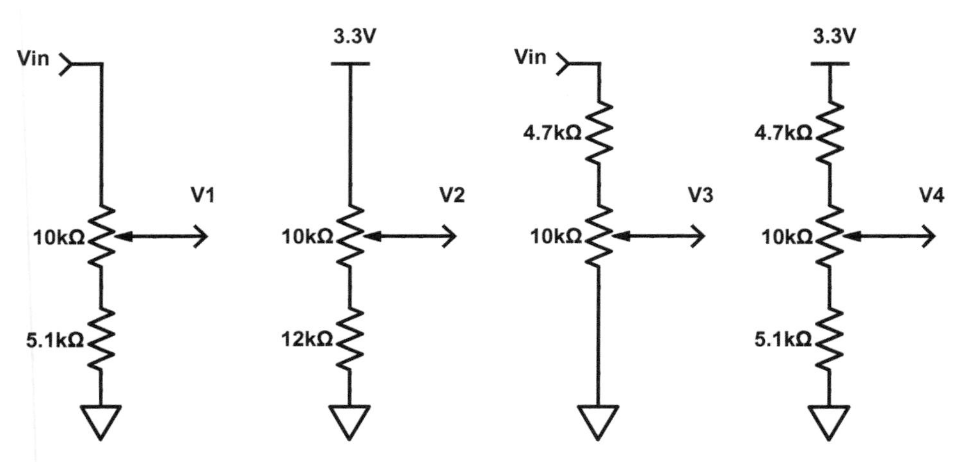

Figure 5.11: *Potentiometers can be used to make adjustable voltage dividers that have a gain range less than the full 0–1 range of a voltage divider made with just a single potentiometer by adding series resistors. See Exercise 5.8.*
Figure drawn with Digi-Key's Scheme-it [18].

Exercise 5.8

For each of the potentiometer circuits in Figure 5.11, figure out the voltage range for the outputs (V_1 through V_4), assuming that there is no current through the output. V_1 and V_3 will be functions of V_{in}, but V_2 and V_4 will have absolute voltage ranges.

5.1.4 *Summary of voltage dividers*

A *voltage divider* consists of a pair of resistances connected in series. An input voltage is applied across the pair, and the voltage across one of the resistors is used as the output.

As long as the current through the output is negligible compared to the current through the resistors, the ratio of the output voltage to the input voltage is the same as the ratio of the resistances that the voltage is across:

$$\text{gain} = \frac{V_{\text{out}}}{V_{\text{in}}} = \frac{R_{\text{d}}}{R_{\text{d}} + R_{\text{u}}} .$$

We'll use the voltage divider concept again in Chapter 11, using impedances rather than resistances.

5.2 Thermistors

A thermistor is a semiconductor device whose resistance varies with temperature, drawn in schematics as shown in Figure 5.12. Most thermistors are *negative thermal coefficient* devices, which means that the resistance decreases as the temperature increases (unlike most metals, which have a positive thermal coefficient of resistance, increasing resistance as temperature increases).

There are various formulas used for estimating the resistance at a particular temperature (or, equivalently, the temperature for a particular resistance).

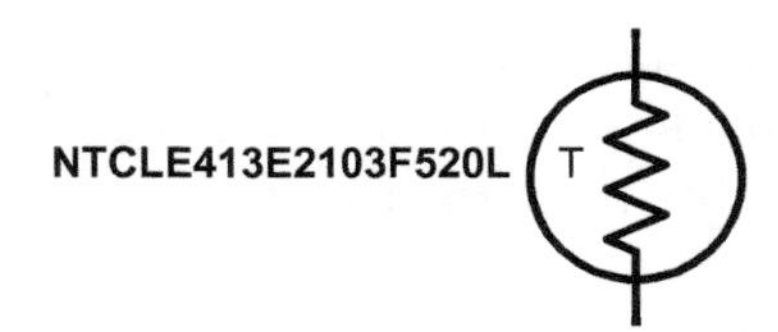

Figure 5.12: *In a schematic, the symbol for a thermistor (or RTD) is a resistor with a "T" next to it for "temperature", usually included in a circle. Put a part number with the symbol, as different thermistors have quite different characteristics, even if they have the same nominal resistance at 25°C.*
Figure drawn with Digi-Key's Scheme-it [18].

The simplest such formula, used on many specification sheets for thermistors, is the B equation:

$$\frac{1}{T} = \frac{1}{T_0} + \frac{1}{B} \ln\left(\frac{R}{R_{T_0}}\right),$$

where T is the temperature in kelvin, R is the resistance, B is a parameter that depends on the material used in the thermistor, and R_{T_0} is the resistance at some calibration temperature T_0. We can also write this as

$$R = R_{T_0} e^{B(1/T - 1/T_0)}$$

or

$$R = R_\infty e^{B/T},$$

where R_∞ is the projected "resistance at infinity", $R_\infty = R_{T_0} e^{-B/T_0}$.

For example, the thermistor we will use in this lab (NTCLE413E2103F520L) has a data sheet at https://www.vishay.com/docs/29078/ntcle413.pdf. (In future labs we won't be providing URLs for the data sheets—you'll be expected to look for, download, and read the data sheets without being specifically instructed to do so. We will help you try to understand the most important parts of each data sheet.)

The thermistor data sheet unpacks the part number into several different fields:

NTC A negative temperature coefficient thermistor
L Leaded—that is, the device has wires coming out of it (unlike some surface mount devices that are soldered directly to a printed-circuit board, for example)
E413 describes the wires and encapsulation: 30-gauge wires with the thermistor encapsulated in PVC and epoxy, rated for a maximum of 105°C. So this thermistor would not be suitable for measuring the temperature in an autoclave.
E2 Tin-alloy on the wires.
103 A nominal resistance of 10 kΩ at 25°C. We'll see a lot of parts labeling where a number like 103 should be interpreted as 10.E3 or 10×10^3.
F The nominal resistance should be accurate to ±1%
520 The leads are 52 mm long.
L The B-value is "low", between 3000 K and 3500 K.

Further down in the data sheet, they give the $B_{25/85}$ value as 3435 K. This is the B-value that they got from a pair of measurements: the resistance at 25°C and the resistance at 85°C. They also provide a table of measured or calculated resistances (they don't say which) for the 10 kΩ 1% low-B device we are using. You are not to use this table in the lab, but to make your own independent measurements—though you can check against this table to see if the measurements you are making are close to the specs.

The B equation is an adequate approximation if you are keeping fairly close to the reference temperature, but using just 2 measurements to estimate B is a bad idea, particularly if you plan to use the thermistor at lower temperatures than either measurement. If you want to use the B equation, you need to fit the best values for B and R_{T_0} (or R_∞, the extrapolated resistance at infinite temperature) for a number of measurements across the temperature range you are interested in. When I did this lab at home, I made 63 measurements, but that may be overkill—10 measurements should be plenty, as long as they span the full range of interest and are made fairly accurately.

There is a better formula for approximating the relationship between thermistor resistance and temperature: the Steinhart-Hart equation [161]:

$$\frac{1}{T} = A_0 + A_1 \ln(R) + A_3(\ln(R))^3 \ .$$

It is much messier to invert the Steinhart-Hart equation to get R in terms of T, if you need to do that, but having an extra parameter allows much closer fitting of the temperature vs. resistance curve. Read the Wikipedia articles on thermistors and the Steinhart-Hart equation for more background [161, 164].

In Figure 5.13, I have plotted several responses for possible implementations of the thermistor circuit, using different resistance values for the fixed resistor that allows us to convert the thermistor reading into a voltage.

The *sensitivity* of a sensor circuit is the change in output corresponding to a change in input—that is, the derivative of the output with respect to the input. The curves of Figure 5.13 are *sigmoidal*, with very little slope at low or high temperatures (so sensitivity goes to 0 at low and high temperatures), but

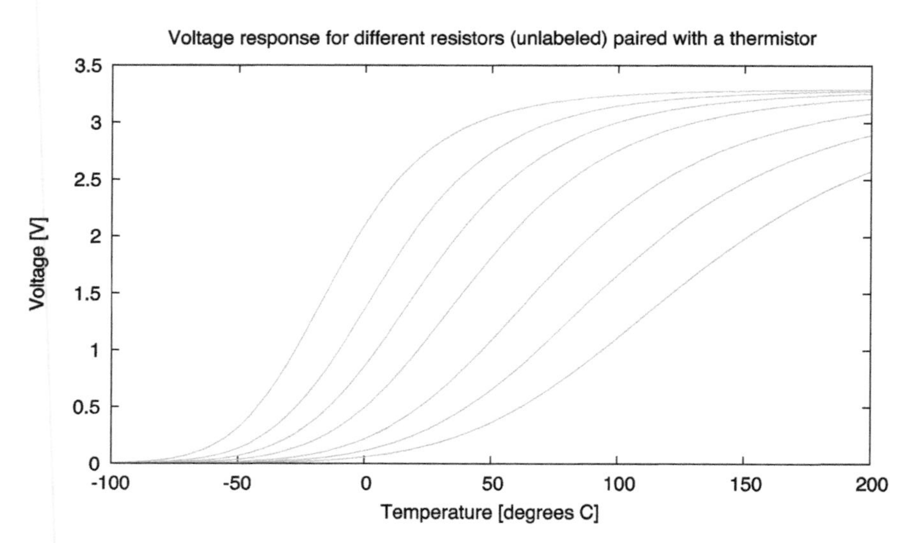

Figure 5.13: *The voltage as a function of temperature is shown here for a circuit with a 3.3 V power supply and different resistors paired with a thermistor in a voltage divider (see Figure 5.1).*
Figure drawn with gnuplot [33].

a fairly steep slope in the middle. Changing the resistor in the design changes where that steepest part occurs. Picking the wrong resistor can result in a very low sensitivity at the operating temperature.

To make the circuit maximally sensitive at some operating temperature, T_{op}, we would want to maximize the formula for sensitivity (derived in Pre-lab 2.5). For a given thermistor, we have only one parameter in the circuit that we can change: the fixed resistor R that we use to convert the thermistor resistance to a voltage. Remember that to maximize a function $f(T)$ we can take the derivative and set it equal to 0: $\frac{\mathrm{d}f}{\mathrm{d}T} = 0$. Because our function is the sensitivity $f(T) = \frac{\mathrm{d}V_{\text{out}}}{\mathrm{d}T}$, the function we want to set equal to 0 is $\frac{\mathrm{d}f}{\mathrm{d}T} = \frac{\mathrm{d}^2V_{\text{out}}}{\mathrm{d}T^2}$.

The second derivative is a measure of how nonlinear the function is at a point (how concave or convex the curve), so that setting the second derivative to 0 means not only that the derivative is maximized, but also that the formula for voltage is well approximated as a straight line as a function of T at T_{op}.

5.3 Other temperature sensors

Thermistors are the most popular electronic temperature sensors, but they are not the only ones. Three other sensor types are fairly common: Resistance-Temperature Detectors (RTDs), diode-voltage temperature sensors, and thermocouples.

> RTDs are temperature-dependent resistors, like thermistors, but with a positive thermal coefficient (increasing resistance with temperature), not a negative thermal coefficient. Their main advantages are that they can withstand higher temperatures (up to 500°C, depending on the materials used to encase them), they have a very linear response, and they are available with fairly tight tolerances, providing accurate temperature readings.

Moderately priced RTDs can have an accuracy of 0.1% and a fairly wide temperature range, making them popular for lab instruments. They are not as common as thermistors—as of 2015 March 23, Digi-Key distributed 5027 different thermistors and only 60 RTDs—and are less sensitive than thermistors (that is, their output changes less for a given change in temperature), more expensive, and slower to respond to temperature changes.

The resistance of RTDs increases as temperature rises: $R_{\text{T}} = (1+\alpha T)R_0$, where T is the temperature in °C, R_{T} is the resistance at T, R_0 is the resistance at 0°C, and α is the temperature coefficient. The current standard calls for $\alpha = 0.00385/°\text{C}$, which is achieved by adding carefully controlled impurities to platinum wire. A nominal 100 Ω RTD would have a change in resistance of 0.385 Ω/°C.

When higher accuracy is needed, RTD resistance can be better approximated by a higher-order polynomial in T, rather than a linear one. The Callendar-Van Dusen equation is one such better approximation (a cubic approximation), and international standards organizations use 9th- or 12th-order polynomials [110].

> When wider temperature range is needed than an RTD can provide, then thermocouples are a popular choice. Properly speaking, a thermocouple does not measure *temperature*, but *difference in temperature* between two junctions of dissimilar materials.

A K-type thermocouple has a junction between chromel and alumel as the sensing junction, and junctions between the chromel and copper wires and alumel and copper wires as reference junctions (see Figure 5.14).

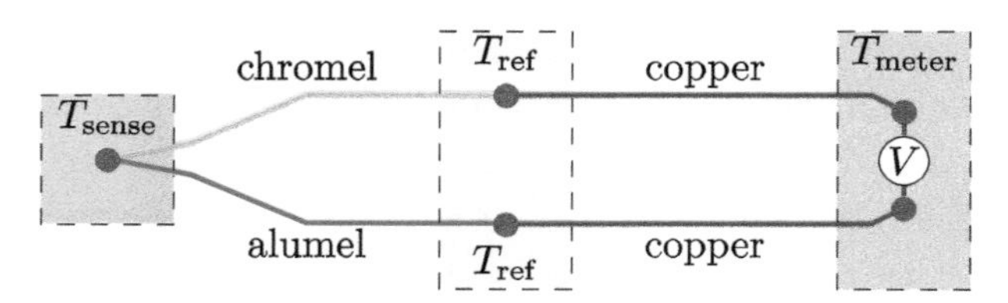

Figure 5.14: *Diagram of a type K thermocouple, showing the sensing and reference junctions. The voltage at the output is produced by integrating the voltages produced by the Seebeck effect in the alumel and chromel wires, which are in turn proportional to the gradient of the temperature along the wires. The difference in material of the wires, hence differences in the Seebeck effects, results in a voltage difference that is proportional to the difference in temperature between the sense and reference junctions. The Seebeck effects in the copper wires between the meter and the reference junctions cancel each other, since the wires are made of the same material and have the same temperature differences.*

Figure copied from https://en.wikipedia.org/wiki/File:Thermocouple_circuit_Ktype_including_voltmeter_temperature.svg distributed under Creative Commons 0 version 1.0 Universal Public Domain license.

The *Seebeck effect* causes a voltage to develop along the alumel and chromel wires based on the gradient of the temperature. Because the alumel and chromel wires are made of different materials with different Seebeck coefficients, the accumulated voltage differences from the sensing junction to the reference junctions don't cancel, and we can measure a voltage between the two reference junctions.

The voltages produced by thermocouples are small (a few millivolts), so care must be taken to use a very high impedance voltmeter to measure the voltage, lest the variation in resistance of the wires due to temperature swamp out the signal due to the change in voltage.

The two reference junctions (alumel-copper and chromel-copper) have to be at the same known temperature, which is usually managed by having them very close to each other and using another sensor, perhaps an RTD, to measure the temperature there.

Although thermocouples have a wide temperature range (generally up to about 1000°C), they are not particularly accurate—it is difficult to measure temperatures to better than ±1°C with thermocouples. Bioengineers rarely have a need for the wide temperature range of thermocouples and usually need more accuracy than they provide, so we won't discuss thermocouples further in this book.

Another common temperature sensor relies on the temperature sensitivity of the voltage across a semiconductor diode at a fixed current. (Semiconductor diodes are explained in Section 22.1—basically they are parts that conduct in one direction when the voltage is sufficiently large, and don't conduct in the other direction.) These *diode temperature sensors* are commonly included in electronic parts such as microcontrollers, because they are cheap to add and can be used to monitor the temperature of the electronics.

The sensors are sometimes referred to as *silicon band gap temperature sensors*, because the formula for the voltage across the diode is a function of the band-gap voltage—a quantum-mechanical property of the semiconductor material. Quantum mechanics is way beyond the scope of this class, and we won't need the concept of band gaps for any of the electronics we do here.

The "band-gap" nomenclature for diode temperature sensors is a little misleading, because one method for using the effect uses two diodes with different amounts of current through them and measures the difference between their voltages, which cancels out the band-gap voltage. The difference in voltages is $\frac{KT}{q} \ln\left(\frac{I_1}{I_2}\right)$, where K is Boltzmann's constant (81.73324(78) μeV/K), T is temperature in K, and q is the charge of an electron. For reasonable ratios of currents, the voltage difference is around $100\,\mu$V/K. This is usually amplified to get a more useful voltage range.

When less accuracy is needed, it is more common just to pass a known current through the diode and measure the voltage across the diode, relying on the approximately -2 mV/K temperature dependence of that voltage. The PteroDAQ data-acquisition system provides access to the temperature sensor built into the microcontroller on the Teensy boards—the temperature read is the temperature of the processor chip (which may be much warmer than the surrounding air). The temperature sensors in the Teensy LC (KL26 processor) and the Teensy 3.1 and 3.2 boards (K20 processor) claim (716 ± 10) mV at 25°C and (1.62 ± 0.07) mV/°C [28, 29]. At the maximum junction temperature of 125°C, the voltage is nominally 878 mV, but could be as high as 895 mV (a 2% variation in voltage or a 10.5°C error in temperature). Cheap standalone diode temperature sensors are usually specified as ±2°C and expensive ones as ±0.2°C.

Although diode-based temperature sensors are fairly cheap, they are not particularly suitable for most bioengineering applications—thermistors and RTDs remain the most suitable devices, with the choice depending on whether sensitivity or accuracy is the most important property.

5.4 Other resistance sensors

There are several other types of resistance-based sensors besides thermistors. Some of the most important ones for bioengineers include

Photoresistors As the name implies, *photoresistors* are sensitive to light, decreasing in resistance as the light intensity increases. (It might be better to have named them "photoconductors", to get the relationship between light and conductivity correct.) Photoresistors are commonly used as ambient light sensors, because they can be made with a spectral sensitivity that matches human perception, but they are not as sensitive or as fast to respond as photodiodes or phototransistors, so don't see as wide-spread use. We will look at phototransistors in Chapter 22 and Lab 6.

Piezoresistors Piezoresistors change resistance due to mechanical strain on the material. Almost all metals will exhibit some change in resistance due to simple geometric considerations: $R = \rho l/A$, where R is the resistance of the device, ρ is the resistivity of the material, l is the length, and A is the cross-sectional area, but semiconductors may have a much larger change in resistance than any simple geometric effect. Piezoresistance is generally expressed as the change in resistivity as strain increases:

$$\rho_\sigma = \frac{\left(\frac{\partial \rho}{\rho}\right)}{\left(\frac{\partial l}{l}\right)} .$$

For this class, the only piezoresistors we will use are silicon ones built into the strain-gauge pressure sensors of Chapter 20 and Lab 5.

Chemiresistors A device whose resistance changes according to the chemical environment it is in is called a *chemiresistor*. Some popular ones for hobbyists include alcohol sensors, carbon monoxide sensors, methane sensors, hydrogen sensors, ... Many chemiresistors rely on adsorption to a ceramic surface, and require heating to drive off moisture and contaminants before making a measurement.

We will not be using chemiresistors in this course.

5.5 Example: Alcohol sensor

Cheap sensors for detecting breath alcohol are resistance-based sensors. The chemiresistors in alcohol sensors usually use a tin-oxide (SnO_2) semiconductor film on a ceramic (like Al_2O_3). The tin-oxide material behaves as an n-type semiconductor, because of a slight deficiency of oxygen atoms in the oxide, resulting in tin atoms acting as electron donors. When the material is heated, oxygen molecules are adsorbed on the surface binding some of the free electrons in the tin oxide, and making its resistance higher. If a reducing gas is then adsorbed, it will either interact with the oxygen or release electrons itself, making more charge carriers available and reducing the resistance [103].

The sensors consist of two parts:

(1) a heater resistor to raise the temperature of the sensor and
(2) a chemiresistor whose resistance varies with the alcohol concentration in the air.

Most sensors are designed now to reach the correct temperature when a specified voltage is applied to the heater resistor. The heater generally has to run for a minute or two before the sensor is up to temperature to take a measurement. A longer "burn-in" time is often specified—this is only required on the first use or when a sensor has been stored for a while, to remove any contaminants from the surface—not before every measurement.

The chemiresistor's resistance can be measured with a voltage divider, as we use for thermistors, but you can get somewhat better linearity if you put a constant voltage across the chemiresistor and measure the current with a transimpedance amplifier (see Chapter 23).

The data sheets for MQ3 [40] and MQ303A [75] alcohol sensors give calibration curves in ppm (parts per million by volume), which is a good unit for measuring gases, as the ratio of the number of gas molecules does not vary with temperature or pressure (unlike units like g/L, which are very dependent on pressure).

The chemiresistors are also sensitive to temperature and humidity and their characteristics change with age and storage conditions, so this style of alcohol detector is usually used only for "toy" breathalyzers, not for professional ones. Professional breathalyzers use fuel-cell technology, which is more expensive, but more accurate.

The drunk-driving laws in the USA are mainly based on blood alcohol concentration (BAC) defined as weight/volume ratios, so that a BAC of 0.04% means 400 mg/L, which with ethanol's molecular mass of 46.06844 g/mol is 8.683 mM. Even if a good measurement is made of alcohol on the breath, the blood alcohol concentration can vary quite a bit.

Standard calibration gases for breathalyzers in the USA use 104.2 ppm (ethanol in nitrogen) as the standard for 0.04% blood alcohol, which is based on a legal standard that assumes that there are 2100 μg/L of alcohol in the blood for each μg/L of alcohol in the breath. (The 2100:1 ratio is known as the *partition ratio*.)

We can check that this calibration is reasonable, by assuming that exhaled breath is about 35°C. At 35°C, there are $\frac{273.15}{(273.15+35)\cdot 22.4}$ mol/L, so at 104.2 ppm, there would be 4.126 μmol/L of ethanol or 190 μg/L in the breath. With the assumed partition ratio of 2100, this gives 398.9 μg/L—the 0.04% BAC.

Breath measurement of alcohol is not a very precise measurement of blood alcohol, as the partition ratio for individuals varies by as much as 20%, but it is a cheap, non-invasive measurement.

5.6 Block diagram

A big part of engineering design is dividing problems into subproblems that are more easily solved. It often helps to have graphical ways of explaining this subdivision, and of succinctly communicating how the subproblems are related. *Block diagrams* are a popular way of conveying this information in electronics—they will be covered in more detail in Section 7.10.2.

Figure 5.15 shows a simple two-block block diagram for the system you will design in Lab 2. Each rectangle corresponds to a different subsystem (or *block*) of the overall system; each line to a wire

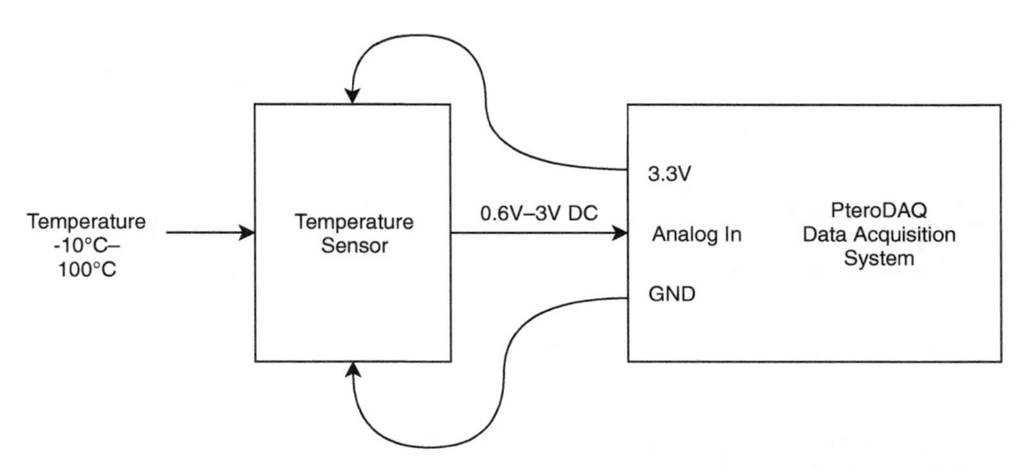

Figure 5.15: *Simple block diagram for a temperature-measuring system. The first block is the sensor, which converts the temperatures that are input to voltages, which are measured and recorded by the data-acquisition system (the second block).*
Figure drawn with draw.io [21].

or other connection carrying information (a *signal*) or power between blocks. A single arrow may represent several wires or no wires, as it is intended to show information transfer, not flow of electric current.

Information flow is generally left-to-right in a block diagram, with inputs on the left and outputs on the right. Power connections are often put on the tops or bottoms of blocks, rather than as inputs. The 3.3 V power from the data-acquisition system (DAQ) is provided to the temperature sensor, and an arrow is used to show that this connection is the reverse of the normal left-to-right flow.

I have also shown the ground connection, which is the reference node for both the 3.3 V power signal and the voltage output from the sensor, but it is common practice to omit such low-level details in block diagrams. (In schematic diagrams, which are at a finer level of detail, omitting the ground connection would be a serious error.)

Connections between blocks may be named and may have additional information about how information is encoded in the connection. In Figure 5.15, the input signal is given as a temperature in the range −10°C to 100°C, and the connection between the blocks as a voltage approximately in the range 0.6 V–3 V.

The temperature range is a design constraint on the system—we want to design a temperature-recording system that can handle those temperatures. This particular temperature range is one that works well for thermistors—if we needed to measure much higher temperatures, we would probably need to use an RTD or even a thermocouple, as thermistors generally are limited to −40°C to 125°C.

The voltage range was selected based on the data-acquisition system we use, the PteroDAQ system with a Teensy microcontroller board. These boards use a 3.3 V processor, and the analog inputs for analog-to-digital conversion are limited to being in the range 0 V–3.3 V (with the voltage measured relative to the ground pin on the processor).

The information in the block diagram is not a complete specification for the sensor block, the DAQ block, or the signal between them. We have not specified the precise mapping between temperature and voltage for the sensor, nor specified how much precision the DAQ needs to measure the voltage with nor how often measurements need to be made.

A block diagram is not a complete specification of the whole system, nor of the interfaces between the subsystems, but gives the engineer an idea how to decompose the problem in separate subproblems: here separating the conversion from temperature to voltage from the recording of the signal in a computer.

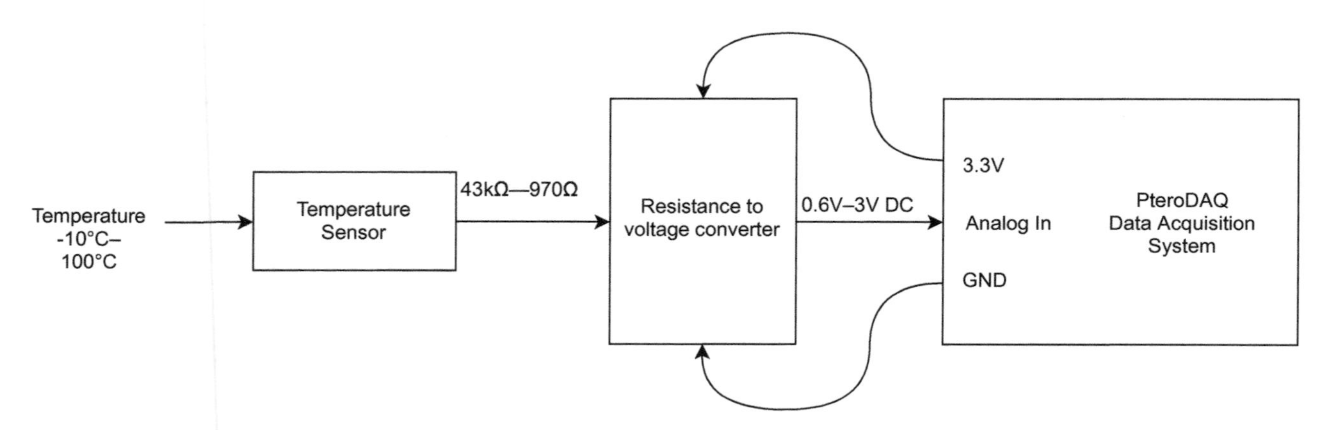

Figure 5.16: *A more detailed block diagram of the system in Figure 5.15, expanding the temperature-to-voltage sensor into a temperature-to-resistance sensor, followed by a resistance-to-voltage converter. Each information-carrying arrow is labeled with information about the type of data and the range, so that each block can be designed looking just at its connections.*
Figure drawn with draw.io [21].

We can subdivide the block diagram further: the temperature sensor we have does not produce a voltage directly, but changes the resistance of a resistor. So we can break up the sensor block in Figure 5.15 into two blocks: a temperature-to-resistance sensor block and a resistance-to-voltage conversion block, as shown in Figure 5.16.

Chapter 6

Signals

6.1 Signals

A *signal* is any physical value that varies with time. An *electrical signal* is a signal in which the physical property is an electrical property like voltage. Sensors are devices that convert signals in some physical property to electrical signals (generally voltage, current, resistance, capacitance, or inductance).

Most of this book is concerned with further processing of the electrical signals from sensors into voltage signals that can be converted into sequences of numbers by an analog-to-digital converter. (For more information on interpreting the sequences of numbers, see Chapter 9.)

We often describe signal in various summary forms, to provide concise descriptions of the important aspects of the signal. Try not to confuse the signal with its concise description—the average value, amplitude, or frequency of a signal V_{sens} is not the same thing as the signal itself. Often, when we want to be clear that we are talking about the signal, and not some property of the signal, we write it as a function of time: $V_{\text{sens}}(t)$.

Section 6.2 talks about measuring voltage, Section 6.3 talks about properties of time-varying voltage signals (though most of the properties can be applied to any signal), and Section 6.5 talks about data-acquisition systems used for recording signals in digital form.

6.2 Measuring voltage

Let's review what voltage is. A physicist might say that voltage is the work done to move a positive charge between two points in an electric field, with units of joules per coulomb. To simplify notation, we use the term *volt* for a joule per coulomb.

In this class, we're mainly interested in charges moving through wires, resistors, semiconductors, and other electronic circuitry, so we won't be looking much at electric fields in three dimensions. Instead, we'll be looking at more abstract models represented by *schematic diagrams*. These models have two types of objects: *nodes* and *components*. The nodes are represented by lines in the schematic diagram—every part of the line is modeled as having exactly the same electrical potential. Components have two or more connections to nodes (sometimes referred to as the *ports* of the component).

Voltage in circuits is the difference in electrical potential between two nodes of the circuit. It is always a difference, and in this book I'll try to speak of it that way, but there is a convention among electronics engineers to talk about the voltage *of* a node or *at* a node. This convention relies on defining some node of the circuit as 0 V (usually called *ground* and abbreviated *gnd*). So if someone talks about the voltage at the output node as being V_{out}, what they mean is that voltage difference between the output node and the ground node is V_{out}. Initially, I'll write this as $V_{\text{out}} - V_{\text{gnd}}$, but after a while I'll simplify and leave out the $-V_{\text{gnd}}$.

You can think of nodes as wires connecting components, but remember that the nodes are a modeling convenience, not a real-world object. The same piece of wire may sometimes be modeled as a node (everything on the wire is treated as being at the same potential), as a resistor (one end of the wire may be at a different voltage from the other), an inductor, or even a more complicated circuit with distributed inductance, resistance, and capacitance.

Simple linear circuit elements (resistors, capacitors, inductors, voltage sources, and current sources) are always two-port components. They must have both ports connected to nodes to be part of a circuit. We often talk about the *voltage across a component*, meaning the difference in voltage between the two ports.

Writing hint: be careful with your prepositions. We can talk about the voltage difference *between* two nodes, the voltage *at* or *of* a node (meaning the voltage difference between that node and ground), or the voltage *across* a component (meaning the voltage difference between the two nodes that are the ports of the component). It is never correct to talk about the voltage *through* a component.

When we make voltage measurements with a meter or data-acquisition system, there must always be two connections between the circuit being measured and the measuring device, so that we can take the difference in electrical potential between two nodes of the circuit. Figure 6.1 shows how we draw schematics to represent the voltmeter connection.

It is quite common when taking multiple measurements to designate one node as a common ground for the measuring instrument, and to take all voltage measurements with respect to that node. This is referred to as *single-ended* measurement, and it is the most common method for oscilloscopes and data-acquisition systems.

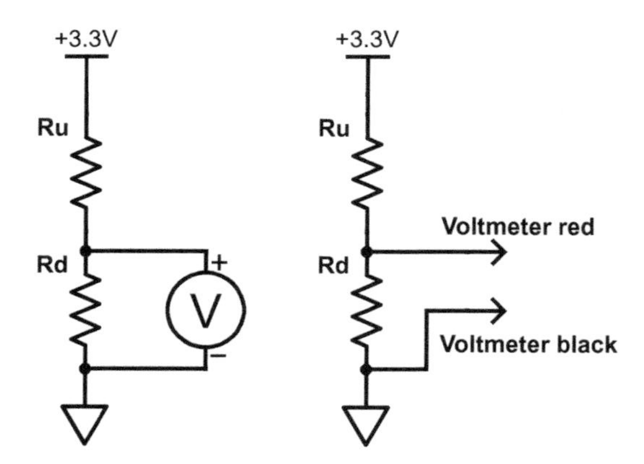

Figure 6.1: *There are two common ways to show how a voltmeter is connected to a circuit in a schematic. On the left, the voltmeter symbol is explicitly included in the schematic—both ports of the voltmeter must be connected to nodes. On the right, the connections are show as output ports of the circuit—I've labeled them here with the standard color codes for voltmeter leads.*

When using ports on the schematic, it is common to have just a single port drawn for the red lead of the voltmeter, with the assumption that the black lead is connected to ground. The PteroDAQ data-acquisition system almost always measures voltages relative to ground, so we commonly use a single port symbol for each measurement channel, and label them with the pin numbers on the microcontroller board.

Figure drawn with Digi-Key's Scheme-it [18].

It is also possible to do *differential measurements*, where the voltage difference between two nodes, neither designated as ground, is measured. This is common in multimeters and instrumentation amplifiers (see Chapter 40) and is sometimes available in oscilloscopes and data-acquisition systems, though it is more frequent for all the channels of an oscilloscope or DAQ to share a common ground node. PteroDAQ supports measuring some differential voltages, but generally only for one or two pairs of pins, depending on the microcontroller board used.

6.3 Time-varying voltage

When we talk about voltages and currents, we generally distinguish between two types of signals: ones that are constant with respect to time and ones that vary with time. Constant signals are referred to as DC (short for *direct current*) and time-varying signals are referred to as AC (short for *alternating current*).

DC voltage can be represented by just a single number—the voltage, but AC voltage is more complicated, and there are several different ways that AC voltage can be represented:

waveform A time-varying signal can be represented as a function of time, $V(t)$. This is the most detailed representation of the AC voltage and is what data-acquisition systems and oscilloscopes are designed to capture.

average The average (arithmetic mean) of the waveform,

$$V_{\mathrm{DC}} = \int_{t_{\mathrm{lo}}}^{t_{\mathrm{hi}}} V(t)\, dt/(t_{\mathrm{hi}} - t_{\mathrm{lo}}) \ ,$$

can be thought of as the DC voltage around which a time-varying voltage is varying. It is quite common to separate time-varying signals into DC and AC parts:

$$V(t) = V_{\mathrm{DC}} + Af(t) \ ,$$

where $f(t)$ has 0 as its average value. The DC part is referred to as the *DC bias*, *DC offset*, or *DC component* of the signal.

AC RMS The size of the AC component of a waveform can be summarized by the standard deviation of the waveform,

$$V_{\mathrm{RMS(AC)}} = \sqrt{\int_{t_{\mathrm{lo}}}^{t_{\mathrm{hi}}} (V(t) - V_{\mathrm{DC}})^2\, dt/(t_{\mathrm{hi}} - t_{\mathrm{lo}})} = A\sqrt{\int_{t_{\mathrm{lo}}}^{t_{\mathrm{hi}}} f^2(t)\, dt/(t_{\mathrm{hi}} - t_{\mathrm{lo}})} \ ,$$

which is the square *root* of the *mean* of the *square* of the difference from the DC offset, and is commonly referred to as the RMS (short for root-mean-square) voltage of the waveform. AC RMS voltage provides a single number that measures how large the AC component of the signal is—it is exactly the same as the (population) standard deviation of the signal.

AC+DC RMS Sometimes a slightly different definition of RMS is used, where the DC component is not subtracted off:

$$V_{\mathrm{RMS(AC+DC)}} = \sqrt{\int_{t_{\mathrm{lo}}}^{t_{\mathrm{hi}}} V(t)^2\, dt/(t_{\mathrm{hi}} - t_{\mathrm{lo}})} \ .$$

This version of RMS voltage captures an important property of the signal—how much power is lost in a resistor. See Section 29.3 for more discussion of RMS voltage, RMS current, and power.

In this course, we will usually be using the AC RMS definition of RMS voltage, rather than the AC+DC definition, because the DC offset will usually not be the information-carrying part of our signal.

We can expand the definition of $V_{\text{RMS(AC)}}$ to get a relationship between $V_{\text{RMS(AC)}}$ and $V_{\text{RMS(AC+DC)}}$:

$$\begin{aligned}
V_{\text{RMS(AC)}}^2 &= \int_{t_{\text{lo}}}^{t_{\text{hi}}} (V(t) - V_{\text{DC}})^2 \, dt/(t_{\text{hi}} - t_{\text{lo}}) \\
&= \int_{t_{\text{lo}}}^{t_{\text{hi}}} V(t)^2 - 2V(t)V_{\text{DC}} + V_{\text{DC}}^2 \, dt/(t_{\text{hi}} - t_{\text{lo}}) \\
&= V_{\text{RMS(AC+DC)}}^2 - 2V_{\text{DC}} \int_{t_{\text{lo}}}^{t_{\text{hi}}} V(t) dt/(t_{\text{hi}} - t_{\text{lo}}) + V_{\text{DC}}^2 \\
&= V_{\text{RMS(AC+DC)}}^2 - V_{\text{DC}}^2 \, .
\end{aligned}$$

We can also express this relationship as

$$V_{\text{RMS(AC+DC)}} = \sqrt{V_{\text{RMS(AC)}}^2 + V_{\text{DC}}^2} \, .$$

amplitude The *amplitude* of an AC signal is the largest deviation that it has from the DC offset. That is, we can represent the waveform as

$$V(t) = V_{\text{DC}} + Af(t) \, ,$$

where A is the amplitude and $f(t)$ is a time-varying function that has everywhere $|f(t)| \leq 1$ and somewhere the magnitude of $f(t)$ is 1. The amplitude of an AC signal is most useful when the shape of the signal is known, for example, if $f(t) = \cos(\omega t)$.

peak-to-peak The *peak-to-peak voltage* is the maximum of the waveform minus the minimum of the waveform. This is a particularly easy measurement to make and interpret, no matter what shape the waveform has. If $V(t) = V_{\text{DC}} + Af(t)$, then the peak-to-peak voltage is

$$V_{\text{pp}} = A(\max_t f(t) - \min_t f(t)) \, ,$$

independent of the DC offset, but scaling linearly with the amplitude.

If $V(t) = V_{\text{DC}} + Af(t)$, where $f(t)$ has an average of 0 and a maximum absolute value of 1, then

- the DC offset (or DC bias) is V_{DC},
- the amplitude is A,
- the peak-to-peak voltage is $V_{\text{pp}} = A(\max_t f(t) - \min_t f(t))$,
- the AC RMS voltage is $A\sqrt{\int f^2(t)\, dt / \int 1\, dt}$, and
- the AC+DC RMS voltage is $\sqrt{V_{\text{DC}}^2 + A^2 \int f^2(t)\, dt / \int 1\, dt}$.

Figure 6.2 illustrates these relationships for a sine wave.

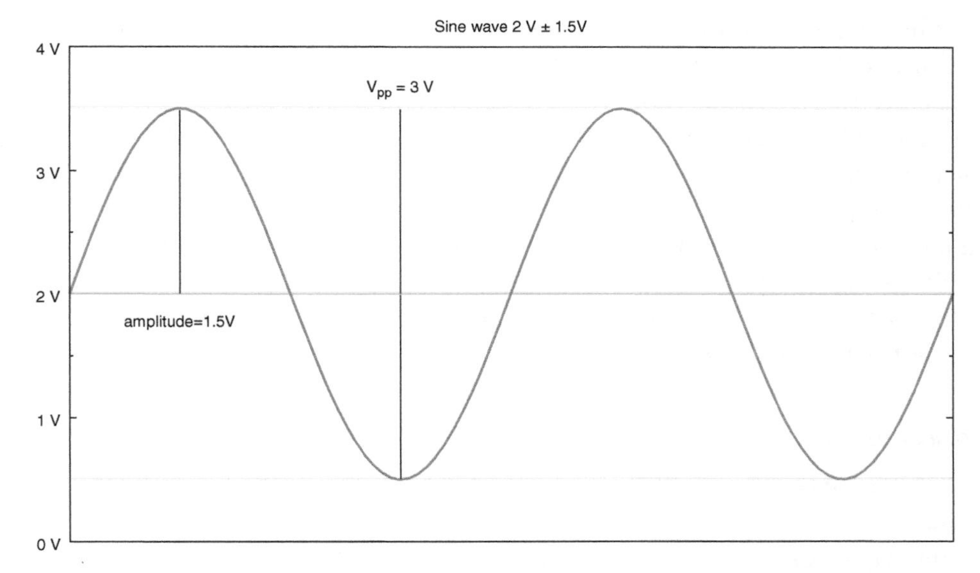

Figure 6.2: *This sine wave has a DC offset of 2 V and an amplitude of 1.5 V. The peak-to-peak voltage,* V_{pp}, *is 3 V (twice the amplitude). The AC RMS voltage is* $0.75\sqrt{2}\,\text{V} \approx 1.061$ *V and the AC+DC RMS voltage is* $\sqrt{2^2 + 2(0.75)^2}\,\text{V} \approx 2.264$ *V.*
Figure drawn with gnuplot [33].

When we have a signal that is approximately a sinusoidal wave, $V(t) = \cos(2\pi f t + \phi)$, we use some other summary numbers:

phase $2\pi f t + \phi$ The *phase* of a sinusoid is a time-varying signal in its own right. Because differences of 2π in the phase result in the same value after taking sine or cosine, we usually reduce the phase to the range $(-\pi, \pi]$ or $[0, 2\pi)$. Sometimes the term *phase* is used as shorthand for the phase at time zero—that is, ϕ.

frequency f . The frequency of a sinusoid is important for designing filters and amplifiers, as the information-bearing signal often has a different frequency range from the noise or interference that contaminates our signal. Frequency is measured in *hertz* (abbreviated "Hz"), which represents the number of full cycles of the signal in one second.

angular frequency $\omega = 2\pi f$ The angular frequency ω is an alternative way to represent the frequency of a signal. It is often mathematically more convenient than frequency, as the angular frequency is the derivative of phase with respect to time.

> When documenting an AC voltage, always indicate which voltage definition is being used—amplitude: $\pm 2\,\text{mV}$, peak-to-peak: $V_{\text{pp}} = 4\,\text{mV}$, or RMS $V_{\text{RMS}} = 1.4\,\text{mV}$. When V_{RMS} is given alone, it is assumed to be AC RMS, not AC+DC RMS. If there is a DC offset, then the amplitude representation is simplest: $1.5\,\text{V} \pm 2\,\text{mV}$.

The PteroDAQ data-acquisition system reports the DC offset and the AC RMS value (the standard deviation) for each channel that it records a waveform for. The averaging is done over the full length of the recording, which is a good summary for a steady-state signal (one that is not changing its properties), but not always the most useful when the properties of the signal are changing—then you might want to take averages over a much shorter time, which is generally done by voltmeters and oscilloscopes.

Data-acquisition systems and digital oscilloscopes, because they have a complete record of the waveform, often give users the choice of several different summary values for an AC signal. When RMS is reported, it is usually the AC RMS (not AC+DC).

Interpreting the AC voltage from a voltmeter can be a bit tricky as different voltmeters report different numbers for the same signal, and what the voltmeter reports is generally hard-wired into the design of the meter, not subject to user selection. A voltmeter may report either AC RMS or AC+DC RMS, for example. In addition, voltmeters may do a "true-RMS" computation or may just measure the peak-to-peak voltage V_{pp} and rescale it to be the AC RMS voltage that a sine-wave with that V_{pp} would have. True-RMS voltmeters are generally much more expensive than peak-finding ones, and the extra expense is often not worth the small improvement in the accuracy of the result.

6.4 Function generators

Engineers used to use sine-wave oscillators as signal sources for testing, but now most use *arbitrary-waveform function generators*, at least for low and medium frequencies.

These function generators are implemented primarily digitally. They consist of a digital clock signal, a memory for a lookup table, some control circuitry (usually a microcontroller or FPGA), a digital-to-analog converter, and an output amplifier to get the desired current and voltage range.

The lookup table L contains one period of the periodic function $f(\phi)$. The table size depends on how precise we need the function to be, as the hardware generally just looks up a number in the table, without doing any interpolation. A typical size is $P = 2^{12} = 4096$ values for the table (the default setting for a function generator in the Analog Discovery 2).

The control logic keeps track of the phase of the signal, $\phi(t)$. More precisely, it keeps track of a scaled version of the phase $\Phi(t) = P\phi(t)/(2\pi)$, using the range $[0, P)$, so that the output value is just $L[\Phi(t)]$.

The phase is maintained on each tick of the digital clock by incrementing $\phi(t)$ by $f\Delta t$—more precisely, by incrementing the scaled phase value $\Phi(t)$ by $Pf\Delta t/(2\pi)$, which is a precomputed constant. If the result of incrementing is out of the range ($[0, P)$, then the function generator subtracts P (or takes the remainder modulo P) to keep the scaled phase value within the legal range for the table lookup. A register that adds a (scaled) frequency to a (scaled) phase on every clock tick is called a *phase oscillator*.

The output from the lookup table, $L[\lfloor\Phi(t)\rfloor]$, is given to the digital-to-analog converter, which changes it to a voltage, which is then amplified and filtered by the output stage of the function generator.

The phase oscillator is not limited to having positive frequencies—a negative frequency decreases the phase and moves backwards through the table, producing a signal that is time-reversed from the one that would be produced with a positive frequency with the same absolute value. We'll revisit the notion of negative frequency in Chapter 9.

More details on function generators will be presented in Chapter 12.

6.5 Data-acquisition systems

A *data-acquisition system* (DAQ) is a hardware and software system that measures electronic signals and records them in data files that can be used in a computer. Perhaps the most widely used DAQ software is LabView, which provides ways to communicate with many different types of sophisticated equipment, but LabView is expensive ($999 and up), so we will be using a much cheaper system that you can continue to use as a hobbyist.

The PteroDAQ system used in this class has two hardware components: a microcontroller board and a laptop or desktop computer, connected by a USB cable (see Figure 6.3). The microcontroller measures the electronic signals, keeps track of the time the measurements were made, and communicates the

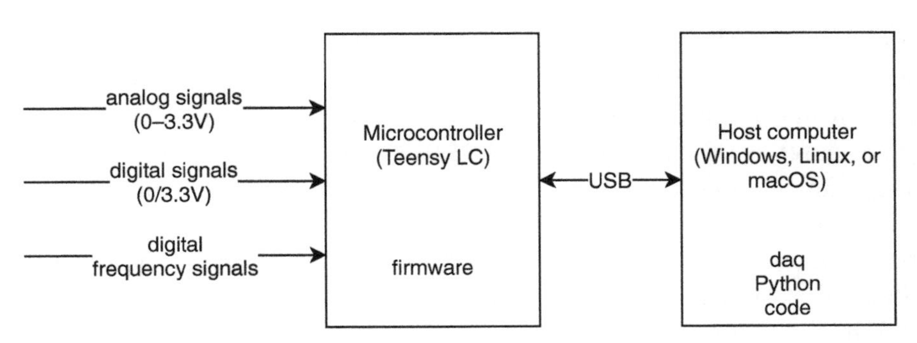

Figure 6.3: *A data-acquisition system (DAQ) is often shown as a single block in a block diagram (as in Figure 7.2), but for the PteroDAQ data-acquisition system, we can split that block into two parts: the microcontroller and the host.*
The microcontroller board handles the timing and the analog-to-digital conversion, and communicates the measured values to a Python program on the host computer which handles the graphical user interface, configuring the software on the microcontroller board, and file input and output. The host computer provides power to the microcontroller and communicates with it via a standard USB interface.
Figure drawn with draw.io [21].

measurements over the USB cable to the host computer. The host computer provides a graphical user interface for configuring the microcontroller, for starting and stopping recording, for adding notes about what is being recorded, and for saving the recorded data (and metadata) to files.

PteroDAQ works with several different microcontroller boards (see Table A.1), but only the Teensy LC will be discussed here, as it is the one used in the UCSC courses. The only electrical measurement that these boards can make is voltage, with a limited range (0 V–3.3 V), so all signals have to be converted to voltage before the DAQ can convert them into computer-usable data.

> Much of bioelectronics is in figuring out how to convert the physical signals we are interested in into voltages measurable by analog-to-digital converters for communication to computers for further analysis and processing. The PteroDAQ system presents a fairly typical electronic interface, and the circuits we'll design for connecting sensors to PteroDAQ are fairly typical of the designs needed for other sensor interfacing.

The microcontrollers used in this course contain an *analog-to-digital converter* (*ADC*) that converts voltages into numbers.

The board we use most often for the course, the Teensy LC, has a 16-bit ADC with a 3.3 V full-scale value. That means that it can report measurements between 0 V and 3.3 V in steps of $3.3\,\text{V}/65536 \approx 50\,\mu\text{V}$.

It can make the voltage measurements on any of 13 pins (A0 through A12). It can only make one measurement at a time, but it quickly cycles through the different measurements that it is asked to make so that the effect is almost the same as looking at all the measurements at the same time.

With the Teensy LC, PteroDAQ can make one differential voltage measurement ($\text{A10} - \text{A11}$), which has a range of values of -32768 through 32767, with steps of size $3.3\,\text{V}/32768 \approx 100\,\mu\text{V}$.

All ADC pins must have voltages between the power rails, in the range 0 V–3.3 V, in order to be read. The negative voltages for the differential channel just mean that A10 has a lower voltage than A11—both A10 and A11 still need to be positive with respect to ground. (In fact, because the differential measurement uses a differential amplifier, A10 and A11 need to be kept to a somewhat narrower range for accuracy—perhaps 0.025 V–3.275 V.)

In addition to analog inputs, PteroDAQ can also record digital inputs (which have only one-bit each, 0 or 1). The Teensy LC board has 14 digital input channels (D0 through D13).

The Teensy LC firmware also supports up to three digital frequency channels, though each channel has only a few digital pins it can be used on. A *frequency channel* counts the number of pulses that are seen on a digital input in each sampling time interval. The counter is a 20-bit counter, so the upper limit of the frequency measurement is the smaller of 3.9 MHz and a million times the sampling frequency.

When reporting analog values, PteroDAQ usually converts the ADC values (0 to 65535 for single-sided analog channels or −32768 to 32767 for differential channels) to volts, using a measurement that it made of the power supply voltage when initialized. (This conversion can be turned off by unchecking the "Supply Voltage" box on the PteroDAQ control panel.)

To reduce noise in the measurements, the Teensy LC analog-to-digital converter can average 1, 4, 8, 16, or 32 measurements at each sample time. The default is to average 32 measurements, which is the slowest, but least noisy. If you need to use a fast sampling rate or sample many analog channels, you can reduce the amount of averaging.

For each channel, PteroDAQ provides a *sparkline* that plots the last few samples recorded (updated 10 times a second). PteroDAQ also displays the last sample measured, the average of all the samples (labeled "DC"), and the standard deviation of all the samples (labeled "RMS").

The lines below each channel can be dragged to change the size of the pane the channel is in. If you have only one or two channels, then giving more space to each one gives a higher resolution to the sparkline.

The PteroDAQ control panel also has a large area for notes—this should always be used to record who collected the data and provide some information about the experimental setup being measured. (The channels can also be renamed to be descriptive of what they are measuring.) The notes do not need to record any of the PteroDAQ settings nor the date, as these are automatically included in the metadata when the recording is saved.

Chapter 7

Design Report Guidelines

7.1 How to write up a lab or design

Reports will emphasize experimental work and concisely summarize that work through an informal reporting style that expresses your individual grasp and understanding. These reports are engineering design reports, as you would write if hired as a consultant on an engineering project. They are **not** just proof that you successfully followed a cookbook procedure, the way high-school chem-lab reports are.

The four "Cs" of technical writing depicted in Figure 7.1 are the adjectives Clear, Correct, Complete, and Concise. There is an obvious tension between completeness and conciseness, and a similar tension between clarity and correctness. Good technical writing requires finding the right balance of these opposing ideals for the audience and purpose of any particular document.

> Every report should go through multiple drafts—the first draft is due before the lab starts. I will provide some pre-lab questions to give you some guidance on this first draft.

Because the pre-lab report is a draft of the final report, it should follow the same guidelines. Of course, it will be less complete than the final draft, with no information about debugging or the final results, but it should lay out clearly what you plan to do in the lab. The more complete the pre-lab report, the more feedback you can get to guide you on doing the lab, and the less time you will spend writing up your work after the lab.

7.2 Audience

The instructor is **not** the audience for the lab report—indeed the professor is rarely the right audience for anything you write in college. You are not writing a textbook (providing chapters full of background) nor a graduate thesis (providing lots of background as well as huge amounts of new material), but a short design report for a busy engineering manager who is not intimately familiar with the problem you are solving or for a new engineer who will need to take over the project.

> Assume that your reader has no access to the textbook or to any pre-lab or homework assignments. Give them all the data, pictures, and explanation they need to read the report without having to turn to other sources.

Your audience is engineering students, engineering managers, or engineers—not technicians. That means that you are providing the information that would allow someone to debug, adapt, or modify your design for a somewhat different set of constraints—you are not providing a step-by-step protocol

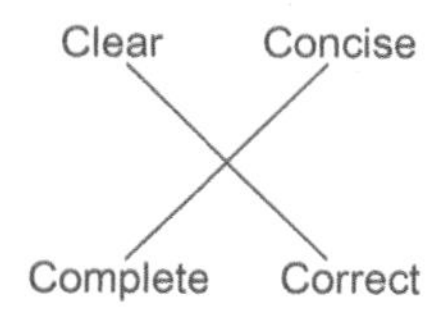

Figure 7.1: *The four Cs of technical writing are in obvious tension. The desired balance between clarity and correctness and between completeness and conciseness depends on who the audience for the document is.*
Figure drawn with Inkscape [45].

for someone to robotically copy your design. There is no need to explain standard tools and parts (like breadboards and alligator clips), nor is there any reason to provide a textual description of the wiring in a schematic. Assume that the reader can wire a circuit from a schematic diagram and concentrate on making sure that the diagram is correct. The reasoning behind your design choices is as important as the choices themselves.

The easiest audience to write to is someone like yourself, before you did the design being reported—what essential information did you need to learn in order to get the design right? What choices did you have to make in creating the design? Why did you choose the way that you did? What parameters are easy to adjust for different goals?

Make the report detailed enough that a student reading it could duplicate your work without having access to the original assignment—though they might have to look a few things up on the web or in text books. (Provide pointers to appropriate readings, when possible.) Explain not just what you did, but *why*, and provide warnings to help your reader avoid mistakes that you made.

To get the level of detail right, imagine that you are writing to a fellow engineering student who has not yet taken this course or who is a few weeks behind in the course. You don't have to teach the course, but you do have to explain exactly what your design is and how you developed it.

You *cannot* assume that the reader of the report has read the assignment—the design report is a standalone document.

7.3 Length

The conciseness goal of Figure 7.1 often gives students the most trouble—there are two parts to it:

Clear, direct sentences: Too often, students use bad academic writing as a model. They imagine that academic writing *requires* vague, passive constructions full of inflated diction and complex sentence structures. This obfuscation (deliberate or accidental) is generally a writing problem and not a reasoning problem, though consistent use of vague phrasing may indicate an underlying lack of comprehension.

Putting in just what is relevant: Often students do a dump of all their notes on a subject, most of which is irrelevant to the audience and the paper that they are writing.

In the past, students have started the quarter writing far too little, not explaining how they did their design or measurements. By the end of the course, students were writing too much, putting in extraneous information that was not relevant to the design, in the vain hope that longer reports would get higher grades.

There is no minimum nor maximum length requirement. The scope and depth of what you report on depends on what you were asked to do, learn, or become familiar with.

Getting the right level of detail in a report is difficult and takes practice—you need to think about what the audience you are writing for really needs to know. Some things are universal for almost all audiences (schematics, part numbers, component values, experimental results, etc.), while others will vary (background material, detailed mathematical derivations, justifications for design choices, alternative designs, future work, etc.).

7.4 Structure

Each report should be complete, thorough, understandable, and literate.

Your report should start with a brief explanation of the problem to be solved, then explain how the problem was solved—what measurements were taken, what calculations were done, what design decisions were made, and what the final result was.

A report will normally consist of the following items:

- A cover sheet or title block on the first page with the usual information (layout is up to you): the name and number of the lab (for example, *Lab 2: thermistors*); class (*BME 51A*); students' names; and the date the report was written. For all engineering documents, use the **full** date, not just month and year, and use the *actual* date of writing, not the due date. (Due date and instructors' names can also be provided as additional information.)
  If you worked with a second student (as most will in this class), both names should appear on the cover. If you turn in separate reports, one name should be as author, the other should be listed as the lab partner. If reports are turned in on paper, including the number of the lab section you are in can make returning the reports easier.
- The report itself, consisting of
  - A statement of the problem—what is the design goal for the lab? The goal should be very specific, including the magnitudes of inputs and outputs and any constraints on the design. For an amplifier design, the goal should include such specifications as gain, power-supply voltage, and frequency range. The goal should include all the specifications that would be needed for someone to use the designed circuit.
  - A statement of the solution to the problem. This solution will often include a block diagram showing how the problem is decomposed into subproblems.
  - Convincing evidence that the solution is a valid, even a good, solution to the original problem. This evidence will usually involve discussion of design constraints, possibly calculations, probably measurements, and almost certainly some citations to data sheets, application notes, web pages, or other sources of inspiration.
  - Schematics for any circuits that are made for the solution or for testing along the way.
  - Plots of any data collected, often with models fitted to the data.

The design reports for this course are not required to have an abstract, though one is standard for most other forms of technical writing. Students often misunderstand the purpose of an abstract—it is intended to be a standalone summary of the paper that can be indexed and distributed separately from the paper itself. People use abstracts to find papers to read (and often read only the abstract, and nothing else of the paper, so key results must be in the abstract).

> There should be nothing in the abstract that is not in the main body of the paper, and every important result in the paper should be in the abstract.
>
> An abstract is not an introduction—people may read the paper without ever reading the abstract. Treat an abstract as a combination of an advertisement for the paper and as a quick summary for those who just need the results and not all the reasoning of the paper.

Many students (and some professional scientists) feel impelled to provide a lot of background before getting to the question that the paper is addressing. This delay is a very bad practice—journalists refer to it as "burying the lede". You should start all papers with a concise statement of the point of the paper—for a design report, start with the specific design goal; for a thesis, start with the specific research question being answered in the thesis; for a grant proposal, start with the goal of the grant. After you've given the specific question or goal, then you can fill in background as needed to explain exactly why that goal is interesting and how you will achieve it.

> There is no "background" section in the structure outlined above—design reports don't generally provide much background, and when some is needed, it is worked into the specific section where it is needed, not separated out into its own section.

Other types of document may have quite different structures, of course—theses, for example, almost always have a chapter on what others in the field have done.

> Section headings should be descriptive, not generic.

Someone picking up the report and reading just the title and section headings should have a fair idea what the report contains. If you can use the same section headings in two different reports, your headings are too generic. This course is not the *Journal of Molecular Biology*, and I don't accept their belief that all articles must have exactly the same section headings in exactly the same order.

> Sections and subsections are ways of splitting higher-level units (such as reports or chapters), so there should never be just one section or subsection—there should either be none (no subdivision) or at least two.

> Every section or subsection must have textual content—generally at least two paragraphs. Having just a figure or table in a section makes no sense, because every figure or table must be referred to in the main body and belongs in the same section as where it is first referred to.

I prefer a writing style where description of techniques, results, and discussion all flow naturally and are not separated into widely different parts of the report. I realize that many biology journals favor or enforce a style in which Methods, Results, and Discussion are all separate sections (with Methods often tucked away at the end as an afterthought). Although this style is common in biology papers, it really is much harder to read than a report that talks about the results of a test as soon as the test is described.

Because the design reports you are writing are describing the results of an iterative design process, you often need to report on measurements that were made that guided the design process. It makes no sense to the reader to talk about the design decisions in one place, the measurement technique in another, the results of the measurements in a third place, and the connections between the measurement and the design in yet a fourth place. A much easier-to-read flow would explain the need for the measurement, the measurement technique, the results of the measurements, and the consequences of those measurements for the design in that order. Try to make sure that your reader has all the information they need for understanding what you are saying without having to jump forward in the report to find it.

7.5 Paragraphs

At a finer level of structure than Section 7.4, we have paragraphs. Just as the whole paper is broken up into meaningful sections, each introduced by a section title, each section is broken up into paragraphs introduced by topic sentences.

> The first (topic) sentence of a paragraph should contain the main point of the paragraph, with subsequent sentences closely connected to it, generally providing supporting detail.

Although some forms of writing allow topic sentences in the final position of a paragraph, this topic-last style is very rare in technical writing and should be avoided.

Don't allow your paragraphs to wander away from the topic introduced by the topic sentence—if you need to change topics, start a new paragraph. Having many short paragraphs is definitely preferable to a long rambling stream of consciousness for technical reports—you aren't trying to write James Joyce's *Ulysses*.

7.6 Flow

Organizing your thoughts into sections and paragraphs helps your reader keep track of what you are talking about, but is not enough to make reading effortless. You also have to guide your reader from topic to topic, so that they can follow along as they read, creating smooth *flow*. Good flow is essential for making reading comfortable, and its opposite is *choppy* writing.

> The main heuristic for establishing good flow is "old information to new information".

Each unit of writing should start with something the reader already knows—usually something introduced in the previous unit—then segue to something new that the reader needs to know. The units of writing can be sentences, paragraphs, or sections—the flow heuristic applies across all levels of structure.

> **Exercise 7.1**
>
> Look at the sentences and paragraphs of Section 7.6. Do the paragraphs start with topic sentences? How are the sentences within each paragraph connected? How are the paragraphs connected? How is this section connected to the previous one?
> The "how" questions are asking about the engineering of the writing to create connections in the readers' heads, not asking for a summary of the content of the paragraphs.

7.7 Tense, voice, and mood

For most of your reports, you will be reporting on something already completed. Therefore, most of what you write will be expressed in the past tense. To aid you in resolving questions about tenses, the following summary should be helpful:

- Experimental results should always be given in the past tense.
- Things that are continuously true should generally be given in the present tense (like what parameters the data sheet claims).

- Discussions or remarks about the presentation of data should mainly be in the present tense.
- Suggestions for actions to take as a result of the study should generally be in the future tense.
- Discussions of results can be in both the present and past tenses, shifting back and forth from experimental results (past tense) to the conclusions and statements that are continuously true.

Your pre-lab drafts will include a fair amount of future tense, about what you are proposing to do, but the tense should be changed to the past tense when revising the document during and after the lab.

Avoid using too much passive voice in your writing.

Passive voice is often overused in scientific writing, partly out of a misguided attempt to sound formal and partly to remove the people who did the experiment from the description of the experiment. The result often sounds like the authors are trying very hard to disassociate themselves from the project. Nick Falkner describes this use of passive well in *The Blame Game: Things were done, mistakes were made* [25]:

> *The error is regretted? By whom? This is a delightful example of the passive voice, frequently used because people wish to avoid associating the problem with themselves.*

I realize that many students have been taught to use passive voice in chemistry and biology lab reports—hiding who did things and talking only about what was done (by unidentified secret agents). A good design report uses "I" (for single authors) or "we" (for multiple authors) when talking about actions: "We measured the impedance in 400 steps from 10 Hz to 1 MHz" rather than "The impedance was measured ... ." Don't use "we" as a single author, unless you mean "you (the reader) and I". You can often avoid passive voice by emphasizing the results of an action, rather than the action: "The impedance varies from 100 kΩ at 1 Hz down to 1 Ω at 1 MHz, as seen in Figure ... ."

There are times when passive voice is appropriate:

- When you don't know who did something—this reason should not come up often in design reports, as you should either have done it all yourself, or provide explicit written acknowledgment and citation for anything done by others.
- When you need to flip a sentence around to put what would normally be its object in the first position, in order to get better flow between sentences by putting old information before new information.
  This flipping of sentences is best shown with some schematic sentences: we start with the choppy sentences

  > A creates B. C modifies B. D controls C.

  and we can improve the flow by modifying to

  > A creates B. B is then modified by C, which is controlled by D.

Students worry that if they avoid passive, then they'll end up starting every sentence with "I". Certainly, starting every sentence identically would be a problem, but avoiding that problem is fairly easy, particularly if students talk about the goals and purposes of experiments, rather than just giving technician-level protocol dumps of what they did. If you re-read this paragraph, you may notice that I did not use passive at all in it, and only this last sentence has "I"—forming gerunds is one good way to create alternative subjects for sentences.

Be careful with your use of the subjunctive mood, signaled by the word *would.* A lot of students seem to think that "would be" is some formal form of past tense—they've seen it in writing, but never understood what it means. In technical writing, the subjunctive mood is generally used only for *contrary-to-fact* hypotheses, though it has some other uses in general writing. For example, *the sound would be picked up by the microphone* implies that it was not picked up, because something was preventing it. I always want to see a reason: *the sound would be picked up, but there is an inch of lead keeping the microphone diaphragm from moving.* "The inverter that we would be using" says that you didn't use that inverter and are about to say why.

> Never is *would be* just a formal form of *was*. If you don't know how to use *would* correctly, then never use it—that approach is safer than using it wrong and won't cramp your style much.

7.8 Formatting with LaTeX

Reports must be organized, neat, and legible; computer-generated typed work is expected. For my courses, I require formatting with LaTeX [62], which is still the best program for typesetting text with mathematics, despite having been released in 1984 and having been built on top of TeX [59], which was released in 1978.

If you don't want to install LaTeX on your computer, there are online implementations on the web that allow simultaneous editing by multiple authors (in much the same manner that Google Docs do, but with LaTeX for formatting). The two main ones merged in 2017, becoming https://www.overleaf.com/.

I have used the online LaTeX, and it seems to work well enough, though it is far slower than running pdflatex on your own machine. For the small documents you will be writing for this class (about 10 pages), the slowness hardly matters, and many users find the syntax highlighting and spelling checking online useful, as well as their keeping up to date on the standard packages. As long as you have good internet access and are not doing huge documents (like this book), the online LaTeX implementations are a good choice.

> To aid students in getting started, I've put a dummy lab report on the web as https://tinyurl.com/overleaf-lab-report-template. To use the dummy report, go to the menu and select either "Copy Project" (to use in Overleaf) or "Download source" (to get the files for use offline or with other LaTeX implementations).

There are incomplete LaTeX implementations as well, which only handle math mode (such as https://www.latex4technics.com/ and https://www.macupdate.com/app/mac/17889/latexit). These can be useful for creating equations to paste into non-LaTeX documents, but are not a substitute for full LaTeX implementations.

On Mac OS X, I installed just the small BasicTeX package from https://tug.org/mactex/morepackages.html, but a lot of students prefer to have a GUI front end such as TeXWorks https://www.tug.org/texworks or TexShop https://pages.uoregon.edu/koch/texshop/. I did end up using `tlmgr`, the TeXLive manager, to add a few more fonts and packages to my installation, so students may prefer just to get the full TeXLive package.

Students on Windows machines have found MiKTeX (https://miktex.org/) to be a good package for running LaTeX without needing internet access.

Because there are so many tutorials for LaTeX readily available, I'm not going to provide one here. The LaTeX manual [62] is a pretty good tutorial, as is the Wikibook guide to LaTeX [105]. There is a very basic tutorial, intended for middle-school and high-school math students, on the Art of Programming web site [5]. Some students have found the LaTeX "cheat sheet" at https://wch.github.io/latexsheet/ to be handy for reminding them of the names of major commands in LaTeX. The online documentation for Overleaf [82] is also pretty good.

I expect students to use sections, citations with bibtex or biblatex, figures and tables, and math mode fluently.

7.9 Math

7.9.1 *Number format*

Engineering reports use a lot of numbers, and the accuracy of these numbers is usually critical to the content of the report, so it is worth our while as writers to be particularly careful with numbers, to make sure that they are read correctly by our readers.

> Never start a number with a period. If you have a number between 0 and 1 to report, start with a leading zero: 0.5, not .5, because periods at the beginning or end of a number are easily lost. Losing one at the end doesn't change the value, but losing one at the beginning is an enormous change.

Students have often been taught *scientific notation*, in which numbers are represented as the product of a number between 1 and 10 and a power of 10. There are two standard ways of writing scientific notation:

- with explicit exponentiation: $2.2\;10^{-12}$, for example, and
- as floating-point numbers as used in many programming languages: 2.2E-12, for example.

When you are using explicit exponentiation, do not use the $\times$ symbol, which is for vector cross products, not for scalar multiplication (at least, not after seventh grade). Instead use a thick space (`\;` in math mode), or, if a punctuation mark is absolutely required, a center dot (`\cdot` to get $\cdot$ in math mode). The thick space is preferred, as the center dot should be reserved for vector dot products.

> When you are using floating-point notation, do not use superscripts! The notation $1e^{-12}$ means 6.144E-6 or $6.144\;10^{-6}$, not 1E-12.

If you want to use floating-point notation in math mode, then enclose it in a box, so that the number does not come out looking like a subtraction when you meant a negative power of 10. Look at the difference between `\(5.1E-6F\)`, which appears as $5.1E-6F$ and `\(\mbox{5.1E-6}\)\,F`, which appears as 5.1E-6 F.

In the example of the previous paragraph, you may have noticed that the space before the unit symbol is given as `\,` rather than as a space. This `\,` is LaTeX's command for a *thin space*, which provides a thinner than usual space without line breaking, so the number and the units won't be separated.

Engineers generally use a slight variant of scientific notation, known as *engineering notation*, in which the exponents are all multiples of three, and the number in front is in the range $1 \leq n < 1000$ [128]. Many calculators include engineering notation as display choice, and it is worth setting your calculator to use it. Engineering notation is popular because it makes conversion to metric prefixes easy, and using the power-of-a-thousand prefixes is how numbers are most commonly written in engineering reports.

> Whenever possible, do not use scientific notation, floating-point notation, or engineering notation, but use units with metric prefixes. When using metric prefixes, the number should be in the range $1 \leq x < 1000$. If the number is outside that range, then switch to a different metric prefix. That is, numbers like $2.2\ 10^{-5}\,$F, 2.2E-5 F, or 22.E-6 F are best written as $22\,\mu$F.
>
> Unitless numbers (like gain) should use engineering notation.
>
> The unit abbreviations (like "nA") should only be used with numbers—if you want to talk about the unit itself, write it out ("nanoamps").

The prefix for 10^{-6} is "μ" not "u", though one occasionally sees "u" used in old documents or programs, from before the days of Unicode. For example, gnuplot uses "u" when it is generating prefixes with its %c code, because it wants to stick with ASCII characters. In your reports, you should use the correct prefix, μ, which in LaTeX is in math mode: `\(\mu\)`.

If you have to exponentiate a unit, be sure to include any metric prefixes attached to the unit in parentheses—for example, report area in $(\text{mm})^2$ rather than mm^2, because of the ambiguity of the latter—is "milli-" squared or not?

> Avoid using commas in numbers, as they can be mistaken for decimal points (in fact, the European convention uses a comma in place of a decimal point).

7.9.2 *Math formulas*

You will have a lot of math formulas in your reports, and common word-processing tools like MSWord do a terrible job of typesetting math. I require LaTeX for producing your reports, as it is the best tool for producing reports that contain math (see Section 7.8).

Whatever tool you use, you will have to find a way to enter math formulas properly, as it is not a good idea to cut and paste formulas as images from web sites. The resulting images are often hard to read, even harder to edit, and more often than not use a different notation than the rest of your report.

> No naked formulas!
>
> Math formulas and equations are grammatically parts of sentences and are treated as noun phrases.

In formal writing, we expect to see full sentences, not just fragments, so equations should *never* appear by themselves, randomly dropped in the text. Because they are parts of sentences, they should also not appear as floating insertions, the way that figures and tables do. If you need examples of how to include math formulas in sentences, there are about 3500 such examples scattered throughout this book.

Because math formulas are parts of sentences, they often have punctuation right after them (commas, periods, em-dashes, colons, etc.). If the math is inline math, not displayed, then nothing special needs to be done—just put the punctuation after the `\)` that ends the formula. For display math, which puts the formula on a line by itself, putting the punctuation outside the formula would result in starting the next line with the punctuation, which is almost always wrong. Instead, put the punctuation inside the formula, but precede it with a tilde to make a space between the formula and the punctuation. `\[\frac{1}{3}~.\]` displays as

$$\frac{1}{3}~.$$

You can have LaTeX number your equations and refer to the equations by number *after* you have introduced the equation in a sentence. Don't refer to the equation number in the same sentence that contains the equation—the number is only for later references back to the equation. If you number equations, use `\label{eqn:name}` to label equations and `Eq.~\ref{eqn:name}` to refer to them, as editing may change the number or order of the equations. (The example I provided on Overleaf defines a command `\eqnref` to make the cross-referencing more consistent.)

It is not adequate to use just the name of a formula in your writing (like *the Steinhart-Hart equation*)—you must include the actual formula (like $\frac{1}{T} = A_0 + A_1 \ln(R) + A_3 (\ln(R))^3$) at the first reference to the formula. Once you have introduced the formula and named it, you can refer to it by name.

> Any variables you use should be defined before being used (or, at worst, in the same sentence as they are first used). For component values, that means the schematic must be presented **before** the variables for component values can be used in formulas.

One common construction in mathematical writing is to introduce a formula by describing the meaning, giving the formula, then adding "where ... " after the formula to define all the variables in the formula. For example, we could write

> One formula for modeling thermistors is the Steinhart-Hart equation,
>
> $$\frac{1}{T} = A_0 + A_1 \ln(R) + A_3 (\ln(R))^3~,$$
>
> where T is temperature in kelvin, R is resistance in ohms at T, and A_0, A_1, and A_3 are so-called Steinhart-Hart coefficients, which are used to fit the equation to the measured behavior of a thermistor.

> Displaying the math formula on a line by itself does not change any of the punctuation rules for a sentence.

In the example above, the math formula is an appositive (mirroring the noun phrase "the Steinhart-Hart equation"), so should be set off with commas.

Because math formulas often have punctuation after them (most often commas or periods), you need to learn how to place the punctuation correctly. For inline math mode, the punctuation goes outside math mode, with `\(a^2+b^2=c^2\).` displaying as $a^2 + b^2 = c^2$.

For display math mode, where the formula ends up on a line by itself, this approach does not work, as the punctuation mark would appear inappropriately at the beginning of the next line. Instead the punctuation mark is put at the end of the formula, inside math mode, generally with some forced spacing to separate it from the formula itself. For example, `\[a^2+b^2=c^2~.\]` displays as

$$a^2 + b^2 = c^2~.$$

(The tilde symbol $\sim$ is used to make an unbreakable space in LaTeX.)

One minor point: the "times" symbol $\times$ is not the same as the variable x—please do not use x as a mathematical operator, but only as a variable. For that matter, there is almost no reason to use $\times$ as a binary operator in this course, as it is used for cross-products of vectors (which we don't need here) and should not be used for scalar multiplication (at least, not after seventh grade).

The $\times$ symbol is OK to use for expressing unitless gain, generally as a postfix operator: *the gain is* 10$\times$. Do not use the letter "x" for $\times$ in this context.

Scalar multiplication is shown by juxtaposition: $\omega = 2\pi f$. When juxtaposition runs things together too much, you can use `\;` for a thick space or `\,` for a thin space. For example, "`100\ohm{}20pF=100\,\ohm{}\;20\,pF=2\,ns~.`" displays as

$$100\Omega 20pF = 100\,\Omega\;20\,pF = 2\,ns~.$$

The command `\ohm` is one I defined to get `\Omega` for Ω in either math mode or text mode. The Overleaf example I provided has the definition.

You can get even better appearance for numbers with units by putting the units in an ordinary font, rather than a math font: "`100\,\ohm{}\;20\,{\rm pF}=2\,{\rm ns}.`" displays as

$$100\,\Omega\;20\,{\rm pF} = 2\,{\rm ns}.$$

Don't use an asterisk ("*") for multiplication in math mode either—that convention is not used for typeset math, but only in computer programming languages (Python, C, Matlab, gnuplot, etc.).

Because juxtaposition is used for multiplication, using multi-letter variable names is problematic in math mode: $Vref$ is the multiplication of four different variables, not a single variable. Using subscripts is a common way around the problem: V_{ref} is written as `\(V_{ref}\)`. If you want to be even more clear that the subscript is a word, not distinct variables, you can take it out of math font with `\rm`: $V_{\rm ref}$ is written as `\(V_{\rm ref}\)`.

Although you are limited to single-letter variables (potentially with subscripts) in math mode, you are not limited that way in scripts and programming languages—in those contexts, it is better to use full-word names. In either context, don't rely on case to distinguish between different variables—although some programming languages easily distinguish between V and v, people do not.

We do a lot of fractions in this book—sometimes complicated ones like

$$\frac{1}{\frac{1}{R_1} + \frac{1}{R_2 + \frac{1}{j\omega C}}}~.$$

If we need to put parentheses around such big fractions, it can look pretty silly:

$$(\frac{1}{\frac{1}{R_1} + \frac{1}{R_2 + \frac{1}{j\omega C}}})~,$$

so LaTeX provides a way to make scalable left and right grouping symbols, with `\left` and `\right`:

```
\[ \left( \frac{1}{\frac{1}{R_1} + \frac{1}{R_2 + \frac{1}{j\omega C}}} \right)~.\]
```

produces

$$\left(\frac{1}{\frac{1}{R_1}+\frac{1}{R_2+\frac{1}{j\omega C}}}\right).$$

We use a lot of logarithms and some trigonometric functions in this course, so it is worth knowing the LaTeX commands for showing them in math mode:

(1) `\ln(x) \log(x) \log_{10}(x)` for $\ln(x)\log(x)\log_{10}(x)$.
(2) `\sin(\omega t) \cos(\omega t) \tan(\omega t)` for $\sin(\omega t)\cos(\omega t)\tan(\omega t)$.

Another special symbol we use a lot is the $\pm$ symbol, which is `\pm` in math mode.

For derivatives, use a Roman-font "d"—that is to get $\frac{{\rm d}V}{{\rm d}t}$, use `\frac{{\rm d}V}{{\rm d}t}`, and if you need partial derivatives, use the `\partial` symbol: the formula $\frac{\partial V}{\partial C}$ is written `\(\frac{\partial V}{\partial C}\)`.

To find other LaTeX codes for symbols, you can use the handy Detexify tool at https://detexify.kirelabs.org/classify.html, which lets you draw the symbol and suggests possible LaTeX commands for getting similar shapes. There are also PDF files on the web with 1000s of LaTeX names and what glyphs they represent—one of the best is http://tug.ctan.org/info/symbols/comprehensive/symbols-a4.pdf.

7.10 Graphical elements

Use good drafting practice when producing figures, graphs, drawings, or schematics and label them for easy reference. No picture, table, schematic, or graph should appear without a name (generally of the form "Figure 1" or "Table 3"), and none should appear without a reference to them **by name** in the main body of the writing. The names, being names, should be capitalized.

Space must be provided in the flow of your discussion for any tables or figures—do not collect figures and drawings in a single appendix at the end of the report. The figures should be referred to in the order of their appearance, with Figure 1 being the first one referred to, Figure 2 next, and so forth. Ideally, figures are on the same page as the discussion of them, or shortly afterwards.

Both figures and tables are *floating insertions*, which means that they do not have a fixed location relative to the text. That means that one cannot refer to figures with words like "above" and "below"—the figure may move if the text is edited or margins are changed. Instead, figures must always be referred to by name, like Figure 7.2.

Every figure or table must have a caption. A caption is not just a title for a figure, but a full paragraph below or beside the graphics telling the reader the point of the figure. The caption does not recite a description of the figure, but tells the reader what is important to notice in the figure—why is the reader being shown this particular image?

Most readers of technical writing flip through the writing quickly, looking at the pictures and reading the captions. If they can't get the main ideas of the paper from the figures, they don't bother reading the whole thing—either it doesn't have what they are looking for or they aren't going to understand it.

The figure captions are in some sense the most important part of the paper—together with the abstract, they are all that most readers ever read of the paper. The first part of the caption should convey the *point* of the figure.

Figure credits, acknowledgments, authorship, or citations are the least important part of the caption, and so they should be left to the end.

If you need examples of the sort of details I expect in captions, there are over 300 of them in this book.

In LaTeX, figures should be handled with the "figure" environment and tables with the "table" environment. These environments create *floating insertions*, sometimes shortened to *floats*, which keep the figure (or table) and its caption together, moving them as a unit if necessary. The usual placement of a figure or table in a technical report is at the top of a page or column—usually the page on which the first reference to the figure or table is made. If there is no room for the float on the same page, it is delayed to the top of the next page.

Some technical reports have a large ratio of figures to text and the floating figures may be delayed for several pages from the reference to them. To avoid having figures drifting into later sections of the report, you can put up barriers in the text to prevent later sections from starting until all the floating insertions have been output. I've found the LaTeX `placeins` package to be particularly useful. In this book I used

```
\usepackage[section,above,below]{placeins}
```

to keep figures and tables from crossing section boundaries, except for allowing the floats to appear either at the top or bottom of the page that contains the section, even if the top or bottom of the page is in a different section.

In LaTeX, don't use the `[h]` or `[h!]` options for `\begin{figure}` or `\begin{table}`, as they prevent the figures and tables from floating properly and often result in awkward placement of text above figures.

As text is edited or margins modified, the positions of floats can change, so figures and tables should always be referred to by name, not by location. The references to figures and tables should be done with `\label` and `\ref`, not with manually entered numbers. The Overleaf example I provided shows how to include a PDF file as a figure and how to do figure references simply. The line `\newcommand\figref[1]{Figure~\ref{fig:#1}}` defines a new command `\figref` for generating figure references. For example, in this book `\figref{4Cs}` generates Figure 7.1 because that figure has `\label{fig:4Cs}` in its caption. The tilde after the word "Figure" is TeX's way of requesting an unbreakable space, so that the word "Figure" and the figure number will always appear on the same line, avoiding the awkwardness for the reader that happens when "Figure" appears alone at the end of the line and the number on the next line (or even next page).

I recommend putting the figure and table labels in the captions of the figures, so that they get the right number associated with them. This placement of the labels is particularly important when you have subfigures, which each have their own caption and label.

Never include a figure or table that is not referred to in the main body—figures are an addition to the text not a replacement for it.

Figures and tables should be referred to in the order they appear. The easiest way to get the right order is to insert the `\begin{figure}` block right after the paragraph in which the first reference to the figure appears.

Although both figures and tables are floating insertions, they serve different functions in a report and are named with separate numbering schemes. Figures are for graphical elements (photos, block diagrams, schematics, graphs, and so forth), while tables are for columnar or gridded arrangements of text, often created with LaTeX's "tabular" environment. (Scanned or screenshot images of tabular data are never appropriate.)

7.10.1 *Vector and raster graphics*

There are two different types of image formats: *raster* and *vector* graphics. Raster formats represent images as arrays of pixels, while vector formats represent images as collections of lines and curves. The raster format is best for photographs, but is a poor format for schematics and graphs, which are better represented in a vector format. The vector format is generally much more compact than a raster format and you can zoom in to get a clearer view without images getting fuzzy or "pixelly".

Whenever possible, you should use a vector format rather than a raster format for your images.

If you have to include raster graphics (for a photo, for example), crop the image and compress it with jpeg or png format. A resolution of 600 pixels by 400 pixels is plenty for almost all photographs in a technical report—that is the most common size for photos in this book.

Photographs should not be included directly from a camera or phone, as the initial images take up too much memory, bloating the file size of the report. Direct-from-camera photos nearly always have too much extraneous material around the main object of the photo, also. All photos should be cropped and resampled before being included in a report—there are many free and low-cost tools for doing so, as well as for other commonly needed fixes, like adjusting the brightness and contrast of the image.

Some file formats (such as PDF and EPS) allow both vector and raster representations, depending on how the image is created. Almost all the figures included in this book are PDF files, most created with Scheme-it for schematics [18], draw.io [21] for block diagrams, and gnuplot [33] for graphs. The few photographs in the book are included as jpeg images. The tools listed above (Scheme-it, draw.io, and gnuplot) all create vector graphics when the output file format permits it. Scheme-it has two different vector-graphics formats: PDF and SVG. For some reason, the PDF output format from Scheme-it often has problems with incorrect cropping around the text, so for this book I've used the SVG format, and converted it to PDF using Inkscape [45] using

```
inkscape --export-area-drawing --export-type=pdf --export-margin=3 foo.svg
```

where "foo.svg" is replaced by the appropriate file name. On my Macintosh the "inkscape" executable is called `/Applications/Inkscape.app/Contents/MacOS/Inkscape`, which is a rather awkward name, so I use an "alias" command in my initialization file to make "inkscape" an alias for the long name. MacOS versions of

Inkscape before 1.0 used a different location, `/Applications/Inkscape.app/Contents/Resources/script`, for the executable.

I use the `graphicsx` package for including graphics, which can handle PDF, jpeg, and png files. The example I provided in Overleaf uses

```
\centering
\includegraphics[width=0.8\textwidth]{generic-block-diagram}
```

inside a figure environment to include a file called `generic-block-diagram.pdf`. The optional argument in square brackets sets the width of the figure to 80% of the width of the text column that it is in. You can also use absolute sizes—many of the figures in this book use absolute heights like `[height=5cm]`, instead of giving a width as a multiple of the text width.

There is no point to converting a JPEG or PNG raster image to PDF raster format before giving it to LaTeX—the resulting image is no smaller and often is harder for PDF viewers to render than if the converters invoked by the graphicsx package are used. Use PDF for vector-format images, and PNG or JPEG for raster graphics.

Exercise 7.2

Crop, downsample, and compress the image at https://users.soe.ucsc.edu/~karplus/bme51/book-supplements/large-image-to-crop.jpg to no more than 80 kbytes. Provide the image and report the size of the image file and the resolution (dimensions in pixels) of the image. Say what tools you used to do the image manipulation.

7.10.2 *Block diagrams*

One of the most common tasks in any form of engineering is to decompose a problem into subproblems that can be solved independently or almost independently. In electronics, the problem is often to design a circuit with a particular function, and the subproblems are designs of subcircuits that interact with each other to perform the desired function.

In electronics, a common way of representing this decomposition is the *block diagram* (introduced in Section 5.6). A typical block diagram consists of two types of objects:

blocks: A block is a subsystem with a well-defined function. The block typically has *inputs* and *outputs*, though some ports may serve for bidirectional information flow.

> A block generally (though not always) has one input on the left and one output on the right. Non-information-bearing connections (like power supplies and reference voltages) are either not drawn or are connected to the top or bottom of the block.
>
> Most often, blocks are drawn as rectangles, with inputs on the left and outputs on the right, though amplifiers (a particularly common sort of block) are drawn as isosceles triangles with the base on the left and the apex vertex on the right. For amplifiers, inputs are on the base and the apex is the output.

The "side inputs" to blocks are often power-supply connections or control parameters that do things like set the voltage that the output is centered at or adjust the gain of the circuit. When deciding whether to put an input as a main input on the left of the block or as a side input on the top and bottom, think about what information it carries. If it is carrying the main information of the system (such as a sensor reading), it goes on the left. If it is constant, it goes on the top or bottom. Blocks are not necessarily circuits—they can also be software functions. For example, band-pass filtering can be achieved either in analog circuitry or digitally in software—the same symbol describing the function of the filter can be used in either case.

> The blocks are always functional elements, and so they should be labeled with their function, rather than with part numbers or component names.

For an amplifier in a block diagram, I want to know things like how much gain it has, or what its bandwidth needs to be, not which chip will be used to implement it—that level of detail should be reserved for the schematic diagram.

connections Connections are wires or other means for communicating information or power between blocks, and are usually drawn as lines connecting the blocks. If a connection is carrying information as a time-varying voltage or current, the information is often referred to as the *signal.*

> Signals in block diagrams are always lines, not blocks. If you have an input from outside the system (for example, sound coming into a microphone), the arrow for the connection will have no block at its base.

> If a block has only a single output, then only one connection should be drawn for it, even if it is distributed to multiple places. Observe that the V_{ref} generator in Figure 7.2 has only a single output, even though that output is used in three other blocks.

One tool that I find useful for drawing block diagrams is the free web tool at https://www.draw.io, though there are many other drawing tools for boxes and arrows that can be used. I like draw.io for several reasons: it is easy to use, it is free, it lets me keep my drawings on my own computer, and it provides conversion to several formats, including PDF.

> Block diagrams are usually drawn with a left-to-right information flow, with outputs on the right of each block connecting up to inputs on the left side of subsequent blocks. When the information flow is different, arrowheads are often drawn on the connections, to make it visually clear how the information flows. If a block diagram is too wide to fit on the page, draw it as multiple left-to-right rows, with arrows from the end of one row to the beginning of the next.

A block diagram can be as simple as a *pipeline* in which each block has one input and one output and is connected only to the one before and the one after. But block diagrams can be more complicated, with a particular output going to the inputs of several blocks or one block taking several inputs.

Block diagrams are often *hierarchical*, with each block expanded in another figure to a more detailed block diagram or to a *schematic*, which shows electrical connections of components. The lines in a block diagram or schematic that connect to things outside the diagram (whether the real world or other blocks) are called *ports*, and these ports are usually named. Because of the hierarchical nature of block diagrams, the points on a block that can be connected to are also referred to as ports of that block.

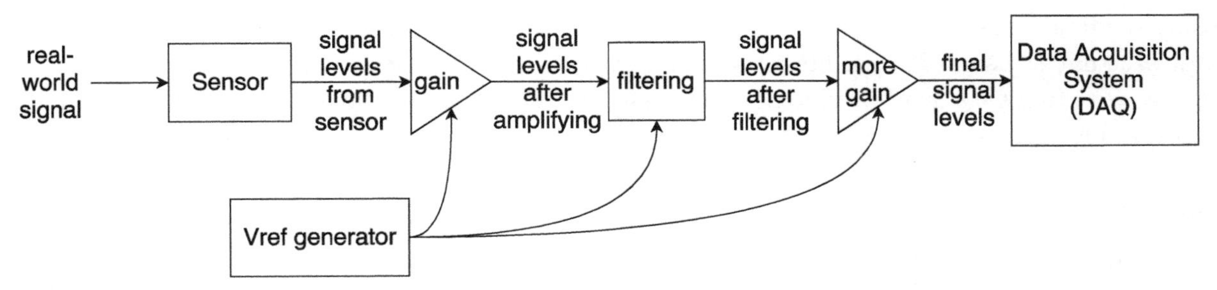

Figure 7.2: *This generic block diagram is typical of the block diagrams that we'll create in this course. There is some real-world signal coming in from the left, a sensor to convert it to voltage, current, resistance, or some other electrical parameter, an amplifier to increase the size of the signal, possibly some filtering to remove high-frequency noise or to block DC bias, some more amplification, and a data-acquisition system to read and record the signal. The block for generating a reference voltage,* V_{ref}*, is commonly needed for setting voltage levels for the amplifiers and filters.*
Figure drawn with draw.io [21].

In a good block diagram, both the blocks and the connections are labeled with extra information. Sometimes the information is just a name (like *preamp* or V_{mic}), so that the block or connection can be referred to in the documentation, but more information is useful.

For blocks, additional information is given about the function of the block—for example, *high-pass filter, 30 Hz* or *preamp, gain=100* or *voltage reference, 1.6 V.*

> The crucial information for the block is the *function* of the block, not its implementation. Part numbers and component values are generally inappropriate on block diagrams—those details are for a lower-level of abstraction (schematic diagrams).
>
> For connections, the crucial information about the signal that the connection carries should be put in a label for the signal. Is the signal voltage, current, or some other physical property? Is there a DC offset? What is the range of the signal? What frequency range is relevant? Is it a single-ended or differential signal?

Detailed information about the signals allows designing the blocks almost independently. If we know the voltage and frequency range of both the input and the output of an amplifier, we can determine the amplifier's gain and any filtering it might need to do. We can check the consistency of a block diagram just by doing local checks: does each block convert the described input to the described output?

Many of the signals we will be looking at in this course will consist of a constant direct-current (DC) voltage with a small time-varying alternating-current (AC) voltage added to it, where the AC voltage is the signal we are interested in. The constant DC voltage is often referred to as a *DC bias*, and may be needed to make sure transistors or other components operate correctly.

For example, we might have a wire from a microphone with the signal from the sound waves causing the voltage to fluctuate from 1.4 V to 1.6 V (centered at 1.5 V). It is handy to label the connection on the block diagram as 1.5 V DC, ±0.1 V AC.

Another sort of signal information that is often attached to connections is the frequency range. For example, music may have a 20 Hz–15 kHz range, while a pulse monitor may be looking at 0.2 Hz–30 Hz. The frequency ranges can be very important in designing filters and amplifiers, so it is convenient to have the information directly on the block diagram.

As another example, the input to a pressure sensor may be labeled as "±25 MPa, DC–20 Hz" to indicate the range of pressures and that we are interested in low-frequency fluctuations, down to constant pressure readings. The output of the sensor may be labeled "differential ±5 mV, DC–20 Hz, $V_{CM} \approx 1.5$ V" to indicate that the signal is a differential signal carried on two wires (but still a single line of a block diagram) with up to a 5 mV difference between the wires, but an average of the two signals (the common-mode voltage) of about 1.5 V. The frequency range is the same as at the input to the sensor.

7.10.3 *Schematics*

Include well-drawn and labeled engineering schematics for each circuit investigated. Don't try to describe wiring in words—draw a schematic diagram! Schematics should be provided not just for the final design, but for prototypes you rejected (and why they were rejected), for test circuitry you set up to understand or measure components, and for models you fit to components (such as the loudspeaker models of Lab 8).

> Provide schematics to name components before you refer to the component values in the main body of the report!

There is a big difference between block diagrams and schematics, both in what information they represent and what stage in the design process they are used:

> - Block diagrams are used early in the design process, to break a large design problem up into smaller, almost independent design problems. The connections between blocks are single lines representing signals that carry information, generally in a left-to-right information flow. Blocks are labeled with the function they perform, including critical parameters like gain or corner frequencies, while signals are labeled with parameters like DC bias, AC voltage range, and frequency.
> - Schematic diagrams are the next step after the block diagram, showing how to implement each block with components. Each line on the schematic is a wire, and components generally need two or three wires to be properly connected. Components are labeled with part numbers or component values, and wires may be named or left unlabeled.

Because the meanings of both the symbols and the lines are different in block diagrams and schematics diagrams, the conventions for the two should not be mixed. The boxes of a block diagram represent functions, and can have various implementations (hardware, software, chemical systems, etc.), but the symbols on an electronic schematic should be only for electronic components. If you need to indicate a connection of a wire to something not part of the schematic diagram, use an input or output port, rather than putting in a block diagram symbol.

> Both the block diagram and the schematic diagram should be drawn before building anything in the lab—one of the first responses to any request for help in the lab is "Show me your schematic!" Without an accurate schematic diagram, debugging is extremely difficult, as it is not possible to separate design errors (wrong schematic) from implementation errors (circuit built doesn't match the schematic).

For this course, and generally for professional reports, schematics must be computer-generated, using a tool like the free web-based Scheme-it provided by Digi-Key [18].

Almost every component should have a value or part number displayed on the schematic. Some may have only a name, with the values for that component discussed in the main body of the text—but that style should only be used if the component has several different values or an unknown value. The value for a component should be consistent throughout the report—any discrepancy between the schematic and the main body or between the schematic and the block diagram should be noted in the caption of the schematic.

Figure 7.3 shows a few of the symbols we commonly use. If you need a symbol for a component not provided by your drawing tool, the standard practice is to create a rectangular box with connections corresponding to the pins of the component. In Scheme-it [18], there is a "Build A Symbol" menu for creating such symbols, adding pins on the left (for inputs), right (for outputs), top (for power and some control inputs), and bottom (for ground and some control inputs). These custom symbols can have wires connected to them like any of the built-in symbols.

Every pin on a component of a schematic should either have a wire connected to it or be labeled with "NC", indicating that there is no connection for that pin. Pins that have no wire and are not labeled are ambiguous—is there a wire missing or is the pin deliberately not connected? For example, if you use a function generator symbol to represent the function generator of an Analog Discovery 2, there are two pins on the symbol: the pin labeled "+" is the output of the function generator and the other pin needs to be connected to ground, which is the internal connection provided by the Analog Discovery 2.

> Two wires that cross at right angles are interpreted as *unconnected.* To show that the wires are connected, you need a *connection dot.* Scheme-it adds the connection dots automatically when you drag a wire to touch another wire.

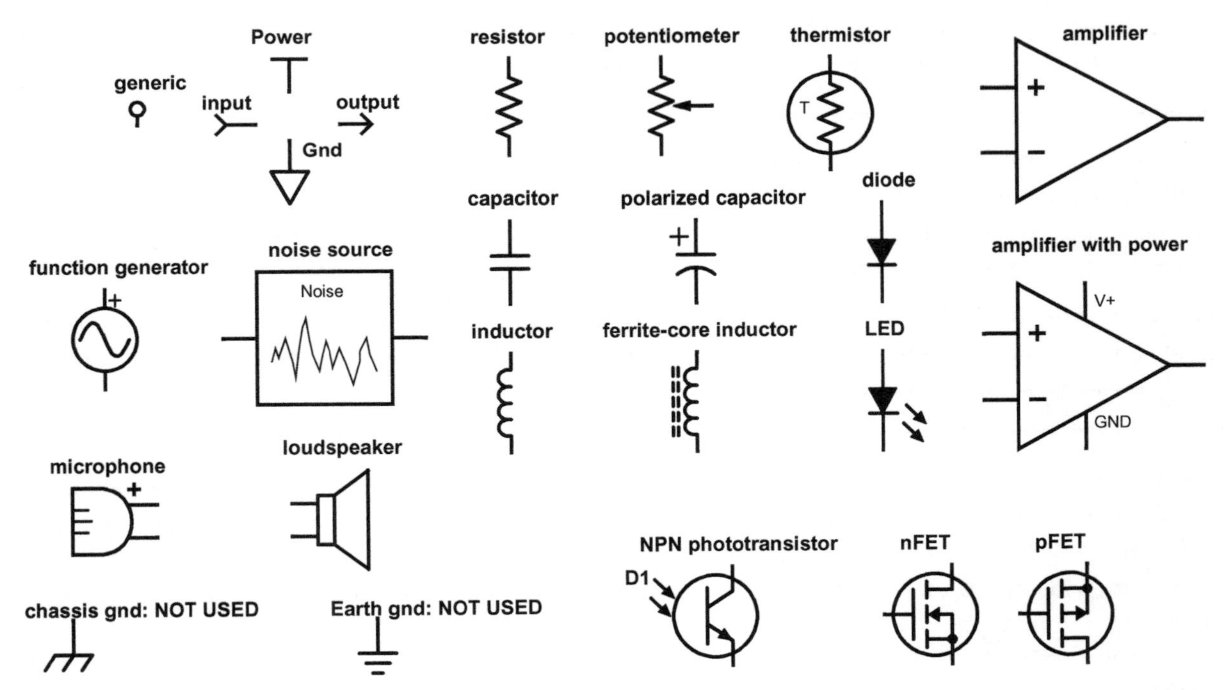

Figure 7.3: *Here are a few of the standard schematic symbols used in this book. We use the American symbols for resistors and capacitors, not the European ones, as the American symbols are visually more distinctive.*
We do not use the battery symbol for voltage sources—that symbol should be reserved for chemical-powered batteries, which we do not use in this course. Use the power-port symbol instead. The symbols for power ports and ground ports cannot be rotated—the ground triangle must always point downward and the power port must always have the crossbar on the top.
Figure drawn with Digi-Key's Scheme-it [18].

Connections to parts of the circuit outside the current schematic diagram are shown with *ports*. There are several different port symbols used for different purposes:

input Input ports are bare arrowheads with a wire coming out of the point of the arrowhead into a wire of the circuit. They are used to indicate an information-bearing signal coming into the circuit. The input port should be labeled with the name of the input signal. **Do not use the input-port symbols for power to a circuit or chip—use the power-port symbols.**

output Output ports are arrows with wires coming into the shaft of the arrow. They are used to indicate an information-bearing signal leaving the circuit. The output port should be labeled with the name of the output signal.

power Constant voltages coming from outside the circuit are drawn as a flat horizontal line, with the wire from the circuit connected to the middle of the line from the bottom. **The T-shaped connector cannot be rotated without losing its meaning.** The power port should be labeled with the constant voltage, either as a number (like `+5V`) or a name (like `Vref`).

ground The ground line has a special symbol, different from other power ports. It is shown as an equilateral triangle, point down, with the wire connected to the center of the top edge. **Like power ports, the ground symbol cannot be rotated without losing its meaning.**
There are other ground symbols (see Figure 7.3) representing true earth ground (important in power and radio electronics) and chassis ground (no longer commonly used, as sheet metal has not been commonly used as the physical support for electronic elements since the end of the vacuum-tube era). We have no use for these symbols in this course.

generic A small circle at the end of a wire is a generic port symbol, used for indicating connections that are not easily classified as input, output, power, or ground. Generic ports are sometimes also used for indicating connections within a schematic without having to draw long connecting lines—all ports with the same name are assumed to be connected together.

> Providing a power supply to a circuit is **not** considered an input to the circuit—use a power-port symbol, not an input-port symbol. If a block diagram has power connections, they should be on the tops and bottoms of boxes, not left inputs nor right outputs (except for the output from the power supply itself, if it is included in the block diagram).
>
> The schematic symbols for power ports and ground ports cannot be rotated—the ground triangle must always point downward and the power port must always have the crossbar on the top.

> When your schematic diagram for your whole circuit is broken up into different figures (a common way to show a complicated circuit), the port names should match between the different schematics. For example, all ports labeled V_{band} should be considered to be connected together. That means that names like V_{in} and V_{out} are very bad names, as the output of one block is the input for the next.

Although block diagrams can be created with Scheme-it, the tool is not well-designed for the task—you would be better off with https://www.draw.io or with Inkscape.

Screen shots don't work well in reports and are often completely unreadable. Scheme-it is capable of exporting SVG, PDF, and PNG formats—PDF is generally better for inclusion in a text document, as it retains its crispness even when scaled, while PNG is better for blog posts and other web-only documents, as PDF is not generally rendered by browsers when embedded in an HTML document.

I have sometimes had problems with PDF generated by Scheme-it having font-spacing problems, bounding-box errors, or color-fill problems when viewed with Preview on a Macintosh. I have gotten around these problems by exporting as SVG, and converting to PDF with Inkscape [45], which seems to produce more robust PDF. Using both `--export-area-drawing` and `--export-margin=3` in the Inkscape command line fixes the bounding-box errors, by determining the true bounding box for the items in the drawing.

> Schematics, part numbers, component values, and other highly technical information have very low redundancy—it is generally not possible for a person who sees an error or inconsistency to figure out what was intended. (I may be able to make up a correct circuit that bears some relationship to what you drew, but it may not be at all what you built.)
>
> Lab reports with incorrect schematics are automatic failures—incorrect schematics are simply not acceptable in this course.

If you are giving a circuit schematic, every wire must be correctly connected, every component must have the correct value, and pin numbers should be correct. Pin numbers are generally placed as labels near the wire where it enters the symbol for the part. Power and ground wires should be included, but it is quite common (even in this book) to omit power connections to chips when all the chips in the system have the same power supply. This shortcut is dangerous for beginning designers—one of the most common errors in lab is forgetting to connect the power to amplifier or logic chips!

Students must check their own and each other's work for accuracy, and obvious errors (like power-ground shorts, components connected on only one end, missing component values, or component values inconsistent with the main body of text) must be corrected. For a biotech student, the equivalent sorts of errors would be getting the wrong reagent in a protocol, putting ice in an autoclave, or replacing μg with mg.

7.10.4 *Graphs*

Transcribe all relevant data from your engineering notes into your report. You do not need to provide tables of the 1000s of data points you collect—but you should graph them. Since you may be collecting 1000s of data points in one experiment using a microcontroller, it will not be acceptable to hand-plot the data. You must use a computerized graphing tool (such as gnuplot).

When using the Analog Discovery 2, there are two options for exporting information: images and data.

> Always export **data** not **images** from software tools, when you have a choice.

Turning data into images is fairly easy using tools like gnuplot, and you can easily rescale the data, add appropriate axis labels, fit models to it, plot just a narrow range of interesting points, add arrows and labels to point out features, and so forth; but changing images into numerical data is much more difficult. Fixing an image that doesn't highlight the point you are trying to make can be nearly impossible. The xkcd cartoon ".NORM normal file format" (https://xkcd.com/2116/) is relevant here [76].

Excel is not an acceptable graphing tool after elementary-school science fairs—learn to use one that works (there are 100s available, but gnuplot will be the one demonstrated in class). You will often need to fit nonlinear models to your data, so make sure that the tool you choose can do that easily, and that you can use log scales for the axes. See Section 2.3.3 for an explanation of when to use log scales.

A plot should always have a meaningful title (usually at the top) and the axes should always be labeled. The axis labels should include the appropriate units (usually in square brackets: *thermistor resistance* [*ohms*] or *temperature* [°C]).

When talking about a plot, remember that the "versus" notation always puts the y-axis first: "voltage vs. time" means that time is on the x-axis (horizontal) and voltage is on the y-axis (vertical).

Don't mix unrelated things or things with different units on the same axis (like resistance vs. temperature and voltage-vs.-temperature on the same plot)—that practice just leads to confusion and hard-to-interpret graphs. Try to keep each axis having a single meaning. You'll see some places where I violate this rule slightly, such as Figure 20.2, where I have one y-axis labeled with pressure in kPa, and the other y-axis labeled in cm H_2O—both y-axes have the same meaning (breath pressure), but one is in the nonstandard units used in medical texts.

However, do collect all *related* data into a single figure. For example, you may have collected current vs. voltage data for different voltage ranges in different experiments, but combining them into a single plot is useful, even if different measurement methods were used for low voltages and high voltages.

Don't use graphics with black backgrounds—they don't print well nor project well.

Don't use PNG and other raster image formats for graphs, as they tend to look fuzzy and don't scale well—use PDF or some other format that supports vector graphics (see Section 7.10 for more information about incorporating graphical elements in your reports).

Instead of using WaveForms 3 screen shots, export the *data* rather than the *image* and replot it with a tool that lets you label each axis and the curves better (such as gnuplot).

Many of our graphs in this course will consist of collected data and curves for models that have been fit to that data. Generally, we use separate points for the data, not connecting them with lines, so that we can see at a glance where the measurements are. If the points are so close together that they all overlap on the plot, then it may be better to use lines connecting the data points rather than separate points. The models we use are generally continuous and should be drawn with smooth curves, rather than as separate points.

When you are creating models in gnuplot, use good programming practices. Any calculations that are needed on the data should be done in the plotting script, not with a hand calculator.

Introduce meaningful variable names, rather than using numeric constants, and use functions that capture the meaning of algebraic expressions, rather than large piles of unreadable hand-created algebra.

Using names for values and setting each value in a single location in the script makes changing component values much easier, as there is only one place to change. Starting each variable name with a mnemonic letter to remind the reader of the type of value is helpful: R for resistance, C for capacitance, L for inductance, f for frequency, G for gain, V for voltage, I for current, and so forth.

If there are several components that happen to have the same value, having a different name for each component (rather than using just numbers in a formula) makes it easier to check whether the meaning of the algebraic expressions matches the structure of the circuit.

Using functions that encapsulate common formulas (like `Zc(f,C)` for the impedance of a capacitor C at frequency f) makes the script much easier to read and debug, because the parts of the formula can reflect the structure of the circuit directly, without the reader having to do pattern matching on the algebraic expressions. Using such functions also reduces the chance of typographic and algebraic errors.

7.10.4.1 *Basic gnuplot commands*

You can run gnuplot in two different ways: interactively or by running a script you have edited in a text editor. To run gnuplot interactively, you just type the command `gnuplot` in a terminal window—after that anything you type in that terminal window will be interpreted as a gnuplot command. You can run from an existing script file by starting an interactive session, then typing `load 'foo.gnuplot'`, where `foo.gnuplot` is replaced by the name of the file.

Including the gnuplot scripts you used as an appendix to your report is a good idea, as it allows the grader to help you debug your gnuplot scripts.

The commands in gnuplot are used to plot data, plot mathematical functions, annotate the plots, control various aspects of the resulting drawing, and fit mathematical models to data. This book is not a full tutorial on gnuplot, but only gives a few of the more essential commands—much more information can be found online and from gnuplot itself using the `help` command. Figure 7.4 has examples of some of the standard commands that I use in many of my gnuplot scripts, and Figure 7.5 gives an example of a short script for plotting data.

Some of the key points of `definitions.gnuplot` are that it sets up log-log plotting (so that functions that are frequency raised to a power plot as straight lines), labels the x-axis, defines standard impedance functions for capacitors and inductors, and gives some utility functions for voltage dividers and parallel

```
# definitions.gnuplot script for amplitude or impedance as function of frequency
set logscale xy
set xlabel "frequency [Hz]"
set xrange[10:1e6]
set format x "%.1s%cHz"
set mxtics 10    # to get standard logarithmic tics at 10,20,30,...,90
set mytics 10    # to get standard logarithmic tics at 10,20,30,...,90
set key bottom left Left
set samples 10000
set grid

divider(Zup,Zdown) = Zdown/(Zup+Zdown)

j=sqrt(-1)
conjugate(a) = real(a) - j*imag(a)
phase(v) = imag(log(v))
real_power(V,Z)= real(V*conjugate(V/Z)) # V is RMS voltage phasor

Zc(f,C) = 1/(j*2*pi*f*C)  # f= frequency [Hz], C=cap [farads]
Zl(f,L) = j*2*pi*f*L      # f=frequency {Hz], L=inductance [henries]
Znonstandard(f,M,alpha) = (2*pi*j*f)**alpha*M # generalization of L, R, and C

Zpar(z1,z2) = z1*z2/(z1+z2)
Zpar3(z1,z2,z3)= 1./(1./z1 + 1./z2 + 1./z3)
```

Figure 7.4: *Gnuplot script to set up functions and x-axis for Bode plots and other functions of frequency. This script is available at https://users.soe.ucsc.edu/~karplus/bme51/book-supplements/definitions.gnuplot. Copying scripts from the PDF file for the book doesn't work, because of character substitutions made in the typesetting.*

```
set title "1 second of 8 Hz 70% ramp wave at different sampling rates" font "Helvetica,14"
set xlabel "time [s]" font "Helvetica,14"
set ylabel "voltage [V]" font "Helvetica,14"
set margins 9,4,4,4

set key top center
set style data lines
set xrange [1.:2.]
set yrange [0:3.9]

plot "8Hz-at-200Hz-70.txt" using 1:2 title "sampled at 200Hz" linecolor "red" , \
     "8Hz-at-200Hz-70.txt" using 1:3 title "sampled at 10Hz" linecolor "blue"
```

Figure 7.5: *This simple gnuplot script reads data from the file "8Hz-at-200Hz-70.txt" and plots the data as a continuous curve. The data file is assumed to have three columns of data: first a time in seconds, then two voltages in volts.*

impedances. Bode plots should always be log-log plots (unless the y-axis is in decibels, which is already a logarithmic transformation—decibels should be shown on a linear scale).

You don't need to copy these definitions into your own file (though you can)—you can just put `load 'definitions.gnuplot'` as the first line of your gnuplot file, then redefine in your script any part of the definitions file that doesn't do what you want.

More explanations of the use of `definitions.gnuplot` are given in Section 11.3.

Each command in gnuplot takes one line. If you have a long command, you can break it into multiple lines by putting a backslash at the end of a line to indicate that the command continues onto the next line. I find continuation lines particularly useful for `plot` commands, which often get quite long.

I generally start a gnuplot script with commands for the framework of the plot, then define any mathematical functions needed, do model fitting (if relevant), and end with a `plot` command to plot the data and any models fitted to the data. Here are some of the most commonly used commands:

- `set title "Title for the plot"` sets the title for subsequent plot commands. Every plot should have a descriptive title.
- `set xlabel "Label for x-axis"` sets the name for the information on the x-axis.
- `set ylabel "Label for y-axis"` sets the name for the y-axis. Both axes should always be labeled.
- `set xrange [10:100]` sets the range of values for the x-axis. You can use an asterisk instead of a number for either the low or the high value, to get gnuplot to select a value automatically, based on the data: `set xrange [*:*]` sets both endpoints automatically.
- `set yrange [10:100]` sets the range of values for the y-axis.
- `set logscale x` chooses a logarithmic scale for the x-axis, while `unset logscale x` chooses a linear scale.
- `set logscale y` chooses a logarithmic scale for the y-axis.
- `set key top right` puts the legend (or key) in the top right corner. There are many options for placing and formatting the key—use `help set key` to get information about the options.
- `set xtics` and `set mxtics` can be used to control the way that tick marks are made on the x-axis—use `help set xtics` and `help set mxtics` for more information.
- `set format x` is used to control the format of the labels for tick marks on the x-axis. The default format is often acceptable, but there are some fancy things that you can do (like automatically including the metric prefixes, to get axes labeled in Hz, kHz, and MHz, for example).

Constants can be set and functions defined using the "=" sign:

```
j = sqrt(-1)
tau = 2.*pi
Zc(f,C) = 1./(2.*pi*j*f*C)
```

It is a good idea to use a decimal point with any numbers, as gnuplot distinguishes between integers and floating-point numbers, and integer arithmetic may give unexpected results. For example, "1/3" is zero, but "1./3." is 0.3333 .... You can have string variables as well as numeric ones, which can be useful for creating labels or file names.

Functions can take up to 12 arguments. As a matter of good style, the first argument should usually be what varies as you plot the function, with subsequent variables used as parameters to control the exact shape of the function. For example, in the definition of "Zc" above, which is the definition of the impedance of a capacitor (see Section 10.2.2), the first variable is the frequency and the second one, the capacitance—we often plot the magnitude of the impedance as a function of frequency.

For more information about defining constants and functions, use the `help user-defined` command.

The main command in gnuplot is `plot`, which is used to plot both data and mathematical functions. Several different data sets or functions can be plotted with a single plot command—the different specifications are separated by commas.

To plot data from a file, we give the file name as a string, then say which columns of data to take from the file. Each line of the file is one data point, though data files may have comments in them, by starting a line with `#`. Each line is assumed to have multiple columns, which are defined by white space (any number of spaces or tabs) between the columns.

For example,

```
plot "datafile.txt" using 3:5 title "raw data"
```

will plot points where the x value is taken from the third column and the y value is taken from the fifth column. The title for the data is put in the key (you can use `notitle` to suppress the entry in the key).

We can do math on the data values in the "using" fields, by putting that field in parentheses and use a dollar sign before any column numbers. For example,

```
plot "datafile.txt" using 3:(3*($5-$6)) title "scaled difference"
```

takes the x values from column 3, as before, but the y value is three times the difference between columns 5 and 6.

Mathematical functions of x can be plotted just by giving the expression with x as a free variable. We often plot data that we've collected and models fitted to that data on the same plot. For example, if we have collected data about the magnitude of impedance of a 1 μF capacitor as a function of frequency, we might plot it with

```
plot \
    "capacitor-data.txt" using 1:2 title "measured", \
    abs(Zc(x,1e-6)) title "predicted"
```

using the previously defined definitions of `Zc` and `j`. I've used backslashes so that I could put each of the elements being plotted onto a separate line, to make editing the script easier. The frequency is assumed to be in column 1 and the measured magnitude of impedance in column 2. The argument `f` of `Zc` is taken from the x-axis, and the argument `C` is set to the value 10^{-6}.

The order in which you plot data files or curves matters, as the key is constructed in the order in which the plots are mentioned in the command. For maximum readability, the curves should be ordered with the highest curve mentioned first, so that the key lists the curves in visual order from top to bottom.

In the `plot` gnuplot command, we can choose `with points` or `with lines` for each data file or function that we plot. By using the command `set style data points` before invoking `plot`, we can have the default be points for data files and lines for functions, which is most often what we want. Occasionally, it is clearer to show both the explicit data points and a continuous curve connecting them, which can be done with the modifier `with linespoints`.

We can construct the title for one trace of a plot using the `sprintf` function, borrowed from the C programming language:

```
Cfit=1e-6
plot \
    "capacitor-data.txt" using 1:2 title "measured", \
    abs(Zc(x,Cfit)) title sprintf("fitted C=%.2f uF",Cfit*1e6)
```

The `help sprintf` command in gnuplot does not give much information, but you can easily find documentation for sprintf on the web.

We often want to fit a mathematical model to the data we have collected. The model should have some theoretical justification, if possible, as fooling yourself by fitting an inappropriate model is very easy (particularly with noisy data) [77].

What model fitting does is to adjust the parameters of a mathematical function so that the difference between the function and the data points is small. Technically, what is minimized is the sum of the squares of the differences between the function and the data points—for a set of points x_i, y_i, model fitting minimizes

$$\sum_i |f(x_i) - y_i|^2 \ .$$

Visually, this balances errors when the data is plotted on linear y axis, but does a poor job when the data is plotted on a log y axis—for data and models that should be plotted on a log-log plot or a semilogy plot, one should fit the log of the model to the log of the data, minimizing

$$\sum_i |\log(f(x_i)) - \log(y_i)|^2 \ .$$

For simple scalar functions, gnuplot provides a fairly simple interface. Continuing the example of having data that measured the magnitude of impedance, we could fit a model of a capacitor to that data:

```
Cfit=1e-6
fit abs(Zc(x,Cfit)) "capacitor-data.txt" using 1:2 via Cfit
plot \
    "capacitor-data.txt" using 1:2 title "measured", \
    abs(Zc(x,Cfit)) title sprintf("fitted C=%.2f uF",Cfit)
```

The `fit` command takes a mathematical expression (with a free variable x), a data file specification, and then the `via` argument that specifies which variables in the mathematical expression need to be adjusted to fit the model.

Because of inherent limitations of model-fitting algorithms, it is important to give reasonable initial values to any parameters you are trying to adjust in the fit. If no values have ever been given for a parameter, gnuplot will start with the value 1, which may not result in convergence for the fitting procedure. Section 29.2.2 provides more examples of fitting models to data.

A lot of students have asked for advice on how to provide different scales on the left and right sides of the plot. You can do this by turning of the y-axis ticks on the right side (`set ytics nomirror`) and turning on $y2$ ticks and ranges. If the right-hand axis is for different sorts of values than the left-hand axis, then you can plot using it by adding `axes x1y2` to a plot command. If the right-hand axis is a rescaling of the left-hand axis (pressure in different units, for example), you can generate the scaling automatically by providing a pair of functions to convert y to $y2$ and vice versa. For example,

```
mmHg_per_kPa = 7.50061683
set link y2 via y*mmHg_per_kPa inverse y/mmHg_per_kPa
set ytics nomirror autofreq
set y2tics nomirror autofreq
```

could be used if the y-axis is in kPa and the $y2$-axis is in mm Hg. Both the "via" and the "inverse" functions are required, and both use "y" as the free variable.

You can do similar things for the bottom and top axes with x and $x2$.

7.10.5 *Color in graphs*

Colors are often used in graphs to identify different curves or sets of data, and the default color choices made by plotting programs are often not helpful.

For example, if you have multiple sets of data and a model fitted for each data set, it may be useful to use the same color for the model as for the data it is fitted to, distinguishing the two by using lines for the model and points for the data.

In gnuplot, you can add the modifier `linecolor` to a plot command, with the color specified either by name or with a hexadecimal number giving the red, green, and blue components of the color. The gnuplot command `show colornames` lists the names that gnuplot understands and gives their translation to RGB color values.

The command

```
plot 'data.txt' using 1:3 with points title 'measured' linecolor 'red', \
    model_RC(x,R,C) with lines title \
          sprintf("%.2f ohms + %.2f %nF",R, 10e9*C) linecolor 'red'
```

will plot both data and model in red, with the data as points and the model as lines. The parameters of the model, R and C, are reported in the legend for the plot.

Avoid using very light colors (such as yellow or grey90), as they will not be easy to see on a white background.

Exercise 7.3

Create a plain text file (fake-data.txt) with the following data in it:

```
# data for homework exercise
0  100
1   50
2  200
30  30
```

and write a gnuplot script that plots the fake data. Include both the script and the plot in the homework. What problems (if any) do you see with the plot, and how might they be fixed?

If multiple exercises from Exercises 7.4–7.7 are assigned, only one example gnuplot script needs to be included in the turned-in assignment, as all the scripts are quite similar.

Exercise 7.4

Write a gnuplot script to plot all four of the following functions on the same log-log plot, from $x = 10$ to $x = 1000$:

- $5x + 500$,
- $x^2/100$,
- $2^{x/100}$, and
- $1000 \log_{10}(x)$.

Include the plot and the gnuplot script in your answer. Which function plots as a straight line?

Exercise 7.5

Plot all four of the functions of Exercise 7.4 on a semilogx plot (that is, with a log scale for the x-axis and a linear scale for the y-axis). Provide your gnuplot script as part of the solution. Which function plots as a straight line?

Exercise 7.6

Plot all four of the functions of Exercise 7.4 on a semilogy plot (that is, with a log scale for the y-axis and a linear scale for the x-axis). Provide your gnuplot script as part of the solution. Which function plots as a straight line?

Exercise 7.7

Plot all four of the functions of Exercise 7.4 with linear scales for both axes. Provide your gnuplot script as part of the solution. Which function plots as a straight line?

7.10.5.1 *PDF from gnuplot*

Many students will create their PDF output interactively from the "qt" or "wx" windows, which have an export button for creating an output file corresponding to the current display in the window. Be sure to export as PDF, not as an image, to get scalable, high-resolution plots.

You can do the creation of a PDF file without having to start an interactive window with gnuplot in several ways. For example, you can create a file (called "pdfscript.gnuplot" here) containing two lines:

```
set terminal pdfcairo color
load 'script-name.gnuplot'
```

and use the command

```
gnuplot < pdfscript.gnuplot > script-name.pdf
```

to produce PDF vector images that can be included in documents with no loss of resolution. The `set terminal pdfcairo color` command tells gnuplot to generate PDF-formatted output for color printing, rather than displaying the graph on the screen.

Because I have around a hundred gnuplot-created figures in this book, I use a Makefile with the rule

```
%.pdf: %.gnuplot
	echo "set terminal pdfcairo color; load '$<'" \
	| gnuplot >$@
```

to create pdf files automatically from gnuplot files.

The command `help set terminal` in gnuplot will explain what output formats and options are available—it can vary from installation to installation, depending on what optional features were included when the installation was done.

7.10.6 *Listing programs and scripts*

There are three ways to show computer programs or scripts in a design report:

- Short snippets can be displayed much like math formulas, though they usually are not grammatically part of a sentence, except as the restatement after a colon.
- Larger programs that are still less than a page long can be put into figures, as in Figure 7.4.
- Still larger programs should be stuck in an appendix, with an explanatory paragraph before the script.

In all three cases, the program has to be formatted correctly. There are different packages for LaTeX catering to different people's taste in formatting. For this book, I use the `fancyvrb` package.

To do a short snippet, I use the `\begin{Verbatim}` environment, but for blocks that are complete programs or files, I keep the program in its own file and use `\VerbatimInput` to grab the file and format it. For example, to set Figure 7.4, I used

```
\begin{figure}
\centering
\VerbatimInput{definitions.gnuplot}
\caption[Script definitions.gnuplot to set up Bode
  plots]{\label{fig:definitions-gnuplot}Gnuplot script to set up
  functions and x-axis for Bode plots and other functions of frequency.


  This\index{definitions.gnuplot} script is available at
  \url{https://users.soe.ucsc.edu/~karplus/bme51/book-supplements/definitions.gnuplot}.
  Copying scripts from the PDF file for the book doesn't work, because of
  character substitutions made in the typesetting.
}
\end{figure}
```

The snippet that I just included above was done with a `\begin{Verbatim}[numbers=none]` environment.

To control the listing format for the book, I used

```
\fvset{numbers=left,fontsize=\footnotesize,obeytabs,frame=single}
```

in the preamble, though I often override the default numbering of lines by specifying `\begin{Verbatim}[numbers=none]`.

7.11 Word usage

Words are very important to engineers, and the words need to be used very precisely. Here are a number of somewhat random usage notes, in alphabetical order:

able to Students often write reports saying things like "We were then able to plot . . . . " Your reader is interested in knowing what you *did*, not what you *were capable of doing*. Use simple past: "We then plotted . . . . " (See Section 7.7.)

above and below In Section 7.10, I mentioned that figures and tables are floating insertions and so should be referred to by name, not with the words "above" or "below," but the same warning applies to other cross-references within a document. Refer to equation numbers, section numbers, or page numbers (all of which LaTeX can generate automatically from `\label` commands), so that rearrangements of the text do not mess up the pointers.

accommodate The verb "accommodate" requires a direct object, not a prepositional phrase beginning with "for". It means to make room for something or to provide someone with something desired. Example: *The PCB accommodates one MCP6004 chip and ten resistors, though more resistors can be added in undedicated spaces.*

actually In informal American English, "actually" is used as a rather meaningless intensifier. In formal written English, "actually" is rarely used—most often to contrast what was planned from what actually occurred. Simply removing most occurrences of "actually" will improve most technical writing.

affect and effect Students often confuse the words "affect" and "effect". There are four different words with two different spellings:

> **affect (transitive verb)** To say that A affects B means that there is some causative relationship between A and B, so that changes to A result in changes to B. Example: *The voltage across a resistor affects the current through the resistor.*
>
> **effect (noun)** An *effect* is the result of a cause. Example: *Increased power usage is an effect of raising the power-supply voltage.*
>
> **affect (noun)** The noun "affect" is specialized jargon in psychology for the experience of feeling or emotion—it has no use in the design reports for this course.
>
> **effect (transitive verb)** The verb "effect" means to create something or cause it to be created. Because it is rarely used correctly and sounds stilted even when correct, avoid the use of "effect" as a verb.

allow vs. allow for The verb "allow" changes its meaning subtly when you add the preposition "for". Generally, "allow" by itself means to permit or enable something: "Zoom allows up to 49 participants per screen in Gallery View, if your computer is powerful enough." "Allow for" expresses a design goal or modeling technique to handle a possible contingency: "Having three identical processors allows for the failure of any one processor without error" or "If you allow for inflation, there has been no real growth in faculty salaries."

although Although we often start sentences with "although", it is not a sentence adverb—it needs to start a clause.

amount Students overuse "amount of" for nouns that can be quantified without the phrase. We talk about "current", not "amount of current", and "gain" not "amount of gain". Students also use "amount of" incorrectly with countable nouns—"amount of" can only be used with uncountable nouns. See also *countable and uncountable*, page 102.

as for The usage "As for x, . . ." sounds like a bad translation of the Japanese topic marker. In English, the standard way to indicate the topic for a sentence is to make the topic the subject. So the construct "As for x, it is . . ." should be tightened to "X is . . .".

as such Some students throw in the phrase "as such" without any apparent awareness of its meaning. The phrase means something like "using the precise definition", so you might say something like "there is no resistor as such, but impedance of the skin acts as a 1 MΩ resistance". If you are not using "as such" this precisely, then just remove it from your writing, as it rarely adds much to the meaning.

based on The correct expression is "based *on* something"—"based off of something" is simply incorrect. Not even the Google Books Ngram Viewer finds support for "based off".

before and after Because block diagrams have directed information flow (from left to right by default, or explicitly indicated by arrows), it makes sense in a block diagram to refer to one block being before another or to talk about signals before and after being processed by a block.

For example, it makes sense to say something like "The signal is buried in 60 Hz noise, but after the notch filter, the signal-to-noise ratio is over 40 dB".

Schematics, however, are undirected graphs of connections between components, with no inherent direction. It usually does not make sense to say things like "before the resistor". Instead, label the nodes in the schematic and refer to them by name. Also avoid "first", "above", and "below" when talking about circuits—instead talk about what connections are made. For example, "the resistor between V_{out} and ground" is a less ambiguous reference than "the resistor below V_{out}".

being that The phrase "being that" is a rather informal usage that should be replaced with "because" in technical writing.

between The preposition "between" requires two objects, separated by "and". It cannot be used with a single object (not even if that object is a range), nor can the objects be separated by a dash.

The phrase "in between" is almost always wrong in technical writing—it is properly used when "between" is a preposition without an object: *The gain is high at each of the corner frequencies and in between.* But it is almost always better to use "between" as a preposition with two objects.

breath and breathe The word "breath" is a noun—the corresponding verb is "breathe".

calculate for The verb "calculate" takes a direct object without a preposition—the usage "calculate for" is a confusion of "calculate" and "solve for". We can solve an equation or inequality for a particular variable, changing it to an equation or inequality that has just the single variable on the left-hand side. We can then use that equation (or inequality) to calculate the value of the variable (or constraint on the value), given values of other variables in the equation.

calibrate *Data* is not calibrated—devices are. Data can be measured, recorded, or analyzed, and the result of all that can be the calibration of a device or system—though most often data is collected for some other purpose.

cMOS The acronym "cMOS", which stands for "complementary metal-oxide semiconductor", is only used as a modifier, not as the head noun of a phrase. We can talk about cMOS logic or a cMOS output stage, but it would be wrong to try to make it into a countable noun and talk about "a cMOS". See also *countable and uncountable*, page 102.

coil vs. twist Wires can be coiled or twisted, both of which refer to helical configurations, but different ones.

A wire is *coiled* when it is in a helical or spiral conformation intended to *increase* its inductance. Generally, the diameter of the helix is substantially larger than the center-to-center separation of the wires.

Wires are *twisted together* when they are in a helical conformation to *reduce* inductance or to match capacitive coupling. The helix generally has a very small diameter (about the diameter of the wire), and the pitch is usually fairly large. The purpose of twisting wires together is to make the wires follow approximately the same path so that they are subject to the same electrical and magnetic fields and thus do not pick up much noise as differential voltage (from capacitive coupling) or current (from magnetic coupling).

compound nouns Some words are created by joining other words together—a classic example is "doghouse". One such word in this book is "loudspeaker", which should not be confused with "loud speaker". I am a loud speaker, because I can make myself heard without needing a loudspeaker.

Compound nouns often evolve in English, starting as separate words, going through a phase of being hyphenated, and eventually becoming single words. If you are not sure where along this continuum a particular compound word is, using Google Ngram Viewer [34] may be a good way to check. For example, "text-book" changed to "textbook" around 1920 [36].

As another example, comparing the n-grams "band pass", "band-pass", and "bandpass" reveals that the two-word phrase started out as the most common, but was quickly overtaken by the hyphenated phrase (probably because of hyphenation rules—see Section 7.12.7), but in the mid-1950s was overtaken by the single word "bandpass", which is now 2.6 times as popular as "band-pass" [35].

Because "low-pass" and "high-pass" are still the preferred forms, I have chosen to use "band-pass" in this book for consistency, but I will certainly accept "bandpass" in your writing—but try not to mix "band-pass" and "bandpass" in the same document.

A pair of compound nouns that are currently making the transition from separate words to hyphenated words are "x-axis" and "y-axis". According to Google ngrams, "x axis" and "x-axis" were about equally common up to 1980, but the hyphenated form has been gradually taking over since then [38]. I have chosen to use the hyphenated form in this textbook, but will accept the unhyphenated form (if used consistently). When used as a modifier (x-axis label), only the hyphenated form is acceptable—see Section 7.12.7.

I have also changed from using "down-sample" to using "downsample", as Google ngrams reports the unhyphenated form as being about twice as frequent.

constraint vs. restraint A *constraint* is an externally imposed restriction on something and is usually expressed as an inequality: $I_{\text{out}} \leq 20\,\text{mA}$. In our applications, we get constraints on voltage, current, power, gain, resistance, and so forth.

The term *restraint* is a forcible limitation on a person's actions (either self-imposed or by others), and so it does not have much use in technical writing. People can be restrained, but voltage, current, and other numeric values can only be constrained.

countable and uncountable English divides nouns into two groups: *countable* and *uncountable*. Countable nouns can be made plural and generally use the "a/an" indefinite article when singular: "I have an apple and two pears". Uncountable nouns are grammatically singular and do not take an indefinite article: "I have bread in the oven". If you need to count something that is uncountable, you need to add a counting word, which is specific to the noun: "three loaves of bread", "fifteen bits of information", or "five bottles of wine".

Demonstrative adjectives, the definite article, and possessives can be used with either countable or uncountable nouns—only counting adjectives and the indefinite article are restricted to countable nouns.

In addition to whether or not you can use the indefinite article "a/an", countability controls whether you use "much" or "many", "less" or "fewer", and "amount of" or "number of". Countable nouns use "many", "fewer", and "number of", while uncountable ones use "much", "less", and "amount of".

There isn't much logic to what nouns are countable—it is an arbitrary property of the word and needs to be learned along with the word. Unfortunately, few dictionaries mark countability of nouns—one that does is the online *Oxford Learner's Dictionary* [83]. The words "software" and "code" (referring to programs) are both uncountable, but "program" is countable.

Some nouns are already numeric, such as "frequency" and "voltage"—such nouns are not usually used with "amount of" or "number of", but are otherwise generally treated as singular nouns: "At a frequency higher than the corner frequency of 37 Hz ... ".

dangling modifier A *dangling modifier* is usually a phrase or adjective at the beginning of a sentence, before the subject of the sentence, which implicitly is grammatically attached to the subject, but which semantically should be attached to something else. For example, "looking at the schematic, the microphone is biased to 2 V" is incorrect because the microphone is not looking at the schematic. The problem can often be fixed by introducing the missing subject: "We can tell by looking at the schematic that the microphone is biased to 2 V". Even better, we can often reword the sentence so that there is no missing subject to pick up the wrong referent: "According to the schematic, the microphone is biased to 2 V".

Quite often, the dangling modifier results from converting an active sentence to passive, eliminating the original subject that the modifier should be attached to. Converting back to an active sentence can resolve the dangling modifier. For example, "Measuring with a handheld ohmmeter, the resistor was 9.98 kΩ" would be better written as "We measured the resistor with a handheld ohmmeter to be 9.98 kΩ" or, if you must keep the passive voice, "Measured with a handheld ohmmeter, the resistor was 9.98 kΩ".

data I don't care whether you consider "data" to be a plural noun (plural of "datum") or an uncountable noun, which behaves like a singular noun in English—just be consistent. A hundred years ago, "data" was about 4–5 times more likely to be treated as a plural noun, but usage has been shifting, and now plural and uncountable uses are about equally common. See also *countable and uncountable*, page 102.

defining context The first time you introduce a new term you should define it. When a word occurs in such a *defining context*, then it is a courtesy to the reader to italicize the word there, so that the reader can quickly scan the page to find the definition of an unfamiliar word.

demonstrative adjectives Never use "this", "that", "these", and "those" as pronouns, because readers may find it hard to figure out exactly what the pronouns' antecedents are. Instead use them as *demonstrative adjectives*, which are words that occur in place of the definite article "the".

Search your report for every occurrence of "this", "that", "these", and "those"—where they are being used as pronouns replace them with noun phrases like "this method", "that approach", "these models", and "those resistors". If you can't figure out the right noun to use, then your reader certainly won't be able to figure out what you mean by the pronoun!

Quite often, you can remove the word "this" entirely by using a relative clause starting with "which". For example, you can replace "*We used a 1 kΩ resistor. This takes 1.65 mA.*" with the single sentence "*We used a 1 kΩ resistor, which takes 1.65 mA.*" You should almost always replace". This is because" with ", because".

Particularly objectionable are phrases like "This is where ... " and "This is when ... ", which generally are just fillers that contribute no content (see also *when and where* on page 109 and *fluff* on page 104).

For more information on demonstrative adjectives, see *the and this* on page 108.

derive The verb "derive" does not mean "take the derivative of". If you do need a single-word replacement for "take the derivative of", you can use "differentiate". The word "derive" in mathematics means to obtain a result by a series of mathematically correct steps—differentiation may have nothing to with the derivation.

each The word "each" is singular, whether used as a noun or as an adjective. You can say "each resistor is ... " (adjective) or "each of the resistors is ... " (noun), but either way you need a singular verb.

efficiency In technical writing, "efficiency" is a ratio of output to input (generally of power or energy) and is unitless—often expressed as a percentage. When applied metaphorically to human endeavors, it is a measure of how much desired work is produced divided by the amount of effort expended. Do not use it as a generic term for effectiveness or goodness.

equates The verb "equates" takes a subject as an actor and two objects as the things being made equal. It is rarely used correctly in student writing, and "equates to" is definitely incorrect usage. What students usually mean is "equals", and most of the time the verb is not needed at all, but is just part of padding to make their writing more verbose.

exponential The term "exponential" is a specific technical term, which refers to functions of the form $f(t) = A^{t/\lambda}$. It does *not* mean just any rapidly growing function, and it can't be intensified with modifiers like "very" or "highly". See also Figure 2.1 to see the difference between exponential and power-law functions.

extrapolate and interpolate Both "extrapolation" and "interpolation" are methods for estimating values of $f(x)$ from a set of $(a, f(a))$ data points. They do have slightly different meanings, though—extrapolation is the process of estimating $f(x)$ for a value of x that lies outside the range of all the data, while interpolation involves estimating for a value of x that lies between known data points.

the fact that Many people use the mouth-filling phrase "due to the fact that" when the simpler, more direct "because" would be better. I don't think I've ever seen an instance where "the fact" really needed to be highlighted like this. Avoid using "fact" with "that".

fitting We fit *models* to the *data*, never *data* to the *models*. Fudging your data to fit your theories is the worst sort of scientific fraud.

The best-fit model is not necessarily a straight line—the expression "line of best fit" only makes sense when the model is a straight line on the graph.

Always give the formula for the model you are fitting and report the values of its parameters—preferably in the same figure that shows the data and the curve for the fitted model.

fluff Students often insert low-content filler phrases like "it is important to note that", "as we can notice", or "because of the fact that". These phrases add a lot of words without adding much, if any, meaning—they are noise that interferes with the message you are trying to get across.

If you want to indicate that something is important, put it in the topic sentence of a paragraph, put it in bold type, or otherwise make it prominent—don't bury it after a load of mush.

gerunds We sometimes want to take a sentence and turn it into a noun phrase to be used as the subject or object of another sentence. One way to do this is to turn the main verb into a *gerund*, using the *-ing* suffix. What was previously the subject of the sentence has to be converted to possessive form (or sometimes omitted).

For example, the sentence "we fit a model to the data" becomes the noun phrase "our fitting the model to the data", and "I learn" becomes "my learning".

In the acknowledgments for the book, I turned two sentences into objects this way then did the same to the resulting sentence: "I thank my wife, who tolerated my turning our bedroom into an electronics lab and my spending many long hours playing with the labs for the course and writing this book."

however The word "however" is a sentence adverb, not a conjunction. It cannot be used to join two sentences, the way conjunctions like "and" and "but" do. It can, however, be used to indicate that a new sentence is to be contrasted with a previous one.

ideal, maximal, negligible, optimal, perfect, and **unique:** Words that already express an extreme condition, like "ideal", "maximal", "optimal", and "perfect", shouldn't be intensified with "more", "most", or "very". Doing so is equivalent to saying "more best"—a mistake rarely made by native speakers.

The word "unique" means "one-of-a-kind, unlike any other", and so cannot be intensified any more than the extreme-condition words can be. Similarly, "negligible" means "small enough that it can be neglected", and so also cannot be intensified.

implement To implement something means to make it—for policies, "implement" can also mean "put into practice". The word does not mean "use", which is a perfectly good word that should be used more often by students.

input and output The compound words "input" and "output" are both nouns and verbs.

As verbs they are irregular, following the pattern of the underlying verb "put", which means that they don't change to form the past tense or the past participle—there is no need for an extra "-ed".

As nouns, "input" and "output" are usually uncountable in normal English, but in engineering writing, they are usually countable—we talk about a differential amplifier having two inputs and one output, meaning the number of different signals that are connected to the amplifier. See also *countable and uncountable*, page 102.

its vs. it's For most singular possessives in English, we add an apostrophe and "s" to the underlying noun, but "it" is an exception to that rule, because "it's" is the contraction of "it is" or "it has". The possessive form of "it" is "its", without any apostrophe.

linear vs. constant A linear relationship between two variables A and B means that there is some constant such that $A = cB$, and we say that A *is linear with respect to* B (or sometimes A *is linear in* B). If A is constant ($A = c$), then it is incorrect to say that it is linear (unless it is a constant 0, which is a very special case). If you have a straight-line relationship, $A = cB + d$, then we say that A is *affine* with respect to B.

-ly There are no words "firstly," "secondly," and "lastly." The words "first," "second," and "last" are already adverbs and don't need an "-ly" ending. Strangely, I never see the corresponding problem with "next", though it is in the same class of words that are simultaneously adjectives and adverbs.

multiple and divisor When talking about digitized signals, we are often concerned about the ratio of the frequency of our signals and the sampling frequency (see Chapter 9 particularly Section 9.2). If we have periodic interference (say at 60 Hz), we can sometimes hide it by sampling at a frequency that is a *divisor* of 60 Hz (like 15 Hz, 20 Hz, 30 Hz, or 60 Hz)—not at a *multiple* of 60 Hz (like 120 Hz, 180 Hz, and so forth).

noise "Noise" is not a synonym for "sound," despite what school disciplinarians may have lead you to believe. In engineering writing, "noise" is undesired interference with a signal and may have nothing to do with sound. In technical writing, "noise" is almost always an uncountable noun, and so does not use the indefinite article "a" and is not made plural. See also *countable and uncountable*, page 102.

You will often see expressions like "white noise," "pink noise," "1/f noise," or "60 Hz noise", which describe the type of interfering signal. *White noise* is a random signal with a broad spectrum, having equal energy at all frequencies; *pink noise* is a random signal that has more energy at low frequencies; *1/f noise* is pink noise in which the energy is inversely proportional to the frequency; and *60 Hz noise* is a periodic signal with a period of 1/60th of a second—often the result of capacitive or inductive coupling from the power lines in the walls.

Some authors prefer to use "interference" when the noise is not random but an undesired signal from a different source—in fact, I more often refer to "60 Hz interference" than to "60 Hz noise."

Noise is usually a varying signal—an undesired DC voltage is referred to as an "offset" rather than as noise. A DC voltage that is needed for the operation of a part is referred to as a "bias voltage" or "DC bias", and we often have to add or remove bias voltages to amplify the desired signal, but the bias is not referred to as "noise".

There are several different sorts of problems with amplifiers that students lump together as "noise", but using the right terminology helps with debugging.

hum Amplifiers often pick up 60 Hz interference (in the US, 50 Hz in some other countries) through capacitive or inductive coupling to power lines. This low-frequency interference results in a low-pitch hum. Hum is easily diagnosed by looking at the frequency of the waveform. Getting rid of hum is often difficult, requiring that wires be kept short and sometimes that they be shielded.

static Static is broad-spectrum noise—often nearly equal energy across the entire spectrum (*white noise*), though sometimes more heavily weighted in one range. Static is often heard on old AM radios that are not tuned to a station. In our circuits, static mainly comes from the electronics itself (thermal noise and shot noise), but may also come from the power supplies.

distortion Distortion is not added to signals (like hum and static), but a change in shape of the waveform due to nonlinear response of the amplifier. Clipping is the most common form of distortion we see in our circuits. Deliberate distortion is often used as an "effect" for electric guitars.

nominal Students often ask whether their schematics should use the component values marked on the parts (the *nominal* values) or component values as measured with a multimeter (the *measured* values). The answer depends on what the schematic is being used for—if it is documenting a design, then the nominal values are the correct ones to report, as that is what would be used to duplicate the design. If the schematic is being used to show the fits of a model to a set of measurements, then the measured or fitted values are the ones to report.

Students often mistakenly refer to the nominal values as "theoretical" values, but this usage is incorrect. A *theoretical* value is one that is derived from some theory, and that theory should be presented. For example, if you have two conductors separated by an insulator, you can compute a theoretical capacitance from the area and separation of the conductors and the dielectric constant of the insulator, which may not agree exactly with the measured capacitance. In contrast, if you put a 1 μF capacitor in your design, but measure the part to have only 0.8 μF, those are the nominal and measured values—neither is a theoretical value, as there is no theory for determining the value.

offset Like the compound words "input" and "output", the compound word "offset" is its own past tense, because the underlying verb "set" does not change form in past tense, just as "put" does not.

operational amplifier An operational amplifier (op amp) is a component in a larger system and has a nearly infinite gain. Amplifiers that use operational amplifiers as components should not be referred to as operational amplifiers—most often the correct term for the circuit is *negative-feedback amplifier*.

optimal The word "optimal" is a formal way of saying "best", and it should be reserved for situations where there is an explicit objective function defining what is being optimized.

prepositions Use the correct prepositions for talking about voltage, current, or resistance.

- We have voltage *between* two nodes, or *across* a component.
- We have current *into* or *out of* a node, or *through* a component.
- We have the resistance or impedance *of* a component.
- We have power *into* a component or dissipated *in* a component. We can also get power *out of* a power supply or battery.

pronouns You can use pronouns to reduce repetition, but make very sure that the antecedent for each pronoun is very clear, and double-check that clarity when editing your writing, as minor rearrangements of text can completely mess up the connections between the antecedent and the pronoun.

Whenever you use a pronoun such as "we" or "it", you need to be crystal clear what the referent for the pronoun is. English does have a referentless-it construction (in phrases like "it became clear that ... "), but most uses of "it" should have an obvious referent. Avoiding referentless-it constructions will generally improve your writing.

proportional The phrases "f is proportional to d" and "f is inversely proportional to d" mean $f = kd$ and $f = k/d$, respectively, for some constant k. They are much more specific than some vague notion of f going up or down when d does. You can use "f increases with d" or "f decreases with d" for the vaguer concept.

rail A "rail" in an electrical circuit is a node carrying a power-supply voltage. In this course, we generally have two power rails in our circuits: ground and a positive power rail at 3.3 V or 5 V.

Some students use "rail" as a verb, to refer to an amplifier saturating or a signal clipping at one of the power-supply voltages, but this usage is too informal for lab reports. Use "saturate" for amplifiers or "clip" for signals.

reason I find the construction "the reason is because ... " particularly grating, because it mixes up two different correct constructions: "the reason for x is y" and "x, because y". The awkward construction is often used with an implicit "x" part, making the construct rather vague as well as ungrammatical.

relatively Some writers are fond of using the adverb "relatively", talking about a value as being relatively small, for example. It is wrong to use "relatively" just to weaken an assertion—that isn't its function. The adverb is a perfectly good one for making comparisons, but the reader should always be clear what the comparison is being made to. Noise may be relatively small compared to the signal it is contaminating, or a parasitic capacitance may be relatively large when compared to the designed-in capacitance. But if you tell me that you want to use a relatively small resistor, I want to know what it is relative to.

replace, substitute, swap, and switch There are two expressions that students often merge, resulting in a confusing hybrid. The expressions "replace x with y" and "substitute y for x" have the same meaning, though the objects of the verb and the preposition have changed places. When students say "substitute x with y", I'm never quite sure whether they've got the wrong verb or the wrong preposition—the meanings are different. Keep the verbs and the prepositions properly paired!

Occasionally, students will use "swap x for y" or "swap x with y", which is even more confusing, as "swapping x and y" is a symmetric operation, in which x and y change places with each other. In the asymmetric case, where one is replacing the other, use "replace with" or "substitute for". Similarly, "switching x and y" is symmetric.

seeing as The phrase "seeing as" is too informal for technical writing. Use the formal equivalent "because".

separable verbs There are a lot of words that are compound words as nouns, but separable verb-plus-particle pairs as verbs. For example, "setup" is a noun, but "set up" is a verb. Other examples include layout, drop-off, turnaround, pickup, put-down, cleanup, close-up, stowaway, flyover, breakdown, breakout, and setback. Sometimes the order is reversed in the noun, which can result in the noun being verbed to become a new verb: "set off" became "offset", which is now both a noun and a verb. Spelling checkers rarely understand whether you mean the noun or the verb, and so they cannot be expected to catch the missing or extra space.

significant The term "significant" should be used in technical writing only for talking about statistical significance. If you are just trying to talk about how big something is, use a different adjective, like "substantial" or "large".

If you use the term "significant", I expect to see a p-value or other statistical test to establish significance.

so that and such that Some students confuse the usages of "so that" and "such that". The conjunction "so that" joins a main clause to a subordinate clause, making the subordinate clause be a consequence of the main clause: "I'm studying bioelectronics, so that I can make neural interfaces." In technical writing, "such that" is a specialized piece of jargon used in mathematics to express a condition that defines a subset or a constraint, and it is often written with a | or : inside set notation. For example,

$$\{n \in \mathbb{Z} | n > 3\}$$

can be read as the set of all integers n, such that n is greater than three. Although there are other uses for "such that" in ordinary writing, this mathematical meaning is so dominant in technical writing that those other uses of "such that" should be avoided.

solve To solve for a variable means to use algebra to rearrange an equation to isolate that variable on one side. Once that rearrangement is done, using the resulting equation is not "solving for" the variable. Use other verbs, like "calculate" when you are just plugging in numbers to an existing formula.

span The verb "span" requires a direct object, not a prepositional phrase. A recording may span several seconds, but it doesn't "span across" them.

straight line vs. constant Students often use the expression "straight line" to refer to a constant function, but this usage is highly misleading, as a straight line (on a linear graph) is any affine function, $y = mx + b$. Most straight lines are not constant functions.

If you can't remember the term "constant" or don't want to use it for some reason, you can refer to the graph of a constant as a "flat line"—though that has come to mean that someone's heart has stopped beating, from the constant output on a cardiac monitor, or brain death, from the constant output on an electroencephalogram.

See also *linear vs. constant* on page 105.

sufficient The term "sufficient" means "enough"—there is no call for the redundant usage "sufficient enough".

synonyms If you have a concept or component you refer to repeatedly, then choose one word for it and use that word consistently. Don't use synonyms to avoid repetition—introducing a new term implies to the reader that a distinction is being made.

the and this If you have previously introduced a noun, you can refer back to it by using the definite article "the": "the noun", "the function $f(x)$ from the paragraph on extrapolation", or "the data points". Whenever you use "the", you are asserting that the reader *already knows* what you are referring to.

If a noun was introduced in the immediately preceding sentence, then you can use a demonstrative adjective such as "this" or "that" to refer back to it. If a sentence mentions three loaves of bread, then the next one can refer to "that bread" or "those loaves". How far back demonstrative adjectives can be applied varies a bit with context—in some cases you can go back a paragraph, but further than that usually requires reintroducing the object being discussed. See *demonstrative adjectives* on page 103 for more information about them.

Don't try to mix articles and possessives—you can say "the Nyquist Theorem" or "Nyquist's theorem", but not "the Nyquist's theorem", as that would attach "the" to "Nyquist", not to "theorem".

thus The word "thus" means "in this manner"—it is *not* synonymous with "therefore".

try and The expression "try and" is very colloquial. In formal writing, you should use "try to", which is more widely accepted (about 10 times as common, according to Google Ngram Viewer [37]).

upstream and downstream The terms "upstream" and "downstream" behave exactly like "before" and "after" (see *before and after* on page 101). They can make sense in the information flow of a block diagram, but not in a schematic.

utilize I have a hatred of the modern usage of "utilize". Originally, "utilize" meant to make something useful—to add utility, but it has now degenerated to being a pompous replacement for the much better word "use". Search your writing for any occurrence of "utilize" and replace it with the corresponding form of "use".

vs. When describing plots, the convention is to refer to "y vs. x" or plotting "y against x". Don't swap the x and y names!

If you must abbreviate *versus*, don't forget the period at the end of the abbreviation. Because it is not a sentence-ending period, you should let LaTeX know not to add extra stretchiness there. I generally use `y vs.~x` to put an unbreakable space after the period.

When you want to modify a noun with the "y vs. x" phrase, it should be hyphenated: *a current-vs.-voltage plot.* See Section 7.12.7.

when and where The word "when" should be used only to refer to times, and "where" should be used only to refer to places. The construction "X is when ... " or "X is where ... " to introduce a definition is considered a childish construction, and so it is not suitable for formal writing.

which vs. that Writing teachers often have difficulty getting students to understand the subtle, but important distinction between "which" and "that" used as relative pronouns, because informal English allows "which" in contexts where formal written English requires "that".

The distinction is often expressed as being between *restrictive* and *non-restrictive* clauses. If the clause starting with "which" or "that" changes the meaning of the preceding noun, restricting it to a smaller class of objects, then the clause is restrictive. If the clause just provides additional information about an already well-defined thing, then the clause is non-restrictive.

Restrictive clauses should start with "that" and not have a comma before them. Non-restrictive clauses should start with "which" and have comma before them.

For example, if we have one microphone in a design, then providing more information about the microphone does not change what we're talking about, and so such additional information would be a non-restrictive clause and use "which": *the microphone, which responds to frequencies up to 20 kHz ...* .

In contrast, we generally have many resistors in a design, and identifying which resistor is usually restrictive, so needs "that": *the resistor that converts the current to voltage ...* . If we name the resistor, then we have no need for further identification, and the same information becomes non-restrictive: *Resistor R3, which converts the current to voltage, ...* .

7.12 Punctuation

7.12.1 *Commas*

The most common punctuation problems in student writing is the misuse of commas. There are three common comma problems:

Missing commas Commas are needed in many places in English—for example, to separate items of a list, to separate noun phrases from non-restrictive clauses, to separate appositive noun phrases from the noun phrase that they are equivalent to, and to separate full sentences that have been joined by conjunctions. It is beyond the scope of these brief guidelines to list the many uses of commas and the rules governing them, but there are many good punctuation handbooks in print and online that can be consulted (the Purdue OWL *Extended Rules for Using Commas* is a good place to start [89]).

Extraneous commas Just as commas are required in many places in English, they are prohibited elsewhere. Don't insert them randomly to break up a sentence! In particular, there should not be commas separating the subject of a sentence from its verb phrase, nor separating a verb from its object.

Commas in place of other punctuation Perhaps the most common error in punctuation with commas is the *comma splice*, which is the use of a comma to join two full sentences without a conjunction. You can't do that in formal written English—if you must join two sentences without a conjunction, then use an em-dash or a semicolon. A better approach is to join the sentences with a conjunction, because the conjunction can explicitly express the relationship between the sentences, which is only implicit if they are stuck together with just a punctuation mark.

> No comma splices! Don't join two full sentences with just a comma.

Lists are extremely common in technical writing, but there are several common errors associated with them. Here are a few basic rules:

- Elements of a list should all be parallel, both grammatically and semantically. If one element is a noun, they should all be nouns—if one is a sentence, they should all be sentences. If one is a property of an object, they should all be properties.
- Displaying the list with bullets does not change any punctuation rules—commas, periods, and semicolons should all be used as usual. White space is not punctuation.
- If an element of the list contains a comma, then the element-separating punctuation should be promoted from commas to semicolons. If an element of the list contains both commas and semicolons, you should probably rewrite the list, as there is no good way to punctuate it.
- A list of two elements uses a conjunction, but no commas: "ham and cheese", "fish or cut bait", "inverting amplifiers or non-inverting amplifiers", ...
- A list of three or more elements has a conjunction before the last element with commas separating the elements. The comma is required between the last two elements, before the conjunction: "resistors, capacitors, and inductors". This usage, having a comma before the final conjunction in a list, is referred to as a *serial comma* or *Oxford comma* and is generally expected in technical writing.

In this class, I will expect use of the *serial comma*, also known as the *Oxford comma*, which comes before the last element of a list of three or more items. That is, I expect to see lists punctuated as "1, 2, and 3", not as "1, 2 and 3".

The detailed rules for the use of commas are beyond the scope of this electronics textbook, but can be found in almost any writer's manual or style guide. See *which vs. that* on page 109 for one specific comma rule involving the words "which" and "that".

7.12.2 *Colons*

Colons are also frequently misused, generally by inserting them where no punctuation at all would be best.

The colon is normally used between a noun phrase and a restatement of the noun phrase. A common noun phrase before a colon is *the following*—consider the following: thing one, thing two, and thing three. This usage is so common that a lot of people try to put colons before every list, which is simply wrong. Having the list displayed as bullet points doesn't change any of the punctuation rules. There are no colons unless you are separating a noun phrase from its restatement.
OK: ... include the following: a resistor, a capacitor, and a transistor.
No colon: ... include a resistor, a capacitor, and a transistor.

Don't use a colon between a verb and its object, nor between a preposition and its object, even if the object is a displayed list or a math formula.

Neither colons nor other punctuation marks are normally used at the end of section titles.

7.12.3 *Periods*

Periods are used for three different purposes: ending sentences, marking abbreviations, and in numbers. They are *not* used with standard unit abbreviations, like Hz, F, A, V, s, ...

LaTeX assumes that any period at the end of a word is intended as a sentence-ending period, and treats the spacing after the period accordingly. If the period is *not* at the end of a sentence, it is a good idea to change the spacing, by using a tilde for an unbreakable space (see Section 7.12.6).

Even if a sentence ends with a math formula, it should still have a period at the end. If the math formula is in-line math, nothing special needs to be done—just put the period after the closing `\)`, but if the math is displayed, then the closing period needs to be inside the displayed math, to avoid a lone period at the beginning of the next line. Separate the period from the rest of the formula with a tilde (unbreakable space)—for example, `\[\frac{1}{3}~.\]` displays as

$$\frac{1}{3}~.$$

Periods in numbers should always be *in* the number, and not at the beginning of the number, as it is too easy to lose a decimal point if it is before the first digit. See Section 7.9.1 for more details on the formatting of numbers.

7.12.4 *Apostrophes*

Apostrophes get used for two purposes in English: to form possessives and to stand for the missing letters in a contraction.

For most singular possessives in English, we add an apostrophe and "s" to the underlying noun, while for plurals that already end in "s", we just add an apostrophe. If we talk about "the transistor's characteristics", we are referring to a specific transistor that has been previously introduced, but "the transistors' characteristics" refers to multiple transistors.

Irregular plurals that don't end in "s" add both apostrophe and "s" to form the possessive: "the children's toys".

There are a few exceptions to the above general rules for apostrophes. One important one is that we use "its" as a possessive with no apostrophe, because "it's" is used for the contraction of "it is" (see *its vs. it's* on page 105). Another is that names that end in the sound [zʌs] or [zəs] get only an extra apostrophe and not another "s" when forming the possessive, to avoid the sound [zəsəs]. The most common such possessives are biblical: *Jesus'* and *Moses'*. Not every name ending with "s" gets this special treatment—the possessive of my last name is correctly *Karplus's*.

In formal written English, we generally avoid using contractions, preferring to write out the longer phrase, so you might find it useful to search your document for apostrophes, replacing any contractions with the phrase they stand for and any mistaken possessives of "it" with "its". In this book, I've deliberately used contractions in places to try to get a more informal tone to reduce the rather dry, pedantic voice that dominates my writing.

Some older style guides recommend using apostrophes for plurals of acronyms, but this usage is almost completely gone—the plural for FET is FETs, and FET's is the possessive form.

7.12.5 *Capitalization*

In recent years, I've noticed a growing tendency for students to overuse capital letters in their writing. Only the beginnings of sentences, proper nouns (names), and acronyms should be capitalized—nothing else. Titles (like doctor and professor) are capitalized when they are parts of names, but not otherwise. Internal names like Figure 7.1 and Section 7.12.5 are still names and do need to be capitalized.

Some names of principles (like "Ohm's law" and "Kirchhoff's current law") are sometimes done with capitals for each of the important words and sometimes with just the person's name capitalized. Both are acceptable, but try to be consistent (Google ngrams shows that capitalization of just the person's name is 3–4 times as common as capitalizing the whole law name [34]).

Many acronyms have very specific capitalization that should be preserved: AC, ADC, cMOS, DAC, DAQ, DC, EKG, nFET, PCB, pFET, RMS, USB, ...

Some names have weird capitalization uses (like LaTeX, PteroDAQ, and LEGO®), which should be preserved intact. Other names, like book and article titles, may have different capitalization rules in different contexts: some journals require *headline capitals*, which capitalize the first word of the title and all the content-containing words (but not prepositions, articles, and conjunctions), while other journals use *sentence capitals*, capitalizing only the first word of a title.

One word we use that has strange capitalization is "H-bridge"—it is always capitalized, because the "H" comes from the shape of the schematic, and "h" does not have the same shape.

> Unit names are not capitalized (hertz, volts, amps, . . .), but symbols for units from people's names are (Hz, V, A). Capitals are very important in unit names and metric prefixes: "M" means *mega-*, but "m" means *milli-*, a factor of 10^9 difference. Similarly "S" means *siemens*, but "s" means *seconds*. You may see both μS and μs used in a report and should not confuse them. We will use "k" for *kilo-* and "K" for *kelvin*—don't mix them up either!

Don't use capitalization for emphasis. If you need to emphasize something, use *italics*, and if you need really strong emphasis, use **bold**.

> Because sentences must start with a capital letter, you should never start sentences with uncapitalizable symbols (like numbers or ω). You can often use a noun phrase that defines the symbol to start the sentence with proper capitalization: "The angular frequency ω ... ".

7.12.6 *Spaces*

There are many debates on the internet about whether to use one space or two after a period. Both sides are wrong, but LaTeX makes the whole debate irrelevant, as it usually handles sentence-ending spaces correctly no matter how much white space you type. The sentence-ending space is slightly larger and much stretchier than a word-separating space, so that justification of text will separate sentences more than it separates words.

The spacing between words and between sentences is handled automatically in LaTeX, and the spaces that you type in your `.tex` files are pretty much ignored. LaTeX uses heuristics to guess when a period is ending a sentence rather than being part of an abbreviation, and it sometimes guesses wrong. Occasionally, LaTeX gets confused by an abbreviation that ends with a period that isn't a sentence-ending period, such as at the end of *Dr.*, *Mr.*, or *vs.*, in which case LaTeX needs to be informed not to use sentence-ending spacing. There are two easy ways to prevent a sentence-ending space, both of which introduce an explicit space:

- Putting a backslash before the space makes it an explicit word-separating space, rather than a sentence-ending spacing.
- Replacing the space with a tilde (`~`) makes the space not only word-separating but also non-breakable, so that no line or page break can occur there. This approach is probably the best way to handle titles at the beginnings of names: `Prof.~Karplus` produces "Prof. Karplus".

You can force a sentence-ending space with `\@`, but I have never had reason to use this feature.

Another place where unbreakable spaces should be used is between numbers and the units associated with the number. To make the grouping of the number and the unit clearer, I use thin spaces, which are a little narrower than the usual word-separating spaces. The thin space in LaTeX is requested with `\,` (a backslash and comma)—note the subtle difference between 10 pF (using `10~pF`) and 10 pF (using `10\,pF`).

> Numbers and unit abbreviations are usually separated by a space: 10 Ω, 50 pF, ... , but with no space between a metric prefix and the unit it modifies. Having that space turn into a line break or page break can be very irritating to the reader. A similar irritant can occur when there is a line break between the first and second part of a name like "Figure 3", "Section 6.1", or "Table 2". Even separating a person's title from their name can be annoying.
>
> To avoid all these undesirable line breaks, you can specify to LaTeX that a space is *unbreakable* by replacing the space with a tilde (`~`). It looks a little better to use a thin space (specified in LaTeX with `\,`) rather than a full-width space before unit names.

Here's one subtle point: spaces are ignored after a LaTeX command, so if you want spaces, you need to end the command with a pair of braces. The place I see this become a problem most often in student writing is after the Ω symbol, for which I provided a command `\ohm`. To use this command correctly, without losing a space, you want to write something like `10\,k\ohm{}`, in order to get 10 kΩ and not lose the space after the Ω.

7.12.7 *Dashes and hyphens*

In typesetting there are at least four different types of dashes:

hyphen (-) a very short mark used inside compound words, to turn a noun phrase into a modifier of another noun, or to mark the end of a line where the word continues onto the next line. Represented in LaTeX as `-`.

en-dash (–) a somewhat wider mark (about the width of a lower-case "n") that is used to represent ranges, such as 1–10 or Jan–Jun. Represented in LaTeX as `--`.

> The range expression "x–y" should be read as "from x to y"—it is incorrect to write "from x–y," as that translates to "from from x to y." Similarly, "x–y" can be expressed as "between x and y", but not "between x–y."

em-dash (—) a much wider mark, used for sentence-level punctuation—somewhat like a semicolon or parentheses. Represented in LaTeX as `---`.

minus sign (−) used only in mathematics, the minus sign is usually about the same size as the en-dash, but has different spacing rules. The text marks (hyphen, en-dash, and em-dash) have *no space* around them (though a few typographers will put thin spaces around em-dashes), but the minus sign has the same spacing rules as the plus sign (with different rules depending on whether it represents a unary or binary operator)—LaTeX handles the spacing around minus signs automatically. The minus sign in LaTeX is represented as `-`, but only while in math mode.
Negative numbers should always be set in math mode, to avoid getting the too-short hyphen instead of a minus sign and to avoid bad line breaks (LaTeX *likes* breaking a line after a hyphen).

There are many prefixes and suffixes that get added to words, sometimes with hyphens and sometimes directly. Most prefixes and suffixes are not standalone words and must be attached to words. Some common prefixes in technical writing are *multi-*, *non-*, *pre-*, *proto-*, *pseudo-*, and *re-*. One common suffix is *-like*.

Although there are some rough rules of thumb for determining whether to hyphenate or not (generally hyphenating when two vowels would be put together or other times where parsing the unhyphenated word would be difficult), there are many exceptions, and hyphens tend to get used less as a word with prefixes or suffixes becomes more common. A pair of exceptions to the two-vowel rule is "preamplifier" and "preamplification", which are most commonly *not* hyphenated, though hyphenation is common enough that you can do so, as long as you are consistent.

When I need to check popular hyphenation usage, I often use Google Ngram Viewer [34]. For example, "noninverting" is now about 1.6 times as common as "non-inverting", the unhyphenated "nonlinear" is about 3.9 times as common as "non-linear", and "pseudocode" about 3 times as common as "pseudo-code". Those ratios are all small enough that authors can choose whichever form they prefer, as long as they are consistent, but "recycle" is about 300 times as common as "re-cycle" so the hyphenated form should not be used. Similarly, "microcontroller" is 40 times as common as "micro-controller", so the hyphen should not be used. Other words that we use in this course that have no hyphen are "transimpedance" and "phototransistor".

The Google Ngram Viewer [34] can also be used to determine whether a particular usage is American or British, as they have separate corpuses for American English and British English. For example, the hyphenated form "nonstandard" is more common in British English, but the unhyphenated form "nonstandard" is more common in American English.

LaTeX attempts to hyphenate words automatically at line breaks, to provide consistent line lengths. The heuristic algorithms it uses are pretty good, but they are not perfect, so sometimes you need to help it out by providing the correct hyphenation for words in the preamble of your document, using the `\hyphenation` command. For example, in the preamble for this textbook, I included

```
\hyphenation{trans-im-ped-ance}
\hyphenation{trans-con-duct-ance}
\hyphenation{Tek-tron-ix}
\hyphenation{Free-scale}
```

to correct LaTeX on some words that it had mis-hyphenated.

If a word already has a hyphen or a dash, LaTeX is not willing to add another one automatically. If I see a need for extra hyphenation, I will sometimes add "discretionary hyphens" to tell LaTeX where it can add hyphens if needed for a line break:

```
volt\-age-con\-troll\-ed switches
negatively-doped metal-oxide-semi\-con\-ductor
oxi\-da\-tion---this
```

Often, rewriting the sentence to avoid so much hyphenation is a better choice than telling LaTeX where it can add more hyphens.

When you use a noun phrase to modify another noun, you should hyphenate the whole modifying noun phrase. Here are some examples:

- The process of synthesizing amino acids is called *amino-acid synthesis*, and the pathway that does it is the *amino-acid-synthesis pathway.*
- An interval of seven seconds on a graph is a *seven-second window*, but if you use a unit abbreviation, it is a 7 s window, without a hyphen.
- A system that acquires data is a *data-acquisition system.*
- Chapter 16 talks about Schmitt triggers and more specifically about *Schmitt-trigger inverters.*
- A system with three ports is a *three-port system.*
- In several places, the book mentions *high-pass, low-pass, and band-pass filters.*
- When the gain of an amplifier is one, we refer to it as a *unity-gain buffer*, and the circuit for it is a *unity-gain-buffer circuit.*
- The voltage difference from the most positive peak to the most negative peak is the *peak-to-peak voltage.*
- A common plot for characterizing components is a *current-vs.-voltage plot.*
- We may need to do *high-pass filtering* in the *first- and second-stage amplifiers* of a *multi-stage amplifier.*
- When we compensate for the impedance of our test fixture with an open circuit, we are doing *open-circuit compensation*; when compensating for the impedance with a short circuit, we are doing *short-circuit compensation*; and when we do both, we are doing *open- and short-circuit compensation.*

Some other noun modifiers also get hyphenated. For example, a physics course that is based on calculus can be described as a calculus-based physics course; when current shoots through an nFET and pFET in series, we refer to it as *shoot-through current*; and PteroDAQ is a USB-based data-acquisition system that uses an analog-to-digital converter to convert analog voltage signals into digital numbers.

If you find a clunky noun cluster in your writing that is hard to hyphenate, you can usually rearrange it using prepositions, making it more understandable in the process. Sometimes words in the noun cluster are just padding, adding no meaning, and should be removed for clarity. Here are a few examples from student lab reports:

- "Transimpedance amplifier gain resistor limitations" can be rewritten as "limitations on the gain resistor of the transimpedance amplifier".
- "Stainless steel electrode impedance spectroscopy data" can be rewritten in several ways, including "impedance-spectroscopy data of stainless-steel electrodes", "data from impedance spectroscopy of stainless-steel electrodes", or "data from impedance spectroscopy of electrodes made from stainless steel". (Of course, "data" may not be needed here, and "impedance spectroscopy of stainless-steel electrodes" may be the best choice).
- "Maximum power supply voltage limit range" probably means "the maximum power-supply voltage", "the limits of the voltage for the power supply", or "the range of legal voltages for the power supply"—I never did figure out exactly what the writer meant.
- "The passive high pass filtered second stage amplifier output signal range" can be made a bit more comprehensible with hyphens: "the passive-high-pass-filtered second-stage-amplifier-output signal range", but prepositions help: "the signal range after filtering the output of the second-stage amplifier through a passive, high-pass filter". The phrase might be clearer if "signal range" were replaced either with "voltage range" or "signal", depending on what was meant.

7.12.8 *Fonts*

Technical reports use font changes (bold, italics, small caps, etc.) sparingly. To get italics in LaTeX, use the `\emph` command: `\emph{put this in italics}` creates *put this in italics*. To get bold face in LaTeX, use the `\textbf` command: `\textbf{put this in bold}` creates **put this in bold**.

Here are a few of the more common uses:

- Use italics for emphasis and bold for very strong emphasis. Other than for safety warnings and section headers, the strong emphasis of bold fonts is rarely appropriate for technical writing. Formatting of section headers is handled by the style file in LaTeX, so you should rarely need to specify bold face.

  Do not use capitals (neither all-caps nor small caps) for emphasis.
- Use italics for book and journal titles in the main body and usually in the list of references (though the reference formatting is controlled by the bibliography style—some styles do not use italics). Some students have been taught to use underlining for book titles, but this advice is obsolete—underlines were used in typescripts to indicate that typesetters should set the corresponding words in italics. Because authors can now set their words in italics directly, underlining has essentially no use.

  Although book and journal titles are set in italics, article titles within journals are not, but put in quotation marks.
- Use italics when introducing jargon words, in the defining context where the meaning of the jargon is made clear. Subsequent uses of the jargon should not be in italics. This slight emphasis on the defining context makes it easier for the reader to look back to find the definition.
- Some style files set the captions of figures and tables in italics, to make a bigger contrast between the captions and the body text. In this book, I used the `subcaption` package with the command

`\captionsetup{font={footnotesize,it},labelfont=rm}` `\subcaptionsetup{width=0.7 \textwidth}` to set up captions in a smaller italic font, slightly narrower than the body text. When a caption is set in italics, the `\emph` command reverts to standard Roman fonts, which is the correct treatment for both emphasis and book titles when the text is in italics.

- In LaTeX, the italics fonts for text emphasis and for mathematics are usually different—when you want to refer to a single-letter variable, always enter math mode to do so, so that you get the correct math font. The variable x is different from the text italic *x*, which is very different from the standard Roman letter x.
- *Monospace fonts*, in which all characters are the same width, are often used to indicate text that has to be typed by a user. The *verbatim mode* used in this text uses a typewriter-style monospace font:

```
This verbatim text is in a monospace font.
```

Monospace fonts are generally much harder to read than fonts with variable character widths and so should not be used for large chunks of text.

7.13 Citation

Cite any external work that you used (data sheets, text books, Wikipedia articles, etc.). You should use and cite data sheets for *every* major component in your design, especially amplifiers and sensors, whose properties are essential to the design. Citing tools used for testing the design or creating graphics for the report is also good practice.

When you cite a data sheet, try to find an online copy from the manufacturer of the part—not just a copy made by a distributor such as Digi-Key or Mouser. Be sure to include the publication date and version number of the data sheet, as data sheets often go through many revisions. Commodity parts such as logic chips, transistors, or diodes often have many manufacturers making parts with the same part number, but their data sheets may differ substantially, both in the level of detail and the specific properties of the part.

Although you don't need to cite references for "common knowledge", such as Ohm's law, if you need to look up a formula, say from a Wikipedia article, then you must cite the article, giving the title, the URL, and the date you accessed the article as a minimum.

Do **not** include anything in your reference list that is not cited in the report—padding your reference list with anything not explicitly cited in the report is a form of academic dishonesty.

Tools like biblatex make honest reference lists easy, as they include precisely those references that are explicitly cited. Citations should be made with the `\cite` macro of LaTeX, with the reference list created automatically by biblatex. There are many different citation styles in common use, and biblatex supports almost all of them, just by changing options in the `\usepackage{biblatex}` command in the preamble of your LaTeX file.

Any citations that are made are wholly parenthetical and are *not* part of the sentence. Citations are *not* noun phrases.

Some authorities disagree with me on the parenthetical nature of citations and permit citations to be used as noun phrases that stand for the thing being cited. That usage is not accepted for this course. The problem with using citations as nouns becomes obvious if you use a style that formats the citations as superscripts.

For example, I can say that the BibTeX system is described in an appendix of the LaTeX manual [62], but it would be completely incorrect to say that it is described in [62], because "in" has no object in the second phrasing.

Many of your figures will be created using software tools (gnuplot, Scheme-it, draw.io, or other graphing and drawing tools). It is considered good form to cite such tools, either in an acknowledgments section at the end of the report or in the figure captions themselves, as was done in this textbook. If you do add such citations to a figure caption, they should be at the end of the caption, not the beginning, as the citation is the least important part of the caption to the reader. The bibliography entry for a software or hardware tool is generally for a web page where the tool can be found, though some academic tool authors request citations to particular papers that describe the tool, as academics get some benefit from citation of their published papers, but much less benefit from citation of web sites. If you use or adapt source code that you find online (such as the `bandpass-filter.py` program provided for Lab 5), you need to cite it. You should provide the author, title, URL, and the date that you accessed the file. If there is no obvious title for the program, use the name of the file. If the document has a date associated with it, use that in addition to the date of access.

> If you copy a figure, not only must you cite the article you copied from, but you must give explicit figure credit in the caption for the figure: *This image copied from ....*

If you modify a figure or base your figure on one that has been published elsewhere, you still need to give credit in the caption: *This image adapted from ....*

> In general, you should not be copying anything from this textbook, Wikipedia, or other sources in your design reports. Close paraphrasing is still plagiarism—so don't have a copy of your source open in front of you when you write the report.
>
> Never cut and paste from another source to make a starting point for a paragraph of your writing—that almost always results in impermissible plagiarism.

In this course, it is acceptable to get other people to help you with your design, your debugging, and your writing, but you must be very careful not to claim any of their work as yours.

> Provide *explicit written acknowledgment* of any human help you get.

It is fine to get help, but claiming that work as your own (which is what failing to acknowledge means) is the definition of the most serious of academic sins: *plagiarism.* You will get charged with an academic integrity violation for failure to acknowledge help.

Citations and acknowledgments are slightly different classes of things—citations are used for saying where specific, identifiable concepts or facts were found, but acknowledgments are used for more general help. There are two places where acknowledgments are generally found: either right at where the help was used, such as a figure caption ending with something like "Mary Smith helped us prepare this figure in gnuplot", or in a separate *Acknowledgments* section near the end of the report, just before the reference list.

In this book, citations have been done with `biblatex`, which is the recommended bibliography style file for new documents in LaTeX, though the older `bibtex` bibliography style files still work. To use `biblatex`, you need two things in the preamble of your LaTeX file: `\usepackage{biblatex}`, perhaps specifying options, and `\addbibresource{lab3.bib}` to specify that your bibliography entries can be found in the file `lab3.bib` (substitute whatever bibliography file name you use). Where you want the bibliography to appear in your document, use the command `\printbibliography`. To cite something from your bibliography file, you use the command `\cite{name-of-entry}`, where `name-of-entry` is the name given to one of the items in the bibliography file.

If you are using a citation format that uses parentheses or brackets to surround the citation, then it is a good idea to put an unbreakable space before the `\cite` command, by using a tilde, to keep the citation from being moved to a new line or a new page by the line-breaking algorithm. The space is *not* desirable if you are using a citation format that calls for superscripts—luckily, most technical journals now call for a format that does not use superscripts.

> To cite multiple sources at a single location in your paper, use a single `\cite` command, separating the names of the entries with commas. Do not put multiple `\cite` commands together.
>
> Do not put a citation right after a math formula—many citation formats will be hard to separate from the formula.

The bibliography file itself is a plain-text file, usually with the extension `.bib`. Each entry in the file begins with an at-sign and contains several fields of data about the item being cited.

Here is an example of an entry for a data sheet found on the web:

```
@online(wp3dp3bt,
    author="Kingbright",
    title="Phototransistor {WP3DP3BT} Datasheet",
    version="5",
    url={https://www.kingbrightusa.com/images/catalog/SPEC/wp3dp3bt.pdf},
    date="2016-10-07",
    urldate="2018-11-24"
)
```

The "@online" specifies that this is an entry for something found on the web—there are many other, more specific types of entries available: check the biblatex documentation [65].

- The first word within the parentheses is the name of the item—this is what you put inside a `\cite` command. For example,

  ```
  The dark current is 100\,nA, according to the data sheet~\cite{wp3dp3bt}.
  ```

  produces "The dark current is 100 nA, according to the data sheet [55]". The tilde makes an unbreakable space between the sentence and the citation.
- The "author" field is used here for a corporate author, as this document did not have an individual author. When you have multiple authors, be sure to separate their names with "and" rather than with commas, as the author field treats a comma as separating the family name from the personal name.

- The "title" field is the title taken from the document itself—many web pages have terrible titles, but this PDF file has a reasonable title. The extra set of braces in the title field indicate that the capitalization of the included word should not be changed when bibliography is typeset. The bibliography entries should be created in headline capitals (all important words capitalized), which biblatex will change to sentence capitals (just the first word capitalized) as needed. Words that shouldn't have their capitalization changed (such as DNA) should be put in braces.
- The "version" field is optional, but many documents (especially data sheets) go through multiple versions, and your reader needs to know which version you are relying on—they may not notice the date.
- The "url" field is where the document was found—you should also use a `doi=` field entry for documents with document object identifiers. The URL you use should be to the specific web page you used and should not contain any tracking junk that is unique to a single visitor. Whenever possible, you should cite primary locations: manufacturer's websites for data sheets, for example, rather than distributor's websites or collections of data sheets.
- The "date" field is the date from the document itself, and "urldate" is the date you found the document (particularly useful for Wikipedia pages, which change frequently, or manufacturer's web sites, which get rearranged frequently). All entries in the bibliography database for online references should include a urldate field.

> The quotation marks in .bib files are simple ASCII quote marks (`"`) not "smart quotes" that are different for left and right quotation marks. Make sure that the editor you are using is not changing the quote characters you type into different left and right quotes.

Wikipedia provides a navigation link "Cite this page" in their left-hand navigation panel for each page—currently, the entry created is suitable for bibtex, but not really for biblatex—you'll need to edit the entry somewhat before including it in your bibliography.

Exercise 7.8

Create a short LaTeX document using biblatex to cite three sources: a book, an electronic data sheet, and a Wikipedia page. Turn in both the final PDF file for the document and the `.bib` file used to create the citations.

Chapter 8

Lab 2: Measuring Temperature

Bench equipment: ohmmeter, voltmeter, thermometer, water baths, secondary containment, ice water, hot water, tape

Student parts: thermistor, resistors, breadboard (optional), clip leads, PteroDAQ

8.1 Design goal

The design goal for this lab is to build a temperature monitor that can record temperature measurements at frequent time intervals (say, 10 times a second) using a microcontroller. You will need to devise a circuit to convert the temperature-varying resistance of the thermistor to a temperature-varying voltage that can be measured and recorded by PteroDAQ (see Prelab 2.3).

This lab consists mainly of measurements, but you do have one design task: to pick an appropriate resistor value for the voltage divider.

> The design should use a single resistor, not a series chain of resistors or a parallel arrangement of resistors. One of the goals of the course is to get students to design around standard parts, rather than custom ones.

I have heard experienced electrical engineers railing about their inexperienced co-workers designing around unavailable, nonstandard parts and the manufacturing delays and increased costs this causes. Using multiple parts where one would do also increases manufacturing costs unnecessarily. For all designs in the course, you should assume that each resistor or capacitor in your design is to be implemented with a single part that you have, not kluged together out of multiple parts. (If there are exceptions, they will be discussed explicitly.)

We want linearity and maximum sensitivity of the monitor at a specified operating temperature T_{op}, but the range should be at least 0°C to 100°C. T_{op} will not be announced until the lab period starts, so do your calculations symbolically—you'll need some numbers you won't have until after analyzing the data from the first part of the lab anyway.

8.2 Pre-lab assignment

Write up your pre-lab report as a draft of your final report—you will correct and add to the report as you do the lab. The pre-lab draft, like the final document, is a standalone report that should be readable by someone with no access to the textbook or the homework exercises.

Design reports do not consist just of the final design, but include the assumptions and thinking that lead to that design, so that an engineer reading the report can figure out why the design is the way it is, and what they can change to adapt it to a different need. Writing up as much of this as possible before building circuits and measuring them makes lab time much more efficient. (That is, the pre-lab exercises are an essential part of the final design report.)

Reading the documentation for lab equipment and components is a big part of an engineer's continuing education. Trying to wing it with unfamiliar equipment or components can result in very expensive mistakes. Even simple equipment can cause you unexpected misery if you aren't aware of the specific limitations of the equipment.

Read the multimeter user's guide to figure out how to use the meter to make resistance and voltage readings.

Learn the differences in function for

- a *voltmeter*, which has a high resistance between the probes and measures the voltage between two nodes in a circuit without taking much current;
- an *ammeter*, which has a very low resistance between the probes and measures the current flowing through the meter without much voltage drop; and
- an *ohmmeter*, which supplies a current through the probes and measures the voltage between them, dividing the voltage by the current to get the resistance of the component being measured.

Be sure you know where to plug in the probes or clips for measuring voltage and resistance, and how to select AC or DC voltage measurement and resistance measurement.

The accuracy specifications for meters are often given in two parts: a percentage of the reading and the number of least-significant digits. For example, if one had a 4.5-digit meter (capable of displaying numbers from -19999 to $+19999$), on a 200 V range, the least-significant digit would represent a step of 10 mV, so a specification of $1\% \pm 2$ would mean that the error on a 10 V measurement could be as much as $10\,\text{V}\ 1\% + 20\,\text{mV} = 120\,\text{mV}$. That means that a 10 V signal could be reported as anywhere from 9.880 V to 10.120 V. Choosing the smallest range that includes the measured voltage usually results in the most accurate measurement.

Pre-lab 2.1

What is the model number for the multimeter you will be using? If you are measuring a DC voltage of approximately 3.3 V, how much error can you expect (using the most appropriate voltage range)? What about measuring a $10\,\text{k}\Omega$ resistor?

There are several approaches to finding data sheets for parts. The two I've found most useful are to

- use Google search with the part number and the word "datasheet" (or "data sheet", Google seems to accept the one-word and two-word forms as synonyms) and
- look for a data sheet on the web site of a distributor who sells the part. I've found the Digi-Key web site, https://www.digikey.com/, particularly useful for finding parts and data sheets.

Most data sheets are now provided as PDF files—I've found it very handy to collect the data sheets for parts that I have used or will use in folders on my laptop (generally sorted by type of part, though it may be even better to collect all the data sheets from a particular project into one folder). This organization makes it easier for me to find the parts again, even if I forget the precise part number, and it allows me to look at the data sheets even when internet access is slow or unavailable.

It takes time to learn to read data sheets, as they often provide far more information than you need for a particular application, or are missing information that you would really like to have. You also have to expect a certain level of sloppiness from some manufacturers, whose data sheets are riddled with typographical errors—apply sanity checks to anything you read from a data sheet, and be very dubious about components whose values are factors of 1000 or 1,000,000 different from other similar components. Check critical values with your own measurements.

> Learn which manufacturers produce clear, reliable data sheets and specify their components in your designs. Carelessness in the data sheets often indicates a corporate culture of sloppy design and manufacturing, which can result in serious problems if you use their components in your designs.

Pre-lab 2.2

Find and read the data sheet for the thermistor you will be using. (We usually use NTCLE413-E2103F520L.) What is the resistance of the thermistor supposed to be at 25°C? Does the resistance increase or decrease with temperature? What is the expected resistance at 0°C? At 60°C?

Make sure that your pre-lab report answers each of the following questions:

Pre-lab 2.3

Draw a schematic of a circuit that converts the thermistor resistance to a voltage measurable by PteroDAQ using the microcontroller board. I suggest basing it on the only circuit you have studied so far: the voltage divider.
Assuming that you have a negative-thermal coefficient (NTC) thermistor and that you want higher temperature to result in higher voltage, which of the resistors in a voltage divider is the thermistor and which a fixed resistor?
Draw the schematic using the proper thermistor symbol (see Figure 5.12), and put the thermistor part number on the schematic.
(For the final report, update this schematic with the resistor value chosen in the lab.)

Pre-lab 2.4

For the schematic of Pre-lab 2.3, do the algebra to figure out the formula for the voltage V_{out} as a function of temperature, using $R_{\text{T}} = R_{\infty}e^{B/T}$, where T is in kelvins, and the voltage-divider formula. Be sure that you are setting up the equation consistent with the schematic you provided. This should be done entirely symbolically, with no numbers plugged in yet. Feel free to use symbolic tools like Mathematica or WolframAlpha.
Plot your voltage as a function of temperature using gnuplot (or other plotting program) from -50°C to 150°C for values appropriate for a typical thermistor: $R_{\text{T}} = 10\,\text{k}\Omega$ at 25°C, $B = 3977\,\text{K}$, $R = 2\,\text{k}\Omega$. Use 3.3 V as the input voltage, as that is the regulated voltage available on a Teensy board. (For the final report, redo the graph with the parameters and voltages measured in the lab.)
You'll have to figure out what value to give R_{∞}, as it is not the same as R_{T}. Be sure to label your axes, and watch out for the conversion between temperature in °C and in K.

Pre-lab 2.5

Using the formula for V_{out} in Pre-lab 2.4, what is the sensitivity of the thermistor circuit, $\frac{dV_{\text{out}}}{dT}$, in V/°C? Feel free to use symbolic tools like Mathematica or WolframAlpha.
Plot your sensitivity as a function of temperature using gnuplot (or other plotting program) from −50°C to 150°C for values appropriate for a typical thermistor: $R_{\text{T}} = 10\,\text{k}\Omega$ at 25°C, $B = 3977\,\text{K}$, $R = 2\,\text{k}\Omega$, $V_{\text{in}} = 3.3\,\text{V}$. Be sure to label your axes with the correct units, and watch out for the conversion between temperature in °C and in K.

You will want to choose the resistor in your voltage divider to make the voltage response of the temperature monitor be linear with temperature and to maximize the sensitivity (the change in voltage for a change in temperature) at some temperature T_{op} (see Pre-lab 2.6). You are not trying to maximize the *voltage*, but the *derivative of the voltage with respect to temperature.*

We can't quite achieve maximum sensitivity and linearity at T_{op} with the same resistor value, as these are slightly different optimizations.

To make the thermometer linear at some operating temperature T_{op} we need to have the curvature of the voltage-vs.-temperature curve be zero at that point,

$$\left.\frac{d^2V}{dT^2}\right|_{T=T_{\text{op}}} = 0\,\text{V}/\text{K}^2 \,,$$

but to maximize the sensitivity at that temperature we need to find the resistor value R in Pre-lab 2.3 that maximizes the sensitivity $\frac{dV}{dT}$, which would be when

$$\left.\frac{d\frac{dV}{dT}}{dR}\right|_{T=T_{\text{op}}} = 0\,\text{V}/(\Omega\text{K}) \,.$$

Pre-lab 2.6

Determine the formula for the second-derivative of voltage with respect to temperature for your circuit in Pre-lab 2.3. Find a formula for the resistance R that sets this second derivative to 0 in terms of T_{op} and the parameters for the thermistor (making the voltage linear with the temperature around the operating temperature T_{op}).
Remember to check your units: B is in K, R_∞ is in Ω—your formula should cancel any K units and end up with just Ω. Also, be careful about not mixing up Ω and kΩ. This should be done entirely symbolically, with no numbers plugged in yet. Feel free to use symbolic tools like Mathematica or WolframAlpha.
Plot your nonlinearity (the derivative of sensitivity) as a function of temperature using gnuplot (or other plotting program) from −50°C to 150°C for values appropriate for a typical thermistor: $R_{\text{T}} = 10\,\text{k}\Omega$ at 25°C, $B = 3977\,\text{K}$, $R = 2\,\text{k}\Omega$, and $V_{\text{in}} = 3.3\,\text{V}$. Be sure to label your axes with the correct units, and watch out for the conversion between temperature in °C and in K.
(Again, this plot should be updated in the final report to use the values measured in the lab.)

Pre-lab 2.7

Determine the formula to choose R to maximize sensitivity at T_{op} rather than aiming for linearity. What value of R maximizes sensitivity at $T_{\text{op}} = 35°\text{C}$?
Plot the sensitivity as a function of temperature from $-50°\text{C}$ to $150°\text{C}$ for values appropriate for a typical thermistor: $R_{\text{T}} = 10\,\text{k}\Omega$ at 25°C, $B = 3977\,\text{K}$, $V_{\text{in}} = 3.3\,\text{V}$ for the value of R that you chose. Be sure to label your axes with the correct units, and watch out for the conversion between temperature in °C and in K.
Bonus: on the same graph, plot the sensitivity for two values of R: the one that maximizes sensitivity and the one that achieves linearity (see Pre-lab 2.6).

Pre-lab 2.8

Invert the function $V_{\text{out}}(T)$ from Pre-lab 2.4 so that you get temperature (in °C) as a function of voltage: $T_c(V)$. This should be done entirely symbolically, with no numbers plugged in yet. Feel free to use symbolic tools like Mathematica or WolframAlpha.

8.3 Setting up the thermistor

The thermistor has two leads that are very close together, which can be handy if you are soldering the thermistor to a printed-circuit board, but which is inconvenient for this lab. Separate the two leads about 1 cm by pulling the wires apart. It may be necessary to make a small cut between the two wires at the end, to start the split. Although a pocket knife or razor is good for this initial cut, it should be doable with the diagonal cutters in the toolkit.

To connect up to the thermistor, it will be handy to have clip leads that can attach to the rather small and fragile wires on the thermistor.

Figure 8.1 shows how to connect an alligator clip to a wire.

(a) *When using a screw to attach a wire to an alligator clip, you need to strip enough wire to wrap around the screw—about 12 mm.*

(b) *The wire is inserted through the base of the clip, up through the hole, and pulled through until the insulation of the wire reaches the hole. A pair of long-nose pliers can help in pulling the wire through.*

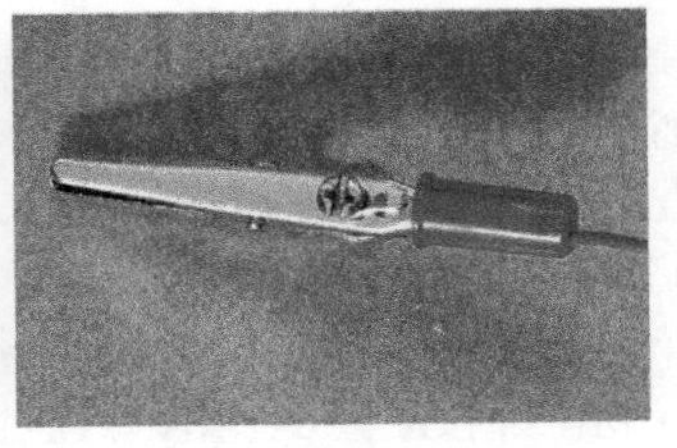

(c) *The wire is wrapped clockwise around the screw, using long-nose pliers, then the screw is tightened to hold the wire firmly in place.*

Figure 8.1: *Wires plugged into a breadboard on one end and with alligator clips on the other end can be used to connect the thermistor in this lab to a breadboard for measuring voltage. If the wires have alligator clips on both ends, they can be used for connecting the thermistor to an ohmmeter. These pictures show how to connect an alligator clip to a wire.*

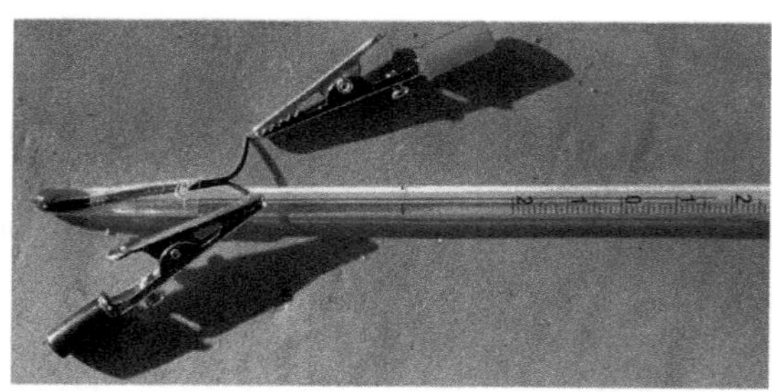

(a) *Thermistor taped to glass thermometer, so that it is in contact with the bulb. The alligator clips are not attached to wires in this picture—that wiring should be done* before *the clips are attached to the thermistor.*

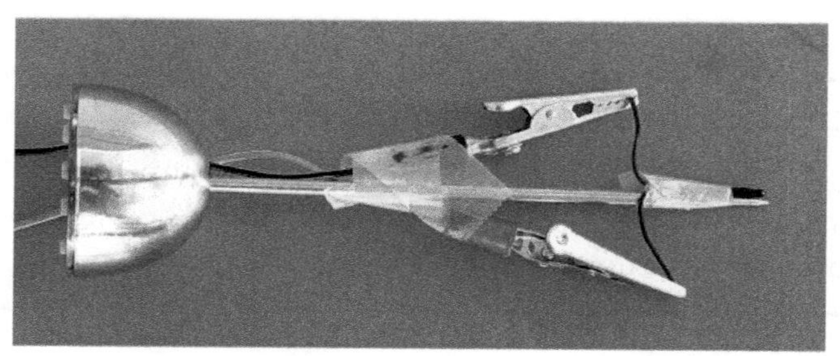

(b) *Thermistor taped to metal thermometer. The clip leads need to be kept from shorting to the metal probe.*

Figure 8.2: *The thermistor should have the leads split for about 1 cm, attached with clip leads to the ohmmeter or test circuitry, then taped to the most sensitive part of a thermometer for calibration. The metal part of the leads should not get wet, as the conductivity of the water could throw off the measurements.*

> Cut a couple of pieces of wire about 2 feet (60 cm) long. Strip the insulation off the last 10 mm of each end, thread the wire through the end of the alligator clip and out the hole next to the screw, wrap the bare wire clockwise around the screw, and tighten the screw to hold the wire. When loosening the screw, don't remove it all the way from the clip, as it can be difficult to line it up to get it back in.

You need to strip a little more from the end of a wire for wrapping around a screw than is needed for connecting to breadboards (10 mm rather than 5 mm).

Aside: in the past some students have had trouble with cheap alligator clips not holding the tiny 30-gauge wires reliably. You can get around this problem by soldering sturdier 22-gauge or 24-gauge wires onto the ends of the thermistor wires, but the extra time to set up soldering irons and do the soldering in lab is not generally worthwhile.

We'll provide water at three or more temperatures (most likely 80°C, ice water, and roughly room temperature), which you can mix to get other temperatures. You'll make your water bath in coffee cups—to get the greatest temperature range, it may be necessary to preheat or prechill the cup with hot or cold water.

> Because water and electronics don't mix well, you must keep the coffee cup inside the secondary containment tub at all times. Move the secondary containment tub, not the cup by itself.

In order to calibrate your thermistor, you will need to measure its temperature and resistance simultaneously. To measure the temperature, we'll use a student thermometer calibrated in °C. Fastening the thermometer and the thermistor together so that the bulbs are in contact will minimize the temperature difference between them, as shown in Figure 8.2.

In your pre-lab you looked up or computed some expected resistance values at a few temperatures. Do sanity checks—are the resistance values at those temperatures near what you expect? If not, is there some error in your measurement technique?

8.4 Measuring resistance

You will need to use a multimeter to measure the resistance of the thermistor. Make sure that the probes are plugged into the high and low jacks labeled for voltage and resistance, not current. By convention, black probes are used for the *ground* and red probes for measured voltage. Although this polarity does not matter for measuring the thermistor resistance, it is good to get in the habit of connecting the probes up correctly.

Make sure you set the meter to measure resistance continuously—with handheld meters, this may be done by either turning a dial or pressing buttons, depending on the interface design. Bench meters have an even wider variety of interfaces, often with special settings for taking measurements only when manually triggered to do so. If you are not familiar with the meter you will be using, read the manual!

Depending on what sort of probe is connected to the multimeter, you may be able to connect them directly to the thermistor wires, or you may need to use alligator clip leads to connect them.

For making precise measurements, it helps to subtract off the measurement of the wires connecting up to the device being measured, by shorting them together and recording that measurement. This is a two-step process: first short the probes together and record the measured resistance (which should be near zero), then measure the device. The improved resistance estimate is the second measurement minus the first one. Some ohmmeters have a "null" or "zero" button that you press for recording the first measurement, with subsequent measurements having that value subtracted automatically.

There is another technique that does an even better job of compensating for wiring resistance: using 4-wire measurement and Kelvin clips. For an explanation, see
https://www.allaboutcircuits.com/textbook/direct-current/chpt-8/kelvin-resistance-measurement/.

Although the Agilent 34401A in some of UCSC's labs is capable of 4-wire measurement, we do not have the Kelvin clip probes for simultaneously measuring current and voltage, so this technique is not available to us. In any case, the resistances we're looking at on the thermistors are large enough compared to the wire resistance that we don't really need the extra accuracy obtainable with 4-wire measurement.

You will want to measure the temperature and resistance of the thermistor at the same time (since the uninsulated water baths will equilibrate to room temperature fairly quickly). Having one person hold and read the temperature, while the other records the temperature and resistance makes the recording easier.

> Although you *can* record the measurements in a lab notebook and transfer to a computer later, I strongly advise that you type them directly into a computer file as you collect them. This reduces the chance of transcription errors (one less copy operation), and allows you to plot the data as you collect it.

In gnuplot, you can just use `plot 'foo.txt'`, replacing foo.txt with the file name of the data file, without bothering to label the axes or title the plot—the idea is to get a quick look at the data while collecting it, not to produce journal-ready figures. Be sure to plot the data as separate points (not lines), so that you can see where you have gaps between measurements.

> Plotting the data as you collect it is a powerful tool for noticing problems with the data collection. If you see anomalous data, you can adjust your lab technique and collect more data.

For example, a sudden drop in resistance may result from getting the leads wet or from having non-uniform temperatures in your water bath, and large deviations from a smooth curve may result from bad contacts to the thermistor or from inconsistent reading of the thermometer.

It is useful to have one window open editing the data file, and another open displaying the data graphically. If you are using gnuplot, you can type "a" in the plot window to redraw the entire graph with automatic scaling—remember to save the data in the file before replotting.

However you collect the data, you will need to have your data in a computer file before the class in which we can analyze the recorded data.

The computer file should start with a number of comment lines, giving the names of the people making the measurements, the date, the part being measured, and the headings for columns (PteroDAQ automatically records much of this metadata, especially if you use the *notes* field, but for manual recording, you will need to remember to record the metadata yourself.) Here is an example of the sort of heading I have in mind:

```
1  # Data collected Sat 23 Jun 2012
2  # by Kevin Karplus
3  # calibration for Vishay BC Components NTCLE413E2103F520L thermistor
4  # using Fluke 8060A multimeter
5  # degrees_F     kohm
6  192     1.30
7  190     1.37
8  186     1.452
9  180     1.65
10 176     1.718
11 ...
```

Your measurements will be in °C not °F, and you will be measuring with a different meter. The # characters are there to hide the comments from the program gnuplot, which you will use for plotting data and fitting models.

8.5 Fitting parameters with gnuplot

Gnuplot is a handy, free program for plotting data and fitting models. There are many other programs available with similar capabilities (some free, some cheap, some expensive), but gnuplot has been around for a long time and is likely to continue to be available for a long time to come.

Gnuplot can be used interactively to explore data, but can also be used with a script file that does everything consistently. I find it more useful, usually, to use a script file from the beginning, and edit it to make changes, rather than trying to remember all the commands I have typed in.

Because this is the first time most of you have used gnuplot, we'll spend the lecture between the halves of the lab developing a gnuplot script to produce plots like the one in Figure 8.3, which fits the Steinhart-Hart model and a simpler 2-parameter model to the data. Bring a computer with gnuplot installed and your data already in a computer file to class, so that you can work along with the class discussion.

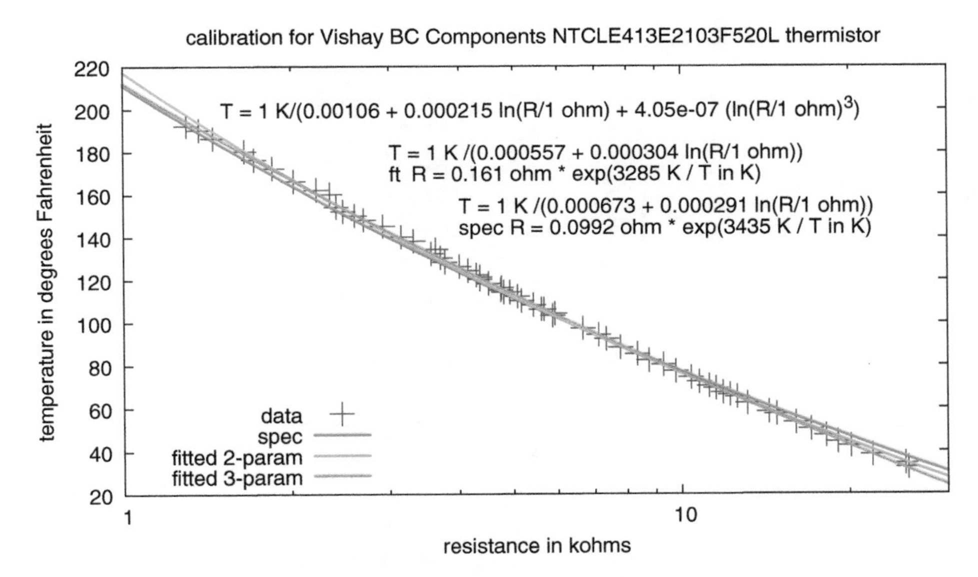

Figure 8.3: *Plot of temperature vs. resistance with a model fitted using the Steinhart-Hart equation (3 parameters) or B-model (2 parameters). This plot was produced from measurements with a low-quality thermometer using* °F*—you should use* °C*. Figure drawn with gnuplot [33]. The gnuplot script for this plot can be found at https://users.soe.ucsc.edu/~karplus/bme51/book-supplements/steinhart-hart.gnuplot and the data at https://users.soe.ucsc.edu/~karplus/bme51/book-supplements/therm_data2.txt. The fits are in* Ω *and K, while the plot is in k*Ω *and* °F.

Before fitting the data, we'll first see how to plot functions (like the B equation), then how to plot data, and finally how to fit a function to the data.

We'll plot and fit the data with temperature as a function of resistance, because the Steinhart-Hart equation is simpler in that form than as resistance as a function of temperature (as one might wish to plot it for a physics experiment, as temperature is clearly the independent variable). The T vs. R plot is also the more useful plot for a calibration curve for using the thermistor later, where we want to change a measured resistance into the appropriate temperature value.

You need to plot temperature in °C as a function of the voltage V_{out} for your thermistor and resistor voltage divider circuit. You may wish to do two plots: for the voltage ratio $V_{\text{out}}/V_{\text{in}}$ and for V_{out} with a fixed $V_{\text{in}} = 3.3\,\text{V}$.

In future assignments, you'll have to do more of the scripting from scratch, so learn what each of the commands in the script do. Gnuplot has a `help` command that you can use interactively to look at the documentation, and there is online documentation at http://www.gnuplot.info/documentation.html

8.6 Using a breadboard

Breadboards are commonly used for prototyping circuits before assembling them more permanently by soldering. Figure 8.4 shows a typical small breadboard, with 30 rows of holes on a 0.1″ (2.54 mm) grid. Each row of five holes is connected together inside the breadboard, so that inserting wires or other components into the holes makes connections.

In addition to the 5-hole rows, many breadboards also have two columns of connected holes running the length of the board, most often used for distributing power and ground connections. If the columns are marked with colors, as in Figure 8.4, then red should be reserved for the positive power rail and the other color (often blue) for ground.

A group of long connections for distributing a signal or power to multiple points is often called a *bus*. In analog systems, generally only the power lines are distributed on buses, but in digital systems buses

Figure 8.4: *Bare breadboard with 30 rows. Each row consists of two nodes: the left node (connecting columns "a" through "e") and the right node (connecting columns "f" through "j"). There are no internal connections between the right and left sides of the breadboard.*

In addition there are two columns on each side of the breadboard connecting holes vertically, making four more nodes. These columns are usually used for distributing power to the circuits on the central part of the breadboard.

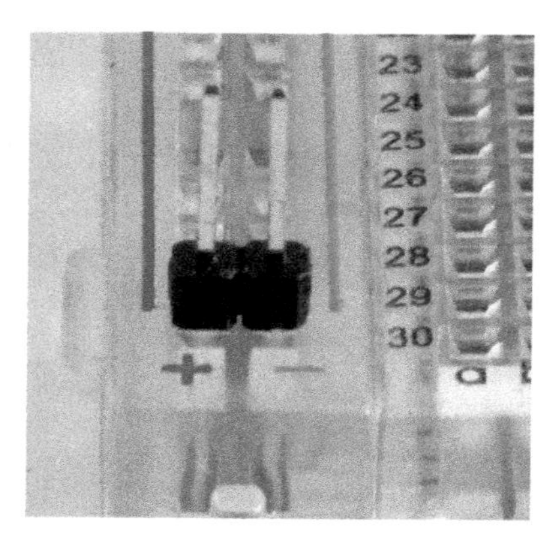

Figure 8.5: *Using double-length male headers in breadboard bus for connecting power and ground from an external source such as the Analog Discovery 2. The male header pins are best left in groups of 2–4 pins, as single pins can be rather wobbly in the breadboard. The shorter header pins intended for soldering (see Figure 3.1) are not suitable for making connections between female headers and the breadboard, as they do not have sufficient length for two header connections.*

are used for most information-carrying signals as well. Figure 8.5 shows how to use the double-headed male headers to make connections to the power and ground buses on a breadboard, which can easily be connected to using the female headers of the Analog Discovery 2.

> One of the most common errors in breadboard use, and the first to check, is whether chips are connected to the power and ground buses, and whether the buses are in turn connected to a power supply (generally either the microcontroller running PteroDAQ or the Analog Discovery 2 in this course).

Figure 8.6: *This picture shows a 0.1 µF capacitor (labeled 104) and a 620 Ω resistor connected in series. The block of three header pins behind them provides oscilloscope or function-generator access to the three rows: 15L for one end of the resistor, 16L for the node where the resistor and capacitor are connected, and 17L for the other end of the capacitor.*
There is also a wire connecting rows 11L and 12L, though it serves no apparent function here, as nothing else is connected to those rows. The wire is inserted far enough that no bare wire is exposed.

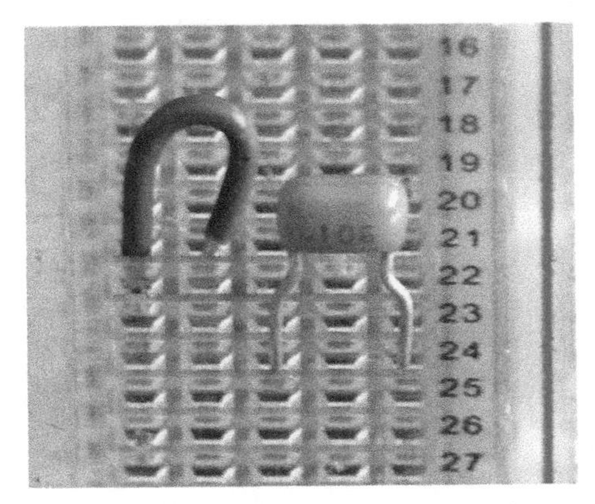

Figure 8.7: *Here we have incorrectly inserted components—both ends of the 10 µF capacitor are in row 24R, so the capacitor is shorted out, and both ends of the wire are in row 21R, so it is duplicating the internal connections of the row.*

Figure 8.6 shows correct use of the breadboard for wires, header pins, and components. Figure 8.7 shows incorrect use of the breadboard, with the capacitor shorted out by the breadboard, and the wire duplicating the connection already made internally.

Figure 8.8 shows a portion of a breadboard with two resistors inserted. The red lines on the picture show which other holes are connected to each end of the resistors. Only the five holes on one side of the central channel are connected—the holes on the other side of the channel are a separate group.

There are two ways to insert resistors into the breadboard: the vertical, *flying-resistor* approach allows very compact wiring, but is somewhat fragile; while the horizontal approach takes up more space on the breadboard, but is sturdier and more reliable. Both approaches are suitable for this class. When using flying resistors, it is a good idea to make the longer wire come from the most-significant color band, so that the color code can be read consistently downward on all resistors. (See Table 4.1 for an explanation of the resistor color code.)

The flying-resistor approach for layout results in compact breadboard layouts, but the long lever arm makes it easy to accidentally brush the resistors out of their holes, and the long bare wires can accidentally short to other resistors in different rows. I use flying resistors a lot in breadboarding, but for more permanent designs with printed-circuit boards, I lay the resistors flat against the board and

Figure 8.8: *Two ways of inserting resistors in a breadboard. The red lines show which holes on the breadboard can be used to connect wires or other components to the resistors.*
On the left is a 330 Ω resistor mounted vertically (a flying resistor*), which allows the ends of the resistor to be in adjacent 5-hole rows. The longer wire is on the end with the most-significant digit of the color code (see Table 4.1).*
On the right is a 270 Ω resistor mounted horizontally, which skips two rows, in order for the resistor to sit flush against the breadboard.

Figure 8.9: *The 15 kΩ (brown-green-black-red is* $150 \cdot 10^2$*) resistor in this picture shows a resistor connecting rows 21L and 24L. There are four other holes in these rows that can be used for other components or wires that need to be connected to the resistor. The wires on this resistor are longer than they should be (compare with Figure 8.8), but still acceptable.*

bend the leads at right angles 0.3″ or 0.4″ apart. That approach works well for breadboards also, but requires a little more care in laying out the design, as the resistors take up more room.

Wires on a breadboard should be cut to the correct length and run close to the breadboard (like the horizontal resistor in Figure 8.8). Generally, 22-gauge solid wire works best in breadboards, but thinner 24-gauge wire can be used if necessary. Larger wires can damage the spring contacts in the breadboard, and thinner wires may not make good contacts. The breadboard contacts wear out after many uses (or a few uses with over-size wires), so breadboards should be regarded as consumable items, not bench lab equipment.

Many breadboards have numbered rows, which aids in talking about where wires go. For debugging circuits on breadboards, it is a good idea to annotate the schematic diagram with the row number for each node. For example, the resistor in Figure 8.9 has one end in row 21L and the other in 24L (the L is for "left"—remember that there is no connection across the chasm down the middle of the board).

One very common error in using breadboards is to connect a wire to the wrong row—often off by one hole.

All the wires that carry the same signal should have the same color insulation. That way, if we see two different color wires plugged into the same row, we can immediately see that there is a wiring error and try to figure out which of the wires is incorrect. Keeping the wiring short and neat helps in getting the wires into the right holes and in seeing when there is a mistake.

When using a breadboard, you need to strip the wires to fit into the holes and make contact with the spring contacts in the board. If you strip too little, you risk not making contact, while if you strip too much, you'll have bare wire hanging out of the board, which may accidentally short to neighboring wires. The breadboards are typically 1 cm deep, so the correct wire stripping length is half that (5 mm). Figure 8.10 shows the correct 5 mm stripping, along with 3 mm (too short) and 1 cm (too long). Wires that are too short are particularly difficult to debug, as they appear to be correctly wired, while not actually making contact.

Wires that have been stripped too much, as in Figure 8.11, can cause short circuits to other wires nearby, resulting in non-functioning circuits and possibly component damage. If you accidentally strip too much insulation off a wire, trim the end down to 5 mm.

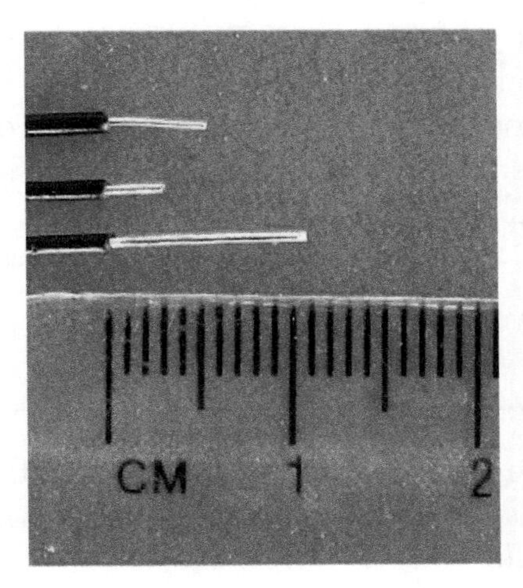

Figure 8.10: *The three wires stripped here illustrate just right (5 mm), too short (3 mm), and too long (1 cm) stripping. These wires are 22-gauge (0.64516 mm diameter), but one can also use 24-gauge (0.51054 mm diameter). The thicker 22-gauge wires hold better in the breadboard and are less likely to cause debugging problems by coming loose.*

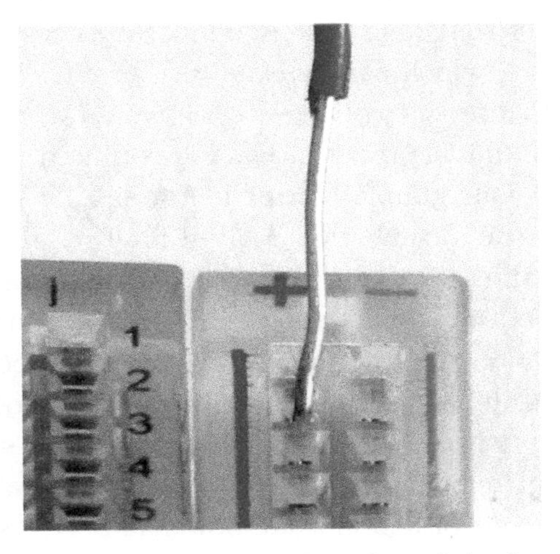

Figure 8.11: *This red wire has had too much insulation stripped from the end, leaving a large stretch of bare wire exposed. This exposed wire can easily short to another wire (like a nearby ground wire), resulting in difficult-to-debug circuit errors and possibly resulting in damage to components.*

8.7 Measuring voltage

We have three instruments available for measuring voltage:

- the multimeter we used for measuring resistance in Section 8.4,
- the PteroDAQ data-acquisition system, and
- the Analog Discovery 2.

Each has different advantages. The multimeter is the quickest to set up if you need to keep changing what pair of nodes you are measuring or just need to make single measurement. PteroDAQ allows recording many voltages over a long time (as well as reporting the most recent measurement), and the Analog Discovery 2 allows recording voltages that change very quickly. Cheap hand-held multimeters tend not to be very accurate—the Analog Discovery 2 and PteroDAQ measurements are likely to be somewhat more accurate. (Expensive bench multimeters are likely to be much more accurate.)

For this lab, it is probably best to use the microcontroller board for PteroDAQ to power your circuit, do quick verification measurements with a handheld multimeter, and use PteroDAQ to do a series of voltage measurements. You can, if you choose, use PteroDAQ for all the voltage measurements—read Section 8.8 for more information on using PteroDAQ.

Choose an appropriate series resistor (usually a single resistor from your parts kit) for the operating point T_{op} specified during the lab. Be sure to record T_{op}, the resistance chosen, and the schematic diagram for your circuit in your lab notebook.

> Come to lab ready to wire up your circuit with a resistor formula that you can plug T_{op} into to get the resistor value.

In the past many students have come to lab not knowing what resistor to use and wasted most of the lab time doing calculations that can be done ahead of time. Lab time is for building and testing circuits—come to lab ready to build.

Once you have selected your resistor, measure it to make sure you have the right resistance. Even with the cheap hand-held meters, this measurement is likely to be more accurate than the nominal value of the resistor—we generally use resistors in class with 1% tolerances, but we get them in cheap assortments, and they may be ones that failed the manufacturers tests and may be 2% off. Some other lab courses use 5% carbon resistors, which can be even further off.

For the lab report, plot the theoretical voltage V vs. temperature T for both the resistance value that was calculated to be optimal and the resistance value you used. Make sure that the region around T_{op} is nearly linear. You can run your gnuplot script before lab for some reasonable choice of resistor value—use the plot to check that the formula you derived for Pre-lab 2.6 is correct. In lab, you should only need to change one value in the script to check your resistance calculation.

Measure and record the temperature and voltage for several different temperatures, both close to and far from T_{op}. We don't need a lot of data points, since we won't be fitting model parameters to the data—this is just a sanity check to make sure that our theoretical model fits the device actually built. Half a dozen well spread-out temperature and voltage measurements should be enough.

If you are using your microcontroller board to provide a 3.3 V power supply to the voltage divider, make sure you record the *measured* voltage of that power supply, as it is not exactly 3.3 V. Report this measurement in your final report! You can use the PteroDAQ software to record single voltage measurements as well as the time course needed for Section 8.8. See Section 8.8 for instructions on running PteroDAQ.

Verify that the voltages are what you expect from your previous calibrations, by plotting both the data and the model on the same plot. Be sure to use the measured voltage of the power supply, and not just the nominal 3.3 V.

> Make the gnuplot script for this validation plot at home, before coming to the lab to make the measurements for it.

Your script should have the B, R_∞, and R values from the data fits for the resistance vs. temperature measurements and the optimization you did. But use the nominal resistance value for the resistor, rather than the theoretically optimal resistor.

In lab, as soon as you measure some voltage/temperature pairs, plot them on your validation plot. If the measurements seem way off from the model, then fix the model, redo the measurements, or both. The problem may be that the resistance you wired on the board does not match the resistance you intended to use—verify it by remeasuring it!

> You can't measure a resistor properly when it is in a circuit—the ohmmeter needs to provide a current and measure the resulting voltage. If there are other paths for the current to take, or other sources of voltage, the ohmmeter measurement will be wrong.

Use wires with alligator clips on one end to connect your thermistor to a breadboard that has both your series resistor and the microcontroller board for PteroDAQ. Use the 3.3 V output and GND of the microcontroller board as your power supply for your voltage divider.

8.8 Recording voltage measurements

For this section of the lab, we will record a time series of voltage measurements using PteroDAQ. Because the temperature vs. time plot does not require any special equipment, you can do this part of the lab at home, but more help will be available if you do it in the lab.

Because the voltages need to stay within the 0 V–3.3 V range of the microcontroller (0 V–5 V for some Arduino boards), the voltage should be supplied by the microcontroller. Using power from the microcontroller makes it much less likely that you will accidentally damage the microcontroller by providing an out-of-range voltage.

Replace the voltmeter connection to the middle of your voltage divider with a wire to an analog input of the microcontroller board (say, A0 on a Teensy board).

On Windows machines the Python program is started from a command line with

```
python C:\ProgramFiles\PteroDAQ\daq
```

and on Mac and Linux machines the program is started with

```
python Documents/BME51/PteroDAQ/daq
```

(in both cases, use whatever directory the program is installed in, if you install the files elsewhere). Python looks for `__main__.py` in that directory.

PteroDAQ should work with any version of Python from version 2.6 on. If you have an older version of Python installed, install either version 2.7 (the final, stable version of the Python2 family) or a recent release of Python3. The anaconda package for installing Python is probably your best choice, as it installs many useful Python modules that are not part of the standard python.org distribution (see Section 3.4).

The first thing to do is to configure what information you want recorded and when.

Use the "Add channel" button to add a channel that records the voltage using the analog pin that you connected to the output of your voltage divider. You can change the name of the channel so that the data file reports what the signal is called, as well as its values.

You probably want to record the analog input every 50 ms (20 Hz) using the "Timed" radio button and typing the time or frequency, and you probably want to record the voltage, rather than the raw analog-to-digital converter readings, by checking the "Supply voltage" box.

If you need to make a single measurement, rather than continuous readings, you can click the "1" button.

If you are using a Teensy board, you have a choice of hardware averaging also. For very slow signals, like we are using in this lab, 32× averaging provides the lowest noise. When we are trying to capture high-speed signals, we may need to use less averaging.

Record a few minutes worth of data (hot water cooling down toward room temperature works well) and save it in a file. Because water changes temperature slowly, recording at 10 Hz or 20 Hz is a fast enough sampling rate. Because you are sampling so infrequently, there is plenty of time to use the 32× hardware averaging option on the boards that support it. Averaging should reduce the noise of the measurement substantially.

You should be able to plot the file easily with gnuplot, though getting proper scaling and labeling of the plot may take some effort. You want to plot temperature (in °C) on the y-axis vs. time (in seconds) on the x-axis.

PteroDAQ just records voltage, not temperature, but no one wants their recording thermometer to report temperature in volts (quick, doctor, they're running a fever of 3.2 V!). So you need to write a function $T_c(V)$ that provides the temperature in °C for a given voltage—the inverse of the function you have already created to compute the voltage as a function of temperature $V(T_c)$. (We are talking about the *functional inverse*, $T_c(V(T)) = T$, not the *multiplicative inverse* $1/x$ nor the *additive inverse* $-x$.)

Try to plot the data before leaving lab, so that you can clear up any mistakes in the inverted function or the plotting and detect any anomalies. If the data shows poor lab technique (wires coming loose, wires shorting together), it may be worth fixing the problem and taking a new data set right away.

8.9 Demo and write-up

The point of the design report is to document your design and measurements—the pre-lab questions are guidance in what sort of analysis you should be doing. The final draft of the report might not contain *any* of the answers to the pre-lab questions, unless those questions are directly relevant to *your* design.

In the lab, the students need to show the instructor that they can correctly measure resistance and temperature, fit parameters of a model to the data, select an appropriate resistor for a given T_{op}, correctly measure voltage from a voltage divider, and record a time series of measurements with the data-acquisition system.

In addition to all the usual stuff described in the lab report guidelines in Chapter 7, the lab report should contain at least the following (though not necessarily in this order):

- A statement of the problem.
- Any pre-lab analysis and design that was done.

- A plot of measurements made and any modeling done (for example, the resistance vs. temperature calibration curve). The plot should use the units measured (Ω or kΩ and °C, for example). The data points measured should be plotted as points, not lines, so that the distinct points can be easily seen.
- A schematic for the voltage divider, including part number for the thermistor and value for the resistor.
- A validation plot of voltage (or voltage ratio) vs. temperature for the voltage divider, showing both measured data and the model built using the equations for thermistor behavior and voltage dividers.

  The purpose of this plot is to make sure that your design is correct (the straightest part of the curve is near the temperature T_{op}) and that what was built matches the design (the data points are on the curve). The parameters for the function (R, R_∞, B) should come from the design you did and the model fitting you did for calibration of your thermistor. **You are not refitting the parameters based on the voltage measurements.**

  This plot will require you to do some scripting in gnuplot or a similar plotting tool. Again, the data points should be plotted as distinct points, but the model as a continuous curve.
- A plot of temperature vs. time from the output recording. Again, this plot will require some scripting. The temperature vs. time plot is a different data set from the voltage vs. temperature data.

  Because we are interested in a continuous-time signal and believe that we are sampling fast enough to catch any real phenomena, the plot will look better if we plot the data with lines, rather than with points. In general, we use lines rather than points for data when the points are so close together on the plot that the size of the points obscures the data. When the data points are separated enough to have space between them, connecting them with lines can be misleading, implying measurement of intermediate values that were not observed.

Chapter 9

Sampling and Aliasing

9.1 Sampling

When we record signals on the computer, we cannot store infinitely many data points to represent continuous functions of time. Instead we store data at fixed time intervals (the *sampling period*). Because it is often easier to deal with frequencies rather than times, we refer to the inverse of the sampling period as the *sampling frequency*, f_s, or *sampling rate*.

Figure 9.1 shows several examples of sampling with different sampling frequencies on a sawtooth-wave signal. If our signals are low frequency compared to our sampling frequency, the computer representation of them is quite good, and little, if any, information is lost in storing them at discrete times (Figures 9.1(a)–9.1(b)). But when the sampling frequency is low, information is lost (Figures 9.1(c)–9.1(f)).

> The Nyquist Theorem says that signals that contain no energy at or above half the sampling frequency, $0.5 f_s$, can be accurately reconstructed from the sampled data—the frequency $0.5 f_s$ is referred to as the *Nyquist frequency* or the *Nyquist limit* [141].

Warning: there is a related concept *Nyquist rate*, which attempts to define the minimum usable sampling frequency in terms of the signal frequency. We do not use that term in this course, as it is much less useful (hence, less commonly used) than Nyquist frequency. Nyquist frequency is always precisely half the sampling rate, no matter what your signal is—it is a property of the sampling, and not of the signal being sampled. The Nyquist frequency is useful for determining the apparent frequency of a sampled signal from the actual frequency, while the Nyquist rate is not.

The math for the Nyquist Theorem is beyond the scope of this course, but should be covered in a *Signals and Systems* course, as it involves Fourier transforms and the sinc function: $\mathrm{sinc}(a) = \sin(\pi a)/(\pi a)$.

When the sampling frequency is an exact multiple of the frequency of the sampled signal, as in Figure 9.1(d), the sampling happens at the same place in each period of the waveform, and so produces a simple periodic pattern.

When the sampling frequency is a little off from an exact multiple, as in Figures 9.1(c)–9.1(e), the samples come from different points in each period and the pattern changes from cycle to cycle. The result can be quite striking—see Figure 9.2(a), which shows a 1 Hz sine-wave beat pattern when a 19 Hz sine wave is sampled at 40 Hz. The *beat frequency* is the difference in frequency between the signal and the Nyquist frequency.

Despite the apparent fluctuation in the amplitude, the original wave form can be recovered with a suitable low-pass filter. Figure 9.2(b) shows the reconstruction of the original 19 Hz sine wave from the

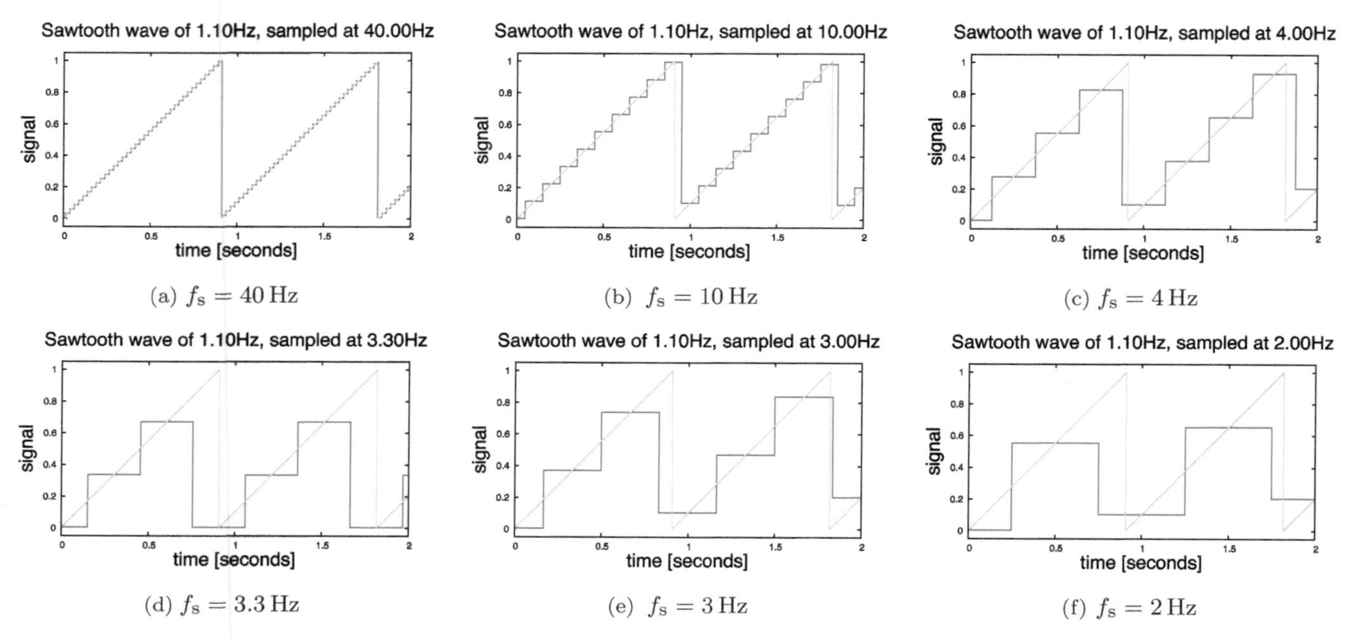

(a) $f_s = 40\,\text{Hz}$ (b) $f_s = 10\,\text{Hz}$ (c) $f_s = 4\,\text{Hz}$

(d) $f_s = 3.3\,\text{Hz}$ (e) $f_s = 3\,\text{Hz}$ (f) $f_s = 2\,\text{Hz}$

Figure 9.1: *This figure shows how changing the sampling frequency changes how well a discrete-time representation of a signal matches the signal. For each part, the continuous-time signal is a 1.1 Hz sawtooth wave, drawn in blue. The discrete-time signal, drawn in red with a step centered at the time the sample was taken, has sampling rates of 40 Hz, 10 Hz, 4 Hz, 3.3 Hz, 3 Hz, and 2 Hz. Figure drawn with gnuplot [33].*

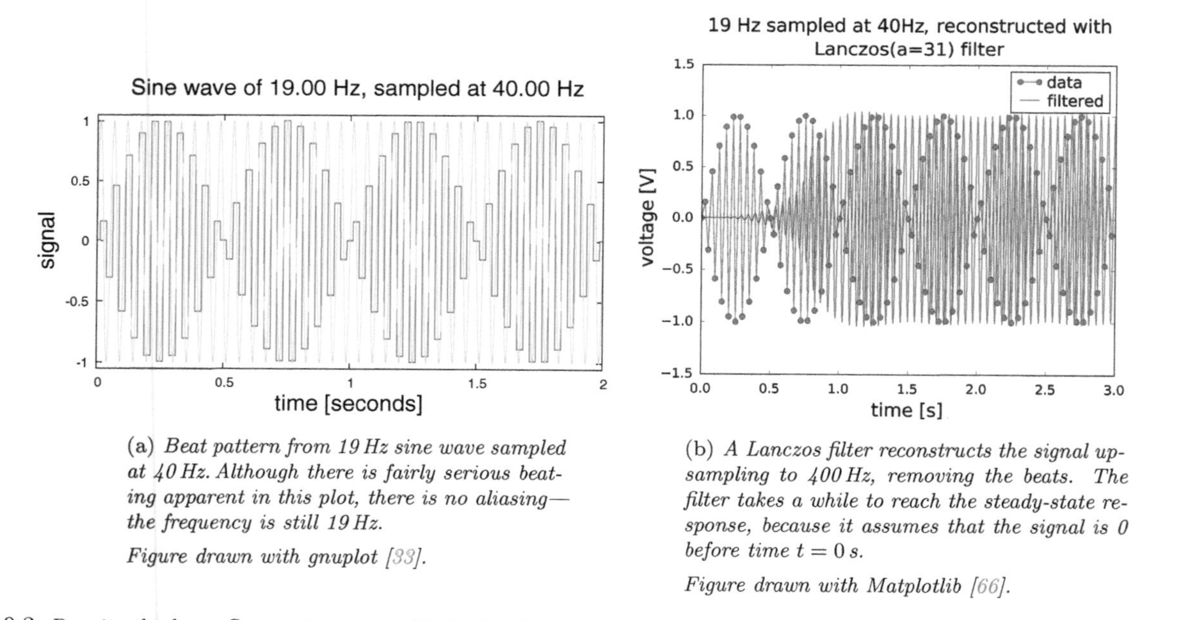

(a) *Beat pattern from 19 Hz sine wave sampled at 40 Hz. Although there is fairly serious beating apparent in this plot, there is no aliasing— the frequency is still 19 Hz.*

Figure drawn with gnuplot [33].

(b) *A Lanczos filter reconstructs the signal upsampling to 400 Hz, removing the beats. The filter takes a while to reach the steady-state response, because it assumes that the signal is 0 before time $t = 0$ s.*

Figure drawn with Matplotlib [66].

Figure 9.2: *Despite the large fluctuation in amplitude for the samples, the original signal can be recovered with a suitable low-pass filter, because all the signal energy is at 19 Hz, which is less than the 20 Hz Nyquist frequency.*

The amplitude fluctuation follows a 1 Hz sine wave—the beat frequency. Visually it looks like a 2 Hz pattern, because the positive and negative parts of the sine wave look much the same when used as amplitude modulation.

sampled waveform using a digital Lanczos filter [133] for the low-pass filter. The Lanczos filter is a finite approximation to the theoretically ideal sinc filter [155]. A high-order filter is needed, because the 19 Hz in the passband of the filter has to be separated from the −21 Hz alias in the stop band.

9.2 Aliasing

Sampling a signal at a fixed sampling rate results in many interesting effects, including aliasing and beating. Be careful to use the correct terminology for different effects.

One important effect is *aliasing*, in which a frequency higher than the Nyquist frequency appears indistinguishable from a frequency lower than the Nyquist frequency (see Figure 9.5). The term *aliasing* refers only to this change in apparent frequency, not to changes in amplitude (the *beating* observed in Figure 9.2(a)).

> Aliasing is an effect in which a signal with frequency higher than the Nyquist frequency is indistinguishable from one lower than the Nyquist frequency. Aliasing (unlike beating) results in loss of information, and the effects of aliasing cannot be removed by filtering.

Aliasing is commonly observed with stroboscopes (which give you visual samples at fixed time intervals) and Moiré patterns, which sample images at a fixed spatial frequency (see Figure 9.3).

The common illusion in movies and videos of wheels or fan blades turning backwards is a result of this aliasing—if the wheel turns just a little less than a full revolution between frames of the video, then it will look the same to the viewer as the wheel turning just a little bit in the other direction. Figure 9.4 shows an example of a wheel turning at 3 Hz and being sampled at 24 frames per second. The speed of turning is well under the Nyquist frequency of 12 Hz, so no aliasing occurs. But if we sample less frequently (say at 4 Hz), we do expect aliasing. Looking down a column shows us what we would see at 4 Hz, and down a diagonal shows us 3.4286 Hz.

Figure 9.5(b) also shows the wheel-turning effect, converting an upward-sloping ramp at 39 Hz into a slow ramp in the other direction. The ramp used here is an asymmetric waveform, clipped from below at 0, so that you can see that the effect is a time reversal, not a negation of the voltage of the signal.

We can compute the effective frequency of an aliased signal fairly easily. For example, consider a wheel with radius 1 rotating counterclockwise at $60f$ rpm (that is, f revolutions per second). We can describe the motion of a point on the wheel in the complex plane as

$$z(t) = e^{j2\pi ft},$$

or, equivalently, $x(t) = \cos(2\pi ft)$ and $y(t) = \sin(2\pi ft)$.

Figure 9.3: *This* Moiré pattern *is formed from two sets of parallel lines with different spatial frequencies. The black ones in back have a period of 40, and yellow ones in front have a period of 44. The gaps between the yellow lines sample the background image, and we see darkness proportional to how much of the black line is visible in the sampling interval. The result looks like a triangular intensity wave with a period of 440, the least common multiple of 40 and 44. If viewed at sufficiently high resolution, there should be three periods visible in the image.*
Scaling the image on a screen may result in changes to number and placement of the dark bars, due to changing the number of pixels available for each line. At some small sizes, the period of 40 and period of 44 can round to the same number of pixels, and the Moiré pattern disappears.

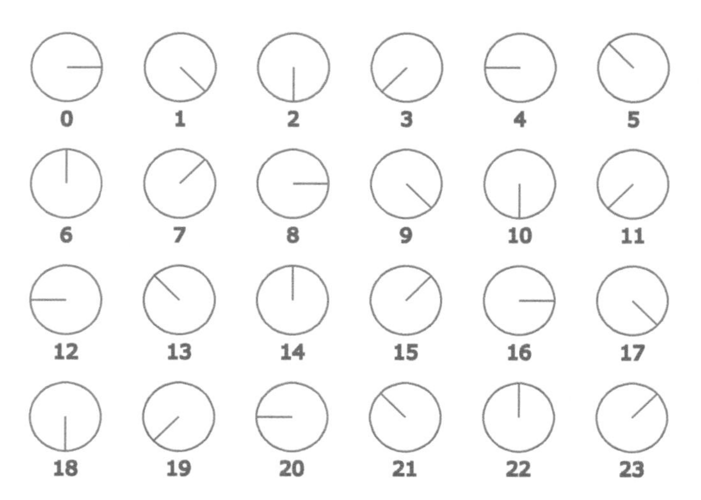

Figure 9.4: *This figure shows a wheel turning at 3 Hz (180 RPM) with a sampling rate of 24 frames per second. In each frame the wheel turns one-eighth of a turn clockwise. At 24 frames per second, this would look like* $24/8 = 3$ *revolutions per second, and the apparent frequency is the same as the actual frequency.*
If we sample at 4 frames per second and look at only every sixth frame (moving down instead of across the figure), we see the wheel rotating counter *clockwise by a quarter turn on each frame—the apparent speed for the 3 Hz rotation is* -1 *Hz.*
Similarly, if we look down the diagonal (sampling at $24/7 \approx 3.4286$ *Hz), we see the wheel turning one-eighth turn backwards on each sample, an apparent speed of* -0.4286 *Hz for an actual speed of 3 Hz.*

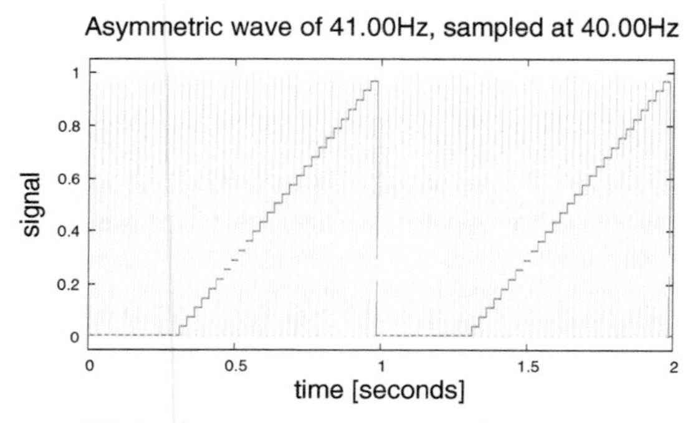

(a) *41 Hz asymmetric ramp waveform, sampled at 40 Hz, looks identical to a 1 Hz waveform with the same asymmetric shape.*

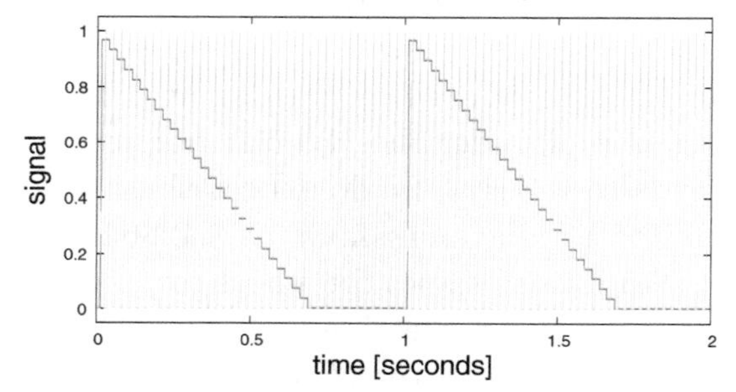

(b) *39 Hz asymmetric ramp waveform, sampled at 40 Hz, looks identical to a* -1 *Hz waveform—that is, it looks like a 1 Hz waveform running backwards in time.*

Figure 9.5: *When sampling a waveform with frequency higher than the Nyquist frequency* $f_s/2$*, the signal appears like one in the range* $[-f_s/2, f_s/2]$*. Here, with a sampling frequency of 40 Hz, signals of 41 Hz and 39 Hz look like 1 Hz and* -1 *Hz, respectively. Figure drawn with gnuplot [33].*

If we sample this every $\Delta t = 1/f_s$ seconds, we have the series of points

$$z_0 = e^0, \quad z_1 = e^{j2\pi f \Delta t}, \quad z_2 = e^{j2\pi f 2\Delta t}, \quad \ldots .$$

If $f < f_s/2$, then $2\Delta t < 1/f$, and $2\pi f \Delta t < \pi$, which means we are advancing by less than π radians (half a revolution) on each sample. With sampling this fast, the wheel's appearance matches its actual speed. If the wheel were turning in the other direction, we would have a negative frequency, but as long as $-f_s/2 < f < f_s/2$, the wheel's apparent motion would match its actual motion.

If $f_s/2 < f < f_s$, then $\pi < 2\pi f \Delta t < 2\pi$ and we are moving more than half a revolution but less than a full revolution on each sample. But the angle $2\pi f \Delta t$ is equivalent to $2\pi f \Delta t - 2\pi$, or $2\pi(f - f_s)\Delta t$. So, the frequency f looks the same as $f - f_s$, which would be a backward rotation.

If $f_s < f$, then the wheel is turning more than a full turn between frames, and so it looks like it is going much slower than it actually is. In fact, if $f_s = f$, then the wheel will look motionless. Figure 9.6 shows the apparent rotation speed as a function of the actual rotation speed.

To avoid artifacts in data recording, it is usually necessary to filter the input signal to remove all components that have frequencies above the Nyquist frequency. You may have already experienced anti-aliasing filters on your computer—characters displayed on the screen are often deliberately blurred (low-pass filtered) to remove the *jaggies* that are discretization artifacts (in space rather than in time). We'll do some simple filter design in this course (see Chapter 11), but more sophisticated filter design is left for the courses *Signals and Systems* and *Digital Signal Processing*.

One important use we'll make of aliasing in this class is to use sampling frequencies that are factors of 60 Hz: 60 Hz, 30 Hz, 20 Hz, 15 Hz, 12 Hz, 10 Hz, and so forth. Using a sampling frequency that is a factor of 60 Hz converts any 60 Hz noise added to our signals into 0 Hz (DC), as shown in Figure 9.7.

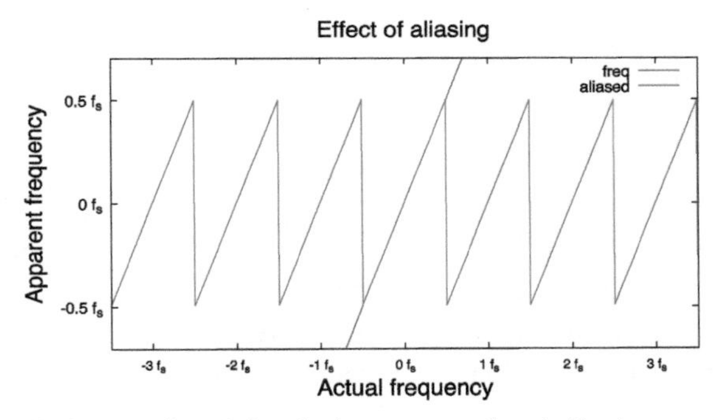

(a) *For a turning wheel in a video, if the wheel turns more than half a turn per frame it will look like a wheel turning at a lower speed (and possibly in the opposite direction, represented here as a negative frequency).*

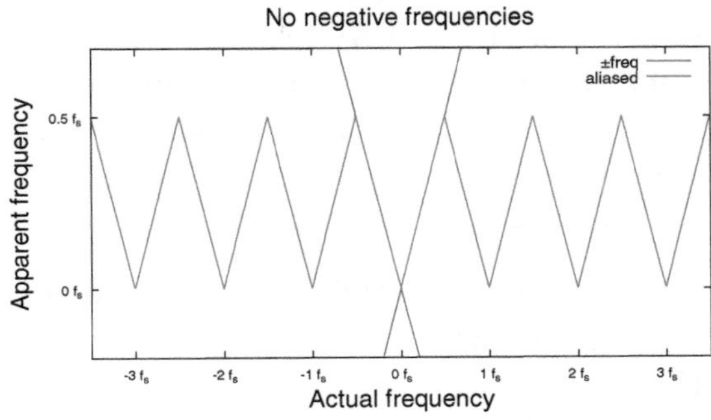

(b) *For sinusoidal waves, we can't distinguish negative frequencies from positive ones with a phase shift, so the negative frequencies "fold over", and all apparent frequencies are between 0 and the Nyquist frequency $f_S/2$.*

Figure 9.6: *For low frequencies, the apparent frequency is the same as the actual frequency, but once the frequency f increases beyond half the sampling frequency $0.5f_S$, the apparent frequency differs, as shown in Figure 9.6(a).*
Figure drawn with gnuplot [33].

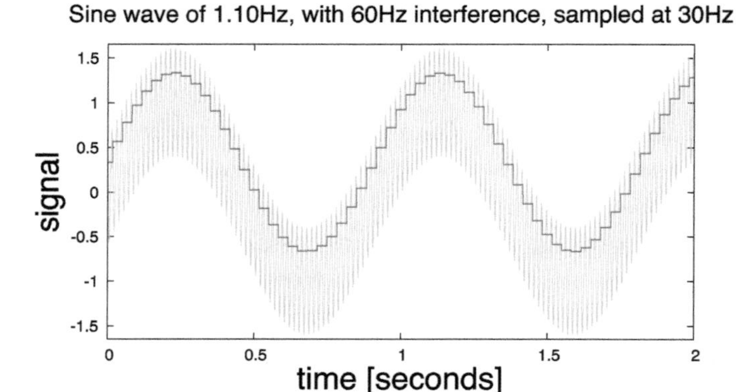

Figure 9.7: *Many signals we want to look at are contaminated by electrical interference at 60 Hz or multiples of 60 Hz (50 Hz in Europe) from capacitive or inductive coupling from the ubiquitous power wiring in our buildings. When we are interested in low-frequency signals, we can suppress this noise by sampling at periods that are multiples of $\frac{1}{60}$th of a second. Because the interference repeats exactly each time we sample, it disappears from our view, becoming just a DC offset to the signal.*
Figure drawn with gnuplot [33].

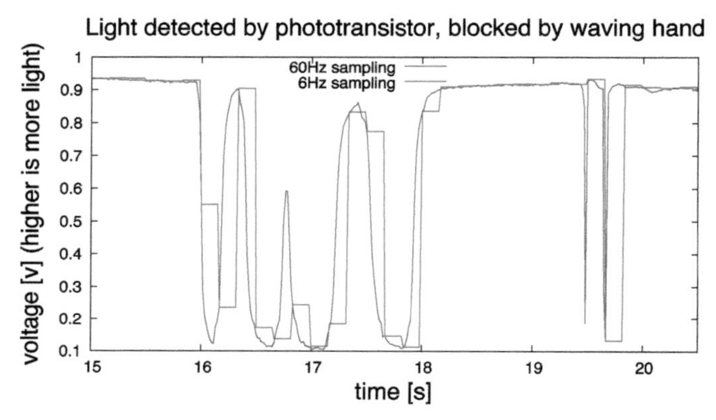

Figure 9.8: *Waving a hand in front of a phototransistor blocks the light as each finger comes between the sensor and the light source. At 60 Hz sampling, we can see the light blocked by individual fingers, but at only 6 Hz sampling we almost miss the gap around 16.8 s, and completely miss the fast-moving finger at 19.5 s.*
Data collected with PteroDAQ using a Teensy LC board [46]. Figure drawn with gnuplot [33].

There is some danger in using low sampling frequencies—we can miss important signals that have short duration. For example, the R spike in an EKG signal may be only 10 ms long, and sampling at 60 Hz or lower frequency may get the wrong value for the spike's amplitude or may miss the spike entirely. Figure 9.8 shows the effect of too low a sampling frequency when looking at light detected by a phototransistor being blocked by a hand waving in front of the phototransistor.

> Slowly changing (low-frequency) signals are fairly well captured even at a low sampling rate, but rapidly changing (high-frequency) signals are badly distorted at low sampling rates, and narrow pulses may be missed entirely.

Chapter 10

Impedance: Capacitors

10.1 Capacitors

As you may have learned in a physics class, a *capacitor* is a pair of conductors separated by an insulator. If charge is added to one of the conductors, there is an electric field created, and the voltage between the conductors is the charge divided by the capacitance: $V = Q/C$. Capacitance is measured in farads, which are coulombs per volt, $1\,\mathrm{F} = 1\,\mathrm{C}/1\,\mathrm{V}$.

We can continue the hydraulic analogy of Section 4.7, in which voltage was pressure, charge was volume of water, and current was flow rate of water, by making two hydraulic analogies for capacitors:

- If one plate of a capacitor is connected to ground, we can model the other plate of the capacitor as a vertical, cylindrical tank with a pipe connected to the bottom of the tank. As water flows into the tank, the water level in the tank rises, with the depth of the water proportional to the total volume of water in the tank (the charge). The pressure in the pipe (the voltage) increases proportional to the depth of the water, which is in turn proportional to the volume of water in the tank. A large cross-sectional area for the tank corresponds to a large capacitance—the pressure only increases gradually with the volume of water in the tank.

  This analogy also works for describing one of the limits of the capacitors—we can only fill the tank so full before it ruptures or overflows, and capacitors can only be charged to a certain voltage before they fail (sometimes in dramatic ways).
- If we want to make an analogy for a capacitor in which neither plate is connected to ground, we'll need a 2-port hydraulic analog: something with a pipe on either side. One such model is a wide cylinder lying on its side, with pipes connected at the left and right ends and a piston on a spring dividing the left half from the right half.

  Water cannot flow through the pipe, because the piston blocks continuous flow, but if we increase the pressure on the left, the piston will move to the right, until the force of the spring divided by the area of the piston matches the difference in pressure between the two sides. If the spring has a linear force per distance, then the difference in the amount of water on each side (corresponding to the charge on the capacitor) is proportional to the pressure difference (corresponding to the voltage).

  If we fluctuate the pressure on one side, the piston will swing back and forth, and we will see water flowing in and out on the other side, even though there is no continuous flow in one direction. This corresponds to a capacitor allowing AC current, but not DC current, through it.

The capacitance of a pair of parallel plates is directly proportional to the area of the plates and the dielectric constant of the insulator, and inversely proportional to the separation between the plates:

$$C = \frac{k\epsilon_0 A}{d} \,, \tag{10.1}$$

where k is the unitless (relative) dielectric constant, $\epsilon_0 = 8.854187817\,\text{pF/m}$ is the permittivity of free space, A is the area of each plate (in m^2), and d is the separation between the plates (in m).

You can have a capacitor with two different dielectrics, as we do in Lab 4, where we have a thin layer of polypropylene and a thicker layer of air when we use the touch sensor as a proximity sensor. You can use the simple geometric formula of Equation 10.1 in this case also, by treating the capacitor as having a plate at the transition between the two dielectric layers. That then gives us two capacitors in series—one with one dielectric, one with the other.

There is an excellent tutorial on capacitors published by Kemet (a manufacturer of capacitors) that explains the basics of capacitors and what the various types are [53]. The Wikipedia article on capacitance is also worth reading, though much of it is too complicated for a first course in electronics [111].

Physics classes also usually cover the formula for discharging a capacitor through a resistor, in which case the voltage follows an exponential decay with time constant RC. (We'll re-derive that formula in Section 11.4.)

Although we are sometimes interested in the charge, Q, we are much more often interested in the current, i, which is the derivative of the charge with respect to time.

Because $Q = CV$, for a capacitance of C, the current through the capacitor is

$$i(t) = \frac{\mathrm{d}Q}{\mathrm{d}t} = C\frac{\mathrm{d}V}{\mathrm{d}t} + V\frac{\mathrm{d}C}{\mathrm{d}t} \,,$$

where V is the voltage across the capacitor.

We are most often dealing with situations where the geometry and dielectric constant of the capacitor are fixed, so the capacitance does not change, which means that $\frac{\mathrm{d}C}{\mathrm{d}t} = 0\,\text{F/s}$ and we can simplify the formula to

$$i(t) = C\frac{\mathrm{d}V}{\mathrm{d}t} \,.$$

Warning: the more general form that allows for fluctuations in capacitance is sometimes needed—for example, capacitance microphones rely on changes in the geometry of a capacitor to measure fluctuations in air pressure.

Capacitors are made deliberately by sandwiching metal foil and various insulators, but they also occur naturally. For example, the lipid bilayer of a cell membrane is a good insulator (charged ions don't easily pass through the membrane) and the salty solutions on either side of the membrane are pretty good conductors, so cell membranes can be regarded as capacitors. Indeed much of neuroscience relies on interpreting electrical signals that result from charging and discharging the capacitor that is the cell membrane of neurons.

The capacitors that we have in our parts kits may be of two types: *ceramic* and *electrolytic*. Ceramic capacitors are generally preferred for dealing with signals (high-pass and low-pass filters, for example), while electrolytics are preferred for energy storage.

10.1.1 *Ceramic capacitors*

Ceramic capacitors have an insulator with a very high dielectric constant (1000–25,000, compared to values around 2–10 for common insulators like plastic, glass, rubber, paper, and so forth). The high dielectric constant allows the capacitors to be made with small areas for the conductors, without having to make the insulating layer extremely thin.

Ceramic capacitors are usually quite small, so marking the size on them requires a rather compressed code.

One common code (Figure 10.1) is to use three digits—the first two are the capacitance and the third an exponent for a power of 10 to multiply by pF to scale the first two. For example, the code "224" means 22×10^4 pF, which is more often written as $0.22\,\mu$F or 220 nF. When there are only 2 digits on the capacitor (or 1), then the multiplier is 1, and the number can be read directly as capacitance in picofarads.

Capacitances are often not very precise, so capacitors are most commonly sold in sizes from the E6 series (see Table 4.2). The collection of ceramic capacitors used in this class don't follow E6 exactly—see Table 10.1 for the values usually included.

There are several different types of ceramic dielectric used in ceramic capacitors, with different temperature and voltage characteristics, as well as different dielectric constants. Because we are buying super-cheap assortments of capacitors, we do not have full specification of the capacitors, but they are likely to be either Y5V or X5R ceramics, as those are the most popular choices for low-cost capacitors. The Y5V capacitors can lose a lot of their capacitance if the temperature changes much from 25°C, while the X5R capacitors will remain within 15% over a much wider range of temperature. Incidentally, the X5R code is technically not a specification of what dielectric is used, but the temperature range

Figure 10.1: *Nonpolarized capacitors showing the markings. The numbers on ceramic disk capacitors usually indicate the capacitance in picofarads. The "20" is a 20 pF capacitor, but "104" is* not *a 104 pF capacitor, it is 10E4 pF, which is more commonly expressed as 0.1 μF or 100 nF. The large capacitor is 56E4 pF or 0.56 μF. The "J" is the tolerance of the capacitor, encoded as in Table 10.2; J is ±5%. The "100" after the J is the voltage rating. Given the 100 V rating and the 5% tolerance, this capacitor is probably a film capacitor rather than a ceramic one (I got a bunch of them a long time ago in a surplus grab bag, so I don't know for sure). The small blue capacitor has pale grey markings that are barely visible in this photo or any other way—they can only be read in bright light with a magnifying lens. The numbers on the blue capacitor are 475, which stands for 47E5 pF, or 4.7 μF.*

Marking	1							5	6	
value	1 pF							5 pF	6 pF	
Marking	10	15	20	22	30	33	47			68
value	10 pF	15 pF	20 pF	22 pF	30 pF	33 pF	47 pF			68 pF
Marking	101			221		331	471			681
value	100 pF			220 pF		330 pF	470 pF			680 pF
Marking	102			222		332	472			682
value	1 nF			2.2 nF		3.3 nF	4.7 nF			6.8 nF
Marking	103			223		333	473			683
value	10 nF			22 nF		33 nF	47 nF			68 nF
Marking	104	154		224		334	474			684
value	100 nF	150 nF		220 nF		330 nF	470 nF			680 nF
Marking	105			225			475			
value	1 μF			2.2 μF			4.7 μF			
Marking	106									
value	10 μF									

Table 10.1: *This table lists the ceramic capacitors usually included in the lab kit. The values may change from year to year, depending on which assortment of ceramic capacitors is cheapest.*

Marking	Tolerance	Marking	Tolerance
B	±0.1 pF	F	±1%
C	±0.25 pF	G	±2%
D	±0.5 pF	J	±5%
		K	±10%
		M	±20%
		Z	−20%, +80%

Table 10.2: *Tolerance values for capacitors can be expressed by a letter code after the value.*

and amount the capacitance changes over that range. There are three parts to the temperature code for capacitors, as explained in Table 10.3.

Ceramic capacitors can also lose effective capacitance when there is a large DC voltage across the capacitor. What happens is that many of the dipoles in the insulator are locked into position by the DC field, and can't rotate with any AC field that is added. With fewer dipoles rotating, the insulating material is less polarizable, and so the dielectric constant is lower, reducing the capacitance. To avoid this problem, it is a good idea to use capacitors at less than about 30% of their rated voltage (so a 5 V design would use ceramic capacitors rated at 16 V or higher).

There is an excellent explanation of the many reasons why the capacitance written on a capacitor is not always what you get, written by engineers at Kemet Electronics (a capacitor manufacturer) [88].

Because of the variation in capacitance with applied voltage and temperature for X5R and X7R capacitors, engineers often choose more expensive C0G capacitors on the signal path. The C0G dielectrics have a much smaller dielectric constant, but much more constant specifications. The more variable X5R and X75 capacitors get used where a 20% variation in the capacitance will not affect the performance of the circuit (as bypass capacitors, for example).

Low temp	High temp	Change from 25°C
X = −55°C	5 = +85°C	F = ±7.5%
Y = −30°C	6 = +105°C	P = ±10%
Z = +10°C	7 = +125°C	R = ±15%
	8 = +150°C	S = ±22%
		T = +22–33%
		U = +22–56%
		V = +22–82%

Table 10.3: *The three characters of the capacitor temperature code encode the low temperature, the high temperature, and the variation in capacitance allowed over that temperature range. These three-letter codes are used for Class II dielectrics, whose dielectric constant is very temperature-dependent.*
One special case is "NP0" or "C0G", a Class I ceramic dielectric, which is ±30ppm/°*C over* −55°*C to* +125°*C, with a maximum change of 0.3% over that range—much more constant than any of the Class II ceramic capacitors. The disadvantage of Class I dielectrics is their much lower dielectric constant, which makes large capacitance values impractical. X5R and X7R are the popular choices for most projects.*

10.1.2 *Electrolytic capacitors*

The low frequencies in several labs in this course require large RC time constants, which in turn require large resistance, large capacitance, or both. Large ceramic capacitors are not generally available, so electrolytic capacitors are often used when a large capacitance is needed.

Electrolytic capacitors get their high capacitance not from materials with high dielectric constants, but from very thin layers of oxides on rough-textured metal (usually aluminum with Al_2O_3, but sometimes tantalum with Ta_2O_5) [127]. The texturing provides a very large area (up to 300 times the area of a smooth-surfaced metal foil [88]), and the thinness of the oxide provides a very small separation, so the capacitance can be large despite a modest dielectric constant.

Because the surface is rough, the other electrode has to conform to it—for the electrolytic capacitors we use, this is done by having a wet electrolyte solution as the other electrode (the cathode). Another layer of aluminum foil without surface texturing or oxide coating is used to connect a wire to the electrolyte. The whole Al-Al_2O_3-electrolyte-Al sandwich is coiled up and stuffed into a cylindrical can.

The electrolyte solution also helps maintain the oxide film—if there is a pinhole flaw in the film, the current there increases considerably, and the metal surface is oxidized to repair the defect. This repair mechanism results in a small leakage current, which increases with the size of the capacitor and with the voltage. A typical leakage current for a 470 μF capacitor rated at 10 V is 47 μA, computed as $\max(0.01CV/\mathrm{s}, 3\,\mu\mathrm{A})$ [54], where C is the capacitance and V is the rated voltage. If the leakage current for a wet-electrolytic capacitor is too large, charging the capacitor at its rated voltage through a 1 kΩ resistor for several minutes can restore the oxide layer and reduce the leakage current.

This repair mechanism relies on the current flowing in the correct direction—a reverse flow would reduce the oxide coating and cause the pinhole defect to get larger, increasing the current and accelerating the process. Indeed, hooking up an electrolytic capacitor backwards can cause the current to go up so fast that the wet electrolyte turns to steam and the capacitor explodes.

To prevent these explosions, electrolytic capacitors are labeled as having a positive and a negative lead and circuits must be designed so that the voltage across the capacitor always has $V_+ - V_- \geq 0$. Schematics indicate which lead is which by putting a +-sign next to one of the plates. The capacitor

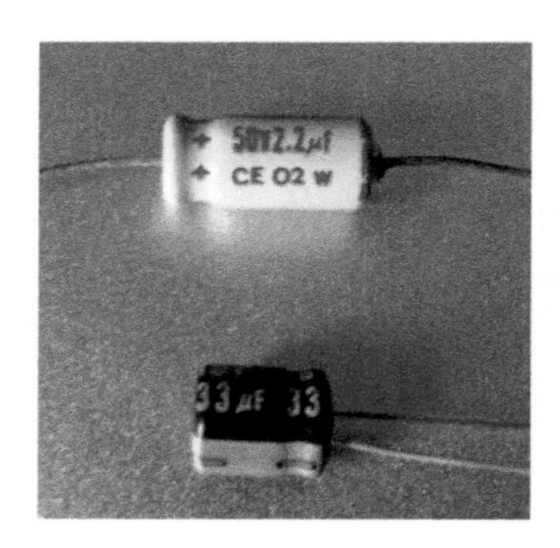

Figure 10.2: *Electrolytic capacitors are usually marked with a clear + or − to indicate polarity. They are usually big enough to have the capacitance and voltage ratings printed in plain text: here the upper picture is an* axial *package for a 2.2 µF, 50 V capacitor, and the lower is a* radial *package for a 33 µF, 16 V capacitor. The radial packages are more popular now, because they are easier for pick-and-place machines to assemble. However, even they are being replaced by surface-mount capacitors.*

cans have markings on the side to indicate which lead is the negative lead, and the negative lead is usually shorter (Figure 10.2).

> Almost every year that we use electrolytic capacitors some student manages to hook one up backwards and blow it up—luckily the cases on the small capacitors we use are designed to fail gracefully, so there isn't much danger of shrapnel in the lab. Still, it is not a bad idea to wear glasses, to avoid the risk of getting something in your eye. The electrolyte is fairly caustic material, so you don't want it getting in your eye or on your skin.

The electrolytic capacitors described above are *polarized*, the most common type. There are also *bipolar* electrolytic capacitors, which have the texturing and oxide coating on both metal foils. These are generally about twice the price of polarized capacitors, and they have a lower capacitance for the same size package (because the dielectric thickness is effectively doubled), but can be used when the voltage across the capacitor may be either negative or positive. They can still be exploded by exceeding their voltage rating.

The electrolyte in the capacitors is not a perfect conductor—it adds a resistor in series with the capacitor. The wet electrolyte has a further effect on the capacitors—the series resistance of the capacitor depends on ionic conduction in the wet electrolyte, so the resistance goes way up at low temperatures. The *ESR (Effective Series Resistance)* of an aluminum electrolytic capacitor may be $2\,\Omega$ at 25°C but $50\,\Omega$ at −55°C [88]. Some electrolytic capacitors use a polymer electrolyte instead of a wet one, which gives them a wider temperature range.

The high effective series resistance of aluminum electrolytic capacitors makes them only suitable for low frequency use. For electrolytic capacitors, the ESR is often not reported—manufacturers sometimes report a unitless *dissipation factor* $RC\omega$ at a particular frequency and temperature instead. That's because the capacitance scales up with the plate area of the capacitor, but the resistance scales down, so the RC constant is more nearly constant than the ESR across a family of capacitors of different sizes.

Worked Example:
How much series resistance should we expect with a cheap electrolytic capacitor?
For example, a 10-cent $10\,\mu$F 16 V electrolytic is specified as having a dissipation factor of 0.16 at 120 Hz and 20°C [54, ESK106M016AC3AA], which would mean an RC time constant of $210\,\mu$s, or an equivalent series resistance of about $21\,\Omega$.

The dissipation factor is a worst-case specification though, and the typical values of ESR may be less than a third of the spec. ESR for wet-electrolytic capacitors is extremely temperature-dependent, going way down at high temperature and way up at low temperature—even fairly small changes in temperature can affect wet-electrolyte capacitors a lot.

I tried measuring the ESR of a nominal $10\,\mu$F 50 V capacitor (part of a cheap assortment bought from ITEADstudio) using the techniques of Lab 8 and Lab 12. The results are shown in Figure 10.3. Since cheap electrolytic capacitors are usually specified as ±20%, the capacitance is within specifications. The dissipation factor at 120 Hz would be only $8.67\,\mu\text{F}\ 2.50\,\Omega\ 2\pi\ 120\,\text{Hz} = 0.016$, well below the usual 0.10 spec for cheap 50 V electrolytic capacitors.

Actually, electrolytic capacitors are more complicated than just a capacitance and a resistance—at higher frequencies the capacitance of the electrolyte matters more than its resistance, and at still higher frequencies the inductance of the wires matters. The Kemet data sheet for their cheap ESK series capacitors provides a good explanation of the behavior of the capacitors at different frequencies [54, pp. 13–15], though without giving precise numbers (which may vary from batch to batch that they manufacture).

For this class, we'll mostly be staying away from the higher frequencies where the more complicated models of electrolytic capacitors are needed, switching to ceramic capacitors for anything over about

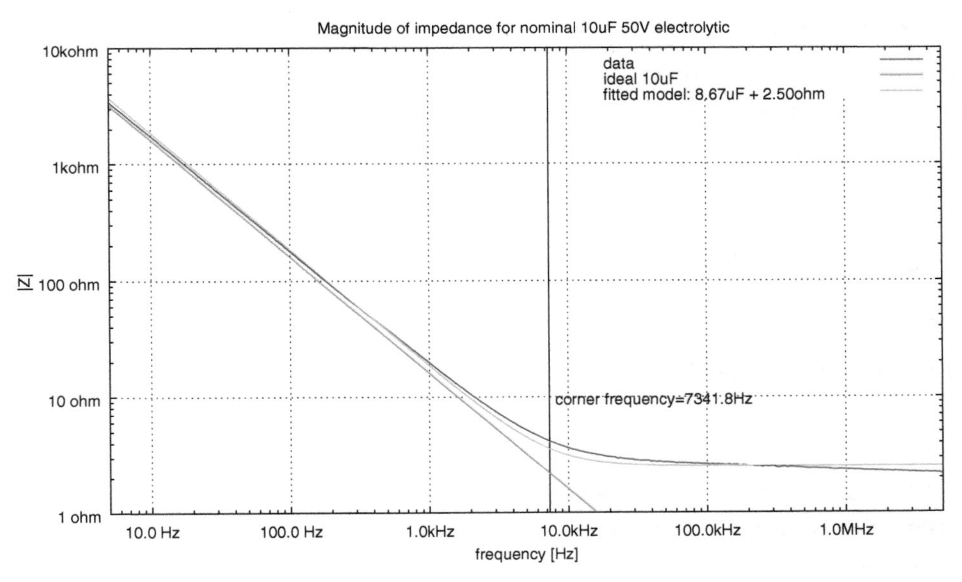

Figure 10.3: *Plot of the magnitude of impedance for a nominally $10\,\mu$F 50 V electrolytic capacitor, measured at 21°C. The capacitor is better modeled as an $8.67\,\mu$F capacitor with a $2.50\,\Omega$ resistor in series (its "effective series resistance" or ESR) than as a simple $10\,\mu$F capacitor. See Section 10.2.2 for an explanation of the impedance of a capacitor.*

Measurements done with $100\,\Omega$ reference resistor on Digilent's impedance-analyzer adapter board.

Data collected with Analog Discovery 2 [19]. Figure drawn with gnuplot [33]. With 50 data points per decade of frequency, the points are close enough together to justify using a line instead of separate points for the data.

1 kHz. In fact, to reduce parts cost in the kits and simplify designs, we eliminated electrolytic capacitors from the labs starting Spring 2016, adding a few larger ceramic capacitors instead.

If you need large capacitors (larger than are cheaply available in ceramic capacitors) with lower ESR than the standard aluminum electrolytic capacitors, the more expensive tantalum electrolytic capacitors have both lower ESR and much less temperature dependence. Aluminum capacitors with polymer electrolytes are now replacing tantalum capacitors, as they have even lower ESR and greater temperature stability.

10.2 Complex impedance

The second big concept of the course, promised in Section 1.1, is impedance.

> *Impedance* is a generalization of the notion of resistance and is defined as the ratio of voltage to current. For DC (direct current) voltages and currents, impedance is just resistance, but for AC (alternating current) voltages that vary with time, the voltage and the current waveforms may not always be in sync, so a more complex representation is needed.

It turns out that it is mathematically easiest to work with voltages and currents that are sums of sine waves. Because we are interested in different phases of sine waves (not just $\sin(\omega t)$, but $\sin(\omega t+\phi)$), we will talk about them as *sinusoids*, meaning any function shaped like a sine wave.

> Since trigonometry gets a bit confusing, I like to work always with complex numbers and the unit circle instead, using Euler's formula:
>
> $$e^{j\theta} = \cos(\theta) + j\sin(\theta) ,$$
>
> where $j = \sqrt{-1}$.
>
> We can represent a sinusoidal voltage
>
> $$v(t) = \cos(\omega t) = \Re\left(e^{j\omega t}\right) ,$$
>
> where $\Re(z)$ means the real part of z, or (less obviously, but more convenient mathematically) as
>
> $$v(t) = \cos(\omega t) = \frac{e^{j\omega t} + e^{-j\omega t}}{2} .$$
>
> The parameter ω is the *angular frequency* in radians per second, but we usually prefer to talk about the frequency f in cycles per second (hertz), with $\omega = 2\pi f$.

It turns out to be very convenient not to take the real part of the exponential formula or add the negative frequency term, but to pretend that the voltage is a complex number: $v(t) = e^{j\omega t}$. If you are paranoid about this engineering trick, you can always do all the math twice, using $\left(e^{j\omega t} + e^{-j\omega t}\right)/2$.

You can think of any complex number as a amplitude A (how far from 0) and an angle from the real axis ϕ: $Ae^{j\phi}$. We can equate these points with different sinusoids (all with the same angular frequency): $A\cos(\omega t + \phi) = \Re(Ae^{j(\omega t+\phi)}) = \Re(Ae^{j\phi}e^{j\omega t})$. The number $Ae^{j\phi}$ is referred to as a *phasor* when it is representing a sinusoid like this.

The real number A is the *amplitude* (often written with a capital letter), and the angle ϕ is the *phase* of the sinusoid (often written with a lower-case Greek letter).

The phase formulas given here all express the phase in radians, as that is the natural representation for exponentials, but sometimes the phase is expressed in degrees rather than radians ($360° = 2\pi$ radians), to get nice round numbers like 90°.

The Wikipedia article on phasors [145] provides a good tutorial introduction to doing algebraic operations on sinusoidal signals, and what the result is when interpreted as phasors.

Summing sinusoidal signals with the same angular frequency ω results in sinusoidal signals with the same frequency, but whose phasor representation is the sum of the phasors of the input signals:

$$Ae^{j(\omega t+\phi)} + Be^{j(\omega t+\psi)} = (Ae^{j\phi} + Be^{j\psi})e^{j\omega t} .$$

If we put a sinusoidal voltage $v(t) = Ae^{j(\omega t+\phi)}$ across a linear component, then the current through the component will be sinusoidal with the same frequency, but the amplitude and phase can be different: $i(t) = Be^{j(\omega t+\psi)}$. The ratio of the voltage to the current is the impedance of the component:

$$Z = \frac{Ae^{j(\omega t+\phi)}}{Be^{j(\omega t+\psi)}} = \frac{Ae^{j\phi}}{Be^{j\psi}} = \frac{A}{B}e^{j(\phi-\psi)} ,$$

which is invariant with time, and is just the ratio of the phasors for the voltage and current. The phase angle of the impedance, $\phi - \psi$, is the *phase difference* between the voltage and the current.

It is a common convention for electrical engineers to use upper-case letters V and I when talking about a DC voltage or current, and lower-case letters v and i when talking about a time-varying voltage or current, often omitting the explicit function notation (that is, using v instead of $v(t)$). I'll try to use the explicit function notation in this book, but you should be aware of the conventional abbreviation of the notation.

Although ω does not appear explicitly in the formula for impedance, the amplitudes and phases of the voltage and current waveforms generally change as the frequency changes, and so the impedance of a component is usually frequency-dependent.

For a simple resistor following Ohm's law, $v(t) = i(t)R$, the impedance is simply the resistance R—the voltage and current waveforms remain *in phase*, that is, with a phase difference of zero.

In Section 10.2.2, we'll examine the impedance of capacitors and in Section 28.3, inductors. Capacitors and inductors are the two ideal linear devices most often used for modeling frequency-dependent behavior of circuits, together with resistors, which are frequency-independent.

10.2.1 *Impedances in series and parallel*

In Section 5.1, we saw that resistances in series add, because the current through them is the same, but the voltages of two components in series add and $R = V/I$. Exactly the same argument applies to impedances, so impedances in series also add.

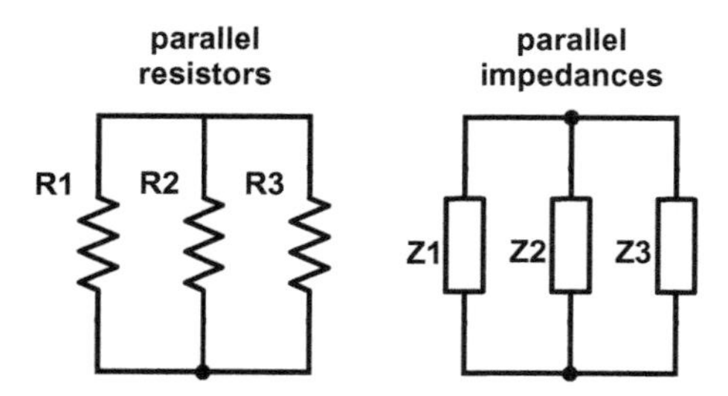

Figure 10.4: *Three resistors in parallel and three impedances in parallel have the same mathematical relationships. Incidentally, the symbols used here for impedances are the standard European symbols for resistors. I prefer the American notation, used on the left, for resistors, because the shape is less generic, and so provides quicker visual recognition as a resistor.*
Figure drawn with Digi-Key's Scheme-it [18].

What about resistances or impedances wired in parallel (as in Figure 10.4)? We can use a very similar argument to see that the components will all have the same voltages across them (they are all connected to the same two nodes), and that the total current from one node to the other is the sum of the currents through the individual components. That is, $I = I_1 + I_2 + I_3 = V/R_1 + V/R_2 + V/R_3$. We can rearrange this equation to get $V(1/R_1 + 1/R_2 + 1/R_3) = I$, or

$$R = V/I = \frac{1}{1/R_1 + 1/R_2 + 1/R_3} .$$

Exactly the same argument can be applied to impedances in parallel, with

$$Z = \frac{1}{1/Z_1 + 1/Z_2 + 1/Z_3} .$$

You will often see the formula for two impedances in parallel written as

$$Z = \frac{Z_1 Z_2}{Z_1 + Z_2} ,$$

which is just an algebraic rearrangement of the general formula. It has the advantage of not giving a divide-by-zero error if one of the impedances is zero, but otherwise has nothing to recommend it. Way too many students try to generalize it (incorrectly) to formulas with three or more impedances in parallel.

I recommend always using the

$$Z = \frac{1}{1/Z_1 + 1/Z_2 + 1/Z_3 + \cdots}$$

formula for impedances in parallel.

10.2.2 *Impedance of capacitor*

Impedance is voltage divided by current, and current is movement of charge. We know from physics classes that the charge on a capacitor is the product of the capacitance and the voltage across the capacitor: $Q = CV$, and any current into one plate of the capacitor is going to increase the charge there. That means that the current into a capacitor plate is the derivative of the charge on the capacitor: $i(t) = \frac{\mathrm{d}Q(t)}{\mathrm{d}t}$, just as velocity is the derivative of position.

If we are dealing with a constant capacitance (a good approximate model for most of the components we'll deal with—except capacitive or electret microphones), then the current is proportional to the derivative of voltage: $i(t) = C\frac{\mathrm{d}v(t)}{\mathrm{d}t}$.

Exponentials are very nice for differentiation: if $v(t) = e^{j\omega t}$, then $\frac{\mathrm{d}v(t)}{\mathrm{d}t} = j\omega e^{j\omega t}$, and

$$\begin{aligned} i(t) &= C\frac{\mathrm{d}v(t)}{\mathrm{d}t} \\ &= Cj\omega e^{j\omega t} \\ &= Cj\omega v(t) \ . \end{aligned}$$

> That means that the impedance for a capacitor is
> $$Z_{\mathrm{C}}(\omega) = \frac{v(t)}{i(t)} = \frac{1}{j\omega C} \ .$$

Although the formula for $Z_{\mathrm{C}}(\omega)$ is easy to derive, it is one of the few formulas worth memorizing in this book (see Appendix B for a complete list of the formulas worth memorizing), as you will use it over and over again.

At DC (a frequency of 0 Hz), $\omega = 0$ rad/s and $Z_{\mathrm{C}}(0) = \infty\,\Omega$, so the capacitor is an open circuit. At very high frequency $Z_{\mathrm{C}}(\infty) = 0\,\Omega$, and the capacitor is effectively a short circuit.

What happens if we put three capacitors in parallel? We have a formula for the impedance from Section 10.2.1:

$$\begin{aligned} Z_{\|}(\omega) &= \frac{1}{1/Z_{C_1} + 1/Z_{C_2} + 1/Z_{C_3}} \\ &= \frac{1}{j\omega C_1 + j\omega C_2 + j\omega C_3} \\ &= \frac{1}{j\omega(C_1 + C_2 + C_3)} \\ &= Z_{C_1+C_2+C_3} \ . \end{aligned}$$

So capacitors in parallel combine the way resistors in series do:

> Capacitors in parallel add.

What happens if we put three capacitors in series?

$$\begin{aligned} Z_{\mathrm{series}}(\omega) &= Z_{C_1} + Z_{C_2} + Z_{C_3} \\ &= \frac{1}{j\omega C_1} + \frac{1}{j\omega C_2} + \frac{1}{j\omega C_3} \\ &= \frac{1}{j\omega}\left(\frac{1}{C_1} + \frac{1}{C_2} + \frac{1}{C_3}\right) \\ &= \frac{1}{j\omega\left(1/\left(\frac{1}{C_1} + \frac{1}{C_2} + \frac{1}{C_3}\right)\right)} \ . \end{aligned}$$

So capacitances in series combine like resistors in parallel do:

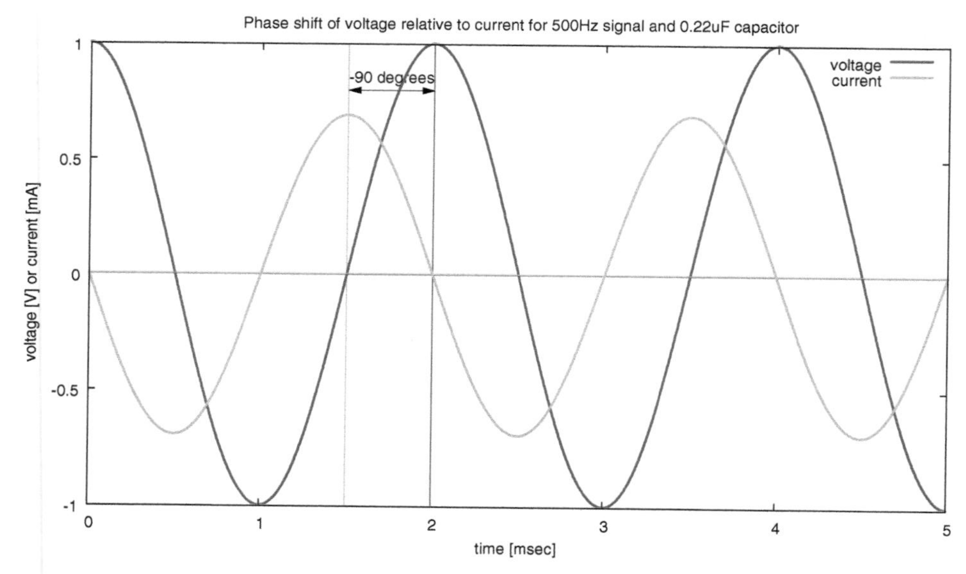

Figure 10.5: *Simulating applying a sinusoidal voltage across a capacitor causes a sinusoidal current through the capacitor. The voltage waveform lags the current waveform by 90° (a phase shift of $-\pi/2$), consistent with an impedance for the capacitor that is a negative imaginary value). For the values given here, $Z_C = 1/(j\ 2\pi\ 500\text{Hz}\ 0.22\,\mu\text{F}) = -1447j\,\Omega$. (The sign change occurs because multiplying top and bottom of the fraction by j results in a $j^2 = -1$ in the denominator.)*
Figure drawn with gnuplot [33].

Capacitors C_1, C_2, ... in series have a combined capacitance of

$$\frac{1}{1/C_1 + 1/C_2 + \cdots}.$$

Note that the impedance of a capacitor is not a real number, but an imaginary one. What does that mean? Basically, it means that the phase of the sinusoid is changed between the current and the voltage (see Figure 10.5). Because the current is proportional to the derivative of the voltage across a capacitor, it is 0 when the voltage is at its maximum and minimum (the derivative of a cosine is the negative of the sine), and at its maximum when the voltage is increasing through its zero crossing. The voltage lags the current with a $-90°$ phase shift. This $-90°$ angle is also the angle between the real and negative imaginary axes of the complex plane, and Z_C is a negative imaginary number for capacitors.

We'll use capacitors to make filters that change the amplitudes of signals depending on their frequencies (see Chapter 11). In more advanced electronics or signal processing courses, you'll look at the effect on the phase of the signals as well, but for this basic introduction, we'll just concentrate on amplitude.

In Section 10.1.2 on page 150, we introduced the notion of *effective series resistance (ESR)* for a capacitor. If we model a capacitor as an ideal resistor in series with an ideal capacitor, $Z = R + 1/(j\omega C)$, and plot the magnitude of impedance, $|Z|$, we get a plot like that of Figure 10.3, in which the magnitude of the impedance is inversely proportional to frequency at low frequencies ($\frac{1}{j\omega C}$ dominates), and constant at high frequencies (R dominates).

Exercise 10.1

If you wire two 680 pF capacitors in parallel, what is the capacitance of the combined circuit?

Exercise 10.2

What is the impedance (as a function of frequency) of a 25 pF capacitor in parallel with a 1 MΩ resistor?

Exercise 10.3

If you wire a 1 nF capacitor in parallel with a 25 pF capacitor, what is the capacitance of the combined circuit?

Exercise 10.4

If you wire two 470 pF capacitors in series, what is the capacitance of the combined circuit?

Exercise 10.5

What is the impedance (as a function of frequency) of a 25 nF capacitor in series with a 100 Ω resistor?

Exercise 10.6

If you wire a 1 nF capacitor in series with a 100 pF capacitor, what is the capacitance of the combined circuit?

Exercise 10.7

At what frequency is the impedance of a 2.2 μF capacitor equal to $-1000j\,\Omega$?

Exercise 10.8

Plot the magnitude of the impedance of a 10 kΩ resistor in parallel with a 0.47 μF capacitor for frequencies from 1 Hz to 100 kHz, using log scales for both frequency and impedance. You may find the gnuplot script in Figure 28.5 useful as a model for your work. Turn in both the plot and the script used to produce it.

Chapter 11

Passive RC Filters

11.1 RC filters

An RC filter is simply a voltage divider which uses two impedances, at least one of which changes with frequency. The simplest filters use a resistor for one of the two impedances and a capacitor for the other (see Figure 11.1).

> A filter is a three-terminal device, having an input node, an output node, and a reference node from which voltages of the input and output are measured. A two-terminal device, such as a resistor and capacitor in series, is **not** a filter, but just an impedance.

In most amplifier designs, you will have to design filters with specific properties, so you need a formal model of how they work. This model is often referred to as the *transfer function*, which is simply the gain of a filter or amplifier as a function of frequency (the ratio of the output to the input). The transfer function is often expressed in terms of its transform into another domain, using the Laplace, Fourier, or z-transform, as these transforms allow easier analysis and design of filters and control systems, but we will not need the sophisticated mathematics of transforms for analyzing simple RC filters. It is enough for us to know that transfer functions provide a complete characterization of a linear system, and that RC filters are linear systems.

The basic idea for our analysis of passive RC filters is to combine two formulas: the voltage divider formula, which we use in almost every design, and the formula for the impedance of a capacitor, which we derived in Section 10.2.2 from basic physics concepts.

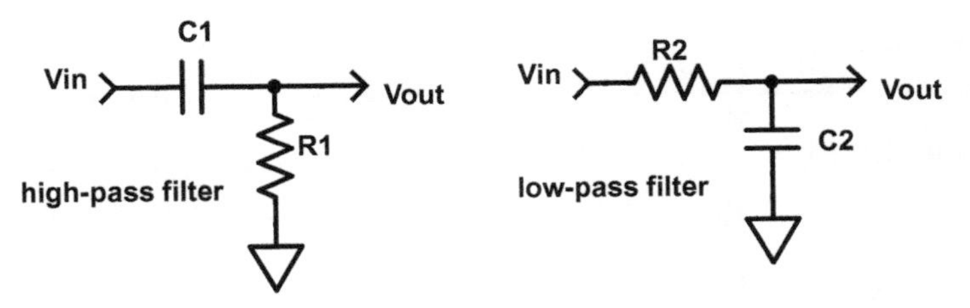

Figure 11.1: *Voltage dividers that act as high-pass and low-pass filters. There is no need to memorize which is which, as it is easy to derive from the behavior at ∞ Hz and 0 Hz: at very high frequency, the capacitors are almost short circuits, and at very low frequency they are almost open circuits.*
Figure drawn with Digi-Key's Scheme-it [18].

11.2 RC voltage divider

Using the formula for impedance of a capacitor, we can get formulas for the behavior of an RC filter by using our voltage divider formula.

For the high-pass filter in Figure 11.1, we have

$$\begin{aligned}\frac{V_\text{out}}{V_\text{in}} &= \frac{R_1}{R_1 + Z_{C_1}(\omega)}\\ &= \frac{R_1}{R_1 + 1/(j\omega C_1)}\\ &= \frac{j\omega R_1 C_1}{1 + j\omega R_1 C_1} \;.\end{aligned}$$

For the low-pass filter in Figure 11.1, we have

$$\begin{aligned}\frac{V_\text{out}}{V_\text{in}} &= \frac{Z_{C_2}(\omega)}{R_2 + Z_{C_2}(\omega)}\\ &= \frac{1/(j\omega C_2)}{R_2 + 1/(j\omega C_2)}\\ &= \frac{1}{1 + j\omega R_2 C_2} \;.\end{aligned}$$

> You can see the high-pass and low-pass nature of the filters by looking at what the voltage divider ratio would be at $\omega = 0$ and $\omega = \infty$.

For the high-pass filter with gain $j\omega R_1C_1/(1 + j\omega R_1C_1)$, as $\omega \to \infty$, the 1+ in the denominator becomes irrelevant and the magnitude of the gain approaches 1. As $\omega \to 0$, the term added to 1 in the denominator becomes negligible, and the magnitude of the gain approaches $\omega R_1C_1 = 2\pi f R_1C_1$, which goes to zero. If we define $f_\text{c} = 1/(2\pi R_1C_1)$, then the magnitude of the gain is asymptotically f/f_c.

For the low-pass filter with gain $1/(1 + j\omega R_2C_2)$, at $\omega = 0$ (DC) the gain is just 1, while as $\omega \to \infty$, the 1+ in the denominator becomes irrelevant and the magnitude of the gain approaches $1/(\omega R_2C_2) = f_\text{c}/f$, which goes to zero. (Here, $f_\text{c} = 1/(2\pi R_2C_2)$.)

Even better is to plot the magnitude of the voltage-divider ratio on a log-log scale as in Figure 11.2. A straight-line approximation to the log-log plot is referred to as a *Bode plot*, after Hendrik Wade Bode, who developed quick ways to plot amplitude and phase response of filters without extensive calculation. The Wikipedia article https://en.wikipedia.org/wiki/Bode_plot will allow you to explore the concept further, but it goes well beyond what you need for this course.

In Figure 11.2, you can see that the gain of both the low-pass and the high-pass filters can be approximated by a pair of straight lines on a log-log plot: one at a constant gain of 1 and the other either f/f_c (for the high-pass filter) or f_c/f (for the low-pass filter).

> The frequency where the filter switches from roughly constant gain to gain that is linearly or inversely proportional to frequency is called the *corner frequency* and is at $f_\text{c} = \frac{1}{2\pi RC}$.
>
> At the corner frequency, we have $\omega RC = 1$, so the magnitude of the gain is $\left|\frac{1}{1+j}\right| = \sqrt{2}/2$. If we specify gain in decibels, we have $20\log_{10}(\sqrt{2}/2) = -3.010\,\text{dB}$ at the corner frequency.
>
> Engineers like to specify the *cutoff frequency* of any filter as the point where the gain drops to $\sqrt{2}/2$ of the peak gain ($-3\,\text{dB}$).

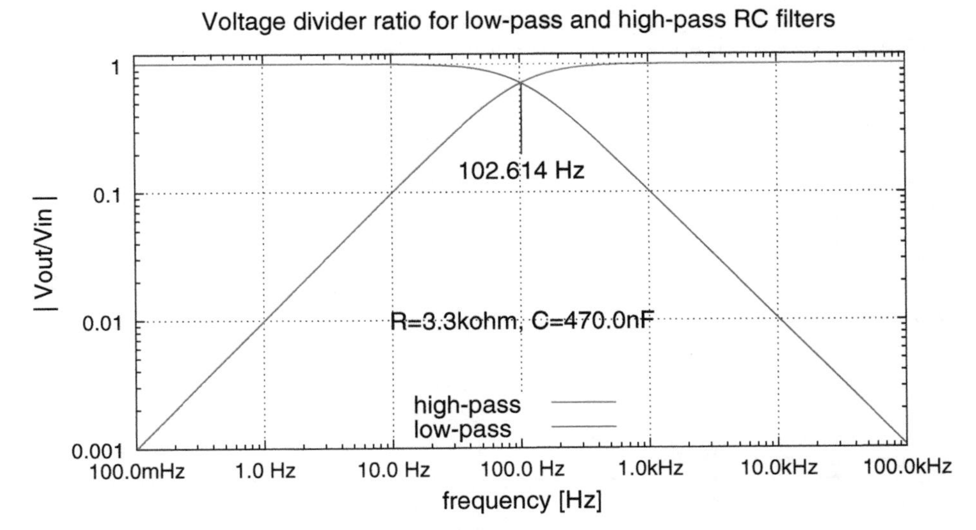

Figure 11.2: *Plot of the effect on amplitude of sinusoids of simple RC filters, such as those in Figure 11.1. The* corner frequency *where* $\omega = \frac{1}{RC}$ *is an important point. (This is often expressed as* $f = \frac{1}{2\pi RC}$*.) The log-log plot allows the behavior of the filter to be approximated by straight lines, which is handy for quick designs.*
Figure drawn with gnuplot [33].

For the simple RC filters in this chapter, the corner frequency (where the asymptotic lines meet) and the cutoff frequency (where gain drops 3 dB) are the same, but for more complicated filters they may be different. Some engineers are sloppy in their terminology, and use the term *corner frequency* for both concepts—when you read other sources, check to see whether they really mean cutoff frequency.

A simple high-pass filter has gain

$$\frac{j\omega R_1 C_1}{1 + j\omega R_1 C_1},$$

with corner frequency $f_c = 1/(2\pi R_1 C_1)$. The gain can be approximated as 1 for frequencies more than three times the corner frequency, as f/f_c for frequencies less than a third the corner frequency, and as $\sqrt{2}/2$ at the corner frequency.

A simple low-pass filter has gain

$$\frac{1}{1 + j\omega R_2 C_2},$$

with corner frequency $f_c = 1/(2\pi R_2 C_2)$. The gain can be approximated as 1 for frequencies less than a third of the corner frequency, as f_c/f for frequencies more than three times the corner frequency, and as $\sqrt{2}/2$ at the corner frequency.

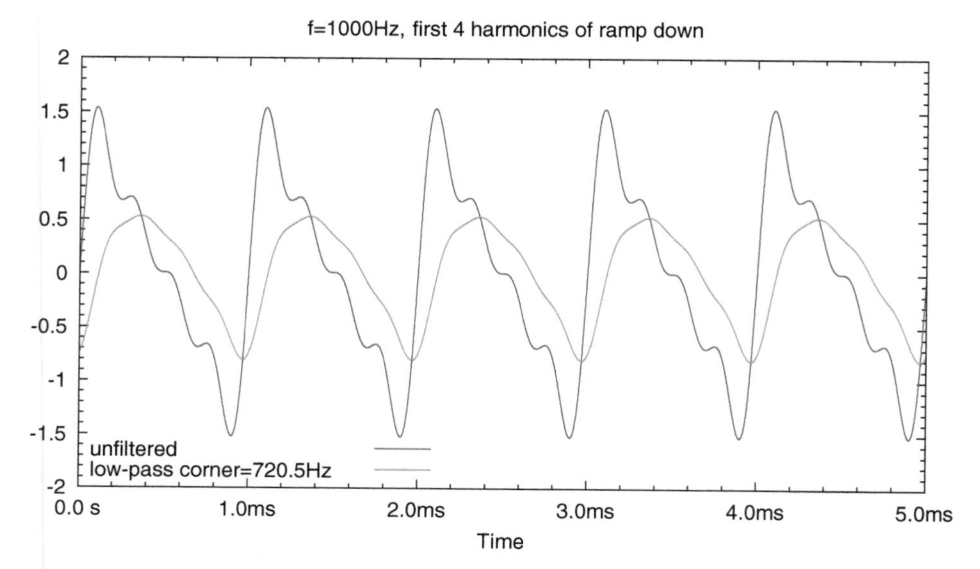

Figure 11.3: *The unfiltered signal is* $\sin(\omega t) + \frac{1}{2}\sin(2\omega t) + \frac{1}{3}\sin(3\omega t) + \frac{1}{4}\sin(4\omega t)$, *where* $\omega = 2\pi f$ *and the frequency* f *is 1 kHz. The filtering is done with a (simulated) 470 Ω resistor and 470 nF capacitor, resulting in a low-pass filter with a corner frequency of 720.5 Hz. Although the corner frequency is lower than the lowest frequency in the waveform, the output still has a substantial signal, though the higher-frequency components have been attenuated more than the lower ones. Changes in phase to the sine-wave components also result in a delay, so that the maximum and minimum of the filtered waveform are later than in the input waveform. Figure drawn with gnuplot [33].*

Just as Section 5.1 defined the input resistance as the sum of the two resistances in a voltage divider, the *input impedance* of a passive RC filter is just the sum of the two impedances: $Z_{\text{in}1} = R_1 + \frac{1}{j\omega C_1}$ or $Z_{\text{in}2} = R_2 + \frac{1}{j\omega C_2}$.

The input impedance expresses the relationship between the input voltage of the current and the current through the input port.

We'll look more at input impedance in Section 11.5.

The formulas just given explain what happens to sinusoidal waveforms, with the complex gain expressing the change of amplitude and phase for the waveform, which remains a sinusoidal waveform with the same frequency. But what happens with other waveforms? Answering this question well requires the more sophisticated math of Fourier series and Fourier transforms, but we can get a hint of it by looking at the sum of several sinusoidal waves of different frequencies.

Consider the function $f(t) = \sin(\omega t) + \frac{1}{2}\sin(2\omega t) + \frac{1}{3}\sin(3\omega t) + \frac{1}{4}\sin(4\omega t)$, shown in Figure 11.3, where $\omega = 2\pi f$ and the frequency f is 1 kHz. This function is an approximation to a 1 kHz downward ramp function, being the first four terms of the Fourier series for that function. Each term on the summation can be filtered separately, and the terms added back together. If we filter the function through a low-pass RC filter, the higher frequencies are attenuated more than the lower frequencies and the resulting waveform looks more like a sine wave than the initial waveform does.

11.3 Simple filters—worked examples

Let's use both the simplified approximations and gnuplot to analyze and design some simple low-pass and high-pass filters. To use gnuplot to produce Bode plots of filter functions, it helps to have a number

```
load '../definitions.gnuplot'

set title "Voltage divider ratio for low-pass and high-pass RC filters" font "Helvetica,14"
set xlabel 'frequency [Hz]' font "Helvetica,14"
set xrange [1e-1:1e5]
set ylabel '| Vout/Vin |' font "Helvetica,14"
set yrange [1e-3:1.2]
set format y "%.2g"
set margins 10,6,4,4
set key bottom center font "Helvetica,14"

r = 3.3e3;       r_name=gprintf("%.1s%cohm",r)
c = 470e-9;      c_name=gprintf("%.1s%cF",c)
corner = 1/(2*pi*r*c)    ; corner_name=gprintf("%.3s%cHz",corner)

unset label
set label corner_name at corner,0.15 center font "Helvetica,14"
set label sprintf("R=%s, C=%s", r_name,c_name) at corner,0.01 center font "Helvetica,14"

unset arrow
set arrow nohead from corner,0.2 to corner,0.7

plot abs(divider(Zc(x,c),r)) title 'high-pass' lt rgb "red", \
     abs(divider(r,Zc(x,c))) title 'low-pass'  lt rgb "blue"
```

Figure 11.4: *Gnuplot script to create Figure 11.2. It starts by loading the* `definitions.gnuplot` *script of Figure 7.4, which happened to be in the directory above the directory that this script was in (hence, the ../).*
It then overrides some of the standard settings and makes some fancy labels and lines. The plot command itself is simple, plotting the magnitude of the gain.
Remember, copying scripts from the PDF file for the book doesn't work, because of character substitutions made in the typesetting.

of functions already defined and a number of gnuplot's options set. Because I do a lot of Bode plots, I created a set of commonly used definitions and put them in a file (Figure 7.4).

To use `definitions.gnuplot`, download the file and put it in the same directory as the gnuplot script you are writing. Start your script with `load 'definitions.gnuplot'`, so that all the defined functions are available. All the frequency-dependent functions are written with frequency as the first argument—keeping this convention in building your models will make it easier to share and debug your scripts.

> When you write your script, build up models for your circuits using the impedance and gain functions from `definitions.gnuplot` and combining them with addition for series connections, `Zpar` for parallel connections, and `divider` for the gain of voltage dividers. Don't do a lot of algebra by hand—all that does is obscure the structure of your model and make debugging almost impossible.

Keep your functions as complex-valued impedances or gains, converting to the magnitude only when doing the final plotting with `plot abs(function_name(x, ...))`.

> Use names for all component values, so that each numeric value occurs exactly once in the file, making it easy to change the file as you change your design.

For example, the script in Figure 11.4 was used to produce the plot in Figure 11.2. Your own scripts need not be as elaborate—setting the title and y-axis label, defining the functions you want to plot, and plotting them is all you really need.

> Don't use Pythagoras's Rule to convert the magnitude of a complex impedance or gain to a real number, because that hides the origin of the formula, making debugging hard. Use $|G|$ in math formulas, and the `abs` function in gnuplot.

The `abs` function is needed in gnuplot to plot the magnitude of an impedance or gain, because gnuplot can only plot real-valued functions, not complex-valued ones. Some other plotting programs will plot complex numbers as the real part of the complex number, rather than the magnitude, which is rarely what we really want.

Worked Example:

Let's work out what a voltage divider with a 10 kΩ resistor on the top (connected to the input) and a 68 nF capacitor on the bottom (connected to ground) would do. Then let's make a plot to confirm our understanding.

First, we can figure out whether this is a high-pass or low-pass configuration by looking at what happens when $\omega = 0$ (DC) and when $\omega \to \infty$. At DC, the capacitor is just an open circuit, so we have a resistor to the input. As long as we take no current, the output is equal to the input $G = 1$. At infinitely high frequency, the capacitor is a short circuit, and the output is connected to ground $G = 0$. This tells us that the circuit is a low-pass filter—it will be approximately 1 up to the corner frequency f_c and approximately f_c/f above that.

We can compute the corner frequency as

$$f_c = \frac{1}{2\pi RC} = \frac{1}{2\pi 10\,\mathrm{k}\Omega 68\,\mathrm{nF}} = 234.1\,\mathrm{Hz}\ .$$

The complex gain of the filter can be written using the definitions in `definitions.gnuplot` as `divider(10E3, Zc(f,68E-9))`, and the Bode plot can be created with a script as simple as

```
load 'definitions.gnuplot'
plot abs(divider(10E3,Zc(x,68E-9))) notitle
```

though adding a title and a y-axis label would be better.

If we want to plot the straight-line approximation, we can increase the script to

```
load 'definitions.gnuplot'
set yrange [*:1.5]     # to make room to see the top of the function
R = 10E3; C=68E-9      # define the resistance in ohms and cap in farads
fcorner = 1/(2*pi*R*C)  # compute the corner frequency in Hz
plot abs(divider(R,Zc(x,C))) notitle, x<fcorner?1:fcorner/x title 'approx'
```

Exercise 11.1

Figure out the corner frequency for an RC filter that has a 22 pF capacitor between the input and the output, and a 47 kΩ resistor to ground. Use gnuplot to plot the magnitude of the gain as a function of frequency on a log-log plot.

Worked Example:
Design a filter that allows blood pulse frequencies (0.5 Hz–3 Hz) through with little change, but blocks DC (gain 0 at 0 Hz).

Because we want to block DC, we need a high-pass filter, which means that the capacitor goes between the input and the output, and the resistor between the output and a voltage reference at whatever voltage we want the output to be centered. (A high-pass filter eliminates the DC bias of the input, but creates a new DC bias wherever we want it to be.)

Because we want to pass 0.5 Hz–3 Hz through with little change, that range needs to be higher than the corner frequency, $3\,\text{Hz} > 0.5\,\text{Hz} > f_c$. We may want our signal to be as much as five times higher than the corner frequency, depending on how much attenuation of the signal we can tolerate.

If we pick the five-times-higher-than-corner guideline for where a high-pass filter is essentially one, then our corner frequency needs to be $f_c \leq 0.5\,\text{Hz}/5 = 0.1\,\text{Hz}$. That gives us $\omega_c = 2\pi f_c \leq 0.628\,\text{radians/s}$, and $RC \geq 1.592\,\text{s}$. The rough rule of thumb does not need to be followed exactly, so $RC > 1.4\,\text{s}$ is probably good enough.

We are free to choose any resistor and capacitor values that have the appropriate RC time constant, but we usually want to pick moderately large resistors and moderately small capacitors, to avoid having a low input impedance for the filter. (By "low impedance" I mean that the *magnitude* of the impedance is small, as complex numbers are not orderable.) A low input impedance for a filter would require high currents from V_{in}, which may not be available. Large capacitors are usually more expensive and often less precisely manufactured than small capacitors, so keeping capacitors small is generally a good idea.

Very large R and very small C, however, can cause problems also—large impedances are more sensitive to noise and parasitic capacitance from conductors close to our wires becomes more of a problem when the design capacitance is small.

As a rough rule of thumb, we try to pick resistors in the 1 kΩ–1 MΩ range and capacitors in the 100 pF–10 μF range.

So for our 1.4 s RC time constant, we could use 10 μF (the largest reasonable value for C) and 140 kΩ, through 1.4 μF and 1 MΩ (the largest reasonable value for R). We should select a capacitor from the E6 series (or even E3: 1, 2.2, 4.7), then pick the closest available resistor to get the RC time constant we want:

C	R	RC
1 μF	E12: 1.5 MΩ	1.500 s
1 μF	E24: 1.6 MΩ	1.600 s
2.2 μF	E12: 680 kΩ	1.496 s
2.2 μF	E24: 750 kΩ	1.650 s
4.7 μF	E6: 330 kΩ	1.551 s
10 μF	E6: 150 kΩ	1.500 s
10 μF	E24: 160 kΩ	1.600 s

All of these are acceptable answers (though the 1 μF ones are pushing above our rule of thumb for resistance a little). Because our corner frequency was not tightly specified, it might be best to go with resistors from the E12 series, and of those the $C = 4.7\,\mu$F and $R = 330\,$kΩ yielding 1.551 s seems best, since all the values are from the very common E6 series. (The $C = 2.2\,\mu$F and $R = 680\,$kΩ choice is also good, as the smaller capacitor is likely to be cheaper and closer to specification, though the time constant is a little smaller.)

The achieved corner frequency for 1.551 s, $f_c = 1/(2\pi RC) = 0.103$ Hz is sufficiently close to our chosen 0.1 Hz, and the magnitude of the input impedance is always at least 330 kΩ, so the filter will impose very little load on (that is, require very little current from) previous stages. On the other hand, the high impedance also means that very little current can be delivered to subsequent stages, so we will probably need to follow the filter with an amplifier stage (to be covered in Chapter 18).

In order to treat subcircuits as independent blocks, we often need them to have a high input impedance (so that they don't affect the preceding stage in the block diagram) and a low output impedance (so that the output voltage is not affected by the next stage in the block diagram). Achieving this pair of design goals simultaneously is often not possible with passive RC filters—Chapter 24 will discuss a different filter design that addresses this problem.

Exercise 11.2

Design an RC low-pass filter with a corner frequency of approximately 40 Hz. Pick component values that are reasonable for ceramic capacitors (not needing electrolytic ones). Round your component values to values on the E6 scale for capacitors and the E12 scale for resistors (Table 4.2). Don't cheat by using multiple resistors or capacitors—you only get one of each. (Hint: select the capacitor value before computing the resistor value.) Report the actual corner frequency.

11.4 RC time constant

Both high-pass and low-pass filters depend not on R and C separately, but only on the product RC, which is measured in seconds (an ohm-farad is a second). You have encountered the RC product before in physics class: if you discharge a capacitor through a resistance, the voltage drops exponentially with an RC time constant. Consider the schematic in Figure 11.5: the capacitor is initially charged to V_0, but when the switch is pressed it decays exponentially to V_∞.

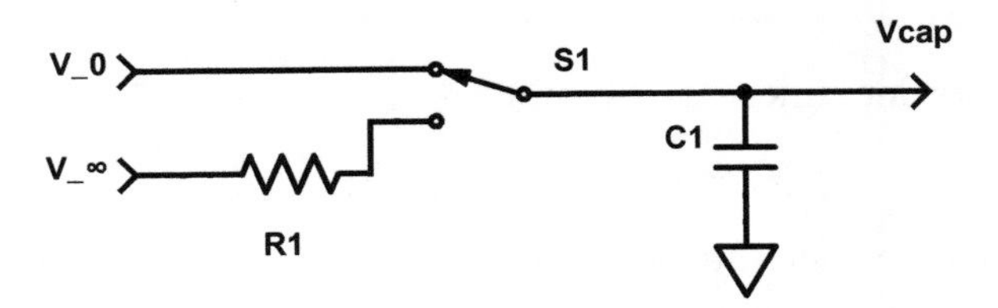

Figure 11.5: *Schematic for a capacitor that is initially charged to* V_0. *When the switch is pressed, the capacitor discharges or charges towards* V_∞, *following the exponential decay* $v_{cap}(t) = V_\infty + (V_0 - V_\infty)e^{-t/(R_1C_1)}$.
Figure drawn with Digi-Key's Scheme-it [18].

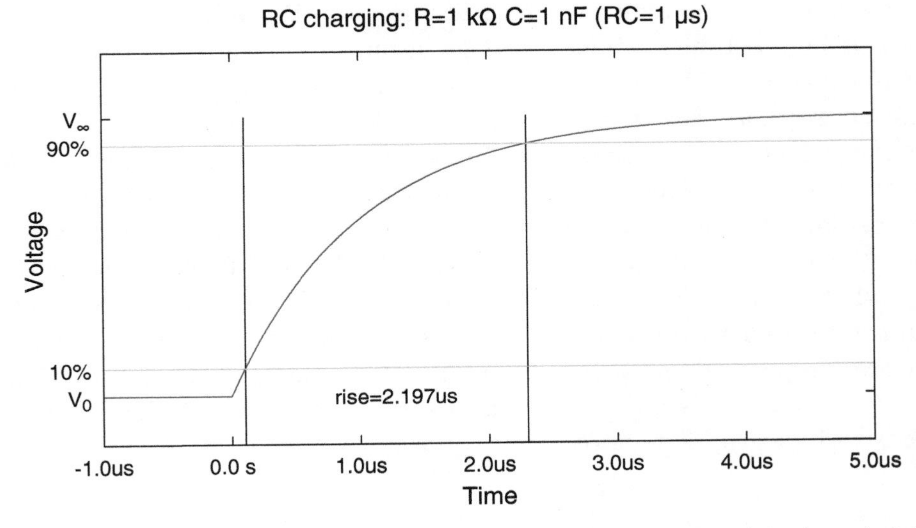

Figure 11.6: *This graph shows the voltage behavior of the circuit of Figure 11.5 if the switch is changed at time 0 s. The voltage starts at* V_0 *and asymptotically approaches* V_∞. *After* $5RC$ *time constants, 99.3% of the transition has been completed. The rise time is about* $2.197RC$ *time constants.*

You can get this exponential decay as the solution to a first-order differential equation based on the following fundamental properties:

- the charge on the capacitor is the product of the capacitance and the voltage $Q = CV$,
- the voltage across the resistor is the product of the current through the resistor and the resistance $V_\infty - V = I_\mathrm{r}R$,
- the current into the capacitor is the derivative of the charge on the capacitor $I_\mathrm{c} = \frac{\mathrm{d}Q}{\mathrm{d}t}$, and
- the current through the resistor is the current into the capacitor $I_\mathrm{r} = I_\mathrm{c}$.

Expressing these concepts in mathematical formulas gives us

$$\begin{aligned} i_r(t) &= (V_\infty - v(t))/R, \\ i_c(t) &= C\frac{\mathrm{d}v(t)}{\mathrm{d}t}, \\ v(t) - V_\infty = -i_r(t)R &= -i_c(t)R = -RC\frac{\mathrm{d}v(t)}{\mathrm{d}t} \,. \end{aligned} \tag{11.1}$$

The solution to the first-order differential equation in Equation 11.1 is

$$v(t) - V_\infty = (V_0 - V_\infty)e^{-t/(RC)} , \tag{11.2}$$

where V_0 is the initial voltage on the capacitor and V_∞ is the voltage on the other side of the resistor (the asymptotic value as time goes to infinity).

This charging curve is shown in Figure 11.6. Equation 11.2 applies equally well for charging and for discharging—it doesn't matter whether $V_0 < V_\infty$ (charging) or $V_0 > V_\infty$ (discharging).

Another way to think of the circuit in Figure 11.5 is as an RC voltage divider, with an input that is a square wave switching between V_0 and V_∞. The output of the voltage divider is no longer a square wave, but has rounded corners. We are often interested in how fast the signal rises or falls, or how long it takes the distorted waveform to settle to close to the ideal square wave.

In analyzing the behavior of a circuit when given a square wave, we can simplify the problem by looking at what the circuit does on a single transition or *edge*. A function that makes a single, instantaneous transition at time 0 (say from V_0 to V_∞) is called a *step function* and the difference $V_\infty - V_0$ is called the *step size*. If $V_\infty > V_0$, we say that the step function has a *rising edge*, and if $V_\infty < V_0$, we say that the step function has a *falling edge*. The output waveform of a circuit whose input is a step function from $V_0 = 0\,\text{V}$ to $V_\infty = 1\,\text{V}$ is referred to as its *step response*. Figure 11.6 can be viewed as the step response of a low-pass filter with $Z_\text{u} = 1\,\text{k}\Omega$ and $Z_\text{d} = 1\,\text{nF}$.

Both rise and fall times and settling times are expressed in terms of the fraction of the step size (100% corresponds to $V_\infty - V_0$). The normalized voltage,

$$\frac{V_0 - v(t)}{V_0 - V_\infty} = 1 - e^{-t/(RC)} ,$$

is 0 at time 0, and 1 at infinite time. We can turn this around to get the time to go a fraction f of the step:

$$t = -RC \ln(1 - f) .$$

Rise and fall times are usually defined as the time it takes to go from 10% of the step to 90% of the step (or vice versa):

$$t_\text{rise} = -RC \ln(1 - 0.9) - (-RC \ln(1 - 0.1)) \approx 2.197RC .$$

See Figure 11.6 for an illustration of rise time.

Settling time is often defined to being 99% of the way (or, more generally for signals that have ringing rather than simple exponential decay) until the remaining fluctuation is less than 1% of the step size:

$$t_\text{settle} = -RC \ln(1 - 0.99) = 4.605RC .$$

As a rough rule of thumb, engineers often approximate signals as being stable after 5 RC time constants.

Exercise 11.3

If we have a passive low-pass filter with a $22\,\text{k}\Omega$ resistor and a $100\,\text{nF}$ capacitor, what is the RC time constant? If we provide a $3.3\,\text{V}$ $100\,\text{Hz}$ square wave as an input, what are the rise and fall times for the edges in the output?

Exercise 11.4

If you have a low-pass filter with a 47 kΩ resistor and a 10 μF capacitor, what is the RC time constant, and how long does it take for the filter to settle after a step?
If you have a 3.3 V step up from 0 V on the input, how long does it take for the signal to rise to 2.9 V?

When a system is first powered up, many of its capacitors in power supplies and filters need to charge to the appropriate DC values, which takes about 5 RC time constants. If some of the RC time constants are large, this can take a while. For example, some of the high-pass filters in an EKG may have corner frequencies of 0.01 Hz, which is an RC time constant of almost 16 seconds, and so the EKG may take over a minute to start responding normally. EKGs often have a *reset* circuit which temporarily reduces the resistance in these filters, to charge the capacitors quickly.

11.5 Input and output impedance of RC filter

In Section 11.2, RC filters were treated as voltage dividers, assuming that

- the input signal is a perfect voltage source, able to produce whatever current is needed by the filter without change to V_{in}, and
- the circuit connected to the output (called the *load*) takes no current from the filter, so that the voltage-divider criterion (page 46) is met.

As with all idealizations, these assumptions are never exactly met, but we often come close enough for the formulas in Section 11.2 to be useful. To check whether the assumptions are reasonable, it helps to compute the input and output impedance of the voltage divider, to compare with the circuits connected to the input and output.

If the magnitude of the impedance of the source of a signal is much smaller than the magnitude of the impedance of the load, then the assumptions that the source is a perfect voltage source and that the load doesn't affect the source are reasonable simplifications.

For checking the input assumption, we need to know (the magnitude of) the output impedance of the source that is driving V_{in}, compared to (the magnitude of) the input impedance of the filter, which is the load being seen by the source.

The *input impedance* Z_{in} of an RC filter is the impedance seen by whatever is driving V_{in}. The current needed from whatever is driving the input is $V_{\text{in}}/Z_{\text{in}}$.

The impedance seen at the input for the passive filters in Figure 11.1 is just the sum of the impedances of the resistor and capacitor, since the two components are in series: $Z_{\text{in}} = R + 1/(j\omega C)$.

We are mainly interested in showing that the magnitude of this impedance is large, and we are guaranteed that $|Z_{\text{in}}| \geq R$. So as long as R is much larger than the impedance of whatever is driving V_{in}, the first assumption (that the input voltage is not changed much by the load of the filter) will be met.

The *output* impedance of a voltage divider measures how much the voltage V_{out} drops as we take current from the voltage divider. Ideally, we would like an output impedance of 0 Ω, so that the voltage does not change, no matter how much current is taken. But this concept is exactly that of Thévenin equivalence, introduced in Section 5.1.2. If V_{in} is connected to a perfect voltage source, we can determine the behavior by replacing the voltage divider with a different voltage source in series with an impedance Z_{out}.

The Thévenin-equivalent voltage with a perfect voltage source as an input is computed by the voltage-divider formula,

$$V_{\text{thev}} = V_{\text{in}} \frac{Z_{\text{d}}}{Z_{\text{u}} + Z_{\text{d}}} ,$$

as in Section 11.2, and the Thévenin-equivalent impedance is the impedance of the resistor and capacitor in parallel:

$$Z_{\text{out}} = \frac{1}{1/Z_{\text{u}} + 1/Z_{\text{d}}} = \frac{1}{1/R + j\omega C} = \frac{R}{1 + j\omega RC} .$$

Another way to look at a voltage divider with impedances Z_{u} and Z_{d} is too look at the open-circuit voltage

$$V_{\text{thev}} = V_{\text{in}} \frac{Z_{\text{d}}}{Z_{\text{u}} + Z_{\text{d}}} ,$$

and the short-circuit current

$$I_{\text{sc}} = V_{\text{in}}/Z_{\text{u}} .$$

By the definition of Thévenin equivalence, $I_{\text{sc}} = V_{\text{thev}}/Z_{\text{out}}$, so we have

$$Z_{\text{out}} = Z_{\text{u}} \frac{V_{\text{thev}}}{V_{\text{in}}} ,$$

or, if we define the gain $G = V_{\text{thev}}/V_{\text{in}} = Z_{\text{d}}/(Z_{\text{u}} + Z_{\text{d}})$, we get that the output impedance is $Z_{\text{out}} = GZ_{\text{u}}$.

We are mainly interested in showing that the magnitude of the output impedance is much smaller than the impedance of whatever load is put on the output, and we are guaranteed that $|Z_{\text{out}}| \leq R$. So as long as the load on V_{out} is much larger than R, the second assumption (that the current through the load is negligible) will be met.

For both assumptions to be met, we need $|Z_{\text{in}}| \ll R \ll |Z_{\text{load}}|$, which can put some pretty strict constraints on the use of a passive RC filter.

Exactly how big "much larger" needs to be depends on how much error we can tolerate in the modeling. As a rough rule of thumb, an impedance 100 times larger will result in only about 1% error, which is usually good enough. We can often get away with a 10% error, which limits our impedance ratios to 10 or more.

Exercise 11.5

What are the magnitudes of the input and output impedances of a passive, high-pass filter with $R = 4.7\,\text{k}\Omega$ and $C = 220\,\text{nF}$ at 15 Hz, 150 Hz, and 1.5 kHz?

Exercise 11.6

Design a passive, low-pass filter with corner frequency around 400 Hz and input impedance of at least $1\,\text{k}\Omega$. Give the component values and compute the magnitude of the output impedance at 100 Hz.

11.6 Recentering a signal

A lot of the high-pass filters you will need to design are primarily for changing the DC voltage level that a signal has—which I refer to variously as *recentering*, *changing the DC offset*, or *level shifting*.[1] For example, in Lab 3, you will recenter the output at 1.65 V, so that it is 1.65 V ± 1.5 V at the output, when it is 0 V ± 1.5 V at the input.

For amplifier labs coming up, you'll often want to center the AC voltage inputs at the mid-point between the two power rails. The term "power rails" can refer to the physical wires that carry the power, the nodes in the schematic that the wires correspond to, or the voltages of those nodes. The term "rails" came from the days when the power supplies had to deliver huge currents, and so the wires were physically quite large—like copper railroad tracks. A symmetric power supply is a pair of power supplies having the same voltage, hooked up in series with ground at the point between them, so that one power supply provides a positive voltage and the other a negative voltage. Symmetric ±15 V supplies are common in old analog electronics books, because that was a popular voltage choice 40 years ago.

The reference voltage around which you center the inputs and outputs of an amplifier is often called *virtual ground*, because amplifier designs traditionally used a symmetric power supply with a positive and a negative voltage source around ground, so the designs used ground as the reference voltage.

Figure 11.7 shows two simple circuits for recentering a signal with an unknown DC bias to be centered at 1.65 V. The first one uses a power supply set to 1.65 V, which is not generally available (though we'll see how to fake one in Section 19.3). The second one takes advantage of the notion of *Thévenin equivalence*, which was introduced in Section 5.1.2.

Thévenin's theorem says that any circuit made up of passive components (resistors, capacitors, and inductors) and sources (voltage sources and current sources) that has only 2 ports (connections to the outside circuitry) can be replaced by a single voltage source in series with a single impedance.

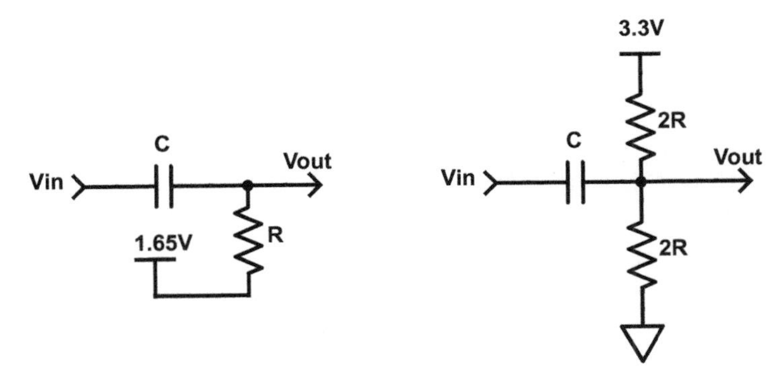

Figure 11.7: *These two circuits show how to change the DC offset of a signal to be centered at 1.65 V. In both circuits, the capacitor blocks any DC current. With a constant* V_{in}, *the capacitor maintains a constant charge, and the output is a constant 1.65 V, as long as no current is taken through the* V_{out} *port.*

Both circuits act as high-pass filters, with corner frequency $1/(2\pi RC)$. *The resistances in the second circuit are effectively in parallel, so that we need to make them twice as big to get the same RC time constant.*

Figure drawn with Digi-Key's Scheme-it [18].

[1] In digital design, *level shifting* may refer to a change in amplitude of a signal as well as a DC shift. Digital level-shifting circuits are often used to change 0 V–5 V signals into 0 V–3.3 V signals or 0 V–1.8 V signals, but for this book, I'm only talking about changing DC levels without changing the amplitude of the AC signal.

We can analyze the circuit on the right in Figure 11.7 by replacing the pair of resistors with their Thévenin equivalent, which gives the circuit on the left. That circuit then can be analyzed as a high-pass filter.

Worked Example:
Design a high-pass filter that attenuates (reduces) 1 Hz signals by a factor of 5, and whose output is centered at 3 V, using a 5 V power supply.

The basic design is like the right-hand design in Figure 11.7 with a 5 V power supply in place of the 3.3 V supply, but the two resistors can't be the same, as that would make the DC output 2.5 V, rather than 3 V. If the bottom resistor is $3R$ and the top one $2R$, then the Thévenin equivalent of the voltage divider is 3 V voltage source in series with a $3R||2R = \frac{1}{1/2+1/3}R = 1.2R$ resistor. That takes care of the centering of the output at 3 V, because the capacitor blocks the DC from the input.

The gain of the filter is

$$G = \frac{1.2R}{1.2R + 1/(j2\pi fC)} = \frac{j2.4\pi fRC}{j2.4\pi fRC + 1},$$

and for the magnitude of the gain to be 0.2 (reducing the signal by a factor of 5) at 1 Hz, we need

$$\begin{aligned}
0.2 &= |G| \\
&= \left|\frac{j2.4\pi 1\,\text{Hz}\ RC}{j2.4\pi 1\,\text{Hz}\ RC + 1}\right| \\
&= \frac{7.54\,\text{Hz}\ RC}{|j7.54\,\text{Hz}\ RC + 1|}, \\
0.2\,|j7.54\,\text{Hz}\ RC + 1| &= 7.54\,\text{Hz}\ RC, \\
0.04((56.85\,(\text{Hz})^2(RC)^2 + 1) &= 56.85\,(\text{Hz})^2(RC)^2, \\
0.04 &= 0.96(56.85\,(\text{Hz})^2(RC)^2), \\
732.9\text{E-}6\,\text{s}^2 &= (RC)^2, \\
RC &= 27.07\,\text{ms}\ .
\end{aligned}$$

If we pick $C = 560\,\text{nF}$, we get $R = 48.3\,\text{k}\Omega$, so our two resistors would be 96.7 kΩ and 145 kΩ, which we can round to 100 kΩ for the resistor to 5 V and 150 kΩ for the resistor to ground.

The Thévenin-equivalent resistance would be 60 kΩ, so the time constant is 33.6 ms. (This time is $1.2RC$ in the above computation, because $2R||3R = 1.2R$. The gain at 1 Hz is

$$G = \left|\frac{j2\pi 1\,\text{Hz}\ 33.6\,\text{ms}}{j2\pi 1\,\text{Hz}\ 33.6\,\text{ms} + 1}\right| = 0.2066\ ,$$

which is very close to the desired five-fold attenuation.

You can save some fuss with calculators by using gnuplot with the definitions.gnuplot file to compute the gain:

```
load 'definitions.gnuplot'
print abs(divider(Zc(1,560e-9), 60e3))
```

Exercise 11.7

Design an RC high-pass filter with a corner frequency of approximately 110 Hz, which centers the output at 1 V given a 3.3 V power supply. There is no other power supply—in particular, not a 1 V one. Hint: remember Thévenin equivalence—what circuit behaves like a resistor to a DC voltage?
Do not dissipate more than 1 mW in the circuit from the DC power supply (we won't worry about AC power dissipation until much later in the course). Round your component values to values on the E6 scale for capacitors and the E12 scale for resistors. Don't use extra capacitors or resistors to duck the E6 or E12 requirement. What is the actual DC power consumption? The actual corner frequency?

11.7 Band-pass filters

A *band-pass filter* combines the properties of a high-pass filter and a low-pass filter, allowing a range of frequencies to pass through (the *passband*), but attenuating frequencies outside that range.

A simple passive band-pass filter can be made with two resistors and two capacitors using a voltage divider with two complex impedances. The subcircuit that connects the input to the output needs to block DC, so its impedance at 0 Hz must be infinite—it consists of a resistor and capacitor in series. The subcircuit that the output voltage is across needs to short out very high frequencies (0 Ω impedance at ∞ Hz), so it consists of a capacitor in parallel with a resistor. The combined voltage divider is shown in Figure 11.8.

We can compute the gain of the band-pass filter by using the complex impedance of the subcircuits and the voltage-divider equation:

$$\begin{aligned}
Z_{\mathrm{u}} &= R_{\mathrm{u}} + \frac{1}{j\omega C_{\mathrm{u}}} \\
&= \frac{1 + j\omega R_{\mathrm{u}} C_{\mathrm{u}}}{j\omega C_{\mathrm{u}}}, \\
Z_{\mathrm{d}} &= \frac{1}{1/R_{\mathrm{d}} + j\omega C_{\mathrm{d}}} \\
&= \frac{R_{\mathrm{d}}}{1 + j\omega R_{\mathrm{d}} C_{\mathrm{d}}}, \\
V_{\mathrm{out}}/V_{\mathrm{in}} &= \frac{Z_{\mathrm{d}}}{Z_{\mathrm{d}} + Z_{\mathrm{u}}}
\end{aligned}$$

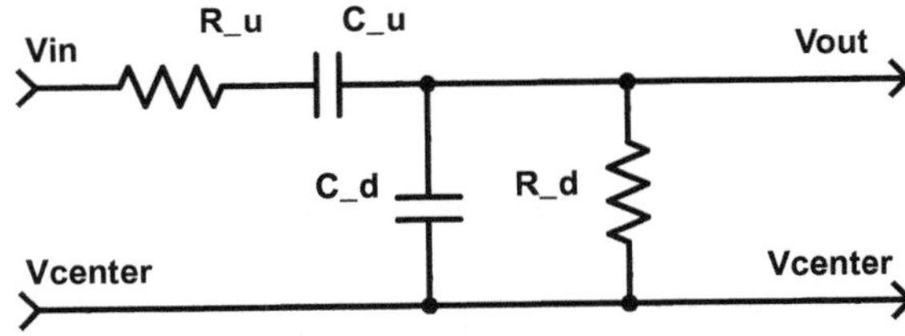

Figure 11.8: *A passive band-pass filter has C_{u} to block DC and C_{d} to short out high frequencies. The peak gain is $\frac{R_{\mathrm{d}} C_{\mathrm{u}}}{R_{\mathrm{d}} C_{\mathrm{u}} + R_{\mathrm{u}} C_{\mathrm{u}} + R_{\mathrm{d}} C_{\mathrm{d}}}$ at frequency $1/(2\pi\sqrt{R_{\mathrm{d}} R_{\mathrm{u}} C_{\mathrm{d}} C_{\mathrm{u}}})$, and the corner angular frequencies are $\omega_{lo} = \frac{1}{R_{\mathrm{d}} C_{\mathrm{u}} + R_{\mathrm{u}} C_{\mathrm{u}} + R_{\mathrm{d}} C_{\mathrm{d}}}$ and $\omega_{hi} = \frac{1}{R_{\mathrm{u}} C_{\mathrm{d}}} + \frac{1}{R_{\mathrm{d}} C_{\mathrm{d}}} + \frac{1}{R_{\mathrm{u}} C_{\mathrm{u}}}$.*
Figure drawn with Digi-Key's Scheme-it [18].

$$= \frac{\frac{R_d}{1+j\omega R_d C_d}}{\frac{R_d}{1+j\omega R_d C_d} + \frac{1+j\omega R_u C_u}{j\omega C_u}}$$
$$= \frac{j\omega R_d C_u}{j\omega R_d C_u + (1 + j\omega R_d C_d)(1 + j\omega R_u C_u)} .$$

We want to approximate the band-pass gain with three straight lines on a log-log plot: an upward sloping line at low frequency, a horizontal line in the passband, and a downward sloping line at high frequency.

At very low frequencies (as $\omega \to 0$), we end up with a gain of approximately $j\omega R_d C_u$, because the denominator approaches 1. At very high frequencies (as $\omega \to \infty$), we end up with a gain of approximately $1/(j\omega R_u C_d)$, because the ω^2 term in the denominator dominates everything else in the denominator. These are the two asymptotic lines that give us the behavior outside the passband.

To find the angular frequency for the peak of the passband, we need to solve

$$0 = \frac{d}{d\omega} \left| \frac{j\omega R_d C_u}{j\omega R_d C_u + (1 + j\omega R_d C_d)(1 + j\omega R_u C_u)} \right|$$

for ω. I tried giving this formula to WolframAlpha [167] (replacing each variable with a single letter and using "i" for $\sqrt{-1}$), but it found only non-useful imaginary solutions for ω. Manually applying Pythagoras's rule to the denominator to get

$$\frac{d}{d\omega} \frac{\omega R_d C_u}{\sqrt{(1 - \omega^2 R_d C_d R_u C_u)^2 + \omega^2 (R_d C_u + R_d C_d + R_u C_u)^2}}$$

simplifies the formula to one involving only real variables, and WolframAlpha can then find the correct solution, which has the peak gain at $\omega = 1/\sqrt{R_u C_u R_d C_d}$.

At that angular frequency, we have a gain of

$$V_{out}/V_{in} = \frac{j\sqrt{\frac{R_d C_u}{R_u C_d}}}{j\sqrt{\frac{R_d C_u}{R_u C_d}} + \left(1 + j\sqrt{\frac{R_d C_d}{R_u C_u}}\right)\left(1 + j\sqrt{\frac{R_u C_u}{R_d C_d}}\right)}$$
$$= \frac{j\sqrt{R_d C_u}}{j\sqrt{R_d C_u} + j\sqrt{R_u C_d}\left(\sqrt{\frac{R_u C_u}{R_d C_d}} + \sqrt{\frac{R_d C_d}{R_u C_u}}\right)}$$
$$= \frac{R_d C_u}{R_d C_u + R_u C_u + R_d C_d} .$$

The peak gain at $\omega = 1/\sqrt{R_u C_u R_d C_d} = \sqrt{\omega_{lo}\omega_{hi}}$ is

$$\frac{R_d C_u}{R_d C_u + R_u C_u + R_d C_d} ,$$

and the corner angular frequencies where the approximation lines intersect are

$$\omega_{lo} = \frac{1}{R_d C_u + R_u C_u + R_d C_d} , \tag{11.3}$$
$$\omega_{hi} = \frac{1}{R_u C_d} + \frac{1}{R_d C_d} + \frac{1}{R_u C_u} . \tag{11.4}$$

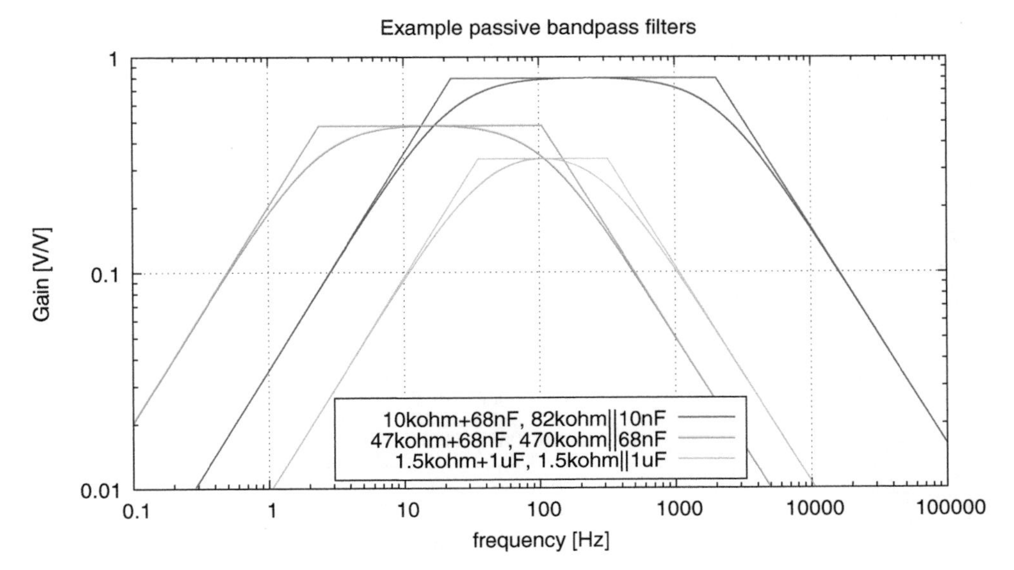

Figure 11.9: *Examples of passive band-pass filter designs, showing both the magnitude of the gain and the straight-line approximations derived in Section 11.7. The first example shows that the gain can be made close to 1, if $R_d \gg R_u$ and $C_u \gg C_d$. The second example has a typical design with $C_u = C_d$ and $R_u \ll R_d$, getting a gain close to 1/2 (0.476, to be more precise). The third example shows the 1/3 gain peak if $R_d = R_u$ and $C_d = C_u$.*
Figure drawn with gnuplot [33].

The relative bandwidth ω_{hi}/ω_{lo} is minimized when $R_u C_u = R_d C_d$ and $R_d C_u \to 0$, but the gain is then arbitrarily small.

11.7.1 *Special cases*

Solving Equations (11.3) and (11.4) to get particular corner frequencies and peak gain is fairly easy for programs like WolframAlpha and Mathematica, but can be a little difficult to do with just a pocket calculator. We can look at some special cases that allow quicker approximate solutions.

For example, if $C_u = C_d$, then the gain in the middle of the passband is $R_d/(R_u + 2R_d)$. If the corner frequencies are well separated, then $R_d \gg R_u$ and the gain is approximately 0.5, the lower corner angular frequency is $\omega \approx \frac{1}{2R_d C}$, and the upper corner angular frequency is $\omega \approx \frac{2}{R_u C}$. A design of this type is shown as the second example in Figure 11.9.

If $R_u = R_d$, then the gain in the middle of the passband is $C_u/(C_d + 2C_u)$. Again, if the corner frequencies are well separated, then $C_u \gg C_d$ and the gain is approximately 0.5, the lower corner angular frequency is $\omega \approx \frac{1}{2RC_u}$, and the upper corner angular frequency is $\omega \approx \frac{2}{RC_d}$. If you swap the capacitors, the frequency for the center of the passband stays the same, but the gain drops substantially, as shown in Figure 11.10.

If $C_u = C_d$ and $R_u = R_d$, then the peak gain is 1/3 and the corner angular frequencies are $\omega = \frac{1}{3RC}$ and $\omega = \frac{3}{RC}$. This design is shown as the third example in Figure 11.9.

> If we want a high gain (near 1), then we need $R_d C_u \gg R_u C_u + R_d C_d$, which in turn requires $R_d \gg R_u$ and $C_u \gg C_d$. Under those conditions, the corner frequencies are approximately $\omega = 1/(R_d C_u)$ and $\omega = 1/(R_u C_d)$.

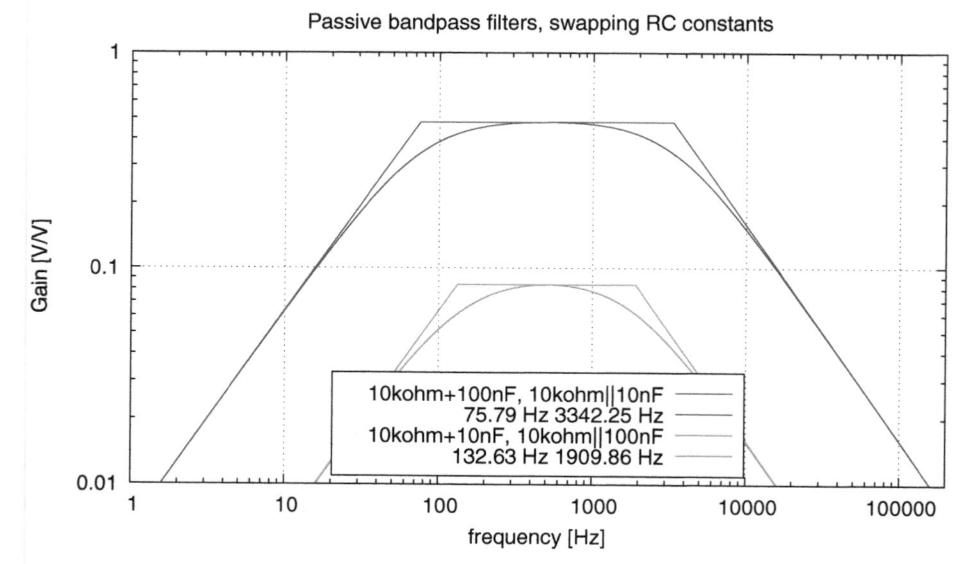

Figure 11.10: *Examples of passive band-pass filter designs, showing both the magnitude of the gain and the straight-line approximations derived in Section 11.7. In this example, the resistors are the same ($R_d = R_u$, but the capacitors differ by a factor of 10. If $C_u \gg C_d$, then the peak gain is close to 0.5 (0.4762), but if the capacitors are swapped, the peak gain drops to 0.0833. The peak of the passband remains the same at* $1/(2\pi\sqrt{R_d R_u C_d C_u}) = 503.3$ *Hz.*
Figure drawn with gnuplot [33].

11.7.2 *Examples and exercises*

Worked Example:
Design a passive band-pass filter with a peak gain over 0.9*, a low corner around* 20 *Hz, and a high corner around* 10 *kHz, using only standard resistor and capacitor values.*

We can set the peak of the passband halfway between the two corners (in log frequency), at $f = \sqrt{20\,\text{Hz}\ 10000\,\text{Hz}} = 447.21\,\text{Hz}$. That means that the product of the RC time constants will be

$$R_u C_u R_d C_d = (2\pi\ 447.21\,\text{Hz})^{-2} = 126.651\ 10^{-9}\ \text{s}^2\ .$$

At that frequency, we want the gain to be greater than 0.9:

$$\frac{R_d C_u}{R_d C_u + R_u C_u + R_d C_d} > 0.9\ ,$$
$$\frac{R_d C_u + R_u C_u + R_d C_d}{R_d C_u} < 10/9\ ,$$
$$\frac{R_u}{R_d} + \frac{C_d}{C_u} < 1/9\ .$$

This means that $R_u \ll R_d$ and $C_d \ll C_u$. If the resistance and capacitance ratios are about the same, we would have $18R_u < R_d$ and $18C_d < C_u$, meaning that we have $324 R_u C_d < R_d C_u$.

For the lower corner, we have

$$R_d C_u + R_u C_u + R_d C_d = 1/(2\pi\ 20\,\text{Hz}) = 7.95877\,\text{ms}\ ,$$

which we can simplify (because of the constraints imposed by the high-gain requirement) to $R_d C_u \approx$ 7.95 ms.

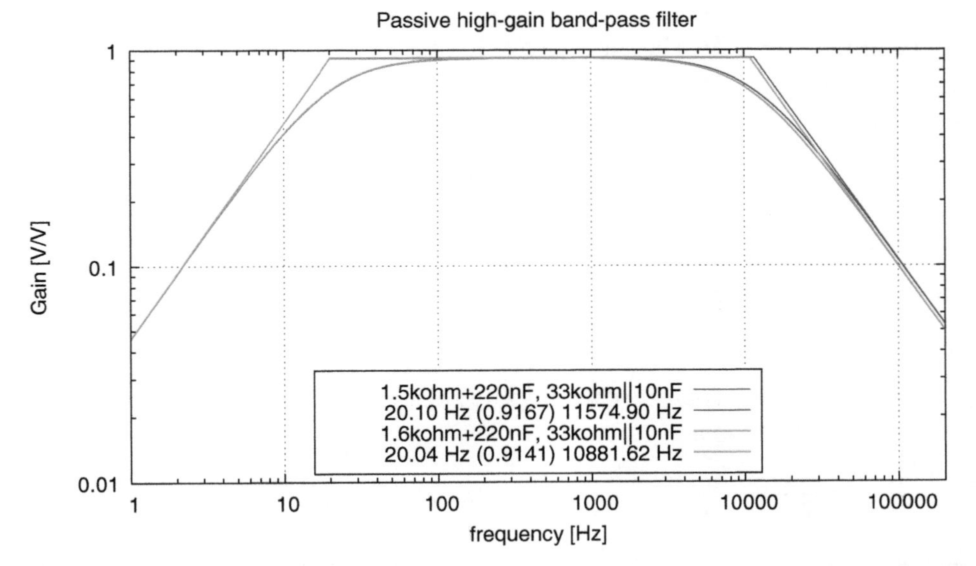

Figure 11.11: *Two solutions for the design challenge of getting a peak gain higher than 0.9 with corner frequencies of 20 Hz and 10 kHz, using standard components. High gain in the passive band-pass filter design of Figure 11.8 requires that the corner frequencies be widely separated.*
Figure drawn with gnuplot [33].

For the upper corner, we have

$$\frac{1}{R_u C_d} + \frac{1}{R_d C_d} + \frac{1}{R_u C_u} = 2\pi\ 10000\,\text{Hz} = 62831.85\,\text{s}^{-1}\ ,$$

which we can simplify because of the gain constraints to $\frac{1}{R_u C_d} \approx 62831\,\text{s}^{-1}$ or $R_u C_d \approx 15.92\,\mu\text{s}$.

The high-gain requirement forces the corner frequencies to be widely separated.

We could pick $R_u = 1.5\,\text{k}\Omega$, $C_u = 220\,\text{nF}$, $R_d = 33\,\text{k}\Omega$, and $C_d = 10\,\text{nF}$, using E6 values for both resistance and capacitance. The achieved peak gain is 0.9167, the lower corner frequency is 20.10 Hz, and the upper corner frequency is 11.57 kHz.

We can lower the upper-corner frequency a little by using $R_u = 1.6\,\text{k}\Omega$ from the E24 series, getting a peak gain of 0.9141 and corner frequencies of 20.04 Hz and 10.882 kHz. The Bode plot for the two filter designs is given in Figure 11.11.

Worked Example:
Design a passive band-pass filter with a peak gain between 0.4 *and* 0.6*, a low corner around* 150 *Hz, and a high corner around* 20 *kHz, using only standard resistor and capacitor values.*

We can set the peak of the passband halfway between the two corners (in log frequency), at $f = \sqrt{150\,\text{Hz}\ 20000\,\text{Hz}} = 1732.05\,\text{Hz}$. That means that the product of the RC time constants will be

$$R_u C_u R_d C_d = (2\pi 1732.05)^{-2} = 8.44343\ 10^{-9}\ \text{s}^2\ .$$

At that frequency, we can set the gain to approximately 0.5:

$$\frac{R_\text{d}C_\text{u}}{R_\text{d}C_\text{u}+R_\text{u}C_\text{u}+R_\text{d}C_\text{d}} \approx 0.5$$

$$R_\text{d}C_\text{u} \approx R_\text{u}C_\text{u}+R_\text{d}C_\text{d}$$

$$1 \approx R_\text{u}/R_\text{d}+C_\text{d}/C_\text{u} \ .$$

We could achieve this, for example, with $C_\text{d} = C_\text{u}$ and $R_\text{u} \ll R_\text{d}$.

For the lower corner, we have

$$R_\text{d}C_\text{u}+R_\text{u}C_\text{u}+R_\text{d}C_\text{d} = 1/(2\pi 150\,\text{Hz}) = 1.061033\,\text{ms} \ .$$

If $R_\text{u} \ll R_\text{d}$ and $C_\text{d} = C_\text{u} = C$, then we can simplify the formula to $2R_\text{d}C \approx 1.06\,\text{ms}$ or $R_\text{d}C \approx 530\,\mu\text{s}$.

For the upper corner, we have

$$\frac{1}{R_\text{u}C_\text{d}} + \frac{1}{R_\text{d}C_\text{d}} + \frac{1}{R_\text{u}C_\text{u}} = 2\pi 20000\,\text{Hz} = 125663.7\,\text{s}^{-1} \ .$$

If $R_\text{u} \ll R_\text{d}$ and $C_\text{d} = C_\text{u} = C$, then we can simplify the formula to $\frac{2}{R_\text{u}C} \approx 125663.7\,\text{s}^{-1}$ or $R_\text{u}C \approx 15.92\,\mu\text{s}$.

These RC time constants also meet the constraint on their product that we started with. We still have an arbitrary choice of the trade-off between the resistance and the capacitance, which affects the total impedance of the filter (and so the load that it imposes on whatever is driving the input). Because we were not given a constraint here, let's pick a reasonable range for the resistors (say $1\,\text{k}\Omega$ to $1\,\text{M}\Omega$) that will work with low-impedance, but low-power, drivers like the output of an op amp.

We could, for example, choose $C = 10\,\text{nF}$, making $R_\text{d} = 51\,\text{k}\Omega$ and $R_\text{u} = 1.6\,\text{k}\Omega$. The peak gain would then be 0.4923 at 1762 Hz, and the corner frequencies would be 153.6 Hz and 20206 Hz. I picked a capacitor from the E6 series and resistors from the E24 series—by giving up a little precision on the frequency, I could use resistors from the E12 series. This solution is not unique—as is often the case in electronics design, the design goal is under-constrained, and we need to use engineering judgment to pick reasonable points in the solution space.

Exercise 11.8

Compute the peak gain and corner frequencies for a passive band-pass filter which has a $1.5\,\text{k}\Omega$ resistor and 220 nF in series for the subcircuit connecting the input to the output, and $15\,\text{k}\Omega$ and 4.7 nF in parallel for the subcircuit connecting the reference voltage to the output. Express the corner frequencies in Hz (f), not radians/s ($\omega = 2\pi f$).

Use gnuplot to create a Bode plot (log-log plot of gain as a function of frequency) for 1 Hz to 1 MHz. Use the functions in definitions.gnuplot (Figure 7.4) to make a readable function (that is, one that expresses the serial and parallel connections) for the gain of a passive band-pass filter, and plot the absolute value of that gain function, passing in the specific values for the components.

(Sanity check: is the peak of the plot at the frequency and gain you expect?)

Bonus: write an approximation function using the if-then-else expression "i?t:e" to plot the three straight lines, changing which line is plotted at each corner: `f<Lc? Lline : f<Hc? Const: Hline`, where each of the names other than f is appropriately replaced with the proper expression for the corner frequency or the straight-line approximation.

Exercise 11.9

Design a passive band-pass filter with a peak gain around 0.5, a low corner around 10 Hz, and a high corner around 200 Hz, with the output voltage centered at V_{ref}.
Choose capacitors from the E6 series, and resistors from the E12 series, even if this means changing the corner frequencies slightly. The corner frequencies should not be more than 20% off, however.
Draw the schematic for the filter, and use gnuplot to plot a Bode plot confirming the design.

11.7.3 *Cascaded high-pass and low-pass filter*

Students are often tempted to design a passive band-pass filter by designing high-pass and low-pass filters separately and feeding the output of one into the input of the other—an arrangement known as *cascading* the filters. Unfortunately, this method does not work well, as the voltage-divider assumption that essentially no current be taken from the output of the voltage divider is violated. (Active filters, introduced in Chapter 24, can be safely cascaded.)

It is possible to analyze a cascaded passive filter, using only the voltage divider formula, but the formulas get a bit messy, as the first voltage divider doesn't have a single component on its lower subcircuit, but the entire second filter in parallel with the component.

Let's look at what can go wrong if we pretend the filters are independent by considering an example. Let's say we want a band-pass filter with corner frequencies 1 Hz and 3.4 Hz, and that we try to design it as a high-pass filter followed by a low-pass filter.

The high-pass filter needs corner frequency 1 Hz or $RC = 159.2$ ms. We can achieve this with a 1 μF capacitor and 160 kΩ resistor. The low-pass filter needs a corner frequency of 3.4 Hz or $RC = 46.8$ ms, which can be achieved with a 10 μF capacitor and a 4.7 kΩ resistor. The schematic for the resulting design is shown in Figure 11.12.

We can compute the gain of the cascaded band-pass filter fairly easily in gnuplot and compare it to the gain we would expect from having treated the filter blocks as independent:

```
high_pass(f) = Rhi / (Rhi + Zc(f,Chi))
low_pass(f) = Zc(f,Clo) / (Rlo + Zc(f, Clo))
intended(f) = high_pass(f)*low_pass(f)
cascaded(f) =  Rhi / (Rhi + Zpar(Zc(f,Chi), Rlo+Zc(f,Clo)) ) * low_pass(f)
```

The magnitudes of the intended and cascaded filters are plotted in Figure 11.13.

The problem with the cascaded design can be fixed by ensuring that the impedance for the second-stage filter is much higher than the impedances in the first-stage filter, so that the paralleling of the impedances does not change the first stage much. For this design, we could leave the time constants

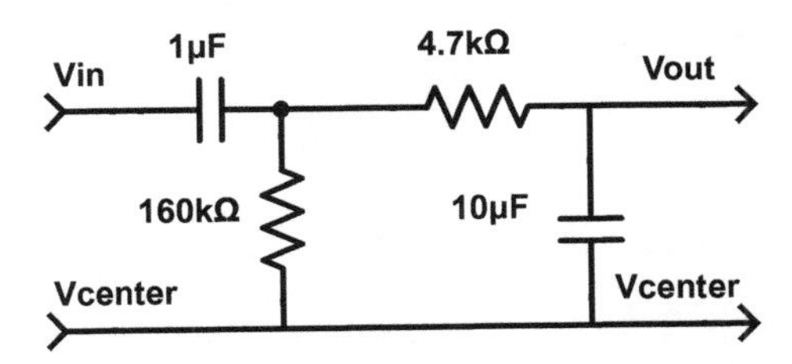

Figure 11.12: *This band-pass filter is a deliberately bad cascaded design, in which the load of the low-pass filter changes the corner frequency of the high-pass filter substantially.*
Figure drawn with Digi-Key's Scheme-it [18].

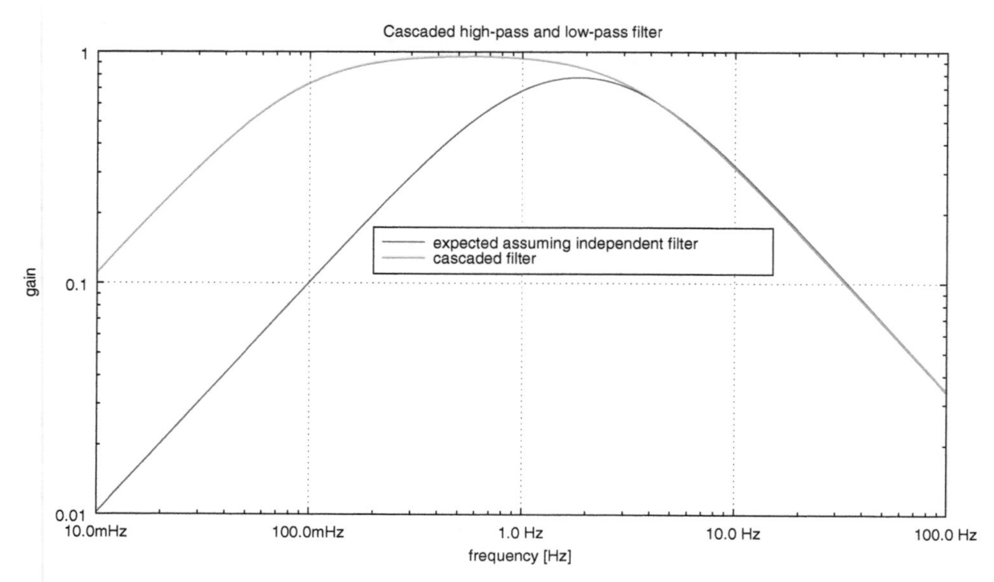

Figure 11.13: *Bode plot for the cascaded filter of Figure 11.12. The corner frequency for the low-pass filter is about where we would expect from treating the filters as independent, but the lower corner frequency is about a factor of 10 too low.*
Figure drawn with gnuplot [33].

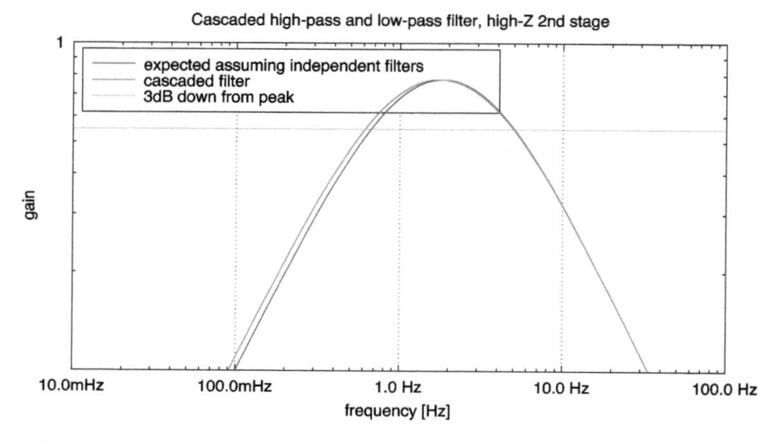

Figure 11.14: *By decreasing the impedances in the high-pass first stage of the filter, and increasing them in the low-pass second stage, we can make a cascaded filter behave much more like we would expect from independent filter blocks.*
The corner frequencies for the cascaded filter, however, are not at the 1 Hz and 3.4 Hz design points, because those corner frequencies are too close to be achieved with a passive band-pass filter.
Figure drawn with gnuplot [33].

alone, but swap the capacitors, getting $10\,\mu$F and $16\,$kΩ for the high-pass and $1\,\mu$F and $47\,$kΩ for the low-pass, resulting in the Bode plot of Figure 11.14.

11.8 Band-stop filters

This section is optional, as we will not be designing passive band-stop filters in this course. The section explains why the simple RC voltage divider does not make a very useful band-stop filter.

A *band-stop* filter is the opposite of a band-pass filter—it tries to pass all frequencies except those in a specified range. A common use for band-stop filters is to exclude a source of interference with the desired signal. Often the stop band is quite narrow, centered around a single source of interference (such as 60 Hz), in which case the band-stop filter may be called a *notch filter.*

It is difficult to make a good notch filter with just resistors and capacitors—usual designs use LC filters (inductors and capacitors), active filters (resistors, capacitors, and amplifiers), or digital filters. But let us look at how much we can do with just RC voltage dividers, like we used for band-pass filters in Section 11.7.

To make a passive, RC, band-stop filter we again use a voltage divider with a parallel RC circuit and a series RC circuit as the two subcircuits, but we swap the subcircuits so that the parallel circuit is between the input and the output and the series circuit is between the output and ground, as shown in Figure 11.15.

The big problem with this filter design is that the sides of the passband have slopes no steeper than ±20 dB/decade (see Section 2.3.2 for an explanation of dB and dB/decade), so we can't get a deep, narrow notch—the deeper the notch, the wider the stop band.

The voltage-divider equation gives us the output voltage as

$$\begin{aligned} Z_{\mathrm{d}} &= R_{\mathrm{d}} + \frac{1}{j\omega C_{\mathrm{d}}} \\ &= \frac{1 + j\omega R_{\mathrm{d}} C_{\mathrm{d}}}{j\omega C_{\mathrm{d}}} , \\ Z_{\mathrm{u}} &= \frac{1}{1/R_{\mathrm{u}} + j\omega C_{\mathrm{u}}} \\ &= \frac{R_{\mathrm{u}}}{1 + j\omega R_{\mathrm{u}} C_{\mathrm{u}}} , \\ V_{\mathrm{out}}/V_{\mathrm{in}} &= \frac{Z_{\mathrm{d}}}{Z_{\mathrm{d}} + Z_{\mathrm{u}}} \end{aligned}$$

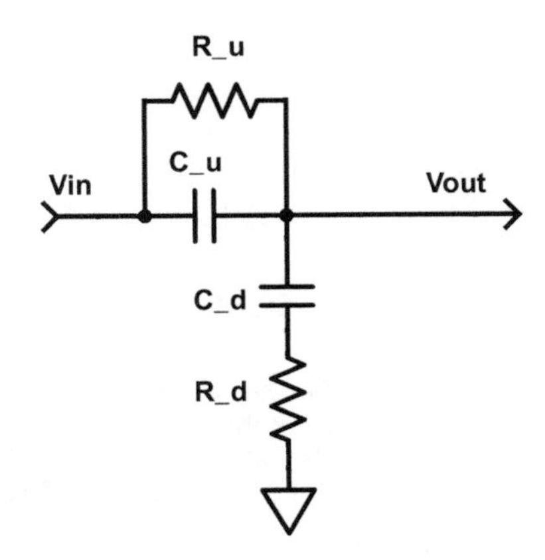

Figure 11.15: *A passive band-stop filter has* C_{u} *to block low frequencies and* C_{d} *to ensure that high frequencies are attenuated. The minimum gain is* $1 - \dfrac{R_{\mathrm{u}} C_{\mathrm{d}}}{R_{\mathrm{u}} C_{\mathrm{d}} + R_{\mathrm{u}} C_{\mathrm{u}} + R_{\mathrm{d}} C_{\mathrm{d}}}$ *at frequency* $\dfrac{1}{2\pi \sqrt{R_{\mathrm{d}} R_{\mathrm{u}} C_{\mathrm{d}} C_{\mathrm{u}}}}$. *Ground can be replaced by any DC voltage source in the system, as there is no DC path from it to the output.*
Figure drawn with Digi-Key's Scheme-it [18].

$$= \frac{1}{1 + Z_u/Z_d}$$

$$= \frac{1}{1 + \frac{j\omega R_u C_d}{(1+j\omega R_d C_d)(1+j\omega R_u C_u)}}$$

$$= \frac{1 - \omega^2 R_d C_d R_u C_u + j\omega(R_d C_d + R_u C_u)}{1 - \omega^2 R_d C_d R_u C_u + j\omega(R_d C_d + R_u C_u + R_u C_d)} \,.$$

At the extremes $\omega = 0$ and $\omega = \infty$, the gain of the band-stop filter is 1, as desired. When $\omega = 1/\sqrt{R_u C_u R_d C_d}$, the gain of the filter is minimized with the value

$$1 - \frac{R_u C_d}{R_u C_u + R_d C_d + R_u C_d} \,.$$

Although it is possible to come up with straight-line approximations to the band-stop filter, they are no simpler to work with than the voltage-divider equation itself, and so are not worth the trouble of deriving them. As a rough rule of thumb, the notch is narrowest for a given depth if $R_u C_u = R_d C_d$. Figure 11.16 shows notch filters for this case, illustrating the increase in width for the filters as the notch gets deeper.

Because the notches get very wide before they get a useful depth, simple RC band-stop filters like these are rarely used—it is much easier to make narrow notch filters digitally.

Figure 11.17 has passive band-stop filters with the same center frequency as in Figure 11.16, but with $R_u C_u \leq R_d C_d$. When $R_u C_u \ll R_d C_d$, then the gain in the stop band is approximately $R_u/(R_u + R_d)$. There does not seem to be any advantage to having this constant gain in the stop band, rather than having a deeper notch at the center frequency with the same width to the stop band.

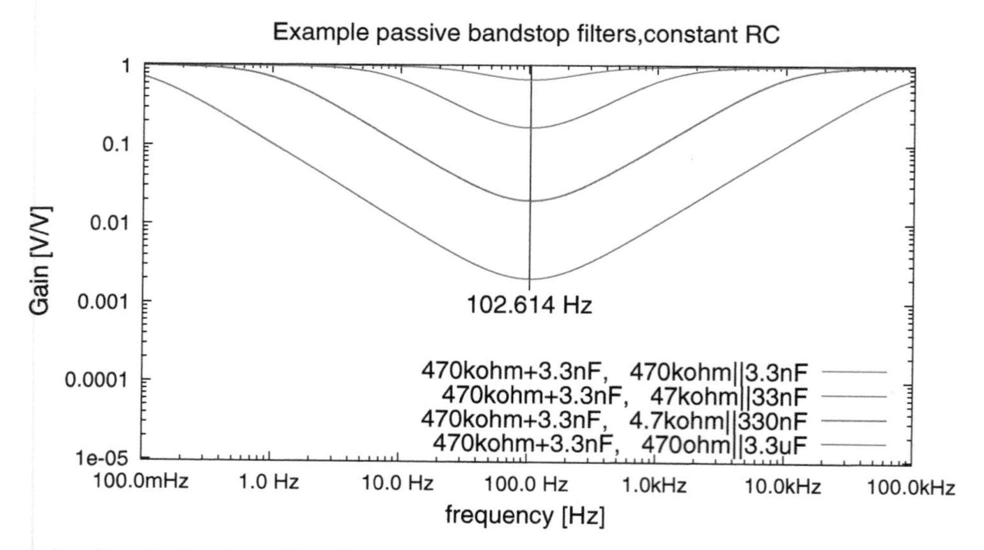

Figure 11.16: *Examples of passive band-stop filter designs with $R_u C_u = R_d C_d = 1.551$ ms, setting the center frequency to 102.614 Hz. Increasing $R_u C_d$ by increasing C_d and decreasing R_d increases the depth of the notch, simultaneously increasing its width. Figure drawn with gnuplot [33].*

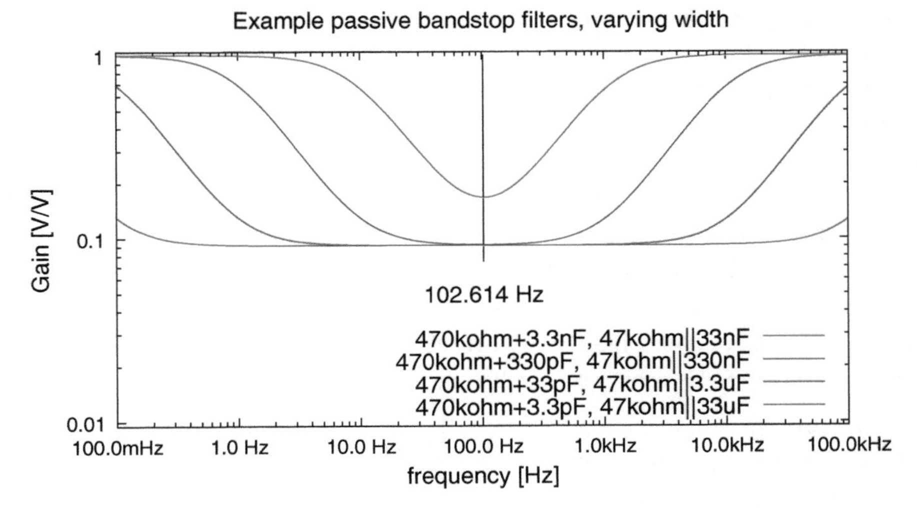

Figure 11.17: *Examples of passive band-stop filter designs with a fixed center frequency (the same as in Figure 11.16) but with varying bandwidth. The top curve here is the same as the second highest curve in Figure 11.16. Figure drawn with gnuplot [33].*

11.9 Component tolerance

All the filter designs in this chapter have assumed that the resistors and capacitors had exactly the values that they were labeled with, but this ideal situation does not happen in the real world. Component values differ from their nominal values for several reasons: manufacturing variation, temperature dependence, and voltage dependence, for example.

The manufacturing variation is usually given as percentage, with the expectation that essentially all the components shipped will be within the specified range. For example, a 1 kΩ resistor with a 1% tolerance will be between 990 Ω and 1010 Ω.

The tolerance may be specified for a specific temperature or over a temperature range. Resistors are usually specified at a particular temperature (25°C is common) and a thermal coefficient given. For example, a 0.1% resistor may have a thermal coefficient of $\pm$50 ppm/°C, which could result in being 0.65% off at the high end of its temperature range, rather than just 0.1% off.

Resistors come in three major classes these days: metal film, carbon, and wire-wound. The metal-film resistors can have fairly precise manufacturing tolerances—the cheap ones are rated at 1%, and ones rated at 0.1% are easily available. Carbon resistors have mostly gone out of favor now except in very cheap toys, as they generally have a tolerance 5% and aren't substantially cheaper than metal-film resistors.

One big difference between carbon resistors and metal-film resistors is that the carbon resistors usually have a negative temperature coefficient (resistance decreases as temperature increases), but metal-film resistors usually have a positive temperature coefficient (resistance increases as temperature increases). In thermal-compensation circuits where the change in resistance with temperature is important for the function of the circuit, carbon and metal-film resistors are not interchangeable.

Wire-wound resistors, like metal-film resistors, use a metal as the resistance element and have a positive temperature coefficient, but are designed for high-power applications—in those applications, precision is not really required, and most wire-wound resistors have 5% tolerance.

The tolerances for capacitors are usually much looser than for resistors, particularly for cheap ceramic capacitors. Moderately good ceramic capacitors (C0G or NP0) generally have tolerances of ±5% over a wide temperature range, capacitors with X7R dielectrics have ±10%, and Y5V capacitors have a tolerance of +80%/−20% over a narrower temperature range. The asymmetric tolerance range is acceptable for bypass capacitors (see Section 11.10), where extra capacitance does not cause any problems.

The capacitance and tolerance of ceramic capacitors is given for small signals with 0 V DC bias—large DC biases saturate the dielectric, reducing the capacitance substantially, and this effect is not usually included in the specified tolerances. It is a good idea to keep DC biases to less than 20% of the maximum DC rating of the capacitor. The high-dielectric-constant ceramics (like X75 and Y5V) suffer much more from this problem than C0G/NP0 ceramics.

> If you need fairly precise capacitance values, C0G/NP0 ceramics, which are available with 1% tolerance, are usually the best deal, though high capacitance values are not available.

The variation in resistance and capacitance means that filter parameters using the components will vary. Engineers usually have to do a worst-case analysis—how far off are the filter frequencies if the resistors and capacitors are at the edge of their tolerance ranges?

Worked Example:
If we have a low-pass filter design using a 10 kΩ resistor and a 100 nF capacitor, what is the corner frequency if the actual values are the nominal values? What is the range of corner frequencies if the resistor is ±1% and the capacitor is +80%/−20%?

The corner frequency with nominal values is $1/(2\pi RC) = 159.2$ Hz. The largest values for R and C are 10 kΩ 1.01 = 10.1 kΩ and 100 nF 1.8 = 180 nF, giving a corner frequency of 87.54 Hz. The smallest values for R and C are 10 kΩ 0.99 = 9.9 kΩ and 100 nF 0.8 = 80 nF, giving a corner frequency of 201 Hz.

Exercise 11.10

Design a passive high-pass filter with a corner frequency of approximately 340 Hz, using capacitors no larger than 1 nF. If the capacitors have a ±5% tolerance and the resistors, ±1% tolerance, what is the range of corner frequencies for your design?

11.10 Bypass capacitors

> If you look at modern electronics, you will find a lot of capacitors that don't seem to be doing anything—they are just connected across a power supply and so should be charged to a constant voltage. These capacitors are called *bypass capacitors* or *decoupling capacitors* and are particularly common in high-speed digital circuits and in analog circuits that handle small signals.

You may remember that on page 64 we pointed out that wires may sometimes be modeled as nodes and sometimes as more complicated circuits involving resistance or inductance. It turns out that the resistance and inductance of power wiring often matters. If we have a changing current through the power wiring, then the voltage across the wire will change, and the power supply voltage seen by the circuit will fluctuate, no matter how good our voltage regulation is at the power supply itself. Rapid changes in current produce particularly large voltage spikes from the inductance of the power wiring.

> Putting bypass capacitors right next to chips that have fluctuations in their current draw provides a way to smooth out the current through the power wires (sudden changes in current charge or discharge the capacitor), greatly reducing the production of voltage spikes, which could otherwise affect circuits throughout the system. For this noise reduction to work well, we want very little resistance or inductance between the capacitor and the power pin on the chip, so the bypass capacitor is usually placed immediately next to the chip.

Not only are bypass capacitors put next to chips that generate noise on the power lines, they are also put next to chips that are sensitive to noise. The resistance and inductance of the power wiring, combined with the capacitance of the bypass capacitor, form a low-pass filter that removes high-frequency noise from the power supply.

For the sorts of circuits we'll be doing in this class, it is not necessary to model the power wiring very precisely. Instead, we'll use bypass capacitors that are much bigger than we probably need and not worry about it further.

Ceramic capacitors are used for bypass capacitors because of their low series resistance and inductance. Where large currents are expected (such as in motor drivers), electrolytic capacitors are used as bypass capacitors to get much larger capacitances, but the higher internal resistance and inductance of electrolytic capacitors limits them to rather low frequencies, so they are almost always used in parallel with a smaller ceramic capacitor, which handles the higher frequency components.

The most common size of bypass capacitor appears to be a $0.1\,\mu$F ceramic capacitor with X5R or X7R dielectric (based on what Digi-Key stocks the most of). This is the most stocked size both in surface-mount and through-hole packages. I'm partial to $4.7\,\mu$F ceramic bypass capacitors, though they are 2-to-10 times as expensive as $0.1\,\mu$F capacitors. I should probably break myself of this habit.

The Wikipedia article on bypass capacitors has a good list of sources for more information about the use of bypass capacitors [119].

Chapter 12

Function Generator

A *function generator* is a piece of bench equipment used for generating electrical signals for testing analog electronics. There are many on the market, ranging from hobbyist ones like the $50 FG085 kit from JYE Tech to ones costing over $6,000 like the Keysight 33622A. Many of the hobbyist ones are nearly useless, having high distortion, awkward user interfaces, and low frequency ranges.

The Analog Discovery 2 USB oscilloscope contains two function generators that are quite good for signals of moderate frequency (up to about 1 MHz). They are of much higher quality than the under-$100 hobbyist function generators, but not as good as the very expensive professional ones, which can produce higher-quality signals to much higher frequencies. For the applications in this book, the function generators in the Analog Discovery 2 are an excellent trade-off of price and performance.

A user generally specifies a wave shape, a frequency, an amplitude, and a DC offset for the function generator. These specifications may be by knobs on the instrument, by menu selections, or by typing in the numbers.

Almost all function generators provide a choice of a variety of different signals: sine waves, square waves, triangle waves, and ramps (also known as sawtooth waves). Many also provide arbitrary waveform generation, where the shape of the waveform is defined by a table of numbers—the user interfaces for specifying the arbitrary waveform vary considerably and are sometimes quite complicated, but we won't need that feature for this course—sine waves, ramps, and triangle waves are all we really need.

The Analog Discovery 2 also provides random noise (useful for some types of testing as in Lab 4, but not really essential for this course) and "pulses", which are just a shorthand way of providing a square wave that has 0 V as one of its value and the amplitude as its other value. A 3.3 V pulse is the same thing as a square wave with an amplitude of ± 1.65 V and a DC offset of 1.65 V, but the pulse is easier to specify.

A function generator can be fairly modeled as an ideal voltage source in series with a resistor as shown in Figure 12.1. This resistance is referred to as the *output impedance* of the function generator.

At high frequencies or with long wires, we need to think about reflections of our signal from any place in the wiring where the impedance changes. The reflections don't matter right at the impedance change, but at some distance from it, where the original signal and the reflection get out of phase and some cancellation occurs.

Most function generators are designed with a 50 Ω output impedance and are usually used with 50 Ω coaxial cable connected to a 50 Ω load. By having 50 Ω impedance throughout, we have no reflections at the connectors.

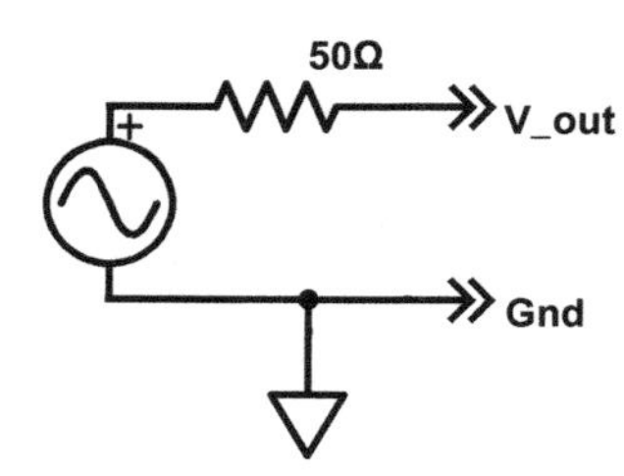

Figure 12.1: *Schematic diagram for modeling a function generator. The ideal voltage source is in series with an* output impedance *consisting of a 50 Ω resistor. The Analog Discovery 2 does not include this 50 Ω resistance—its output impedance is effectively 0 Ω, though the optional BNC adapter board has a jumper for adding a 50 Ω series resistor.*

If the function generator settings are known, they should be used to label the voltage source in the schematic (something like "105 Hz, (1.3 ± 0.7) V").

Figure drawn with Digi-Key's Scheme-it [18].

Because the Analog Discovery 2 uses a feedback amplifier (see Section 19.2) to set the voltage, the output is best modeled as a voltage source with 0 Ω in series, but with a current limit that clips the output. The current limit is approximately ±30 mA, though I have seen larger currents when driving very-low-impedance loads.

The optional BNC adapter board has a jumper for adding a 50 Ω series resistor, to provide the conventional interface for 50 Ω coaxial cables.

In this course, we are generally working at low enough frequencies and with short enough wiring that impedance matching is not a concern—the difference in phase between the reflection and the original signal is too small to matter. We don't bother with 50 Ω coaxial cable and 50 Ω loads for the function generator, but connect it directly to high-impedance inputs like the oscilloscope or amplifier inputs.

The Analog Discovery 2, like most recent function generators, specifies the voltage at the voltage source—if you use a low-impedance load and the optional 50 Ω output impedance on the BNC Adapter Board, then you need to use the voltage-divider formula (see Section 5.1) to figure out the voltage at the load.

12.1 Agilent 33120A function generators

For several years, the course at UCSC used Agilent 33120A function generators, which are now obsolete and can be obtained used for around $500–$1000. These function generators were hard for beginners to use, as the controls specified the voltage assuming that a 50 Ω load had been applied to the output. Given the output impedance of 50 Ω, this meant that the internal voltage generated was twice what the controls specified, as the 50 Ω output impedance and the 50 Ω load were expected to form a voltage divider, cutting the internal voltage in half. If the function generator is connected to a high-impedance load (like an oscilloscope, an op-amp input, or an analog-to-digital converter), the voltage observed is close to what the internal voltage source provides, doubling the voltage set on the control panel.

Warning: the Agilent 33120A function generator claims to be reporting peak-to-peak voltage, but it actually delivers twice as much voltage as it reports. The designers of that function generator assumed that there would be a 50 Ω load on the output of the function generator, which we are not providing. In Section 5.1, we saw how to compute the voltage provided by circuits like this one, with a 50 Ω internal impedance for the function generator in series with the 50 Ω load.

For the Agilent 33120A, you'll want a setting of $1.6V_{pp}$ to get a peak-to-peak voltage of 3.2 V. Offset voltages are also twice what you input on the controls. (There is a menu option to get the function generator to report the voltage for a high-impedance load, so that the voltages it reports correspond to the ones you get, but it is turned off by default and takes some effort to set up.)

12.2 Analog Discovery 2 function generator

The Analog Discovery 2 provides two independent function generators, which can be controlled by the WaveForms 3 software. Each of the function generators provides a single-wire output, with the other side of the function generator connected to ground.

With the standard "flywire" connector for the Analog Discovery 2, Channel 1 is provided on a yellow wire and Channel 2 on a yellow wire with a white stripe. The ground connection is provided by any of the four black wires.

Each function generator works by providing a 14-bit number to a digital-to-analog converter every 10 ns (a 100 MHz sampling rate). The size of the voltage steps depends on the amplitude of the signal being generated—if it is less than 1.25 V (2.5 V peak-to-peak), then the step size is 166 μV, but if the amplitude is larger, then the step size is 665 μV. There is a separately generated DC offset (with a 672 μV step size) added to the AC signal, but the combined AC+DC signal has to stay within the range $-5\text{ V} \le V_{\text{out}} \le +5\text{ V}$.

The function generators provide excellent waveforms at audio frequencies for amplitudes around 1 V, but the waveforms are not as clean at very high frequencies or at very low amplitudes. If you need a very low amplitude signal, you may want to generate a larger signal and use a voltage divider (see, for example, Figure 42.1). If you need clean signals with a frequency above 5 MHz, you probably need a different function generator.

One unusual feature of the Analog Discovery 2 is that you can listen to the two channels with a stereo headphone plugged into a jack at the back of the device. Channel 1 is played to the left earpiece, and Channel 2, to the right. The audio output uses AC coupling, to avoid presenting a DC offset to the headphone.

We will mostly be using the function generator to create sine waves, ramps, triangle waves, and pulses, all of which can be run from the "Simple" function-generator type, as shown in Figure 12.2.

The function generators on the Analog Discovery 2 provide lots of fancy features that we don't need for this course: AM and FM modulation, sending bursts by using the "synchronized" option with a wait time and run time, sweeping the frequency continuously, creating arbitrary custom waveforms, playing waveforms imported from files, and so forth.

In addition, the Analog Discovery 2 can treat the power-supply outputs as low-frequency function generators, so that one can have 700 mA function generators, rather than just 30 mA ones. To activate this feature, you need to use the "Device Manager" and select the configuration that includes 4 Wavegen channels. The new channels (3 and 4) correspond to the positive and negative power supply. They can be given only very low frequency waveforms (up to 1 Hz), because the power supplies are designed to change voltage only slowly. You also need to take a minimum of about 1 mA of current, in order for the power supply to provide the right voltage (at lower loads the voltage may be off by quite a bit). The power supplies are also fairly low resolution, with only a 10-bit DAC, for a step size of about 5 mV.

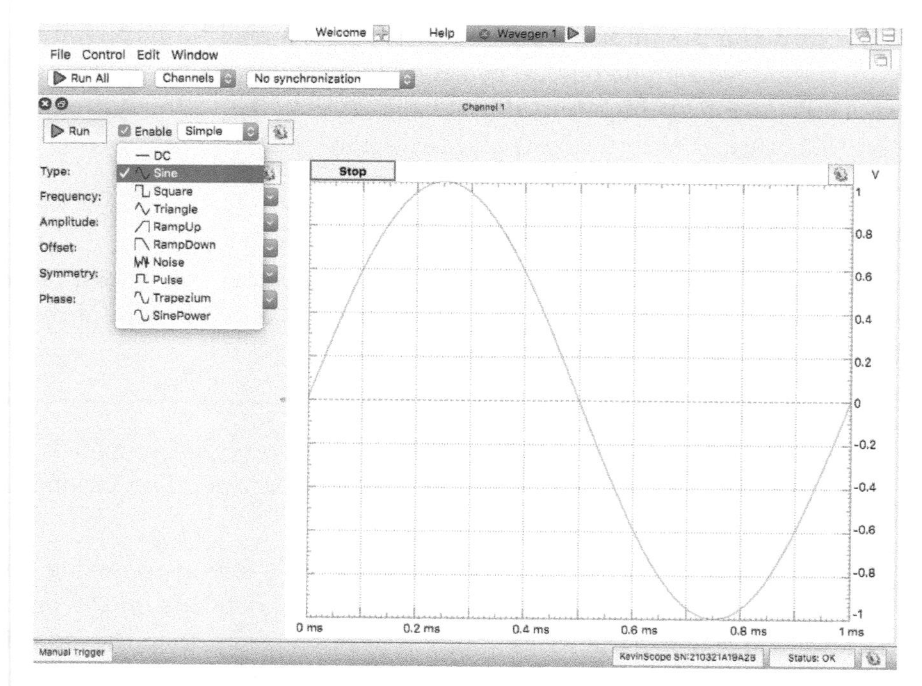

Figure 12.2: *This screenshot of the WaveForms 3 Wavegen window shows how to select a sine wave for Channel 1. To get controls for both channels, click on the "Channels" button and select "2". To start the function generator, either click on the "Run" button in this window or click on the green triangle in the "Wavegen1" tab at the top of the window (which allows turning the function generator on and off while looking at the oscilloscope window, for example).*

Exercise 12.1

Given a ±30 mA current limit for a function generator, what is the smallest resistor you can put across the function generator and still get a ±3 V signal? (Reminder: putting a component *across* the output of a function generator means that it connects the two output nodes of the generator.)

Exercise 12.2

Given a ±30 mA current limit for a function generator, what is the largest voltage you can specify while driving a 50 Ω load? If you have a 1 V DC offset, what is the largest amplitude AC signal you can specify?

Chapter 13

Debugging

Debugging is one of the most important skills in engineering spanning all engineering fields. Details of the debugging techniques will change from field to field, but the basics are universal.

One of the main reasons for teaching electronics to bioengineers is to practice debugging. Unlike fields like protein engineering, electronics design has fast enough turnaround time to allow several iterations of debugging in an ordinary lab session. But electronics also has enough real-world complexity that actual implementations do not match the simplified models we use for design (unlike, for example, most computer programming, where the models match the implementation).

Debugging generally consists of a repetition of several steps:

1. Set up an expectation about how the system should behave under specific conditions.
2. Observe the system under those conditions.
3. Note discrepancies between expected and observed behavior, and form hypotheses about causes for the differences.
4. Modify the system based on the hypothesized causes of discrepancies.

13.1 Expectation vs. observation

Building up clear expectations about how a system behaves and observing the behavior are key to debugging. There are several parts to this:

model It isn't really possible to debug a system unless you know in detail what it is supposed to do. Many students copy circuits from the web without understanding them—this is very risky, as the specific constraints of our design problems may be different from the solutions to even very similar problems one may find on the web.

Understanding a design means that you need a model that predicts its behavior under conditions that you can test. This model can be a simple mental model (for example, an amplifier with a gain of 50 will convert a $\pm 20\,\text{mV}$ input to a $\pm 1\,\text{V}$ output) or a more complicated computational model (like the ones used in Chapter 11 to give the gain of an RC filter as a function of frequency).

simulate or compute You will often need to use simulation or plots of theoretical behavior to confirm that your design should do what you want. For example, in Lab 6, students often have trouble with a passive band-pass filter having a lower gain than they expected. Using gnuplot with the functions in the `definitions.gnuplot` file (see Figure 7.4) to plot the gain of the filter lets students see if their design matches their mental models.

check assumptions of model It is important in building your model to check any assumptions of the model—for example, are the signal levels and frequencies in a range where the gain computation you are doing is applicable, or are you in a range where the output range would be limited?

> Because so many of our circuits involve voltage dividers, you should always check the voltage-divider assumption that no current is taken from the output of the voltage divider. If you do take current from something you've treated as a voltage divider, then you need to use a more complicated model of the circuit to figure out how it should behave.

To avoid this complexity, we often put a unity-gain buffer between the voltage divider and the node that needs current—the input of the unity-gain buffer takes essentially no current but can provide substantial current at the same voltage at its output (see Section 19.3 for explanation of unity-gain buffers).

In this course, we use single power-supply designs, but many designs on the web are based on the assumption of a dual power supply. When trying to adapt a design, think about the assumptions of the design: are your signals centered at the voltages that the design expects? What do you need to change in either the design or your signals to make them compatible?

If the output of an amplifier is stuck at one of the power rails, it is quite often because the designer forgot to take into account what center voltage the input and output of the amplifier are referenced to or neglected to compute the effect of the gain of the amplifier on a DC offset at the input.

> We often design as if our components were ideal mathematical objects, rather than real-world implementations that only approximate the ideal. It helps to be aware of some of the limitations of our components—for example, that op amps have limited output current and limited gain at high frequencies—and check that the limitations will not affect the expected behavior too much.

block diagram Because complicated systems are harder to understand than simple ones, you are often best off testing one subsystem at a time. A block diagram that breaks the complicated system into simpler subsystems is a great aid in debugging, especially if you can observe the signals at the connections between the blocks.

If you have explicit expectations for what the signals should look like at each connection, based on your mental (or computational) models of the subsystems, you can quickly localize problems.

For example, when debugging a multi-stage amplifier, measure the voltages at the outputs of each amplifier stage, so that the difference between expectation and observation can narrow down where the problem is. If the signal is as expected out of the first stage of an amplifier but not out of the second stage, then the problem is most likely in the second stage.

> For this debugging technique to work, you need explicit prior expectations for each of the signals, which is why I require that a block diagram have voltage information written for each signal.

build a little at a time Some students try to design an entire system on paper, build it, then power it up, and hope it works—it usually doesn't. A better approach is to build one block of the system at a time, and test each block separately before adding it to the system. By adding just one block at a time to a tested system, debugging is simplified—if something doesn't work, it is most likely in the block just added or the interactions between that block and the already tested parts.

measure and record Measuring the voltages, gain, corner frequencies, and other properties of signals and blocks of the system makes it easier to build and check mental models of how the system works. Careful recording of measurements makes it easier to see if behavior changes, perhaps as a result of adding a new block to the system.

correct your homework The homework, particularly the pre-lab assignments, are designed to help you build models, block diagrams, and schematics that can be used to debug your physical implementation. If there is a mistake in the homework, you need to correct your mental or computational model, or you have nothing to compare the physical implementation to, making debugging impossible. You should correct any errors you made in your homework, not just shrug your shoulders at lost points.

> The point of homework is to help you build robust mental models useful for debugging hardware.
>
> Doing homework just for points is a waste of everyone's time.

ask for help intelligently Debugging is a difficult skill to learn, and it is perfectly proper to ask for help when you get stuck. But don't let someone else do your debugging for you—you won't learn much that way.

Instead, when asking for help, start by explaining the difference between what you see and what you expect to see. Be prepared to explain why your expectations are what they are—our job as teachers is to help you build correct mental models, but we don't have access to the insides of your mind, so you need to tell us what you are thinking.

design or implementation? One of the most difficult things to figure out when a new system doesn't work as expected is whether the problem is with the design or with the implementation.

When you explain your expectations and the reasons for them to someone else (particularly a teacher or lab assistant), they can help you narrow down whether the design or the implementation is the problem.

If they find that your expectations don't make sense to them, they help you re-examine your design to uncover the problems in your model. The problems here may be minor misunderstandings or fundamental ones.

If your expectations make sense, but the hardware is not behaving as expected, they can help you locate the difference between what you intended to build and what you actually built.

When you are building a multi-block system, wire up your circuit one block of the block diagram at a time, checking the function of each block after you build it. It is often easiest to start at the sensor end of an amplifier chain, and work your way forward, because you may find an error in your assumptions about the sensor after you connect it to the first stage. Doing a redesign is less painful if you find errors in your assumptions early, before you have built a system based on the mistaken model.

13.2 Show me your schematic!

One of the most frequent things I say in the lab (so much so that it was put on the class T-shirts in 2017) is "Show me your schematic!" The schematic diagram is the detailed description of exactly what was to be built and thus provides the dividing line between the design and the implementation.

When you are explaining how you expect the system to behave, the schematic diagram provides the system description that you can point to in your explanations. You should be able to circle parts of the schematic to indicate where the blocks of your block diagram are—where is the first stage of the amplifier, where is V_{ref} generated, where is the output, where are you making measurements with your oscilloscope, ...?

When checking whether the design makes sense, we can compare the schematic to the block diagram, checking that the connections are the same, and that each block of the diagram is built with a reasonable circuit that should have the function. We can also check that the assumptions used to design the individual blocks are met when they are connected together.

One of the most common problems in electronic labs is incorrect wiring from a correct design. With a clear schematic, it is fairly easy to check that each node is correctly wired, especially if the schematic has pin numbers on it.

> Every connection to an integrated circuit on a schematic should be labeled with a pin number, so that the wiring can be easily checked.

Before asking a teacher or lab assistant for help, first check that the wiring matches the schematic.

One trick that helps in quickly spotting wiring errors is to note for each node how many components are connected together by the node, then check the corresponding place on the breadboard and count how many connections there are. Many of the problems that I see can be very quickly spotted this way—0 connections to an input instead of 1, no resistor connected to an op-amp output, when the schematic shows one between the negative input and the output, and so forth.

One of the most common wiring errors is forgetting to provide power and ground to each chip—check your power and ground connections first.

> The first question that many call center staff are trained to ask is "Is it plugged in?" Checking power connections back to the power source (Teensy board, USB cable, Analog Discovery 2, wall wart, etc.) yourself can save a lot of embarrassment.
>
> Also, check that the power is the right way around (reversing the power connections can damage integrated circuits and electrolytic capacitors).

On many data sheets, the positive power pin is labeled V_{dd} and the negative one V_{ss}, based on conventions from the days when nMOS was the most common integrated-circuit technology. The "d" comes from the *drain* of the nMOS transistors and the "s" from the *source*—see Chapter 32.

When you change your design during debugging (changing a resistor value to adjust gain, for example), update the schematic—it is very difficult to debug a design if it exists only in your head as a series of modifications to an earlier design.

13.3 Color code for wires

The most common errors in this course are wiring errors, with wires connected differently on the breadboard or soldered protoboard than was intended. Debugging wiring errors can be very difficult, especially if all the wires look alike and are in tangled messes that are hard to trace.

A little bit of planning before building the circuit can save an enormous amount of debugging time, both by reducing the number of wiring errors and by making the errors that do occur much easier to find and fix. One of the most powerful tools for reducing wiring errors is color coding the wires.

The idea is simple: on the schematic diagram, each node is assigned a color and that color is used for all wires that implement that node.

There are a few rules that make color coding more useful:

- Use black wires for your ground node, and don't use black wires for anything else.
- Use red wires for your positive power supply, and don't use red wires for anything else.
- All wires belonging to a node must be the same color.
- Wires for adjacent rows on a breadboard or PC board that are not part of the same node should be different colors.

If you follow these rules, then debugging becomes much faster—any place where wires of different colors connect must be an error. Any place where wires of the same color go into adjacent rows of a breadboard is likely to be an off-by-one wiring error (unless the two rows are the same node, which does happen in some of our designs).

The color coding of black for ground and red for positive power is highly standardized—you will encounter that convention in almost every electrical engineering or computer engineering lab course, as well as car batteries and other DC systems. Warning: the color codes for *house* wiring are different—black is used for the "hot" AC wire, white for neutral, and green for true ground.

To assign your wire colors, start with your schematic. Put pin numbers on the schematic so you know which nodes are adjacent on the breadboard or PC board. Pick a node, assign it a color that doesn't conflict with any of its neighbors, and color the entire node. If you start with the nodes that have the most connections (V_{ref} in many of our designs), you can usually color every node following the rules with only three or four colors, without having to backtrack and reassign colors.

Exercise 13.1

Color each node of the schematic in Figure 13.1 following the color-coding convention described in this section.

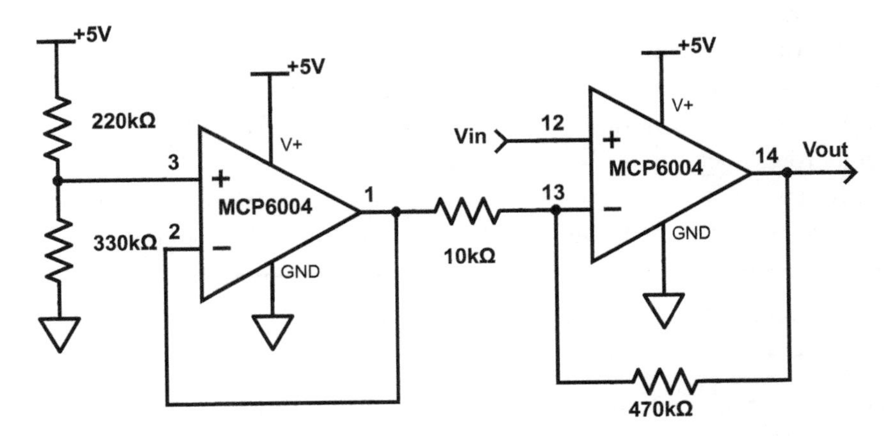

Figure 13.1: *This amplifier design will be explained in Chapter 19, but you can color-code the wires without understanding how the amplifier works. If you are unfamiliar with the symbols in this schematic, check Figure 7.3.*
Figure drawn with Digi-Key's Scheme-it [18].

13.4 Good breadboard practice

Section 8.6 describes how to use a breadboard correctly. This section will just repeat some of the highlights to keep in mind for debugging.

- Use short wires: they are easier to trace and debug, and are less likely to pick up stray signals (like 60 Hz interference). Make connections to the closest point that has the correct node. (Important on PC boards used for amplifiers in Lab 9 and Lab 13 also.)
- Strip the correct length so that there are no exposed bare wires (see Figures 8.10 and 8.6).
- Keep resistors close to breadboard or as a "flying resistor" between adjacent rows (see Figure 8.8).
- Use double-ended header pins (in groups of 2 or more) as test points for oscilloscope and voltmeter probes.
- Color code your wires: see Section 13.3.

Here are some quick debugging checks that will uncover about 80% of student errors on breadboards:

- Are power and ground connected to a power supply that is turned on?
- Are power and ground connected correctly to each chip? The red and black color coding really helps here.
- Are orientable components (such as diodes, phototransistors, and microphones) the right way around?
- Check each node on the schematic—are the number of components connected there the same as on the breadboard?
- Are the components the right ones (resistor values particularly)?
- Is there exposed bare wire that may be shorting?
- Are any of the wires making a poor connection (not inserted far enough or broken)? This test is very difficult to do visually, but the continuity test on the multimeter can help find missing connections that are visually there but not electrically connected.

13.5 Limitations of test equipment

Most debugging of hardware requires looking at the voltage or current waveforms of inputs, outputs, or intermediate nodes of the circuit. The test equipment may be an oscilloscope, a voltmeter, a logic analyzer (for digital signals), or other, more esoteric equipment.

The point of probing various nodes is to compare the observed signal to the expected signal—if you don't know what to expect, then observing the signal will not provide you with much information. Unexpected behavior is what we're looking for when debugging, but not all unexpected behavior helps—sometimes the unexpected behavior is the *result* of our looking at the signal.

Inserting test equipment into a circuit changes the circuit!

When we measure a voltage, whether with a voltmeter or an oscilloscope, we are connecting the test equipment in parallel with whatever we are measuring. The impedance of the test equipment is high but is not infinite, and so the modified branch of the circuit (original in parallel with test equipment) has a lower impedance than the original. This change in impedance can change the behavior of the circuit. It is frustrating, but fairly common, to have a circuit work when it is being debugged, but not when the test equipment is removed.

The input impedance of any piece of test equipment is an important parameter, and so is usually reported on the data sheet for the equipment. Section 15.7 discusses the input impedance of the Analog Discovery 2 in some detail.

To measure current, we usually introduce a small resistor (referred to as a *sense* resistor) in place of the wire we wish to measure the current in, and measure the voltage across the resistor. The added resistance is an obvious change to the circuit, but the meter or oscilloscope in parallel with the resistor may also have an effect—particularly at high frequencies, where the capacitance of the test equipment may result in a lower impedance than the sense resistor.

If you don't have an oscilloscope, you can do much of your testing with the PteroDAQ data-acquisition system, but you'll want to limit your power supply to 3.3 V, so as not to have signals that exceed the voltage range that the Teensy boards can handle.

In addition to the changes in the circuit due to the input impedance of the test equipment, the test equipment can inject noise into the circuit. For example, the analog input channels of the Teensy board are connected one at a time to the shared analog-to-digital converter in the microcontroller. There is a small capacitance coupling the control signal for this connection to the input pin. The result is a voltage divider merging the noise with the desired signal, as shown in Figure 13.2.

There are two ways to reduce the noise injection:

- Use a relatively small impedance for R_{circuit}. The manufacturer of the KL26, the microcontroller on the Teensy LC board, recommends keeping the resistance below 5 kΩ.
- Use an amplifier to separate the signal being measured from the analog input pin. The amplifier output has a very low impedance, making the noise injection into the measured signal small, and any noise injected into the amplifier output is not coupled back to the input. The Analog Discovery 2 incorporates such amplifiers before its analog-to-digital converters.

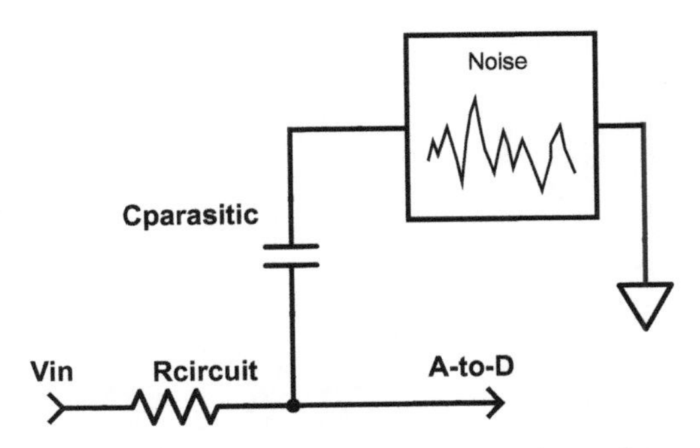

Figure 13.2: *This circuit shows how the analog channels of the Teensy board (and of most microcontrollers) can inject noise into the system being measured. If the Thévenin-equivalent impedance of the device being measured,* R_{circuit}, *is large, then the noise will not be reduced much by the voltage divider, but if* R_{circuit} *is small, the noise will not affect the measurement much.*

You can see the noise clearly by recording an open analog channel—one that has nothing connected to the corresponding pin. In fact, seeing a very noisy channel in a PteroDAQ recording is usually a sign that the corresponding input pin has not been correctly connected.

Figure drawn with Digi-Key's Scheme-it [18].

Another strange phenomenon that you can observe with PteroDAQ is memory in the analog-to-digital converter. Try connecting a function generator to pin A0, then recording pins A0, A1, A2, and A3 with PteroDAQ. The first channel should faithfully report the signal being presented, but the subsequent channels also report that signal, with gradually increasing noise.

The problem here is that the analog-to-digital converter takes a sample of an input voltage by connecting the input briefly to a *sample-and-hold* capacitor, and then measures the voltage on that plate of the capacitor. If the impedance driving the input is very large, then the sample-and-hold capacitor is neither charged nor discharged, but retains its previous value. The noise injection shown in Figure 13.2 gradually corrupts this value.

Chapter 14

Lab 3: Sampling and Aliasing

Bench equipment: function generator (Analog Discovery 2)
Student parts: breadboard, PteroDAQ, double-headed male headers, resistors, and capacitors

14.1 Design goal

This lab is more of a demo than a design lab. Your real goal is to write a clear tutorial explaining aliasing and to illustrate it with figures from the data you collect. Your audience for this tutorial is students coming into the course you are currently taking—create the explanation you would have liked to have two weeks ago. *Do not plagiarize your explanation from the textbook, Wikipedia, or any other source—write your own explanation of aliasing.*

The main point is to look at sampling frequency and how it affects the representation of signals stored on computers.

In addition, you will learn how to set up a function generator, record data and plot it, and design a high-pass filter to change the DC bias of an analog signal.

The filter design is not the main point of this lab, but I try to include some design elements in every lab. The original version of this lab was done using a function generator that made setting the DC offset difficult, so the filter was an essential part of the lab. With the ease of setting the offset on the Analog Discovery 2, the filter design is now more of a side project, though it is a reusable building block as similar filter designs will be useful in future labs. Only a short write-up of the filter design is needed, explaining the goal of the filter, showing the schematic, and computing the corner frequency achieved.

14.2 Pre-lab assignment

Remember from Section 7.1 that the pre-lab report is a draft of your final design report and should follow the guidelines for the final report.

In order to record data at different sampling frequencies, you first need to have a signal to sample that you can record and display nicely with PteroDAQ. You will want the signal on one of the analog inputs—say, pin A0 on a Teensy board.

The default setting of the function generator produces sine waves that are centered at 0 V, but the analog-to-digital converter on our microcontroller board for PteroDAQ requires signals between 0 V and 3.3 V. Any signals below 0 V or above 3.3 V would be *clipped*, so that negative voltages would appear as 0 V and too high voltages as 3.3 V.

You should never deliberately apply voltages outside the operating range of 0 V–3.3 V, but very small excursions outside that range (due to noise, for example) are not necessarily harmful. But applying voltages well outside the intended range can damage the microcontroller. The data sheet for the Kinetis

KL26 subfamily (used on the Teensy LC board) says that the *absolute maximum* range for voltages is from −0.3 V to 3.6 V (0.3 V above the positive power rail), but that negative voltages with current less than 3 mA will not cause damage—there is no such safety for voltages above 3.6 V, though [29, p. 6].

We will use two different approaches to getting the appropriate voltage range for the signal:

Setting the offset voltage of the function generator. If we set the offset to 1.6 V and the amplitude to 1.5 V, we would have a range from 0.1 V to 3.1 V, which stays comfortably inside the desired range. On block diagrams and schematics, we would annotate such a signal with the notation "1.6 V DC ±1.5 V AC", giving both the DC bias voltage and the AC amplitude of the signal.

Building a circuit that recenters the voltage into the correct range. Centering a signal means adjusting the DC offset to be a particular value—for example, if our analog-to-digital converter has a range of 0 V to 3.3 V, and we have a ±1 V AC signal, we want to add a 1.65 V DC offset, to make a signal that is 1.65 V DC ±1 V AC.

For many of our later designs, we will not have a function generator as an input, but the output of a sensor that has some DC bias different from what we want. So it is useful for us to have a building block in our design toolkit that can change the DC voltage of a signal without appreciably affecting the signal.

Much of this book is about reusable building blocks, so each lab should be examined by the student to find the parts that can be re-used in future designs.

To do the design, we need to break the problem into subproblems, labeling what we know about the inputs and outputs, as shown in the block diagram in Figure 14.1.

For the first block, we will use a capacitor to prevent any direct current flow (see Section 10.1 for more information about capacitors), and for the second block we'll use a pair of resistors to set the DC voltage to midway between the power rails. The resulting circuit is shown in Figure 14.2.

The design process illustrated here will be used in almost all design assignments: divide the design up into subparts, draw a block diagram that shows the relationship between the subparts (with information about what each block does and what is known about the interfaces between the blocks), then refine the blocks into schematics that include the wires and components.

Remember from Section 5.1.2 that the voltage divider with two equal resistors is equivalent to having half the voltage and half the resistance, so we can model the two-470 kΩ-resistor voltage dividers as a single resistor of 235 kΩ to 1.65 V.

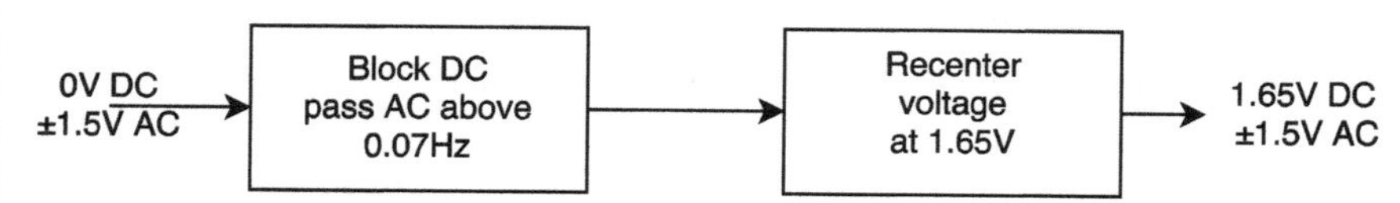

Figure 14.1: *This block diagram shows how we can go about designing a circuit by breaking the problem into subproblems. The input signal is centered at 0 V, but swings ±1.5 V, and we want the output to have the same voltage swing, but centered at 1.65 V. The first block has to prevent any DC current from flowing through it but allow AC (time-varying "alternating current") to pass through. The second block has to recenter the signal at 1.65 V, but with only 3.3 and 0 V available as power supplies.*

We'll expand this block diagram into a more detailed schematic in Figure 14.2—also look at Section 11.6 for more details on how the filter works.

Figure drawn with draw.io [21].

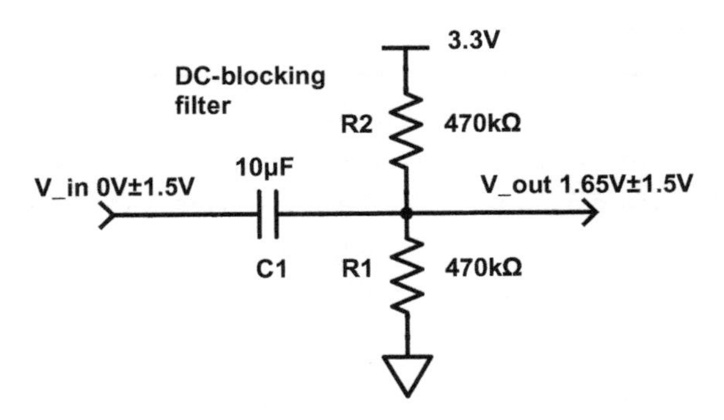

Figure 14.2: *This schematic expands the block diagram of Figure 14.1. The capacitor prevents any direct current flow, while allowing time-varying current to flow. The pair of equal resistors centers the output voltage halfway between the power rails (as long as the output takes no current).*

The design shows an unpolarized capacitor (such as a ceramic capacitor). If a polarized (electrolytic) capacitor is used, then the positive side has to be toward the output, as that side has the higher DC voltage, and electrolytic capacitors can explode if hooked up backwards (see Section 10.1.2).

Figure drawn with Digi-Key's Scheme-it [18].

Because we purchase cheap resistor assortments, in some years there have been no 470 kΩ resistors in the parts kits! There are several work-arounds for this situation:

- Choose a resistor size that isn't too far off, without changing the capacitor, to get a different corner frequency.
- Choose a different resistor, but change the capacitor to keep the RC time constant about the same.
- Construct approximately 470 kΩ resistances by putting other resistors in series or parallel. We will almost never use this solution in this course—I want students to design around available parts, rather than kluging together multiple parts to fit a pre-existing design.

Each of these solutions could be appropriate in different circumstances. For the current lab, the total resistance does not matter much—anything from 1 kΩ to 1 MΩ would probably work OK—but we want a low corner frequency for our high-pass filter (between 0.05 Hz and 0.5 Hz), so changing both the resistor and capacitor to keep the RC constant about the same is probably best.

This DC-blocking filter of Figure 14.2 passes even fairly slowly changing waveforms, but that means that filter takes a long time to settle down after any change to the characteristics of the input. Trade-offs like that are an inherent part of any design process—a lot of design consists of deciding which trade-offs are most suitable for the particular product being designed.

We'll see circuits like this one throughout the course, as voltage dividers (Section 5.1), RC filters (Chapter 11), and Thévenin equivalents (see Section 5.1.2).

Pre-lab 3.1

Analyze the DC-blocking filter of Figure 14.2. What is the RC time constant? What is the corner frequency? Use the Thévenin equivalent for the voltage divider that sets the center voltage.

Pre-lab 3.2

Plot the magnitude of the gain of the filter in Figure 14.2 ($V_{\text{out}}/V_{\text{in}}$) as a function of frequency on a log-log plot.

Pre-lab 3.3

Redesign the DC-blocking filter to use different resistor or capacitor sizes, making sure that your lab kit includes the sizes in your design. You should have roughly the same corner frequency as in Pre-lab 3.1, but the specification is not tight—because your ceramic capacitors have a tolerance of about ±25%, you are not going to get an exact match anyway.
Draw the schematic. What is the RC time constant and corner frequency of your new design?

Pre-lab 3.4

Draw a schematic diagram of your setup for the lab, including the function generator, the DC-blocking filter, and a custom symbol for the PteroDAQ data-acquisition system.

The pre-lab questions are intended as warm-up exercises to give you an idea of the sort of analysis and design you should be doing in your design report. You do not need to include these warm-ups in your final report, but you should include *full* explanation for your design.

14.3 Using function generator with offset

You will use a function generator to generate waveforms of known frequency and look at them using the PteroDAQ data-acquisition system.

The instructions in the book assume that you will be using the "Wavegen" tool built into the Analog Discovery 2 USB oscilloscope, but a different function generator can also be used. For example, for 2013 through 2017, we used Agilent 33120A bench-top arbitrary-waveform generators [1], which are much more expensive instruments, though no better for our purposes, than the ones in the Analog Discovery 2.

You won't need most of the capability of the function generators (which are really overkill for a beginning electronics course)—you just need to learn how to generate a signal of known shape, frequency, amplitude, and voltage offset.

If you select the Wavegen tool from the Welcome tab in WaveForms 3, it defaults to the "Simple" mode, where you can select the shape of the waveform, the frequency, the amplitude, and the offset.

I recommend using a "RampUp" signal with the "Symmetry" field set to 50%, which ramps up for half the period then remains constant at the highest value for the rest of the period. This signal makes it easy to tell if the waveform has been reversed in the time domain (it will alternate between ramping down and staying at the highest value), negated in amplitude (it will alternate between ramping down and staying at the lowest value), or both (it will alternate between ramping up and staying at the lowest value).

After you select the amplitude and offset in "Simple" mode, confirm that the waveform and voltage range are correct on the plot provided by the Wavegen tool—the sine wave should remain between 0 V and 3.3 V if you are feeding the output to a Teensy analog input. If you are testing your DC-blocking filter, you will want to change the offset, to test that the filter is properly changing the DC voltage.

Clicking the "Run" button starts running the function generator, providing an output voltage on the yellow wire labeled W1 on the Analog Discovery 2.

> Before clicking "Run" make sure that the voltage range displayed in the Wavegen window is the voltage range you want.

Once you have the function generator running, plug the Teensy LC board into the end of the breadboard, so that there are two spare holes on the left side and three on the right and use a USB cable to connect the Teensy board to your laptop, as shown in Figure 14.3. Connect the ground pin on the Teensy to the blue bus on the right side with a black 22-gauge wire. Use double-headed male headers to connect the Wavegen output of the Analog Discovery 2 to an analog input and ground on the Teensy Board. Two wires are needed for this connection: the yellow wire labeled W1 for the function generator and a black wire for ground.

> In PteroDAQ, set the timed trigger to sample at 200 Hz. Add **two** channels, both looking at the same pin, and use the *downsampling* option for the second channel to downsample by a factor of 20, so that it is sampling at $200\,\mathrm{Hz}/20 = 10\,\mathrm{Hz}$.

The downsampling option is a rarely used feature, so it is hidden under the >> button on the channel.

Set the function generator for a very low frequency (say 1 Hz) and observe the waveforms with PteroDAQ. Make notes in your lab notebook of what you observe. Make sure your notes include the frequency of the sine wave and the sampling frequency of each channel, so that you can reproduce the effect later. Save the recording to files, so that you can plot them in your report.

Use the notes field of PteroDAQ to record who worked on the lab, the frequency of the signal, and any other *metadata* that you might need when plotting the data. Get in the habit of always keeping notes of experiments along with the raw data. You'll find that the design reports are much easier to write if you keep careful notes—reconstructing your experiments and your thoughts is amazingly difficult without notes, as what you later think overwrites what you were thinking at the time.

Figure 14.3: *Breadboard with Teensy LC at the top end and function generator of Analog Discovery 2 connected to pin A0. Note the use of double-headed male headers to make connections to the breadboard. I find that the double-headed male headers work best if I leave them in groups of 2 or 3 pins—single pins tend to get wobbly.*
Don't forget to connect the blue bus to the ground pin on the Teensy board with a black wire (from Row 2, Column J in this picture).

Record the waveforms with the function generator set to asymmetric upward-sloping ramps at frequencies of 1 Hz, 5 Hz, 9 Hz, 10 Hz, 11 Hz, 15 Hz, 19 Hz, 20 Hz, and 21 Hz, all with the sampling frequency of 200 Hz and downsampled by a factor of 20 to 10 Hz.

Don't mix together different frequencies in the same file—saving just a few seconds of input with a constant frequency, with notes in each file saying what the input is, makes it much easier to analyze and plot the data. Naming your files so that you can keep track of which one is which is also very important.

Find interesting behavior at other frequencies or with other waveforms. Describe the patterns you see and record them for inclusion in your report. Again, save any data you wish to talk about in files.

14.4 Wiring high-pass filter

Wire up your high-pass filter from Pre-lab 3.3, using the microcontroller board's 0 V and 3.3 V lines for power. We use a strict color-coding convention in this class: the 0 V wire is always black, and the positive power wire (3.3 V for the Teensy) is always red. Those colors of wire are not used for any other signals. So you need at least three colors of wire for this lab: red, black, and another color for the output of the filter.

Connect the function generator to the input of your DC-blocking filter, starting out with the same settings as used in Section 14.3. Use PteroDAQ to record a waveform as before.

Check that the signal from your circuit remains in the desired range (low peak between 0 V and 0.5 V, high peak between 2.8 V and 3.3 V) at frequencies above the corner frequency: 1 Hz, 3 Hz, 10 Hz, and 30 Hz.

If the waveform at the output of the filter is not centered at 1.65 V, then debug your design and wiring.

Once the filter seems to be working, change the DC offset a small amount in the Wavegen tool, and verify that the output of the filter does not change. If everything is working correctly, you should be able to use any offset voltage within the range of the function generator and still record the correct waveform.

Be sure to include notes in your PteroDAQ files that indicate what the Wavegen settings are and what filter you are using, as looking at the signals themselves will not tell you this information—the whole point of a DC-blocking filter is to throw away the DC offset of the input!

The very large RC-time constant for your high-pass filter means that there is a slow power-on transition when you first apply power to the filter, as the capacitor slowly charges up. Wait for that to settle before you start trying to adjust the design.

14.5 Using gnuplot

In this class, we'll use gnuplot to plot data and to fit parameters of models to the data. There are many other fine tools for plotting and for data modeling, but gnuplot is the one I'm most comfortable with, and it provides a reasonable balance between ease of use, power, and formats for the output.

Gnuplot can be used in two ways: interactively and with scripts written ahead of time. I generally use a blend of the two, where I write and edit scripts that I first debug interactively, then use the script non-interactively to create PDF files for inclusion in reports (like the approximately 86 gnuplot-generated figures in this book).

Figure 14.4 provides a simple script for viewing a data file recorded for this lab. It assumes that the first channel that you added in PteroDAQ is the one sampled at 200 Hz, and that the second one is the one downsampled by a factor of 20 to get a sampling frequency of 10 Hz.

The data file may contain comment lines (beginning with "#") that will be ignored by gnuplot—PteroDAQ includes a lot of metadata as comments at the beginning of each file. Blank lines (but not comment lines) will cause a gap in the plotted curve.

The `title`, `xlabel`, and `ylabel` commands set up the labeling of the graph—the font options are used here to make the labels larger than the default, for greater readability. Increasing the size of the title and axis labels requires playing with the margins, to make sure that the created PDF file does not clip off some of the text.

> Make sure that quotes that you use are simple ASCII quotes, not "smart quotes" that are different on the left and right, as gnuplot only understands the simple quotes. You may have to change a preference setting on your text editor to turn off smart quotes.

The command `set key top center` tells gnuplot where to put the legend. The command `set style data lines` tells gnuplot to plot the data with lines from each point to the next. Without that command, gnuplot defaults to doing scatter diagrams for data (`set style data points`) and curves for formulas, which is useful for showing data and the model fit to the data on the same plot.

Setting the xrange limits the points plotted to those between 1 and 2 seconds. Using `*` would tell gnuplot to pick a value that includes the bottom (before the :) or top (after the :) of the observed range.

The `using 1:2` modifier tells gnuplot to use the first column of data for the x-axis and the second column for the y-axis, which is the default it would use if not instructed. The titles for the plots are used to make a key for the different lines, and `linecolor "red"` and `linecolor "blue"` choose red and blue for

```
set title "1 second of 8 Hz 70% ramp wave at different sampling rates" font "Helvetica,14"
set xlabel "time [s]" font "Helvetica,14"
set ylabel "voltage [V]" font "Helvetica,14"
set margins 9,4,4,4

set key top center
set style data lines
set xrange [1.:2.]
set yrange [0:3.9]

plot "8Hz-at-200Hz-70.txt" using 1:2 title "sampled at 200Hz" linecolor "red" , \
     "8Hz-at-200Hz-70.txt" using 1:3 title "sampled at 10Hz" linecolor "blue"
```

Figure 14.4: *This simple gnuplot script reads data from the file "8Hz-at-200Hz-70.txt" and plots the data as a continuous curve. The data file is assumed to have three columns of data: first a time in seconds, then two voltages in volts.*

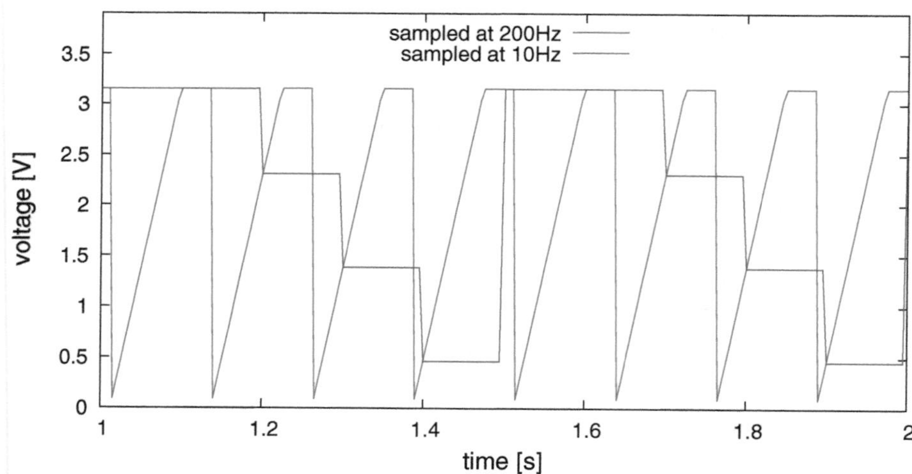

Figure 14.5: *Example plot of sampling an 8 Hz 3.2 V peak-to-peak upward ramp, with "symmetry" set at 70% at 200 Hz and 10 Hz. The 200 Hz sampling catches the shape of the waveform, but the 10 Hz sampling makes the signal look like a −2 Hz signal, with 2 peaks every second, because of aliasing.*

The negative sign on the frequency is shown by the alternation of a constant portion at the highest voltage and a downward-sloping ramp—the time-reversal of the input, alternation of a constant portion at the highest voltage and an upward-sloping ramp.

This waveform was generated using DC offset by the Analog Discovery 2 Wavegen tool and plotted with the script of Figure 14.4.

Data collected with PteroDAQ using a Teensy LC board [46]. Figure drawn with gnuplot [33].

the two lines. The default colors for the plots depend on the version of gnuplot and the terminal, so it is best to be explicit.

I specified red and blue as the two colors for the plot, as they are easily distinguished even by (red-green) colorblind people. Avoiding red-green color distinctions is important, because about 7% of the male population in the U.S. is red-green colorblind [116].

To make PDF files for inclusion in your report, you can use the "qt" or "wx" window and save the images as PDF files with a button click, or use the other methods described in Section 7.10.5.

If you have the appropriate data, you should get a plot like Figure 14.5

14.6 Demo and write-up

For the demo, show the lab assistant some of the interesting waveforms on PteroDAQ, to make sure you are recording useful information.

The point of the write-up is to present a brief tutorial for someone who doesn't yet understand aliasing and to illustrate the important points with the waveforms you gathered. Remember that your audience for lab reports is not your professor or teaching assistant, but students who are a few weeks behind you in the class (see Section 7.2).

> You should present a quantitative view of aliasing—something that allows the reader to *predict* the apparent frequency from the actual frequency and the sampling frequency. I want to see a general rule that says what apparent frequency to expect for any input frequency and sampling frequency. Vague assertions about distortions or lower frequency are not sufficient.

You may find the *floor function*, $\lfloor x \rfloor$, useful for writing your quantitative function. The floor $\lfloor x \rfloor$ is defined as the largest integer n such that $n \leq x$.

Superimposing the higher sampling frequency plot from one channel and the lower sampling frequency plot from the other channel on the same graph (as in Figure 14.5) helps to explain what is going on. You may need to specify a small xrange (only a few seconds) to show what is going on clearly.

Your report should have schematics for your entire test setup. I recommend using Scheme-it [18] to draw the circuit, exporting as a PDF file for inclusion in the document. Other schematic drawing tools are acceptable, as long as they support the American symbol for a resistor, and not just the European one.

> Warning: the word "frequency" is not a synonym for "signal". A frequency is a number with units of Hz—it is not a time-varying voltage.

Chapter 15

Oscilloscopes

An oscilloscope [144] is a device for visualizing a time-varying voltage, like the data acquisition systems described in Chapter 6, but with a greater emphasis on visualizing the signals as they occur, rather than saving them for later analysis. Oscilloscopes are optimized for displaying high-frequency signals of short duration or repetitive signals, while data acquisition systems are optimized for lower frequency signals and longer durations. There is, however, considerable overlap between the two types of systems, and for many of the applications in this book, either can be used.

There are two main types of oscilloscopes, analog and digital, as explained in Sections 15.1 and 15.2.

15.1 Analog oscilloscopes

Analog oscilloscopes are an older design—they use a cathode-ray tube, which shoots an electron beam at a phosphor-coated screen to make a bright dot. The electron beam is deflected horizontally by a sawtooth waveform, so that the spot moves repeatedly across the screen from left to right at a constant speed. The beam is deflected vertically by the input voltage, so that what appears on the screen is the voltage as a function of time. An analog oscilloscope can use very simple circuits (when I was a kid, we built one from a kit that used a handful of vacuum tubes), so can be made fairly cheaply, but the cathode-ray tubes are bulky, power-hungry, and delicate, so analog oscilloscopes are disappearing from labs.

Although analog oscilloscopes have much simpler interfaces than digital scopes, their fragility and inability to record or measure waveforms exactly has lead to their replacement by digital scopes in most applications.

Analog scopes are easier to learn to use and would be sufficient for most of this course, though digital scopes offer considerably more capability if you can figure out the controls. In the first few offerings of the course, I wavered between teaching mainly the use of analog scopes and mainly the use of digital scopes. The extra capabilities of the digital scopes and their ubiquity in modern labs has led me to teaching using only digital oscilloscopes.

15.2 Digital oscilloscopes

Digital oscilloscopes record the input voltage for a chunk of time, then display it on a screen. The recorded data can be manipulated in various ways that are not feasible with the analog scope, and the data can be saved on a computer for further analysis and display.

When using a digital oscilloscope, you want to export the *data* from the scope, not just an *image* of the display screen. It is easy to create a plot from data, but it is much harder to recreate the data from a picture—a lot of information is lost.

The most popular dedicated oscilloscopes for home use are currently the Rigol oscilloscopes—the Rigol DS1054Z is a 4-channel 50 MHz oscilloscope with 1 Gsample/s (shared among the four channels) for under $400. The new Keysight EDUX1002A oscilloscopes are also quite good 2-channel 50 MHz scopes for about $450, with a somewhat better user interface than the Rigol scopes at a slightly higher price.

There are two expensive parts to a digital oscilloscope: the data capture system (analog electronics and analog-to-digital converter) and the user interface (display and knobs). Because almost all engineers now have laptops and similar devices with very high quality displays, you can save a lot of money by making graphical user interfaces on a laptop or tablet computer and building digital scopes with just the front end, connected to the laptop with a USB connection, much like the way we use the Teensy boards for the PteroDAQ data-acquisition system.

Most of these USB scopes are currently less sophisticated than dedicated oscilloscopes, but also considerably cheaper for similar levels of capability. I have three USB oscilloscopes: a BitScope Pocket Analyzer (BS10 [10]), an Analog Discovery 2 by Digilent, and an OpenScope MZ by Digilent

The BitScope BS10 costs about $245 and is not very featureful (fairly low resolution in both time and voltage, and a rather awkward user interface). I bought it several years ago, because it was one of the first USB oscilloscopes to run under Mac OS X.

The Analog Discovery 2 costs about $399 (but only $279 with academic discount), and is a much better instrument—it combines a good two-channel digital storage scope, two-channel function generator, and regulated dual power supply in a single package, and has an easy-to-use, fairly intuitive graphical user interface. The function generator and oscilloscope are well-integrated—the impedance analyzer tool, for example, makes Lab 8 and Lab 12 much easier.

The 14-bit, 100 Msample/sec resolution for the oscilloscope is very good for such a low-cost product, as is the 14-bit, 100 Msample/sec resolution for the function generator. Measuring low voltages is inherently inaccurate, as you are limited by the resolution of the analog-to-digital converter. The oscilloscope channels have two gain settings: at low gain the step size (based on information from the hardware reference manual [3]) is

$$1\,\mathrm{V}/(20/1040\,3.9/2.2)/8192 = 3.581\,\mathrm{mV/step},$$

and at high gain the step size is

$$1\,\mathrm{V}/(220/1040\,3.9/2.2)/8192 = 325.5\,\mu\mathrm{V/step}.$$

The low-gain setting is used if the oscilloscope channel is set to $> 500\,\mathrm{mV}$/division and the high-gain setting is used if the channel is set to $\leq 500\,\mathrm{mV}$/division. Averaging many measurements can improve the precision.

The triggering for the oscilloscope channels on the Analog Discovery 2 is very stable, with much less jitter than I've seen in other low-cost scopes. The FFT analysis of the function generator output shows very low distortion.

The power supplies are programmable up to 5 V and −5 V, but are limited to 250 mW each when powered from the USB cord, or 700 mA each when powered with an external 5 V power supply. I have found that powering from the USB cable is unreliable, and so I nearly always use an external "wall-wart" power supply with the Analog Discovery 2.

The Analog Discovery 2 has a software development kit (SDK) that lets it be controlled from C or Python programs, making it extensible to much more sophisticated uses than what is built into the provided software. The documentation of the hardware and application programmer's interface is excellent, but there does not seem to be a good user's manual for the WaveForms 3 software that is the main user interface—the online documentation is only adequate [20].

Digilent has introduced a lower-cost instrument, the OpenScope MZ (priced at $89), but it has only hobbyist-level specifications—the Analog Discovery 2 is a better buy, even at the higher price. I purchased an OpenScope MZ to see what it is good for—the main attraction seems to be that it communicates with the host computer wirelessly, so that with a battery power supply it can be electrically isolated from the computer or other equipment. (It can also be used from a phone app, and not just a laptop.) The OpenScope MZ is probably adequate for the labs in this course, but barely.

Overall, the Analog Discovery 2 looks like the best investment that a student could make in a "bench" instrument.

15.3 Differential channels

Most oscilloscopes have multiple channels, so that more than one signal can be viewed at a time. The channels usually share a common ground connection, rather than allowing arbitrary differential voltage measurements. It is usual to hook up the ground of the oscilloscope leads to the ground node in the circuit being examined, but that is not always required. Whatever node the ground wire of an oscilloscope probe is connected to, make sure that *all* the scope grounds connect to the *same* node.

> Some oscilloscopes, including the Analog Discovery 2 without the BNC adapter board, have differential inputs. *Differential measurement* allows you to measure the difference between two nodes, neither of which needs to be ground. With differential measurement, you always need to connect *both* wires to your circuit for each channel you use.

The signal displayed on an oscilloscope with differential input is just the difference in voltage between the signal wires, but the individual wires are allowed only a limited range of voltages relative to the oscilloscope ground. Because of this range limitation, we nearly always need to connect a third wire (a ground wire) between the oscilloscope and the circuit we are testing, even when we are using a differential input channel.

15.4 DC and AC coupling

Traditionally, oscilloscopes have a display that is 10 divisions wide and 8 divisions high. The oscilloscopes have controls for setting the scaling of the inputs (vertical controls) and of the time base (horizontal controls), in terms of how many volts or seconds there are per division. The inputs usually also have a choice of *DC-coupled* or *AC-coupled* input.

DC coupling [121] connects the input to the display via an amplifier that can amplify DC signals. If you have a 3 V DC input and have set the display scale to 1 V/division, the output trace on the display will be 3 divisions higher than a 0 V signal would be.

AC coupling (also called *capacitive coupling* [112]) connects the input through a *DC-blocking capacitor*, like the one used in Lab 3, to allow only varying signals to appear on the display—all DC signals will appear at the same vertical position on the screen. Of course, this removal of the DC offset means that you can't see very slowly changing signals—they look the same as constant signals.

The designers of AC coupling for an oscilloscope have to trade off how quickly the DC bias is removed when first connecting to a signal versus how slow a signal can be seen. The specification is usually given as a high-pass filter frequency (see Chapter 11). For example, the digital oscilloscopes in the TDS3000 series have a lower frequency limit of 7 Hz when AC-coupled with a 1× probe. When the probe is switched to 10×, the lower frequency limit is 0.7 Hz (see Section 15.7 for more information about oscilloscope probes).

The Analog Discovery 2 has only DC coupling, unless you purchase the optional BNC adapter board and a couple of oscilloscope probes, which adds about $40 to the price. The BNC adapter board allows AC or DC coupling, but switching between them requires moving a shorting jumper on the board, so is not very convenient. The AC coupling of the BNC adapter board has a corner frequency of about 0.15 Hz for a 10× probe and about 1.5 Hz for a 1× probe.

The main advantage of the BNC adapter board is the ability to use standard 10× oscilloscope probes, which reduce the load on the circuit (from 1 MΩ||24 pF with the wires to about 10 MΩ||15 pF). The 10× probe should also increase the bandwidth allowing higher frequencies to be displayed without attenuation, but this is difficult to measure with just the Analog Discovery 2, as the function generator drops to half power (−3 dB) at about 9 MHz, which is lower than the oscilloscope bandwidth.

The main disadvantage of the BNC adapter board is that both channels become referenced to ground, rather than having full differential inputs as on the wired inputs without the adapter board.

The way the Analog Discovery 2 compensates for having only DC coupling is to *add* a DC offset to the input signal before amplifying and digitizing it.

> The oscilloscope offset has to be the *negative* of the DC offset that the input signal has.

Using the offset allows viewing a small AC signal with a large DC bias, but if the DC bias changes, the offset setting on the oscilloscope needs to be manually changed—unlike with AC coupling, where the circuitry automatically adjusts for changes in DC bias after a few seconds.

15.5 Triggering an oscilloscope

Both analog and digital scopes have *trigger* conditions, which determine when a sweep across the scope display starts. These trigger conditions can be based on any of the inputs to the scope (there is often an additional input to the scope available purely for triggering). On analog scopes, the trigger is usually a voltage level and whether the input signal is rising or falling as it crosses that voltage level. Digital scopes may have the ability to do more complicated triggering logic. Because digital scopes have memory, they can allow you to see what happened before the trigger, as well as afterward.

The triggers can be done once (single triggering), producing one sweep across the screen. On digital scopes, this gives you a record of a one-time event that can be carefully analyzed, but on an analog scope, the fluorescence quickly fades, unless the trace is captured photographically. The triggers can also be done in *normal* mode, where each occurrence of the trigger starts a new trace. If you have a repetitive input signal, this allows the analog display to show the same portion of the signal repeatedly, giving a stable image (like looking at a rotating object with a stroboscope). Finally, there is usually an *auto* mode that starts a new trace after a while even if there is no trigger signal, which allows you to see the input without having to set proper trigger conditions. Generally, you use the auto mode initially to look at the signal and choose appropriate trigger conditions.

On the Analog Discovery 2, using the WaveForms 3 software, the triggering modes are "auto", "normal", and "none", with the last mode starting a new trace as soon as the previous one has been recorded. Single triggering is possible in all three modes. The trigger level can be set by typing a voltage level, selecting from a menu, or using the mouse on the right-hand side of the oscilloscope plot.

> The trigger point is usually displayed in the center of the oscilloscope screen for WaveForms 3, but the "Time position" control allows capturing the 8192-sample waveform with an arbitrary delay after the trigger (centered at the specified delay), and up to 4096 samples before the trigger (using a negative time for the position).

When a repetitive signal is being captured, the WaveForms 3 software allows averaging the different traces, which can improve the signal-to-noise ratio enormously if the signal is identical in each trace and the noise varies. (As a rough rule of thumb, the signal-to-noise ratio improves as the square-root of the number of traces averaged.)

For more information about triggering, look at Hobby Projects' online tutorial [43].

15.6 Autoset

Many digital oscilloscopes have an "autoset" button that attempts to choose automatically the vertical scaling and offset, the horizontal time base, and the triggering level to produce a meaningful display.

> The "autoset" button on an oscilloscope rarely results in good choices for the low-frequency signals that we are interested in. Instead it often chooses to zoom in on high-frequency noise. You nearly always need to correct *all* the settings after pressing "autoset", so only use it when you aren't getting any trace on the screen and can't see why not.

Although the algorithms used for choosing the automatic settings are proprietary, they generally follow the same outline:

- Look at the signal for a while to determine the maximum and minimum voltage.
- Pick a vertical scaling and offset to make the observed max and min fill a substantial part of the display without being clipped. Generally round numbers are chosen, though not always—you may need to correct them.
- Set the triggering level half way between the maximum and minimum voltage. On the Analog Discovery 2 with WaveForms 3, the trigger level is not rounded, and autoset always chooses a rising-edge trigger, even when a falling-edge trigger would give a more consistent signal.
- Try to determine the frequency of the signal and pick a time base that shows 2–4 periods of the fundamental. Generally, round numbers are chosen.

The autoset button for WaveForms 3 is hidden in the extended triggering options (click the down-arrow at the right-hand end of the triggering control bar).

Although autoset is often useful for getting a quick look at a signal, the choices it makes are not always the ones you want to use. You may want to zoom in to a particular part of the waveform by using less time per division or trigger on something other than the halfway point of the voltage swing. You may want to zoom out to see the overall envelope of the waveform, rather than the individual periods.

> Learn to adjust the controls of the oscilloscope to get the most informative display you can, rather than relying on autoset to guess what you want.

15.7 Oscilloscope input impedance and probes

The oscilloscope inputs typically have a fairly high impedance, so that they can measure voltages without changing the voltages being measured much. But for bioelectronics, where we often have very small currents available, the impedance of the oscilloscope is often still too low, necessitating amplification before we can use the oscilloscope.

What is the input impedance of an oscilloscope?

A typical oscilloscope has an input impedance that is equivalent to a 25 pF capacitor in parallel with a 1 MΩ resistor between the input node and the reference node that define the voltage measured by the oscilloscope. The cable that connects the scope to the circuit being tested usually adds more capacitance.

To reduce the effect on the circuit being tested, oscilloscopes are often used with a probe that can be switched between being a 1× probe and a 10× probe, connected to the oscilloscope with coaxial cable and a BNC connector. (Oscilloscopes designed for very fast signals often use higher-bandwidth SMA connectors instead, as the BNC connector bandwidth is about 3 GHz, and the SMA connector bandwidth is about 18 GHz.)

The 10× probe is a voltage divider (see Figure 15.1) that reduces the signal seen by the oscilloscope to one-tenth the signal at the tip of the probe, but also increases the impedance of the probe tip by a factor of 10, reducing the current diverted from the tested circuit.

> You usually have to tell the oscilloscope whether your probe is switched to 1× (shorting out the top half of the voltage divider) or 10×, as the standard BNC connector has no way of conveying this information to the scope, though some expensive oscilloscopes use custom connectors for their probes to tell the scope about the probe settings.

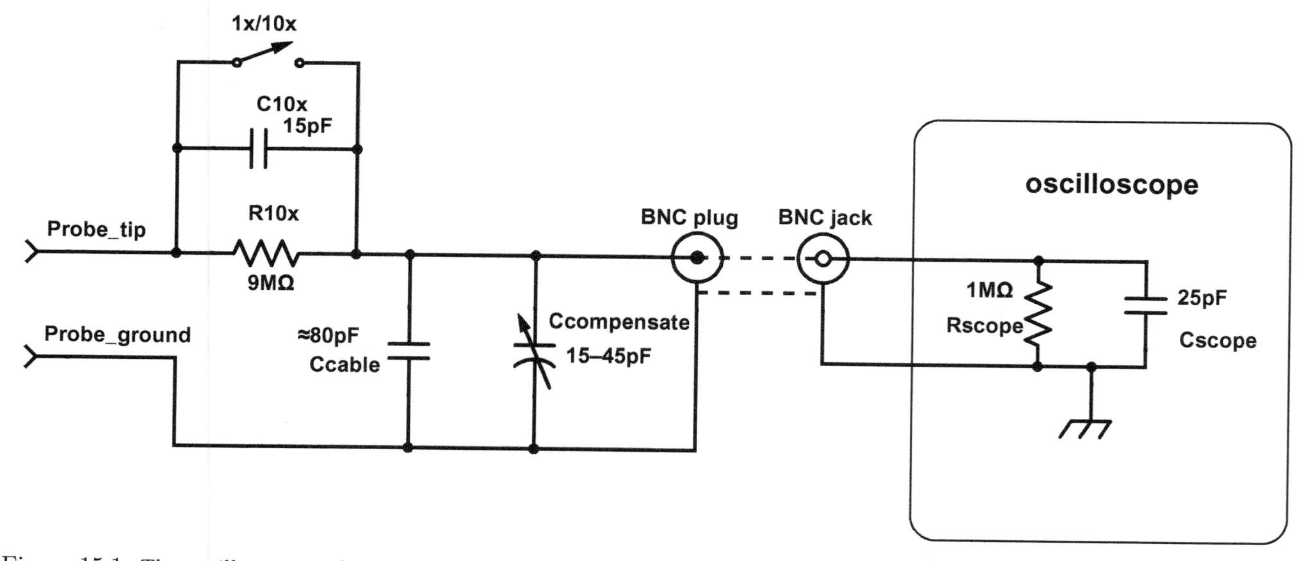

Figure 15.1: *The oscilloscope probe in the 10× setting serves as a voltage divider, dividing both DC and AC voltages by a factor of 10, but presenting a 10 times higher impedance at the probe tip than would be present with the switch in the 1× setting (shorting out R_{10x} and C_{10x}). The compensation capacitor is used to adjust the capacitance of the lower subcircuit of the voltage divider, so that it is 9 times the C_{10x} capacitance. Exercises 15.3–15.5 derive that factor of nine (don't just assume it for Exercise 15.5). Figure drawn with Digi-Key's Scheme-it [18].*

Without the BNC adapter board, the wiring of the Analog Discovery 2 adds about 5 pF in parallel with the scope input impedance. With the adapter board, the input impedance of a scope probe depends on what probe is connected and the setting of the probe to 1× or 10×. With the 1× setting, the probe just adds some capacitance in parallel with the oscilloscope input, resulting in an input impedance around $1\,\text{M}\Omega \| 100\,\text{pF}$. With the 10× setting, the probe acts as a voltage divider providing an input impedance around $10\,\text{M}\Omega \| 15\,\text{pF}$, but reducing the voltage that the oscilloscope sees by a factor of 10.

Exercise 15.1

Write the formula for the complex impedance of the Analog Discovery 2 without the BNC adapter board, assuming that the wiring adds 5 pF capacitance in parallel with the oscilloscope, which has an input impedance of 1 MΩ in parallel with 24 pF. Plot the magnitude of this impedance as a function of frequency on a log-log scale from 1 Hz to 1 MHz, using gnuplot (or other computer plotting program). What is the corner frequency?

Exercise 15.2

Write the formula for the complex impedance of the Analog Discovery 2 with the BNC adapter board and a 1× probe assuming that the probe adds 75 pF capacitance in parallel with the oscilloscope, which has an input impedance of 1 MΩ in parallel with 24 pF. Plot the magnitude of this impedance as a function of frequency on a log-log scale from 1 Hz to 1 MHz, using gnuplot (or other computer plotting program). What is the corner frequency?

Exercise 15.3

With the switch in the 10× position, what components are between the tip of the probe and the input to the oscilloscope? What components are between the input to the oscilloscope and the probe ground, counting both the components in the probe and the components in the oscilloscope? Use the impedance formulas for resistors and capacitors and for impedances in parallel to get the impedances of the two subcircuits of the voltage divider of a 10× scope probe and oscilloscope in Figure 15.1, using symbolic names ($C_{10\text{x}}$, C_{cable}, etc).

Exercise 15.4

Use the impedance formulas from Exercise 15.3 to determine the gain of the oscilloscope probe (the voltage at the oscilloscope divided by the voltage at the probe tip) as a function of frequency.

Exercise 15.5

The probe gain in Exercise 15.4 can be made constant (independent of frequency) for one value of $C_{\text{compensate}}$. Determine that value and show that it results in constant probe gain. What is the gain of the probe when $C_{\text{compensate}}$ is set correctly? Hint: the gain is less than 1—the scope probe *attenuates* the input signal.

Exercise 15.6

Use gnuplot (or other plotting program) to plot the gain of the scope probe (voltage at the oscilloscope divided by the voltage at the probe tip) versus frequency from 1 Hz to 1 MHz for a scope probe with the correct setting of the compensation capacitor as determined in Exercise 15.5. Plot the gain also with the compensation capacitor set 5 pF too large.
Use log-log plots for both (you may put both on the same plot, as long as they are clearly labeled).

If you have done Exercise 15.6 correctly, you should see a nearly constant gain as a function of frequency for the oscilloscope. A constant gain with respect to frequency is referred to as a *flat response*, and is a desirable property for many test instruments and amplifiers, as signals of different frequencies are treated the same, not favoring one over another.

To set the compensation capacitor in a 10× oscilloscope probe correctly, a simple visual technique is usually used.

1. Switch the probe to the 10× setting.
2. Connect the probe to a 1 kHz square-wave source. Most oscilloscopes provide one specifically for adjusting the probes—the Analog Discovery 2 can use one of the function generators for this purpose. The ground lead of the probe should be connected to the ground of the square-wave source.
3. Set the oscilloscope to display the waveform clearly. The autoset button, if your scope has one, is useful here.
4. Use a small screwdriver to adjust the capacitor built into the base of the probe. A metal screwdriver can add some extra capacitance to ground, which makes the compensation different when the tool is in place than once the tool has been removed. Using a screwdriver with a nonmetallic body avoids this problem of added capacitance from the metal of the screwdriver.
5. When $C_{\text{compensate}}$ is correctly adjusted, the oscilloscope should display a crisp square wave with sharp corners. If the corners are rounded, then $C_{\text{compensate}}$ is too large, and the probe is acting as a low-pass filter. If the waveform shows overshoot on the rising and falling edges, then $C_{\text{compensate}}$ is too small, and the probe is acting as a high-pass filter.

The compensation correction described above applies only to the 10× setting on the probe—the response of the 1× setting will be only slightly affected.

Chapter 16

Hysteresis

16.1 What is hysteresis, and why do we need it?

When we put a signal into a digital input of a computer, we want it to be treated as a simple on or off. There are lots of inputs we might want to treat this way (detecting a human input on a control panel, detecting the lack of water in a water bath, and so forth).

The electronics will deliver a voltage to a digital input pin, which the computer will interpret as a 0 or 1. You can think of this as a 1-bit analog-to-digital converter.

> The simplest model of a digital input is a simple step function: below a *threshold* voltage V_t the input is interpreted as 0, and above that, it is interpreted as 1. Because threshold voltages are input voltages at which the output *changes*, output voltage values are not thresholds.

The term *threshold* is used to emphasize that crossing the threshold changes the state of the system (like crossing the threshold of a house changes you from being outside to being inside). This model is a bit too simple sometimes, as voltages near the threshold may not result in a clean 0-or-1 decision and may even damage some circuits if held for too long. A more detailed model is shown in Figure 16.1.

If we have noisy inputs (the usual case in the real world), the interpretation of the input as a digital signal can get difficult. See, for example, Figure 16.2, which shows what happens when we have a digital input using the transfer function of Figure 16.1 interpreting a noisy input (a *transfer function* is just a function relating the output of a system to the input of the system—here the function is a nonlinear one giving the output voltage as a function of the input voltage). Instead of a single transition from 0 to 1 or from 1 to 0, we see a number of pulses at the transition. This behavior is particularly undesirable if we are using the digital signal to count events (cells in a cell sorter, people passing through a door, etc.). Increasing the gain of the input circuit would not help by itself, as we would still see a series of pulses as the input voltage crossed back and forth across the threshold.

> One solution to this problem is *hysteresis*: have two thresholds and let the conversion from a noisy analog input to a digital output have one bit of memory. If you are currently in the high state, you have to cross the lower threshold before you go to the low state, and if you are in the low state, you have to cross the higher threshold to go to the high state.

The transfer function is shown in Figure 16.3, and how such a digital input circuit would handle the same noisy input as Figure 16.2 is shown in Figure 16.4.

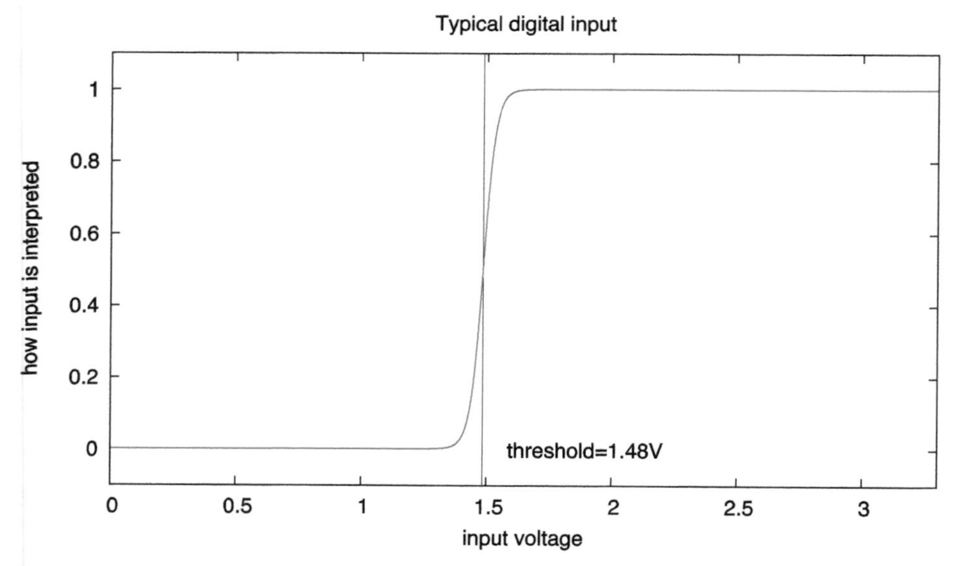

Figure 16.1: *Representative transfer function for a digital input. The gain of a digital amplifier is the slope of the curve at its steepest point. Increasing the gain makes the decision between off (0) and on (1) crisper, but often at the cost of other desirable properties (such as speed or power consumption).*

The comparator chips that we'll look at in Chapter 35 use this approach of having a very high gain (and an extra input for setting the threshold) for converting analog signals to one-bit digital values.

Figure drawn with gnuplot [33].

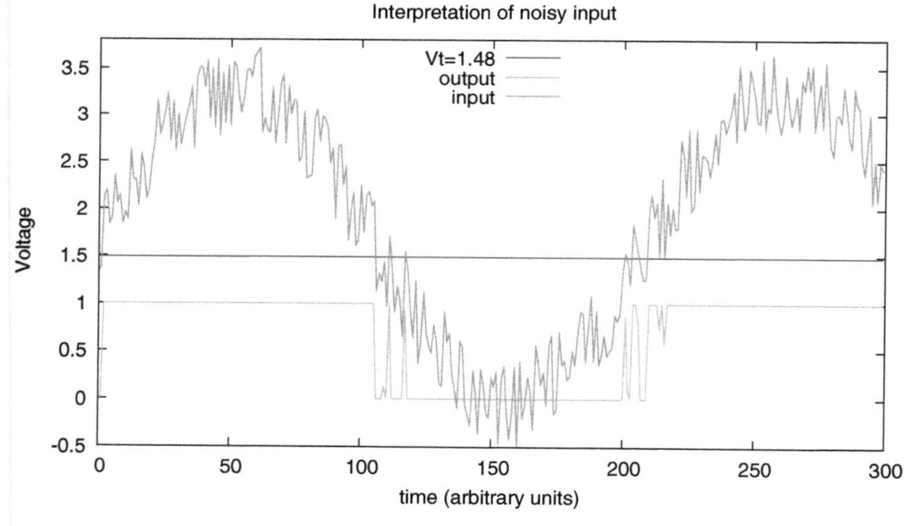

Figure 16.2: *The results of interpreting a noisy input with the simple function of Figure 16.1. When the input is near the threshold, the output can fluctuate wildly. If you were trying to count the events (button presses, cells in a flow sorter, etc.), you could end up counting many more than you should.*

Figure drawn with gnuplot [33].

Although the transfer function of Figure 16.1 is fairly representative of normal digital inputs, which expect their inputs to stay away from the threshold as much as possible, the circuits for *Schmitt triggers*, which implement hysteresis for digital inputs, have much higher gain (steeper slopes), and often have a larger separation between the two thresholds V_{IL} (input low voltage) and V_{IH} (input high voltage). The separation $V_{\mathrm{IH}} - V_{\mathrm{IL}}$ is called the *hysteresis voltage.* The gain (the slope of the curve) is not specified

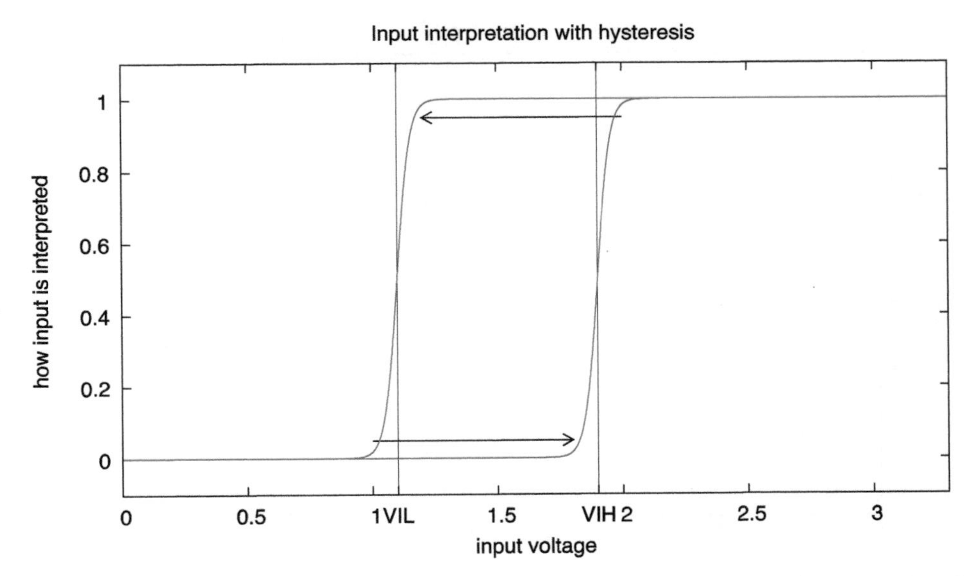

Figure 16.3: *Representative transfer function for a digital input with hysteresis. The output is not a function of just the input, but of the input and the previous state. The hysteresis curve here is drawn with the same gain as Figure 16.1. The horizontal separation between the curves is the* hysteresis voltage $V_{\text{IH}} - V_{\text{IL}}$.
Figure drawn with gnuplot [33].

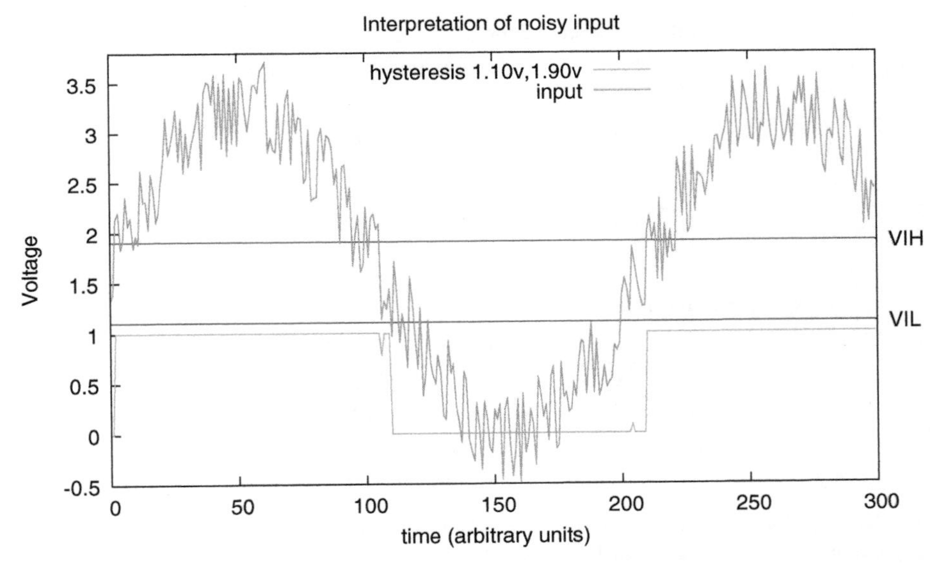

Figure 16.4: *The results of interpreting a noisy input with the hysteresis of Figure 16.3. The output does not wobble wildly when the input is near the threshold. We can get even cleaner transitions by increasing the gain (making the slope at the transition points steeper), as is done in the design of Schmitt-trigger inputs, which are often used for converting such noisy inputs into clean digital signals.*
Figure drawn with gnuplot [33].

on the data sheets for the Schmitt triggers we use (74HC14N) and was too high for me to measure easily—more than 1000, compared to a gain of only 10 in Figure 16.3.

The hysteresis curve for the 74HC14N does not look exactly like the curve in Figure 16.3. Not only are the thresholds different and the gain higher, but the 74HC14N is an *inverter*, which means that

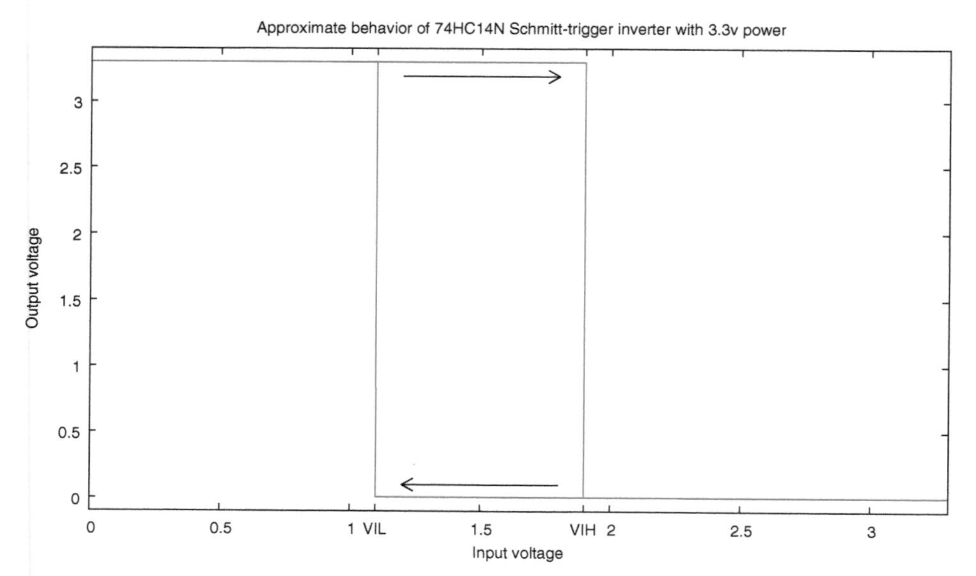

Figure 16.5: *An approximate transfer function for the 74HC14N operating with a power supply of 3.3 V (estimated, not measured). On real devices the thresholds may be at somewhat different voltages.*
Figure drawn with gnuplot [33].

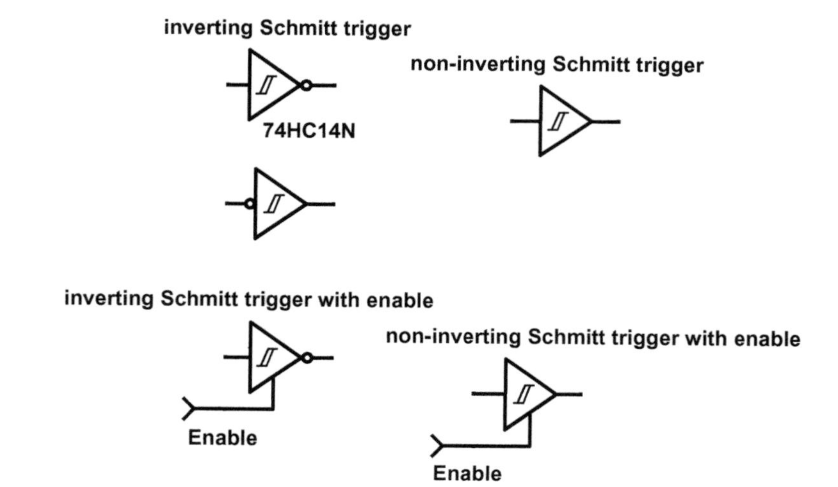

Figure 16.6: *The schematic symbol for an inverting Schmitt trigger is an inverter symbol (an amplifier triangle with an inversion bubble on the input or output) with a superimposed hysteresis symbol.*
We do not use non-inverting Schmitt triggers in this course, nor Schmitt triggers with an enable input. Do not use those symbols by mistake.
Figure drawn with Digi-Key's Scheme-it [18].

the output goes low when the input goes high and vice versa. A more realistic transfer function for the 74HC14N operating on a 3.3 V power supply is shown in Figure 16.5.

The schematic symbols for Schmitt triggers are shown in Figure 16.6—we use only the inverting Schmitt-trigger symbols, as those are the ones for the 74HC14N part that we use. The 74HC14N is a cheap, fast part, but the threshold voltages V_{IL} and V_{IH} are not exactly constants—the spec for them allows a pretty wide range of values.

16.2 How a hysteresis oscillator works

Although the main use of Schmitt triggers (circuits that implement a hysteresis transfer function) is to convert noisy analog inputs to clean digital inputs, they have another minor use as oscillators, which we will take advantage of in Lab 4 to make a touch sensor that is sensitive to changes in capacitance.

Figure 16.7 shows a typical circuit for a *hysteresis oscillator* (also called a *relaxation oscillator* [150]). The principle of operation is simple: when the output of the Schmitt-trigger inverter is high, the capacitor C_1 charges through resistance R_1, until the voltage across C_1 is larger than V_{IH} for the Schmitt-trigger inverter, then the output goes low and C_1 discharges through R_1 until its voltage drops below V_{IL} and the cycle repeats. The time spent with the output high depends on how long it takes to charge C_1 from V_{IL} to V_{IH}, with the output voltage at V_{OH} (output high voltage). The time spent with the output low depends on how long it takes to discharge C_1 from V_{IH} to V_{IL}, with the output voltage V_{OL} (output low voltage).

The output voltages V_{OH} and V_{OL} are not *thresholds*, as there is no change of state associated with crossing them. They are better thought of as the states that the Schmitt trigger is in, with the state changing when the input voltage crosses the appropriate one of the thresholds V_{IL} or V_{IH}.

The oscillator of Figure 16.7 has a fixed period, proportional to R_1C_1, but we can use a very similar design to create a variable frequency from a variable capacitor (Figure 16.8). The capacitances C_1 and C_2

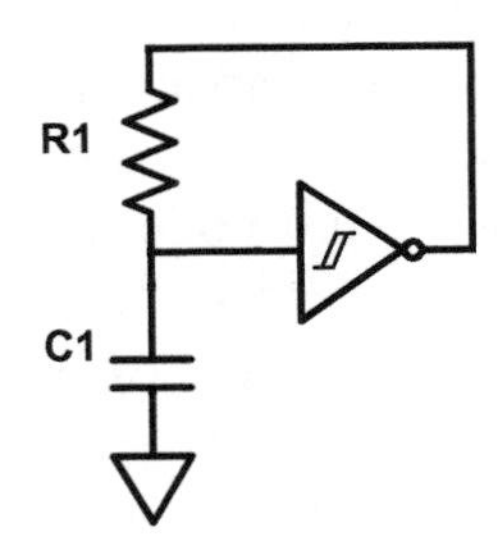

Figure 16.7: *This is the simplest form of the hysteresis oscillator, consisting of a Schmitt trigger and a feedback resistor R_1 that charges and discharges a capacitor C_1. When the output of the inverter is high, the capacitor C_1 charges through resistance R_1, until the voltage across C_1 is larger than V_{IH} for the Schmitt-trigger inverter, then the output goes low and C_1 discharges through R_1 until its voltage drops below V_{IL} and the cycle repeats.*
The period of the oscillator is proportional to R_1C_1, with the constant of proportionality depending on the voltages V_{IL}, V_{IH}, V_{OL}, and V_{OH} of the Schmitt trigger.
Figure drawn with Digi-Key's Scheme-it [18].

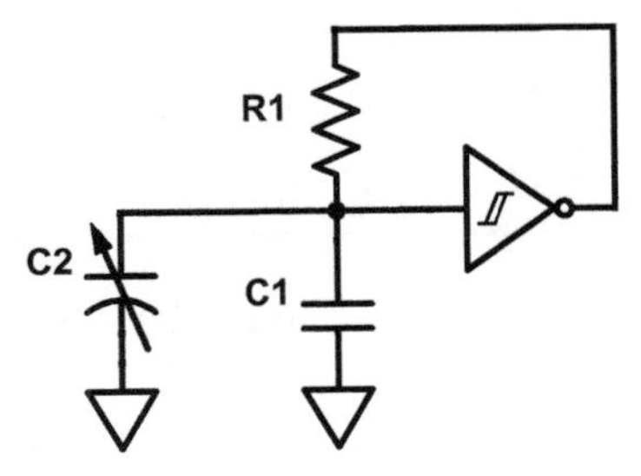

Figure 16.8: *By adding a sensor (C_2) whose capacitance varies with the property we are trying to measure, we can turn a simple hysteresis oscillator into a sensor input circuit that converts capacitance into a digital signal whose period increases with increasing capacitance.*
The fixed capacitor C_1 sets an upper limit on the frequency of the oscillator.
Figure drawn with Digi-Key's Scheme-it [18].

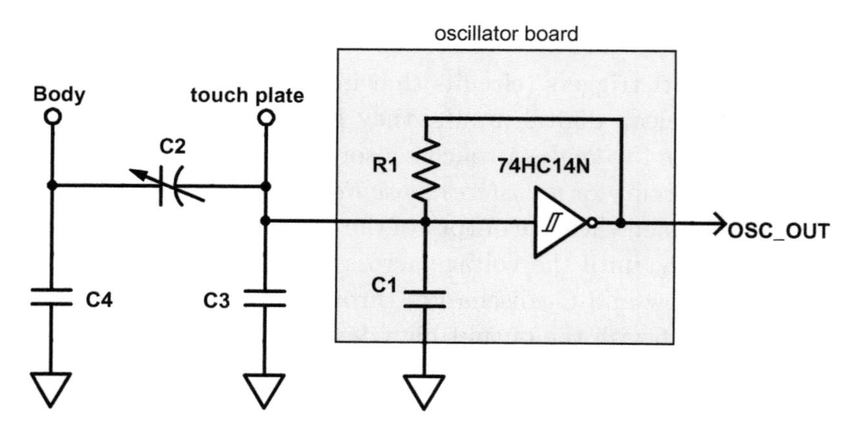

Figure 16.9: *We can make a touch sensor out of a hysteresis oscillator by adding a touch plate (one plate of a capacitor) to the input of the Schmitt trigger. The touch plate adds some more capacitance to ground (C_3), and touching the touch plate adds capacitance to the body (C_2) in series with capacitance of the body to ground (C_4). It is the change in C_2 from touching the touch plate that we want to detect.*

Figure drawn with Digi-Key's Scheme-it [18].

simply add, so the period is proportional to $R_1(C_1+C_2)$. No matter how small C_2 gets, the capacitance C_1 sets a minimum period (hence, maximum frequency) for the oscillator.

To make a touch sensor, we make a variable capacitor with one conductive plate being the finger being sensed and the other plate being connected to the input of the Schmitt trigger. Because our body is not connected directly to the oscillator's ground node, we need to model not just the touch, but also the capacitive connection between the body and ground, see Figure 16.9.

The charge and discharge times depend on the voltages (V_{IL}, V_{IH}, V_{OL}, V_{OH}) and the RC time constant R_1C, where $C = C_1 + C_3 + 1/(1/C_2 + 1/C_4)$. The charge and discharge times do not depend on R_1 and C separately, but only on the product. This means that we can design the hysteresis oscillator for a particular frequency with quite different component values. Because the capacitances from the body (C_2 and C_4) in Figure 16.9 are fairly small, we need to keep C_3 and C_1 small so that R_1C varies substantially as C_2 changes.

Figure 16.10 provides an example of the charge and discharge curves for a low-speed hysteresis oscillator using a 74HC14N chip. When the output is high, the input charges toward the output value, until it reaches V_{IH}, when the output changes, and the input discharges towards 0 V, switching again when the input reaches V_{IL}.

The charge/discharge curves are not straight lines, but portions of exponential RC curves, giving a characteristic "shark's-fin" waveform. We'll see in Chapter 36 how to modify this oscillator into a triangle-wave oscillator with constant slope for the charge and discharge curves.

> The RC time constant for Figure 16.10 is 0.1 s, but the oscillation frequency is *not* 10 Hz—the period is *proportional* to RC, *not equal to* RC.

Also, the 10 μF capacitor used was a cheap ceramic capacitor, and those often have capacitance 20% or more below their nominal capacitance.

Look at Figure 16.11, which shows typical charging and discharging curves for a Schmitt trigger powered at 5 V, and Figure 16.12 shows how switching between charging and discharging curves whenever a threshold is crossed results in a periodic waveform.

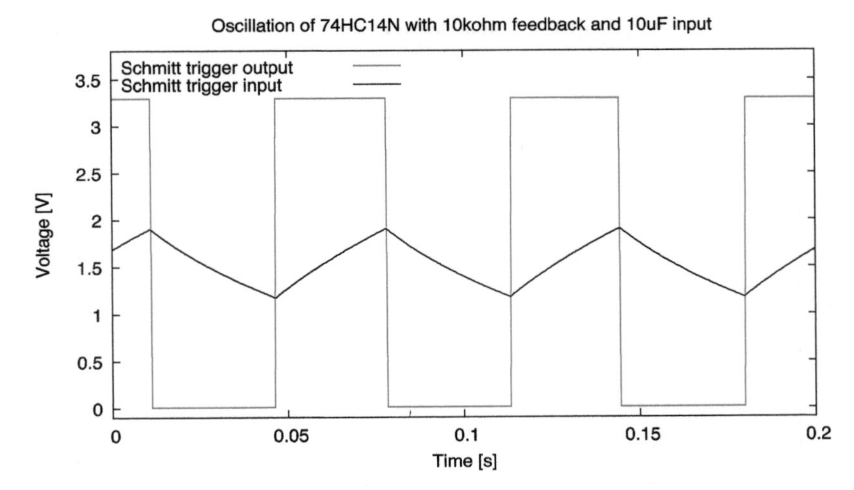

Figure 16.10: *Low-frequency oscillation of a 74HC14N Schmitt-trigger with a 10 kΩ feedback resistor and 10 μF capacitor on the input. The oscillation frequency is approximately 15.00 Hz. This recording was made with PteroDAQ using a Teensy 3.1 board with 32× averaging. The input was recorded on A10−A11, to take advantage of the differential amplifier to reduce the noise that was due to having a high-impedance source. The output was recorded as a digital signal on D0 and multiplied by the recorded power-supply voltage in the gnuplot script.*

Data collected with PteroDAQ using a Teensy LC board [46]. Figure drawn with gnuplot [33].

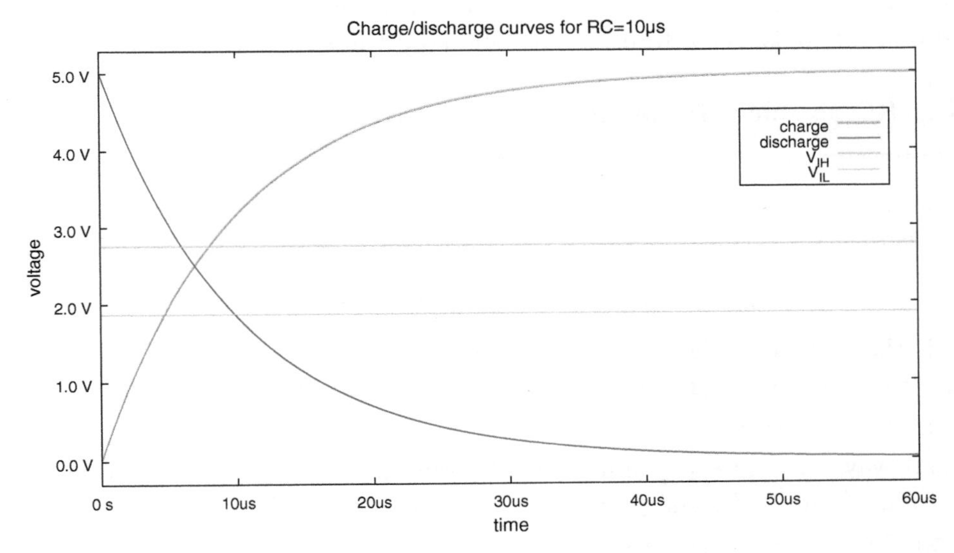

Figure 16.11: *The curves here are the theoretical exponential charge/discharge curves for an RC time constant of 10 μs. Although both curves have the same time constant, the rising and falling time between the hysteresis thresholds are different. The time to rise from* V_{IL} *to* V_{IH} *is about* $7.98\,\mu s - 4.68\,\mu s = 3.30\,\mu s$. *The time to fall from* V_{IH} *to* V_{IL} *is about* $9.83\,\mu s - 5.98\,\mu s = 3.85\,\mu s$.

Figure drawn with gnuplot [33].

The rise and fall times between the thresholds are not the same and changing the thresholds would change the relationship. Moving the thresholds up (keeping the difference between them fixed) would increase the rise time and decrease the fall time, and moving the thresholds down would have the opposite effect. The sum of rise and fall times determines the period, and it is minimized for a given hysteresis voltage when the rise and fall times are equal.

Changes to the high and low output voltages of the inverter also will result in changing charge and discharge times. The threshold voltages of a Schmitt trigger are dependent on the power-supply voltage,

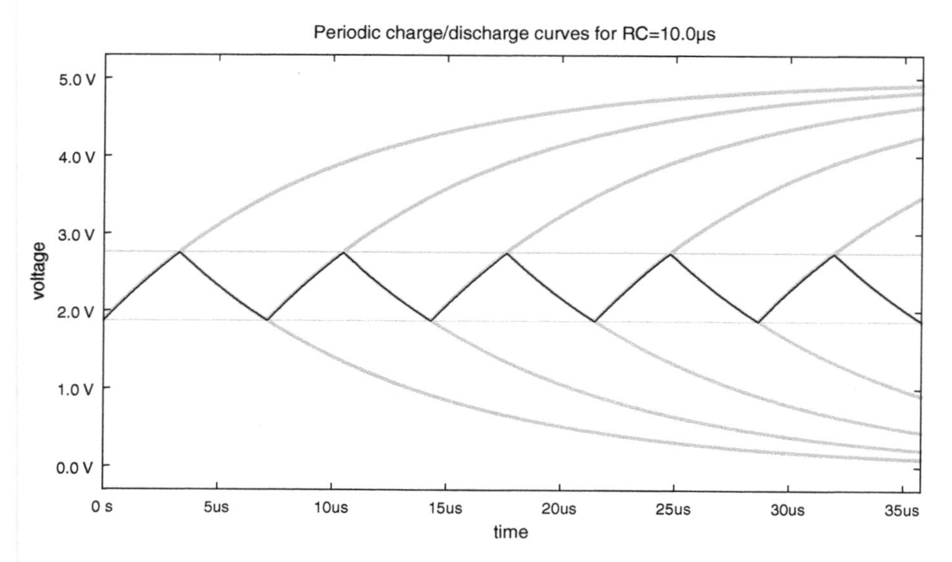

Figure 16.12: *This figure shows how repeating the charge and discharge curves of Figure 16.11 produces a characteristic "shark's fin" periodic waveform.*

Figure drawn with gnuplot [33].

so design for a particular frequency will depend on measuring (or estimating) the threshold voltages. Even when the power supply is constant, I have observed threshold-voltage changes of $\pm 30\,\text{mV}$.

16.3 Choosing RC to select frequency

From Section 11.4, remember that the charging/discharging curve for a capacitor charging through a resistor is

$$V(t) = (V(0) - V_\infty)\, e^{\frac{-t}{R_1 C}} + V_\infty \ ,$$

where $V(0)$ is the voltage at the beginning of the charge or discharge time and V_∞ is the voltage on the other side of the resistor (the asymptotic value as time goes to infinity).

There are two ways to figure out how long charging lasts. For the first method, we can use the exponential charging formula for charging from V_{IL} to V_{IH}, which happens when the other end of the resistor is at V_{OH}, so $V(0) = V_{\mathrm{IL}}$ and $V_\infty = V_{\mathrm{OH}}$:

$$V_{\mathrm{IH}} = V(t) = (V_{\mathrm{IL}} - V_{\mathrm{OH}}) e^{\frac{-t}{R_1 C}} + V_{\mathrm{OH}} \ .$$

That can be rearranged to get

$$\frac{V_{\mathrm{IH}} - V_{\mathrm{OH}}}{V_{\mathrm{IL}} - V_{\mathrm{OH}}} = e^{\frac{-t}{R_1 C}} \ ,$$

or

$$t = -R_1 C \ln \frac{V_{\mathrm{IH}} - V_{\mathrm{OH}}}{V_{\mathrm{IL}} - V_{\mathrm{OH}}} \tag{16.1}$$

$$= R_1 C \ln \frac{V_{\mathrm{IL}} - V_{\mathrm{OH}}}{V_{\mathrm{IH}} - V_{\mathrm{OH}}} \tag{16.2}$$

$$= R_1 C \ln \frac{V_{\mathrm{OH}} - V_{\mathrm{IL}}}{V_{\mathrm{OH}} - V_{\mathrm{IH}}} \ . \tag{16.3}$$

Exercise 16.1

Do a similar exponential analysis for discharging, showing the derivation for the discharge time, and add the charge and discharge times to get the total period.

The other method is not quite as elegant, but can be done without needing the exponential charging formulas. Instead, we can simplify our analysis considerably by pretending that the charging and discharging are done with a constant current—turning the curving charge and discharge curves into straight lines. We can estimate this average current by pretending that the voltage at the input of the Schmitt trigger averages out to $(V_{\text{IH}} + V_{\text{IL}})/2$, so that for charging, the current would be

$$I = \frac{V_{\text{OH}} - (V_{\text{IH}} + V_{\text{IL}})/2}{R_1} ,$$

and looking at how much the charge needs to change on the capacitor:

$$\Delta Q = C\Delta V = C(V_{\text{IH}} - V_{\text{IL}}) .$$

The time to charge the capacitor is

$$t = \Delta Q/I , \tag{16.4}$$

$$= R_1 C \frac{V_{\text{IH}} - V_{\text{IL}}}{V_{\text{OH}} - (V_{\text{IH}} + V_{\text{IL}})/2} . \tag{16.5}$$

Both Equations (16.3) and (16.5) are proportional to $R_1 C$, but the multipliers are somewhat different. In practice the differences are pretty small—much smaller than the variation in the input threshold voltages, so either formula may be used for design purposes.

Exercise 16.2

Do a similar constant-current analysis for discharging, showing the derivation for the discharge time, and add the charge and discharge times to get the total period.

The exponential charging curve model is a more accurate model of the phenomenon being modeled, but the constant-current model is a little easier to think about. This sort of trade-off between accuracy of a model and ease of reasoning with the model is one of the reasons that we have many different models for any given phenomenon.

16.3.1 *Improved model of 74HC14N*

The model of a Schmitt trigger developed in Section 16.1 is not a complete one, as it only talks about the DC behavior of the chip when the output has no load on it. For many purposes, this very simple model is good enough, but sometimes we'll need more sophisticated models.

To pick reasonable R and C values for a hysteresis oscillator, given a desired RC time constant, we need to look at where we have neglected aspects of the 74HC14N to simplify our model—where might our model break down as being too simple? If we pick a very small R_1 what might go wrong? What about if we pick a very large R_1?

Consider the slightly more sophisticated model of Figure 16.13, which includes output resistance, input capacitance, and feedback capacitance.

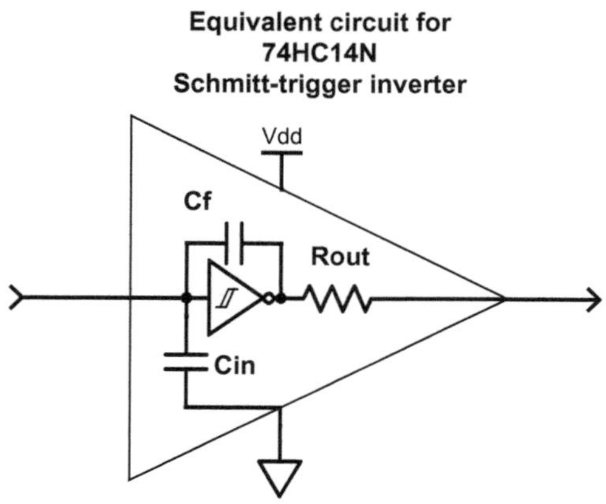

Figure 16.13: *Sometimes we need a slightly more sophisticated model of a Schmitt-trigger inverter than just the transfer function of Figure 16.5. The model here adds an input capacitance C_{in} to ground, an output resistance R_{out}, and a feedback capacitor C_{f} from the output back to the input.*
Figure drawn with Digi-Key's Scheme-it [18].

16.3.2 *Minimum value for R*

For the hysteresis oscillator of Section 16.2, if the resistance R_1 is very small, then we will get a large current when there is a voltage drop across it, but taking a large current from the 74HC14N will cause the output voltage to change. To estimate how much the current will change, we can model the output as a voltage source in series with a resistor (this is a very crude model, but good enough for us here).

The output impedance of digital outputs is not usually given on data sheets, since the outputs have nonlinear behavior. What you can do to get a rough estimate is to look at how much the output voltage changes with a large load—there is usually one or more specifications for how the output voltage changes with specified currents. For example, if an output voltage drops from 5 V to 4.5 V with a 4 mA output current, then we could approximate the output as a 5 V voltage source in series with a $(5\,\text{V} - 4.5\,\text{V})/4\,\text{mA} = 125\,\Omega$ resistor.

The equivalent resistor R_{out} may be different for the high output voltage and the low output voltage, as different transistors in the integrated circuit are used to connect to the two power rails—use the larger resistance for approximation here. A more sophisticated model would not treat R_{out} as a simple resistor, as the output current and voltage are not precisely linearly related, but we can get a good enough approximation with just a resistor.

As long as R_1 is at least $100\times$ the resistance of the output, we can ignore the resistance of the output. If R_1 is smaller, we should probably compute our RC time constant using $R_1 + R_{\text{out}}$ instead of just R_1. For most of the R and C values that students are likely to choose for a hysteresis oscillator, the R_{out} values are too small to matter.

16.3.3 *Maximum value for C*

If we are using a hysteresis oscillator to detect a change in capacitance for C_2, then we need the frequency of the oscillator to change by an easily measured amount for the change in capacitance we expect. If $C_1 \gg C_2$, then changes in C_2 will not affect the overall capacitance much, and touches will be difficult to detect.

Generally, the smaller C_1 is, the more the frequency will change in response to changes in C_2. If we ground the body (shorting out C_4 in Figure 16.9), then C_2 is in parallel with C_1, and the capacitances add. We can then estimate the change in the RC time constant from a change in C_2 as being proportional to the change in the total capacitance.

Including C_4 in the model reduces the effect of changes in C_2, unless $C_4 \gg C_2$.

16.3.4 *Minimum value for C*

The input capacitance C_{in} of the 74HC14N gets added to the external capacitor C_1 that we design in the circuit. If C_1 is large, then C_{in} is unimportant, but if C_1 is small, then C_{in} may be critical. Parasitic capacitance from wiring on the board may also matter when capacitances are small. The input capacitance C_{in} is often provided on the data sheet for a Schmitt trigger or other digital circuits, because the total capacitance affects how fast signals can be.

We have also neglected the effect of 60 Hz interference from capacitive coupling to nearby power lines, as shown in Figure 16.14. If C_1 is small and the resistance R_1 is large, then the current through R_1 will be very small, and capacitive coupling of 60 Hz interference could result in large currents relative to the charging/discharging current through R_1. Moving the signal on C_1 up and down is roughly equivalent to changing the hysteresis thresholds, which changes the time for charging or discharging. These changes result in fluctuations in frequency of the oscillator tied to the 60 Hz interference. This fluctuation in frequency due to a voltage is referred to as *frequency modulation*.

16.3.5 *Maximum value for R*

Another phenomenon to consider is that there can be a small current leakage through the input of the Schmitt trigger—if this is comparable to the current we are using to charge or discharge the capacitor, then our charging time will be way off (or we may not be able to charge the capacitor at all). Make sure that the charging current you expect is at least 3× and preferably 10× the worst-case input current for the Schmitt trigger (found on the data sheet).

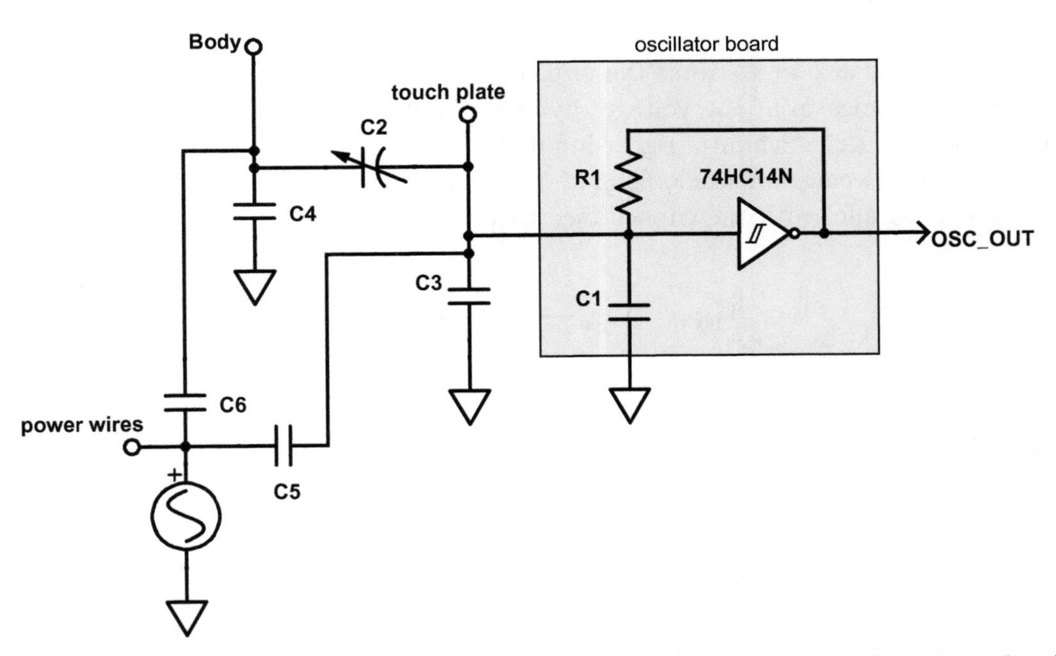

Figure 16.14: *Relaxation oscillator with touch sensor Figure 16.9 with 60 Hz interference coupled in from power wires. The changing voltage at the input to the Schmitt trigger results in some fluctuation in the frequency of the oscillator (*frequency modulation*). Figure drawn with Digi-Key's Scheme-it [18].*

For example, if you expect to put 2 V across the resistor, and the worst-case input current is 4 μA, then you want to have at least 40 μA through the resistor, and a maximum resistance of $2\,V/40\,\mu\text{A} = 50\,\text{k}\Omega$. (All those numbers are just illustrations—look up the input current specs for the chip and compute the voltage you expect to have across your resistor.)

Given a range of reasonable R_1 values, we'll get a range of capacitance values. Small capacitors (0.1 μF or smaller) are generally cheaper than larger ones, but resistors of reasonable sizes are all about the same price, so we usually choose a small capacitor and large resistor, when the choice is feasible.

16.4 Feedback capacitance

We also sometimes need to model a feedback capacitor, that is, a capacitor between the output and the input. If there is a large external capacitance on the input, or a low-impedance source driving the input, then the small feedback capacitor is unimportant. But if the input is driven only by a high-impedance source, then the capacitive coupling of the rapidly changing output back to the input can cause the input voltage to change substantially.

One can think of the feedback circuit as a voltage divider having $R_1 \parallel C_\text{f}$ as the upper subcircuit and $C_\text{i} = C_1 \parallel C_\text{in} \parallel C_\text{wire}$ as the lower subcircuit. For the very high frequencies of a rapidly rising or falling edge, the impedance of the capacitors will be very small, and we have a capacitive voltage divider. If the capacitance between the input and ground is large compared to the feedback capacitor, the voltage jump when the output switches will be small, and we can ignore the feedback capacitor. If the input capacitance is small, then the feedback capacitor may be important to include in our modeling.

To illustrate the phenomenon more clearly, I set up a hysteresis oscillator with and without a deliberately added large C_f and plotted the oscilloscope trace in Figure 16.15. Small resistance and large capacitance values were used, so that the impedance of the oscilloscope probes would not affect the circuit much.

The large jump in the input as a result of the feedback capacitor reduces the period of the oscillator, because less charge has to be transferred through the resistor.

We can estimate the size of C_f from the size of the jump in voltage at the input of the Schmitt trigger. The voltage comes from the voltage divider formula, $V_\text{jump} = V_\text{step}\frac{Z_\text{i}}{Z_\text{i}+Z_\text{f}}$, where V_jump is the voltage jump observed at the Schmitt-trigger input, and V_step is the size of the transition at the output of the Schmitt trigger. Because we are looking at a very rapid change (and hence a high frequency), the large resistor R_1 in parallel with the capacitance C_f does not change the impedance much, so we can simplify to

$$\begin{aligned} V_\text{jump} &= V_\text{step}\frac{1/(j\omega C_\text{i})}{1/(j\omega C_\text{i}) + 1/(j\omega C_\text{f})} \\ &= V_\text{step}\frac{C_\text{f}}{C_\text{f} + C_\text{i}} \,. \end{aligned}$$

We can solve this equation for C_f getting

$$C_\text{f} = \frac{V_\text{jump}}{V_\text{step} - V_\text{jump}} C_\text{i} \,.$$

If you make a measurement like this with an oscilloscope, be sure to include the input capacitance of the oscilloscope probe in C_i.

When C_1 is sufficiently small, the internal parasitic capacitance of the 74HC14N chip is enough to produce a jump in the input without added feedback capacitance. For example, the plot in Figure 16.16 shows the input and output waveforms for an oscillator with a $C_1 = 33\,\text{pF}$ and no added feedback capacitance—there are large jumps in the input voltage when the output changes.

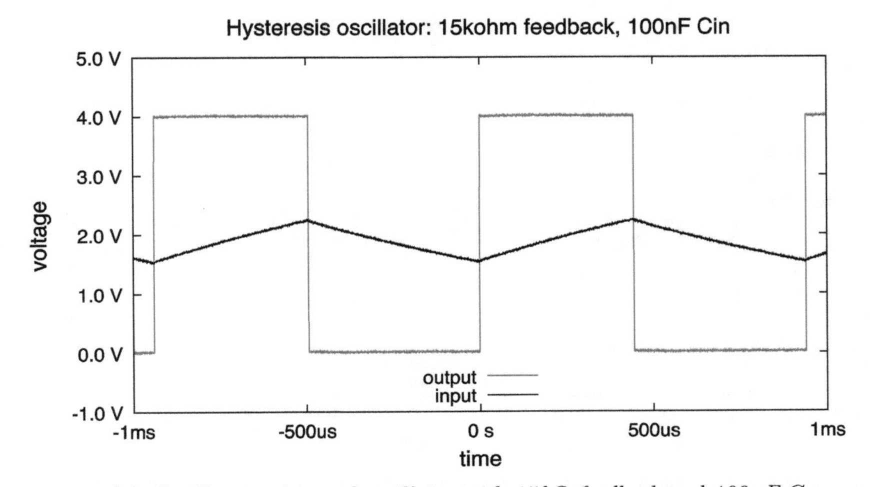

(a) *Oscilloscope trace of oscillator with 15* kΩ *feedback and 100* nF C_1.

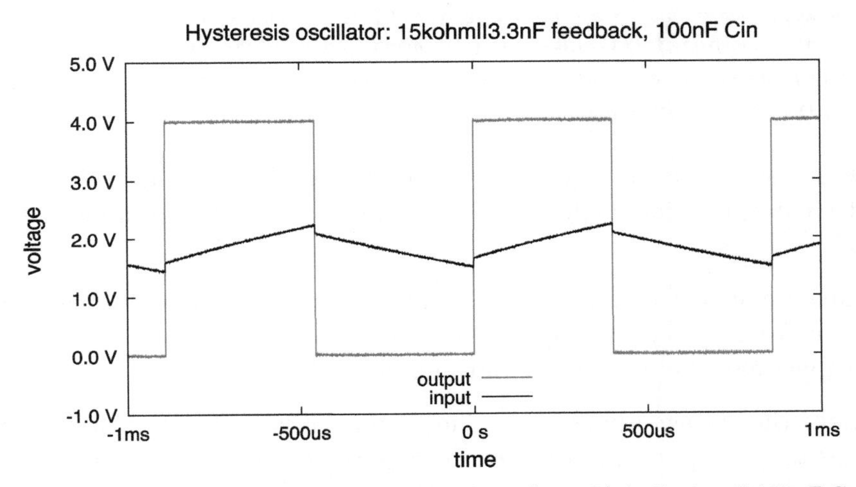

(b) *Oscilloscope trace of oscillator with (15* kΩ||*3.3* nF*) feedback and 100* nF C_1.

Figure 16.15: *Oscilloscope trace for a hysteresis oscillator using 74HC14N Schmitt-trigger inverter showing the effect of an added feedback capacitor in parallel with* R_1*, running at 4 V.*
Data collected with Analog Discovery 2 [19]. Figure drawn with gnuplot [33].

Feedback capacitance can cause serious problems even when using the Schmitt trigger in its normal application of converting analog signals to clean digital signals. Although the approximately 2 pF internal feedback capacitance is not large enough to cause problems in normal use, even small amounts of additional feedback capacitance from wiring can cause ringing that turns a single edge into multiple edges.

What happens is that when the input crosses one of the thresholds (say rising above V_{IH}), it causes the output to rapidly move in the opposite direction a short time later (the *propagation delay*, usually around 20 ns). The edge in the output drives a corresponding edge in the input through the feedback capacitor. If the input changes enough to cross the other input threshold, then the output reverses direction, again after the propagation delay. The result is an oscillation with a period of twice the propagation delay. The oscillation stops when the input doesn't swing far enough to cross both thresholds.

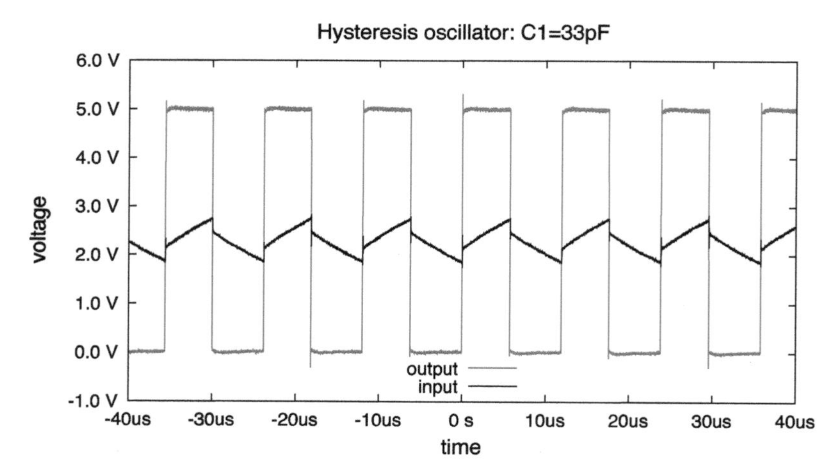

Figure 16.16: *This hysteresis oscillator, with $C_1 = 33\,pF$, shows the effect of capacitive feedback on the waveform of the output. The output of the Schmitt trigger swings from 0 V to 4.9 V in about 20 ns, and the input simultaneously swings from 1.87 V to 2.16 V, as a result of the capacitive voltage divider that has the internal feedback C_f as the upper subcircuit and $C_1 \parallel C_{in} \parallel C_{wiring}$ as the lower subcircuit. The downward transition from 4.98 V to 0 V also takes about 20 ns, and the input voltage swings from 2.75 V to 2.47 V. The relative size of the voltage swings suggests that the feedback capacitance is around 2 pF.*

Data collected with Analog Discovery 2 [19]. Figure drawn with gnuplot [33].

Because the purpose of a Schmitt trigger is to turn noisy analog transitions into single digital edges, having ringing that creates additional edges is highly undesirable. You can prevent the ringing in three ways:

- by keeping the capacitance between the input and the output due to wiring as small as possible,
- by driving the input from a very low-impedance source, and
- by adding some capacitance between the input and ground.

All three of these make the voltage divider from the output back to the input have a low gain, resulting in only small changes to the input—not enough to cross the other threshold and create an extraneous spike.

Adding extra capacitance between the input and ground is probably the most direct solution, but adds some low-pass filtering to the input, which may or may not be desirable. If you have a low-speed signal with high-frequency noise, low-pass filtering cleans up the signal that the Schmitt trigger sees, allowing the circuit to work correctly even with fairly high amounts of noise in the input. On the other hand, the low-pass filter adds extra delay to the input, which may be unacceptable for high-speed circuits that need precise timing of the detected edges.

In Figure 16.17, I show a ringing transition resulting from having 68 pF of additional feedback capacitance, when the input is driven from a 1 kΩ source. The period of the ringing is about twice the 16 ns propagation delay. The ringing is reduced by the 1 MΩ || 25 pF load of the oscilloscope channel monitoring the input—without that load, the ringing can last for a couple of microseconds. I've seen ringing that caused extraneous edges with as little as 5 pF of extra feedback capacitance—an amount that is common with breadboard wiring.

16.5 Capacitance touch sensor

A capacitance touch sensor is used for switches on some lab equipment, because it has no moving parts (other than electrons) and can be put behind an easily-cleaned glass surface. It is also easy to operate with gloves on. Capacitive sensing can also be used in other contexts, to detect the presence

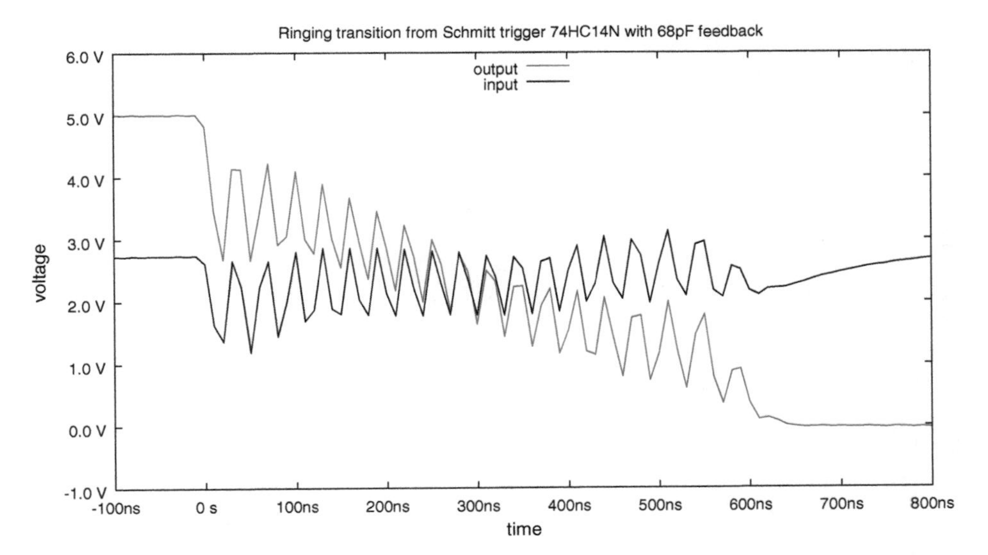

Figure 16.17: *With only an extra 68 pF of feedback capacitance, a 74HC14N Schmitt trigger can have substantial ringing on each transition. The input here is driven through a 1 kΩ resistor. When the input rises above* V_{IH}*, the output drops, and the capacitive feedback causes the input to fall far enough to cross the* V_{IL} *threshold, which causes the output to go high again. This cycle repeats until the input doesn't move enough to cross both thresholds—with somewhat more feedback capacitance, the ringing can be sustained a long time.*
Data collected with Analog Discovery 2 [19]. Figure drawn with gnuplot [33].

of conductive substances without having to make direct electrical contact with them or to measure distances between conductors.

Other common capacitive sensors include proximity sensors (like touch sensors but with a longer range), humidity sensors (detecting the change in dielectric constant of a polymer film), and soil moisture sensors (detecting the change in dielectric constant of the soil—dry soil has a much lower dielectric constant than water).

The basic idea of our capacitance touch sensor is simple:

- A touch plate consists of one plate of a capacitor and an insulator—your finger is the other plate of the capacitor.
- The capacitance of the touch plate varies depending on how close the finger is to the touch plate and how much area of the finger is in contact.
- The varying capacitance is used to change the frequency of an oscillator.
- Because of the 60 Hz energy all around, you act much like a low-voltage 60 Hz voltage source, so the frequency of the oscillator is not constant, but fluctuates with 60 Hz frequency modulation, as shown in Figure 16.18.
- The frequency (or period or duration of one of the pulses) of the oscillator is measured to determine whether the touch plate is currently being touched or not.

Changing the frequency of an oscillator may not be the most energy efficient way to detect a capacitance change, but it is how the built-in capacitive touch sensors work on the Teensy boards. Their hysteresis oscillator uses a Schmitt trigger followed by a constant-current driver, so that they get triangle waves instead of the "shark-fin" charge/discharge curves that we get with resistor feedback, but is otherwise similar [30, p. 863]. They correct for variations in circuit parameters by running another *reference* oscillator without a connection to the touch sensor and comparing the frequencies by counting the number of reference oscillator periods in some number of touch oscillator periods.

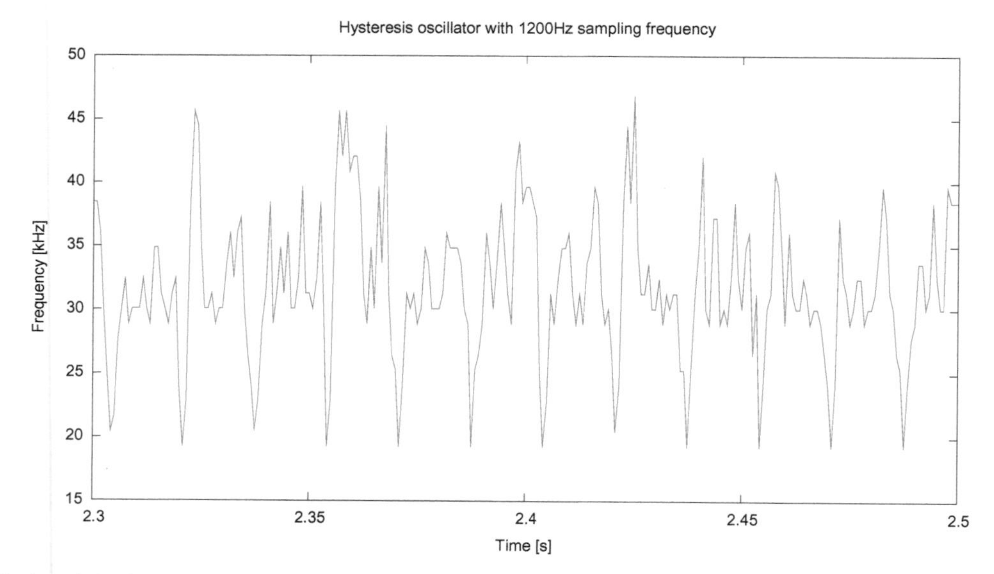

Figure 16.18: *A plot of the frequency of the oscillator in the capacitive touch sensor with a high sampling rate, showing the frequency modulation by 60 Hz line noise,*

If 60 Hz interference is a problem (as it often is on breadboard setups), measuring the frequency by counting pulses over 1/60 s averages out the fluctuation, as was done in Figure 16.19.

Data collected with PteroDAQ using a Teensy LC board [46]. Figure drawn with gnuplot [33].

For Lab 4, you will make a cheap capacitance touch plate out of a piece of aluminum foil (the plate of the capacitor) and a layer of packing tape (the insulator). You can connect this plate of the capacitor to the inverter input of the hysteresis oscillator, effectively putting the capacitor in parallel with C_1. The period of the oscillator will then be proportional to $R_1(C_1 + C_2)$, rather than just R_1C_1, where C_2 is the capacitance of the touch plate and finger.

To detect the change in period easily, we want the change in C_2 to be fairly large relative to $C_1 + C_2$, which means we want to keep C_1 fairly small.

I used frequency channels on PteroDAQ to make a plot of the frequency of the oscillator with a particular set of R_1 and C_1 values (see Figure 16.19).

16.6 Multi-dielectric capacitors

If we use our touch sensor as a proximity sensor, the capacitor whose plates are the finger and the touch plate is a little more complicated than the simple model introduced in Section 10.1, because we have two different dielectrics: the air between the finger and the sensor and the plastic or glass covering the touch plate. These two dielectrics have different thicknesses and different dielectric constants (see Figure 16.20).

We can make an approximate model for the proximity sensor by modeling the pair of dielectrics as two capacitors in series, as if there were a zero-thickness conductor on the top surface of the plastic or glass dielectric. If we use a simplified model that ignores fringe effects that extend out past the edges of the conductors, treating both the finger and the metal plate as having area A, we get two capacitances:

$$C_{\text{air}} = \frac{\epsilon_0 A}{d},$$

$$C_{\text{plastic}} = \frac{k\epsilon_0 A}{t},$$

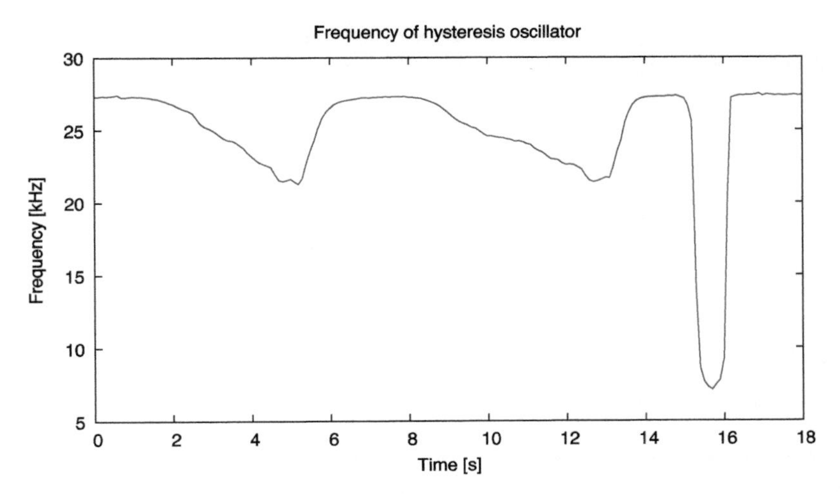

Figure 16.19: *A plot of the frequency of the oscillator in the capacitive touch sensor with respect to time, sampled every 0.1 s, to average out 60 Hz frequency modulation. For the first part of the plot, the sensor is used as a proximity sensor for a hand a centimeter or so from the touch plate. Between 15.2 s and 16.2 s is a finger touch to the sensor.*
The noise in the frequency measurement is quite small, and the change in frequency due to touching and releasing the capacitive sensor is much larger than necessary for reliable detection.
Data collected with PteroDAQ using a Teensy LC board [46]. Figure drawn with gnuplot [33].

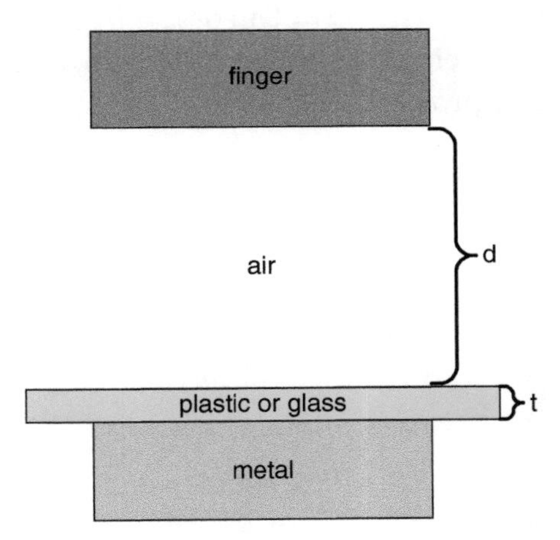

Figure 16.20: *The proximity sensor can be viewed as a multi-dielectric capacitor, with both air and a solid insulator as the dielectric materials. The distance d varies and is what we want to measure, while the thickness t is a constant.*

where k is the relative dielectric constant of the plastic or glass and $\epsilon_0 = 8.854187817\,\mathrm{pF/m}$ is the permittivity of free space. (The relative dielectric constant for air is very close to 1, and modeling it as being 1 is a tiny error compared to other simplifications we are making.) Putting the two capacitors in series gives us a total capacitance of

$$\begin{aligned} C_{\text{sensor}} &= \frac{1}{1/C_{\text{air}} + 1/C_{\text{plastic}}} \\ &= \frac{1}{d/(\epsilon_0 A) + (t/k)/(\epsilon_0 A)} \\ &= \frac{\epsilon_0 A}{d + t/k} \,. \end{aligned}$$

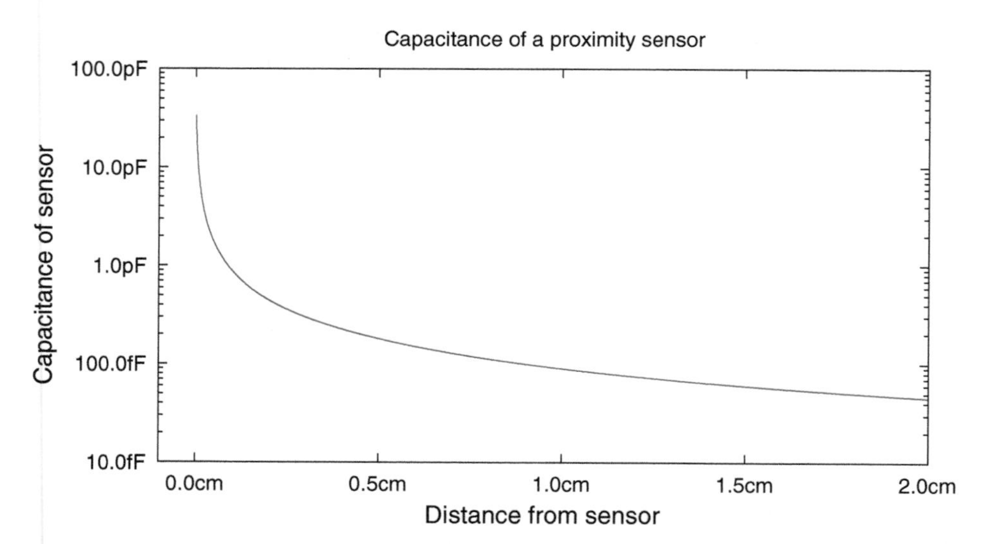

Figure 16.21: *Capacitance versus distance d for a proximity sensor modeled using the simplified formula* $C = \frac{\epsilon_0 A}{d+t/k}$. *The parameters A, t, and k are deliberately not provided here, as you need to determine the values for Lab 4.*
For me, the most interesting part of this plot is how large the capacitance is when the touch plate is touched, and how little it is and how little it changes when only 1 cm–2 cm away.

If we remove one of the dielectric layers (setting $d = 0\,\mathrm{m}$ or $t = 0\,\mathrm{m}$), this formula simplifies to the formula for the remaining single-dielectric capacitor. The plot in Figure 16.21 shows the modeled behavior for a multi-dielectric capacitor used as a proximity sensor.

Chapter 17

Lab 4: Hysteresis

Bench equipment: function generator, oscilloscope, soldering station, fume extractors, micrometer

Student parts: 74HC14N Schmitt trigger, resistors, capacitors, aluminum foil, packing tape, alligator clip, hysteresis-oscillator board, PteroDAQ

17.1 Design goal

There are five parts to this lab:

(1) characterizing a Schmitt-trigger inverter,
(2) demonstrating the use of a Schmitt trigger to convert a noisy analog signal to a clean digital signal,
(3) designing a hysteresis oscillator for use as a capacitance touch sensor,
(4) soldering the oscillator circuit on a PC board, and
(5) testing the capacitance touch sensor as a proximity sensor.

For characterizing the Schmitt trigger, the goal is simply to determine the two input voltage thresholds at which the output switches from low to high and from high to low (see Figure 16.5 for the approximate transfer function).

The oscillator circuit is an extremely simple one, with the Schmitt-trigger inverter, one resistor, and one capacitor. The design problem is to select appropriate parameters for the resistor and capacitor to get a desired period and large enough ratio of periods between the oscillator with a touched and untouched sensor.

> For ease of measuring the frequency with PteroDAQ, you should probably design your oscillator with a frequency between 1 kHz and 1 MHz when the sensor is not touched.

The soldering practice is just that—practice at using the soldering iron on a simple enough circuit that few solder joints are needed. You will also make a simple capacitance touch plate (a conductor with a thin insulator covering it), whose characteristics will affect the design of the hysteresis oscillator. Everybody needs to solder their own oscillator board—one per group is not enough.

The demonstration consists of producing a noisy sine wave using the Analog Discovery 2 function generators, and using its oscilloscope function to observe the input and output.

17.2 Design hints

Designing the oscillator comes down to setting the RC time constants appropriately for the touched and untouched sensor plate. Remember that all the times in this system (pulse widths or periods) are proportional to the RC time constant.

For generating the noisy sine wave, use both Analog Discovery 2 channels: one for the sine wave, one for noise. Average the two signals with a voltage divider (see Exercise 5.3).

17.3 Pre-lab assignment

Read Chapter 16 making sure you understand how the hysteresis oscillator works and how you can use the frequency of an oscillator to detect changes in capacitance. You may also wish to read the Wikipedia article on relaxation oscillators [150], which gives a somewhat different view of the same idea.

You will measure the four voltages V_{IL}, V_{IH}, V_{OL}, and V_{OH} for the 74HC14N chip in the lab, but you should be able to get good estimates of them from the data sheet.

We'll be powering off of a 3.3 V supply (so that we can observe signals with PteroDAQ), but the data sheets may not report the hysteresis voltage or threshold voltages for that power-supply voltage. You may have to scale voltages down from a power-supply voltage that does have specifications or interpolate between provided values.

The data sheets usually provide three values for a parameter: a minimum, a typical value, and a maximum. You can get good guesses for how a system will behave by using typical values, but the parts are only guaranteed to be somewhere between the maximum and the minimum values.

Pre-lab 4.1

What are the typical values for the thresholds of a 74HC14N with a 3.3 V power supply? (Interpolate if needed.) What about output high and low voltages?
Provide a citation for the 74HC14N data sheet used, as different manufacturers may have slightly different specifications. Which manufacturer made the chips in your kit?
Plug the typical voltage values into the formulas from Exercises 16.1–16.2 to get simplified formulas for the period of your oscillator.

Pre-lab 4.2

Instead of getting the period from the RC time constant, turn the formulas from Exercises 16.1–16.2 and Pre-lab 4.1 around to provide formulas for the RC time constant you would need to get a particular frequency of oscillation, f.
What RC time constant do you need to get an oscillation at 100 kHz?

The RC time constant is **not** $\frac{1}{2\pi f}$, as that formula is irrelevant here—we are not looking for corner frequency of an RC filter. Please think about how the RC time constant affects the oscillation time, rather than grabbing a random formula out of memory—this class is not a memorize-and-regurgitate class.

You will need to estimate the capacitance of a finger touching the touch plate. What you are really interested in is the *difference in capacitance* between having a finger touching the plate and not having a finger touching the plate (C_2 in Figure 16.9, not C_3). One plate of the capacitor is the finger, one is the aluminum foil, and the insulator is the packing tape. You can use Equation 10.1 from Section 10.1 to estimate the capacitance from the geometry.

Pre-lab 4.3

Estimate C_2, the capacitance of a finger touch on the packing-tape and foil sensor, by estimating the area of your finger that comes in contact with the tape, and assume that the tape is 2-mil tape (0.002″ thick) made of polypropylene (look up the dielectric constant of polypropylene online). Make sure that your units are consistent!

Warning: the tape that you measure in the lab may not be 2-mil tape—the range is about 1.6 mils to 3 mils, depending on the quality. You will probably need to redo the calculation of Pre-lab 4.3 using the measured thickness for your design report.

Warning: an inch is not a meter, and the area of your finger tip touching a plate is not a square meter—watch your units in your calculations! It is probably easiest to convert all your numbers into standard SI units before doing calculations with them.

If putting your finger on the touch sensor should increase the period of the pulse by a factor m, then $C_1 + C_2 + C_3$ should be about $m(C_1 + C_3)$, if we further assume that the body is grounded or that the body-to-ground capacitance C_4 is large enough that C_2 in series with C_4 is not very different from C_2.

Estimating C_3 is difficult for this touch sensor design, so you can assume that it is too small to matter when doing the preliminary design—once you build the sensor and oscillator, you can estimate the capacitance C_3 from the change in frequency of the oscillator when the touch sensor is attached. The assumption that C_3 is small relative to C_1 turns out to be a pretty good one for most student choices for C_1.

Pre-lab 4.4

Use your estimate of the capacitance of the finger touch to get initial values for R_1 and C_1, so that the period approximately doubles when you touch the sensor.
Make your initial values for capacitance and resistance be in the E6 and E12 series, respectively.

We are interested in how much the frequency changes when the sensor is touched, but we are more interested in the *ratio* of the frequencies than in the *difference*. A change from 1 MHz to 1.001 MHz is so small that it may not be distinguishable from noise, but a change from 2 kHz to 1.8 kHz is a big change, even though the difference in the first case is 1 kHz and only 200 Hz in the second case.

When we are interested in ratios of quantities, then taking the logarithm of the quantities (or plotting them on a log scale) is appropriate, because logarithms turn ratios into differences: $\log\left(\frac{f_1}{f_2}\right) = \log(f_1) - \log(f_2)$.

Pre-lab 4.5

Thinking of the hysteresis oscillator with a touch plate as a sensor—what is the sensitivity of the sensor? Remember: sensitivity is $\frac{\partial \text{out}}{\partial \text{in}}$, your input is the capacitance C_2, and your output is the log of the period of the oscillator, $\ln(T)$. We use $\ln(T)$ rather than T, because we are interested in the ratio of the frequencies, rather than the difference in frequency or period.
(It helps to simplify the model by pretending that the body is grounded or that C_4 is so large that it doesn't matter.)
What can you redesign to change the sensitivity?

Pre-lab 4.6

Design a simple circuit for averaging two signals from time-varying voltage sources. Make sure that looking at the output with an oscilloscope won't change the result much and that you don't take more than 10 mA from the voltage sources, even if one is at 3.3 V and the other is at 0 V. Furthermore, make sure that the output impedance of your circuit is $< 2\,\text{k}\Omega$, so that you can drive the inputs of the Teensy LC running PteroDAQ without too much noise.
Provide the schematic, complete with component values.
Hint: see Exercise 5.3.

17.4 Procedures

17.4.1 *Characterizing the 74HC14N*

To characterize the 74HC14N Schmitt-trigger inverter, we want four voltages: input thresholds V_{IL} and V_{IH} and output voltages V_{OL} and V_{OH}.

When I tried doing this part of the lab, it took me about half an hour to set up the breadboard, make and record the measurements, and analyze the results. The measurements do fluctuate a bit—I saw a standard deviation of 2 mV–3 mV for each input threshold.

Using PteroDAQ makes all four voltages easy to measure. The simplest are the output voltages: just connect +3.3 V from the Teensy board to pin 14 and GND to pin 7 of the chip (providing power to the chip), then connect either +3.3 V or GND (depending on which output voltage you want to measure) to pin 1 (the input of one Schmitt trigger) and measure the output on pin 2 (the output of that Schmitt trigger) with one of the analog pins (say A1) using PteroDAQ.

Figure 17.1 identifies pins of a 14-pin DIP. Consult the spec sheet for the part to determine what the various pins do—don't trust my instructions above without checking them against the data sheet!

Please use red wires for your positive power supply and black wires for your ground throughout this course (and don't use those colors for other signals). Following this convention will make your circuits much easier to check and debug.

Be sure to record the *measured* power-supply voltage as well as the output voltage measurements! PteroDAQ reports the power-supply voltage on both the graphical user interface and in the metadata of files it saves.

The input threshold voltages are a little harder to measure, because we are interested in looking at places where a small change in voltage causes the output to change, but we can't just tweak up a little

Figure 17.1: *The numbering of pins on a dual-inline package is counterclockwise, with a notch at the end of the package between the lowest and highest numbers. Some packages have a small dot by pin 1, rather than a notch.*

or down a little to see the change—once we make a change, we have to go all the way to the other threshold to change back.

The basic idea of the measurement is to slowly sweep the input voltage up and down, noting the voltage when the output changes.

> You can set up PteroDAQ to record values when one of the digital inputs changes, rather than at constant time intervals, by clicking on the checkbox "Pin Change" instead of "Timed". The trigger input can be set to trigger a sample on a rising edge, a falling edge, or either edge. Within about a microsecond after the edge, the microcontroller board starts recording the sample.

Set up a function generator with a very slow triangle wave (say 0.1 Hz) with the voltage sweeping from 0 V to 3.3 V. You want a very slow sweep, so that the voltage does not change in the short time it takes for PteroDAQ to respond to the trigger. Connect the function generator to the input of the Schmitt trigger and to an analog pin on the microcontroller. Connect the output of the Schmitt trigger to a digital pin.

> If PteroDAQ triggers on a rising edge, when the output of the Schmitt-trigger inverter has just changed from low to high, then the input has just dropped below V_{IL}.
>
> If PteroDAQ triggers on a falling edge, when the output of the Schmitt-trigger inverter has just changed from high to low, then the input has just risen above V_{IH}.

You can take many measurements in a couple of minutes, and PteroDAQ records and averages them for you. PteroDAQ also reports the power-supply voltage, which is an important parameter to include in any report, because V_{IL} and V_{IH} vary with the power-supply voltage.

If you don't have a function generator, it is possible to do the sweeps with just a potentiometer as a voltage divider with the 3.3 V power supply from the microcontroller board as the input voltage. You can even make these measurements before lab time, so that you have exact voltages to use in the pre-lab questions. Sweeping back and forth with a multi-turn trimpot is a bit tedious, especially if you don't go far enough in each direction to change the state of the Schmitt trigger, so using a function generator is an easier method.

With the Analog Discovery 2, it is easy to verify the voltage range that your function generator will put out before you enable the generator, as it shows you a plot of one period of the waveform with the voltages clearly marked. With other function generators, you might want to confirm the voltage range on an oscilloscope before applying the signal to the input of the 74HC14N, to make sure you stay in the legal range.

For example, you can set the amplitude and offset on the Agilent 33120A function generators—see the User's Guide [1]—but the settings on the Agilent 33120A usually need to be half what you think they ought to be (see page 188).

Once you have the threshold voltages V_{IH} and V_{IL}, recompute the RC time constant, and check that your starting values for R_1 and C_1 based on your initial estimates of the voltages still seem reasonable. If not, adjust them.

17.4.2 *Breadboarding the hysteresis oscillator*

The hysteresis-oscillator board that you will solder up does not have exactly the same circuit as Figure 16.7. The circuit that the board implements is shown in Figure 17.2.

The circuit on the PC board has one extra component: a *bypass capacitor* (also known as a *decoupling capacitor*) to keep the fluctuations in current from the inverter chip from propagating too much voltage noise into the power lines. Bypass capacitors to keep noise from propagating through power wiring are an important part of both digital and analog design—the Teensy boards have 8 or 9 bypass capacitors. (See Section 11.10 for more information.)

Although your circuit will work fine without the bypass capacitor, as there are no other chips to disturb with the noise except on the microcontroller board, which has more than adequate bypass

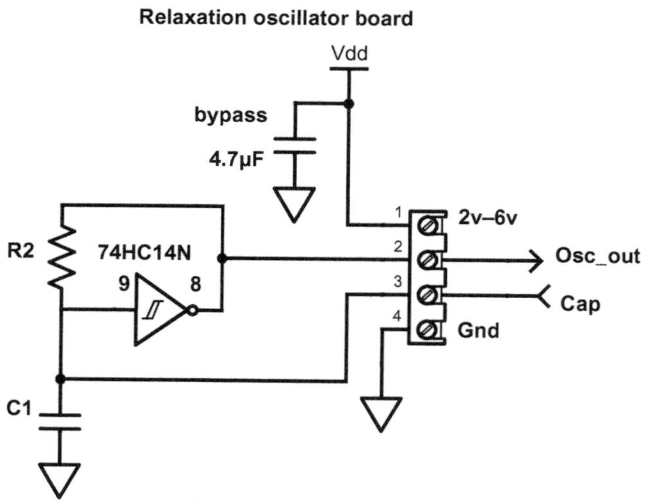

Figure 17.2: *The schematic for the rev0.5 hysteresis oscillator PC board. The bypass capacitor is near pin 14 of the 74HC14N chip, and C1 for the oscillator is near pin 7. Only one of the six Schmitt-trigger inverters in the package is used—which one has been indicated by putting pin numbers on the schematic. Pin numbers on schematics are a great aid when debugging circuits, and you should put them on all your schematics.*
The screw terminal shown in the schematic is a way to connect wires non-permanently to a printed-circuit board—it would not be used on a breadboard, which already provides a way to attach wires non-permanently.
Figure drawn with Digi-Key's Scheme-it [18].

capacitors, you should get into the habit of including them—the 0.1 μF capacitor (the largest of the little disk capacitors in your assortment) is a reasonable size, though I often like to use 4.7 μF ceramic capacitors as bypass capacitors.

Make your touch plate by folding a piece of aluminum foil into a neat rectangle about 2–3 cm × 5–7 cm. There should be four or more layers of aluminum foil throughout and eight or more layers on the end that you will clip an alligator clip to. The extra layers are to keep the foil from tearing when clipped to.

Then fold a 9 cm-to-12 cm-long piece of packing tape over the foil, so that the foil is covered with a single layer of tape everywhere except at one end, where bare foil sticks out to grab with an alligator (see Figure 17.3).

We use an alligator clip to attach the aluminum foil to a wire, rather than attempting to make a solder joint, because soldering to aluminum is very difficult. The aluminum oxide on the surface of the aluminum is not removed by standard fluxes and standard solders do not wet aluminum well. There are

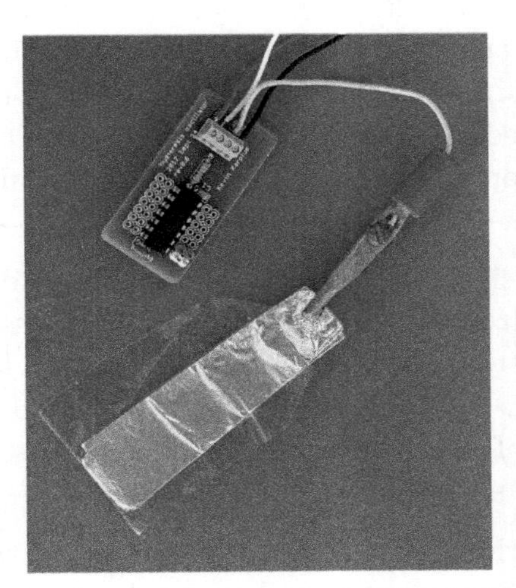

(a) *The red, black, and white wires go to a breadboard, either for frequency measurement with PteroDAQ or for output waveform display with Analog Discovery 2.*

The white and blue header pins on pins 1 and 2 of the 74HC14N chip are for testing a single Schmitt trigger that isn't used in the oscillator. These pins are not needed for the touch sensor.

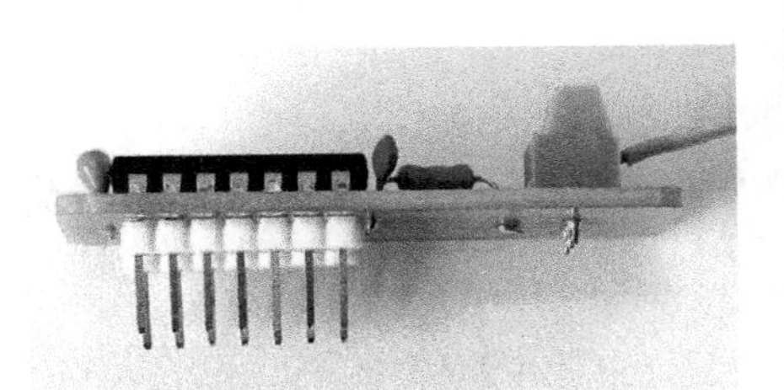

(b) *Alternatively, male headers can be soldered to the bottom of hysteresis-oscillator board.*

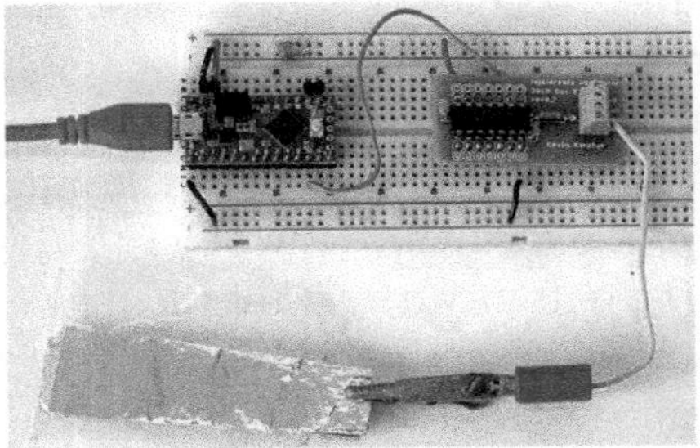

(c) *With header pins soldered to the bottom of the hysteresis-oscillator board, the board can be plugged into a breadboard, with wires connected to either the breadboard rows or the screw terminal. The undedicated Schmitt triggers can be used normally.*

Figure 17.3: *Touch sensor made with aluminum foil and packing tape, connected to the hysteresis-oscillator board.*

specialty fluxes and solders for soldering aluminum, but even with these materials, soldering aluminum is much more difficult than soldering copper [106].

> Measure a double thickness of the packing tape (stuck to itself without aluminum foil in between) with a micrometer to check the thickness estimate used in the pre-lab. Different brands of packing tape have different thicknesses! Remember to divide your reading by two, to get the thickness of a single layer of tape.

You can also make a touch sensor by taking two insulated wires twisted together connecting one end of each wire to the two plates of C_1. Touching the insulation on the pair of wires increases the capacitance between the wires. A paired-wire sensor does not rely on your being grounded, and so is less susceptible to differences in behavior depending on what else you are touching.

Wire up your circuit on a breadboard (keeping the wires fairly short, to reduce stray capacitance), and connect the touch plate to the inverter input with a wire and an alligator clip. Power the circuit either from the microcontroller board or from the bench power supply. For +3.3 V and GND wires, remember that the standard convention is red for +3.3 V and black for GND. Not following this convention will make it much harder for other people to help you debug your circuit.

> Look at the output with an oscilloscope. Is it oscillating? How does the period of the oscillation change as you touch the touch sensor? If it is changing by less than 20%, try using a smaller capacitor for C_1. What is the period? If it is too short, try using a larger R_1 to increase the RC time constant. What happens if you leave the touch plate connected but remove C_1? What happens to the output of the oscillator if you look at the input to the oscillator with the oscilloscope?

Measure the frequencies or periods with different capacitance values: with the touch plate unconnected; connected but not touched; connected and touched; and connected, touched, and with person grounded (to the ground used by the hysteresis oscillator). The different configurations correspond to C_1 only, $C_1 + C_3$, $C_1 + series(C_2, C_4) + C_3$, and $C_1 + C_2 + C_3$. You can also try changing C_1 to see the effect of changing the capacitance on the frequency.

You can measure the frequency using the Analog Discovery 2's "Measurements" options under the "View" menu of the "Oscilloscope" tool, though a frequency channel of PteroDAQ is easier to use. To get accurate measurements, make sure that you have many periods on the screen for the oscilloscope.

If you don't have an oscilloscope, you will have a hard time viewing the signal itself with PteroDAQ, as the intended frequency of operation is much higher than the sampling rate for PteroDAQ (unless you deliberately design the oscillator to work at the low end of the usable range). You can, however, look at the *frequency* of the output with one of the *frequency* channels on PteroDAQ, as the output is a square wave with clean edges—exactly what is needed for the PteroDAQ frequency channels. To monitor the frequency, you want to use a 60 Hz sampling frequency (or a factor of it, like 20 Hz or 30 Hz), so that any frequency modulation of the oscillator output by capacitively coupled line noise is averaged out (see Section 16.5).

The oscillation you get with just the internal capacitance of the 74HC14N is too high to measure with PteroDAQ, which can't handle frequencies above about 4 MHz.

How much does the frequency change with how firmly you press the sensor? The Al foil is not sensitive to pressure, so why does the frequency change? Bonus: measure the frequency of the oscillator that uses just the internal input capacitance of the 74HC14N chip itself (this may exceed the frequency limits of the PteroDAQ frequency channel).

The frequency should be inversely proportional to the capacitance, so you can use the ratio of two frequencies to get the corresponding ratio of capacitances, without relying on all the simplifying assumptions of the calculations in Exercises 16.1–16.2 and Pre-lab 4.1.

Use the ratios of frequencies to estimate the ratios of capacitance, and assume that your labeled capacitance is the value it is labeled (we'll look next quarter at ways that you can measure the capacitance more accurately). Once you have one known capacitance and a bunch of ratios, you should be able to determine the values of the other capacitors in your system.

It is valuable to plug the resulting capacitance values back into your RC model to see how well that model predicts the frequencies. Use the nominal value of your feedback resistor, as the 1% tolerance on the resistor is more accurate than DT9205A multimeters used as ohmmeters. If you have access to bench ohmmeters that are accurate to about 0.1%, then measuring is better than relying on the nominal value.

The simple RC model we used to get the timing is not very accurate when C is small, because of the effect of feedback capacitance (as explained in Section 16.4), so you should not expect the RC timing calculation to provide accurate frequency estimates for the oscillator, nor should you expect the frequency measurements to provide accurate capacitance estimates from the RC model—that is why I recommend using the ratio of frequencies to get the ratio of capacitances for estimating the unknown capacitance values.

> Try turning the touch plate into a proximity sensor—keep the sensor in a fixed location and move your hand to be nearby. Measure the distance to your hand with a ruler, and plot the frequency as a function of distance. At what range does the noise of the measurement exceed the observable signal? Could you make the sensor more sensitive by using a larger area for your touch plate?

You can use the standard equation, Equation 10.1, to approximate the expected capacitance of the proximity sensor by treating the sensor as two capacitors in series: one corresponding to the thin layer of tape and one corresponding to the thicker layer of air.

In previous versions of this lab, I provided students with a program that made the frequency measurements on a FRDM-KL25Z microcontroller board, and changed the color of the LED when the frequency dropped sufficiently below the initial frequency. That code used hysteresis internally to keep the LED from flickering when the finger touch was near the on-off threshold. Based on the recordings you made of frequency with and without the touch, where would you make the two frequency thresholds for turning an LED on or off? How would you adjust those thresholds for changes in the components (for example, the touch plate capacitance changes depending on where it is placed and what conductors are nearby)? Can you think of a way to have such a program set the thresholds automatically?

17.4.3 *Using hysteresis to clean up a noisy analog signal*

This section assumes that you have an Analog Discovery 2 or two other function generators that can generate a sine wave and a noise signal. If you don't have such equipment, this part of the lab should be omitted.

The main use of hysteresis is to clean up noisy analog signals to make clean, single-edge digital signals.

We can create a noisy analog signal using the Analog Discovery 2's function generators. One channel can be set to produce a sine wave (perhaps from 0.5 V to 2.7 V at 100 Hz), while the other produces high-frequency noise (perhaps from 1.6 V to 1.7 V at 40 kHz). Using the circuit you developed in Pre-lab 4.6, average the two signals and provide the output of that circuit to the input of the Schmitt trigger.

Connecting the function generators to the breadboard will require three wires: the two function generator outputs and ground.

The noisy-sine demo is easiest to do if you have a 74HC14N chip on a breadboard, but even if you have already soldered all your 74HC14N chips to the hysteresis printed-circuit boards, you can still connect to the Schmitt-trigger inverters on the chip using the extra soldering holes provided on the board. You can solder in a pair of male header pins for connecting to the oscilloscope, plus a wire to connect the input of the Schmitt trigger to your averaging circuit on a breadboard. Power and ground can be connected through the screw terminals, just as when running the oscillator. Alternatively, you can solder male header pins to the bottom of the board on one side, to plug the board into a breadboard.

You can look at the input and output of the Schmitt trigger with either PteroDAQ or the oscilloscope channels of the Analog Discovery 2. Because PteroDAQ does not support very high frequency sampling, it is best used with 2 kHz sampling, 10 Hz sine, and 400 Hz noise. The Analog Discovery 2 can use higher frequencies, limited mainly by the low-pass filtering caused by the capacitance of the oscilloscope channel as a load on the averaging circuit. The power for your circuit will come from the Teensy board, which in turn is powered from the USB port on your laptop.

One interesting phenomenon when using the Teensy LC is that the microcontroller injects a little noise on the analog pins that it is reading. If the output impedance of the averaging circuit is too high, the voltage of this noise can be large enough to cross the hysteresis thresholds of the Schmitt trigger, causing output of the Schmitt trigger to flip as result of reading the voltage at the input!

If you use the oscilloscope inputs of the Analog Discovery 2, you need to connect two wires for each channel, as they are differential inputs—when looking at single-ended signals, the positive oscilloscope connection goes to the signal, and the negative one to ground.

There also needs to be a ground connection between the ground of the system being measured and the ground of the oscilloscope, to make sure that the average (common-mode) voltage for the differential input stays within range. Because you are using the function generator, you should have already connected the ground.

Your setup should be using a total of 7 wires from the Analog Discovery 2 to your breadboard (and PC board): 2 function generator, 1 ground, and 4 oscilloscope inputs. If you power the 74HC14N chip from the Analog Discovery 2 power supply, that adds an eighth wire.

Do not connect the Analog Discovery 2 power supplies to the +3.3 V connection on the Teensy board. Driving the +3.3 V pins from the outside can damage the Teensy board.

Once you have the oscilloscope wired up, you can set up the user interface. I find the following sequence of actions useful:

- Set the voltage offset and range for each channel so that your signal should be clearly visible on the screen. The left-hand y-axis has tick marks that correspond to the voltages for the channel selected at the top of the plot. The offset is *added* to your signal to make the display, so if you want your display to be centered at 1.5 V, you need to set the offset to −1.5 V.

Set the triggering to be on the rising edge of the output channel, with a voltage in the middle between the power rails. Set the time base so that one period of your waveform is around 2 divisions on the screen. For example, if you are looking at a 1 kHz signal, you will want about 500 μs/division.
There is an "Autoset" button (normally hidden—use the downward arrow at the right of the triggering options to reveal it) that attempts to get reasonable initial values for the vertical and horizontal settings of the two channels. If your signal is not appearing at all on the screen, using autoset may make it visible for you, but you'll usually have to adjust all the parameters manually afterwards, so autoset is often a last-resort option when all the settings are wrong.

- Use the gear icon on the oscilloscope display pane to choose the "light" color option. With this option you can avoid the black background that does not work well in printing or projecting images.
- Use the gear option on each channel to change the name of the channel to something meaningful (like "input" or "output") rather than "Channel 1" and "Channel 2". At the same time, you may want to turn the "noise" display off and perhaps change the colors of your traces—the default orange for Channel 1 is too pale for the "light" background.
- Adjust the offset and range for each channel to make the signals easily visible on the screen. The offset can be adjusted either by changing the value displayed, or by moving the arrow on the left side of the screen up or down. The arrows are color-coded to correspond to the channels.
- Only one set of y-axis tick marks are displayed (color-coded for the channel). The color-coded buttons at the top of the oscilloscope pane select which axis labels are used.
- Adjust the triggering to trigger on what you want. The "Autoset" button usually sets the trigger to Channel 1 rising at the middle voltage for the channel, but you probably want to trigger on your output channel, which has crisper edges and less noise than the input channel.
- You may wish to adjust the horizontal scale with the "Time" subpane on the left. The "position" is what time the center of the horizontal axis corresponds to, expressed as the delay after the trigger event. You can center the screen *before* the trigger, as the oscilloscope can store up to 8192 samples continuously, and just take a snapshot when the trigger event happens. Changing the position can be done by entering a number in the position box, or by using your mouse to drag the signal in the oscilloscope pane left or right.
- Once you have the data you want displayed on the screen, you can stop the oscilloscope and export the data either as a PNG image or as a data file, using the "Export" option under the "File" menu.

> I recommend tab-separated data files, which are easiest to work with in gnuplot. You should add metadata (explaining the experimental setup, such as the Wavegen settings) to the header when you save the file.

The PNG image of the screen is useful for tutorials on how to use the oscilloscope, but is a terrible way to record data, as it is very difficult to recover the data from the image for further analysis.

After you have confirmed that the Schmitt trigger is doing what you want and recorded waveforms to include in your report, do more experiments:

- Try increasing the noise level in the Wavegen tool until the output shows multiple edges where only one is expected. Around what amplitude is the noise?
- Bonus: with the noise level low enough to get single edges, look in more detail at one edge, by changing the time base to 200 ns/division. Are you still seeing only one edge? (If not, try adding an extra 20 pF–100 pF between the input of the Schmitt trigger and ground.)

- Bonus: if you have spare time, you can try forcing ringing on the transitions by adding 20 pF–100 pF of feedback capacitance from the output to the input. You can change the triggering of the oscilloscope to trigger on the output when it has a pulse less than 1 μs long, rather than on every edge—that will catch just the edges that have ringing, even if they are somewhat rare.

Only do bonus steps in the lab after you have completed all the required steps—there is no bonus if you don't complete the required parts!

17.4.4 *Soldering the hysteresis oscillator*

Put the components on the board **in the right orientation**. Follow tutorial instructions on how to solder the components in place—you may want to re-read the instructions in Section 3.3. Be careful not to burn yourself with the iron!

Make sure that all your connections are shiny (not cold-soldered) and that no adjacent pins have been soldered together. If you do accidentally solder two points together, remelt the solder and use the solder sucker to remove the excess.

Use a handheld multimeter to check that the +3.3 V and GND terminals are not shorted together, then hook up wires to PteroDAQ (or to an oscilloscope) and test the soldered board the same way you tested the breadboard version—measuring the frequency without the touch plate, with the touch plate connected but not touched, and with the touch plate being touched.

Bonus: you can listen to the output of your oscillator on an earbud, if you lower its frequency into the audio range by adding capacitance between the "cap" input and ground. See Lab 9 for information about connecting to earbuds.

17.5 Demo and write-up

There are five checkpoints in this lab that should be demonstrated to an instructor:

1. V_{IH} and V_{IL} measured.
2. Noisy sine wave converted to clean digital signal.
3. Breadboard hysteresis oscillator oscillating and displayed on the oscilloscope.
4. Breadboard hysteresis oscillator working with PteroDAQ to measure frequency.
5. Soldered hysteresis oscillator working with microcontroller to measure frequency. The instructor may wish to examine the solder joints on the back of the board.

Turn in a write-up describing what you did. Remember that the instructor is *not* your audience—a bioengineer who has not read the assignment is.

Provide at least the following:

- Write up all the parameters that you measured, explaining briefly how they were determined. Make sure you include the measured voltage of your power supply when you are reporting V_{OH} and V_{OL}—PteroDAQ reports the voltage it is using as a reference in its metadata.
- Provide a plot of a noisy input signal and the corresponding output signal as two curves on the same plot.
- Provide a plot of frequency vs. time for the sensor being not touched and being touched.
- Provide a plot of frequency vs. distance for the sensor being used as a proximity sensor.
- Write up both your initial estimate of C_2 and improved estimates after you have measured the frequency of your oscillator under different conditions.

Chapter 18

Amplifiers

18.1 Why amplifiers?

An *amplifier* is a device that increases the amplitude of electronic signals, taking power from a power supply rather than from the input signal. The most common type of amplifier is a *voltage amplifier*, in which both the input and the output signals are voltages. There are also *current amplifiers*, in which both the input and output are currents, *transimpedance amplifiers*, with current inputs and voltage outputs, and *transconductance amplifiers*, with voltage inputs and current outputs. (You can remember the "trans-" names, by considering the output divided by the input: voltage divided by current is impedance, and current divided by voltage is conductance.)

For most of this chapter, we'll be talking about voltage amplifiers, and that should generally be assumed when the word "amplifier" is used without a qualifying adjective.

Voltage amplifiers amplify a voltage—a potential difference between two nodes in the circuit. We classify the voltages into two broad categories: single-ended and differential.

> A *single-ended signal* is a voltage relative to a fixed node (often ground, but also commonly a node with a reference voltage, V_{ref}). The fixed, reference node is usually shared by several parts of the system.

When we label single-ended signal on a block diagram, we usually give the DC offset (relative to ground) and the amplitude of the signal: $1.65\,\text{V} \pm 0.6\,\text{V}$, for example. Sometimes a range is given rather than a DC offset and amplitude: $1.05\,\text{V}$–$2.25\,\text{V}$ instead of $1.65\,\text{V} \pm 0.6\,\text{V}$.

Which labeling is better depends on the meaning of the signal—if the DC offset is just for the convenience of the electronics and does not carry information, then the offset and amplitude notations is more useful, but if the DC part of the signal carries information, then the range would be more appropriate. In the pressure-sensor lab (Lab 5), the DC part of the signal corresponds to the air pressure and is a major part of what we are measuring, so the range notation would be best. In the pulse-monitor lab (Lab 6) we are not interested in the overall light level, but only in the fluctuation, so the offset and amplitude notation may be better.

> A *differential signal* is the voltage between a pair of nodes, called a *differential pair*, both of which are associated with the signal, not shared with other parts of the system.

When we put differential signals on a block diagram, we use only a single line on the diagram, even though the signal is carried by a pair of wires. The signal is labeled by the amplitude or range of

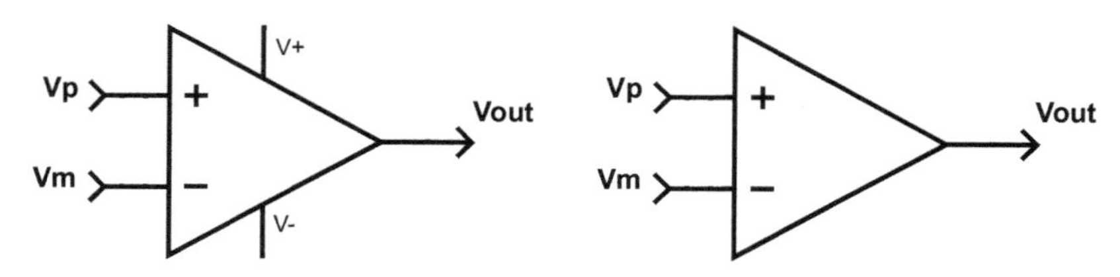

Figure 18.1: *The schematic symbol for a differential amplifier has two inputs on the left (labeled "+" and "-") and a single output on the right. Additional inputs (such as power supplies or the reference voltage for the single-ended output) are generally added to the top or bottom of the symbol, with positive power supplies on the top and negative power supplies on the bottom.*
When all amplifiers in a design use the same power supplies, designers often omit the power-supply connections, showing just the inputs and outputs of the amplifiers.
In a block diagram, rather than a schematic, differential signals are usually shown as just a single line, and so a conventional amplifier symbol with a single input is used—if the differential signaling needs to be noted, it is done in the label for the edge representing the signal. Figure drawn with Digi-Key's Scheme-it [18].

the *difference* voltage, "±100 mV differential", not by the individual voltages of the wires separately. If necessary, the average of the voltages on the two wires (called the *common-mode voltage*) can be included on the label: "±100 mV differential, CM 1 V–3 V".

The distinction between single-ended and differential signals is a subtle one, since a single-ended voltage and a reference voltage can always be treated as a differential pair. The converse, however, is not true—a differential pair cannot usually be handled as if one node were a fixed reference voltage.

> A *differential amplifier* has two inputs that form a differential pair, and outputs a single-ended output relative to a reference voltage that is either provided to the amplifier or derived from the power supply to the amplifier. The schematic symbol for a differential amplifier (show in Figure 18.1) has two connection ports on the left for the two input nodes and one at the tip of the triangle on the right for a single-ended output node.

The main theme of this book is connecting sensors to computers, and amplifiers are often a major part of that connection, because the sensor provides insufficient voltage or current for direct digitization. The design constraints for the amplifier are usually determined by working inward from both ends—from the sensor outputs and from the computer's analog-to-digital inputs.

This chapter will look at amplifier parameters (Section 18.2), at multi-stage amplifiers (Section 18.3), at some examples at the block-diagram level (Section 18.4), and at instrumentation amplifiers (Section 18.5), which are commonly used building blocks. The next chapter, Chapter 19, will go into more detail on operational amplifiers, the most commonly used building block in amplifier design.

18.2 Amplifier parameters

There are many amplifier parameters given on data sheets for amplifiers—this section will cover a few of the ones that affect our designs the most.

18.2.1 *Gain*

The most important characteristic for an amplifier is usually its *gain*, the change in output divided by the change in input. For voltage and current amplifiers, the gain is unitless (V/V or A/A), but for transimpedance amplifiers, the gain is in ohms (V/A = Ω), and for transconductance amplifiers, the gain is in siemens (A/V = S).

Gain is not output over input, $V_{\text{out}}/V_{\text{in}}$,
but the change in output over the change in input, $\frac{dV_{\text{out}}}{dV_{\text{in}}}$.

If you plot the output vs. the input (the *transfer function*), then the gain is the slope of the curve. The notion of gain is the same as the notion of sensitivity in a sensor: how much does the output change for a change in the input.

The general principle for choosing the gain of an amplifier is to start by looking at

- what the inputs are (what signals do we get from the sensor, how big are they, and what frequency range are we interested in?) and
- what the outputs need to be (how much voltage and current do we need?).

For example, if we expect an input signal that is $1\text{V} \pm 5\,\text{mV}$ and we need the output to range from $0.5\,\text{V}$ to $3\,\text{V}$, then the amplifier will have to have a gain of $\frac{3\,\text{V}-0.5\,\text{V}}{1.005\,\text{V}-0.995\,\text{V}} = 250$, and we will have to do something to recenter the voltage at $1.75\,\text{V}$, instead of $1\,\text{V}$.

The gain does not have to be constant for all input voltages. For example, the 74HC14 Schmitt trigger that we looked at in Figure 16.5 has essentially 0 gain for most input voltages, but very large negative gain where the output makes its transition between states—that low gain except at the transition is typical of digital amplifiers.

For most analog designs, though, we want *linear* amplifiers, where the transfer function is essentially a straight line for a certain range of inputs and the gain is constant over that range.

For a differential amplifier, which converts a differential signal into a single-ended one, the output is encoded as $V_{\text{out}} - V_{\text{ref}}$, where V_{ref} is the voltage output when the input difference is zero: $V_{\text{p}} - V_{\text{m}} = 0\,\text{V}$.

The gain of a differential amplifier is

$$G = \frac{V_{\text{out}} - V_{\text{ref}}}{V_{\text{p}} - V_{\text{m}}} ,$$

where V_{p} and V_{m} are the "plus" and "minus" input ports of the differential amplifier, as shown in Figure 18.1.

The reference voltage V_{ref} is usually externally supplied to the amplifier for greater design flexibility, though some amplifier designs may generate it internally.

18.2.2 *Gain-bandwidth product*

The gain discussed in Section 18.2.1 is not necessarily the same for all frequencies of input signals. In fact, the standard amplifier chips that we use as building blocks (*op amps*) have built-in low-pass filters, so that the gain decreases as frequency increases. This low-pass filter is intended to prevent undesired high-frequency oscillations, which can be difficult to avoid in an amplifier that has high gain at high frequencies.

At low frequencies, we can easily get a fairly large gain from a single op amp, but at high frequencies this can be difficult, as for most of the frequency range the maximum gain G is inversely proportional to the frequency f—that is, Gf is a constant, referred to as the *gain-bandwidth product*.

The gain-bandwidth product has many standard abbreviations, including GBWP, GBW, GBP, or GB, but in this book we'll use f_{GB} to emphasize that the units are Hz.

The gain-bandwidth product is typically 1 MHz for the MCP6004 amplifiers, so the gain for the chip at 2 kHz is 500, at 20 kHz is 50, and at 200 kHz is only 5. Gain-bandwidth products for amplifiers vary from 1 kHz to 18 GHz, but larger is not always better, as other properties of an amplifier may need to be sacrificed to get high gain at high frequencies.

Because the gain of op amps is limited by an internal low-pass filter, we can approximate the gain of the op amp, A, as

$$A = \frac{A_0}{1 + j\omega\tau},$$

where A_0 is the DC gain and τ is a time constant that depends on the internal filter design. Because these two parameters are coupled and are not independently carefully controlled in manufacturing, it is standard to report the gain-bandwidth product instead, which is $f_{\mathrm{GB}} = A_0/(2\pi\tau)$, the frequency at which the magnitude of the op amp's gain is approximately 1.

The gain of an op amp at frequency f can then be approximated as

$$A = \frac{1}{1/A_0 + jf/f_{\mathrm{GB}}},$$

which is approximately A_0 for very low frequencies, where $f < f_{\mathrm{GB}}/A_0$ and approximately $-jf_{\mathrm{GB}}/f$ for moderate frequencies, where $f_{\mathrm{GB}}/A_0 < f < f_{\mathrm{GB}}$.

This approximation breaks down when f gets large (somewhere around $f = f_{\mathrm{GB}}$), because the simple approximation to op-amp gain is no longer good at high frequencies—additional filtering is built into the op-amp circuit for higher frequencies.

For a more detailed view of the effect of the limited gain at high frequencies, see Section 19.5.

Data sheets usually give only typical values for the gain-bandwidth product, not worst-case values. For safe designs, it is recommended that you not exceed about 60% of the maximum gain that would be available with the typical gain-bandwidth product.

18.2.3 *Distortion and clipping*

We generally want our amplifiers to be *linear*, which means that if we take a complicated signal $cf(t) + dg(t)$ as input to an amplifier, the effect is the same as if the components of the signals had been run through the amplifier separately:

$$A\left(cf(t) + dg(t)\right) = cA\left(f(t)\right) + dA\left(g(t)\right) .$$

Linear amplifiers make sine wave inputs into sine wave outputs. They may filter the signal (changing the amplitude or the phase of the sine waves differently for different frequencies), but they do not distort a sine wave into a different waveform shape.

Waveforms other than sine waves may change shape as result of filtering—for example, a low-pass filter will round the corners of a square wave, but these changes can be analyzed as the combined effect of changes in amplitude and phase for the underlying sine waves that make up the signal (a method known as *Fourier analysis*).

The term *distortion* is not used for the changes made by linear filtering, but only for nonlinear changes. nonlinear amplifiers distort the waveform, so that an input sine wave does not result in an output sine wave.

Sometimes the nonlinearity is deliberate (such as the logarithmic amplifiers of Section 23.2) and sometimes it is an undesired by-product of limitations of the amplifier.

The most common form of distortion is *clipping*, which results from limitations on the output of an amplifier stage, as shown in Figure 18.2. Amplifiers generally have limitations on both the output voltage range and how much current the outputs can supply. Both the output voltage range and the current limitations usually depend on the power-supply voltage chosen.

Figure 18.3 shows the distortion caused by clipping a simple sine wave. The distortion is commonly observed in amplifiers whose gain is set higher than output range of the amplifier. Signals smaller than clipping limits are not distorted, but those that go outside the clipping limits are truncated, resulting in flat spots that introduce *harmonic distortion*, with substantial energy being transferred from the frequency of the sine wave, f, (the *fundamental frequency*) to higher harmonics ($2f$, $3f$, $4f$, $\ldots$).

Speech and music are distorted in irritating ways by clipping, and clipping signals from sensors can greatly distort measurements being made. To avoid this distortion, we nearly always design our amplifier systems so that clipping does not occur.

Amplifiers are usually classified into two groups based on their output voltage limitations:

(1) *rail-to-rail* amplifiers, whose outputs can come very close to the positive and negative power-supply voltages, and
(2) other amplifiers, in which the output voltage has a much more limited range.

For example, the instrumentation amplifier INA126PA is not a rail-to-rail amplifier—the output cannot go all the way to the positive power rail V_+ nor to the negative power rail V_-. The output

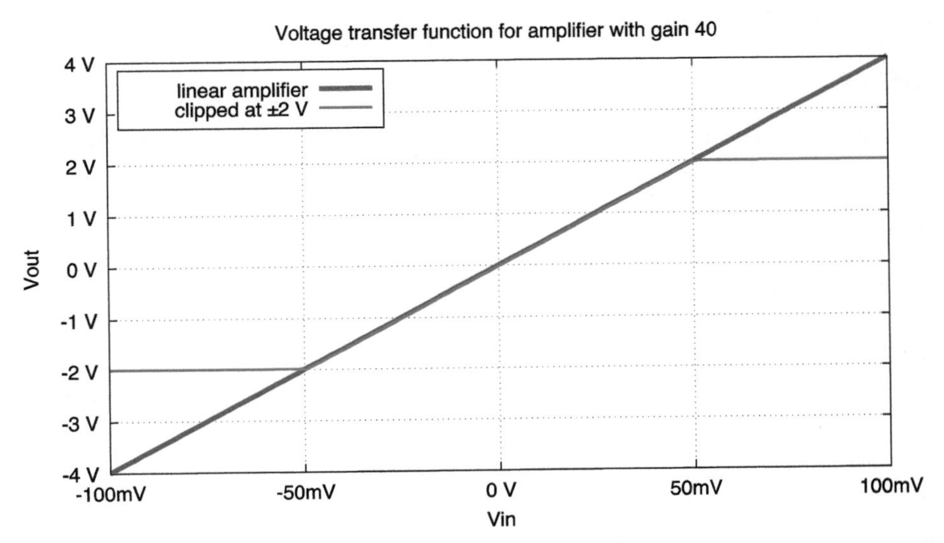

Figure 18.2: *Voltage transfer functions for two amplifiers: one a linear amplifier, the other clipping at* ± 2 *V. Both amplifiers have a gain of 40. As long as the input remains in the range* $\pm 50\,mV$*, the two amplifiers behave identically, but for larger input signals, the clipping amplifier distorts the signal, as shown in Figure 18.3.*

Figure drawn with gnuplot [33].

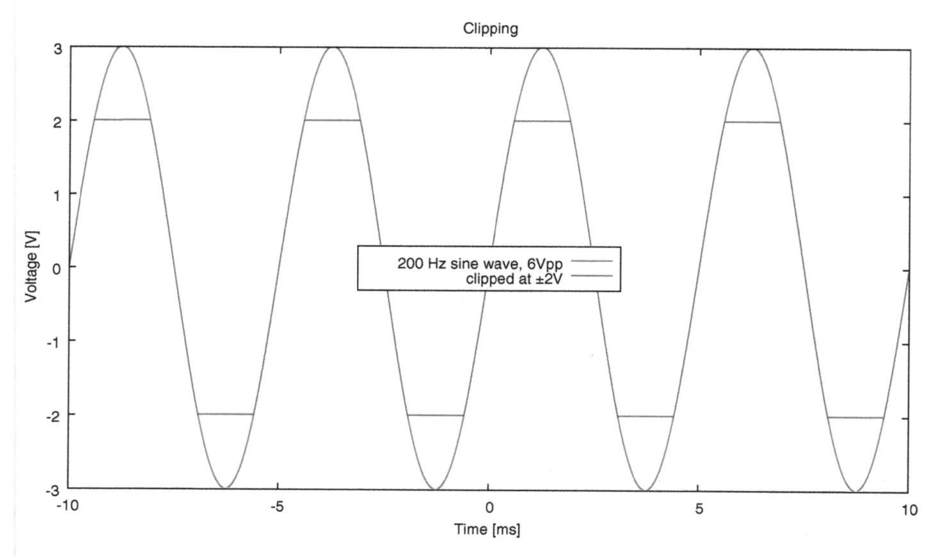

Figure 18.3: *The 6 V peak-to-peak sine wave shown in red changes shape considerably when clipped to ±2 V. V_{PP} is a common notation for peak-to-peak voltage. Signals that stayed within the ±2 V clipping limits would not be affected by the clipping. Figure drawn with gnuplot [33].*

is only guaranteed to work in the range $V_- + 0.95\,\text{V}$ to $V_+ - 0.9\,\text{V}$, though it typically can cover the wider range $V_- + 0.8\,\text{V}$ to $V_+ - 0.75\,\text{V}$. (Look for the "Output Voltage, Positive" and "Output Voltage, Negative" specifications on the INA126 data sheet [97].) Because of this range limitation on the output, we often need to keep the gain of the instrumentation amplifier fairly small and follow it with a rail-to-rail op amp to get full-scale voltage range.

Clipping can also be caused by current limitations of an amplifier, particularly when driving a low-impedance load, like a loudspeaker. For example, if we try to drive an $8\,\Omega$ load whose other end is connected to V_x with an amplifier that has a $\pm 20\,\text{mA}$ output limitation, then our output voltage range is limited to $V_x \pm 20\,\text{mA}\ 8\,\Omega = V_x \pm 160\,\text{mV}$.

18.2.4 *Input offset*

Two other important properties of instrumentation amps (and of op amps, for that matter) that matter to us are the *input offset voltage* and the *input bias current.*

The *input offset voltage* V_{offset} is the differential voltage between the inputs needed to make $V_{\text{out}} - V_{\text{ref}} = 0\,\text{V}$. Ideally it is zero, but amplifier specifications give a range around zero that it is guaranteed to be in. Most of the input offset voltage comes from inherent mismatches of the input transistors and other components of the amplifier [84]. On Digi-Key in 2016, 90% of the instrumentation amps had $|V_{\text{offset}}| \leq 250\,\mu\text{V}$, and the INA126PA instrumentation amps (one of the cheapest available as through-hole parts) are specified to have a typical offset $|V_{\text{offset}}| \leq 150\,\mu\text{V}$ and a maximum input offset of $\pm 500\,\mu\text{V}$.

18.2.5 *Input bias*

Input bias current is the current needed at the input of an amplifier for it to function correctly. There are two different parameters that are relevant—the *bias current,* which may be the current in either input, and the *offset current,* which is the difference in the bias currents and contributes to the offset in the output voltage. The bias current varies in instrumentation amps from about $3\,\text{fA}$ to $2.5\,\mu\text{A}$—on Digi-Key 90% of the instrumentation amps have $\leq 22\,\text{nA}$ of input bias current, and the INA126PA has

$-10\,\text{nA}$ typical input bias current (max $-50\,\text{nA}$). But the INA126PA has only $\pm 0.5\,\text{nA}$ typical ($\pm 5\,\text{nA}$ max) input offset current—the bias currents into the two inputs are well matched.

One can build instrumentation amps out of separate op amps, as explained in Sections 40.1–40.2, but it is difficult to match amplifiers and resistor values precisely when doing so. Integrated circuit manufacturers make single-chip instrumentation amps with laser-trimmed resistors for very precise matching of component values that track each other over a wide temperature range. (Having the components on the same chip helps keep them very close in temperature also.)

The INA126PA instrumentation amp is very cheap for a through-hole part and is one of the worst one-chip instrumentation amps with respect to input voltage offset and input bias current. Still, it would be difficult to build an instrumentation amp as good out of separate op amps and resistors. While 90% of instrumentation amp chips have $\leq 250\,\mu\text{V}$ of input offset voltage, the corresponding level for op-amp chips is $3\,\text{mV}$, and only about 40% have $\leq 250\,\mu\text{V}$. While 90% of instrumentation amps have $\leq 22\,\text{nA}$ of input bias current, the corresponding level for op amps is $11\,\mu\text{A}$, and only 7% have $\leq 22\,\text{nA}$. (Percentages computed from sorting the relevant parts lists on Digi-Key's web site in 2016.)

It is possible to add trimpots and other external circuitry to instrumentation amps made from op amps to correct the imbalances that lead to large input offset voltage and input offset current, and engineers used to have to do that all the time. Nowadays, it is much cheaper and more reliable to buy an integrated instrumentation amp with the desired properties.

18.2.6 *Common-mode and power-supply rejection*

The output of a differential amplifier is supposed to depend only on the difference between the inputs $V_\text{p} - V_\text{m}$, and not on the average of the inputs, $(V_\text{p} + V_\text{m})/2$. The average of the inputs is called the *common-mode voltage*, and data sheets for instrumentation amplifiers and op amps usually specify the *common-mode rejection ratio* (CMRR), as a gain for the amplification of the common-mode voltage. Typical values are in the range $-100\,\text{dB}$ to $-60\,\text{dB}$ (gains of 0.00001 to 0.001—see Section 2.3.2 for definition of dB), meaning that a $1\,\text{V}$ change in the average of the inputs results in only a $10\,\mu\text{V}$ to $1\,\text{mV}$ change in the output.

The range of values allowed for the inputs V_p and V_m varies with the design of the amplifier, but generally they need to stay between the power rails—the limitation is often presented on data sheets as a limitation on the common-mode voltage.

An amplifier should also be good at eliminating noise from the power supply. The *power-supply rejection ratio, PSRR,* is generally also expressed as a gain, with $\text{PSRR} = -60\,\text{dB}$ meaning that a $1\,\text{V}$ change in the power-supply voltage would result in a $1\,\text{mV}$ change in the output (assuming that the inputs and the output stay within the legal range for the amplifier with that power-supply voltage).

18.2.7 *Other amplifier parameters*

In addition to input offset and input bias, there are many other parameters that can be important in some designs: the input voltage noise, the slew rate, the bandwidth, ... In this course, we look mainly at fairly large, slow signals going to an analog-to-digital converter that has fairly coarse steps ($3.3\,V/2^{16} \approx 50\,\mu\text{V}$), and so we can simplify our design process by ignoring most of these parameters. For applications with lower-voltage or higher-frequency signals, these other parameters may become crucial.

18.3 Multi-stage amplifiers

Some designs (such as the one in Figure 18.7 for Section 18.4.2) require an amplifier with a small gain, which can easily be obtained with a single op amp (see Section 19.2 for how to design such an amplifier), but many of our amplifier designs will not be so easy. There are several reasons we might want to use more op amps, for example,

- high gain needed,
- high frequency to amplify,
- DC offset to remove, or
- differential signal to convert to single-ended signal.

If you need a higher gain than is easily obtained in a single amplifier, then you can put multiple amplifiers together as a *multi-stage amplifier* to get the desired gain. Each *stage* of the amplifier consists of one op amp, associated feedback circuitry, and possibly associated filter or gain-control circuitry.

For example, if we wanted a gain of 250 at 20 kHz using MCP6004 op amps, whose gain is limited to 50 at that frequency, we could use two stages: the first one providing a gain of 10 and the second one a gain of 25, to get a combined gain of 250 (Figure 18.4). We could also use two inverting amplifiers with gains that multiply to 250 (say −12.5 and −20, Figure 18.4).

Exercise 18.1

Design an amplifier *at the block-diagram level* to amplify a 40 kHz burst from a piezoelectric sensor (these sensors are used in cheap robots to detect walls and other reflecting surfaces). Assume that the sensor has very high impedance, but produces an output that is about ±40 mV for a strong echo from a close reflector and ±3 mV for a distant reflector—a typical set of echoes might look like those of Figure 18.5.

You want to read the signal from the piezoelectric sensor with the 0 V–3.3 V analog-to-digital converter of a Teensy board.

What gain is needed? Can this be done in one stage with MCP6004 op amps? What other blocks do you need to get usable values for the analog-to-digital converter?

Draw a block diagram for the design.

Bonus: make your design not sensitive to 60 Hz interference.

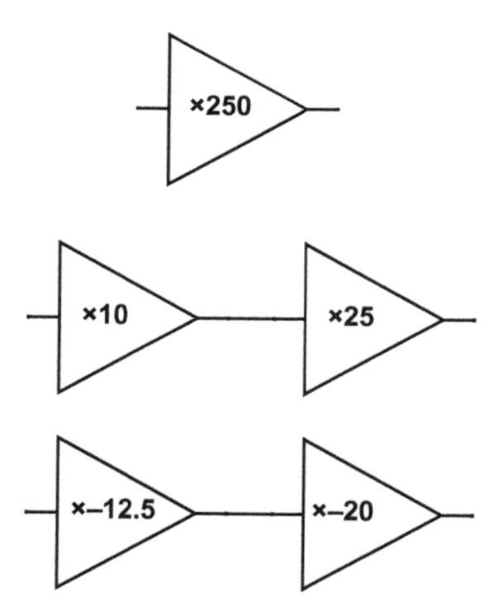

Figure 18.4: *Block diagrams for different implementations of an amplifier with a gain of 250. The top block diagram just specifies the gain of the amplifier—it is ambiguous about whether a single-stage amplifier is intended or a multi-stage one that has not yet been further refined into separate stages. The middle block diagram specifies a two-stage amplifier with non-inverting (positive-gain) stages, giving the gain for each stage. The bottom block diagram specifies a two-stage amplifier with inverting (negative-gain) stages. Figure drawn with Digi-Key's Scheme-it [18].*

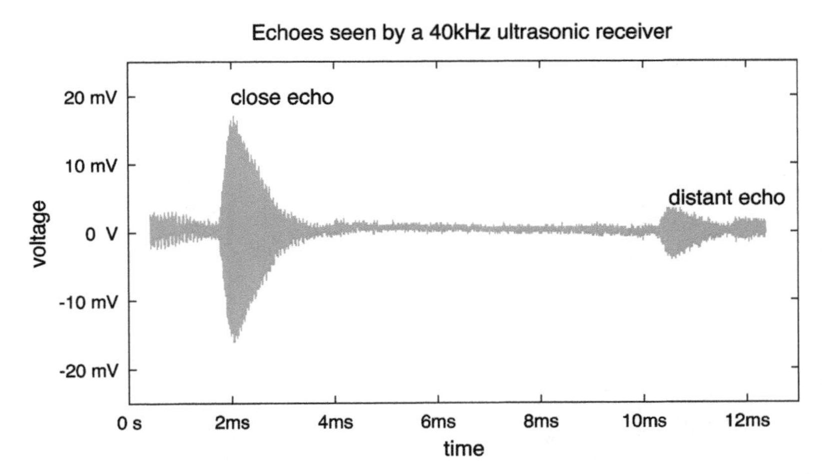

Figure 18.5: *At time 0, four pulses of a 40 kHz square wave were sent to a resonant transmitter. This plot is a recording of the echoes received by an ultrasonic resonant receiver. The delay of the echo tells you how far away the reflector is—the round-trip distance is approximately 343 m/s times the delay, so the 2 ms echo is from a reflector about 34 cm from the transmitter and receiver and the 10.7 ms echo is from a reflector about 1.8 m away.*
Data collected with Analog Discovery 2 [19]. Figure drawn with gnuplot [33].

Multi-stage amplifiers are often used to remove DC bias from a signal. This DC bias may come from the sensor or it may come from the amplifier itself. If you have a DC offset of 10 mV at the input to an amplifier and want a gain of 1000 for your signal of interest, the amplification of the 10 mV DC at the input would result in a 10 V DC offset at the output—more than the output range of the op amps we use in this course. The result would be *saturation* of the amplifier, where the output is as close to one of the power rails as it can get.

To avoid saturation of an amplifier due to DC bias, we can reduce the gain of a single stage, so that stage does not amplify the offset much, then use an RC high-pass filter (see Section 11.6) to remove the DC bias before amplifying further in the next stage. The gain of the first stage should be low enough that even with the largest expected DC offset and largest AC signal, the amplifier will not hit its output limits.

Op-amp designs often introduce a small DC bias, typically the equivalent of less than ±5 mV at the input, which limits the gain you can get from a single stage, even if there is no DC offset in the signal from the sensor or the previous amplifier stage.

For the MCP6004 op amps, the *input offset voltage* is specified to be in the range −4.5 mV to 4.5 mV, so if the output is limited to 0 V to 3.3 V, the gain must be less than $\frac{3.3\,\text{V}-0\,\text{V}}{4.5\,\text{mV}-(-4.5\,\text{mV})} \approx 366.7$ to avoid saturating the amplifier. This limitation applies to any frequency, even down to 0 Hz (DC).

Consider the block diagrams in Figure 18.6, which look at what happens to an input offset voltage in a multistage amplifier. If we break an amplifier with a gain of 1024 into two stages, with a high-pass filter to remove DC offset between the two stages, then the output offset of the amplifier is determined just by the last stage. If we choose a gain of 32 in each stage, then at each stage an input offset of ±4.5 mV becomes an output offset of ±4.5 mV 32 = ±144 mV. If that offset is removed with an RC filter between the stages, we can avoid problems with saturating the amplifier and have only a ±144 mV offset at the output to correct for in processing the digitized output. Using three stages, with gains of 32, 32, and 1, we could reduce the output offset to ±4.5 mV.

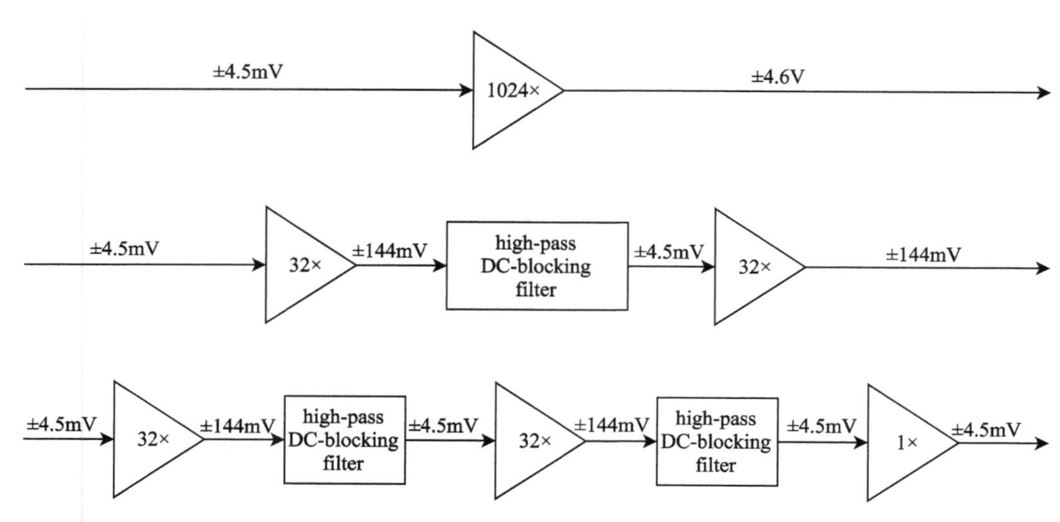

Figure 18.6: *Block diagrams for different implementations of an amplifier with a gain of 1024, illustrating the effect on DC offset. Each amplifier stage is assumed to have an input offset voltage in the range ±4.5 mV. The top block diagram shows the effect of a single stage—the offset is amplified to ±4.6 V, exceeding the voltage range available to the amplifier, and so saturating it at one of the power rails.*
The middle block diagram specifies a two-stage amplifier with two stages with a gain of 32 each, separated by a filter that removes the DC offset. A new offset of ±4.5 mV is assumed at each amplifier stage, so the final output has an offset in the range ±144 mV. The bottom block diagram specifies a three-stage amplifier with gains of 32, 32, and 1, again separated by DC-blocking filters. The low gain of the final stage limits the output voltage offset to ±4.5 mV. The unity-gain buffer also ensures that the second high-pass filter is not affected by whatever load the output is connected to.
Figure drawn with draw.io [21].

Exercise 18.2

Draw a block diagram for an amplifier that needs to achieve a gain of 20,000 for a 1 kHz–10 kHz signal that is 0.5 V DC ±50 μV AC. The output should be 1.65 V DC ±1 V AC. Output offset is allowed to be as much as 0.2 V from the nominal 1.65 V.
Assume you will be using MCP6004 op amps with a single 3.3 V power supply and take into consideration both gain-bandwidth product and input offset for the amplifiers.
How many stages do you need? What is the gain for each stage? What do you need in the block diagram besides just amplification?

The final reason given for multi-stage amplifiers, converting differential signals to single-ended signals, is the topic of Section 18.5.

18.4 Examples of amplifiers at block-diagram level

18.4.1 *Example: Temperature sensor*

Worked Example:
What amplification is needed for an MCP9701A temperature sensor?
According to the data sheet from Microchip, this sensor requires a 3.1 V to 5.5 V power supply, and provides a voltage output that is (nearly) linear over the temperature range from −10°C to +125°C, with a temperature coefficient of 19.5 mV/°C. The linear approximation to the output voltage given in the spec sheet is $V_\mathrm{T} = 400\mathrm{mV} + 19.5\frac{\mathrm{mV}}{^\circ\mathrm{C}}T$ when powered with 5 V (an error of up to 0.1 °C/V is possible if the voltage differs from 5 V).

If we want to measure temperature from 0°C to 100°C, then the output voltage of the sensor will vary from 400 mV to 2.35 V. This is a very convenient range for the 0 V–3.3 V range of the analog-to-digital converter on the Teensy boards, so no gain is needed. The output impedance of the sensor is around 3 Ω, which means that it can provide plenty of current without appreciably changing the output voltage (not that the ADC inputs of the Teensy require much current). In short, the MCP9701A temperature sensor can be connected to the ADC input of the Teensy with no amplifier needed! This ease of connection is not a coincidence—many voltage-output sensors are designed to have a voltage range compatible with typical analog-to-digital ranges.)

18.4.2 *Example: pH meter*

Worked Example:
What amplification is needed in a pH meter?
Let's look at pH meters instead. Since we aren't doing a lab in which we build pH meters, I'll skip the theory and just treat the pH probe as a black-box sensor, which we can model as a voltage source that depends on pH in series with a large resistance (typically about 250 MΩ).

The cell voltage for a pH probe is $V_7(T) - 0.1984\,\text{mV/K}(\text{pH} - 7)T$, where $V_7(T)$ is an offset voltage measured at pH 7, and T is the temperature in kelvin [99, pp. 9–10]. Most pH probes are designed so that the $V_7(T)$ function is constant with temperature and close to 0 V. The probes typically include a temperature sensor so that the pH meter can compute the correct pH from the cell voltage.

If we want to make a pH meter that can measure from pH 0 to pH 14 at 25°C, we can expect an output range of $(\pm 7)0.1984\,\text{mV/K}(273.15\,\text{K} + 25°\text{C}) = \pm 414\,\text{mV}$. Standard pH probes have an output impedance of approximately 250 MΩ—that is, the probe can be electrically modeled as a voltage source (proportional to pH difference from pH 7) in series with a 250 MΩ resistor. Trying to read the pH-probe voltage with the analog-to-digital converter on a Teensy board would have two problems: the voltage goes negative (not in the 0 V–3.3 V range) and the 250 MΩ output impedance of the probe is too large to drive the ADC input—a 10 kΩ impedance is the largest that is recommended.

We can fix the problem of the voltage being centered at the wrong value by connecting the pH probe to a 1.65 V voltage source rather than to ground (see Section 19.3 for how to construct such a voltage source).

The problem of the pH probe having too large an impedance for the ADC can be fixed by using an amplifier—we can use an amplifier that does not require much current from its input (high input impedance) but can provide substantial current at its output (low output impedance).

For any sensor amplifier, the voltage gain needed for the amplifier can be determined by working inward from the ends: what voltage range is needed at the output divided by what voltage range is available at the input.

For the pH probe, if the analog-to-digital converter needs 0.1 V to 3.2 V, but the pH probe provides ±414 mV, then we want a gain of approximately 3.1 V/828 mV = 3.74. The design can be summarized in a block diagram, showing the signal levels and crucial parameters of the different parts of the system as in Figure 18.7.

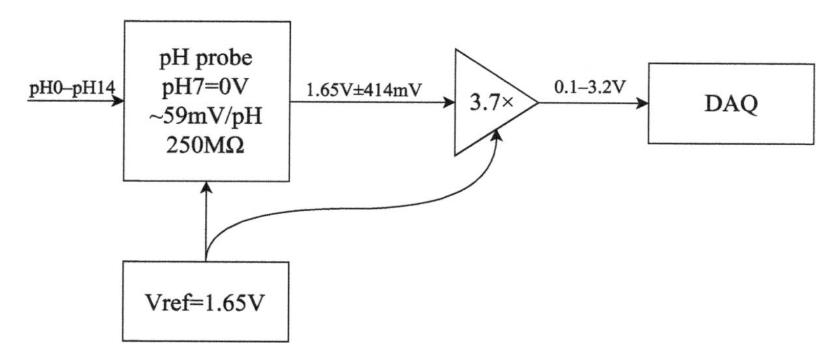

Figure 18.7: *A pH meter consists of three main parts: a pH probe, an amplifier, and conversion from analog to digital form for display or recording. The gain needed for the amplifier is determined by the ratio of the voltage range needed for the analog-to-digital converter and the voltage range provided by the pH probe.*

This block diagram is incomplete, as it does not include the temperature measurement needed to determine the sensitivity of the pH probe.

Figure drawn with draw.io [21].

The high impedance of the pH probe means that there will be very little current from the probe: at pH 7.01, we'd have only $0.01\,\mathrm{pH} \cdot 0.1984\,\mathrm{mV}/(\mathrm{pH\,K})(273.15+25)\mathrm{K} = 590\,\mu\mathrm{V}$ output, which would produce only $590\,\mu\mathrm{V}/250\,M\Omega = 2.4\,\mathrm{pA}$ of current. That is a very small current, and so the design of pH meters requires a lot of care in shielding to avoid picking up stray currents through capacitive or inductive coupling. Many industrial pH meters put an amplifier right at the probe to minimize this problem.

The amplifier block diagram in Figure 18.7 is not a complete pH meter design—the voltage from the pH probe depends not only on the pH, but also on the temperature of the probe, which also needs to be measured to determine the pH. (See Section 5.3 for more information about the RTDs most often used for this temperature measurement.)

18.4.3 *Example: Ultrasound imaging*

Worked Example:
What amplifiers are needed for ultrasound imaging?

A popular medical imaging technique is ultrasound imaging, which emits a burst of high-frequency sound, then listens to the echoes off of structures inside the body.

The echoes are caused by differences in the mass or elasticity of different tissues, causing a mechanical impedance mismatch between adjacent tissues. The mechanical impedance can be computed as the product of the density of the material and the speed of sound in the material.

The speed of sound in liquids is higher than in gases (about 1484 m/s in water and 343 m/s in air) and higher still in solids. Different body tissues have different speeds of sound from 1450 m/s in fat to 4080 m/s in bone, with soft tissue, blood, and muscle around 1560 m/s–1580 m/s [57].

To get good image resolution, you need a short wavelength for the sound, which means a high frequency. Typical frequencies are in the range 2.5 MHz–15 MHz [74], which gives a wavelength around 100 μm–625 μm (shorter in bone, longer in fat). Some ultrasound imagers go as high as 200 MHz. The reason for using such a wide range of frequencies is that high-frequency sounds are attenuated more by tissues, so lower frequencies are used when deep penetration is needed (sacrificing some resolution) and higher frequencies are used for imaging closer to the surface or for ultrasonic probes on catheters, where high resolution matters more than depth of penetration.

Transducers for ultrasound imaging generally have between 128 and 512 transducers in an array [56], each of which has its own driver and amplifier for listening to the echo. Some of the amplifiers can be shared by using a fast multiplexing switch and a high-bandwidth amplifier—just as the microcontroller on the Teensy board uses a multiplexer to share the analog-to-digital converter among many channels, but the need for high speed limits how much sharing is possible.

Although high-frequency amplifier design is outside the scope of this course, let's look a little at what it takes to get a gain of 2000 at 10 MHz. (Real ultrasound imagers adjust the gain with time after a pulse is sent, to compensate for the loss of intensity due to attenuation by the tissue, rather than using a fixed gain.)

For a gain of 2000 at 10 MHz, we'd need a gain-bandwidth product of 20 GHz. Op amps of that speed aren't really available! If we break the gain into two stages (say 50 and 40), then we only need a gain-bandwidth product of 500 MHz (though we should probably specify 800 MHz to be safe, since manufactures only give typical gain-bandwidth product, not worst-case). Op amps with 1 GHz gain-bandwidth product are available for under $3 in single-unit quantities—not as cheap as the low-speed (1 MHz gain-bandwidth product) op amps that we use in this course, but not astronomical.

We could use slightly cheaper op amps by using three stages with a gain around 12.6 each (for a gain-bandwidth product of only 130 MHz), but the increase in the number of amplifiers may exceed the cost savings from using a cheaper amplifier.

18.5 Instrumentation amplifiers

An *instrumentation amplifier* is a differential amplifier with very high input impedance on both inputs (pure voltage input, with essentially no current flow). The output of the amplifier is almost purely a function of the *difference* between the input voltages, with very little influence from the *average* of the two inputs.

The term "instrumentation amplifier" is occasionally abbreviated to *IA*, *in amp*, or even *inst amp*, but these abbreviations are considered very informal and rarely used in formal writing, such as design reports. (The abbreviation "op amp" for operational amplifier is much more widely accepted.)

Most instrumentation amplifiers have three inputs: positive and negative signal inputs and a voltage reference input that defines the output voltage when the difference between the main inputs is 0 V (see Figure 18.8). The signal inputs are typically very high impedance, so as not to disrupt the circuit being measured.

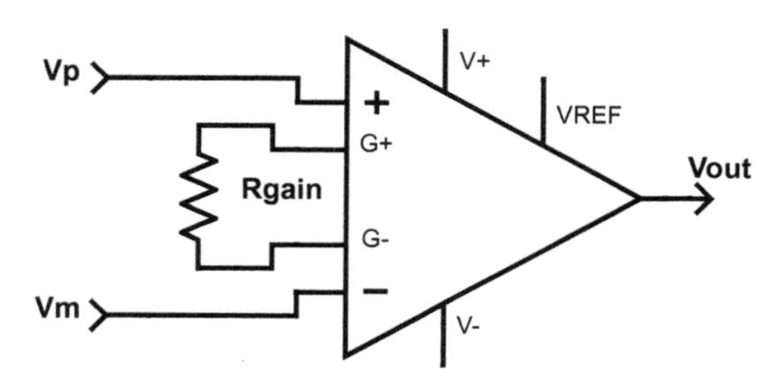

Figure 18.8: *The standard schematic symbol for an instrumentation amplifier has two inputs on the left, surrounding an extra pair of connections that are used for an external resistor that controls the gain. The V_{ref} input appears on either the top or the bottom of the symbol, to show that it is being treated like a power-supply input. The input names V_p and V_m used in this book are selected based on the "plus" and "minus" symbols in the amplifier triangle.*

In a block diagram, a single line represents the differential input, the gain is shown with a label on the amplifier, and the V_{ref} voltage comes in the top or bottom of the amplifier symbol (see Figure 18.9).

Figure drawn with Digi-Key's Scheme-it [18].

The reference voltage V_{ref} is not generated by the instrumentation amp—it should be thought of like a power supply, because the instrumentation amplifier can take current from the V_{ref} pin. Because the V_{ref} pin requires some current, it should be connected to a low-impedance source like a power supply or unity-gain buffer (see Section 19.3 for an explanation of unity-gain buffers).

When digitizing the output of an amplifier, it is often a good idea to connect V_{ref} to an extra analog-to-digital converter channel, so that $V_{out} - V_{ref}$ can be digitized directly, rather than having to guess V_{ref}.

The R_{gain} resistor in Figure 18.8 specifies the gain, but the translation of resistor values to gains depends on the specific amplifier chip—look it up on the data sheet for the chip.

A multi-stage instrumentation amplifier consists of an instrumentation amplifier as a first stage, followed by negative-feedback amplifiers (see Section 19.2) to provide more gain and possibly filtering (see Chapter 24). We often use fairly low gain in the first stage of an instrumentation amplifier, because that allows the amplifier to reject more common-mode noise. Some instrumentation amplifiers have a limited output range, but adding subsequent amplification stages can increase the output range to be rail-to-rail.

Worked Example:
Make a block diagram for a 2-stage instrumentation amplifier with an overall gain of 30, with the first stage having a gain of 5. What is the legal range of inputs for the first stage?
The block diagram is fairly simple—most of the information of the diagram will be in the labels for the signals and the amplifiers.

The first stage is an instrumentation amplifier with a gain of 5, but the instrumentation amplifier also needs a V_{ref} input to set the midpoint of the output voltage range. That means that we need to generate the V_{ref} voltage reference in another block (we'll see how to create that block in Section 19.3). The most common choice of V_{ref} is halfway between the power rails, to allow the largest possible symmetric signal around that voltage. The output of the first stage is $V_1 = 5(V_{in+} - V_{in-}) + V_{ref}$.

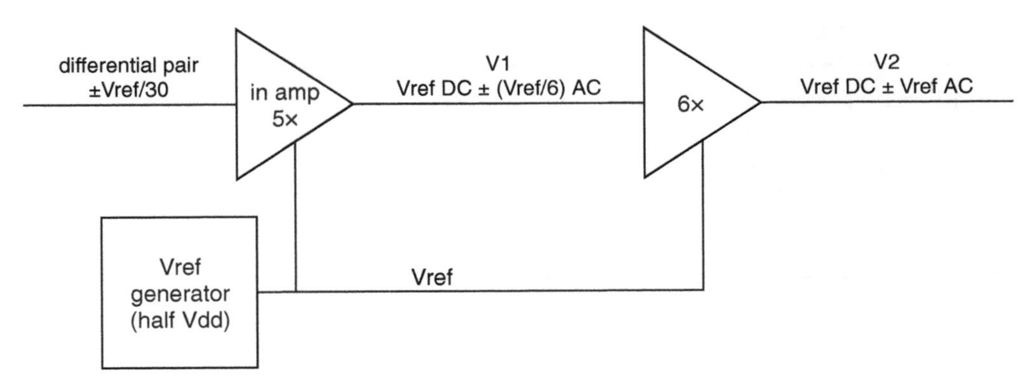

Figure 18.9: *This block diagram for a 2-stage instrumentation amplifier uses only a single line for each signal, even though the input to an instrumentation amplifier consists of a pair of wires. The differential pair carries information only in the difference in voltage between the wires, so it is still a single signal for block diagrams.*
Figure drawn with draw.io [21].

The second stage is a single-ended amplifier with a gain of 6. Its output is

$$V_2 = 6(V_1 - V_{\text{ref}}) + V_{\text{ref}} = 30(V_{\text{in+}} - V_{\text{in-}}) + V_{\text{ref}} \ .$$

If V_{ref} is set to half the power supply, and the output V_2 is a rail-to-rail output from 0 V to $2V_{\text{ref}}$, then the differential input signal is limited to

$$|V_{\text{in+}} - V_{\text{in-}}| < V_{\text{ref}}/30 \ .$$

How much common-mode voltage the pair of inputs can have depends on the specifications of the instrumentation amplifier used as the first stage. Normally, it would be important to put this information into the block diagram as a constraint on the input signals, but we do not have enough information in this example to do so.

The constraint on the input means that the output of first stage is limited to

$$\left(1 - \frac{5}{30}\right) V_{\text{ref}} < V_1 < \left(1 + \frac{5}{30}\right) V_{\text{ref}} \ ,$$

which can be written as $V_{\text{ref}} \pm V_{\text{ref}}/6$. The completed block diagram summarizing the design is in Figure 18.9.

Exercise 18.3

Use the data sheet to determine upper and lower limits for voltage outputs for instrumentation amplifier INA126P powered at 3.3 V. Use worst-case values (those that minimize the output range), rather than typical values.

Exercise 18.4

If we want to amplify a ±100 mV differential signal with an INA126PA chip powered from a 3.3 V power-supply, what is the highest gain you can use without hitting the output voltage limits? What gain resistor would you need to set that gain? If you need to round to the E12 series, which resistance value would you choose? What gain would that produce?

Exercise 18.5

For the amplifier of Exercise 18.4, what does V_{ref} need to be set to in order to keep the output in range without clipping? Consider adding a second stage to the amplifier using a rail-to-rail amplifier for the second stage. If you center the output of the multi-stage amplifier at V_{ref}, what is largest gain you can specify for the second-stage amplifier and still stay within bounds?

Chapter 19

Operational Amplifiers

19.1 What is an op amp?

The building block for our amplifier designs is an *operational amplifier*, usually abbreviated as *op amp*. This device has two voltage inputs (V_{pa} and V_{ma} in Figure 19.2 for the "plus" and "minus" inputs to the amplifier), two power connections (V_{dd} for the positive power rail and V_{ss} for the negative power rail), and one voltage output (V_{out}). It is common in analog-design courses to use two power supplies, symmetric around 0 V, so that $V_{\text{ss}} = -V_{\text{dd}}$, but we'll mostly be using single-supply designs, in which $V_{\text{ss}} = 0\,\text{V}$ and V_{dd} is our only power supply (often 3.3 V, since that is easily provided by the microcontroller board).

The amplifier amplifies the voltage difference between the two inputs, which are usually labeled with a plus and a minus inside the triangle that symbolizes an amplifier, to indicate which one is which (see Figure 18.1).

> Plus and minus symbols on the amplifier refer to the sign of the *derivative* of the output with respect to that input: the + means that the output *increases* when that input increases, and the − means that the output *decreases* when that input increases.

Some authors prefer to use the terms *inverting input* for V_{ma} and *non-inverting input* for V_{pa}, to emphasize that the sign refers to the gain of the amplifier, not the voltage of the input. I find that this notation introduces confusion when we introduce the inverting (negative-gain) and non-inverting (positive-gain) amplifier configurations.

The input voltages themselves have to stay between the two power-supply voltages, and the output voltage is naturally constrained to be between the power-supply voltages. If our power supply rails are $V_{\text{ss}} = 0\,\text{V}$ and $V_{\text{dd}} = 3.3\,\text{V}$, then there are no nodes at a negative voltage with respect to ground—everything has to stay between 0 V and 3.3 V.

The MCP6004 op amps we use are *rail-to-rail* op amps, which can drive the output voltage fairly close to the voltages on the power rails. The data sheet claims that the output can go to about 25 mV from the rails, from $V_{\text{ss}} + 25\,\text{mV}$ to $V_{\text{dd}} - 25\,\text{mV}$ [71]. In other words, with a 3.3 V power supply, the output can go from 0.025 V to 3.275 V. Many older op-amp designs cannot get closer than about 1 V to the power rails—1 V to 2.3 V for a 3.3 V power supply. As modern systems use lower and lower voltages (a lot of digital logic now runs with a 1.8 V supply voltage), the older designs are becoming less and less used—if you had to stay 1 V away from the power rails, you would not be able to do anything with a power supply of 2 V or less.

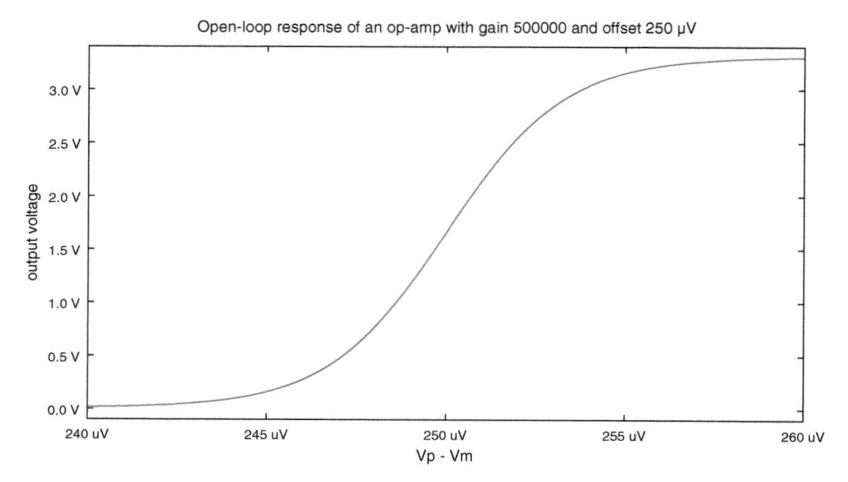

Figure 19.1: *The open-loop gain of an op amp is quite large at DC, which means that the output changes from one voltage rail to the other with only a small change in the differential input. The voltage difference at which that change occurs is the* offset voltage*—see Section 18.2.4.*

This graph is a theoretical curve, not a measured one, as measuring the DC open-loop response of an op amp is somewhat difficult.

Figure drawn with gnuplot [33].

The astute reader will have noticed that I have been very sloppy in the description of the op amp—I said that there was a single output voltage, but voltages are always differences between two nodes. There is no way that the chip can know what node we have arbitrarily labeled ground, so what is the output voltage in reference to? By convention, it is usually referenced to the average of the two power-supply voltages, since analog designers traditionally used symmetric dual power supplies, with the average of the positive and negative supplies then being 0 V. In practice, we try to design circuits using op amps that allow us to insert our own reference voltage, V_{ref}.

In this class we will usually use a single power supply, so the formulas that describe the output voltage of an op-amp chip may have to mention the power supply voltages directly. Because the op-amp output is between the power rails V_{dd} and V_{ss}, the mid-point between the power rails, $V_{mid} = (V_{dd} + V_{ss})/2$, is an important point of symmetry.

The op amp provides a very high gain: that is, $A = \frac{dV_{out}}{d(V_{pa} - V_{ma})}$ is very large, but the gain is not very tightly specified. The output of the amplifier can be expressed in terms of the inputs to the op amp:

$$V_{out} - V_{mid} = A(V_{pa} - V_{ma} - V_{offset}) \ , \qquad (19.1)$$

where A is a large number (called the *open-loop gain* of the op amp) and V_{offset} is the *input offset voltage*, as shown in Figure 19.1, but this expression only holds for a fairly narrow range of $V_{pa} - V_{ma}$.

The gain of the op amps we use is around 200,000–800,000 at DC, but drops considerably at high frequencies (down to 1 at 1 MHz). It is almost always a mistake to design an op-amp circuit where the precise open-loop gain of the amplifier chip matters—we always use op amps in feedback circuits that set the gain.

19.2 Negative-feedback amplifier

The third big idea of this course is negative-feedback amplifiers, using op amps as a building block. The "negative feedback" refers to feedback from the output of the op amp back to the negative input of the op amp. The same circuit is used for both positive-gain (non-inverting) and negative-gain (inverting) amplifiers.

> A negative-feedback amplifier is not an operational amplifier—it may use an operational amplifier as a component. If you want to highlight that, you can call it a "negative-feedback amplifier with an op amp" but *not* a "negative-feedback operational amplifier".

Figure 19.2 shows the standard negative-feedback circuit used for most op-amp amplifier circuits. We'll look at the general formula for the output voltage of the amplifier, then some special cases that are particularly useful.

The op-amp inputs are related to the external inputs:

$$\begin{aligned} V_{\text{pa}} &= V_{\text{p}} , \\ V_{\text{ma}} &= \frac{Z_{\text{i}} V_{\text{out}} + Z_{\text{f}} V_{\text{m}}}{Z_{\text{i}} + Z_{\text{f}}} . \end{aligned} \tag{19.2}$$

The formula for V_{ma} in Equation 19.2 is just another way of writing the voltage-divider formula. We can simplify the formula if we introduce a "feedback gain" $B = Z_{\text{i}}/(Z_{\text{i}} + Z_{\text{f}})$ to get

$$V_{\text{ma}} = BV_{\text{out}} + (1 - B)V_{\text{m}} .$$

We can put the equation for an operational amplifier, Equation 19.1, together with these formulas to get the general form

$$V_{\text{out}} - V_{\text{mid}} = A\left(V_{\text{p}} - ((1 - B)V_{\text{m}} + BV_{\text{out}}) - V_{\text{offset}}\right) ,$$

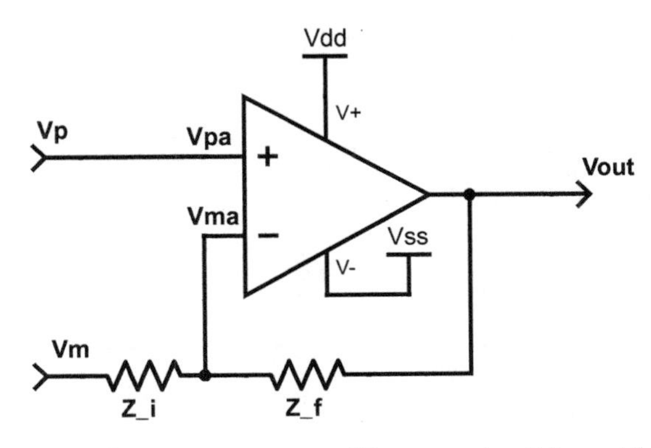

Figure 19.2: *Generic negative-feedback amplifier using an op amp. The same circuit is used for positive gain, negative gain, low-pass active filters, high-pass active filters, and (with nonlinear devices replacing the impedances) logarithmic amplifiers.*

In this schematic, V_{pa} and V_{ma} are the "plus" and "minus" inputs to the op amp, while V_{p} and V_{m} are the "plus" and "minus" inputs for the overall negative-feedback amplifier. Amplifiers are drawn as triangles pointing to the right, with the inputs on the base of the triangle and the output at the point of the triangle. The top and bottom of the triangle are used for the power supplies and any control inputs for the amplifier.

The impedances Z_{i} and Z_{f} are the input impedance and feedback impedance for the circuit. (Technically, Z_{i} is the input impedance for V_{m}, as V_{p} has effectively infinite input impedance.)

Figure drawn with Digi-Key's Scheme-it [18].

which we can simplify and rearrange to get

$$V_{\rm out}\left(1+AB\right)=V_{\rm mid}+A\left(V_{\rm p}-(1-B)V_{\rm m}-V_{\rm offset}\right)\ .$$

The general form for the output of a negative-feedback amplifier is

$$V_{\rm out}=\frac{A}{1+AB}\left(V_{\rm p}-(1-B)V_{\rm m}-V_{\rm offset}\right)+\frac{V_{\rm mid}}{1+AB}\ ,$$

where A is the open-loop gain of the op amp, B is the voltage-divider gain of the feedback, $V_{\rm offset}$ is the input voltage offset of the op amp, and $V_{\rm mid}$ is the central voltage about which the op amp amplifies.

For very large A, where $|AB| \gg 1$, the center voltage $V_{\rm mid}$ contributes little to the output, and so we usually design op-amp circuits without paying much attention to $V_{\rm mid}$.

The general form looks like an awful mess, but see what happens as $A \to \infty$. The terms not multiplied by A become negligible, and we get

$$\begin{aligned} V_{\rm out} &\to \frac{V_{\rm p}-(1-B)V_{\rm m}-V_{\rm offset}}{B} \\ &\to V_{\rm m}+\frac{V_{\rm p}-V_{\rm m}}{B}-\frac{V_{\rm offset}}{B}\ , \end{aligned}$$

or, rearranging to pull out $V_{\rm p}$ instead of $V_{\rm m}$,

$$V_{\rm out}\to V_{\rm p}+\frac{(V_{\rm p}-V_{\rm m})(1-B)}{B}-\frac{V_{\rm offset}}{B}\ .$$

If we plug back in the definition of B and ignore the input voltage offset, we can get a general formula for negative-feedback amplifiers:

The general form for negative-feedback amplifiers (with *ideal op amps* having $A=\infty$ and $V_{\rm offset}=0\,{\rm V}$) is

$$V_{\rm out}=V_{\rm m}+(V_{\rm p}-V_{\rm m})\frac{Z_{\rm i}+Z_{\rm f}}{Z_{\rm i}}\ , \tag{19.3}$$

or

$$V_{\rm out}=V_{\rm p}-(V_{\rm m}-V_{\rm p})\frac{Z_{\rm f}}{Z_{\rm i}}\ . \tag{19.4}$$

If you need to worry about the offset voltage, just replace $V_{\rm p}$ by $V_{\rm p}-V_{\rm offset}$ in the above formulas.

Negative-feedback amplifiers are usually thought of as having single-wire inputs and single-wire outputs, with the input and output voltages both measured with respect to a common reference voltage $V_{\rm ref}$ somewhere between the power rails. The voltage $V_{\rm ref}$ is sometimes called a *virtual ground*, because it is used as the reference voltage against which input and output voltages are measured (instead of measuring relative to ground as we normally would do). It should be thought of like a power-supply input—a constant voltage provided from outside the amplifier.

If we connect V_{ref} to one of the negative-feedback amplifier inputs, the formula for the output can be written as one of the following:

non-inverting If $V_{\mathrm{m}} = V_{\mathrm{ref}}$, we have $V_{\mathrm{out}} - V_{\mathrm{ref}} \approx (V_{\mathrm{p}} - V_{\mathrm{ref}})\left(1 + \frac{Z_{\mathrm{f}}}{Z_{\mathrm{i}}}\right)$, which we can also write as

$$\frac{V_{\mathrm{out}} - V_{\mathrm{ref}}}{V_{\mathrm{p}} - V_{\mathrm{ref}}} \approx \left(1 + \frac{Z_{\mathrm{f}}}{Z_{\mathrm{i}}}\right). \tag{19.5}$$

This is the standard setup for a *positive-gain* or *non-inverting* amplifier.

inverting If $V_{\mathrm{p}} = V_{\mathrm{ref}}$, we have $V_{\mathrm{out}} - V_{\mathrm{ref}} \approx -(V_{\mathrm{m}} - V_{\mathrm{ref}})\left(\frac{Z_{\mathrm{f}}}{Z_{\mathrm{i}}}\right)$, which we can also write as

$$\frac{V_{\mathrm{out}} - V_{\mathrm{ref}}}{V_{\mathrm{m}} - V_{\mathrm{ref}}} \approx -\left(\frac{Z_{\mathrm{f}}}{Z_{\mathrm{i}}}\right). \tag{19.6}$$

This is the standard setup for a *negative-gain* or *inverting* amplifier.

Remember that the gain of an amplifier is defined as $\frac{\mathrm{d}V_{\mathrm{out}}}{\mathrm{d}V_{\mathrm{in}}}$, so the gain of a non-inverting amplifier is approximately $\left(1 + \frac{Z_{\mathrm{f}}}{Z_{\mathrm{i}}}\right)$ and for an inverting amplifier approximately $-\left(\frac{Z_{\mathrm{f}}}{Z_{\mathrm{i}}}\right)$. Both the input and output of the amplifier are centered at V_{ref}. Forgetting that the amplification is happening around V_{ref}, with an input at V_{ref} producing an output at V_{ref}, is a common design mistake.

For easy designing with negative-feedback amplifiers with op amps, we make several simplifying assumptions:

- both the input voltages are within the range accepted by the op amp,
- the output voltage is within the range that the op amp can produce,
- the input current to the op amp through the V_{pa} and V_{ma} ports is negligible,
- the open-loop gain of the op amp is large compared to the gain we want from the negative-feedback amplifier (that is, $|A| \gg 1/|B|$), and
- the output impedance of the op amp is negligible compared to the impedance of the feedback loop.

When these conditions are met, the V_{ma} node is brought to the same voltage as the V_{pa} node by the feedback loop, making analysis simple.

When the conditions are not met, we need to use more complicated models of the op amp to figure out what will happen. Those more complicated models are mostly beyond the scope of this course.

We've used complex impedances (rather than simple resistances) for all the calculations, so we can make the gain vary as a function of frequency, by having Z_{i}, Z_{f}, or both vary with frequency (for example, by using capacitors).

There is a rule of thumb that you can use to help figure out how a negative-feedback amplifier will behave: as long as the open-loop gain A is high enough and the output voltage doesn't hit the power-supply limits, the two inputs to the op amp will be held by the amplifier to be the same: $V_{\mathrm{pa}} = V_{\mathrm{ma}}$. The voltage-divider formula for V_{ma} then tells you what is going on.

One important caveat applies to that rule of thumb, especially when amplifying DC signals—the two inputs of the op amp are *not* brought to exactly the same value, but only close. The difference is referred to as the *input offset voltage*:

$$V_{\mathrm{off}} = V_{\mathrm{pa}} - V_{\mathrm{ma}}, \text{ when the output is in the middle .}$$

For non-inverting amplifiers with gain G, the effect is to center the output at $V_{\mathrm{ref}} - GV_{\mathrm{off}}$, instead of at V_{ref}. Using the formula that assumes infinite gain, $A = \infty$, but non-zero offset (Equation 19.3, with V_{off} subtracted from V_{p}), we get

$$\begin{aligned} V_{\mathrm{out}} &= V_{\mathrm{ref}} + G(V_{\mathrm{p}} - V_{\mathrm{off}} - V_{\mathrm{ref}}) \\ V_{\mathrm{out}} - (V_{\mathrm{ref}} - GV_{\mathrm{off}}) &= G(V_{\mathrm{p}} - V_{\mathrm{ref}}) \ . \end{aligned}$$

Because the offset is amplified by the gain, even the small offsets (say, $\pm 5\,\mathrm{mV}$) of typical op amps can cause problems with clipping when a high gain is needed.

For non-inverting amplifiers, we can go through a similar calculation including the offset voltage (Equation 19.4), with V_{off} subtracted from V_{p}):

$$\begin{aligned} V_{\mathrm{out}} &= V_{\mathrm{ref}} - V_{\mathrm{off}} + G(V_{\mathrm{m}} - V_{\mathrm{ref}} + V_{\mathrm{off}}) \\ V_{\mathrm{out}} - (V_{\mathrm{ref}} - (1 - G)V_{\mathrm{off}}) &= G(V_{\mathrm{m}} - V_{\mathrm{ref}}) \ . \end{aligned}$$

That is, the offset is amplified by $1 - G$, which is exactly the gain of the negative-feedback amplifier when thought of as a non-inverting amplifier.

> The input voltage offset of an op amp is amplified by the gain of non-inverting configuration of the negative-feedback amplifier, no matter which input of the negative-feedback amplifier is the "real" input and which is V_{ref}.

Let's look at a couple of examples: one for an inverting amplifier and one for a non-inverting amplifier.

Worked Example:

Design a single-stage amplifier with a gain of -22*, using a single* $5\,V$ *power supply. The input and output should be centered at* $2\,V$*. If the op amp used has an input offset voltage in the range* $-5\,mV$ *to* $5\,mV$*, what is the range of possible values for the output voltage when the input voltage is* $2\,V$*?*

Because the gain is negative, we need to use an inverting configuration for the negative-feedback amplifier. The ratio of the resistances must be $R_{\mathrm{f}}/Ri = 22$ to get the desired gain. We usually want our resistances to be in the range $1\,\mathrm{k}\Omega \leq R \leq 1\,\mathrm{M}\Omega$, so one reasonable choice is $R_{\mathrm{f}} = 220\,\mathrm{k}\Omega$ and $R_{\mathrm{i}} = 10\,\mathrm{k}\Omega$.

We also need to generate a reference voltage of $2\,\mathrm{V}$ for the positive input of the op amp, but we don't need to provide any current from the reference voltage, so we can use large resistors in a voltage divider. The resistor to ground would have $2\,\mathrm{V}$ across it, and the resistor to the $5\,\mathrm{V}$ power supply would have $3\,\mathrm{V}$ across it, so the ratio of the resistances needs to be 3:2. We can achieve this with a $330\,\mathrm{k}\Omega$ resistor and a $220\,\mathrm{k}\Omega$ resistor.

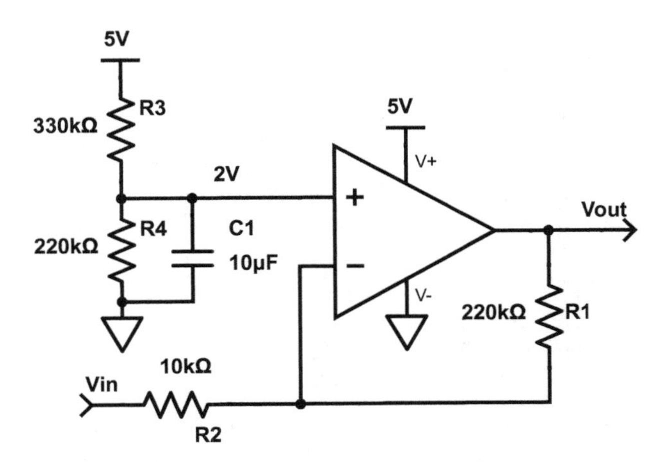

Figure 19.3: *This amplifier has a gain of −22, with the input and output centered at 2 V. The low-pass filter on the voltage divider has a corner frequency of 121 mHz.*

To reduce noise from the power supply appearing on the reference voltage (and so being amplified by the amplifier), we should add a capacitor in parallel with the 220 kΩ resistor to make the voltage divider be a low-pass filter. We want a very low corner frequency for the filter, so that 60 Hz noise is suppressed—something around 0.1 Hz–1 Hz is probably reasonable, for an RC time constant of 160 ms–1.6 s.

The Thévenin-equivalent of the voltage divider is 2 V voltage source and a resistor with resistance $220\,\text{k}\Omega || 330\,\text{k}\Omega = 132\,\text{k}\Omega$.A $10\,\mu$F capacitor gives us a time constant of 1.32 s, for a corner frequency of 121 mHz.

Figure 19.3 shows the schematic diagram for the whole circuit.

When the input voltage is 2 V, the output voltage with an ideal op amp would be 2 V, but the input offset of the amplifier effectively changes the input reference voltage, and the output voltage offset will be 23 times the input offset voltage (not −22 times). So the ±5 mV input offset becomes a ±115 mV output offset, and the voltage at the output could be anywhere in the range 1.885 V to 2.115 V.

Worked Example:

Design a single-stage amplifier with a gain of 34, *using a single* 3.3 *V power supply plus a reference voltage source of* 2 *V. (The reference source can provide up to* ±10 *mA of current.) If the op amp used has an input offset voltage in the range* −5 *mV to* 5 *mV, what is the range of possible values for the output voltage when the input voltage is* 2 *V?*

A single-stage non-inverting amplifier with a gain of 34 requires a resistor ratio of $R_\text{f}/R_\text{i} = 34 - 1 = 33$. We don't want to use a lot of current for the resistors, so $R_\text{f} = 330\,\text{k}\Omega$ and $R_\text{i} = 10\,\text{k}\Omega$ may be a good choice. This design is shown in Figure 19.4.

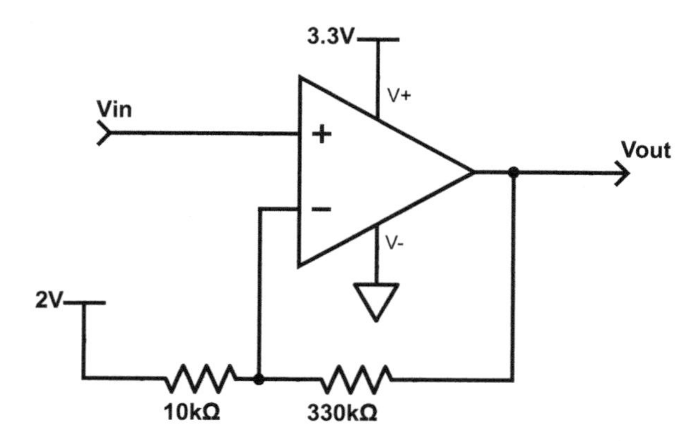

Figure 19.4: *This amplifier has a gain of 34,with the input and output centered at 2 V. The V_{ref} reference voltage needs to be able to supply up to $\pm 200\,\mu A$ of current without changing voltage, so cannot be provided by a simple voltage divider.*

The voltage drop across R_i is in the range $-2\,V = 0\,V - 2\,V \leq V_{Ri} \leq 3.3\,V - 2\,V = 1.3\,V$, and so the current would be in the range $-200\,\mu A \leq I \leq 130\,\mu A$, which is well within the specified range of the reference source. It would be a serious mistake to make the resistors so small that all the available current from the voltage source were used for this one amplifier.

The input offset of $\pm 5\,mV$ is amplified by the non-inverting gain of 34, so the output offset is $\pm 170\,mV$. That means that for an input equal to the reference voltage (2 V), the output is in the range $1.83\,V \leq V_{out} \leq 2.17\,V$.

Exercise 19.1

Design a single-stage negative-feedback amplifier with a gain of −10, using a single 3.3 V power supply. The input and output should be centered at about 1.03 V. All resistors should be from the E12 series.

Exercise 19.2

Design a single-stage negative-feedback amplifier with a gain of approximately 3.14, with the input and output centered at V_{ref}. All resistors should be from the E12 series.

19.3 Unity-gain buffer

There are some interesting special cases in negative-feedback amplifier design. For example, what happens if $Z_i = \infty\,\Omega$? In that case, we have that $V_{out} = V_p$, a circuit known as a *unity-gain buffer* (see Figure 19.5). It might seem strange to have an amplifier whose gain is 1—the input and the output voltage are the same! The important difference is that the amplifier can provide substantial current (several milliamps) without drawing any current (well, maybe a few picoamps) from the input.

A unity-gain buffer is very useful for connecting high-impedance voltage sources to analog-to-digital converters. The input impedance of the op amp may be $10^{13}\,\Omega$ in parallel with 3 pF, requiring essentially

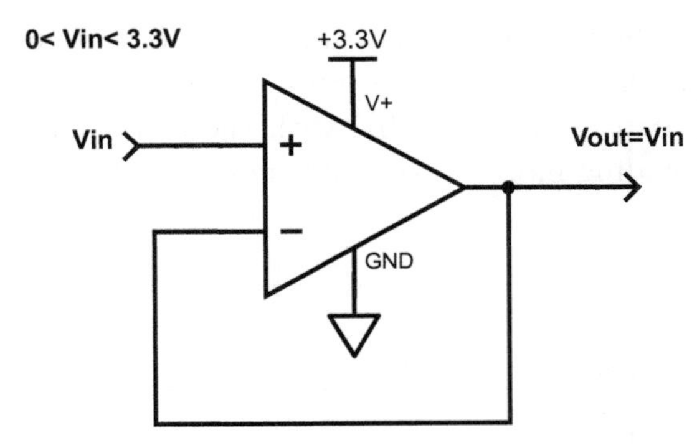

Figure 19.5: *A unity-gain buffer has a gain of 1, so that the output voltage is the same as the input voltage. The input port sees the very high impedance of the input to the op amp, so takes essentially no current from the input, but the output can deliver several milliamps of current without the voltage changing appreciably.*
The input, of course, has to be within the range of outputs that the op amp can deliver. That is, the voltage has to be between the power supply rails, and for some op amps may have to be at least a volt above the lower power rail and at least a volt below the upper power rail.
Figure drawn with Digi-Key's Scheme-it [18].

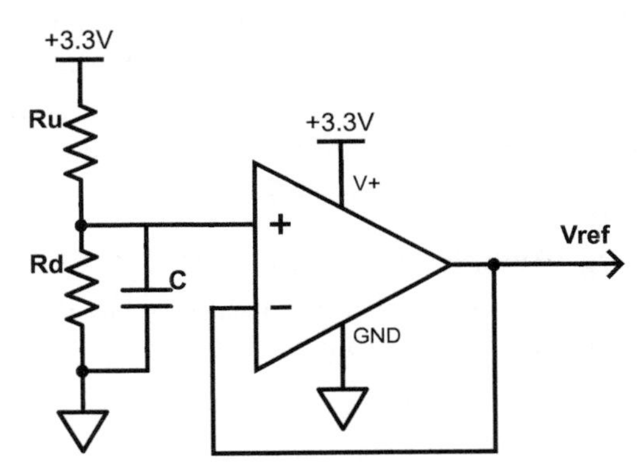

Figure 19.6: *A unity-gain buffer can be used to make reference voltage that can provide small currents to circuits without appreciably changing the voltage. The pair of resistors can be used to set the voltage (relative to the power supply) and the capacitor creates a low-pass filter that reduces noise from the power supply. We usually want the corner frequency of the filter to be well below 60 Hz, to reduce one of the most common noise frequencies.*

no current from the source, while the output impedance of the op amp may be only 40 Ω, able to provide substantial current to the analog-to-digital converter.

A unity-gain buffer lets us create a low-impedance voltage source—one that does not vary much in voltage as current is taken from it.

> If we want to create a *virtual ground* V_{ref}, we can use a voltage divider to set the voltage we want, then use a unity-gain buffer to copy the voltage and provide current, as shown in Figure 19.6. The high input impedance of the op amp meets the voltage divider constraint that there be no current through the output of the voltage divider, and the feedback circuit ensures a faithful copy of the voltage independent of the current (until the limits of the op amp are exceeded).

The voltage-divider approach for creating a virtual ground is adequate for this class, but it has the disadvantage of copying any noise on the power supply to noise on the virtual ground. The noise can be reduced by putting a large capacitor in parallel with the resistor to ground in the voltage divider. There are other circuits for providing more consistent reference voltages, using Zener diodes or voltage-reference integrated circuits.

Worked Example:

Design a V_{ref} circuit to provide 1.65 V from a 3.3 V power supply, with a low-pass filter whose corner frequency is at most 1 Hz.

Because 1.65 V is half the power-supply voltage of 3.3 V, we can make both resistors be the same size, $R_{\text{u}} = R_{\text{d}} = R$. The Thévenin-equivalent of the voltage divider is then a voltage source of 1.65 V in series with a resistance of $R/2$. That makes the time constant for the low-pass filter be $RC/2$.

We want the corner frequency f to be less than 1 Hz, or $\omega = 2\pi f < 2\pi$ radians/s. That gives us $RC/2 = 1/\omega > \frac{1}{2\pi}$ s or $RC > 318$ ms.

Picking a reasonably large capacitor value, say $C = 330$ nF, gives us $R > 965$ kΩ. Using 1 MΩ resistors and a 330 nF capacitor would just barely meet our specs.

A larger capacitor would give more noise reduction, at the cost of slowing down the rise of the V_{ref} source when the power supply is first turned on. Using the heuristic of five time constants for the output of the circuit to settle (see Section 11.4), we get a settling time of $5 \cdot 165\,\text{ms} = 825\,\text{ms}$. Increasing the capacitance to 1 μF would increase the settling time to 2.5 s.

The unity-gain buffer is a good example of the rule of thumb for designing op-amp circuits—since V_{ma} is directly tied to the output, and the two inputs V_{ma} and V_{pa} are made to match by the feedback loop, the output voltage must be equal to the input voltage V_{p}.

19.4 Adjustable gain

We often want to change the gain of an amplifier using a potentiometer (see Section 5.1.3)—think of the volume control knob on an audio amplifier, for example.

One of the simplest ways to adjust gain is to use a potentiometer as a voltage divider between the stages of a multi-stage amplifier, as shown in Figure 19.7. The voltage gain of the voltage divider varies from 0 to 1, and the gain of the amplifier is the maximum gain times the voltage-divider ratio, so the total gain can be varied from 0 to the maximum gain.

This voltage divider between stages works well as long as the input is not so large that the first-stage gain causes clipping, as the gain control cannot compensate for clipping that occurs before the signal gets to the voltage divider.

For some sensors, we can put a potentiometer voltage divider immediately after the sensor, before the first stage of amplification—this approach requires that the sensor works when loaded with the resistance of the voltage divider.

We can also put a potentiometer voltage divider after the last stage, but only if whatever we are driving does not need much current, as the impedance of the voltage divider is much higher than the op-amp output impedance. An analog-to-digital converter may work OK driven through a 10 kΩ potentiometer, but driving an 8 Ω loudspeaker would not work at all.

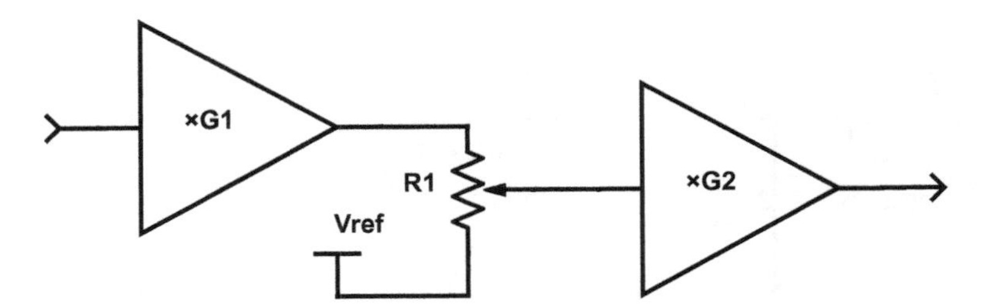

Figure 19.7: *We can get adjustable gain by putting a potentiometer as an adjustable voltage divider between two stages of a multi-stage amplifier. (The voltage-divider constraint is met only if the input impedance of the second amplifier is effectively infinite, as it would be for a non-inverting amplifier.) The gain can be anywhere from 0 to* G_1G_2*, but if the gain from the first stage* G_1 *is too large, then no setting of the gain control can prevent clipping.*

The input impedance of the second stage needs to be large compared with R_1*, so that the voltage-divider constraint is satisfied (negligible current taken from the voltage divider).*

Figure drawn with Digi-Key's Scheme-it [18].

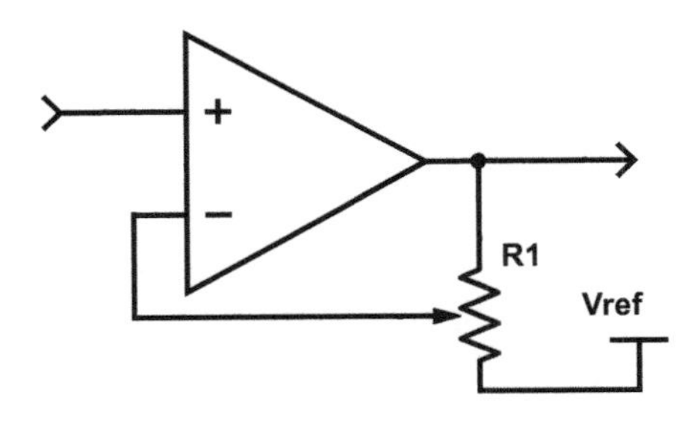

(a) *For this non-inverting amplifier, if the wiper arm for the potentiometer is all the way at the top, the gain of the amplifier is 1. If the wiper arm is all the way at the bottom, the gain of the amplifier is the open-loop gain of the op amp.*

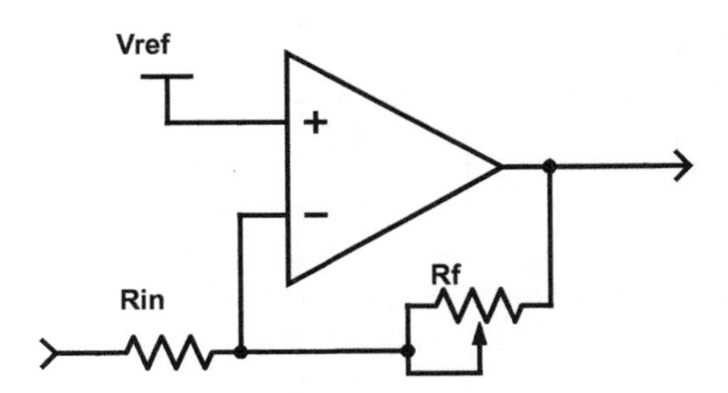

(b) *If the wiper arm for the potentiometer is all the way to the left, the gain of the amplifier is* $-R_f/R_{in}$*. If the wiper arm is all the way to the right, the gain of the amplifier is 0. We usually do not want to make* R_{in} *adjustable, as it determines the input impedance of the amplifier.*

Figure 19.8: *We can adjust the gain of a single-stage amplifier by using a potentiometer as either the entire voltage divider (as in Figure 19.8a) or as just the feedback resistor (as in Figure 19.8b) in the negative-feedback circuit.*

Figure drawn with Digi-Key's Scheme-it [18].

Another approach is to use a potentiometer in the feedback portion of a negative-feedback amplifier, as shown in Figure 19.8. In a non-inverting amplifier the gain can be set anywhere from 1 to the open-loop gain of the op amp, while for an inverting amplifier the range is 0 to $-R_f/R_{in}$.

In any of the gain-control designs, we can limit the range of the gain, so that adjustment is easier, by adding series resistors on either end of the potentiometer, as we saw previously in Figure 5.11.

Exercise 19.3

On the schematic of Figure 19.9, draw a box around each functional block, then redraw the block diagram with appropriate labels for each block and each signal between blocks. Work out the signal levels by assuming a full-scale signal at the output.

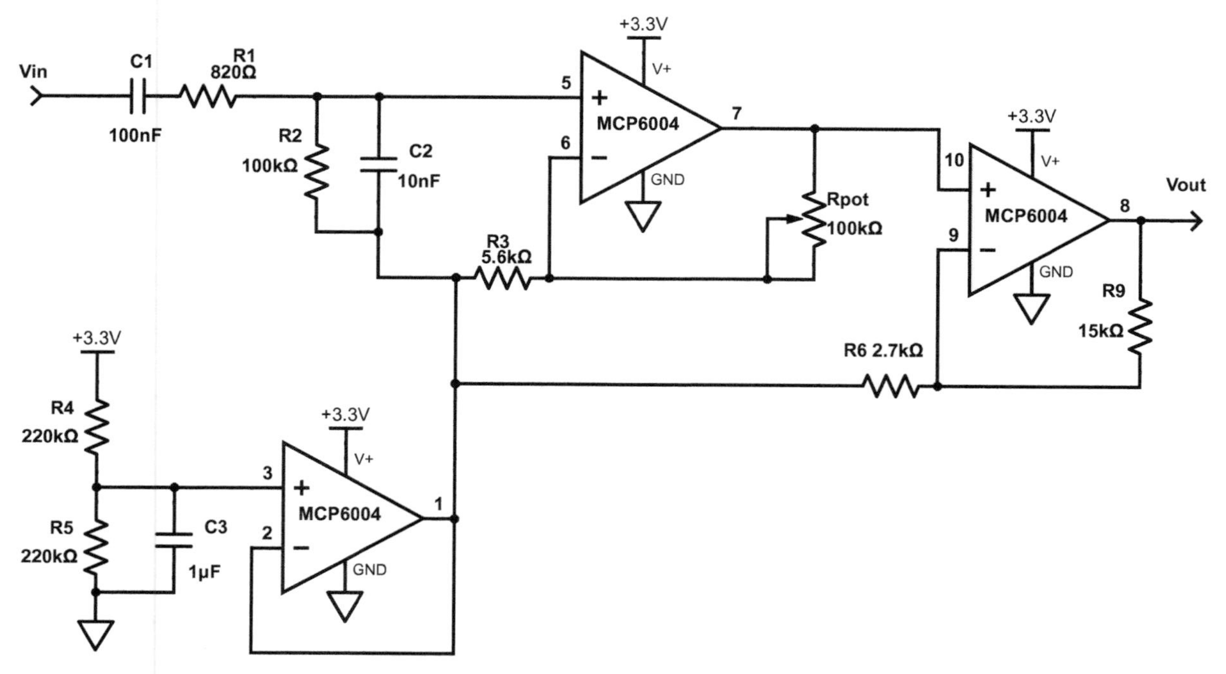

Figure 19.9: *This schematic for a multi-stage amplifier is for the block-diagram exercise in Exercise 19.3. It is available for download at https://users.soe.ucsc.edu/~karplus/bme51/book-supplements/multi-stage-amplifier.pdf.*
Figure drawn with Digi-Key's Scheme-it [18].

19.5 Gain-bandwidth product in negative feedback

So far we have looked mainly at ideal op amps, where we could treat the open-loop gain A as infinite. For most of the work we do in this course, we are dealing with very low frequencies, and the approximation that A is infinite is good enough. But when we amplify high-frequency signals, A may be quite small, and the $A \to \infty$ approximation very bad.

Figure 19.10 shows the effect of this non-infinite open-loop gain—at high frequencies, the gain of the non-inverting amplifier is less than the nominal gain we would get with an ideal op amp. The gain curves can be modeled out to 1 MHz using the gain-bandwidth product introduced in Section 18.2.2.

From Section 19.2 we have the general formula for the output of an op amp:

$$V_{\text{out}} = \frac{A}{1+AB}\left(V_{\text{p}} - (1-B)V_{\text{m}} - V_{\text{offset}}\right) + \frac{V_{\text{mid}}}{1+AB} ,$$

where A is the open-loop gain, $B = Z_{\text{i}}/(Z_{\text{i}} + Z_{\text{f}})$ is the gain of the feedback voltage divider, V_{offset} is the input offset voltage, and V_{mid} is the center voltage about which the op amp amplifies in an open loop.

In non-inverting configurations, we hold V_{m} constant, and we can write

$$V_{\text{out}} = \frac{A}{1+AB}(V_{\text{p}} - V_{\text{m}} - V_{\text{offset}}) + \frac{AB}{1+AB}V_{\text{m}} + \frac{1}{1+AB}V_{\text{mid}} ,$$

or

$$V_{\text{out}} - V_{\text{m}} = \frac{A}{1+AB}(V_{\text{p}} - V_{\text{m}}) + \frac{1}{1+AB}(V_{\text{mid}} - V_{\text{m}} - AV_{\text{offset}}) .$$

The gains A and B for the $V_{\text{p}} - V_{\text{m}}$ term depends on the frequency of V_{p}, but because V_{mid}, V_{m}, and V_{offset} are all constants, the gains for the $V_{\text{mid}} - V_{\text{m}} - AV_{\text{offset}}$ term are just dependent on the DC gains

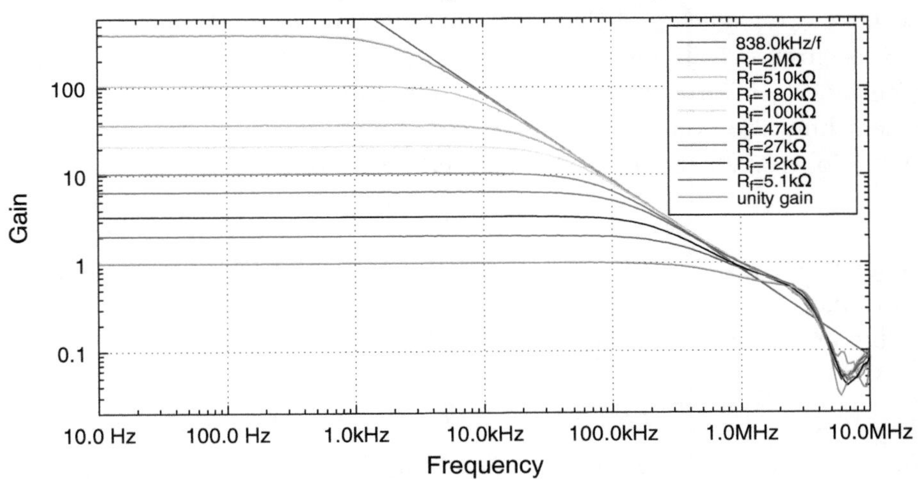

Figure 19.10: *I used a network analyzer to measure the gain of several different non-inverting amplifier configurations. All the configurations had $R_i = 5.1\,k\Omega$, except the unity-gain buffer, whose R_i was the 1 MΩ impedance of Channel 1 of the Analog Discovery 2 (effectively infinite, given the very small resistance of the wire for R_f).*
As the ideal gain of the configuration increases, the corner frequency of the resulting non-inverting amplifier decreases inversely. The measurements here are consistent with a gain-bandwidth product of about 838 kHz, somewhat less than the data sheet's typical value of 1 MHz.
Above 1 MHz, the open-loop gain A drops even faster than the model presented in this section, with a very sharp dip around 6.5 MHz. Data collected with Analog Discovery 2 [19]. Figure drawn with gnuplot [33].

A_0 and B_0,

$$V_{\text{out}} - V_{\text{m}} = \frac{A}{1+AB}(V_{\text{p}} - V_{\text{m}}) + \frac{1}{1+A_0B_0}(V_{\text{mid}} - V_{\text{m}} - A_0V_{\text{offset}}) \,.$$

Similarly, in inverting configurations, we hold V_p constant and treat V_m as input, getting

$$\begin{aligned} V_{\text{out}} &= \frac{A(B-1)}{1+AB}(V_{\text{m}} - V_{\text{p}}) - \frac{AB}{1+AB}V_{\text{p}} - \frac{A}{1+AB}V_{\text{offset}} + \frac{V_{\text{mid}}}{1+AB} \,, \\ V_{\text{out}} - V_{\text{p}} &= \frac{A(B-1)}{1+AB}(V_{\text{m}} - V_{\text{p}}) + \frac{V_{\text{mid}} - V_{\text{p}} - AV_{\text{offset}}}{1+AB} \,, \\ V_{\text{out}} - V_{\text{p}} &= \frac{A(B-1)}{1+AB}(V_{\text{m}} - V_{\text{p}}) + \frac{V_{\text{mid}} - V_{\text{p}} - A_0V_{\text{offset}}}{1+A_0B_0} \,. \end{aligned}$$

We can summarize the gain of the negative-feedback amplifiers as follows:

The gain of a non-inverting amplifier with open-loop gain A and feedback gain B is

$$G = \frac{\mathrm{d}(V_{\text{out}} - V_{\text{m}})}{\mathrm{d}(V_{\text{p}} - V_{\text{m}})} = \frac{A}{1+AB} \,, \tag{19.7}$$

and the gain of an inverting amplifier is

$$G = \frac{\mathrm{d}(V_{\text{out}} - V_{\text{p}})}{\mathrm{d}(V_{\text{m}} - V_{\text{p}})} = \frac{A(B-1)}{1+AB} \,. \tag{19.8}$$

As $A \to \infty$, $G \to 1/B = 1 + Z_f/Z_i$ for non-inverting amplifiers, and $G \to (B-1)/B = -Z_f/Z_i$ for inverting amplifiers, just as we previously derived. The asymptotic approximation only depends on

A having a large magnitude—the phase of A doesn't matter as long as $|AB| \gg 1$, when the 1 in the denominator doesn't matter much and the factor of A on top and bottom cancels.

But when $|AB|$ is close to 1, its phase matters a lot, because $1 + AB$ can result in complex numbers that vary in magnitude from 0 to 2. In Section 18.2.2, gain-bandwidth product was introduced—the frequency-dependent limitation on the open-loop gain.

As we saw in Section 18.2.2, the open-loop gain can be approximated as

$$A = -jf_{\mathrm{GB}}/f \ ,$$

for moderate frequencies $f_{\mathrm{GB}}/A_0 < f < f_{\mathrm{GB}}$, where A_0 is the DC open-loop gain and f_{GB} is the gain-bandwidth product.

For non-inverting amplifiers, the closed-loop gain is then

$$G = \frac{A}{1+AB} = \frac{-jf_{\mathrm{GB}}}{f - jf_{\mathrm{GB}}B} = \frac{1}{B}\,\frac{1}{1+jf/(Bf_{\mathrm{GB}})} \ ,$$

which is the formula for a low-pass filter with corner frequency $f_{\mathrm{GB}}B$ and passband gain $1/B$, at least for real values of B, as we would get if the feedback voltage divider consisted only of resistors. The passband gain $1/B$ times the corner frequency $f_{\mathrm{GB}}B$ is just f_{GB}, which is where the name *gain-bandwidth product* comes from. Above the passband, where $f \gg f_{\mathrm{GB}}B$, we can approximate the gain as $-jf_{\mathrm{GB}}/f$, which is the same as the open-loop gain.

Figure 19.10 shows the f_{GB}/f approximation to the gain and measured data for several different non-inverting amplifiers made using an MCP6004 op amp. Figure 19.11 compares the $\frac{1}{B}\,\frac{1}{1+jf/(Bf_{\mathrm{GB}})}$ model with the measured data for three of the amplifiers. The model is an excellent fit below f_{GB}, but above that frequency the gain of the MCP6004 drops faster than f_{GB}/f.

We can do a similar analysis for inverting amplifiers, with the closed loop gain

$$G = \frac{A(B-1)}{1+AB} = \frac{-jf_{\mathrm{GB}}(B-1)}{f - jf_{\mathrm{GB}}B} = \frac{B-1}{B}\,\frac{1}{1+jf/(Bf_{\mathrm{GB}})} \ ,$$

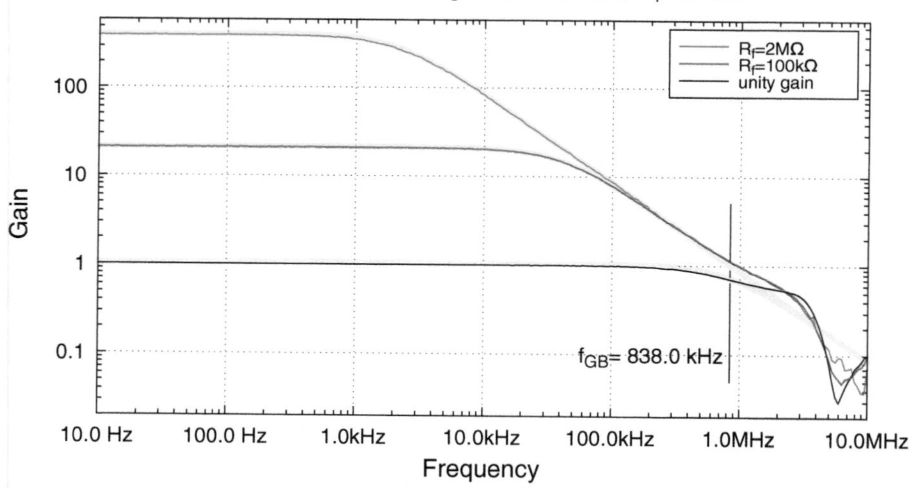

Figure 19.11: *Three of the measurements of Figure 19.10 are compared here to the model* $\frac{1}{B}\,\frac{1}{1+jf/(Bf_{\mathrm{GB}})}$, *with* f_{GB} *fit using the data from the 2 MΩ and 5.1 kΩ feedback network. The model results are shown as thick yellow lines behind the plots of the measured data.*

Data collected with Analog Discovery 2 [19]. Figure drawn with gnuplot [33].

which is the formula for a low-pass filter with corner frequency $f_{\text{GB}}B$ and passband gain $(B-1)/B$, at least for real values of B.

> A non-inverting negative-feedback amplifier behaves as a low-pass filter with passband gain $1/B$ and corner frequency $f_{\text{GB}}B$.
>
> An inverting negative-feedback amplifier behaves as a low-pass filter with passband gain $(B-1)/B$ and corner frequency $f_{\text{GB}}B$.

Worked Example:
What is the bandwidth of an amplifier with an ideal gain of 11, if the gain-bandwidth product of the op-amp chip is 1 MHz?

For a non-inverting amplifier, the gain in the passband is $1/B$, and the corner frequency is just $f_{\text{GB}}B = 1\,\text{MHz}/11 = 90.9\,\text{kHz}$.

Worked Example:
If we use that amplifier with an ideal gain 11 to amplify a 25 kHz signal, what gain do we actually get?

If we are trying to amplify a 25 kHz signal with an op amp that has a gain-bandwidth product $f_{\text{GB}} = 1\,\text{MHz}$, then the open-loop gain is approximately $A = -40j$ at that frequency. If we design a non-inverting amplifier (say with $Z_{\text{f}} = 10\,\text{k}\Omega$ and $Z_{\text{i}} = 1\,\text{k}\Omega$) to get an ideal gain of 11, $B = 1/11$, and our output is

$$\begin{aligned} V_{\text{out}} - V_{\text{m}} &= \frac{A}{1+AB}(V_{\text{p}} - V_{\text{m}} - V_{\text{offset}}) + \frac{1}{1+A_0B}(V_{\text{mid}} - V_{\text{m}}) \\ &\approx \frac{-40j}{1-3.636j}(V_{\text{p}} - V_{\text{m}} - V_{\text{offset}}) \,, \end{aligned}$$

and so the magnitude of the gain is only 10.61 instead of 11, and there is a $-15°$ phase shift of the output relative to the input.

We often do not want to use amplifiers all the way out to their corner frequencies, because the gain there is about 30% less than the ideal passband gain.

For example, if you want the open-loop gain of your amplifier to be within 10% of the ideal gain, you need to have $|A/(1+AB)| > 0.9/B$ or $|1+\frac{1}{AB}| > 10/9$. If AB is purely imaginary, this requires $|AB| < 2.06$, and the highest frequency we can use is $f_{\text{GB}}B/2.06$, which is about half the frequency we could use if we were willing to go all the way to the corner frequency.

Worked Example:
If you have gain-bandwidth product $f_{\text{GB}} = 1$ MHz, and you want a gain of 15, dropping to no lower than 14, what is the highest frequency you can amplify in one stage?

We need $G = \left|\frac{A}{1+AB}\right| > 14$ with $B = 1/15$. We can rearrange the inequality to get

$$\left|\frac{15A}{15+A}\right| > 14,$$
$$\left|\frac{15+A}{15A}\right| < 1/14,$$
$$|1 + 15/A| < 15/14\ .$$

Since A is purely imaginary at the frequencies of interest, we need $|15/A| < \sqrt{(15/14)^2 - 1^2} = 0.385$ or $|A| > 39$, which limits our frequency to $f < 1\,\text{MHz}/39 = 25.64\,\text{kHz}$.

Worked Example:

If we are using op amps with a gain-bandwidth product of 1 *MHz, how many stages do we need to get a ideal gain of* 300 *and a cutoff frequency of* 50 *kHz or more?*

At 50 kHz, the open-loop gain is only $1\,\text{MHz}/50\,\text{kHz} = 20$, so one stage is not enough. Two stages each with a gain of $\sqrt{300} = 17.32$ look plausible, but we need to check that the combined gain of the two stages is at least $\sqrt{2}/2\ 300 = 212.1$ at 50 kHz.

The gain of a single-stage non-inverting amplifier is $\frac{A}{1+AB}$, with $B = 1/\sqrt{300} = 0.05774$ and $A \approx -j1\,\text{MHz}/50\,\text{kHz} = -20j$ at 50 kHz. So the magnitude of the gain per stage at 50 kHz is about 13.093, and the combined gain of both stages is 171.4, which is not enough!

What if we go to three stages? If the stages are identical, we'd need a gain $300^{1/3} = 6.694$ per stage, and we can compute the magnitude of the gain per stage at 50 kHz as 6.348, for a combined gain of 255.8, which meets our constraints.

We even have a little wiggle room here, so we might be able to get the 300 gain with standard resistors: say 68 kΩ and 12 kΩ for the feedback resistors, to get an ideal gain of $1 + 68\,\text{k}\Omega/12\,\text{k}\Omega = 6.6667$ per stage or 296.3 for all three stages combined. The gain at 50 kHz is about 6.325 per stage or 253 over all three stages, comfortably within our specifications.

Exercise 19.4

For an inverting amplifier with gain −10 made using an op amp with $f_{\text{GB}} = 1\,\text{MHz}$, what is the corner frequency?

Exercise 19.5

If we want an amplifier with gain 1000 and passband at least 20 Hz–75 kHz, built from op amps with $f_{\text{GB}} = 1\,\text{MHz}$, how many stages of amplification do we need? If the stages are identical, what is the gain per stage?

Chapter 20

Pressure Sensors

Pressure (force per unit area) is a generally useful measure for fluids. In biomedical engineering, both breath pressure and blood pressure are commonly measured properties.

20.1 Breath pressure

For breath pressure, one may want to know the maximum *expiratory* pressure (how hard one breathes out) and the maximum *inspiratory* pressure (how hard one breathes in). Often doctors want records of the full breath cycle, plotting pressure versus flow rate.

For professional breath measurements, pressure is measured with non-zero flow, to make sure that the pressure is from the lungs, rather than from the cheeks. One way to achieve the controlled flow is by having a calibrated leak (a hole 2 mm in diameter) in the standard instrument. I made some very cheap apparatus for measuring breath pressure by drilling a deliberate 2 mm hole in PVC plumbing fixtures (see Figure 20.1).

Different sources give very different estimates of the maximum expiratory pressure (MEP) and maximum inspiratory pressure (MIP), from very high numbers [44, Table 91, p. 97] to rather low ones [166]. A fairly recent survey article summarizes the estimates for adults with four linear regressions based on age in years, y [23]:

$$\begin{array}{lll} \text{MEP} & \text{male} & 174\,\text{cm}\,\text{H}_2\text{O} - y\,0.83\,\text{cm}\,\text{H}_2\text{O/year}, \\ \text{MEP} & \text{female} & 131\,\text{cm}\,\text{H}_2\text{O} - y\,0.86\,\text{cm}\,\text{H}_2\text{O/year}, \\ \text{MIP} & \text{male} & 120\,\text{cm}\,\text{H}_2\text{O} - y\,0.41\,\text{cm}\,\text{H}_2\text{O/year}, \\ \text{MIP} & \text{female} & 108\,\text{cm}\,\text{H}_2\text{O} - y\,0.61\,\text{cm}\,\text{H}_2\text{O/year}. \end{array}$$

I made two different low-cost mouthpieces for breath-pressure measurement out of PVC plumbing components (see Figure 20.1). The larger one in Figure 20.1(b) allowed me to get to somewhat higher pressures than the smaller one in Figure 20.1(c), but the smaller one is cheaper and easier to make, so we'll use that design in the lab. An alternative approach is to make the Figure 20.1(b) adapters as permanent lab equipment and issue each student their own commercial mouthpiece (Figure 20.1(a)).

Figure 20.2 shows a few traces of exhalation followed by inhalation using the smaller mouthpiece made from a PVC elbow. The maximum expiratory pressure is around 11 kPa to 12 kPa (105 cm H_2O–120 cm H_2O) and maximum inspiratory pressure is around −4 kPa to −7 kPa (−40 cm H_2O to −70 cm H_2O). These five traces were selected for having the most constant pressure during inhalation or exhalation from 12 recorded traces.

If I covered the hole in the breath-pressure apparatus, so that there was no air flow, just static pressure, I could easily get maximum pressures of 13 kPa and −23 kPa, showing the importance of the leak. It would be interesting to experiment with different size holes for the leak, to see how the maximum

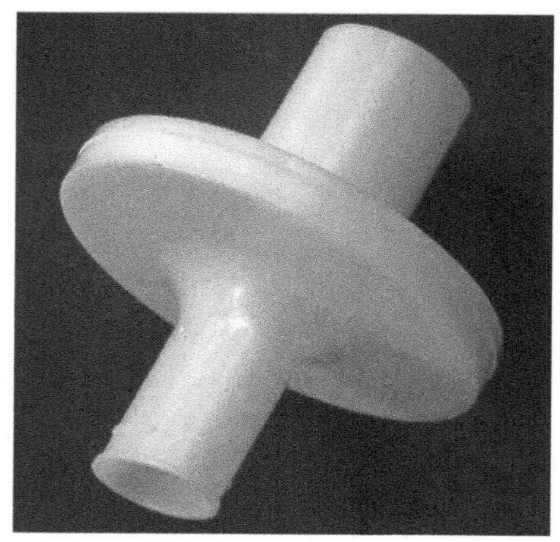

(a) *This MicroGard® II mouthpiece is used for commercial pulmonary function testing and costs about \$1.35 each, if you can get a distributor to sell to you (US distribution limited to hospitals). The mouth goes around the oval end, while the round end connects to the measuring equipment. The large disk contains a filter, to prevent bacteria and viruses from contaminating the equipment or being transferred to other patients. The filter adds a pressure drop of 0.04 kPa/(L/s) [102]. The mouthpiece will fit into the apparatus of Figure 20.1(b).*

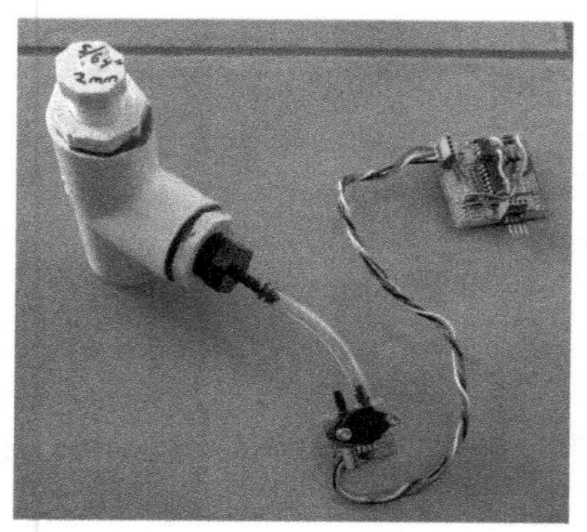

(b) *Breath-pressure apparatus with 1″ PVC tee, with bushings to provide 1/2″ threaded connections on one leg and the base of the tee. One leg of the tee goes around the mouth for breathing, the other accepts a screw-in plug with a 5/64″ (2 mm) hole. Using a screw-in plug allows experimenting with different hole sizes.*

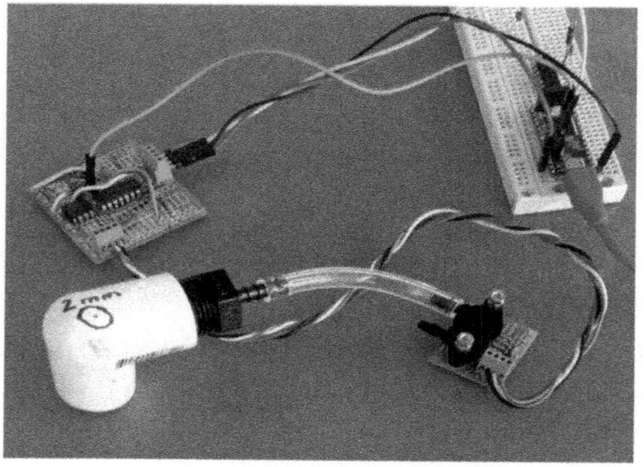

(c) *Breath-pressure apparatus with 1/2″ PVC elbow. The 2 mm hole is drilled into the back of the elbow. The elbow can go either around the mouth or into the mouth, with the lips closing around the outside to seal it.*

Figure 20.1: *Two different low-cost ways for making breath-pressure measurements. Both have a 2 mm hole for an air leak, to prevent measuring pressure generated by the cheeks instead of from the diaphragm. The threaded PVC fitting holds a barbed fitting for connection to the pressure sensor.*

pressure depends on the size of the hole—I expect the air flow to be proportional to $p^2 d^2$, where p is pressure and d is the diameter of the hole, so the pressure drop might be inversely proportional to the diameter of the hole.

The breath-pressure numbers are not particularly close to the ±50 kPa of the full-scale reading of the pressure sensor used in Lab 5, so you might want to use a higher gain for the amplifier than just what it would take to make the full-scale range of the sensor correspond to the full-scale range of the analog-to-digital converter. But consider the pressure range you need for blood-pressure-cuff measurements, before you choose the gain for your amplifier.

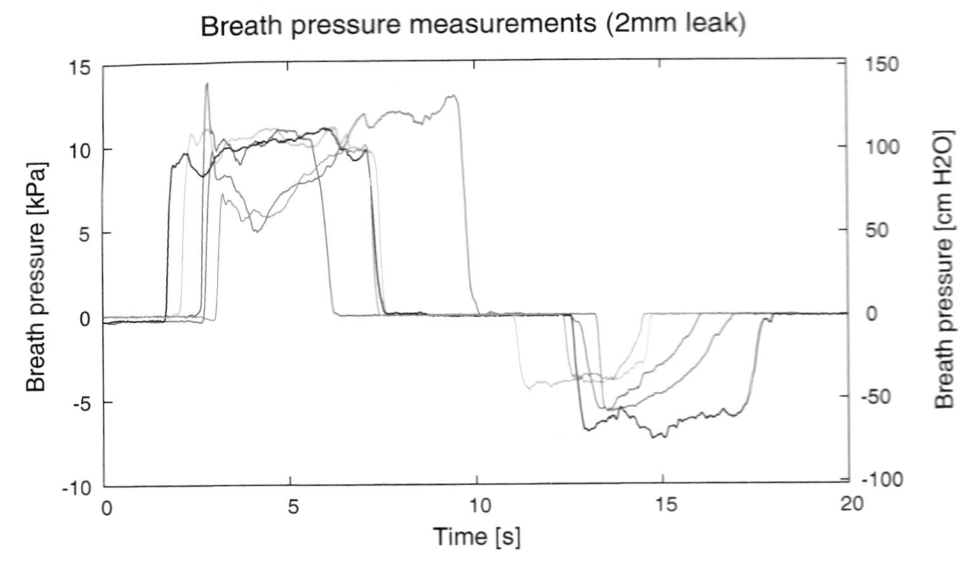

Figure 20.2: *Breath-pressure measurements were made for maximum-effort exhalations or inhalations, with all air flow through a 2 mm diameter hole, with the apparatus of Figure 20.1(c).*
Data collected with PteroDAQ using a Teensy LC board [46]. Figure drawn with gnuplot [33].

20.2 Blood pressure

Blood pressure is much more commonly measured than breath pressure—my blood pressure is measured on every visit to the doctor (no matter what the reason for the visit), and many people measure their blood pressure daily, to monitor their dosage of blood-pressure medication.

> Blood pressure is not a constant, the way pressure in a water main is, but is pulsed. Each time the heart contracts, it squeezes blood out into the arteries, raising the blood pressure. When the heart relaxes, the blood pressure drops again. There are two pressure measurements of interest: the *systolic pressure*, which is the maximum pressure of the cycle and the *diastolic pressure*, which is the minimum pressure of the cycle.

Blood pressure is usually reported in the US as two numbers separated by a slash: 120/80, which are the systolic and diastolic pressures in mm Hg (a rather old-fashioned unit of pressure that is not much used any more except for blood pressure—use online unit converters or Google to translate to more standard units). The typical 120/80 would correspond to about 16 kPa systolic pressure and 10.7 kPa diastolic pressure.

There are two main methods for measuring blood pressure: invasive monitoring with an intravenous (or arterial) needle, causing direct contact between the blood and the pressure sensor, and indirect reading through a blood-pressure cuff, most commonly on the upper arm. For other techniques for noninvasive blood-pressure measurements, see the Wikipedia article on blood pressure [109].

NXP's disposable medical-grade pressure sensor, the MPX2300, is intended for invasive measurements of blood pressure, and so is made of biomedically approved materials that can be sterilized with ethylene oxide. I considered using the MPX2300 for the course, and even designed a PC board for mounting the part, but the package does not protect the chip well, and I decided the lab needed more rugged parts.

Of course, I never had any intention of doing invasive monitoring of blood pressure in the lab—it would be unethical to do invasive procedures without real medical need, and students sticking themselves with needles is likely to result in a bloody mess.

> Blood-pressure cuffs provide a familiar and non-threatening way to measure blood pressure. An inflatable balloon (the *cuff*) is wrapped around the upper arm, the pressure is raised until blood flow is cut off, and then the pressure is lowered at a constant rate through a controlled leak until blood flow resumes.

For cheap blood-pressure cuffs, the "leak" is not built-in—one has to loosen a valve very slightly to produce the leak, which takes a little practice. The pressure in the cuff is measured as the cuff deflates, and the systolic and diastolic pressures are recorded at two specific points on the deflation curve.

It is conventional to measure blood pressure in the arm at the height of the heart, in order to get a standardized measurement. Higher in the body, the blood pressure is lower, and lower down the blood pressure is higher. You can approximate the pressure change due to measuring at a different height by the difference in height of the heart and measurement point, multiplying by the density of blood (about 1.06 g/mL), and converting from cm H_2O to a more convenient pressure unit. The top of a person's head might be 60 cm above their heart, so blood pressures there may be 6.2 kPa (47 mm Hg) lower than at their heart. The feet of a standing person may be 120 cm below the heart, so blood pressures there might be 12.5 kPa (94 mm Hg) higher than at the heart. These differences in pressure are much larger than the differences used to diagnose disease, hence the need for a standardized location on the body for measuring pressure.

To further complicate matters, different arterial pressures may be observed in the ankle arteries and the brachial (arm) arteries, even when the patient is lying down, with ankle pressure being typically about 10%–20% higher than brachial pressure. In fact, peripheral artery disease is diagnosed by the ankle-brachial index (the ratio of ankle systolic pressure to brachial systolic pressure) being less than 1 [68, 94].

There are two main methods for determining the systolic and diastolic pressures during the deflation:

Ausculatory: When the cuff pressure is between the systolic and diastolic pressures, the blood flow in the arteries becomes turbulent and makes characteristic sounds called the *Korotkoff sounds.* Doctors and nurses are trained to listen to the blood flow through a stethoscope and note what pressure these sounds start at (systolic pressure) and stop at (diastolic pressure). This is the standard method in most medical practices.

Oscillometric: If you observe the cuff pressure while deflating the cuff through a controlled leak, you will see that it does not descend smoothly, but varies with the pulse of the person. By monitoring just the pressure in the cuff, you can determine the pulse rate and the blood pressure fairly accurately. This method is the preferred method for most automatic blood pressure meters.

The variation in cuff pressure is strongest at the mean arterial pressure, and drops off quite a bit at the systolic and diastolic pressure. These oscillations can be separated from the smooth descending pressure with a high-pass (or band-pass) filter for the oscillations and a low-pass filter for the deflation. To get good filter characteristics, it is easiest to implement these filters as digital filters in software, rather than in hardware.

Figure 20.3 shows an *amplitude envelope*—a slowly changing signal that acts as a multiplier for a higher-frequency *carrier wave* to produce the overall signal. The amplitude envelope of the

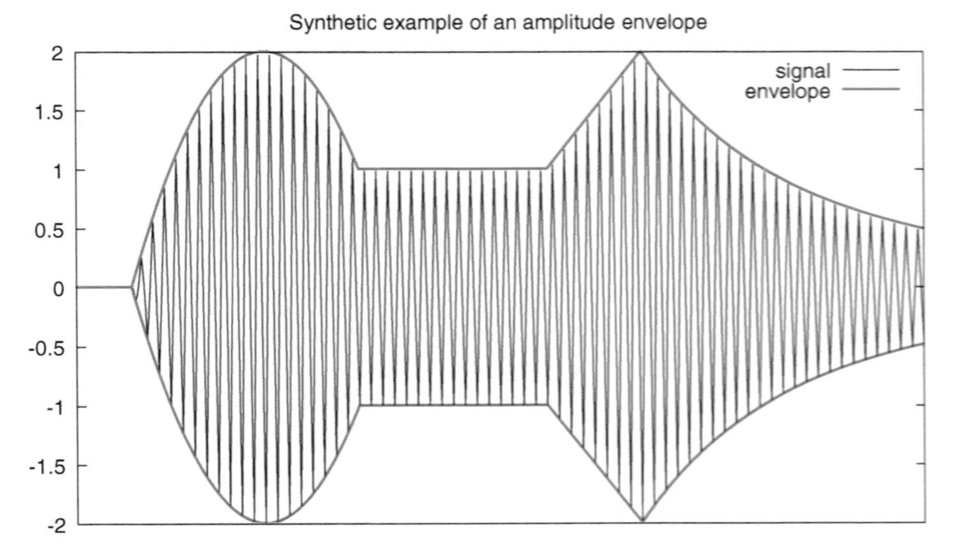

Figure 20.3: *This figure shows an amplitude envelope modifying an underlying signal. You can think of an underlying constant-amplitude triangle wave being multiplied by the (positive) amplitude envelope to get the overall signal.*

oscillations on a blood-pressure measurement can tell you where on the smooth deflation curve to read the systolic and diastolic pressures.

Different manufacturers use different proprietary algorithms to determine where on the amplitude envelope of the oscillations to measure the systolic and diastolic pressures, depending on exactly how the amplitude envelope is determined from the pulsations.

Here is a typical algorithm for determining systolic and diastolic pressures [95]:

(1) Fit an amplitude envelope to the peaks of the oscillations. This is often just a pair of straight lines down from the biggest peak at the mean arterial pressure (MAP). The biggest peak is the one with the largest swing from maximum to minimum pressure, not the one with the largest positive excursion.
(2) Note the time where the amplitude envelope is 40%–50% of the maximum before the peak (so while the pressure is still higher than the MAP). The pressure at this time is the systolic pressure.
(3) Note the time where the amplitude envelope drops to 75%–80% of the maximum after the peak (so the pressure is less than the MAP). The pressure at this time is the diastolic pressure.

The ratio of the blood pressure estimated by these methods and by the traditional ausculatory method varies quite a bit (as much as 10%–20%) [32]. Taking into account the shape of the envelope can improve the accuracy of the measurements [7].

Figure 20.4 shows a recording of pressure in a cuff as it is manually inflated then slowly deflated. The sampling was done at 200 Hz, which means that 60 Hz line-frequency interference needs to be filtered out. The fluctuations due to the pulse are barely visible.

In Figure 20.5, the same recording is shown as in Figure 20.4, zoomed in to the region of slow deflation, after passing separately through two different filters: a low-pass filter (to remove the fluctuations) and

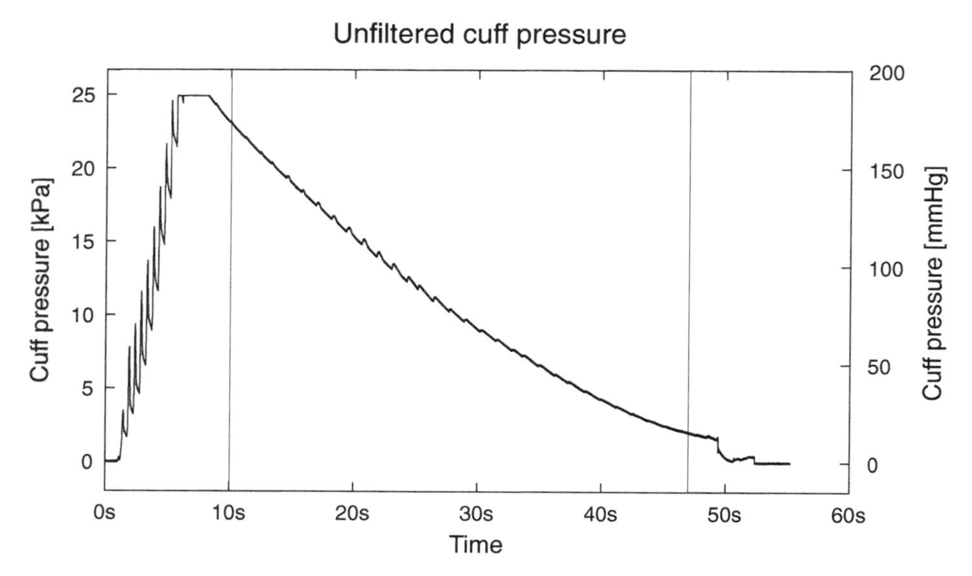

Figure 20.4: *Pressure measurements from a blood-pressure cuff on the upper arm. From about 0s to about 7s, the cuff was inflated to 200 mm Hg using a manual squeeze bulb, but the reading clips at about 187 mm Hg (25 kPa), because of the gain used on the amplifier. The cuff then deflates at about 570 Pa/s (4.2 mm Hg/s), controlled by an engineered leak in the inflation bulb. A slightly slower leak would give a longer time for measuring the blood pressure, but you don't want to leave the blood flow constricted for more than about 100 s.*

The fluctuations due to the pulse are visible (barely) between 11 s and 42 s. Around 49 s a release valve was used to release the remaining pressure. We will focus on the region from 10 s to 47 s.

Data collected with PteroDAQ using a Teensy LC board [46]. Figure drawn with gnuplot [33].

a band-pass filter (to accentuate the fluctuations). Each filter is applied separately to the originally recorded data—not one filter to the output of the other.

The cutoff frequency for the low-pass filter was selected so that the slow release of pressure from the cuff would be kept, but the higher frequency fluctuations from the pulse would be blocked. That requires a cutoff frequency lower than the lowest expected pulse rate—for these plots 0.3 Hz, corresponding to 18 bpm, was chosen.

The band-pass filter has to eliminate the slow shift in baseline due to the gradual decrease in pressure, and show only the pulse. For this plot, the same 0.3 Hz cutoff was used to distinguish between the slow leak and the pulse, and a high-frequency cutoff of 6 Hz was used, to eliminate higher harmonics and noise.

The 6 Hz cutoff was chosen to handle heart rates up to 360 bpm, while substantially reducing any 60 Hz interference. The highest theoretical rate for a human heart beat is about 300 bpm (5 Hz), though a rate of 600 bpm has been reported (based on electrical activity, not pressure fluctuation in the peripheral circulatory system) [12].

The pulse rate can be measured very precisely from the band-pass filtered oscillations shown in Figure 20.5. Look for a feature that occurs once in each period (the tallest spike or an upward zero crossing, for example) in the filtered data file, and compute the time needed for 10 periods (say, from 40.2 s to 52.45 s, or 12.25 s). Use several periods, to average out the natural fluctuations in the heartbeat.

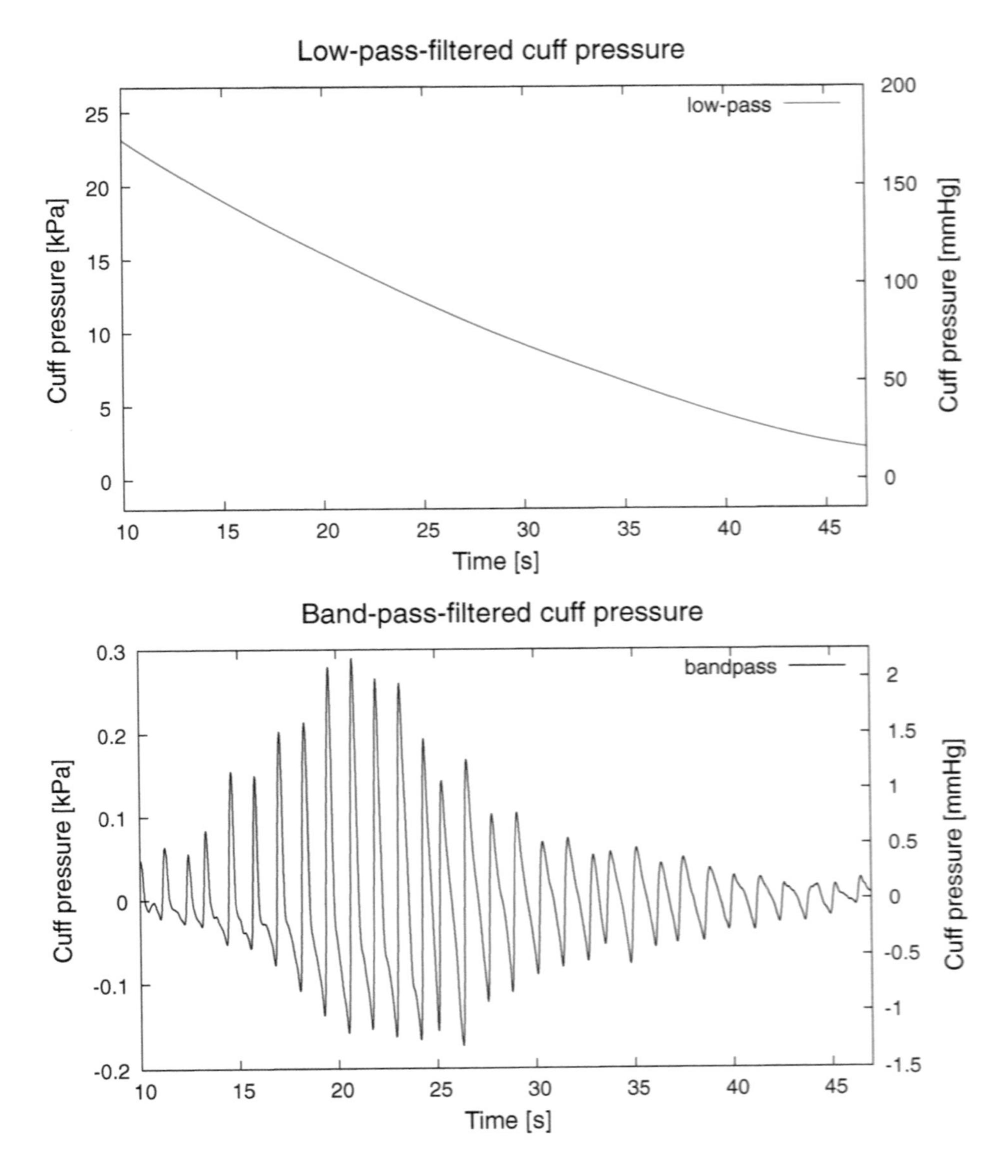

Figure 20.5: *The cuff-pressure measurements of Figure 20.4 during the deflation of the cuff are shown here after low-pass (0.3 Hz) and band-pass (0.3 Hz–6 Hz) filtering. The band-pass-filtered signal can be used to determine when to read the systolic and diastolic pressures from the low-pass signal.*
Data collected with PteroDAQ using a Teensy LC board [46]. Figure drawn with gnuplot [33].

> You want to measure the time for an *integer* number of periods, rather than estimating the number of periods in a fixed time.
>
> When counting features for periods, watch out for *fencepost errors*. You need 11 of the features to get 10 periods, just as you need 11 fenceposts for 10 sections of fence in a row.

The pulse rate in beats/min is 60 times the beats/s: here, 49 bpm. It is conventional to round pulse rates to the nearest bpm, since that is all the precision that is needed and heart rates fluctuate enough that higher precision is likely to be misleading.

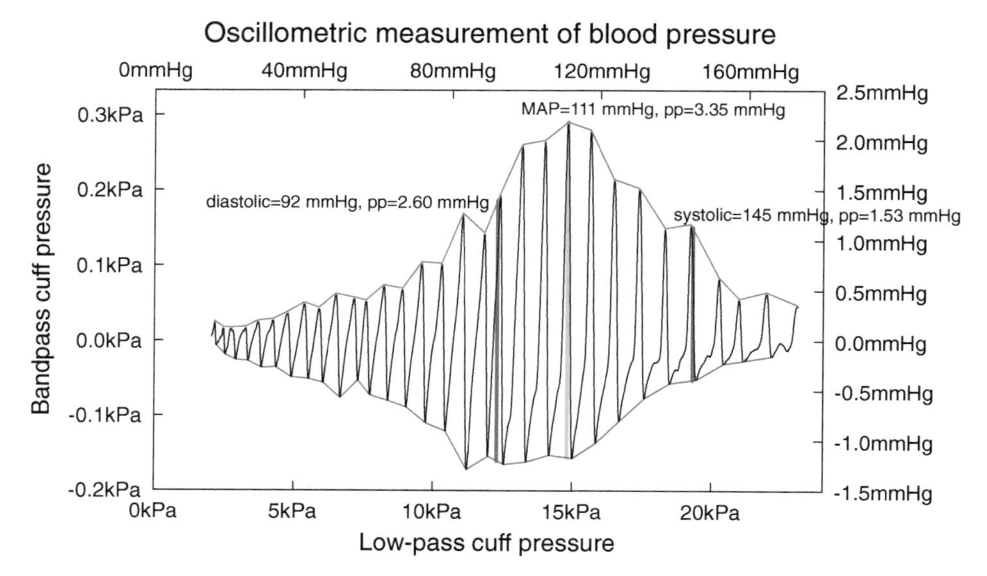

Figure 20.6: *Plotting the cuff deflation as the band-passed signal vs. the low-passed signal. The shape of the curve does not depend (much) on how fast the cuff is deflated, and the mean arterial pressure can be estimated directly from where the biggest fluctuation occurs.*

I drew an envelope around the peaks and picked the point of maximum amplitude as the mean arterial pressure, then chose 77.5% of the maximum amplitude at lower than the mean arterial pressure to mark the diastolic pressure and 45.5% of the maximum amplitude at higher than the mean arterial pressure to get the systolic pressure. The envelope height is reported as "pp=" for diastolic, mean arterial, and systolic pressure. Both systolic and diastolic pressures seem anomalously high for this particular subject—this may be due to the crude method for estimating the two pressures or due to a mistake in the amplifier gain.

Data collected with PteroDAQ using a Teensy LC board [46]. Figure drawn with gnuplot [33].

The normal pulse rate for a human being varies with age, with young children having higher pulse rates than adults. The resting heart rate for an adult is usually 60 bpm–100 bpm, with lower heart rates generally associated with higher physical fitness, though there is a genetic component also (I and some others in my family have resting heart rates under 50 bpm). Exercise or excitement increase the heart rate, and target pulse rates are often used to set desired levels of aerobic exercise.

It can be useful to plot the band-passed signal (just the oscillations) vs. the low-passed signal (without the oscillations) to see what static pressures correspond to various parts of the waveform (Figure 20.6). I used Python programs to select the maxima and minima and to automatically compute and generate the labels for the plot—you can get the same effect with much less effort using a ruler to measure the biggest swing, then sliding the ruler on the plot to find where the swing is reduced to the levels needed for the systolic and diastolic pressures.

The filtering was done digitally after the signal was recorded, using the filter functions of the SciPy signal-processing package—warning: the SciPy package is not routinely installed in all installations of Python, though it is in Anaconda and Enthought.

The first part of the program is standard boilerplate to allow "print" and "zip" to mean the same thing in different versions of Python. Then the signal-processing package is imported from the SciPy package

The "times" and "values" lists are arrays of numbers taken from the first two columns of the data file read from standard input. If there is PteroDAQ metadata that gives the sampling rate, that sampling rate is used, otherwise 20 Hz is assumed.

The cutoff frequencies have to be converted to a standard scaling, in which the Nyquist frequency is 1 and the sampling frequency is 2, as that is the parameterization that SciPy's signal-processing package assumes.

The iirf function computes the parameters for an *infinite-impulse-response filter*, and puts them in an array that is then passed to the sosfiltfilt routine that implements the specified filter and filters the values, returning the result to be stored as "bandpass". The same setup is done to compute the parameters for a low-pass filter, and filter values into "lowpass". For both filters, a 4th-order Bessel filter was applied twice (once forward in time and once backward in time) with the scipy.signal.sosfiltfilt function to eliminate phase delays due to filtering (see Figure 20.7).

Finally, some comments are added to the output, giving metadata about the filters. Time, values, band-passed values, and low-passed values are printed out in four columns.

20.3 Pressure sensors and strain gauges

The pressure sensor we use at UCSC, the MPX2050DP from NXP (see Figure 20.8), is a temperature-compensated strain-gauge pressure sensor which does not include a built-in amplifier. If I were designing an instrument that needed a pressure sensor, I'd most likely choose one that *does* include an integrated amplifier, since the extra cost is small and the integrated amplifier makes design easier. But there are many applications where the integrated amplifier is not available—for example, the MPX2300 disposable medical sensor mentioned in Section 20.2 does not include an amplifier, in order to keep the cost below $5 a piece (about $3.30 each in thousands).

> Strain gauges are also used in applications other than pressure gauges (anywhere that small bending or stretching of solid objects needs to be measured), and most strain gauges don't come with built-in amplifiers. So there are plenty of applications for an amplifier of the type you will design in Lab 5.

A pressure sensor works by having a membrane separating two compartments: one of known pressure, the other of the fluid whose pressure is to be measured.

> There are basically three types of pressure sensors: *differential*, *gauge*, and *absolute* pressure sensors. In a differential sensor, both compartments have accessible connections; in a gauge sensor, one of the compartments is the ambient air pressure and the other the pressure to be measured; and in an absolute sensor, one of the compartments is sealed with a known pressure (often vacuum, since its pressure doesn't change with temperature nor as the membrane moves).

The cheapest pressure sensors are barometric pressure sensors (absolute sensors that measure the pressure of the ambient air), because these are high-volume parts and don't require a sealed package, just the tiny sealed chamber behind the membrane. NXP sells the MPL3115A2 for about $2.35 in hundreds, which can measure 50 kPa–110 kPa (actually down to 20 kPa, but calibration is not guaranteed below 50 kPa), with noise as low as 1.5 Pa RMS. Most of the barometer/altimeter chips, including the MPL3115A2, do not use an analog readout at all, but include an on-chip amplifier and analog-to-digital converter, together with interface circuitry to do I^2C or SPI digital interfaces to microcontrollers.

```
#!/usr/bin/env python

from __future__ import print_function, division
import sys
import numpy as np
from scipy import signal
try: from future_builtins import zip
except ImportError:     # either before Python2.6 or one of the Python 3.*
    try: from itertools import izip as zip
    except ImportError: # not an early Python, so must be Python 3.*
        pass

sampling_freq = 20      # if not read from input
times=[]        # time stamps from first column
values=[]       # data values from second column
for line in sys.stdin:
    if line.startswith('#'):
        print (line.rstrip())
        if line.startswith('# Recording every'):
           # extract from input: # Recording every 0.05 sec (20.0 Hz)
           fields = line.split('(')
           sampling_freq=float(fields[1].split()[0])
        continue
    fields = [x for x in map(float, line.split())]
    if len(fields)>=2:
        times.append(fields[0])
        values.append(fields[1])
values=np.array(values)  # needed because sosfiltfilt omits the cast

# define the corner frequencies
low_end_cutoff = 0.3    # Hz
lo_end_over_Nyquist = low_end_cutoff/(0.5*sampling_freq)
high_end_cutoff = 6.0 # Hz
hi_end_over_Nyquist = high_end_cutoff/(0.5*sampling_freq)

# band-pass filter passes frequencies between low_end_cutoff and high_end_cutoff
sos_band = signal.iirfilter(4,
            Wn=[lo_end_over_Nyquist,hi_end_over_Nyquist],
                            btype="bandpass", ftype='bessel', output='sos')
bandpass = signal.sosfiltfilt(sos_band, values)

# low-pass filter passes frequencies below low_end_cutoff
sos_low = signal.iirfilter(4, Wn=[lo_end_over_Nyquist],
                            btype="lowpass", ftype='bessel', output='sos')
lowpass = signal.sosfiltfilt(sos_low, values)

print("# Bessel bandpass to {:.3g}Hz to {:.3g}Hz".format(low_end_cutoff, high_end_cutoff))
print("# Bessel lowpass {:.3g}Hz".format(low_end_cutoff))

for t,v,b,l in zip(times,values,bandpass,lowpass):
    print("{:.7f}\t{:.6f}\t{:.6f}\t{:.6f}".format(t,v,b,l))
```

Figure 20.7: *Python script for filtering blood-pressure signals, used for creating Figure 20.5. Two new columns are added to the output: one filtered with a band-pass filter, the other with a low-pass filter.*
Get this file from https://users.soe.ucsc.edu/~karplus/bme51/book-supplements/bandpass-filter.py as cut-and-paste from the PDF file will fail due to character replacement.

They are used in weather monitoring and as altimeters, but generally have a fairly slow response (1 update/s for high resolution or up to 16 updates/s at low resolution), and so are often not suitable for biomedical applications.

Our sensor is a differential sensor, with two ports that we can connect tubing to, but we'll generally be using it as a gauge sensor, leaving one port open to the ambient air and connecting a tube to the other port. The sensor is designed to measure positive pressure only (with the positive port having

Figure 20.8: *The MPX2053DP sensor, mounted on a* breakout board *that provides easy connection to the pins of the sensor. The MPX2053DP is no longer made, though the data sheets are still available. When we scaled up the lab section sizes for Winter 2018 we used the similar MPX2050DP for Lab 5, which has better linearity and temperature compensation, but is otherwise identical. (To save money, one could also use the MPX53DP, which lacks temperature compensation, has a non-zero offset, and has poorer linearity.)*

a higher pressure than the negative port), but we will abuse the part somewhat by measuring small negative pressures as well. Too large a negative pressure could damage the sensor—the manufacturer does not give any information about how much negative pressure the sensor can tolerate.

When there is a pressure difference between the two compartments, the membrane flexes, bending or stretching some component. A strain gauge on that component changes resistance as a result of the stretching, and the resulting change in resistance is converted to a differential voltage with a Wheatstone bridge circuit (Figure 20.9).

Warren Schultz writes

> The essence of piezoresistive pressure sensors is the Wheatstone bridge shown in Figure 1 [redrawn as Figure 20.9 in this book]. Bridge resistors RP1, RP2, RV1, and RV2 are arranged on a thin silicon diaphragm such that when pressure is applied, RP1 and RP2 increase in value while RV1 and RV2 decrease a similar amount. Pressure on the diaphragm, therefore, unbalances the bridge and produces a differential output signal. One of the fundamental properties of this structure is that the differential output voltage is directly proportional to bias voltage B+ [93].

I don't believe that the bridge circuit given by Schultz precisely represents the thermal compensation circuitry of the MPX2050DP sensor we are using, since the resistances I measured on the device were not consistent with this simple circuit, but I've not been able to find the exact circuit used, despite looking at several patents on temperature compensation circuits from the company. The basic idea, that the differential voltage between the S+ and S- signals is proportional to the pressure and to the supply voltage, is still correct, and that is all we really need for Lab 5.

There are many different silicon strain-gauge pressure sensors—you will need to look up the specifications from the data sheet for the one you will be using. As an example, I will use the MPX2202 sensor from NXP, which is *not* the one we use in the lab. The MPX2202 data sheet gives the output as $40\,\text{mV} \pm 1.5\,\text{mV}$ at $200\,\text{kPa}$ [27], with an input voltage of $10\,\text{V}$. The device is *ratiometric*, which means that the output voltage is proportional to the input voltage, so it is better to report the sensitivity as a gain ($4\,\text{mV/V}$ full scale, or 20E-6/kPa).

If we power the sensor with about $3.30\,\text{V}$, then we would expect to see about $13.2\,\text{mV}$ for a full-scale reading, or $66\,\mu\text{V/kPa}$. This is the *differential* voltage $V_{\text{S+}} - V_{\text{S-}}$—the individual voltages are both close to $V_{\text{dd}}/2 = 1.65\,\text{V}$, halfway between the power rails for the sensor.

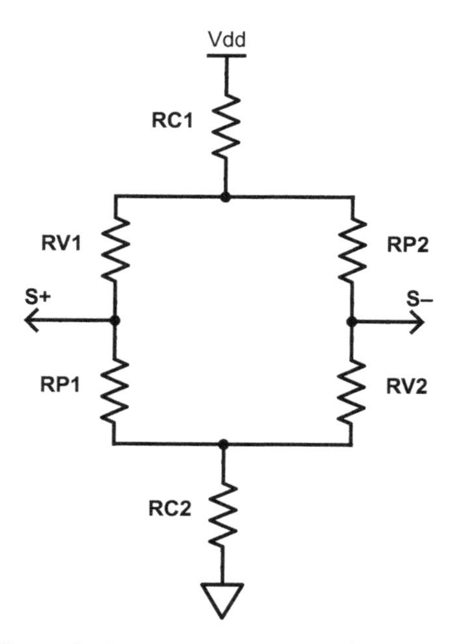

Figure 20.9: *This is the circuit that NXP (formerly Freescale Semiconductor) claims is the equivalent circuit for their temperature-compensated pressure sensors, redrawn from an application note [93]. The R_V and R_P resistors form a Wheatstone bridge—the R_V values decrease with pressure, while the R_P values increase with pressure. The fixed R_C resistors are adjusted so that the positive temperature coefficient of the unstressed piezoresistors can compensate for the negative temperature coefficient of the piezoresistive effect.*
When I tried measuring the resistances of pairs of wires, I did not get values consistent with this equivalent circuit—I think that they are using a more complicated temperature-compensation scheme, though the effective output is still like this bridge circuit.
Figure drawn with Digi-Key's Scheme-it [18].

If you are uncomfortable thinking of pressure in kilopascals, then you may want to use a converter, like the one at http://hyperphysics.phy-astr.gsu.edu/hbase/pman.html to convert to nonstandard units, like pounds per square inch, mm Hg, atmospheres, and inches or cm of water. The Google search box will also do unit conversions (try searching "180 mm Hg to kPa", for example).

Exercise 20.1

For the MPX2202 pressure sensor described above, what would the differential output voltage be with a supply voltage of 5 V and a pressure of 25 kPa?

Exercise 20.2

If you connect an MPX2202 pressure sensor to a 5 V supply and feed its output into an instrumentation amplifier with a gain of 100, what pressure in kPa is indicated by an output of $V_{\text{out}} - V_{\text{ref}} = 1.3\,\text{V}$?

Chapter 21

Lab 5: Strain-Gauge Pressure Sensor

Bench equipment: pressure sensor on breakout board, blood-pressure cuff, drill press

Student parts: INA126PA instrumentation amp, MCP6004 op amp, resistors, breadboard, PteroDAQ, air tubing, PVC elbow, barbed fitting

21.1 Design goal

In this lab, you will design, breadboard, and test an instrumentation amplifier to interface a strain-gauge pressure sensor to a microcontroller board running a data-acquisition system to record breath pressure as a function of time. You will also measure pressure in a blood-pressure cuff and use digital filtering to separate the low-frequency deflation of the cuff from the fluctuations due to pulse. Some students make the mistake of referring to digital filtering as "simulated filtering"—but there is no simulation involved. Digital filtering is just as real as analog filtering, and the digital filtering algorithms are not simulations of analog filters.

21.2 Pre-lab assignment

Look up the data sheets for the parts you will be using: the MPX2050DP pressure sensor from NXP, the INA126PA instrumentation amp, and the MCP6004 quad-op amp. Make sure you have a copy of them in front of you when you do your circuit design and layout.

21.2.1 *Sensor values*

Pre-lab 5.1

With a 3.3 V power supply, what differential voltage do you expect for a high blood pressure of 200 mm Hg using an MPX2050 pressure sensor? (Warning: this is not the same sensor as the MPX2202 in Chapter 20—look up the right data sheet.)

Pre-lab 5.2

What is the worst-case offset voltage you can have with the MPX2050 pressure sensor? The offset voltage is the differential output voltage that you get with zero pressure input.

Pre-lab 5.3

What is the worst-case input offset voltage you can have with the INA126PA instrumentation amplifier?

Pre-lab 5.4

Calculate the differential sensor output voltage for 100 mm Hg with 3.3 V power—you will later compare this computation to a measurement.

Pre-lab 5.5

If you want the output of the complete amplifier to change by 1.5 V when the pressure changes by 200 mm Hg, how much gain do you need? (Still assuming a 3.3 V power supply.)

Pre-lab 5.6

Using your results from Exercise 18.3, Pre-labs 5.1–5.3, figure out the maximum gain that you can request from the INA126PA instrumentation chip, assuming a 3.3 V power supply and a pressure range of −200 mm Hg to 200 mm Hg.

Express your gain as a *constraint*—an inequality, not an equation.

Make sure you used worst-case, rather than typical specs for Exercise 18.3, "worst-case" meaning that the output range allowed by the chip is the smallest that it might be according to the data sheet.

The inputs need to stay between the power rails, which is guaranteed by the bridge circuit for strain gauge sensors.

We also need to have V_{ref} near the middle of our voltage range, so that both positive and negative pressures can be measured. V_{ref} is not generated by the INA126PA chip, but must be provided from external circuitry.

Pre-lab 5.7

If you set V_{ref} to half the 3.3 V power supply, what is the maximum gain for a pressure range of −200 mm Hg to 200 mm Hg that you can request of the INA126PA chip without exceeding its worst-case voltage limits? Keep all your formulas as inequalities—we are discussing constraints here, not design targets.

You do not want to use the full range of the INA126PA output voltage—use a smaller output range, so that the voltages can be centered in the middle of the power-supply range without clipping, even if your inputs exceed the design specs. You can compensate in the second stage.

Remember that the exercises are giving you *constraints*—the most extreme possible values. It is rare in engineering that you want to push up against the constraints, unless you are trying to do something that is only marginally feasible. Keeping your constraints as inequalities helps you remember their purpose.

For an amplifier, keeping the gain to about half the maximum allowed generally provides ample safety margins.

Pre-lab 5.8

If you use a power supply of 3.3 V, $V_{\text{ref}} = 1.2$ V, and an instrumentation-amplifier gain of 122.65, what is the largest input voltage range that you can use while staying within the worst-case INA126PA output range? What pressure range does that correspond to? Is this pressure range wide enough for both breath and blood pressure measurements?

The pre-lab exercises above all assumed that a 0 Pa difference in pressure at the sensor would result in an output that is exactly at V_{ref}, but the sensor and both amplifier stages can have offsets, so that the voltage corresponding to zero pressure is not exactly zero for the differential signals, nor exactly V_{ref} for the single-ended signals. The offsets for the sensor and the first stage are amplified by the same amount as the signal, but the offset for the second stage is amplified only by the second stage.

21.2.2 *Block design*

For this class, we usually label signal levels on a block diagram as "DC ± AC amplitude". For example, the output of the pressure sensor might be designed to be 1.5 V±5 mV (I'm not suggesting that is the right label—you'll have to choose a reasonable label based on Pre-lab 5.4).

Signals in block diagrams use a single line on the diagram, even if the signal is physically represented as a difference in voltage between two wires (a *differential pair*). For a differential signal, the label on the block diagram often just has the difference voltage.

When the DC part of a signal is not just a bias voltage but carries information we care about, we might use "1.495 V–1.505 V" to indicate the voltage range.

Pre-lab 5.9

Draw a block diagram for all the components of the system, from the pressure sensor to the microcontroller. Figure out the characteristics of the signal at the input and output of each block (pressure range, voltage, current, frequency—anything you might need to design the blocks).

21.2.3 *Schematics*

When drawing your schematics, label each node with the color of wire that you will use for connecting up that node. All your ground connections should be black, and all your 3.3 V connections should be red. For other nodes, pick colors of wire we have available, but try to use a different color for each node. If you have more nodes than colors, then try to make sure that wires that go to adjacent pins on a component have different colors (unless they are the same node, of course).

Coloring your schematic before you wire your breadboard makes debugging much easier, as many of the most common mistakes are off-by-one placement of wires. If you are careful with your color coding, then whenever two wires of different colors come together, you know you have a mistake. Also, when two adjacent pins that should be different nodes have wires of the same color, you have a mistake.

This lab is *not* a tinkering lab—the idea is to do the design right the first time, so that you don't have to spend a lot of time debugging.

The INA126PA chip will require current from V_{ref}, so V_{ref} should not come directly from a voltage divider (which would violate the voltage-divider rule), but from a low-impedance source like an op amp. The current needed is $I = (V_{\text{m}} - V_{\text{ref}})/40\,\text{k}\Omega$, which will be only a few microamps, because V_{m} will be close to halfway between the power rails, so the difference from V_{ref} will be small. Because the current needed is so small, we don't need a powerful voltage source.

Pre-lab 5.10

Come up with a circuit to produce V_{ref} and draw the schematic.
Warning: avoid using low resistances for a voltage divider to set V_{ref}—that would waste the scant power available from the voltage regulator on the microcontroller board and could damage the board.

Pre-lab 5.11

After you have a block diagram, come up with a circuit for each block. Draw your schematic very carefully, with all the pin numbers for the chip labeled.
Prepare it for the final report—you want a really clean schematic before you commit to wiring anything. Have both partners check the schematic carefully for errors before going on to the next step.

After drawing the schematic, you need to come up with sizes for the resistors.

Pre-lab 5.12

Compute the needed gain resistor for the gain computed in Pre-lab 5.6. Use the E12 series of resistors, making sure you don't exceed the maximum gain. What gain do you achieve with this resistor?

Once you have settled on a gain for the first stage, you need to choose an appropriate gain for the second stage. The second stage not only amplifies the signal, but also the input offset of the op-amp chip.

Pre-lab 5.13

Compute the gain needed in the second stage to get $V_{\text{ref}} \pm 1.5\,\text{V}$ at the final output, if you use the gain chosen in Pre-lab 5.12 for the first stage.

Pre-lab 5.14

What resistor values will you use to get the gain for the second stage, as computed in Pre-lab 5.13? Remember to use values rounded to the E12 scale (or whatever resistor values you have in your parts kit).
Look up the input offset voltage of the op-amp chip. How much output offset could you have as a result of the gain you chose? Is this large enough to cause clipping in the second stage?

Pre-lab 5.15

What is V_{out} from your complete amplifier (Pre-labs 5.4–5.13) as a function of pressure? Pressure as a function of V_{out}?

Pre-lab 5.16

What sampling rate will you be using on the microcontroller? Will you need any filtering in the amplifier?

Write down all the design questions and decisions you make. Don't count on remembering them later! You should write half your design report before wiring anything.

21.3 Procedures

The building blocks are simple for this lab, but you need to be careful with your gain computation and with your wiring. Sloppy schematics and sloppy layout diagrams will cost you a lot of debugging time.

Start by connecting the sensor up to a 3.3 V power supply and an oscilloscope, to make sure that the signals you get from the sensors are what you were expecting when you did your gain calculations.

The four wires from the pressure sensor to the amplifier board (+3.3 V, ground, S+, S−) should be color-coded (four different colors, with red and black for +3.3 V and ground) and twisted together so that there is little pickup of noise from magnetic fields.

Always use red wires for your positive power supply and black wires for your ground throughout this course (and don't use those colors for other signals). Following this convention will make your circuits much easier to check and debug.

Using the mechanical gauge on the blood-pressure cuff, measure and record the sensor voltage difference for 100 mm Hg with 3.3 V power (also 0 mm Hg and −100 mm Hg). You can get negative pressure readings by connecting the higher pressure to the negative port of the differential sensor. Compare these values with what you computed in the pre-lab.

When making a measurement of 0 mm Hg, disconnect the sensor from the blood-pressure cuff, so that both ports are connected to the ambient air in the room, to make sure that there is no residual pressure in the cuff that would throw off your measurement. The zero-pressure measurement may not result in a differential sensor voltage of 0 V, as the sensor can have a small offset (maximum offset specified on the data sheet).

There may also be an offset at the input of Analog Discovery 2 input channel (or PteroDAQ differential input channel). You can determine how much of the offset comes from the pressure sensor and how much from the amplifiers in the Analog Discovery 2 (or PteroDAQ) by swapping the output wires from the sensor, as that would negate the offset from the sensor, but leave the offset from the channel unchanged. With the sensor hooked up one way, the measurement is $V_O + V_S$ and the other way is $V_O - V_S$, where V_O is the offset of the oscilloscope channel and V_S is the offset of the sensor.

> The comparison with the mechanical gauge is just a sanity check to make sure your gain calculations are correct. The mechanical gauge is less accurate than the pressure sensor, so should not be used as a calibration device!

Wire the virtual ground (V_{ref} as designed in Pre-lab 5.10) and the instrumentation amplifier with the gain resistor selected in Pre-lab 5.12. Measure and record the voltages at the input and output of the instrumentation amplifier ($V_{out} - V_{ref}$) to check whether your gain calculations were correct and whether you have wired the amplifier correctly.

The measured gain is best estimated from a pair of measurements at high pressure and zero pressure (or high positive and high negative pressure), not from a single pair of measurements at one pressure, but the measurement of small voltages with the equipment you have available is not very accurate, so your gain computation from the R_{gain} value will be more accurate than estimates from measurements of the input voltage to the instrumentation amplifier. The gain measurements are a sanity check on your design, not a calibration.

The instrumentation amplifier may have a small input voltage offset, whose worst-case can be determined from the data sheet. Both the sensor offset and the input offset of the amplifier get amplified by the gain of the amplifier. You can distinguish between the sensor offset and the amplifier offset by swapping the wires from the sensor to the amplifier input. Similarly, you can distinguish between amplifier offset and Analog Discovery 2 offset by swapping the wires between the amplifier and the input channel.

The wires from the output of the amplifier to the oscilloscope or analog-to-digital converter are much less sensitive to noise than the wires to the input of the amplifier, but color coding them is essential and twisting them together is probably a good idea. Twisting together the power-supply wiring to the amplifier is definitely good for reducing 60 Hz noise on the power supply (and hence on the output of the sensor).

Wire up an op amp for your second stage and again measure and record the final output voltage at the three pressures used before. The op amp has an input offset also, and the final stage amplifies not only that offset, but the offsets that were already amplified by the instrumentation amplifier. You need to set your system gain small enough that adding all these offsets to your maximum signal does not cause clipping. When you are done, your zero-pressure will not be exactly at V_{ref}, but off by amplified offsets. If you record the value at the beginning of a pressure-sensor recording, you can subtract it off to get accurate pressure readings.

> The amplifier in this lab has a fairly low gain, and so is debuggable on a breadboard, but if you needed more gain, the capacitive and electromagnetic noise you would pick up from long wiring on the breadboard could cause a lot of problems. Keeping your wiring short and close to the breadboard will help a lot in reducing hum and noise pickup.

21.4 Breath pressure

> You will make your own breath-pressure apparatus, like that in Figure 20.1(c), by using a drill press to drill a 2 mm diameter hole in a PVC elbow.

In the lab, you should demonstrate being able to measure breath pressure (both positive and negative) and record the pressure with the PteroDAQ software. You might want to try some rapid in-and-out pressure changes on the tube—how fast can you get the pulsing to go? A higher sampling rate than the default 10 Hz may be needed to get a clear view of these rapid fluctuations.

> Make sure that you record a time when you are neither inhaling nor exhaling, so that you have a recording of the voltage for zero pressure. The signal during this time interval can be fitted in gnuplot with a constant, which can then be subtracted for all plots, measuring and removing the cumulative offsets from the sensor and the amplifiers.

The offsets can vary from sensor to sensor and from amplifier chip to amplifier chip. Make sure that you record which sensor (not just the part number, but the specific instance of the sensor) that you use for each measurement.

21.5 Blood pressure

After measuring breath pressure, connect to the blood-pressure cuff and do a blood-pressure measurement. The American Medical Association and American Heart Association reported several common mistakes made in measuring blood pressure, even with professional equipment [2]:

- Go to the rest room and empty your bladder. A full bladder can add 10 mm Hg.
- Support your feet and back while sitting upright (hard to do with lab stools—maybe make some measurements at home or in the library). Lack of support can add 6 mm Hg.
- Support your arm so that the upper arm is relaxed and the cuff is level with your heart—an unsupported arm can add 10 mm Hg.
- Wrap the cuff around a bare arm—clothing can add 5–50 mm Hg to the reading.
- Use the right size cuff—too small a cuff can increase the reading by 2–10 mm Hg. Our lab has cuffs sized for average adults—there are also cuffs sold for children and for people with unusually large arms.
- Don't cross your legs—put both feet flat on the ground. Crossing your legs or having one leg unsupported can increase the reading by 2–8 mm Hg.
- Stay still and don't talk—talking can add 10 mm Hg.

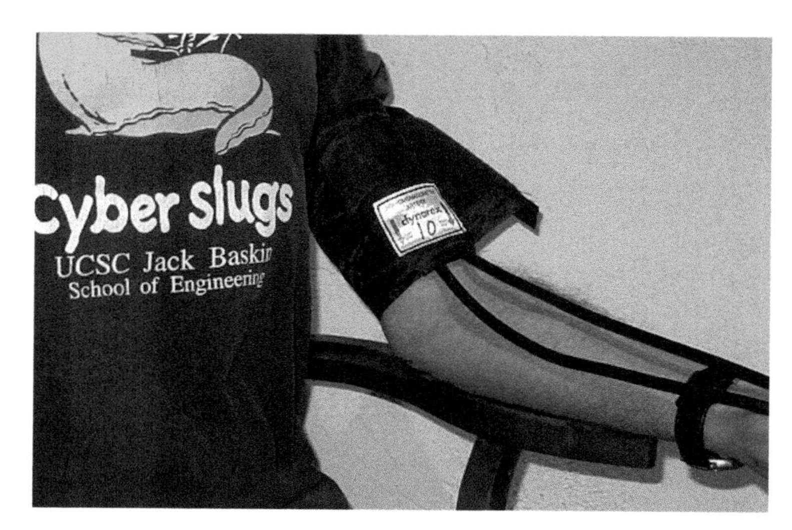

Figure 21.1: *This picture shows the correct positioning of the cuff, with the left-arm arrow pointing to the brachial artery, the arm fully supported, and the cuff at the level of the heart.*

> The cuff should be wrapped around the biceps with the tubes over the inside of the elbow. Most cuffs have an arrow that should point to the brachial artery (on the inside of your elbow, close to your body)—see Figure 21.1.
>
> With your arm comfortably supported, palm-up, so that the blood-pressure cuff is at the level of your heart, inflate the cuff quickly to about 180 mm Hg, then let the air out slowly (about 3–4 mm Hg/s or 0.4–0.5 kPa/s).
>
> Set the slow leak up **before** pumping up the cuff, as fussing with the valve after inflation results in glitches in the pressure.

To record the zero-pressure offset for the sensor, disconnect the cuff—don't rely just on the cuff being deflated, as there may still be a small residual pressure.

Don't inflate the cuff too much, or you might bruise your arm, nor keep it inflated for too long to avoid getting swelling in your hand from too little return blood flow. If you have to repeat a measurement, wait a couple of minutes between trials to let your arm recover.

Check to make sure that your measurements are not being clipped at the highest pressure, but that you are using most of the voltage range of the analog-to-digital converter.

Record the cuff pressure at a sampling frequency that has a period that is an integer multiple of the period of the electrical interference from the 60 Hz line frequency (say 20 Hz or 30 Hz) to reduce interference by aliasing the interference to a DC value. Check that the pulsations are visible when you plot the data (they may be too small to see on PteroDAQ's sparklines while you are recording).

Use the provided bandpass-filter.py Python filter program
https://users.soe.ucsc.edu/~karplus/bme51/book-supplements/bandpass-filter.py
from Figure 20.7 to extract the fluctuations and the steady decrease in pressure. The script can be run from a command line as

```
cd Documents/BME51/pressure-lab
python bandpass-filter.py < pterodaq-data.txt > filtered.txt
```

assuming that the file `bandpass-filter.py` and the data file (here called `pterodaq-data.text`, but you should use the file name you gave your data file) are both in the same directory (`Documents/BME51/pressure-lab`). The input file is assumed to have time stamps in Column 1 and data in Column 2. The output file (called `filtered.txt` in the example) will echo those columns and add two more columns: band-pass-filtered in Column 3 and low-pass-filtered in Column 4. **Do not use the same file name for the input and the output—this script cannot do an update in place.**

Make plots like Figures 20.4–20.6 (though you don't have to be as fancy as Figure 20.6—fitting and drawing the envelope is not necessary). Do mark the systolic, diastolic, and mean arterial pressures, though. Use a gnuplot command like

```
set arrow front nohead from MAP,lowP to MAP,highP lw 2 linecolor "red"
```

to create a red line for mean arterial pressure (after setting values for MAP and the high and low pressure there).

21.6 Demo and write-up

The main purpose of a design report is to present and *explain* your design choices. This is not an "answer-getting" class with one right answer—there are many acceptable designs, with trade-offs between them. You should explain what choices you made and why those choices are reasonable.

Also explain any testing you did to show that your design worked correctly and was properly calibrated. The final demos of breath pressure and blood pressure are just demonstrations, not tests that show that gain and offset have been correctly measured or corrected for.

In the report, in addition to the usual block diagram and schematic, you should include a photo of the breadboard to show the layout you used.

Include a plot of a recording of breath pressure, with the y-axis properly scaled in kPa and with the ambient air pressure at 0. The scaling should be based on the sensor specs and your gain computations—we don't have the time or equipment to do a proper calibration—but the zero can be set by fitting to a part of the curve that you know is at ambient pressure.

Include plots showing the blood-pressure measurements, and report pulse rate, mean arterial pressure, and rough estimates of systolic and diastolic pressures based on these measurements.

> When determining the pulse rate, use precise time measurements for an *integer* number of periods, rather than estimating the number of periods for a fixed time interval, and watch out for fencepost errors (see page 285).

21.7 Bonus activities

You can also try pinching the tube shut and pressing on the tube to increase pressure (this is what the long tubes you see across roads to count traffic do). What other pressure fluctuations can you make and record in a piece of flexible tubing?

If you have more time in the lab, try using the pressure sensor to listen to a loudspeaker or phone playing a low frequency sine wave. What is the highest frequency that you can reliably detect with the pressure sensor? (If you have an Analog Discovery 2 USB oscilloscope and a loudspeaker, you can use the network analyzer to determine the frequency response of the system consisting of the loudspeaker and pressure sensor.)

Chapter 22

Optoelectronics

22.1 Semiconductor diode

A *diode* is a two-terminal device that allows current to flow in one direction (the *forward* direction), but not in the other (the *reverse* direction). Figure 22.1 shows the schematic symbol used for diodes, as well as the symbols for optoelectronic parts based on diodes.

A *semiconductor diode* is a semiconductor with two different regions: each with slightly different materials. One region has a carefully controlled impurity resulting in free electrons in the crystal structure (referred to as *n-doped*, because the charge carriers have negative charge). For example, a silicon semiconductor may be doped with phosphorus. When the crystal structure is formed, four of the five electrons in the outer shell of phosphorus form the crystal bonds and the fifth electron is free to move around in the crystal.

The other region has a different impurity that results in electrons being missing from the crystal structure (referred to as *p-doped*, because the charge carriers, called *holes*, have positive charge). For example, a silicon semiconductor may be doped with boron, which has only three valence electrons. A fourth electron for the crystal structure is borrowed from a nearby atom, resulting in a positively charged hole that can move around, though it is not quite as mobile as the free electrons of n-doped silicon.

The junction between the two types of semiconductor is where everything interesting happens, as shown in Figure 22.2.

When the n-doped semiconductor is connected to a *higher* voltage than the p-doped semiconductor, then the majority charge carriers on each side (electrons on the n-doped side, positively charged holes on the p-doped side) are pulled away from the junction, and a non-conducting *depletion layer* or *depletion region* is formed. This connection *reverse biases* the diode, so that no current flows. When there is no applied voltage from the outside, there is still a small depletion layer formed (see Figure 22.2(a)). A reverse-biased diode acts like a capacitor, with two conducting plates separated by an insulator, but the capacitance varies with the bias voltage.

When the n-doped semiconductor is connected to a *lower* voltage than the p-doped semiconductor, then the majority carriers in each type of semiconductor are pushed toward the junction and cross over, allowing current to flow (*forward biasing*, see Figure 22.2(b)). The voltage has to be large enough to eliminate the normal depletion layer before conduction starts—this minimum forward voltage is often called the *diode voltage* or *threshold voltage*.

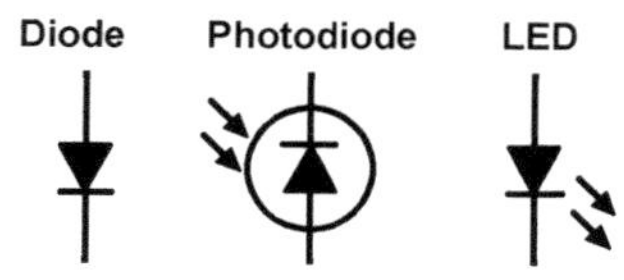

Figure 22.1: *The schematic symbols for diodes, light-emitting diodes (LEDs), and photodiodes are essentially all the same, with the addition of arrows to indicate light coming out of LEDs and going into photodiodes. LEDs sometimes have circles around them, and photodiodes sometimes do not—the arrows are the essential difference, not the circles. Sometimes the arrows are drawn with wavy lines instead of straight ones. The triangles in the diode symbols indicate the direction of conventional current flow (forward biasing). The positive end (the* anode*) points to the negative end (the* cathode*)—the -ode names can be kept straight if you remember that the cathode rays of cathode-ray tubes are electron beams, so the cathode is where the electrons leave the wire.*

The photodiode is normally used reverse biased (with the cathode more positive than the anode), and so I've drawn it with the opposite orientation from the other diodes.

Figure drawn with Digi-Key's Scheme-it [18].

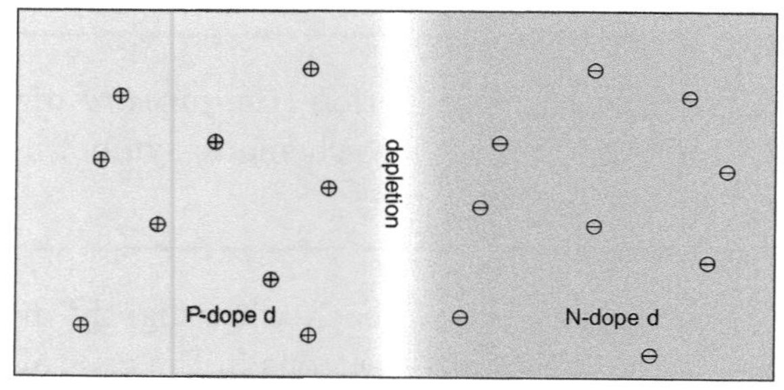

(a) *A diode junction with no applied voltage has a small, non-conducting depletion region, and so acts as a capacitor.*

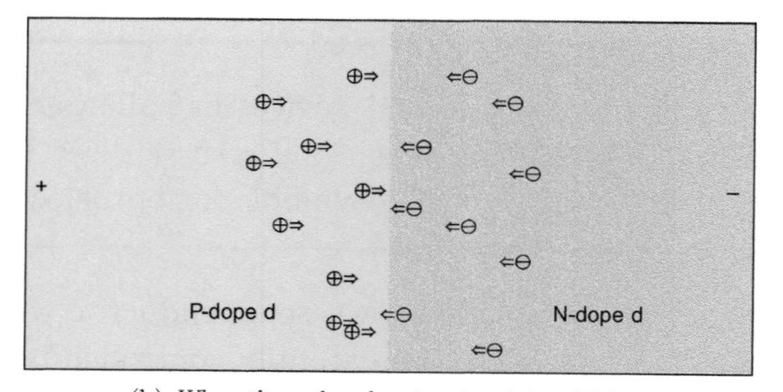

(b) *When the p-doped region is set to a higher voltage than the n-doped region, the charge carriers are pushed towards the junction. When the voltage is high enough, the depletion region is eliminated, and current flows through the diode.*

Figure 22.2: *A semiconductor diode does not conduct until a sufficiently high voltage is applied to the p-doped region (the anode) relative to the n-doped region (the cathode).*

22.2 Light-emitting diodes (LEDs)

LEDs are semiconducting diodes that emit light when there is sufficient forward current through them. LEDs often have a fairly large diode voltage compared to other diodes, and you need to exceed the diode voltage before you get enough current to turn on the LED.

Light is emitted when a free electron from the n-doped side combines with a hole from the p-doped side. The drop in the energy of the electron is released as a photon. Because the electrons vary in energy, the photons do not all have the same wavelength, so LED light is not spectrally pure, but the variation is generally not very large, at least when compared to thermally produced light, like black-body radiation and incandescent light bulbs.

For a no-math view of how LEDs (and semiconductor diodes in general) work, try the *How Stuff Works* website [41].

Data sheets for visible-wavelength LEDs often report two wavelengths: the *peak* wavelength, where the energy from the LED is largest, and the *dominant* wavelength, which takes into account the varying sensitivity of the human eye. The dominant wavelength is closer to the maximum of human sensitivity, around 555 nm [158], than the peak wavelength is. Infrared and ultraviolet LEDs generally report only peak wavelength, because human eye sensitivity is irrelevant for light that is not visible.

When you are trying to look at the LEDs with photodetectors, rather than human eyes, the adjustments for human eye sensitivity are just a nuisance, so you should use only the *peak* wavelength, not the *dominant* wavelength.

The light output of the LED may be reported in different ways. For example, the output luminous intensity of the LED may be reported in candela (lumen per steradian) or watts per steradian, or the total luminous flux of the LED may be reported in lumens or watts. A *steradian* is a measure of solid angle that would cut an area of r^2 out of a sphere of radius r, so a full sphere is an area of 4π sr.

Some of the units are purely physical measures (watts, milliwatts per steradian, watts per square cm), while others have been adjusted for human eye sensitivity (lumens, candelas, lux, foot-candles). The Wikipedia article on photometry [147] gives tables of the SI units for both photometric (adjusted for human eye sensitivity) and radiometric (physical) measurements.

For infrared emitters, only the purely physical units make any sense, as the human-adjusted ones would all report the output as zero. Sometimes a manufacturer stupidly reports the light output of IR emitters in human-adjusted units anyway, relying on the conversion that there are 683.002 lumen/watt at 540 THz (a wavelength of about 555 nm), ignoring the correction for wavelength that is part of the definition of lumens.

The data sheet will also give the forward voltage of the LED, either for a typically used current or as a plot of current vs. forward voltage. The current rises rapidly with voltage (it is well fit by an exponential), and so engineers often use an approximation that the forward voltage is constant over the range of currents they are interested in. The variation from one LED to another from the same batch may be larger than errors introduced by this approximation.

For example, the 151034BS03000 blue LED from Würth Electronics has a peak wavelength of 465 nm and a dominant wavelength of 470 nm [169]. The output at 20 mA is typically 2500 mcd, but could go as low as 850 mcd. (The abbreviation "mcd" stands for "millicandela", a luminous intensity of one millilumen per steradian.) Also at 20 mA, the forward voltage is typically 2.8 V, but could go as high as 3.6 V. We would not want to specify this part for a 3.3 V system, as it might work fine in prototyping, but then fail to light up in production units, because of changes in the characteristics.

The data sheet for an LED often provides a forward-current vs. forward-voltage plot—a typical cheap red LED might have a 7 mA current at 2 V but 35 mA at 2.2 V. Because an LED is a semiconductor diode, the current goes up exponentially with voltage at low voltages, growing by a factor of 10 for every 100 mV. The growth rate slows down for higher voltages, to about a factor of two change in current per 100 mV at the normal operating range of the LED. This means that small changes in the forward voltage can produce large changes in the current, but the forward voltage specification for an LED is often fairly wide. The amount of light produced by an LED is proportional to the current through the LED, not to the voltage.

From a design standpoint, having light proportional to current and only a roughly specified diode voltage means that we usually try to control the current through an LED, rather than trying to control the voltage.

If you connect an LED directly across the power supply, there is a very high probability of burning it out, as the current goes up exponentially with voltage, and the part has an absolute maximum current.

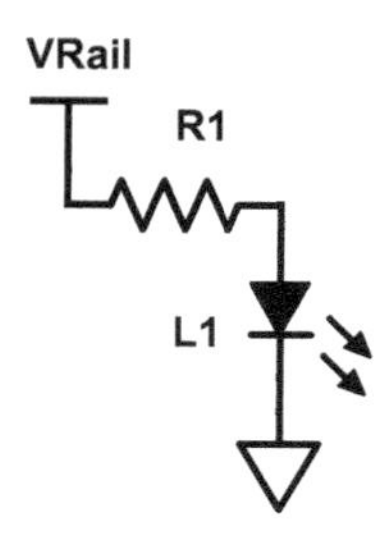

Figure 22.3: *One common circuit for setting the current through an LED is a current-limiting resistor. The resistor value R is chosen so that the voltage drop across the resistor produces the desired current via Ohm's law: $V_{\text{rail}} - V_{\text{f}} = IR$. The forward voltage V_{f} of the LED increases with increasing current, but slowly enough that one can often get away with a constant approximation for V_{f}.*
Figure drawn with Digi-Key's Scheme-it [18].

For the 151034BS03000 blue LED, the maximum continuous current is 30 mA. This maximum current is mainly due to heating the device, so if you only turn the LED on a tenth of the time for 100 μs at a time, you can push the peak current for the 151034BS03000 up to 100 mA through.

To control the current through a small LED, the simplest approach is to use a current-limiting resistor, as shown in Figure 22.3. This approach takes advantage of the forward voltage of the LED being roughly constant over a wide range of currents, so that the voltage across the resistor is just the power-supply voltage minus the forward voltage of the LED. Using Ohm's law, we can pick a resistance to give us the desired current.

For example, for 151034BS03000 with a 5 V power supply, we would choose $R = (5\,\text{V} - 2.8\,\text{V})/I$, so that to get 6 mA, we would choose a resistor around 366 Ω. The closest in the E12 series is either 330 Ω or 390 Ω, giving currents of 6.7 mA or 5.6 mA, respectively.

If the 151034BS03000 LED does not have its typical value for forward voltage, but the maximum (3.6 V), then 330 Ω would provide only 4.2 mA and 390 Ω only 3.6 mA. The data sheet does not give the minimum forward voltage, which is an unfortunate omission, as we cannot compute how high the current could get in the worst case.

Exercise 22.1

Look up the data sheet for the LTL-4234 green LED. Find the peak wavelength, the dominant wavelength, and the forward voltage. Figure out what size current-limiting resistor to use to get approximately a 5 mA current with a 3.3 V power supply—pick a resistor from the E12 series. How much current would there be with the typical forward voltage? What if the forward voltage is the maximum, rather than the typical value?

22.3 Photodiode

A *photodiode* is a semiconductor diode whose junction is exposed to light [146]. To use it as a photodiode, the diode is *reverse-biased*—that is, the positive voltage is connected to the n-doped semiconductor, and the negative voltage to the p-doped semiconductor.

The reverse biasing attracts the majority charge carriers away from the junction between the two types of semiconductor, creating a *depletion* region that does not normally conduct, but when a photon

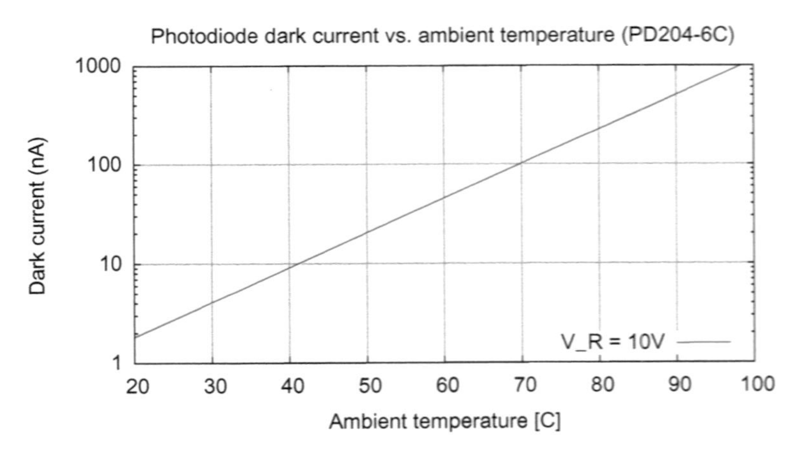

Figure 22.4: *The dark current for a photodiode grows exponentially with temperature. This plot gives approximate typical dark currents for a PD204-6C photodiode reverse-biased by 10 V—drawn by fitting a simple exponential to two points from the photodiode data sheet [24].*

The photodiode has a photocurrent of only 3.5 μA at 1 mW/(cm)^2 [24], and so a hot photodiode is difficult to tell from one that is detecting light.

Figure drawn with gnuplot [33].

is absorbed in the semiconductor in or near the depletion layer, it can knock an electron loose, resulting in a pair of charges (a hole and an electron), that can move under the influence of the electric field to the conducting regions. This movement of charges creates a current which is proportional to the number of photons absorbed (as long as the hole and electron do not recombine).

> There is a very small current due to thermal effects even with no photons knocking loose electrons (referred to as the *dark current*). To avoid noise problems, the dark current should be much smaller than the smallest current you plan to measure, but the dark current typically grows exponentially with temperature, as shown in Figure 22.4.

Photodiodes tend to be used in one of two ways:

photoconductive mode A fixed reverse-bias voltage is used and the photocurrent is measured. The larger the reverse-bias voltage, the thicker the depletion region, and the lower the capacitance of the photodiode. The reduced capacitance makes the photodiode faster at responding to changes in light level (which is particularly important if light is being used to transmit information—the speed of information transfer may be limited by the response speed of the photodiode).
Increasing the reverse-bias voltage, however, also increases the dark current without changing the photocurrent much, reducing somewhat the dynamic range of the sensor.

photovoltaic mode If we don't allow much current to flow, then the photocurrent charges up the capacitance until the voltage is high enough that the diode starts to conduct in the forward direction. This voltage is almost independent of the amount of light, but the current we can take from the photodiode at that voltage is linear with the amount of light. (This is how a solar cell works—a solar cell is just a very large photodiode used in photovoltaic mode.)

22.4 Phototransistor

Figure 22.5 shows the schematic symbols for photodiodes, phototransistors, photodarlingtons, and photoresistors. We will not use photodarlingtons nor photoresistors in this course.

A phototransistor is a two-terminal device whose current is proportional to the illuminance of the transistor. Unlike photodiodes, phototransistors do not *generate* current—instead they *control* the current that passes through them from an external voltage source.

A phototransistor can be thought of as a photodiode combined with an amplifying bipolar transistor. The two terminals are the collector and the emitter (in the schematic symbols in Figure 22.5, the emitter is the terminal with an arrow, and the collector is the other terminal).

There are two types of phototransistors: *NPN* and *PNP*, referring to the sandwiching of the types of semiconductor making up the transistor. The phototransistor we have in our kits is an NPN silicon transistor, which means that the emitter and collector are n-doped silicon and the base is p-doped silicon.

The symbols for bipolar transistors (of which phototransistors are a special case) have the collector and emitter as diagonal lines connecting to a vertical line representing the base. The emitter always has an arrow on it, and the arrow points in the direction of conventional current flow when the transistor is on. For NPN transistors, the arrow points away from the base, while for PNP transistors it points toward the base. In Figure 22.5, all the symbols are drawn so that the lower end would be connected to the more negative voltage in normal usage.

For an NPN phototransistor, the collector and emitter are n-doped semiconductor regions, and the base is a thin layer of p-doped semiconductor between the conductor and emitter. The base in a regular bipolar transistor controls the flow of current between the collector and emitter:

- If the base-emitter junction is reverse-biased, then the transistor is turned off, and essentially no current flows.
- If the base-emitter junction is forward-biased, then current flows from the collector to the emitter proportional to the base-emitter current (the current gain of a bipolar transistor is usually quite large—more than 100).

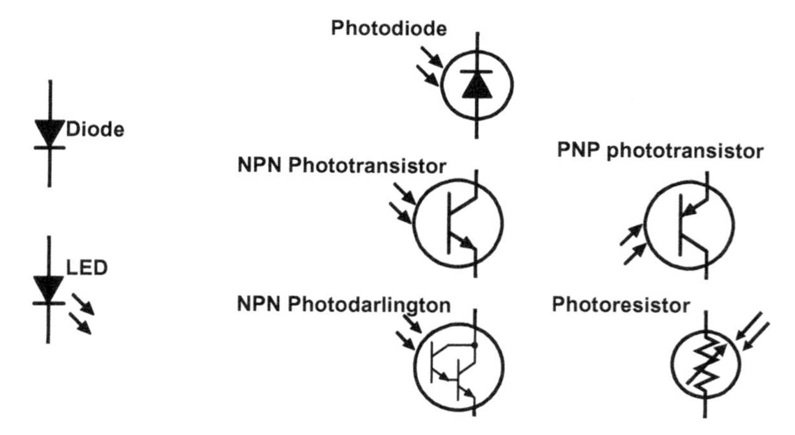

Figure 22.5: *The triangles in the diode symbols and arrows in transistor symbols indicate the direction of conventional current flow (forward biasing). All the symbols in this diagram have the end connected to the more positive voltage on top (remember that photodiodes are reverse-biased).*

Photodarlingtons are basically phototransistors with an extra transistor for more current gain. They provide larger currents than phototransistors (32 mA seems to be a common number) and are even slower and less linear.

Photoresistors provide a spectral response close to human eyes, but have very slow response to changes in light. Like resistors, they have no preferred direction of current flow.

Figure drawn with Digi-Key's Scheme-it [18].

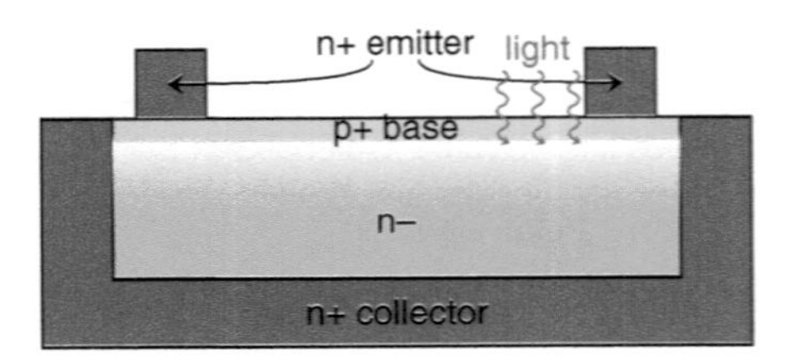

Figure 22.6: *Simplified cross section of an NPN phototransistor. The emitter covers only a small portion of the base, so that photons can easily reach the depletion region at the p+/n- junction that is light sensitive.*

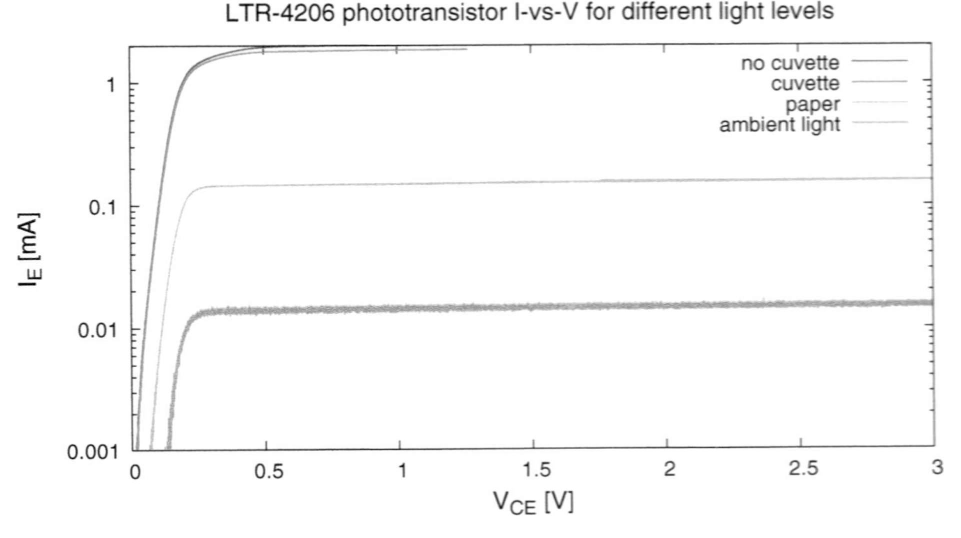

Figure 22.7: *This plot shows the current through a phototransistor (an LTR-4206, which may be a different NPN phototransistor than the one in your parts kit) for different levels of illumination. This phototransistor was being used as part of a homemade colorimeter for measuring OD600, using a 333-2UYC/H3/S400-A6 LED as the light source. The current is roughly constant once the collector-to-emitter voltage is high enough (this is called the* active region *in bipolar transistors).*
Data collected with PteroDAQ using a Teensy LC board [46]. Figure drawn with gnuplot [33].

In a phototransistor (see Figure 22.6), the base is not wired to anything—a configuration known as a *floating base*. The floating base normally results in the base having a voltage intermediate between the collector and the emitter. In order to have the forward and reverse biases of the floating base correct, we need to have the base be more positive than the emitter (forward bias) and less positive than the collector (reverse bias). That means that we want the collector wire to be connected to a more positive voltage than the emitter wire: $V_C > V_E$.

In fact, the difference $V_C - V_E$ (usually called V_{CE}) needs to be bigger than a constant, $V_{CE(sat)}$, called the *saturation voltage* for the phototransistor, in order for the transistor to conduct. V_{CE} generally needs to be much higher than $V_{CE(sat)}$, up in the active region for the bipolar transistor, for the current specifications from the data sheet to be accurate.

> The V_{CE} bias voltage can be arranged either with a bias resistor (like the one in Lab 2) or with a transimpedance amplifier (to be explained in Chapter 23).

The base-emitter current that controls the transistor comes from the photocurrent generated at the reverse-biased base-collector junction. This current then controls the collector-emitter current that we can measure from outside the device.

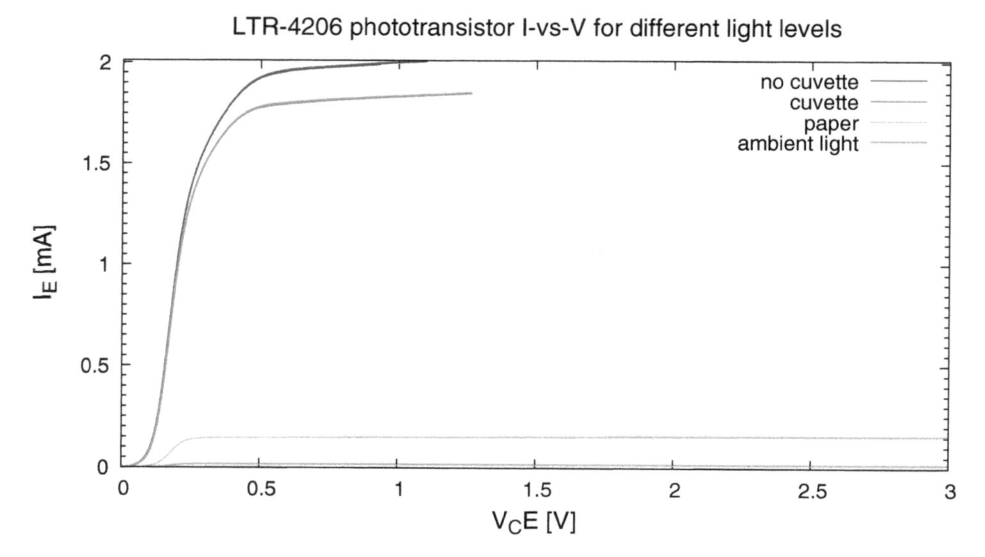

Figure 22.8: *This plot shows the same data as Figure 22.7, but with a linear scale for current. It is much harder on this plot to see what is happening at low currents, but the slope of the current at high currents is now apparent.*
Changing between log and linear scales for plots is commonly used to accentuate different features of the data—here the not-quite-constant nature of current at high current is apparent, while Figure 22.7 makes the low-current behavior clearer.
Data collected with PteroDAQ using a Teensy LC board [46]. Figure drawn with gnuplot [33].

A bipolar transistor acts as a current amplifier (with the collector current proportional to the base current) when the collector-to-emitter voltage V_{CE} is sufficiently high. The gain of the amplifier drops rapidly when the V_{CE} gets too small. The *saturation voltage* $V_{\mathrm{CE(sat)}}$ of a bipolar transistor is the V_{CE} value at which the current gain makes this sudden change.

We want to use phototransistors substantially above $V_{\mathrm{CE(sat)}}$, so that the current gain is roughly independent of the voltage, and the collector current that we can measure is proportional to the base current, which is the photocurrent proportional to the light input.

We can easily measure I-vs.-V plots at different light levels using PteroDAQ or the Analog Discovery 2, as shown in Figure 22.8 measured using the circuit in Figure 22.9. These plots can help us decide what voltage to use even more than $V_{\mathrm{CE(sat)}}$ does. For example, the LTR-4206 data sheet's value for $V_{\mathrm{CE(sat)}}$ at 0.5 mA is only 0.4 V, but at the high currents used in the colorimeter design, we probably want to keep $V_{\mathrm{CE}} > 0.6\,\mathrm{V}$.

Because of the current gain of the bipolar transistor, the current through a phototransistor is much larger than through a photodiode, often 100–1000 times larger. However, the high gain comes at a price—the capacitance of the base-collector junction is also multiplied by the gain (the *Miller Effect*), so phototransistors are about 1000 times slower also. For looking at low-frequency signals like in pulse monitoring, the slowness of the phototransistor is unimportant, and phototransistors are still faster than photoresistors.

A phototransistor also has a narrower dynamic range than a photodiode. The phototransistor has a fairly linear response over about three decades, while a photodiode is fairly linear over seven or more decades. (That is, there is a range of about 1000:1 in irradiance over which the phototransistor can be well modeled as current proportional to irradiance, but there is about a 10,000,000:1 range for a photodiode.) The increased linearity makes photodiodes more popular for measuring instruments, even though the photocurrents are so much smaller.

Both photodiodes and phototransistors provide a current output that is proportional to the light power input, so their sensitivity can be expressed in A/W. The data sheet usually gives the current

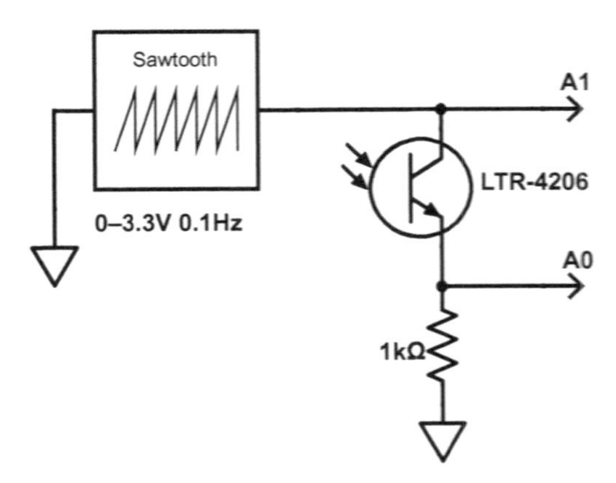

Figure 22.9: *The plots in Figures 22.7–22.8 were created with this simple test fixture, except that a triangle wave was used instead of a saw-tooth wave. PteroDAQ was used to record the voltages at A0 and A1,* V_{CE} *was computed as* $V_{A1} - V_{A0}$*, and the emitter current was* $I_E = V_{A0}/1\,k\Omega$*.*
Figure drawn with Digi-Key's Scheme-it [18].

for bright light—often $1\,\text{mW}/(\text{cm})^2$, calling it the *on current.* If a photodetector collects all the light from a circle 3 mm in diameter ($7.07\,(\text{mm})^2$), then a light intensity of $1\,\text{mW}/(\text{cm})^2$ corresponds to a power input of about $71\,\mu$W. The data sheet also gives the dark current (which is very small at room temperature).

A phototransistor may have a sensitivity of 2.8 A/W, while a photodiode in a similar package might have only 50 mA/W. The larger current is due to gain of the bipolar transistor in the phototransistor, ameliorated somewhat by different areas for the light-sensitive part of the devices that the lens of the package focuses on.

The current-vs.-voltage curves of Figure 22.8 show that the active current is not constant, particularly for larger current values. In order to get more linear measurements from a phototransistor, we need to hold V_{CE} constant and measure the current at that constant voltage. Chapter 23 explains a circuit that does just that: the transimpedance amplifier.

Exercise 22.2

Look up the data sheet for phototransistor SFH325FA.
What is the minimum photocurrent for group 4A with an illumination of $0.1\,\text{mW}/(\text{cm})^2$ at bias voltage of $V_{CE} = 5\,\text{V}$?
What bias voltage V_{CE} do you need across the phototransistor to get at least 0.3 times the photocurrent at 5 V? (Hint: look through the values in the specification table for things in units of voltage.)
What current would you expect with a 5 V bias, if the irradiance (brightness of light falling on the photodetector) is $2\,\text{mW}/(\text{cm})^2$?

22.5 Optical properties of blood

For Lab 6, you will design an optical pulse monitor to detect pulse by shining light through a finger and measuring how much light comes through. The amount of light changes as the amount of blood in the blood vessels changes—when the blood vessels have more blood, they block more of the light and the photocurrent drops.

Each time the heart beats, the surge in the flow causes red blood cells to accumulate where the blood vessels narrow, but in the time between pulses the blood cells redistribute more uniformly. The fingers get rapidly more opaque as the pulse reaches them, then gradually get clearer again.

In this section, we'll look a little at the opacity of blood, to help choose what wavelength(s) to use for a pulse monitor.

Our phototransistor has its maximum sensitivity in the infrared portion of the spectrum (around a wavelength of 940 nm), which is typical for silicon photodiodes and phototransistors. But that doesn't mean that the best LED to use for illumination is an infrared one—we need to balance photodetector sensitivity with the opacity of blood.

> We want to monitor how much blood is in the finger based on how much light is absorbed, so we need to look at the light absorption of hemoglobin. For detection to be easy, a lot of the light needs to make it through the finger, so we need a wavelength at which flesh is not too opaque. As a good first approximation, we could look at the absorption spectrum of hemoglobin, the main coloring agent of blood—see Figure 22.10.

Because we don't want enormously bright lights, we'll want a wavelength at which there is only moderate absorption. A red LED at 627 nm (a common peak wavelength for red LEDs) seems reasonable, but a different LED with a peak around 700 nm (like MT1403-RG-A) might be even better, as there would be less attenuation of the signal.

We can look up typical silicon photodetector sensitivity [151] and see that silicon photodetectors are only about 70% as sensitive at 627 nm as at their peak around 950 nm, and 80% at 700 nm. The three-fold greater transparency of blood at 700 nm more than compensates for the 20% reduction in sensitivity compared to using 950 nm.

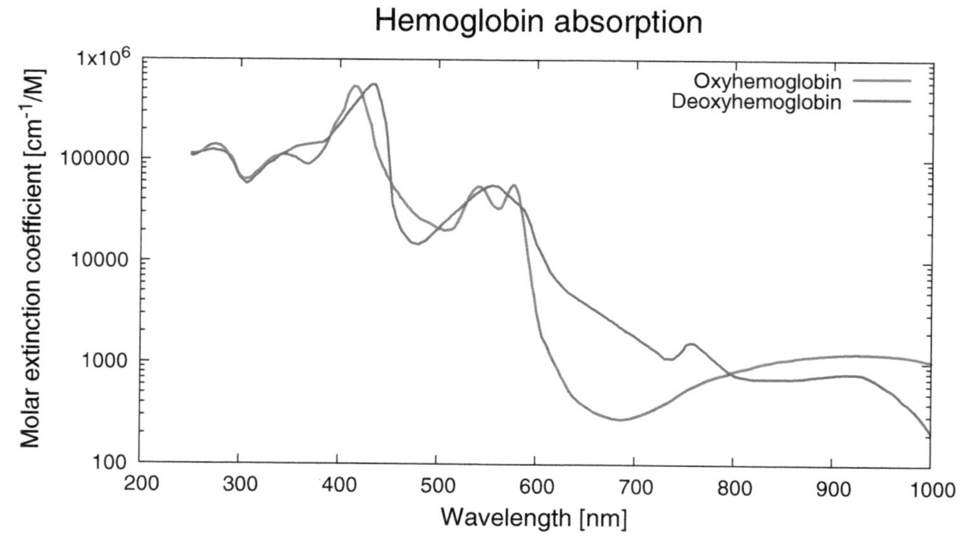

Figure 22.10: *Oxyhemoglobin is most transparent around 686 nm (272.8* (cm)$^{-1}$/M*) and does not get much more opaque for* 700 nm ± 25 nm *(284–368* (cm)$^{-1}$/M*). At a shorter wavelength, which looks brighter to the eye,* 627 nm ± 22.5 nm*, absorption is higher (370–2130* (cm)$^{-1}$/M*). At the peak of our photodetector sensitivity,* 950 nm ± 27.5 nm*, oxyhemoglobin absorbs more (1136–1225* (cm)$^{-1}$/M*). Deoxyhemoglobin is not very important here, since oxyhemoglobin in the blood is usually well over 95% in healthy individuals. Data for figure from https://omlc.org/spectra/hemoglobin/summary.html Figure drawn with gnuplot [33].*

If we were making a pulse oximeter, to measure the ratio of oxyhemoglobin and deoxyhemoglobin, then we'd need to use two wavelengths, choosing ones at which the molar extinction ratios were quite different. The minima of oxy/deoxy absorbance are around 654 nm, 438 nm, and 370 nm, while the maxima are somewhere greater than 1000 nm, 464 nm, 412 nm, 578 nm, and 538 nm. A good pair of wavelengths might be 650 nm (or 660 nm) and 940 nm, which are readily available LED wavelengths that have very different oxy/deoxy ratios.

Of course, hemoglobin is not the only substance in the body, so we should also look at the absorbances of other substances that might interfere with our signal: water, lipids, and melanin. The Wikipedia article *Near-infrared Window in Biological Tissue* [139] has absorption spectra for many of the relevant substances, though the spectra there have different ranges of wavelengths, with melanin's mainly in the visible range, so it may be necessary to look at other sources if you are interested in using infrared illumination.

A lot of wearable pulse monitors rely on light being scattered back to a photodetector next to the LED, not on opposite sides of a finger. The scattering of light by biological tissues decreases with wavelength, favoring short wavelengths for this application, but the opacity of melanin in the skin generally decreases with wavelength, so we need to make trade-offs. Reflection pulse monitors often use green LEDs (peak around 565 nm), where blood is much more opaque than in the red or infrared, and biological tissue scatters light back to the surface, but the melanin in skin does not yet block too much light.

When the Apple Watch first came out with a pulse monitor, it did not work well on people with dark skin or with wrist tattoos—a long-standing problem with reflection-based heart monitors [11]. Picking wavelengths at which melanin absorbs relatively little light compared to hemoglobin could reduce this problem, but one does have to check that the total amount of light scattered back to the sensor is large enough to provide a usable signal.

Tattoo pigments are a harder problem to design around, as they may be opaque at arbitrary wavelengths—either broad-spectrum illumination should be used or the user advised to put the sensor somewhere away from tattoos.

Researchers have developed fairly sophisticated optical models of skin and the flesh underneath it to model optical biosensors more accurately. One of these models has been released by Maxim Integrated, to support the use of their pulse oximeter and heart-rate sensors [67]. This model includes parameters for five different optical layers: stratum corneum, epidermis, papillary dermis, vascularized dermis, and subcutaneous adipose tissue. The model takes into account all the major sources of absorption and scattering, and traces the ray path of the light, so that either transmission or reflection modeling can be done.

An example is given of reflection modeling for someone with moderately light skin (1%–10% melanosomes, when the normal human range is given as 1%–40%). In that model, there is a minimum penetration depth around the wavelength of green LEDs, with the light penetrating about 1 mm, while for the wavelengths we've been considering (650 nm or more), the light penetrates about 3 mm. The choice of green LEDs for cis-illumination may be a deliberate choice to get light mainly from the vascularized dermis layer, where it will be modulated by the pulse, rather than from the underlying adipose layer.

Chapter 23

Transimpedance Amplifier

A *transimpedance amplifier* takes a current input and provides a voltage output. The gain (output/input) is expressed in ohms, hence the name *transimpedance.*

Figure 23.1 shows a typical transimpedance amplifier. It is useful to mentally separate the schematics of Figures 23.1b and 23.1(c) into two blocks: a photosensor and a transimpedance amplifier. The photosensor gets a voltage applied across it and provides a current proportional to the light sensed. The transimpedance amplifier both sets the voltage for the sensor and converts the current to an output voltage. The sensor is *not* part of the amplifier.

Since there is no current flowing through the input of an op amp, any current I that flows downward through the photodiode D_1 in Figure 23.1(b) must flow through the resistor R_1. Remember that an op amp in a negative-feedback loop essentially holds the negative input at the same voltage as the positive input, which means that the voltage across R_1 is $V_{\text{out}} - V_{\text{ref}}$. Ohm's law gives us $V_{\text{out}} = IR_1 + V_{\text{ref}}$.

That is, the gain of the amplifier

$$G = \frac{V_{\text{out}} - V_{\text{ref}}}{I} = R_1$$

is just the impedance of the feedback element, and the output voltage with zero input current is V_{ref}.

A transimpedance amplifier differs from just using a pull-up resistor to convert current to voltage, because the transimpedance amplifier ensures that the voltage across the photodiode, $V_{\text{ref}} - V_{\text{bias}}$, remains constant, independent of the current. This property is used for measuring the properties of ion channels in electrophysiology, where the photodiode is replaced by a pair of electrodes—one inside and one outside the cell. By changing V_{ref} we can change the voltage across the cell membrane, while measuring the resulting current changes.

High-gain transimpedance amplifiers (sometimes called *patch-clamp amplifiers* for their early applications in electrophysiology) are essential to the nanopore and nanopipette work at UCSC. In those projects, the currents involved are very small (10 pA–100 pA) and the interesting changes are also small (as low as 0.5 pA), and so the amplifiers need to be designed for very low noise. This course will not get into the esoteric field of low-noise design, as the currents we will deal with are mostly thousands of times larger (though still pretty small).

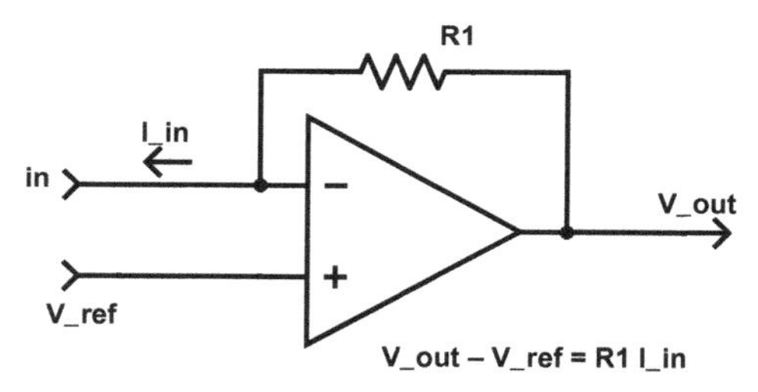

(a) *Bare transimpedance amplifier. The current direction for a positive output is shown.*

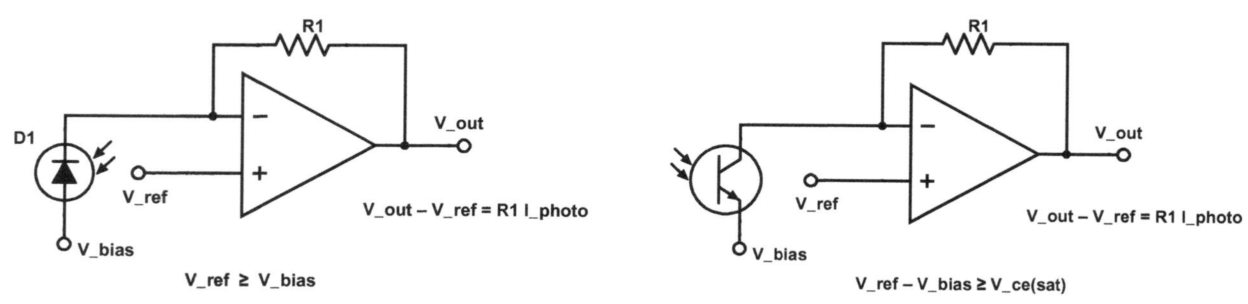

(b) *Transimpedance amplifier with photodiode as current source.*

(c) *Transimpedance amplifier with NPN phototransistor as current source.*

Figure 23.1: *A transimpedance amplifier provides a voltage output for a current input. The feedback loop holds the two inputs of the op amp at the same voltage V_{ref}, and the bias on the photodiode or phototransistor is $V_{\text{ref}} - V_{\text{bias}}$. For both designs $V_{\text{ref}} - V_{\text{bias}}$ should be positive, so that the photodiode is reverse-biased and the NPN phototransistor has its collector at a higher voltage than its emitter. For the phototransistor, the collector voltage should be at least $V_{\text{CE(SAT)}}$ higher than the emitter voltage.*

The photodiode or phototransistor is not *part of the transimpedance amplifier, but is the current-output sensor providing the current that is being amplified.*

Figure drawn with Digi-Key's Scheme-it [18].

23.1 Transimpedance amplifier with complex gain

There is nothing in the design of the transimpedance amplifier that requires that the feedback element be a resistor—any impedance can be used to get a gain that varies with frequency. If there is a DC component to the current, though, we usually want a finite impedance at 0 Hz (so not a capacitor alone or in series with other impedances as a feedback element).

For example, if we replace R_1 with a resistor R and capacitor C in parallel, then the gain equation becomes $V_{\text{out}} - V_{\text{ref}} = ZI$. If the current is sinusoidal (angular frequency $\omega = 2\pi f$), then the gain is

$$Z = \frac{1}{1/R + j\omega C} = \frac{R}{1 + j\omega RC} ,$$

which is essentially R for very low frequency, but $1/(j\omega C)$ at high frequency—a low-pass filter with corner frequency $\frac{1}{2\pi RC}$.

Exercise 23.1

Design a low-pass transimpedance amplifier with a 100 kΩ DC gain and corner frequency of about 5 Hz, using standard component values. What corner frequency is achieved? Provide a schematic and a plot of the magnitude of the gain of the amplifier as a function of frequency. (Remember that Bode plots of gain are always log-log plots.)

23.2 Log-transimpedance amplifier

The transimpedance amplifier design is not limited to linear impedances—we can use nonlinear devices to good effect also. For example, we may want to use a phototransistor or photodiode in a wide range of different lighting conditions. Full sunlight can have an illuminance of 1120 W/m^2, with a lot of that energy in the infrared range (over 50%) [162], while a candle a foot from a surface (bright enough for reading) provides about 0.1–1 W/m^2, again with most of the energy in the infrared.

For applications like optical pulse monitoring, we are not very interested in the absolute light level, but in small fluctuations in the light level—a few percent of the total light. If we try to make a transimpedance amplifier sensitive enough to measure these small fluctuations at very low light levels, then the gain would need to be so high that the amplifier would saturate at high light levels, but if we designed it for the high light levels, we would not be able to measure the fluctuations at low light levels.

Your eyes have the same problem—we need to be able to see both in very bright light and very dim light. Furthermore, what we are interested in seeing is not the overall light level, but the variations in reflectance of different objects, so what we are interested in is the ratio of light from different parts of the scene. Your eye has evolved to have a roughly logarithmic response to light, so that the ratios of light intensity are converted to differences in neural firing rates.

Because our pulse monitor is looking for changes in the fraction of light blocked by our blood (see Section 22.5), we are most interested in looking at the ratios of light intensity for different times in the pulse waveform. Because the phototransistor provides a current proportional to light intensity, we want to have an output voltage that is proportional to the logarithm of current. A transimpedance amplifier design can do that—if we have a nonlinear device that we can put in the feedback path whose current is exponential with the voltage.

It turns out that diodes have exactly the sort of exponential behavior we want.

The *Shockley diode model* [120, 153] gives the forward current through a semiconductor diode as

$$I = I_S(e^{\frac{V_d}{nV_T}} - 1) ,$$

where V_d is the voltage across the diode, I_S is the *scale current* dependent on the size of the diode junction, $V_T = k_B T/q$ is a temperature-dependent *thermal voltage* (about 26 mV for silicon diodes near room temperature), k_B is Boltzmann's constant (81.73324(78) μeV/K), q is the charge of an electron or hole, T is temperature in kelvin, and n is a fudge factor dependent on the semiconductor material that is generally between 1 and 2 for silicon diodes. The fudge factor n is known as the *diode ideality*.

For voltages much bigger than the thermal voltage, the exponential part of the model is much bigger than one, so the current can be approximated as

$$I \approx I_S e^{\frac{V_d}{nV_T}} ,$$

the exponential behavior we were looking for.

Neither the Shockley model nor this simplified exponential model captures the full current-vs.-voltage characteristics of a real diode, and there are various ways to improve the model, but for our purposes, the model is good enough.

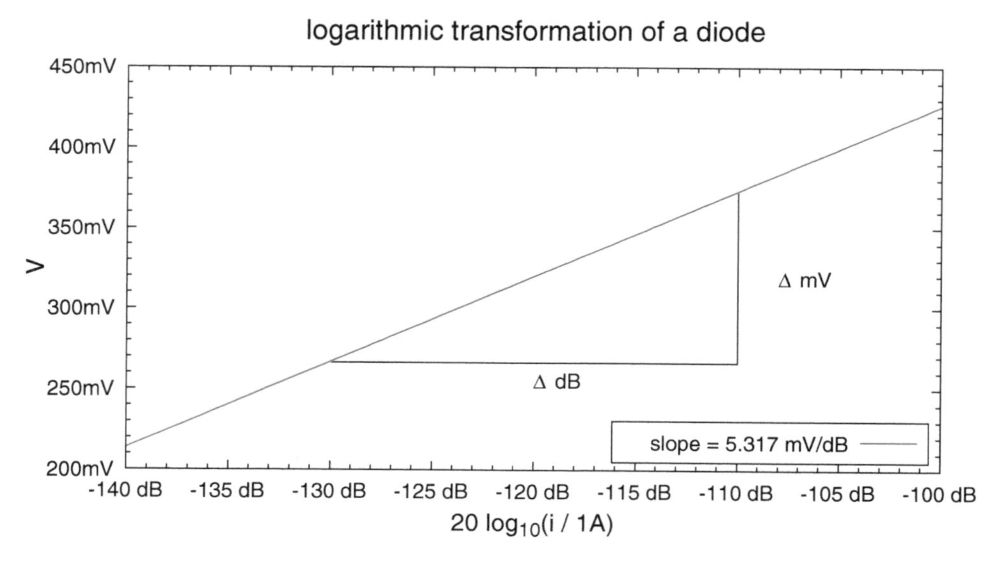

Figure 23.2: *A representative plot of the logarithmic transformation of current to voltage, showing why mV/dB are the correct units for the gain of such a transformation. This plot is* **not** *for any specific diode—look up the specifications for the diode that you are using, rather than treating this plot as canonical.*
Figure drawn with gnuplot [33].

Any semiconductor diode can be used for logarithmic current-to-voltage conversion: a small-signal silicon diode, a Schottky diode, the body diode of an FET (see Section 32.1), or even an LED. Different diodes will have different proportionality constants, as both I_S and n depend on the diode.

To determine the characteristics for a logarithmic transformation, we need to estimate the I-vs.-V curve for the diode that we choose or look it up from the data sheet. If we plot the voltage versus the log of the current (as in Figure 23.2), we can see that the gain of the logarithmic transformation is correctly expressed in mV/dB.

As was discussed in Section 2.3.3 for semilogx graphs, the proper unit of gain for a sensor or amplifier with logarithmic response is V/dB, V/decade, or V/octave. A gain of 1 V/decade means that for every 10-fold increase in the input, there is 1 V more output—that could also be expressed as 50 mV/dB or 0.301 V/octave.

A gain of 5 mV/dB corresponds to a straight line on a semilogx graph, with an equation of the form $V = (5\,\text{mV/dB})\ (20\log_{10}(I/1\,\text{A}) + V_1$ or $V = (5\,\text{mV/dB})\ (20\log_{10}(I/I_0)$, where I_0 is a scaling factor in amps and V_1 is an offset in volts. (You can think of V_1 as the y-intercept, and I_0 as the x-intercept.)

Exercise 23.2

Look up data sheets for a 1N914 diode. There are different manufacturers of 1N914 diodes, and the quality of their data sheets varies enormously. You will need to find a good one, not a minimal data sheet. Report what manufacturer's data sheet you use.

Exercise 23.2 *(continued)*

What is the forward voltage for a forward current of $1\,\mu$A? For $100\,\mu$A? for 10 mA? How many decibels difference is there between $1\,\mu$A and $100\,\mu$A? Between $100\,\mu$A and 10 mA? Give the formula for V_{d} as a function of the forward current. Be sure to use the correct units for any constants (for example, mV or mV/dB).
The 1N914 is not an ideal diode—you need to fit the model to the data from the data sheet to get a useful model.

I have measured the 1N914 diode from 10 nA to 40 mA, and the simple logarithmic fit is pretty good from 10 nA to about 10 mA, as shown on the data sheets (see Figure 23.3).

Exercise 23.3

Using the formula you derived in Exercise 23.2, for a 0.5 dB fluctuation in current, what fluctuation would you expect to see in the output voltage?
What is the exponential dependence of voltage on current in mV/dB? (Compare the estimate from the data sheet with the estimate from Figure 23.3.)

Exercise 23.4

Using the formula you derived in Exercise 23.2 or Figure 23.3, what ratio of current does a 100 mV change in voltage represent?

The generic circuit for a log-transimpedance amplifier is shown in Figure 23.4. The feedback loop holds the voltage on the input I_{in} to V_{ref}, and any current that flows into the I_{in} port must pass through the diode. Because of the orientation of the diode, only negative currents can flow. The voltage across the diode is thus $V_{\mathrm{out}} - V_{\mathrm{ref}}$ and the current through it is

$$-I_{\mathrm{in}} \approx I_{\mathrm{S}} e^{\frac{V_{\mathrm{out}} - V_{\mathrm{ref}}}{n V_T}} ,$$

which gives us

$$V_{\mathrm{out}} - V_{\mathrm{ref}} \approx n V_{\mathrm{T}} \ln\left(\frac{-I_{\mathrm{in}}}{I_{\mathrm{S}}}\right) .$$

The next two exercises ask you to derive the formula for the output voltage of a light sensor combined with a log-transimpedance amplifier. To do them, break the problem into parts: what is the response of the sensor (formula for photocurrent)? what is the voltage across the diode for that current? If you think of these circuits as a block diagram having two blocks, a sensor block and an I-to-V block, the overall response can be seen as the product of the responses of the two blocks.

The phototransistor has to have sufficient voltage across it to work as a light sensor, and the diode of the I-to-V converter needs to be forward-biased to act as a logarithmic converter. Furthermore, the inputs and outputs of the amplifier need to stay within the power rails. These constraints, together with the formula for the output voltage as a function of the light, provide constraints on the possible voltages for V_{ref}.

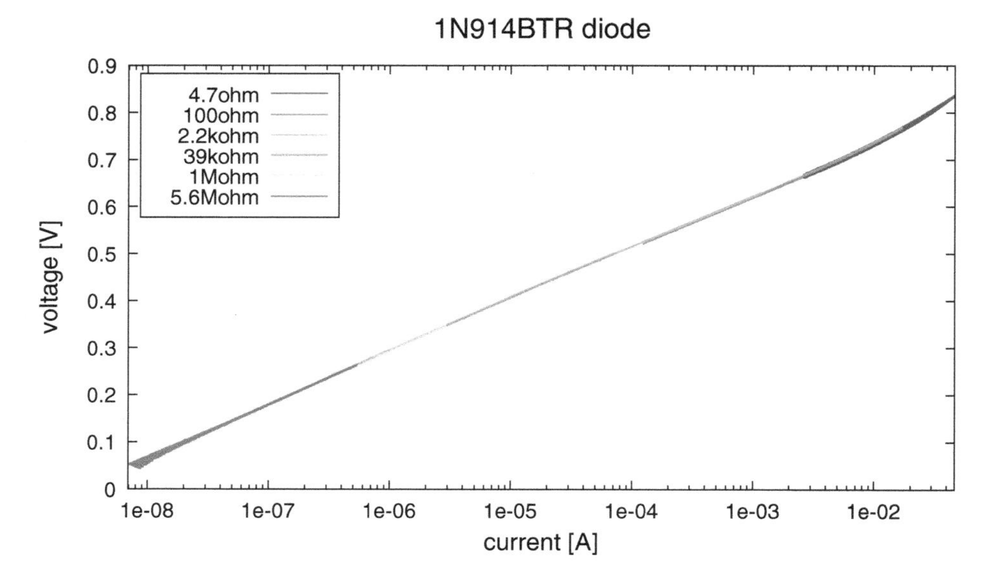

Figure 23.3: *The voltage-vs.-current plot measured for a 1N914BTR diode shows the expected exponential behavior from 10 nA to about 10 mA. The model may still apply at lower currents, but the measurement setup used was already having difficulty with stray currents at 10 nA. At higher currents, the diode voltage is slightly higher than the model predicts.*

The resistance values in the legend refer to the size of the sense resistor used for measuring the current.

Data collected with PteroDAQ using a Teensy LC board [46]. Figure drawn with gnuplot [33].

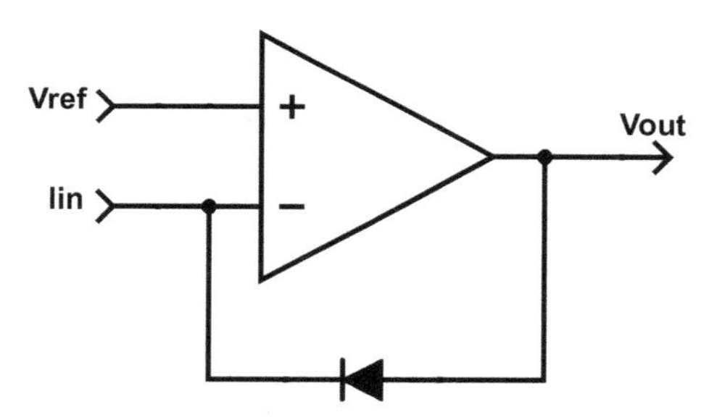

Figure 23.4: *The basic circuit for a logarithmic transimpedance amplifier. The input I_{in} is held to a constant voltage V_{ref}, and the output $V_{out} - V_{ref}$ is proportional to $\log(-I_{in})$. Note the sign, which indicates that the current flows from V_{out} through the diode and the input to ground.*

To get the logarithm of a positive current into the amplifier, the diode would need to be reversed, giving $V_{out} - V_{ref}$ proportional to $\log(I_{in})$.

Figure drawn with Digi-Key's Scheme-it [18].

Exercise 23.5

For the design on the left in Figure 23.5, provide a formula for the output voltage V_{out} in terms of the irradiance E_e in $W/(cm)^2$, sensitivity of the phototransistor s (in units of $A(cm)^2/W$), the scale current I_S for the diode, and nV_T.

What constraints are there on V_{ref}?

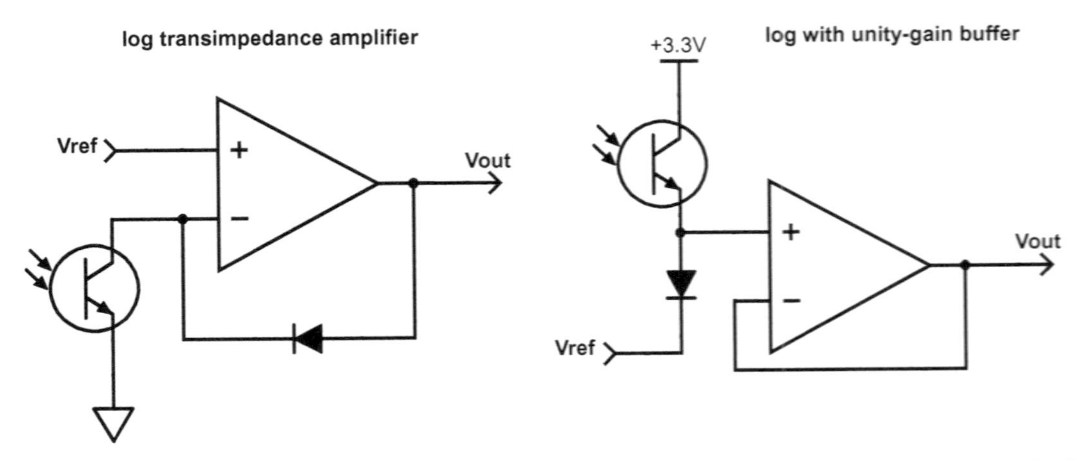

Figure 23.5: *Two different designs for logarithmic light detectors, for Exercises 23.5–23.6. The design on the left is a log-transimpedance amplifier, but the design on the right is not—it uses a simple diode to convert log(current) to voltage, then a unity-gain buffer to duplicate the voltage while providing a low output impedance.*
Figure drawn with Digi-Key's Scheme-it [18].

Exercise 23.6

For the design on the right in Figure 23.5, provide a formula for the output voltage V_{out} in terms of the irradiance E_e in $\mathrm{W/(cm)^2}$, sensitivity of the phototransistor s (in units of $\mathrm{A(cm)^2/W}$), the scale current I_S for the diode, and nV_T.
What constraints are there on V_{ref}?

23.3 Multistage transimpedance amplifier

The simple one-stage transimpedance amplifier (in either incarnation of Figure 23.5) does not produce a large enough signal to see a pulse clearly (as you will see in the first part of Lab 6). But if we just increase the gain, we run into the problem that the DC part of the input signal saturates the amplifier and we can't see any fluctuation at all.

We can also have our signal buried in 60 Hz interference. We can reduce the interference problem by taking advantage of what you learned about aliasing: if we sample with a period that is a multiple of 1/60 s, then we'll be in the same place on a 60 Hz or 120 Hz waveform on each sample and the interference will not be visible in the output. Picking 60 Hz, 30 Hz, 20 Hz, or 15 Hz as our sampling frequency will give us enough resolution to see the pulse waveform, without much problem from 60 Hz interference.

Simply aliasing away the 60 Hz interference with a low sampling rate is not a cure-all. If the interference is large compared to the signal we want, it will limit the amplification we can do before the limits of the op amp are reached and clipping occurs. That clipping can remove the cycle-to-cycle changes that are the pulse signal we want to see, and so removing unwanted interference *before amplification* is generally a better design strategy.

We have to be careful, though, that the impedance of our filter does not interfere with the measurements made by the sensor. For low-current sensors, like the phototransistor in the pulse monitor or the electrodes of an EKG, we generally have one stage of amplification before we do any filtering. This first stage generally has fairly low gain, as its main purpose is to provide a low-impedance voltage (one that can provide moderately large currents) that follows the sensor input, without requiring much current from the sensor.

Relying on aliasing can have other undesired side effects. For example, in 2015, students observed that if they recorded for a long time, the DC shift in their signal seemed to be periodic—they correctly attributed this to a slight difference in frequency between the sampling frequency and the 60 Hz interference, producing aliasing to a very low frequency. The difference was larger than could be accounted for by the ±50 parts-per-million error in the crystal oscillator on the FRDM-KL25Z board they were using, but was within the ±0.02 Hz allowable error in line frequency for the power company.

We can eliminate the DC shift (and reduce the 60 Hz interference problem) by *filtering*—changing our amplifier design so that the gain varies with frequency. We want to make the gain be large for the signal we want to see, but small for interfering signals that we want to get rid of.

We can do simple filtering with analog RC filters, before amplifying the signal, then use digital filters, if necessary, to clean up the signal for further processing or display. The analog filters are just to clean up the signal enough that we can amplify it into a range suitable for our analog-to-digital converter—they do not have to be very tightly specified, as long as whatever modification of the signal they introduce is acceptable.

A filter that allows through a range of frequencies while blocking frequencies outside that range is called a *band-pass filter*, and is usually specified in terms of the lower and upper corner frequencies.

What range of frequencies are we interested in keeping for a pulse monitor? A person's pulse is normally in the range 40 beats/minute to 200 beats/minute, depending on their genetics, age, physical condition, whether they are exercising, and so forth. This translates to 0.6 Hz to 3.3 Hz for the fundamental frequency of the pulse waveform. If we want to see the shape of the pulse, we need to include the first few harmonics—in addition to the fundamental frequency, f, we might want to include $2f$, $3f$, $4f$, and so on. The more harmonics we include, the more information we preserve about the shape of the pulse, but the harder it is to remove interference that is not related to the pulse.

If we use a simple RC filter, then we get little reduction in magnitude for frequencies near the corner frequency, and we can set our low-pass and high-pass corner frequencies near the 0.6 Hz and 3.3 Hz limits we are interested in. There will be some loss of signal near the limits, but for pulse monitoring, we don't need the shape of the signal to be well preserved, but we do want to get rid of 60 Hz interference and DC drift.

After our first stage has converted the logarithm of the photocurrent into a voltage, we can add a high-pass filter to block DC, then add a second-stage amplifier to amplify the small AC pulse signal. The second stage can be inverting or non-inverting, depending on which way up you want the output signal to be.

Low-pass filtering can easily be built into an amplifier, by making the feedback impedance be a complex impedance, as described in Section 19.2.

Worked Example:
How do you determine the output range of a two-stage log-transimpedance amplifier that includes a high-pass filter?

The first stage would be a log-transimpedance amplifier, with reference voltage V_{ref1}, and the second stage would be a voltage amplifier with gain G_2 and reference voltage V_{ref2}, which may or may not be the same as V_{ref1}.

The output of the log-transimpedance amplifier, is $V_1 = A\log_{10}(I_{\text{in}})+B+V_{\text{ref1}}$, where A and B are constants that depend on the diode (though A will depend mostly on the type of semiconductor and B on the size). I used $\log_{10}(I)$ instead of $\ln(I)$, so that I could read the value of A more easily off of the graphs on the data sheet or in the book. Using $\log_{10}$ makes the units for A be mV/decade.

If the log-transimpedance amplifier is used for a phototransistor, the currents will be small enough that $A\log_{10}(I_{\text{in}}) + B$ does not get very large (maybe 300 mV to 800 mV, depending on how bright the light is), but other applications have higher currents from the sensor and so produce higher voltages.

A high-pass (or band-pass) filter would eliminate the $B + V_{\text{ref1}}$ constant term, so that the second stage would not amplify it, recentering the voltage at V_{ref2}. We can assume that the corner frequency of the high-pass filter is low enough that the signal we are interested in is not substantially changed by the filter (either a passive filter with a gain of 1, or an active filter in the second stage, with a gain of G_2 in the passband).

As a result, we are mainly interested in how big the varying signal, $A\log_{10}(I_{\text{in}})$, is. The range is from $A\log_{10} I_{\text{min}}$ to $A\log_{10} I_{\text{max}}$, so the voltage range we are amplifying is $A\log_{10}\left(\frac{I_{\text{max}}}{I_{\text{min}}}\right)$ peak-to-peak.

The final output would then have the range

$$G_2\log_{10}\left(\frac{I_{\text{max}}}{I_{\text{min}}}\right)$$

peak-to-peak, centered at V_{ref2}.

Exercise 23.7

Draw a block diagram for an optical pulse monitor, showing all the necessary conversion, filtering, and amplification stages. Give the corner frequencies of any filters included, and estimate the necessary gains to get a pulse signal that is 2 V peak-to-peak for a 6% fluctuation in the opacity of the finger being monitored.

23.4 Compensating transimpedance amplifiers

Section 23.4 is optional, more advanced material. It looks at the phase response of op amps and discusses how to keep high-gain transimpedance amplifiers from ringing or oscillating.

Transimpedance amplifiers are used for measuring some very small currents in bioengineering, both in biological systems (such as measuring the current flow through ion channels) and in instrumentation (nanopipettes and nanopores).

Digitizing these small currents requires high-gain, low-noise amplifiers, but the simple solution of using a very large feedback resistor in a transimpedance amplifier runs into a problem: *instability*. An unstable amplifier can start spontaneously oscillating, either with a fixed period or chaotically, introducing a spurious signal that can be much larger than the genuine signal being amplified.

One solution is to keep the transimpedance gain low and use subsequent voltage gain stages, as explained in Section 23.3, but that solution generally does not have the best noise performance, because

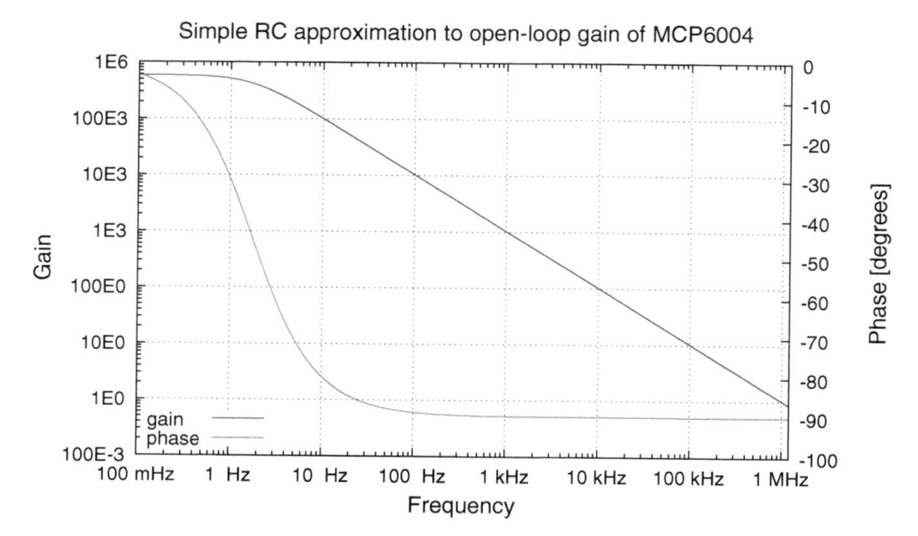

Figure 23.6: *This plot shows a simple RC-filter approximation to the open-loop gain of an MCP6004 op amp, which has a 1 MHz gain-bandwidth product. The gain is fairly accurate up to 1 MHz, but the phase approximation is only good to about 100 kHz. According to the data sheet [71], above 100 kHz, the phase change reduces from −90° to about −80° at 1 MHz, then rapidly increases to −180° around 10 MHz.*

Figure drawn with gnuplot [33].

electrical noise is amplified by the subsequent stages (low-noise design is outside the scope of this book). So we often want to put as much gain as we can into the transimpedance stage without triggering instability.

To understand the instability, we need to look not just at the magnitude of the gain of the amplifier and feedback loop, as we do for most of this book, but also the phase change of the gain, as shown in Figure 23.6.

The key concepts are the following:

- The voltage gain of a negative-feedback amplifier is $\frac{A}{1+AB}$ for a non-inverting amplifier or $\frac{A(B-1)}{1+AB}$ for an inverting amplifier, where A is the open-loop gain of the op amp and B is the gain of the voltage divider used for feedback (see Section 19.2). In either case, instability is a problem if the denominator gets close to zero, that is if AB is close to -1.
- The open-loop gain of most op amps can be approximated as a simple low-pass filter: $A = A_0/(1 + j\omega\tau)$, where $\tau = A_0/(2\pi f_{\mathrm{GB}})$ (see Section 18.2.2). When the frequency is high enough (where the open-loop gain is limited by the gain-bandwidth product) the phase change of the op amp is about $-90°$.
- If we set up a transimpedance amplifier with feedback resistor R, then the feedback consists of a low-pass RC filter: a voltage divider with R on top and the input capacitance of the amplifier and any capacitance in parallel with the current source on the bottom, making a combined capacitance C. That means that $B = \frac{1}{1+j\omega RC}$.
- The phase change of a low-pass RC filter ($B = \frac{1}{1+j\omega RC}$) approaches $-90°$ above the corner frequency.

Having a phase change of $-180°$ and gain 1 for AB at some frequency means $AB = -1$, which results in instability and possible oscillation. The *phase margin* of an amplifier is the angle between the phase change of AB and $-180°$ at the frequency where $|AB| = 1$. To avoid instability, designers try to get a phase margin of 45° or more.

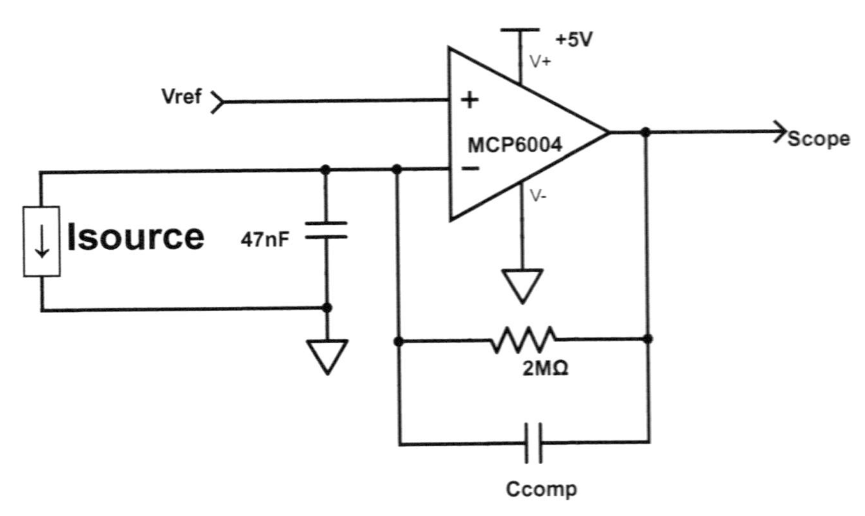

Figure 23.7: *This schematic shows a transimpedance amplifier with a large input capacitance (47 nF) and a moderately large gain (2 MΩ). The feedback voltage divider produces a low-pass feedback filter, which when combined with the internal low-pass filter of the op amp can result in a feedback loop with a phase change of 180° and gain of 1, which makes the amplifier unstable.*
A compensation capacitor can be added, as shown by C_{comp} *to change the phase relationships and remove the instability.*
Figure drawn with Digi-Key's Scheme-it [18].

Those conditions for instability mean that we can have instability at frequencies between $f_c = \frac{1}{2\pi RC}$ and the gain-bandwidth product f_{GB} (though we may have problems only for frequencies at least a factor of 2 or 3 above the low-pass corner frequency f_c, since the phase change of the filter is only asymptotically $-90°$—at the corner frequency it is only $-45°$). If the parasitic capacitances are low and we only request small transimpedance gain R, then RC is small, and the corner frequency of the low-pass feedback is above the gain-bandwidth product, so there are no problems, but if we ask for large gain R and the capacitance of the sensor or amplifier input is also large, then RC is large and we may have instability.

Worked Example:
To demonstrate the problem, I deliberately made an unstable transimpedance amplifier using the MCP6004 op amps that we use in class.

The op amps have a gain-bandwidth product of 1 MHz, so I needed an RC time constant much larger than 160 ns. I chose 2 MΩ and 47 nF for an RC time constant of 94 ms and a corner frequency of 1.69 Hz (see Figure 23.7). For V_{ref}, I used a pair of 10 kΩ resistors in a voltage divider, with a 470 μF electrolytic capacitor and a couple of 10 μF ceramic capacitors to ensure that there is no feedback through the power lines to V_{ref}. We do indeed get chaotic oscillation with no current source connected as the input, as shown in Figure 23.8.

The reasoning about the amplifier instability suggests that the oscillation should be at about the frequency where the gain around the loop is 1, that is where $\frac{f_{\text{GB}}}{f}\frac{1}{2\pi fRC} = 1$ or $f = \sqrt{\frac{f_{\text{GB}}}{2\pi RC}}$. For the circuit of Figure 23.7, that would be around $\sqrt{1\,\text{MHz}\;1.69\,\text{Hz}} = 1.3\,\text{kHz}$. An average of many Fast Fourier Transforms (FFTs) shows that the energy of the oscillation does indeed peak near the expected frequency (see Figure 23.9).

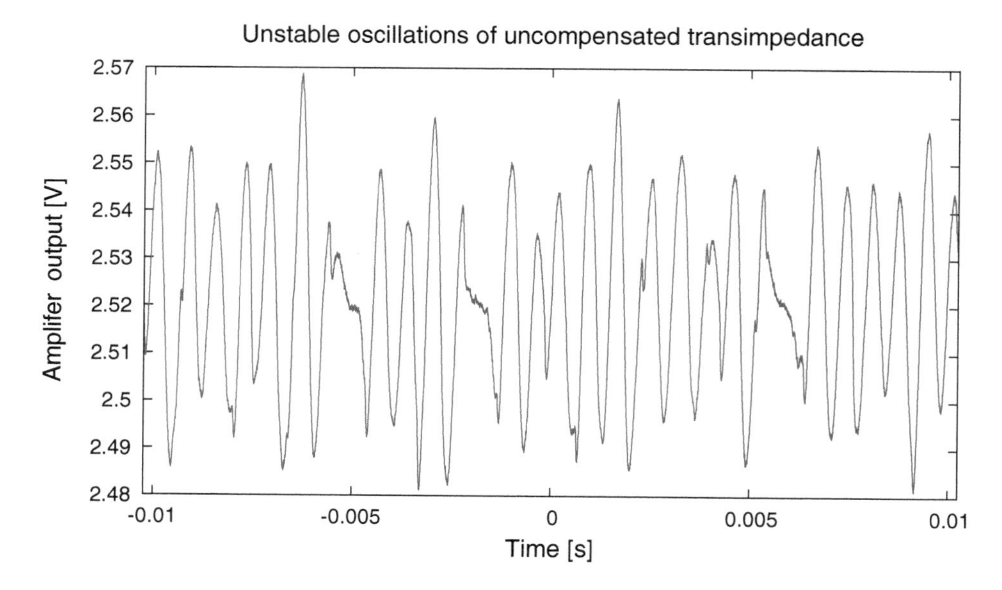

Figure 23.8: *The transimpedance amplifier shown in Figure 23.7 does indeed produce a small, chaotic oscillation. Data collected with Analog Discovery 2 [19]. Figure drawn with gnuplot [33].*

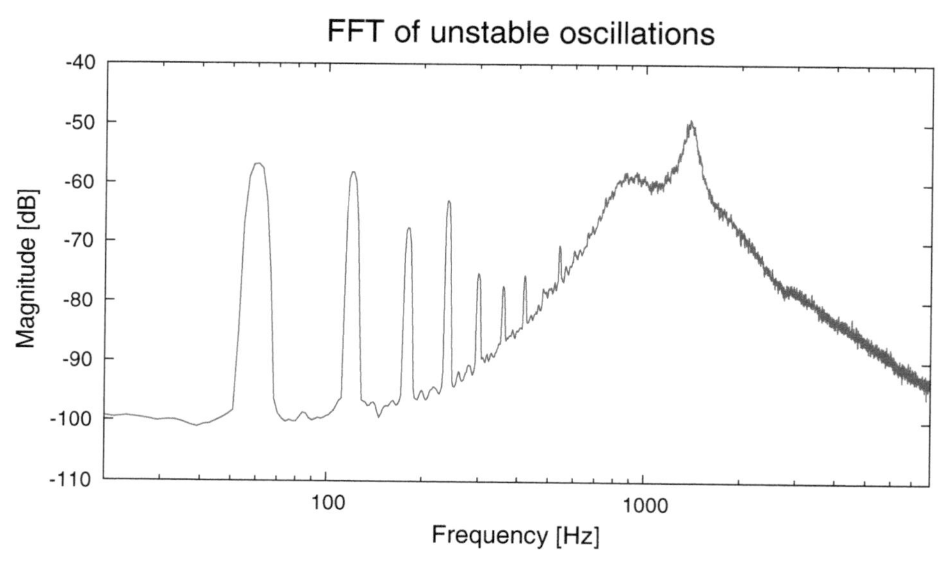

Figure 23.9: *If we average many Fast Fourier Transform (FFT) traces of the output of the transimpedance amplifier shown in Figure 23.7 with no input, we can see that the chaotic oscillation has a peak around 1380 Hz, close to the predicted frequency. Also visible are harmonics of 60 Hz, which are the correct output of the transimpedance amplifier (picking up stray currents by capacitive coupling).*

The data was collected using an exponential average of 150 traces with a weight of 30. Data collected with Analog Discovery 2 [19]. Figure drawn with gnuplot [33].

Eliminating the unintended oscillation is referred to as *compensating* the amplifier. The idea is to reduce the phase change of the feedback circuit at the frequency where $|AB| = 1$, so that $AB \neq -1$. The reduction is accomplished by adding a small capacitor in parallel with the feedback resistor, making the

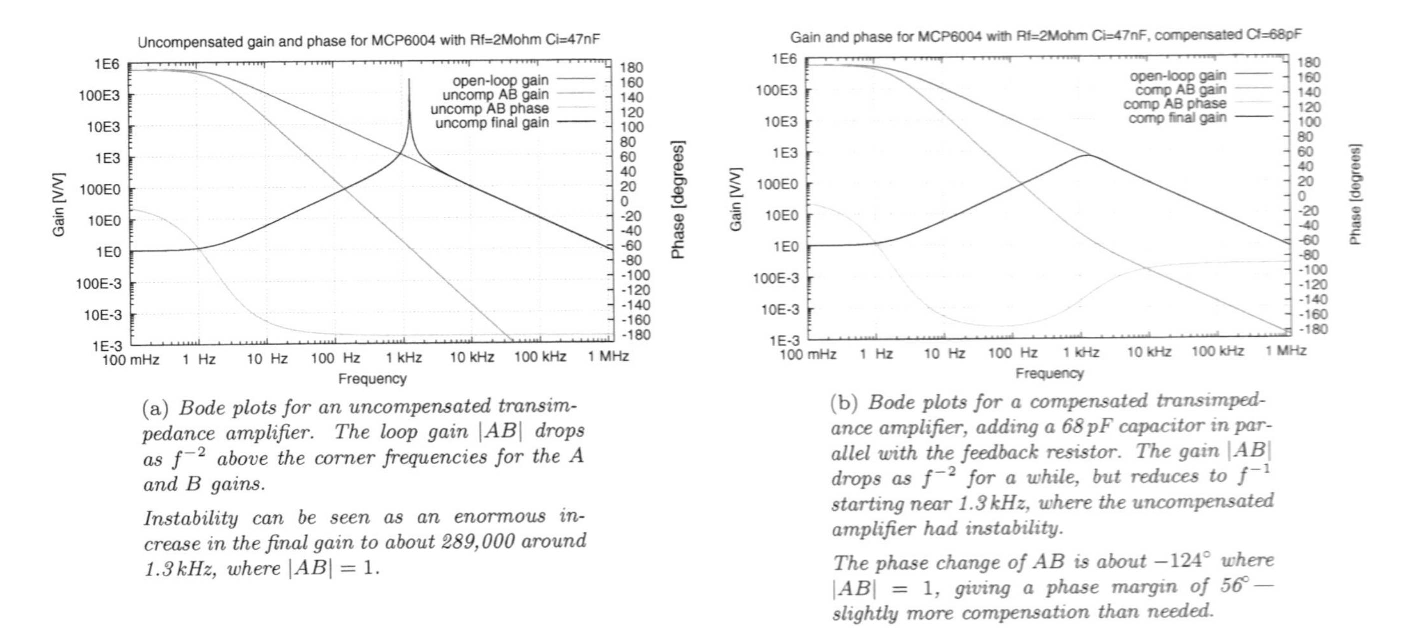

(a) *Bode plots for an uncompensated transimpedance amplifier. The loop gain* $|AB|$ *drops as* f^{-2} *above the corner frequencies for the A and B gains.*

Instability can be seen as an enormous increase in the final gain to about 289,000 around 1.3 kHz, where $|AB| = 1$.

(b) *Bode plots for a compensated transimpedance amplifier, adding a 68 pF capacitor in parallel with the feedback resistor. The gain* $|AB|$ *drops as* f^{-2} *for a while, but reduces to* f^{-1} *starting near 1.3 kHz, where the uncompensated amplifier had instability.*

The phase change of AB is about $-124°$ *where* $|AB| = 1$, *giving a phase margin of 56° — slightly more compensation than needed.*

Figure 23.10: *Theoretical Bode plots for uncompensated and compensated transimpedance amplifiers using an op amp with 1 MHz gain-bandwidth product, a feedback resistance* $R_f = 2\,M\Omega$, *and an input capacitance of 47 nF.*
Figure drawn with gnuplot [33].

gain of the feedback filter $\frac{1+j\omega R_F C_F}{1+j\omega R_F(C_F+C_i)}$, where R_F and C_F are the feedback components and C_i is the input capacitance.

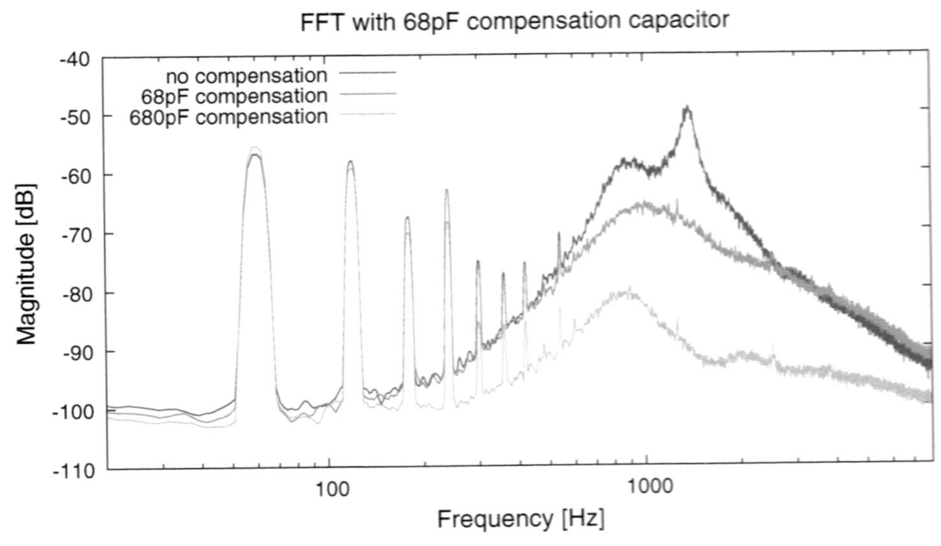

Figure 23.11: *Taking an FFT of the noise from a transimpedance amplifier that has no input (other than capacitively coupled 60 Hz interference) shows the expected behavior. Adding a 68 pF compensation capacitor in parallel with the 2 MΩ cuts out the oscillation peak, but there is still a fair amount of noise around the corner frequency of the amplifier (1.2 kHz), matching the prediction of Figure 23.10b. Overcompensating with a 680 pF capacitor reduces the noise substantially, but the bandwidth is reduced to 120 Hz.*
Data collected with Analog Discovery 2 [19]. Figure drawn with gnuplot [33].

For so-called "optimal" compensation, where the phase change of B at $|AB| = 1$ is $-45°$ instead of $-90°$, we want to set the upper corner frequency $1/(2\pi R_F C_F)$ at the geometric mean of the lower corner frequency $1/(2\pi R_F(C_F + C_i))$ and the gain-bandwidth product f_{GB}:

$$\frac{1}{2\pi R_F C_F} = \sqrt{\frac{f_{GB}}{2\pi R_F(C_F + C_i)}} .$$

We can simplify the formula for the optimal compensation by assuming that $C_i \gg C_F$ to get

$$C_F \approx \sqrt{\frac{C_i}{2\pi R_F f_{GB}}} ,$$

which for my design yields $C_F = 61\,\text{pF}$, which is indeed much smaller than $C_i = 47\,\text{nF}$. Figure 23.10 shows Bode plots for both the uncompensated and the compensated amplifiers. Although the amplifiers are transimpedance amplifiers, so the gain is nominally an impedance, the Bode plots give the gain as a unitless V/V gain, to show the amplification of voltage noise at the input or internal to the amplifier.

Using a larger capacitor (*overcompensating*) increases the phase margin (thus allowing for more variation from specs for the various components) at the cost of reducing the bandwidth of the final amplifier. Figure 23.11 shows that the compensation does work to suppress the oscillation, but overcompensation can reduce noise further, at the expense of bandwidth.

For more details on stabilizing transimpedance amplifiers, you may wish to read the Maxim Integrated application note *Stabilize your transimpedance amplifier* [9]. That application note was the basis for much of the analysis in this section.

Chapter 24

Active Filters

24.1 Active vs. passive filters

Active filters are filters that include an amplifier as part of the design, while passive filters use only resistors, capacitors, and inductors (so-called *passive* components).

The main advantages of active filters over passive filters include

- possible gains greater than one (more precisely, because we use inverting amplifiers, < -1);
- easier decomposition of the design into nearly independent blocks, because the output impedance of the active filter is very low (near $0\,\Omega$);
- easier cascading of multiple stages of filters, because the gains of the active filters just multiply, without further interaction between stages;
- ability to provide substantial current at the output of the filter; and
- simpler design for setting the corner frequencies of band-pass filters.

The main advantages of passive RC filter design (as in Chapter 11) include

- fewer and cheaper components,
- no gain-bandwidth limitations from amplifiers, and
- no power supply needed.

Cascading multiple stages of filtering increases the *order* of the filter, so that transition from the passband to the stop band can be steeper than gain proportional to f or f^{-1}, as we get with the simple RC filters of Chapter 11. An nth-order filter can have lines on a log-log plot with slopes up to $\pm n$, that is, with gains proportional to f^n or f^{-n}. The slopes of these lines are referred to as the *roll-off* of the filter (more precisely, the absolute value of the slope), and are expressed in dB/octave or dB/decade. The simple, first-order RC filters we've looked at so far have a roll-off of 6 dB/octave or 20 dB/decade.

The active-filter designs used in this course are very simple ones that are easy to analyze—the high-pass and low-pass filters are first-order filters and the band-pass filter is a combination of a first-order high-pass and first-order low-pass filter. There are more advanced designs for active filters that provide faster roll-off (40 dB/decade per stage) or narrower bandwidths than the simple ones we use. Optional Section 24.7 describes one such design.

The active-filter designs presented in this chapter all have very low output impedance, for easy interfacing to subsequent stages, but their input impedance is not very high. The relatively low input impedance imposes some constraints on the previous stage For example, we often do not want to use these designs in the first stage of amplification after a sensor (unless the sensor is a voltage-output sensor with a much lower impedance than the input impedance of the filter). We also do not want to put an active filter right after a passive filter, as the input impedance of the active filter will change the behavior of the passive filter.

24.2 Active low-pass filter

You can make a negative-feedback amplifier into a low-pass filter by putting a capacitor C_f and resistor R_f in parallel for Z_f as shown in Figure 24.1. At low frequencies (well below the corner frequency), the feedback impedance is approximately R_f and the gain of a non-inverting amplifier is $1 + R_\mathrm{f}/R_\mathrm{i}$. At high frequencies (well above the corner frequency), the impedance is approximately $1/(j\omega C_\mathrm{f})$, and the gain of a non-inverting amplifier at these high frequencies is approximately $1 + 1/(j\omega C_\mathrm{f} R_\mathrm{i})$, which gets smaller as the frequency goes up, dropping to 1 at high frequencies.

The inverting amplifier configuration, which is the same as Figure 24.1 but with the V_in and V_ref port names swapped, is more popular, because the gain

$$-Z_\mathrm{f}/R_\mathrm{i} = \frac{-R_\mathrm{f}}{R_\mathrm{i}} \frac{1}{j\omega R_\mathrm{f} C_\mathrm{f} + 1}$$

goes to zero as ω goes to ∞, rather than bottoming out at 1 as the non-inverting amplifier does. The negative input of the op amp is held constant by the feedback loop, so the input impedance of the filter is R_i.

The gain of an inverting, active, low-pass filter is

$$\frac{-R_\mathrm{f}}{R_\mathrm{i}} \frac{1}{j\omega R_\mathrm{f} C_\mathrm{f} + 1} ,$$

so the passband gain simplifies to

$$\frac{-R_\mathrm{f}}{R_\mathrm{i}} ,$$

the corner frequency is

$$f_\mathrm{c} = \frac{1}{2\pi R_\mathrm{f} C_\mathrm{f}} ,$$

and the input impedance is R_i.

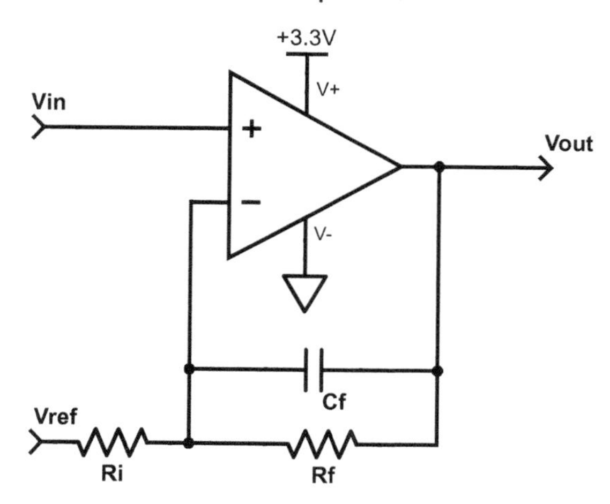

Figure 24.1: *An active low-pass filter is a negative-feedback amplifier with a feedback impedance that decreases as frequency increases, so that the gain of the amplifier also decreases with increasing frequency. Shown here is an active low-pass filter configured as a non-inverting (positive gain) amplifier, though the inverting configuration is far more common.*
Figure drawn with Digi-Key's Scheme-it [18].

One disadvantage of the inverting configuration is that the input impedance is only R_i at DC rather than the effectively infinite input impedance of the non-inverting configuration, so whatever is providing V_{in} needs to have a fairly low impedance. If the input to the active filter is being driven by an amplifier, then the inverting configuration is usually the better choice, but if the input is coming directly from a high-impedance sensor or a high-impedance passive filter, then the non-inverting configuration may be the only feasible one, despite the limited filtering at high frequencies.

One of the big advantages of active filters is that they can be cascaded—that is, the output of one filter can drive the input of the next without affecting the gain of either stage. So cascading multiple active filters simply multiplies the gains of the individual stages.

For example, we can look at the effect of cascading identical active low-pass filters. We can write the gain equation for a simple low-pass filter as $G(f) = \frac{G_p}{1+jf/f_c}$ and take its magnitude:

$$|G(f)| = \frac{G_p}{\sqrt{1+(f/f_c)^2}} .$$

Then the magnitude of the gain off n cascaded filters is

$$|G(f)|^n = (G_p)^n \left(1+(f/f_c)^2\right)^{-n/2} ,$$

with passband gain of G_p^n. We can find the cutoff frequency (where the gain is $\sqrt{2}$ of the passband gain)—the gain is $\frac{G_p^n}{\sqrt{2}}$ when $2 = (1+(f/f_c)^2)^n$. We can simplify that to $2^{1/n}-1 = (f/f_c)^2$ or $f = f_c\sqrt{2^{1/n}-1}$.

The cutoff frequency for cascading n low-pass filters that have the same corner frequency keeps decreasing as the number of stages increases—it is

$$f_c\sqrt{2^{1/n}-1} ,$$

where f_c is the corner frequency for a single filter.

Worked Example:
Design an active low-pass filter with three identical stages that has a cutoff frequency of about 30 Hz. What is the magnitude of the gain at 60 Hz?

Given the number of stages and the cutoff frequency we can compute the corner frequency for the individual stages:

$$\begin{aligned} 30\,\text{Hz} &= f_c\sqrt{2^{1/3}-1} \\ &\approx f_c 0.5098 \\ f_c &\approx 58.84\,\text{Hz} . \end{aligned}$$

That corner frequency gives us a time constant $R_f C_f \approx 2.705\,\text{ms}$. If we select $C_f = 100\,\text{nF}$, we want $R_f \approx 27.05\,\text{k}\Omega$. We could pick $27\,\text{k}\Omega$ from the E12 series, which results in a corner frequency of 58.95 Hz, and a cutoff frequency for the cascaded filter of 30.05 Hz.

The passband gain can be set as needed by setting the input resistors (as long as not too high a gain is required). It might be best not to make the stages identical, but to use a low gain for the first stage (to get a high input impedance) and higher gain for the later stages.

The gain at 60 Hz for one stage is

$$G_{60,\mathrm{Hz}} = \frac{-G}{1 + j\omega 2.7\,\mathrm{ms}\, 2\pi\, 60\,\mathrm{Hz}}$$
$$\approx G(-0.4911 + 0.4999j) \; ,$$

where G is the magnitude of the passband gain for one stage and the gain for all three stages is approximately $G^3(0.24977 + j0.2368)$, which has magnitude $0.3442G^3$.

In professional designs, it is rare to cascade identical filters—instead subtle changes are made to each stage to get somewhat better properties from the resulting multi-stage filters. The design of such filters is the subject of entire courses and beyond the scope of this course, but calculators for various designs with different tradeoffs can easily be found on the web. Some of the popular designs include the Butterworth, Bessel, Cauer, and Chebychev filters.

Exercise 24.1

Plot the gain of the amplifier in Figure 24.1 vs. frequency on a log-log plot from 0.1 kHz to 100 kHz, using the following parameter values: $R_\mathrm{i} = 1\,\mathrm{k}\Omega$, $R_\mathrm{f} = 100\,\mathrm{k}\Omega$, and $C_\mathrm{f} = 22\,\mathrm{nF}$.

Exercise 24.2

Design a low-pass filter with a cutoff frequency of about 10 Hz, a passband gain of about 1000, and a gain at 60 Hz of less than 20. (*Hint*: this will require more than one stage.)

24.3 Active high-pass filter

An active high-pass filter can also be made from an inverting amplifier, as shown in Figure 24.2. The analysis is similar to that for an active low-pass filter, as the gain is just the negative of the ratio of impedances:

$$G = \frac{-Z_\mathrm{f}}{Z_\mathrm{i}}$$
$$= \frac{-R_\mathrm{f}}{R_\mathrm{i} + 1/(j\omega C_\mathrm{i})}$$
$$= \frac{-j\omega R_\mathrm{f} C_\mathrm{i}}{1 + j\omega R_\mathrm{i} C_\mathrm{i}} \; .$$

As ω goes to 0, the gain is asymptotically $-j\omega R_\mathrm{f} C_\mathrm{i}$, and as ω goes to ∞, the gain is asymptotically $-R_\mathrm{f}/R_\mathrm{i}$. The (angular) corner frequency is where the magnitudes of these two functions match: $\omega C_\mathrm{i} = 1/R_\mathrm{i}$ or $\omega = 1/(R_\mathrm{i} C_\mathrm{i})$.

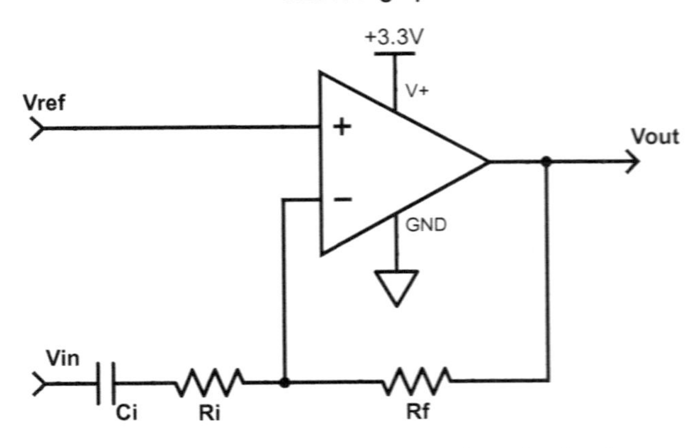

Figure 24.2: *A generic active high-pass filter has a capacitor and resistor in series for the input impedance and a simple resistor for the output impedance. The corner frequency is determined by the* R_iC_i *time constant, and the gain in the passband is determined by the ratio of resistors* $-R_f/R_i$.

The order of the two components in series does not matter.

Figure drawn with Digi-Key's Scheme-it [18].

The gain of an inverting, active, high-pass filter is

$$\frac{-j\omega R_f C_i}{1 + j\omega R_i C_i},$$

so the passband gain simplifies to

$$\frac{-R_f}{R_i},$$

the corner frequency is

$$f_c = \frac{1}{2\pi R_i C_i},$$

and the input impedance is

$$R_i + \frac{1}{j2\pi f C_i}.$$

Worked Example:
We have a $\pm 20\,mV$ *signal that is near the peak of human hearing* ($2\,kHz$), *and we want to amplify it to about* $V_{ref} \pm 1\ V$ *for digitization, removing any DC offset and* 60 *Hz hum from it.*

Removing a DC offset usually means using a high-pass filter, and the 60-Hz interference has a much lower frequency than the 2 kHz signal, so we can use the high-pass filter to remove it also. A corner frequency around 600 Hz would provide 20 dB (10-fold) reduction at 60 Hz.

Amplifying from $\pm 20\,\text{mV}$ to a range of $\pm 1\,\text{V}$ requires a gain of 50 (or -50, as no constraint was given on the polarity of the signal). If we want that large a gain from a single stage, then we'll need a large ratio for the resistors, which requires a fairly small resistor for the input and a fairly large one for the feedback. To keep both between $1\,\text{k}\Omega$ and $1\,\text{M}\Omega$, the smaller one must be no bigger than $20\,\text{k}\Omega$ and the larger one no less than $50\,\text{k}\Omega$.

A 600 Hz corner frequency means an RC time constant of about $265\,\mu\text{s}$, which we can approximate with a 33 nF capacitor and an $8.2\,\text{k}\Omega$ resistor, to get a time constant of $270.6\,\mu\text{s}$ or a corner frequency of 588.2 Hz.

A gain of -50 would require a feedback resistance of $410\,\text{k}\Omega$, which is not a common value. The closest values in the E24 series are $390\,\text{k}\Omega$ and $430\,\text{k}\Omega$, which would give gains in the passband of -47.56 and -52.44, for output voltage ranges of $V_{\text{ref}} \pm 951\,\text{mV}$ and $V_{\text{ref}} \pm 1.049\,\text{V}$. Either of these would be a reasonable solution, if a 5% difference in the output voltage is acceptable.

So for $R_{\text{i}} = 8.2\,\text{k}\Omega$, $C_{\text{i}} = 33\,\text{nF}$, and $R_{\text{f}} = 390\,\text{k}\Omega$, we have an output signal of $V_{\text{ref}} \pm 951\,\text{mV}$ with the 60 Hz interference having about 20 dB less gain than the desired signal.

Active high-pass filters can have a serious problem with instability, just as we analyzed for transimpedance amplifiers in Section 23.4. The problem is that the gain on an inverting amplifier is not just $(B-1)/B$, for a negative feedback of B, but $\frac{A(B-1)}{1+AB}$, where A is the open-loop gain of the op amp. The limited gain-bandwidth product of the op amp can cause problems if $AB \approx -1$, making the denominator near 0.

When can AB get close to -1? We can approximate the open-loop gain A as a low-pass filter $A = \frac{A_0}{1+j\omega\tau}$ with $\tau = A_0/(2\pi f_{\text{GB}})$. We can get B from the voltage divider formula:

$$B = \frac{Z_{\text{i}}}{Z_{\text{i}} + Z_{\text{f}}} = \frac{1 + j\omega R_{\text{i}} C_{\text{i}}}{1 + j\omega (R_{\text{f}} + R_{\text{i}}) C_{\text{i}}} ,$$

which gives us

$$-1 \approx AB = \frac{A_0 (1 + j\omega R_{\text{i}} C_{\text{i}})}{(1 + j\omega\tau)(1 + j\omega (R_{\text{i}} + R_{\text{f}}) C_{\text{i}})} .$$

If we multiply both sides by the denominator and look at just the real part, we get

$$(\omega^2 (R_{\text{i}} + R_{\text{f}}) C_{\text{i}} \tau - 1) \approx A_0 .$$

Because A_0 is a large real number, we get the real part of AB close to -1 if

$$\omega = \sqrt{\frac{A_0}{(R_{\text{i}} + R_{\text{f}}) C_{\text{i}} \tau}} .$$

If that frequency is in our passband, everything is fine, because the phase change for AB in our passband approaches $-90°$, well away from $-180°$ where we could get instability. That is, we won't have instability if

$$\sqrt{\frac{A_0}{(R_{\text{i}} + R_{\text{f}}) C_{\text{i}} \tau}} > \frac{1}{R_{\text{i}} C_{\text{i}}} ,$$

$$\frac{A_0}{(R_{\text{i}} + R_{\text{f}}) C_{\text{i}} \tau} > \frac{1}{R_{\text{i}}^2 C_{\text{i}}^2} ,$$

$$\frac{A_0}{\tau} > \frac{(R_{\text{i}} + R_{\text{f}}) C_{\text{i}}}{R_{\text{i}}^2 C_{\text{i}}^2} ,$$

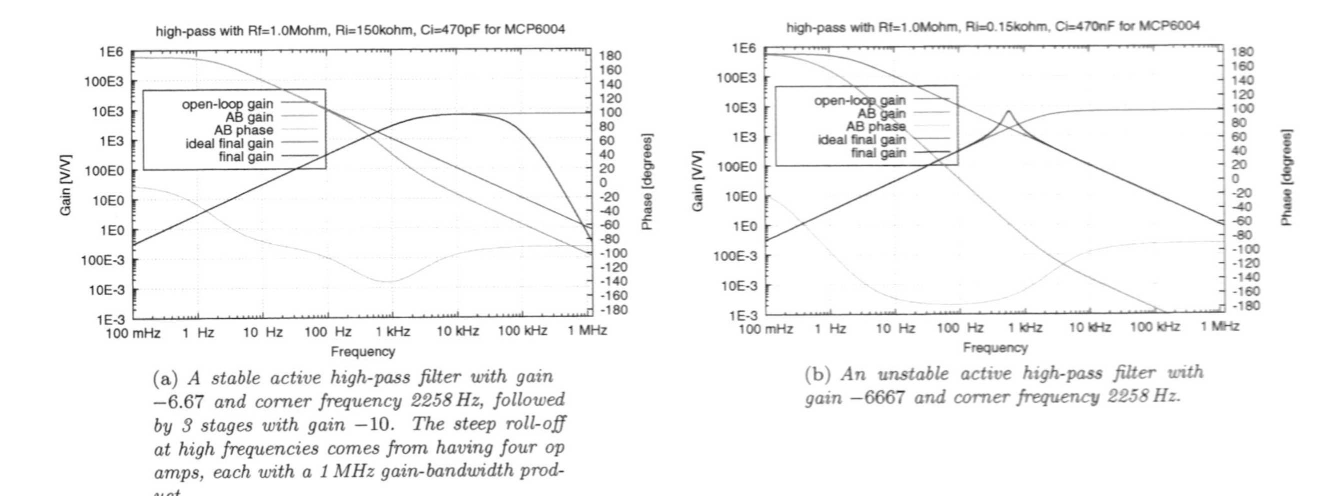

(a) *A stable active high-pass filter with gain −6.67 and corner frequency 2258 Hz, followed by 3 stages with gain −10. The steep roll-off at high frequencies comes from having four op amps, each with a 1 MHz gain-bandwidth product.*

(b) *An unstable active high-pass filter with gain −6667 and corner frequency 2258 Hz.*

Figure 24.3: *This pair of plots shows useful Bode plots for two different active high-pass filters, each with corner frequency 2258 Hz. The low-gain filter in (a) is stable, but the high-gain filter in (b) is unstable. In each plot is an approximation to the open-loop gain A of the MCP6004 op amp, the gain and phase change for AB, and both the ideal high-pass response and the high-pass response expected taking into account the gain-bandwidth limitations of the op amp.*

Figure drawn with gnuplot [33].

$$2\pi f_{\mathrm{GB}} > \frac{R_{\mathrm{i}} + R_{\mathrm{f}}}{R_{\mathrm{i}}} \frac{1}{R_{\mathrm{i}} C_{\mathrm{i}}} ,$$

$$f_{\mathrm{GB}} > \frac{R_{\mathrm{i}} + R_{\mathrm{f}}}{R_{\mathrm{i}}} f_{\mathrm{c}} ,$$

for the corner frequency f_{c}.

The simplest solution for avoiding instability in single-stage high-pass filters is to keep the gain low:

$$\frac{R_{\mathrm{i}} + R_{\mathrm{f}}}{R_{\mathrm{i}}} < \frac{f_{\mathrm{GB}}}{f_{\mathrm{c}}} .$$

Use subsequent stages to increase gain, if necessary.

Worked Example:

Design an active high-pass filter with corner frequency around 2260 *Hz, with a gain of over* 6500 (*or* −6500 *as we don't care about the phase*) *using MCP6004 op amps.*

The $R_{\mathrm{i}}C_{\mathrm{i}}$ time constant needs to be around $70.4\,\mu$s to get the corner frequency, which can be achieved with $R_{\mathrm{i}} = 150\,\Omega$ and $C_{\mathrm{i}} = 470\,\mathrm{nF}$, getting a corner frequency of 2258 Hz.

A first design might try to get the gain in one stage, with $R_{\mathrm{f}} = 1\,\mathrm{M}\Omega$. But $\frac{R_{\mathrm{f}}+R_{\mathrm{i}}}{R_{\mathrm{i}}} f_{\mathrm{c}} = 15.05\,\mathrm{MHz}$, which exceeds the gain-bandwidth product of 1 MHz. The result would be an unstable high-pass filter. We can reduce the gain for the same corner frequency by choosing $R_{\mathrm{i}} = 150\,\mathrm{k}\Omega$ and $C_{\mathrm{i}} = 470\,\mathrm{pF}$, getting a gain of only −6.667, but having a stable design (see Figure 24.3 to compare the two designs). Using three subsequent stages with a gain of −10 each would get us the necessary total gain, without running into problems with instability.

To keep the gain of individual high-pass filter stages low, we may need to use multiple stages. Just as with cascading multiple low-pass filters, we can determine the effect of cascading several high-pass filters that have the same corner frequency. (Also, just like for low-pass filters, professional filter designs usually have different designs for the different stages, to get better filter characteristics, but that sort of filter design is beyond the scope of this course.)

The gain of a single-stage, active, high-pass filter can be written as

$$-G_p \frac{jf/f_c}{1+jf/f_c} ,$$

where G_p is the passband gain, and f_c is the corner frequency of the stage. A chain of n such stages will have its cutoff frequency when the gain is $1/\sqrt{2}$ of the passband gain:

$$1/\sqrt{2} = \left| \frac{jf/f_c}{1+jf/f_c} \right|^n .$$

We can rewrite this as

$$\sqrt{2^{1/n}} = \left| \frac{1+jf/f_c}{jf/f_c} \right| = |1 - jf_c/f| ,$$

and we can apply the definition of the magnitude of complex numbers to get

$$2^{1/n} = 1 + (f_c/f)^2 .$$

Solving for the cutoff frequency f, we get

$$f = f_c/\sqrt{2^{1/n}-1},$$

which is the same scaling factor as for the cutoff frequency of cascaded low-pass filter, but inverted, as the cutoff frequency for cascading high-pass filters increases with more stages.

24.4 Active band-pass filter

An active band-pass filter can be made with an op amp, two resistors, and two capacitors. Like the low-pass filter in Figure 24.1, the feedback element consists of a resistor and capacitor in parallel, but another capacitor is added in series with R_i to block DC.

The feedback impedance and the input impedance are not filters by themselves—it is the combination of the two impedances with the op amp that makes a filter.

The band-pass filter is nearly always used in the inverting configuration (see Figure 24.4), because we don't usually want a gain of 1 at 0 Hz—we want to recenter the output at V_ref.

The gain of the band-pass filter in Figure 24.4 can be determined by combining the generic gain equation for an inverting negative-feedback amplifier, Equation 19.6, with the impedances of series and parallel RC circuits:

$$\begin{aligned} Z_\mathrm{i} &= R_\mathrm{i} + \frac{1}{j\omega C_\mathrm{i}} \\ &= \frac{1+j\omega R_\mathrm{i} C_\mathrm{i}}{j\omega C_\mathrm{i}} , \end{aligned}$$

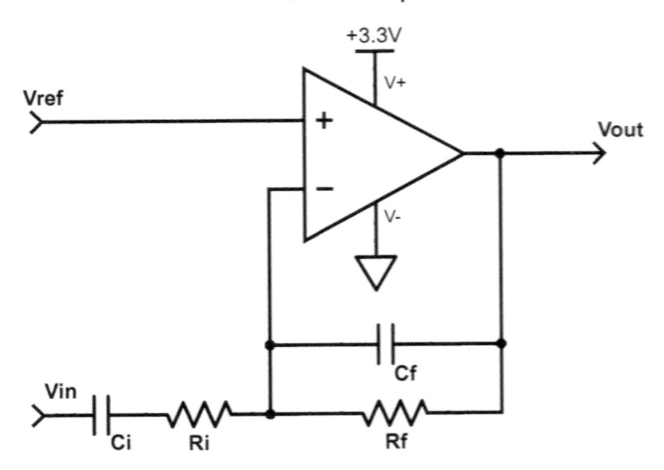

Figure 24.4: *An active band-pass filter is a negative-feedback amplifier with roughly constant gain in a passband between two corner frequencies, but decreasing gain outside the passband, as the frequency increases above the top corner frequency or as the frequency decreases below the bottom corner frequency.*
Shown here is an active band-pass filter configured as an inverting (negative gain) amplifier, so that gain goes to 0 at 0 Hz and ∞ Hz.
Figure drawn with Digi-Key's Scheme-it [18].

$$\begin{aligned} Z_\mathrm{f} &= \frac{1}{1/R_\mathrm{f} + j\omega C_\mathrm{f}} \\ &= \frac{R_\mathrm{f}}{1 + j\omega R_\mathrm{f} C_\mathrm{f}}, \\ \frac{V_\mathrm{out} - V_\mathrm{ref}}{V_\mathrm{in} - V_\mathrm{ref}} &= \frac{-Z_\mathrm{f}}{Z_\mathrm{i}} \\ &= \frac{\frac{-R_\mathrm{f}}{1+j\omega R_\mathrm{f} C_\mathrm{f}}}{\frac{1+j\omega R_\mathrm{i} C_\mathrm{i}}{j\omega C_\mathrm{i}}} \\ &= \frac{-j\omega R_\mathrm{f} C_\mathrm{i}}{(1 + j\omega R_\mathrm{f} C_\mathrm{f})(1 + j\omega R_\mathrm{i} C_\mathrm{i})} . \end{aligned}$$

The gain of the active band-pass filter in Figure 24.4 is

$$\frac{-j\omega R_\mathrm{f} C_\mathrm{i}}{(1 + j\omega R_\mathrm{f} C_\mathrm{f})(1 + j\omega R_\mathrm{i} C_\mathrm{i})} .$$

We want to approximate the band-pass gain with three straight lines on a log-log plot: an upward-sloping line at low frequencies, a horizontal line in the passband, and a downward-sloping line at high frequencies.

At very low frequencies (as $\omega \to 0$), we end up with a gain of approximately $-j\omega R_\mathrm{f} C_\mathrm{i}$, because the denominator approaches 1. At very high frequencies (as $\omega \to \infty$), we end up with a gain of approximately $-1/(j\omega R_\mathrm{i} C_\mathrm{f})$, because the ω^2 term in the denominator dominates everything else in the denominator.

The two asymptotic lines that give us the behavior outside the passband are

$$G \approx -j\omega R_f C_i, \text{ at low frequencies, and}$$

$$G \approx \frac{j}{\omega R_i C_f}, \text{ at high frequencies.}$$

We can approximate what happens in the passband by looking at $\omega = 1/\sqrt{R_i C_i R_f C_f}$, which should be the peak of the passband. At that frequency, we have a gain of

$$\begin{aligned}\frac{V_{out} - V_{ref}}{V_{in} - V_{ref}} &= \frac{-j\sqrt{\frac{R_f C_i}{R_i C_f}}}{\left(1 + j\sqrt{\frac{R_f C_f}{R_i C_i}}\right)\left(1 + j\sqrt{\frac{R_i C_i}{R_f C_f}}\right)} \\ &= \frac{-j\sqrt{\frac{R_f C_i}{R_i C_f}}}{j\left(\sqrt{\frac{R_f C_f}{R_i C_i}} + \sqrt{\frac{R_i C_i}{R_f C_f}}\right)} \\ &= \frac{-R_f C_i}{R_f C_f + R_i C_i}\,.\end{aligned}$$

The corner angular frequencies where the approximating lines meet the peak gain are then

$$\omega_{lo} = \frac{1}{R_f C_f + R_i C_i}\,, \tag{24.1}$$

and

$$\omega_{hi} = \frac{1}{R_f C_f} + \frac{1}{R_i C_i}\,. \tag{24.2}$$

The ratio of the corner frequencies is

$$\frac{\omega_{hi}}{\omega_{lo}} = 2 + \frac{R_f C_f}{R_i C_i} + \frac{R_i C_i}{R_f C_f} \geq 4\,. \tag{24.3}$$

The passband is easily expressed in terms of the lower corner frequency and two of the components:

The gain of an active band-pass filter at the peak of the passband is

$$G_{peak} = \frac{-R_f C_i}{R_f C_f + R_i C_i} = -\omega_{lo}\, R_f C_i\,,$$

and is reached when $\omega = 1/\sqrt{R_i C_i R_f C_f} = \sqrt{\omega_{lo}\omega_{hi}}$. This gain is a purely negative real gain, with no phase shift (or, if you prefer, a positive gain with a phase shift of 180°).

The active band-pass filter has simpler formulas than the passive band-pass filter presented in Section 11.7, which is one of the attractions of the design. The active filter is also capable of gain larger than 1 (unlike the passive design).

The formula for the corner frequencies is symmetric with respect to the feedback and the input time constants, so let's just call the two time constants τ_1 and τ_2.

To design an active filter given the peak gain and the two corner frequencies, we need to determine the two time constants R_iC_i and R_fC_f, choose which time constant to assign to the input and which to the feedback, and then choose resistor and capacitor values that get us the desired gain.

If the two corner frequencies are far apart, $\omega_{lo} \ll \omega_{hi}$, we can approximate $\tau_1 = 1/\omega_{lo}$ and τ_2/ω_{hi}, but in general we need to solve for the two time constants. Our equations for the corner angular frequencies are

$$\omega_{lo} = \frac{1}{\tau_1 + \tau_2} ,$$

$$1/\omega_{lo} = \tau_1 + \tau_2 ,$$

and

$$\omega_{hi} = \frac{1}{\tau_1} + \frac{1}{\tau_2} ,$$

$$\tau_1\tau_2\omega_{hi} = \tau_1 + \tau_2 ,$$

$$\tau_1\tau_2\omega_{hi} = \omega_{lo} ,$$

$$\tau_1\tau_2 = \frac{1}{\omega_{lo}\omega_{hi}} .$$

We can view τ_1 and τ_2 as solutions of the equation $(t - \tau_1)(t - \tau_2) = 0$, or $t^2 - (\tau_1 + \tau_2)t + \tau_1\tau_2 = 0$. In other words, we can find τ_1 and τ_2 by applying the quadratic formula to

$$t^2 - \frac{1}{\omega_{lo}}t + \frac{1}{\omega_{lo}\omega_{hi}} = 0 ,$$

or, avoiding the fractions,

$$\omega_{lo}\omega_{hi}t^2 - \omega_{hi}t + 1 = 0 .$$

Once we have the two time constants, we need to assign them to the input and feedback branches. We can use the peak gain to get another time constant $R_fC_i = |G_{peak}|/\omega_{lo}$. To get a large gain, we need a large value for R_fC_i, but those component values each contribute to a different one of the time constants τ_1 and τ_2. For the low frequencies we deal with in this course, we are often constrained by how large a capacitor we can use for C_i, as large capacitors are either expensive or very imprecise.

At this point, we can make an arbitrary choice of C_i (perhaps as large as we have easily available), get R_f from the desired gain, and use the Equations 24.1–24.3 to get R_i and C_f for each of the two assignments of the time constants to R_iC_i and R_fC_f. If the values for any of the resistors or capacitors fall outside our desired region, we can scale everything by changing C_i.

Putting the larger time constant on the input branch, $R_iC_i \geq R_fC_f$, makes the gain depend more on R_f/R_i than on C_i/C_f, as ω_{lo} depends more on the larger time constant than on the smaller one. Because resistors are much more precise than capacitors (particularly for the cheap ones we use in this class), this assignment gives us better control of the gain. Also, putting the larger time constant on the input usually makes the input impedance larger, which we usually want for a building block that we will be connecting to other circuits. For a given gain and pair of time constants, putting the larger time constant on the input generally results in resistor values that are closer together and capacitor values that are further apart.

If there is a lot of leeway on the gain or the corner frequencies, we may be able to choose the capacitances C_i and C_f to match, as it is easier to match capacitors than to get a well-defined ratio between capacitances of different values. The peak gain if $C_i = C_f$ is $-R_f/(R_i + R_f)$, which never has

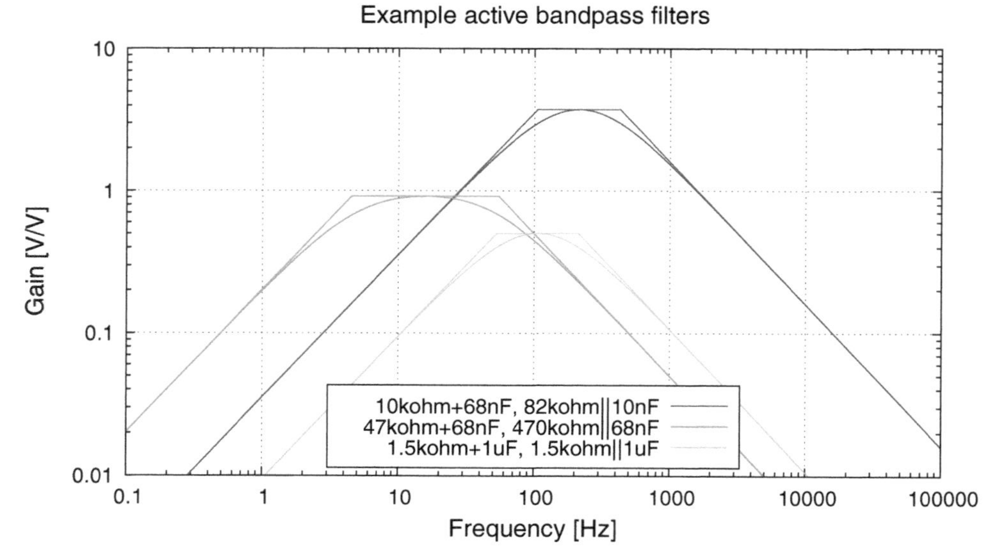

Figure 24.5: *Examples of active band-pass filter designs, showing both the magnitude of the gain and the straight-line approximations derived in Section 24.4. The component values are the same as for the examples for passive band-pass filters in Figure 11.9, but the gains are larger. The first example shows that the gain can be made larger than 1, if $R_f > R_i$ and $C_i > C_f$. The second example has a design with $C_i = C_f$ and $R_i \ll R_f$, getting a gain close to 1 (0.909, to be more precise). The third example shows $R_f = R_i$ and $C_f = C_i$ having a peak gain of 1/2.*
Figure drawn with gnuplot [33].

magnitude bigger than 1, so one of the advantages of active band-pass filters is lost. If we also have $R_f \gg R_i$, then the gain is approximately -1, the lower corner frequency is $\omega_{lo} \approx \frac{1}{R_f C}$, and the upper corner frequency is $\omega_{hi} \approx \frac{1}{R_i C}$. A design of this type is shown as the second example in Figure 24.5.

If $C_i = C_f$ and $R_i = R_f$, then the peak gain is 1/2 and the corner frequencies are $\omega_{lo} = \frac{1}{2RC}$ and $\omega_{hi} = \frac{2}{RC}$. This design is shown as the third example in Figure 24.5. The ratio between the corner frequencies of a factor of 4 is as narrow a passband as the filter design in Figure 24.4 gets—for narrower passbands, different filter designs are needed (see Section 24.7, for one example).

Worked Example:
Design an active band-pass filter with a peak gain of about 5 *in the passband* 100 *Hz*–1 *kHz.*

The ratio of the cutoff frequencies is 10, which is larger than the minimum ratio of 4 for the band-pass filter in Figure 24.4, so this design can be used. The peak gain for this filter design is negative, not positive, but we'll assume that the phase is not important, so a gain of -5 is as good as a gain of $+5$.

We can determine the RC time constants by solving the system

$$\tau_1 + \tau_2 = \frac{1}{2\pi\, 100\, Hz} ,$$
$$1/\tau_1 + 1/\tau_2 = 2\pi\, 1\, kHz .$$

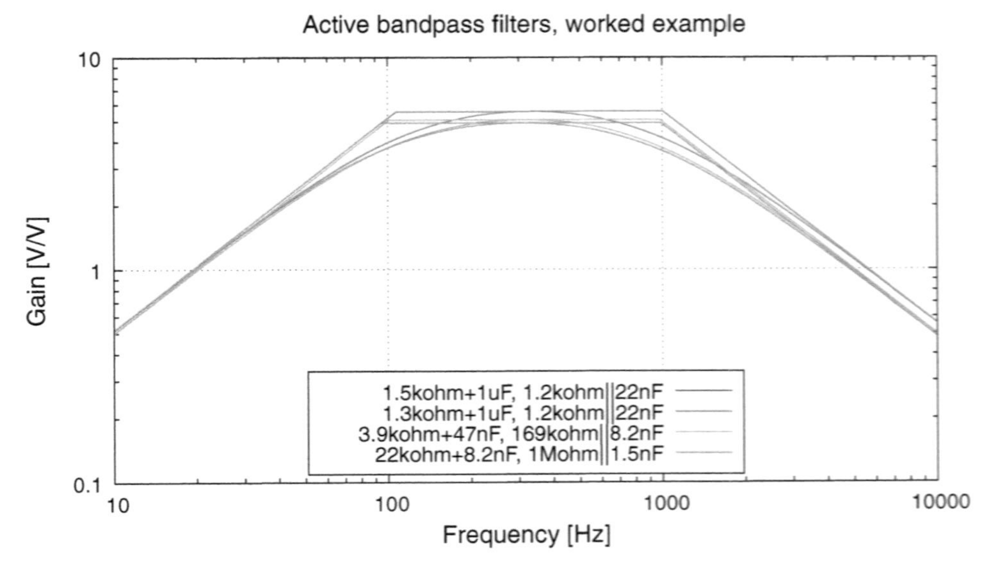

Figure 24.6: *Examples of active band-pass filter designs, showing both the magnitude of the gain and the straight-line approximations derived in Section 24.4. These examples are different solutions to the worked example. Very different component values result in almost the same response from the filters.*
Figure drawn with gnuplot [33].

Using the quadratic formula (or WolframAlpha) to solve the system, the τ values are $\frac{5\pm\sqrt{15}}{2000\pi}$ s, or 1.41218 ms and 179.370 μs, but the equations for the corner frequencies do not give us any guidance on which of these time constants is $R_f C_f$ and which is $R_i C_i$.

From the peak gain of -5, we also have that $R_f C_i = \frac{5}{2\pi 100\,Hz} = 7.95775$ ms.

We now have three time constants, but need to determine four parameter values (two resistors and two capacitors). Designing these active filters always ends up under-constrained like this—we have to make some arbitrary choices to finish the design.

The largest time constant is $R_f C_i$, so let's pick C_i to be fairly large—say 1 μF, which sets $R_f =$ 7.958 kΩ. If we choose $R_i C_i = 1.41218$ ms (the bigger of τ_1 and τ_2, as suggested earlier), we get $R_i = 1.412$ kΩ, getting $C_f = 179.370\,\mu\text{s}/7.958\,\text{k}\Omega = 22.54$ nF. If we had chosen $R_i C_i = 179.370\,\mu$s, we would have gotten $R_i = 179.4\,\Omega$ and $C_f = 177.5$ nF—the capacitance value is reasonable, but the value for R_i is a little low, as it determines the input impedance of the filter, which we would like to keep fairly high.

If we round the larger resistor values to standard E24 values, we get

$$R_i = 1.5\,\text{k}\Omega\ ,$$

$$R_f = 8.2\,\text{k}\Omega\ ,$$

$$C_i = 1\,\mu\text{F}\ ,$$

$$C_f = 22\,\text{nF}\ ,$$

giving us time constants of $R_i C_i = 1.5$ ms and $R_f C_f = 180.4\,\mu$s, corner frequencies of 94.71 Hz and 988.3 Hz, and a peak gain of -4.88. The Bode plot for this filter is shown in Figure 24.6.

If this gain is too small, we could try rounding R_i down instead of up, getting

$$R_i = 1.3\,\mathrm{k}\Omega ,$$
$$R_f = 8.2\,\mathrm{k}\Omega ,$$
$$C_i = 1\,\mu\mathrm{F} ,$$
$$C_f = 22\,\mathrm{nF} ,$$

giving us time constants of $R_iC_i = 1.3\,\mathrm{ms}$ and $R_fC_f = 180.4\,\mu\mathrm{s}$, corner frequencies of 107.5 Hz and 1004.7 Hz, and a peak gain of -5.539.

If E48 resistors are acceptable, then $R_i = 1.4\,\mathrm{k}\Omega$ is available, and we can have time constants of $R_iC_i = 1.4\,\mathrm{ms}$ and $R_fC_f = 180.4\,\mu\mathrm{s}$, corner frequencies of 100.7 Hz and 988.3 Hz, and a peak gain of -5.189.

Although $1\,\mu\mathrm{F}$ ceramic capacitors are cheap, the cheap capacitors do not have very constant capacitance values, varying with both temperature and voltage. A better grade of ceramic capacitors are the C0G (or NP0) capacitors, for which sizes larger than 100 nF are expensive. What happens if we choose $C_i = 47\,\mathrm{nF}$? We then get $R_f = 169\,\mathrm{k}\Omega$, which is a good E48 value. If we pick $R_iC_i = 179.370\,\mu\mathrm{s}$, we get $R_i = 3.816\,\mathrm{k}\Omega$, which rounds to $3.9\,\mathrm{k}\Omega$ and $C_f = 1.41218\,\mathrm{ms}/169\,k\Omega = 8.36\,\mathrm{nF}$, which rounds to 8.2 nF.

With these values we get

$$R_i = 3.9\,k\Omega ,$$
$$R_f = 169\,k\Omega ,$$
$$C_i = 47\,\mathrm{nF} ,$$
$$C_f = 8.2\,\mathrm{nF} ,$$

giving us time constants of $R_iC_i = 183.3\,\mu\mathrm{s}$ and $R_fC_f = 1.3858\,\mathrm{ms}$, corner frequencies of 101.4 Hz and 983.1 Hz, and a peak gain of -5.153.

This design has reasonable resistor and capacitor values (even for C0G capacitors) and comes close to all the design specifications.

If we wanted to maximize the input impedance, we could scale R_i and R_f up, and C_i and C_f down. If we set $C_i = 8.2\,\mathrm{nF}$, then we get $R_f = 970\,\mathrm{k}\Omega$, and we can set the values (rounded) to be

$$R_i = 22\,k\Omega ,$$
$$R_f = 1\,M\Omega ,$$
$$C_i = 8.2\,\mathrm{nF} ,$$
$$C_f = 1.5\,\mathrm{nF} ,$$

giving us time constants of $R_iC_i = 180.4\,\mu\mathrm{s}$ and $R_fC_f = 1.5\,\mathrm{ms}$, corner frequencies of 94.7 Hz and 988.3 Hz, and a peak gain of -4.880.

One important take-away from this example is that there is never a single correct solution to design problems like this—instead, there are a wide range of different designs which have different trade-offs in price and performance.

Of the various designs here, I think that I like the last one best, even though it is not very precise on the corner frequencies and gain, because it can be implemented with high-quality but low-cost capacitors and has a large input impedance, which is less likely to cause problems with previous stages of the system.

Exercise 24.3

Compute the peak gain and corner frequencies for an inverting active band-pass filter which has a 1.5 kΩ resistor and a 220 nF in series for the input subcircuit, and 15 kΩ and 4.7 nF in parallel for the feedback subcircuit. Express the corner frequencies in Hz (f), not radians/s ($\omega = 2\pi f$). Use gnuplot to create a Bode plot (log-log plot of gain as a function of frequency) for 1 Hz to 1 MHz. Use the functions in definitions.gnuplot (Figure 7.4) to make a readable function (that is, one that expresses the serial and parallel connections) for the gain of an active band-pass filter, and plot the absolute value of that gain function, passing in the specific values for the components. (Sanity check: is the peak of the plot at the frequency and gain you expect?)
Bonus: write an approximation function using the if-then-else expression "i?t:e" to plot the three straight lines, changing which line is plotted at each corner: `f<Lc? Lline : f<Hc? Const: Hline`, where each of the names other than f is appropriately replaced with the proper expression for the corner frequency or the straight-line approximation.

Exercise 24.4

Design an inverting active band-pass filter with a peak gain between −15 and −20 and corner frequencies around 0.3 Hz and 3 Hz, using standard components (E6 series for capacitors, E12 series for resistors). Plot the absolute value of the gain as a log-log plot from 0.01 Hz to 100 Hz.

24.5 Voltage offset for high-pass and band-pass filters

We often use high-pass and band-pass filters to remove DC offset from an input signal. But what about the input offset of the amplifier used to create the active filter? What happens to it?

The analysis is surprisingly simple. At DC, we can treat the capacitors as open circuits, so there is nothing connected to the negative input of the op amp except the feedback resistor, which means that the op amp is configured as a unity-gain buffer. The output DC voltage is just the voltage at the positive input plus the input offset of the op amp.

The output voltage offset of an active high-pass or band-pass filter is just the input offset voltage of the op amp used to build the filter.

Low-pass filters are different, of course, because they have gain at DC, and the input offset voltage is amplified by that gain.

24.6 Considering gain-bandwidth product

This section is optional material—careful analysis of the effect of gain-bandwidth product when designing filters is probably not needed in this course. The simple low-pass analysis of Section 19.5 *should suffice.*

In Section 19.5, we looked at the effect of the gain-bandwidth product of op amps on simple negative-feedback amplifiers. Our main conclusion was that the amplifiers act as low-pass filters, rather than amplifying all frequencies equally.

If we use an op amp to make an active filter, the results are more complicated—Figure 24.7 shows the response of a high-pass filter made using an MCP6004 op amp.

The analysis we did before to get the gain of an inverting amplifier (Equation 19.8) is still valid—the gain is

$$G = \frac{A(B-1)}{1+AB} ,$$

and B is still $Z_i/(Z_i + Z_f)$, but for active filters B is no longer a real number but a complex number that depends on frequency.

Rather than working out the details analytically, it is simplest to use a computer to calculate the effect of gain-bandwidth product, using $A = \frac{1}{1/A_0 + jf/f_{GB}}$ as the model for open-loop gain. For a high-pass filter, A_0 is not very critical (I used the minimum value for DC gain given on the MCP6004 data sheet), but f_{GB} is important. The data sheet only gives typical values, so I fit the model to the data by adjusting f_{GB}—the resulting value is quite close to the fit found in Figure 19.11.

I also had to fit values for C_i, since the nominal values for cheap ceramic capacitors are not very accurate. The fitting strategy I used first fit C_i at low frequencies (below 5 kHz), where the gain-bandwidth

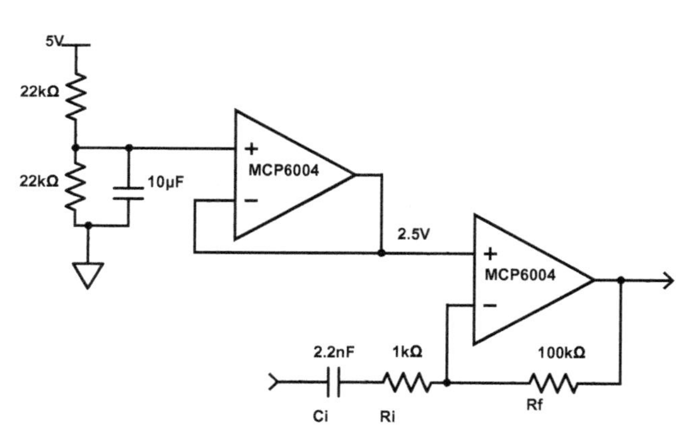

(a) *The schematic for the active high-pass filter tested in Figure 24.7b.*

Figure drawn with Digi-Key's Scheme-it [18].

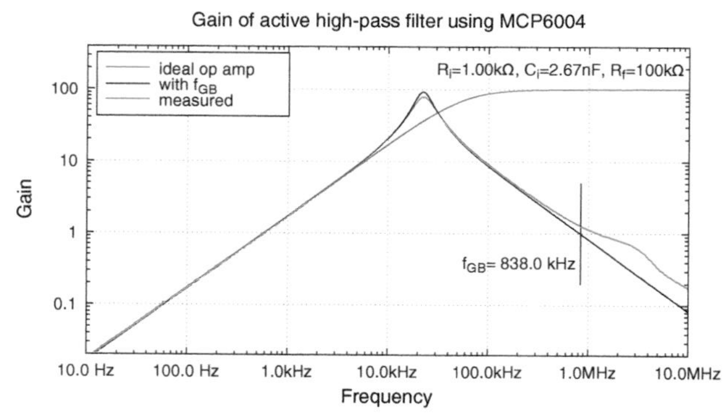

(b) *The model for the active high-pass filter using an ideal op amp is good at low frequencies, but fails at 5 kHz and above. Measurements were made with a ±20 mV input and a 5 V power supply using the network analyzer of the Analog Discovery 2.*

Figure drawn with gnuplot [33].

Figure 24.7: *A high-gain active high-pass filter may not be well-modeled by using an ideal op amp, if insufficient gain is available at the corner frequency of the filter.*

Modeling the gain as $\frac{A(B-1)}{1+AB}$, *with* $A = \frac{1}{1/A_0 + jf/f_{GB}}$ *captures the measured behavior much better.*

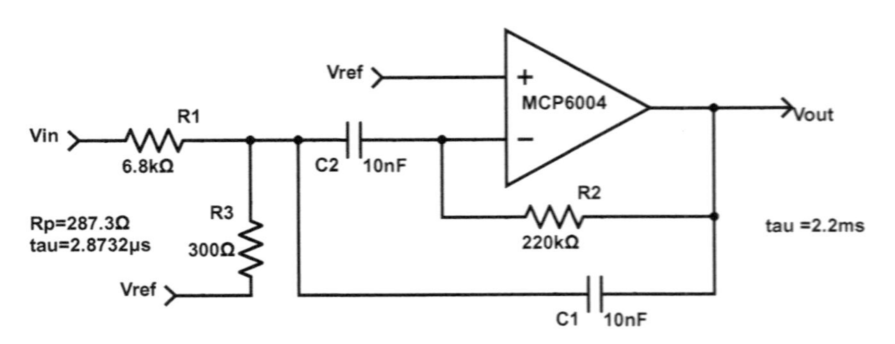

Figure 24.8: *Multiple-feedback active band-pass filter, centered at 2 kHz with a bandwidth of 145 Hz. The choice of component values is explained in the main body text.*
Figure drawn with Digi-Key's Scheme-it [18].

limitations are not important, then fit f_{GB} for frequencies up to 500 kHz. At higher frequencies, the simple model for the open-loop gain A is not sufficient to capture all the open-loop behavior of the op amp.

Interestingly, the limited bandwidth of the op amp does not just provide a low-pass filter cascaded with the desired high-pass filter. Where the two filter lines cross, we get *more* gain than we expect from a high-pass active filter using an ideal op amp.

The model using $G = \frac{A(B-1)}{1+AB}$ and the simple low-pass filter for A captures most of the measured frequency response. The main shortcomings are that frequencies above f_{GB} are not properly modeled and that the gain at the peak is overestimated.

24.7 Multiple-feedback band-pass filter

This section is optional material—it is not necessary to use multiple-feedback active filters for any of the labs in this course.

The simple filters of Sections 24.2–24.4 are not the limit of what one can do with active filters. More sophisticated filters, such as the Sallen-Key topology [152] or the multiple-feedback filter shown in Figure 24.8, are probably more common designs.

To improve my understanding of active filters, I tried deriving the formulas for a band-pass multiple-feedback filter and then designing one. I started with the schematic Figure 24.8, but without any component values.

Worked Example:
Design a multiple-feedback band-pass filter that is centered at 2 kHz, but which firmly rejects 1.5 kHz and 2.5 kHz.

We can derive the formula for the gain of the active filter by working from

- the definitions of the components (the relationship between voltage and current),
- the basic concept of ideal op amps (that the negative feedback holds the negative input to the same voltage as the positive input), and
- Kirchhoff's current law (the sum of currents into a node is 0 A).

To simplify writing the formulas, I will use multiplication by s to indicate differentiation with respect to time. (This is a common shorthand, based on the Laplace transform—we don't need the full power of Laplace transforms here, but the notation is handy.)

Let's look at the current into the node where R_1, R_3, C_1, and C_2 meet, calling the voltage at the node V_m:

$$\begin{aligned} I_1 &= (V_{in} - V_m)/R_1 , \\ I_2 &= s(V_{ref} - V_m)C_2 , \\ I_3 &= (V_{ref} - V_m)/R_3 , \\ I_4 &= s(V_o - V_m)C_1 , \\ I_1 + I_2 + I_3 + I_4 &= 0 . \end{aligned}$$

We also have $I_2 = (V_o - V_{ref})/R_2$, because an ideal op amp takes no current through its inputs, and so all current through C_2 must also go through R_2.

We can simplify a bit, by using the two formulas for I_2 to replace $V_m - V_{ref}$ with $-(V_o - V_{ref})/(\tau_2 s)$, where $\tau_2 = R_2 C_2$. This substitution corresponds to replacing the amplifier, R_2, and C_2 with an inverting high-pass filter.

It is conventional to recenter the voltages so that V_{ref} is zero, which simplifies the algebra. Doing that gives us

$$(V_{in} + V_o/(\tau_2 s))/R_1 + V_o/R_2 + V_o/(\tau_2 s R_3) + V_o(1 + 1/(\tau_2 s))C_1 s = 0 ,$$

and multiplying both sides by $\tau_2 s$ and rearranging gives us

$$\begin{aligned} -V_{in}\tau_2 s/R_1 &= V_o\left(1/R_1 + \tau_2 s/R_2 + 1/R_3 + (\tau_2 s + 1)(C_1 s)\right) , \\ \frac{V_o}{V_{in}} &= -\tau_2 s/\left(1 + \tau_2 s R_1/R_2 + R_1/R_3 + (\tau_2 s + 1)(R_1 C_1 s)\right) . \end{aligned}$$

More correctly, that should be $\frac{V_o - V_{ref}}{V_{in} - V_{ref}} = \ldots$, but I'll just talk about the gain G.

Defining $\tau_1 = R_1 C_1$ we get

$$\begin{aligned} G &= -\tau_2 s/\left(1 + R_1/R_3 + (\tau_2 R_1/R_2 + \tau_1)s + \tau_1\tau_2 s^2\right)) \\ &= -(\tau_2 s/R_1)/\left(1/R_1 + 1/R_3 + (\tau_2/R_2 + \tau_1/R_1)s + \tau_1\tau_2 s^2/R_1\right)) . \end{aligned}$$

If we have $C_1 = C_2 = C$, which is a common design constraint, we can simplify to

$$\begin{aligned} G &= -(R_2 C s/R_1)/\left(1/R_1 + 1/R_3 + 2Cs + C\tau_2 s^2\right)) \\ &= -(R_2/R_1)/\left((1/R_1 + 1/R_3)(Cs)^{-1} + 2 + \tau_2 s\right)) . \end{aligned}$$

We can define $R_p = R_1 \parallel R_3 = R_1 R_3/(R_1 + R_3)$ and $\tau_p = R_p C_1$. Then we have

$$G = -(R_2/R_1)/\left((\tau_p s)^{-1} + 2 + \tau_2 s\right)) .$$

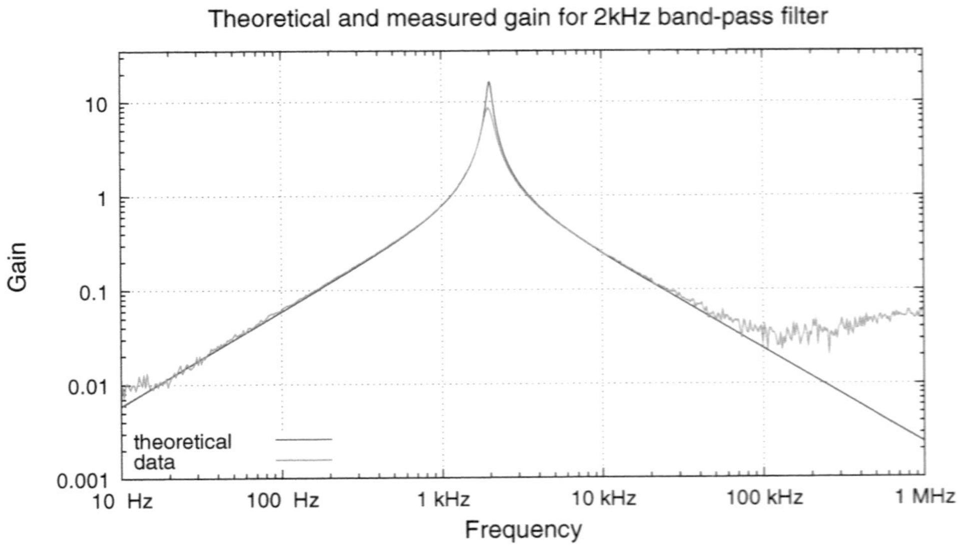

Figure 24.9: *The measured frequency response of the multiple-feedback active filter of Figure 24.8 is fairly close to the theoretical behavior, but the frequency is a bit off (probably because of the low-precision ceramic capacitors I used). The measured peak is not as sharp as the theoretical one. The high-frequency deviation from theoretical behavior is probably due to the gain-bandwidth limitations of the op-amp chip.*
Data collected with Analog Discovery 2 [19]. Figure drawn with gnuplot [33].

One of the magical things about the Laplace transform is that we can get the frequency response just by replacing s with $j\omega$. Intuitively, if s is differentiation, then

$$se^{j\omega t} = \frac{\mathrm{d}}{\mathrm{d}t} e^{j\omega t} = j\omega e^{j\omega t} \; .$$

This allows us to write the gain as

$$G = -(R_2/R_1) / \left(2 + j(\tau_2\omega - (\tau_p\omega)^{-1})\right) \; ,$$

whose magnitude is maximized when the imaginary part of the denominator is zero: $\tau_2\omega - (\tau_p\omega)^{-1} = 0$ or $\omega = 1/\sqrt{\tau_2\tau_p}$, and at that maximum, the gain is $-R_2/(2R_1)$.

The cutoff frequencies, where the gain drops by the square root of 2, occur when the real and imaginary parts of the denominator have the same magnitude: $\tau_2\omega - (\tau_p\omega)^{-1} = \pm 2$. If we take only the positive values of ω, we get $\omega = \left(\pm 1 + \sqrt{(\tau_p + \tau_2)/\tau_p}\right)/\tau_2$, for a bandwidth of $2/\tau_2$ or $1/(\pi R_2 C_2)$.

For $C_1 = C_2 = 10\,\mathrm{nF}$ and $R_2 = 220\,\mathrm{k}\Omega$, we get a bandwidth of 144.7 Hz. With $R_1 = 6.8\,\mathrm{k}\Omega$ and $R_3 = 300\,\Omega$, we have $R_\mathrm{p} = 287.32\,\Omega$, with time constant $2.8732\,\mu\mathrm{s}$, giving a center frequency of 2001.8 kHz. The gain at the center frequency should be $-R_2/(2R_1) = -16.16$. I built this filter and tested it with the Analog Discovery 2, as shown in Figure 24.9.

Chapter 25

Lab 6: Optical Pulse Monitor

Bench equipment: 1/8″ (or 3 mm) drill for LEGO® bricks, jig or vise for holding bricks while drilling, function generator, oscilloscope (optional)

Student parts: 2–3 two-stud black LEGO® bricks, 3 2 × 4 black LEGO® bricks, phototransistor, rubber bands, black electrical tape, MCP6004 op amp, resistors, diode, capacitors, breadboard, PteroDAQ, LED (optional), heat-shrink tubing (optional)

25.1 Design goal

For this lab we'll be designing and building an optical pulse monitor to detect pulse by shining light through a finger and seeing the change in the opacity of the finger as the amount of blood in the blood vessels changes.

The first half of the lab consists of determining characteristics for a phototransistor and building a transimpedance amplifier to convert its current output to a voltage signal for the levels of light that pass through fingers.

The second half of the lab adds a second-stage amplifier and filter to make the signal have an appropriate voltage range to observe the pulse using the PteroDAQ system.

25.2 Design choices

In all the designs we'll be looking at, all the light arriving at the phototransistor has to pass through a finger (or other part of the body with blood pulses).

The first choice to make is to decide where the light to be detected by the pulse monitor is going to come from. There are several possibilities:

trans-illumination By mounting an LED on the opposite side of the finger from the phototransistor, you can shine light through the finger, resulting in a moderately uniform path length for the light. This approach is commonly used in fingertip pulse oximeters and used to be common in ear-clip optical pulse monitors.

ambient light You can use room light or sunlight to illuminate the finger. This has the advantage of not requiring an LED and requiring less mechanical design. It has the disadvantage of losing control over the wavelength of illumination and the intensity of the light. Furthermore, indoor illumination is often modulated (at 60 Hz, 120 Hz, or several kHz for fluorescent or LED bulbs) resulting in large AC signals in the light intensity that we might have to filter out.

cis-illumination Putting an LED next to the phototransistor is mechanically almost as simple as using ambient light, but relies on back-scattering of light from the finger, rather than on light passing all the way through the finger. The LED and the sensor can be placed on almost any exposed patch of skin where there are blood vessels near the surface. Thick layers of fat can cause problems, as fatty tissue has rather little blood flow, and the light can be scattered back to the sensor with little absorption from hemoglobin in the blood.

Ambient light is added to the LED light, so unless the LED is fairly bright or the finger is covered with opaque material, there can still be unwanted modulation of the light from ambient light sources. Making the LED very bright can cause problems with excessive heating of the skin, as all the power delivered to the LED gets dissipated as heat.

The cis-illumination design is probably the most popular now for optical pulse monitors, as the LED and phototransistor can be surface mounted on the same printed-circuit board, resulting in very low manufacturing cost, and bright LEDs can provide large signals for the phototransistor, reducing interference problems from picking up small stray 60 Hz currents.

pulsed light Both the trans-illumination and the cis-illumination designs do not really require continuous light. You can turn on the LED shortly before making the measurement, then turn it off again. Because the LED only needs to be on for 10 μs–100 μs (depending on the response speed of the phototransistor and amplifier) and the measurements only need to be made 30–60 times a second, power consumption can be greatly reduced by pulsing the LED.

For example, my Verily Life Sciences study watch has two green LEDs that it illuminates for 600 μs every 33.33 ms (a 30 Hz sampling rate). This short illumination reduces the average power needed by about a factor of 56, allowing fairly bright illumination without excess battery drain.

Furthermore, ambient light interference can be greatly reduced by measuring the light just before turning the LED off and just before turning it on, taking the difference between the readings as the measurement. This cancels out any ambient light modulation whose frequency is very small compared to the time of the LED pulse. Synchronizing the measurement with the light pulses is easy in a microcontroller program, but is a little trickier if one is using PteroDAQ or an oscilloscope.

Pulsed illumination provides a large fluctuation in light intensity that is not due to the blood pulse, so limits the amplification possible in the analog circuitry.

Look up the data sheets for the phototransistor (WP3DP3BT or LTR4206, for example) and LED (MT1403-RG-A, WP710A10F3C, WP710A10ID, or WP3A8HD, for example) in your parts kit.

Pre-lab 6.1

What current-limiting resistor would you need to place in series with your LED to limit the current to about 80% of the LED's maximum continuous current rating with a 3.3 V power supply? (Be sure to report which LED you are designing around!) Round the resistance to the E24 series and report both the forward voltage across the LED and what current you expect with the selected resistor.

25.3 Procedures

25.3.1 *Try it and see: LEDs*

Hook up your LED and series resistor, then power them from a 3.3 V supply. Does the LED light up? (If not, debug!) If you have an infrared LED, you may not be able to see it lighting up, but looking at it with a digital camera (including a cell phone camera) will probably show it quite visibly, as the blue sensors in the cameras are generally sensitive to near IR light. (Warning: the iPhone cameras supposedly have IR-blocking filters, so may not see the IR LEDs.)

Measure the voltage and current—remember that we always measure current by measuring the voltage across a known resistor, and you can use the current-limiting resistor as your known resistor. Does the LED have the forward voltage and current you computed it should have?

Bonus: you can use the Analog Discovery 2 to measure the I-vs.-V characteristics of the LED. Use a current-sense/current-limiting resistor that is large enough that the current through the LED will not exceed the absolute maximum, even with a 5 V supply. Hook up the function generator as the power supply for the LED and series resistor, and use the two differential oscilloscope channels to measure the voltages across the LED and across the resistor. Set the function generator to have a slow triangle wave (say 1 Hz) with an amplitude of 5 V and record the voltage waveforms. You can plot the current (the voltage across the resistor divided by the resistance) vs. the voltage to get a typical I-vs.-V curve for the LED.

25.3.2 *Set up log amplifier*

Wire your phototransistor and logarithmic current-to-voltage converter (see Figure 23.5) on a breadboard and measure the average DC voltage for your complete sensor in the ambient light of the lab. Shadow the phototransistor by pinching it between your fingers and measure the average DC voltage there. Try illuminating the phototransistor with the LED.

These measurements should give you an idea of the range of DC voltages (and, hence, photocurrents) to expect with the light levels you will be using.

Bonus: you can do the same sort of I-vs.-V plot for the phototransistor as you did for the LED, though you may have to use the offset on the waveform generator to avoid too large a negative voltage. Because the phototransistor current depends on the amount of light (that's the whole point of it!), doing an I-vs.-V plot only makes sense if you can be sure to keep the light level consistent throughout the measurement.

25.3.3 *Extending leads*

You will be putting the phototransistor (and LED, if you use an illuminated design) into a shroud that blocks all light except that which passes through the finger.

Because the shroud would interfere with plugging the LED and photodetector into a breadboard, you will need to solder wires onto the leads of the components to lengthen the leads. The leads need to be long enough to reach the breadboard from a shroud a short distance away (so that you can rest your hand comfortably on the bench top with your finger over the phototransistor, without touching the breadboard). I recommend leads about 10 cm–20 cm long (4″–8″).

If you have flexible (multi-stranded) wire, that can make positioning easier, but you will need to stiffen the free end by tinning it with solder, if you plan to put it into the breadboard. (A more durable solution is to use a crimp-on female header, but that requires a crimping tool, which is probably not worth the expense for just this course.)

Color coding the leads is very useful, so that you can connect the leads correctly, even when the LED and phototransistor are hidden inside the block.

I recommend red and black for the anode and cathode of the LED, respectively, so that you can remember which one gets hooked up to the positive and which to the negative end. I recommend using different colors (say, green for collector and yellow for emitter) on the phototransistor, so that each wire coming from the block will have a unique easily deciphered color.

Document the colors on your schematic!

Figure 25.1 shows how to extend the leads. Be sure to insulate the wires with electrical tape or heat-shrink tubing all the way to the LED or phototransistor, so that the wires don't accidentally short underneath the block—this shorting is hard to see and has been a very frequent failure mode in work by students trying to "save time" by doing a hasty, sloppy job. The time they lost in debugging was orders of magnitude larger than the time it would have taken to do a neat, thorough job of insulating the wires. The wires from the phototransistor should probably be twisted around each other, to minimize stray currents picked up by the wiring.

25.3.4 *Assembling the finger sensor*

In early versions of this lab, I used wooden blocks for holding phototransistors, but I now prefer to use 1×2 black LEGO® bricks, with $1/8''$ or 3 mm holes drilled to accept 3 mm LEDs or phototransistors (one could also drill $13/64''$ holes to accept 5 mm optoelectronics parts), see Figure 25.2. Black LEGO® bricks are more opaque than wood, and it is easier to prototype finger clips with them than with the wooden blocks.

A simple block is not an ideal way to make a pulse sensor, as you need to provide a moderate pressure (between systolic and diastolic blood pressure) to get a really good pulse measurement. The correct pressure provides a slight throbbing in the finger—squeezing too hard cuts off blood flow to the finger tip and too little squeeze results in rather small fluctuations in opacity. If the pressure is too low, movement artifacts easily swamp out the small fluctuations due to pulse.

A gently squeezing clip is better than having to apply the pressure by pressing down with your finger, as the clip can provide a uniform pressure independent of slight movements. Crude clips can be cheaply prototyped with rubber bands and the LEGO® blocks.

Not everyone finds the rubber bands useful, though—a lot of students find it easier to train themselves to provide the correct pressure over a stationary block than to get the right combination of rubber bands for their finger width.

I prefer to have students drill the LEGO® bricks themselves, but the bricks can be predrilled if required. A jig to hold the bricks in the correct place in a drill press can save a lot of time on aligning the bricks for drilling, and is a fairly simple thing to make, as shown in Figure 25.3.

For any of the block designs here, the leads of the LED or phototransistor need to be bent at right angles close to the body of the device. It may be easiest to make those bends before inserting the device into the holder.

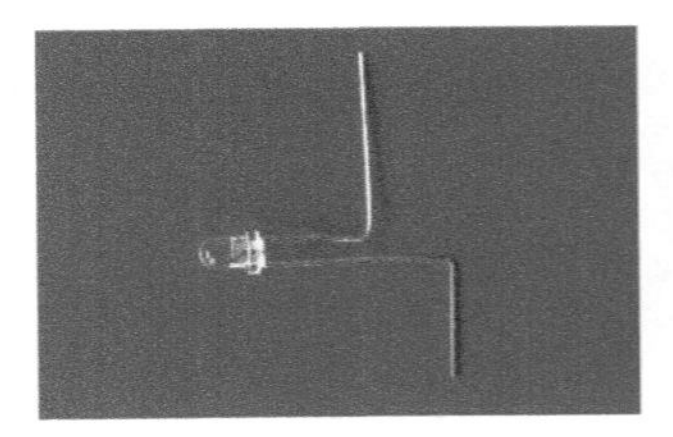

(a) *Bend the leads of the phototransistor in different places, keeping the shorter lead shorter after bending to make it easier to keep track of which lead is which.*

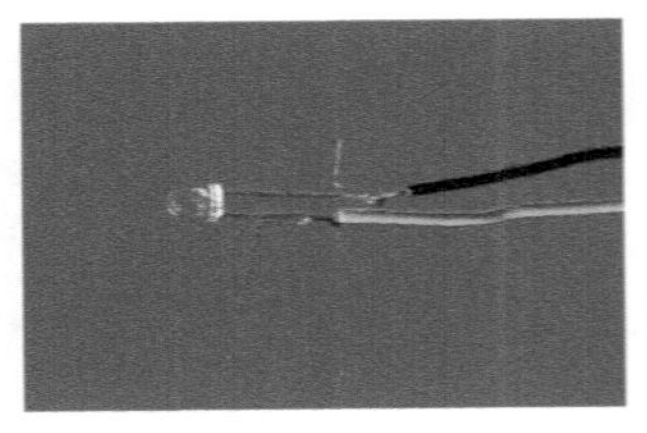

(b) *Mechanically join the wires and the leads by bending the ends of the wires into little hooks with the long-nose pliers, linking them together, and crimping the hooks closed.*

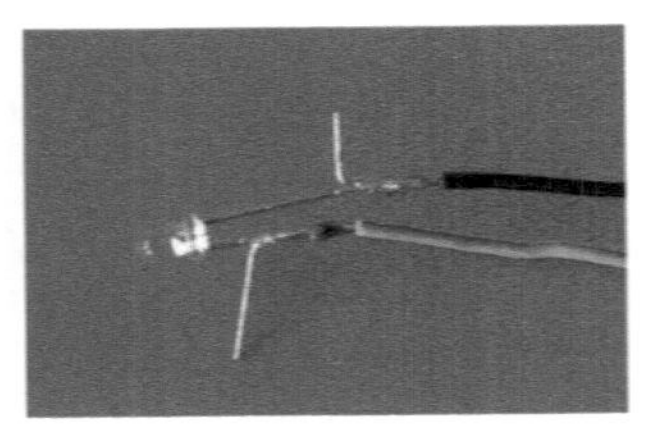

(c) *Solder the joints. This can be done with the wires and phototransistor sitting on the bench top, but it helps to put the lead and the wire both into jaws of a helping hand so that they are held motionless.*

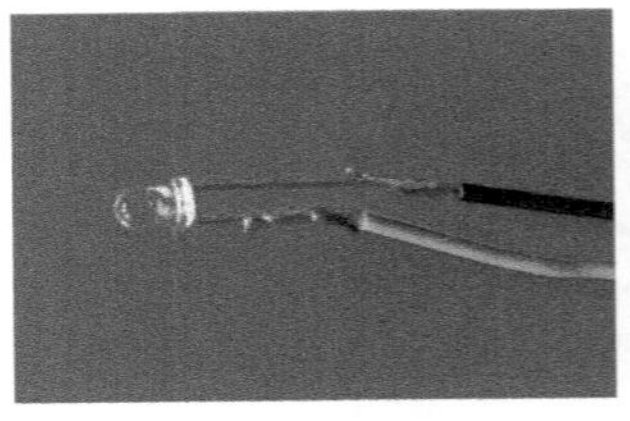

(d) *Clip off the excess lead, to get a smooth connection.*

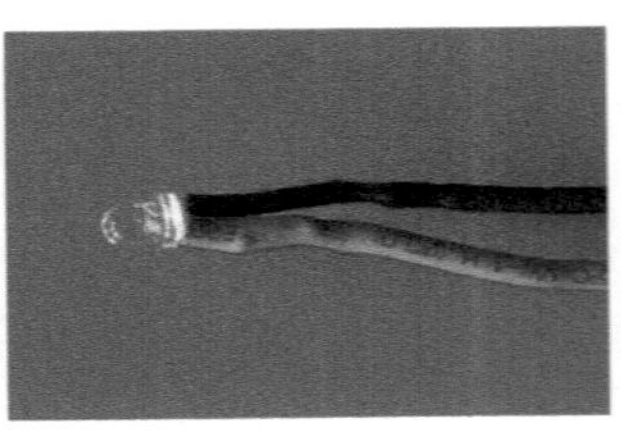

(e) *Use heat-shrink tubing or a little electrical tape around the leads and solder joints to keep the leads from shorting. Cover **all** the bare wire.*

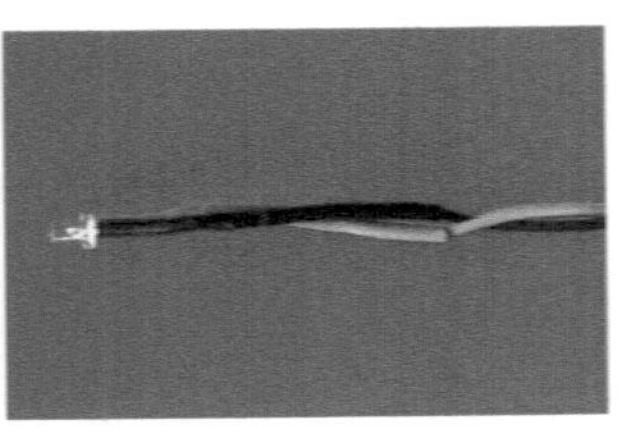

(f) *Twist the pair of wires together, to reduce the pick up of stray signals.*

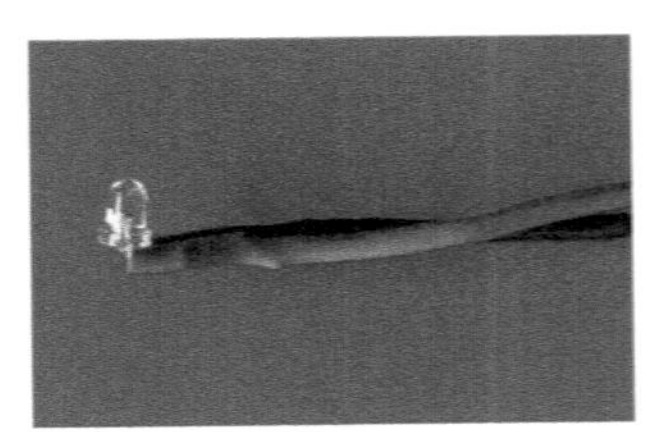

(g) *Bend the phototransistor 90°, so that the wires can be taped flat to the back of a LEGO® brick.*

Figure 25.1: *The leads on the optoelectronic devices need to be lengthened to make connection to the breadboard easier. Use a color code for the wires so that you can tell which lead is which. I chose here to use black for whichever of the wires of the phototransistor had to be connected to the lower voltage. If you wire up an LED as well as a phototransistor, use different colors for them, so that you can tell the devices apart even when they are taped inside opaque holders.*

> Cover the back of the phototransistor and wires with a piece of black electrical tape, both to keep them in place and to prevent stray light from reaching the detector from the back. A second black LEGO® brick snapped onto the bottom of the phototransistor brick blocks light from the bottom.

25.3.5 *Try it and see: Low-gain pulse signal*

After assembling the block, connect the phototransistor up to your logarithmic current-to-voltage converter (see Figure 23.5), rest your finger over the phototransistor, and try recording your pulse with PteroDAQ using a 600 Hz sampling rate. Sample at this high rate so that you can see how much 60 Hz

(a) *Two-stud black LEGO® bricks with 1/8″ holes can be used to hold 3 mm optoelectronics. An extra two-stud brick blocks stray light from entering the opening at the bottom of the phototransistor brick.*

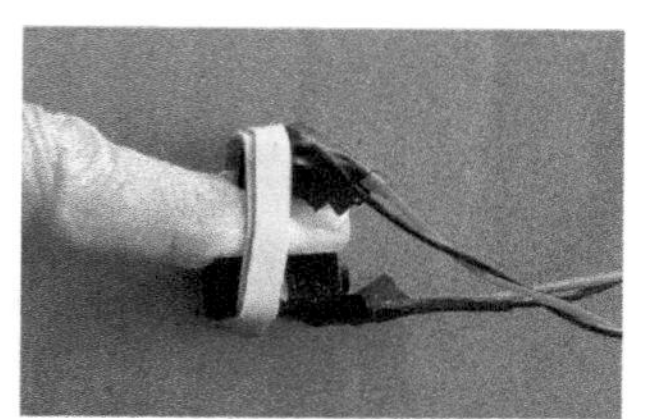

(b) *A rubber band can be used with the LEGO® bricks to hold an LED and phototransistor on opposite sides of the finger (trans-illumination).*

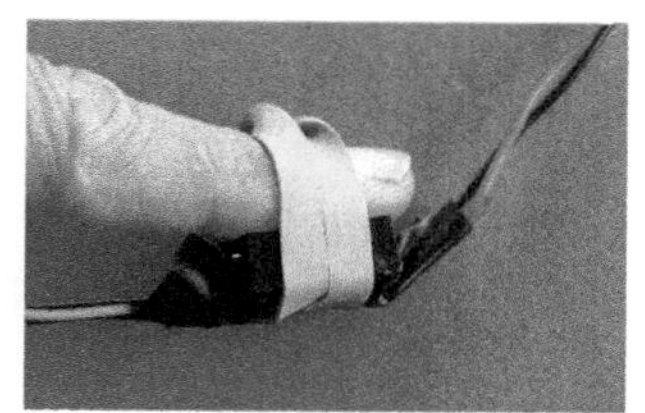

(c) *If the LED brick is snapped onto the bottom of the phototransistor brick, a rubber band can be used to hold both on the same side for cis-illumination.*

(d) *Using some 2×4 LEGO® bricks to support the 1×2 brick makes the lump caused by the wires coming out the back of the phototransistor less of a problem. The brick closing off the bottom of the phototransistor brick should be black, to avoid light leaks. Similarly, the black electrical tape holding the phototransistor to the brick blocks light from reaching the back of the phototransistor.*

(e) *A finger can be placed covering the phototransistor hole while the hand rests on the tabletop, to make keeping the pressure on the fingertip fairly easy.*

Figure 25.2: *The phototransistor brick can be left on the bench top and covered with a finger to use ambient light as the illumination, or strapped to a finger using rubber bands. By changing rubber bands or which finger is used, pressure can be adjusted to be between systolic and diastolic blood pressure, maximizing the signal.*

or 120 Hz interference you are getting—you need to be sure that your amplifier will not start clipping as a result of the interference, which would mask the pulse signal that you want to see. If you are using steady light (sunlight or LED illumination), you should get a signal like the one in Figure 25.4(a), but if you have fluorescent illumination, you will see a lot of interference at 120 Hz, as in Figure 25.4(b).

If you get 60 Hz interference (as is likely), you should try to determine whether the interference is due to modulated light or to capacitive coupling. What simple experiments can you do to distinguish between possible sources of interference?

If the currents are very low, you may also have problems with picking up stray signals from the electrical and magnetic fields that pervade our buildings. These stray signals will mainly be 60 Hz signals (50 Hz in Europe), capacitively coupled to the wires from the phototransistor to the amplifier. You can reduce the problem by twisting the wires together, so that each wire picks up about the same amount of interference; by keeping the wires short; and by grounding yourself, since the finger-to-phototransistor capacitance is probably the largest capacitance contributing to the interference.

You can also minimize the effect of the interference by increasing the intensity of the light, as the small currents induced by capacitive coupling will be a smaller fraction of a large photocurrent than of a small photocurrent.

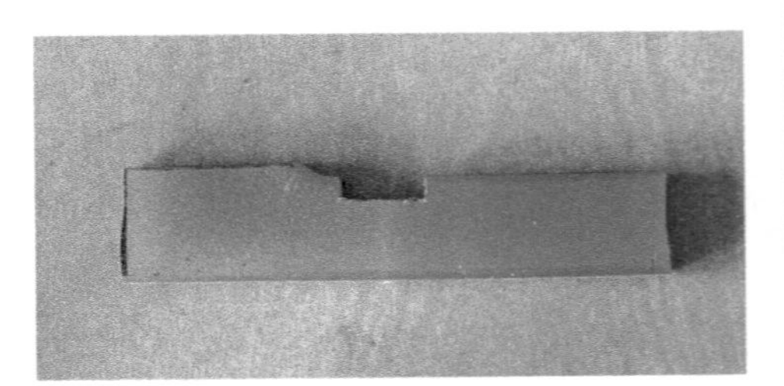

(a) *This drill jig was cut out of an old, 1-cm-thick, HDPE cutting board using a scroll saw. The bottom is flat, and the top has a rectangular notch that is just the width of a 2-stud LEGO® brick (1.57 cm). The notch is only 4 mm deep, about half the thickness of the brick.*

(b) *The jig is the same width as the jaws of the drill-press vice, making alignment easy. In classroom use, it is worthwhile to use some painters' tape to tape the jig to the non-moving jaw of the vice, so that it doesn't move when students insert or remove the LEGO® brick.*

(c) *The LEGO® brick is slightly taller than the thickness of the jig (11 mm vs. 10 mm), so the clamping force of the vice is entirely on the brick—the vice needs to be tightened gently, to avoid crushing the ABS plastic brick.*

Figure 25.3: *A jig for holding LEGO® bricks when drilling the 1/8″ holes for the 3 mm phototransistor (and possibly for a 3 mm LED) makes the drilling much more repeatable. A grey brick is shown here, but black bricks should be used for their greater opacity.*

A source of interference that many people overlook is from the switching power supply powering a laptop—these provide a high-frequency fluctuation in the USB power-supply voltage to the PteroDAQ Teensy board, as well as electromagnetic fields from the inductors in the supply itself. When measuring small signals, it is often better to power the Teensy board from a laptop battery, without the switching power supply connected to the laptop.

> If you sample at 60 Hz (or a divisor like 30 Hz), the line frequency interference will be hidden. It is still present in the signal, though, so you must either limit your gain so that even with the interference your signal is not clipped or filter out the interference before amplifying further.

The V_d signals plotted in Figure 25.4 are only 3 mV–10 mV peak-to-peak, which is too small to be seen on the PteroDAQ sparkline—you have to record them and then plot them with gnuplot.

The DC offset for V_d will vary with the brightness of the overall illuminations—moving your head or your body can shadow your finger and change the DC offset. The pulse signal is much smaller than the DC offset, and the pulse signal in Figure 25.4 is large compared with typical signals seen by students—it took me several attempts adjusting the pressure of my finger on the sensor to get such a large fluctuation in opacity.

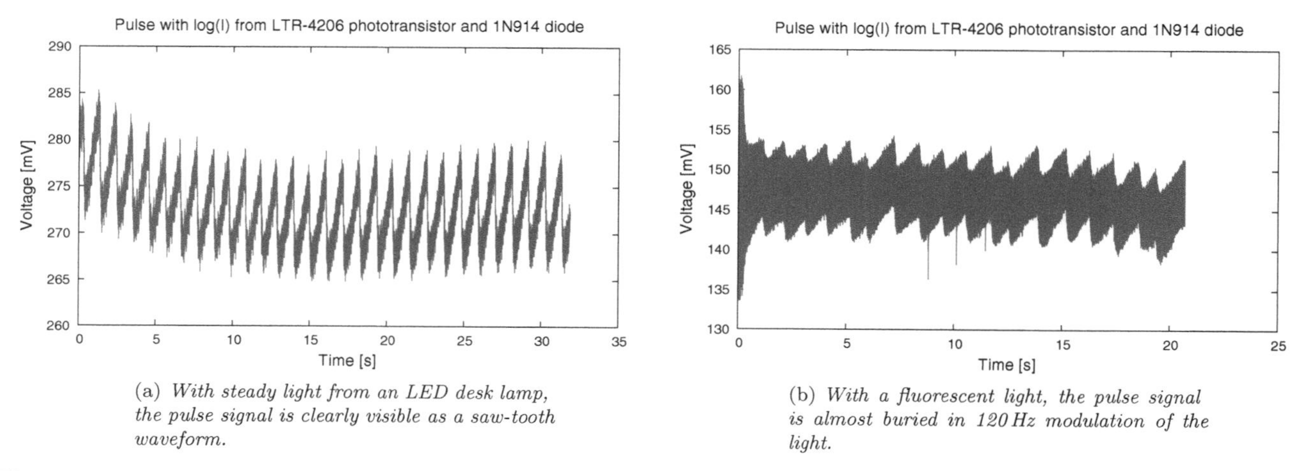

(a) *With steady light from an LED desk lamp, the pulse signal is clearly visible as a saw-tooth waveform.*

(b) *With a fluorescent light, the pulse signal is almost buried in 120 Hz modulation of the light.*

Figure 25.4: *Signals from just the first stage of the pulse monitor. Note the large, but varying DC offset compared to the size of the pulse signal, as well as the size of the higher-frequency interference. The voltage being reported is the difference from* V_{ref}*, so that zero current would appear as zero voltage here.*

Data collected with PteroDAQ using a Teensy LC board [46]. Figure drawn with gnuplot [33].

It may be a little easier to see the fluctuations with the Analog Discovery 2—the resolution is worse than PteroDAQ on the Teensy board (335 μV per step instead of 50 μV), but the user interface makes it easier to zoom in on the signal.

Pre-lab 6.2

For each of the plots of V_{d} in Figure 25.4, what is the range of photocurrent measured? What is the peak-to-peak fluctuation in light passed through the finger due to the pulse? Report fluctuation in dB ($20 \log_{10}(A/B)$) or percent ($100\frac{B-A}{(B+A)/2}$), where A and B are the lowest and highest light amounts in a pulse period.

You can maximize the pulse signal by gently squeezing your finger against the block—when the pressure on your finger is between the diastolic and systolic pressure, you will partially block the blood flow during diastole, increasing the difference in the amount of blood in different phases of the pulse. As with blood pressure measurement, the pulse signal is maximized when the pressure is close to the mean arterial pressure.

In addition to the 60 Hz interference, you may also see a large slow drift in the DC portion of the current. The second design project for this lab is to add filtering and more amplification, so that the 60 Hz interference and the DC drift are removed, while the signal for your pulse is amplified to be large enough to be easily visible.

The slow drift in the DC offset can be caused by changes in pressure of the finger, from changes in illumination (particularly if ambient light is used), or from charging of a high-pass DC-blocking filter.

25.3.6 *Procedures for second stage*

After determining how much additional gain you need, design and construct a second stage to your amplifier to provide this gain, including high-pass and low-pass filtering to remove DC offset and 60 Hz interference.

Pre-lab 6.3

Design a band-pass filter to pass pulses through, but block DC and 60 Hz. You want to have at least 20 dB attenuation of the 60 Hz interference (a factor of 10) relative to the peak gain, and you want to pass pulse signals as low as 20 bpm (0.33 Hz).
The gain for the filter should be based on the observed signal levels out of your first stage.

As mentioned in Section 18.3, you may need more than one additional gain stage, to avoid having the DC offset of the op amps themselves result in clipping of the output.

Test this filter stage independent of the log-transimpedance amplifier. There are two tests worth performing:

- (Required) Use the waveform generator to create a sawtooth waveform (called "ramp up" or "ramp down" by WaveForms 3) at around 1 Hz, to mimic the waveform that should come out of the first stage. The opacity of the blood vessels spikes up sharply when the heart squeezes, then ramps back down as the blood seeps back out through the veins, so the light level drops sharply, then gradually recovers.

  Figure 25.5 shows a recorded pulse waveform illustrating how a ramp-down signal is a reasonable approximation to the opacity of the finger. In recordings for some people, but not this one, you can observe a "notch" in the downward slope about halfway down from the peak, called the *dicrotic notch* [17]. This dicrotic notch seems to be the result of the resistance of peripheral blood vessels, though a similar dip in pressure in the aorta (called the *incisura*) is due to closure of the aortic valve. Use the oscilloscope to observe the output of the filter stage. What do you observe about the shape of the output waveform? You can adjust the amplitude and offset of the waveform generator to

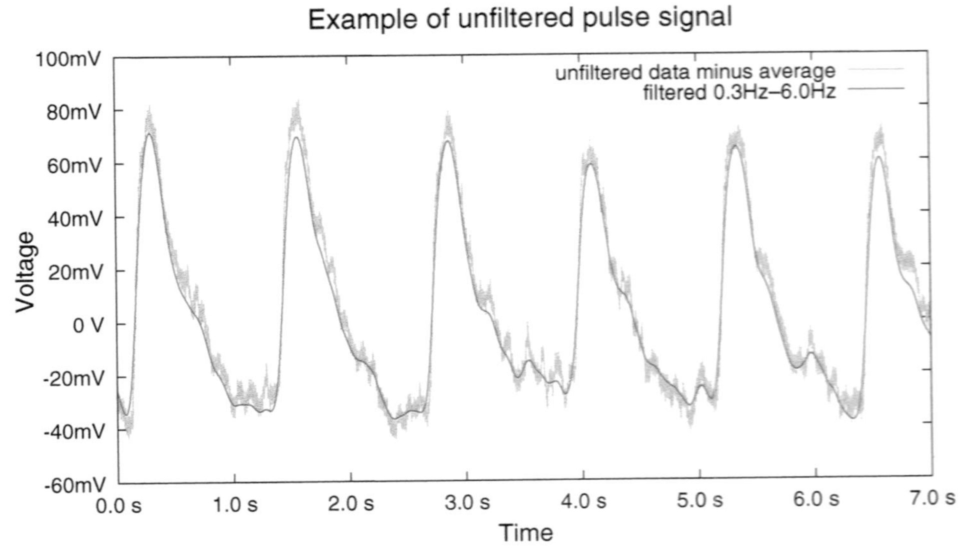

Figure 25.5: *The unfiltered pulse waveform was taken in full sunlight with the patient grounded to minimize 60 Hz interference from capacitive coupling. You will probably not be able to get so little 60 Hz interference in the electrically noisy environment of the lab. The average value of the signal was subtracted off, to center the signal at 0 V.*
Digitally filtering the waveform removes the remaining 60 Hz interference, but rounds the peaks a little. The underlying opacity waveform can be thought of as ramp-down sawtooth waveform, with an initial jump upwards in opacity followed by a steady decline.

investigate whether your filter is blocking DC correctly and has the right gain, but the gain computation is only accurate for sine waves—a ramp waveform has substantial higher frequency components which will have a different gain than the fundamental frequency. If you want to check your peak-gain computation, then use a sine wave at the frequency for which you should get peak gain. The best way to measure the gain of a filter is to measure both the input and the output of the filter, and take the ratio of the measurements. Use sine waves if you are testing a filter, as other waveforms have a combination of many different frequencies.

You don't have to eyeball the measurements on a digital oscilloscope, as it can compute and report the amplitude for you. On the Analog Discovery 2 scope, you can choose "Measurements" from the "View" menu, then use the "Add" button to add "Defined Measurement", "Channel 1", "Vertical", "Amplitude", and the same measurement for "Channel 2". The ratio of the measurements is the gain of the filter for the measured input signal.

- (Bonus) The gain measurement can be fully automated, to get a measured Bode plot for the filter. Connect waveform generator W1 to the input, use the first oscilloscope channel to observe that input, and use the second oscilloscope channel to observe the output. Now use the network analyzer to plot the response of the filter over the frequency range of interest. (Warning: the network analyzer can be quite slow at low frequencies, so you might want to do some quicker checks at the high end of the range of interest first.)

If you are working indoors, you will almost certainly see 60 Hz interference in your pulse waveform (assuming you sample at a high enough frequency to see it, and don't alias the 60 Hz and its harmonics to DC). This interference could come from any of several sources: capacitive coupling between the amplifier input and a conductor with 60 Hz through an electric field, inductive coupling through a magnetic field, or modulation of the light that is illuminating the finger.

Think up some ways that can distinguish between the different sources of interference: changing light sources, grounding the other plate of the capacitor, unplugging the laptop power supply, ... Do some experiments to see what source of interference seems to be the most troublesome and suggest ways to reduce the problem that could be used if you were making a commercial pulse monitor.

Record the pulse output for 20 s with the PteroDAQ data-acquisition system.

Do digital filtering (using the filter program from Lab 5) to plot a further cleaned-up pulse waveform.

25.4 Demo and write-up

For the first half of the lab, show me a single-stage amplifier that produces a DC voltage in a reasonable range when a finger is in the block, and a (small) fluctuation related to the pulse on the oscilloscope or PteroDAQ.

For the second half of the lab, show me a clear PteroDAQ trace of a heartbeat from a finger in the sensor. (That means recording using PteroDAQ and plotting with a program like gnuplot, not a screenshot of PteroDAQ!)

You will probably encounter some problems with the mechanical design of the finger-tip sensor. Describe changes you would make to improve the mechanical design. Incidentally, please refer to the LEGO® bricks as "bricks", not "legos". The Lego company has strong feelings on the matter:

> If the LEGO trademark is used at all, it should always be used as an adjective, not as a noun. For example, say "MODELS BUILT OF LEGO BRICKS". Never say "MODELS BUILT OF LEGOs". [64]

The report should contain the overall gain of the whole system from light in to voltage out. The gain for your system should be reported in mV/dB.

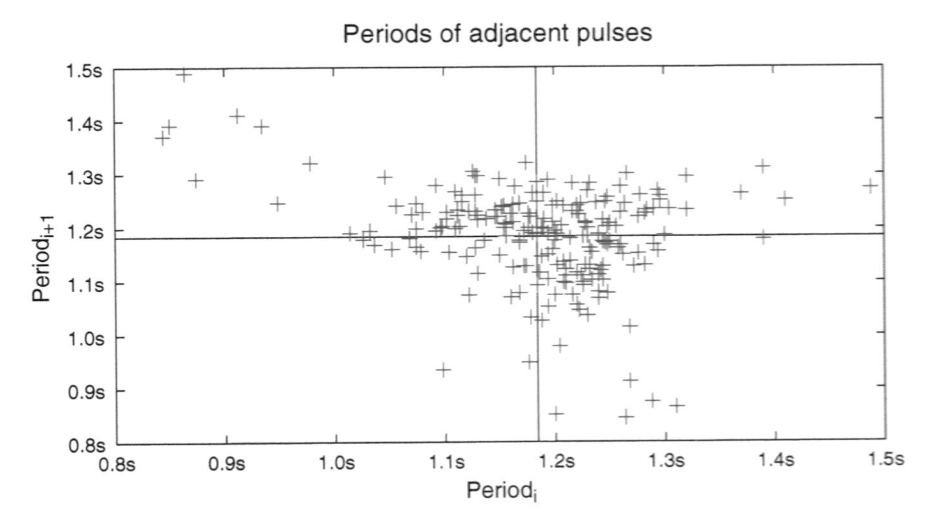

Figure 25.6: *Pulse periods have a wide range and are not independent of each other. The mean period is 1.184 s (for a pulse rate of 50.66 bpm), shown by the horizontal and vertical lines on the plot. Short periods are usually followed by long periods in this data—Pearson's r correlation between adjacent periods is* −0.3068.
Figure drawn with gnuplot [33].

The write-up should include schematics for both the single-stage and multi-stage amplifier, calculations used to get initial component values, adjustments to components that were needed to make the circuit work properly, and a plot of the output from the final amplifier design.

In your schematics, make sure that you use the proper NPN phototransistor symbol—incorrectly using a PNP phototransistor symbol is a serious failure.

Use the amplifier output file to compute the pulse rate for the individual.

Compute the pulse rate by measuring the time t precisely for exactly N periods and dividing the number of periods by the time, N/t. Because the interval between heart beats can fluctuate a lot from one pulse to the next, make the number of periods fairly large (say, $10 \leq N \leq 60$). *Don't get confused by fencepost errors* (see page 285).

Because time is one of the easiest things to measure precisely and accurately, you get much more precise results by taking an integer number of periods and measuring their time than by taking a time interval and estimating the number of periods.

Fluctuations in the period of the pulse can be quite large, and the period of one pulse is not independent of adjacent pulses—Figure 25.6 plots 202 pairs of pulse periods (about 4 minutes of recorded pulse). The mean period is 1.184 s (for a pulse rate of 50.66 bpm), and the standard deviation of the pulse period is 0.0928 s, but the distribution does not look Gaussian, as there are more very short periods than very long periods (skewness −0.8742 s). The range of periods is from 0.84 s to 1.49 s, but short periods are usually followed by long periods.

Averaging pulse periods (taking the time for N periods and dividing by N) results in lower variance for the estimate of the period. There is a trade-off between averaging many periods (to get a good estimate of the average pulse rate) and averaging few periods (to track changes in pulse rate as they occur). An exercise monitor may want to use a shorter window for the averaging than a pulse monitor for resting pulse rate.

Chapter 26

Microphones

A microphone is a device that converts sound (pressure waves in air) to an electrical signal—generally either a current or a voltage. A good microphone has a frequency range a bit wider than human hearing—for example, 10 Hz to 20 kHz, compared to the 20 Hz–15 kHz range of human hearing, but many microphones trade off frequency range for small size, low price, or other desirable properties.

Microphones differ from the pressure sensor we looked at in Section 20.3 and Lab 5, in that microphones are sensitive to much smaller pressure variations and respond to faster changes (higher frequencies), but do not usually respond to very low frequencies. The pressure sensors we looked at in Lab 5 can measure constant pressure levels, but microphones generally don't respond to frequencies below 10 Hz.

There are several different sorts of microphones: dynamic, capacitive, piezoelectric, and so forth [135]. Each microphone type works by a different principle:

dynamic A *dynamic microphone* consists of a diaphragm connected to a coil suspended in a magnetic field. Pressure waves move the diaphragm, which in turn moves the coil, generating a voltage proportional to velocity of the motion. A dynamic microphone is essentially a loudspeaker run backwards, though loudspeakers generally have too much mass to work well as microphones at high frequencies.

capacitor A *capacitor microphone* consists of a conductive diaphragm separated by a small air gap from another conductor. The two conductors separated by an insulator form a capacitor, whose capacitance depends on the area of the conductors and their separation. As pressure waves move the conductive diaphragm closer to or farther from the fixed conductor, the capacitance changes. This change in capacitance can be converted to a more easily communicated electrical change by any of several means [136].

electret An *electret microphone* is a capacitor microphone which has an electret on one plate of the capacitor. An *electret* is an insulator with a permanently charged electrical dipole, which replaced the external voltage source needed with a standard capacitor microphone. With this permanent charge, the fluctuations in capacitance due to sound result in fluctuations in voltage, but the current available is extremely small, so the voltage fluctuations are amplified into current fluctuations via a field-effect transistor, as shown in Figure 26.1 [122, 123].

Electret microphones are very cheap ($0.40–$2 each in 1000s), fairly small (6 mm–9 mm diameter), have a wide frequency range (20 Hz–20 kHz), and run at a wide range of voltages (2 V–10 V). There is a good tear-down of an electret microphone at http://www.openmusiclabs.com/learning/sensors/electret-microphones/ showing the components, including the discrete JFET inside and the metal can used to provide electrostatic shielding so that the JFET amplifies just the microphone signal, and not capacitively coupled noise from the ambient electrical fields.

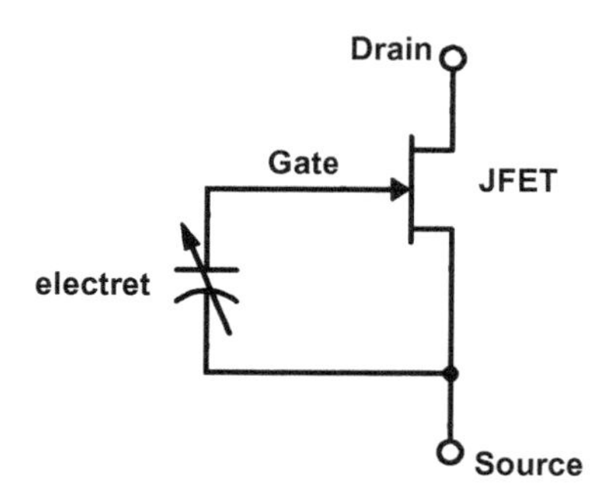

Figure 26.1: *An electret microphone consists of a capacitor that is permanently charged by an electret and whose capacitance varies with air pressure, combined with a junction field-effect transistor (JFET) that amplifies the tiny voltage changes into measurable current changes.*

There is more information on JFETs in Section 26.2, and field effect transistors are explained in more detail in Section 32.1, though the FETs used for electret microphones are usually junction FETs (JFETs) rather than the MOSFETs described there.

MEMS *Microelectromechanical systems (MEMS) microphones* are now the most popular for cell phones, laptops, and other small electronic devices, because they are very small, can be surface-mounted (the standard soldering technique for printed-circuit boards these days), and are fairly cheap ($0.80–$2.70 in 1000s). They typically have a smaller frequency range than electret microphones (100 Hz–10 kHz rather than 20 Hz–20 kHz), but are much smaller (less than 3 mm by 4 mm). MEMS microphones generally run at low voltages (1.5 V–3.6 V).

The MEMS microphones are also a type of capacitor microphone, with the diaphragm micromachined in silicon. They generally include a simple amplifier to provide a voltage output, though some come with a pulse-width-modulation output (see Section 35.2 for more information about pulse-width modulation).

The amplifiers in MEMS microphones result in the mics having fairly low output impedance (around 400 Ω), so that they can provide reasonably large currents for subsequent circuitry.

piezoelectric Piezoelectric substances produce a voltage when squeezed. Connecting a diaphragm to a piezoelectric crystal can make a microphone. Such microphones have a very high impedance, typically 10 MΩ, and so need an amplifier with a high input impedance, as they can't provide much current. Piezoelectric microphones are now mainly used as contact pickups on acoustic guitars.

There are many naturally occurring piezoelectric crystalline materials, such as quartz, sucrose, and Rochelle salt, but most applications now use man-made materials such as zinc oxide (ZnO), lead zirconate titanate (PZT), and aluminum nitride (AlN), generally as ceramics rather than crystals.

26.1 Electret microphones

In this class, we will use only electret microphones, as they are cheap, versatile, and compatible with our breadboards. Although silicon MEMS microphones are more popular these days for applications like cell phones, they do not come in through-hole packages that can be used with breadboards, but only as surface-mount devices. (*Breakout boards*, which solder the MEMS microphone to a small printed-circuit board, are available for prototyping, but cost about $5.)

An electret microphone consists of a permanently charged capacitor (an "electret"). One plate of the capacitor is a diaphragm, moved by the changes in air pressure that make up sound.

Because the charge on the electret is constant, but the capacitance changes with the separation between plates of the capacitor, the voltage across the capacitor changes inversely with the capacitance:

$$V(t) = Q/C(t) \ .$$

Because the capacitance is inversely proportional to the spacing between the plates of the capacitor, the voltage is thus directly proportional to that spacing. If the diaphragm behaves like a linear spring, then the displacement is proportional to the force on the diaphragm, which is in turn proportional to the pressure of the sound wave. Thus, the voltage across the capacitor is directly proportional to the pressure of the sound wave.

The small change in voltage is converted to a change in the saturation current of the field-effect transistor (FET) (see Figure 26.5), which can be measured externally. We will look at field-effect transistors more carefully in Section 32.1 and Lab 11: for this lab we are mainly interested in looking at the electret microphone as a simple sensor, but some of the FET behavior will appear in the DC characterization of the microphone.

26.2 Junction field-efect transistors (JFETs)

The FETs used in electret microphones are junction FETs (JFETs), rather than the more common metal-oxide-semiconductor FETs (MOSFETs). I believe that JFETs are used to provide built-in biasing of the gate electrode, though their higher transconductance gain and lower flicker noise than MOSFETs may also be important in this design choice by the microphone manufacturers.

The self-biasing results from the structure of the JFET—the gate is connected to the source and drain via a back-biased diode, which has a small leakage current that keeps the gate at about 0 V relative to the source, which is a normal operating point for a JFET. A MOSFET would require a biasing circuit to set the gate voltage, making a 2-wire connection to the microphone more difficult (TI makes a MOSFET-based amplifier as a JFET replacement for an electret microphone, the LMV1012 [91], but it costs more than the microphone plus JFET combined do).

The crucial voltages and currents of the JFET are

- V_{gs}, the gate-to-source voltage that is provided by the electret and is not accessible from outside the microphone,
- V_{ds}, the drain-to-source voltage, which needs to be set large enough for the JFET to work well,
- I_{d}, the current through the drain of the JFET, which is what the fluctuations in V_{gs} control.

Figure 26.2 shows a cross section of a junction field-effect transistor. The structure is similar to the diodes that we looked at in Figure 22.2 with the gate being the anode and the source and drain at opposite ends of the cathode. The diode is kept reverse-biased ($V_{\mathrm{gs}} \leq 0$ V), so that there is a thick depletion layer separating the gate from the source and drain and no current flows through the gate. The source and drain are connected by lightly n-doped semiconductor, acting as a resistor, with electrons

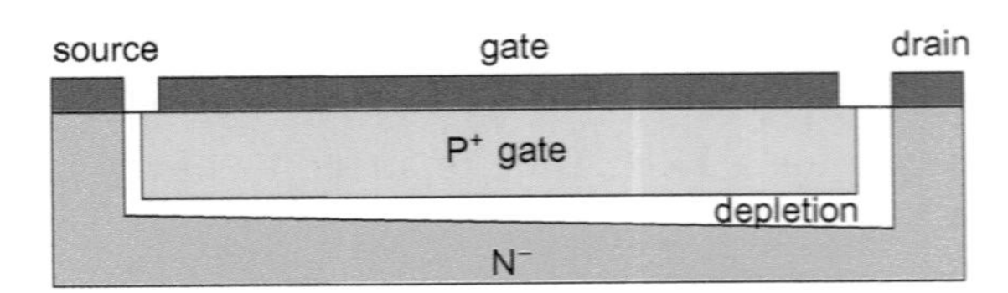

Figure 26.2: *A junction field-effect transistor (JFET) is essentially a reverse-biased diode, with the gate connected to the anode and the source and drain at opposite ends of the cathode. Changing the bias V_{gs} changes the thickness of the depletion region, which changes how much of the negatively doped (N^-) region is left to conduct, thus controlling the current from the drain to the source.*

moving from the source to the drain (so conventional current flow is drain to source). If V_{gs} is made more negative, the depletion region gets thicker, and the n-doped region gets narrower, reducing the current from the drain to the source. In this manner, the voltage V_{gs} controls the current I_d.

Analyzing exactly how a JFET operates and deriving the standard models is beyond the scope of this course, so we'll have to take on faith some of the models of JFETs used elsewhere [132]:

- In the linear region, where the voltage across the microphone is small, the transconductance gain $g_m = \partial I_d / \partial V_{gs}$ is roughly proportional to V_{ds}, the voltage across the microphone.
- In the saturation region, where the voltage across the microphone is larger, $g_m = \partial I_d / \partial V_{gs}$ is roughly proportional to $\sqrt{I_d}$, which does not change much as V_{ds} changes.

We usually use the JFETs in electret microphones in the saturation region, because of the insensitivity of g_m to V_{ds} in that region.

26.3 Loudness

The purpose of an audio amplifier is to take a small electrical signal that represents sound and turn it into a large enough signal to drive headphones or a loudspeaker and produce the sound. We'll be making audio amplifiers to amplify sound directly from a microphone, to make it louder.

In order to design the amplifier, we'll have to know how big our input is and how big we need to make the output. Measuring pressure in pascals is standard scientific practice, but when you are dealing with sound, most people report how loud it is in decibels (or dB(SPL), for "decibels sound pressure level").

The loudness of sound is usually expressed in *decibels*, which are logarithms of ratios:

$$D = 20 \log_{10} \left(\frac{A}{A_{ref}} \right) ,$$

where D is the loudness in decibels, A is the amplitude of the signal, and A_{ref} is the amplitude of the reference being compared to.

I'm being a bit sloppy in my terminology here—technically, *loudness* is a subjective phenomenon, not a physical one. The physical phenomenon is properly termed *sound pressure level.* According to Wikipedia, the most common reference for sound pressure level is $20\,\mu$Pa (RMS), which is close to the threshold of hearing for young adults (old men like me are often somewhat deaf and require higher amplitudes, particularly at higher frequencies) [157]. On this scale, a pressure wave of 1 Pa would be 94 dB(SPL), which is pretty loud—normal conversation is only 40–60 dB(SPL). See Section 29.3 for a brief explanation of RMS for signals.

Exercise 26.1

What is the RMS pressure level for a 1 kHz tone reported as having 70 dB(SPL)?

In some design reports, I've seen students using "noise" and "sound" interchangeably, but they are not synonymous, despite what your parents may think of your choice of music.

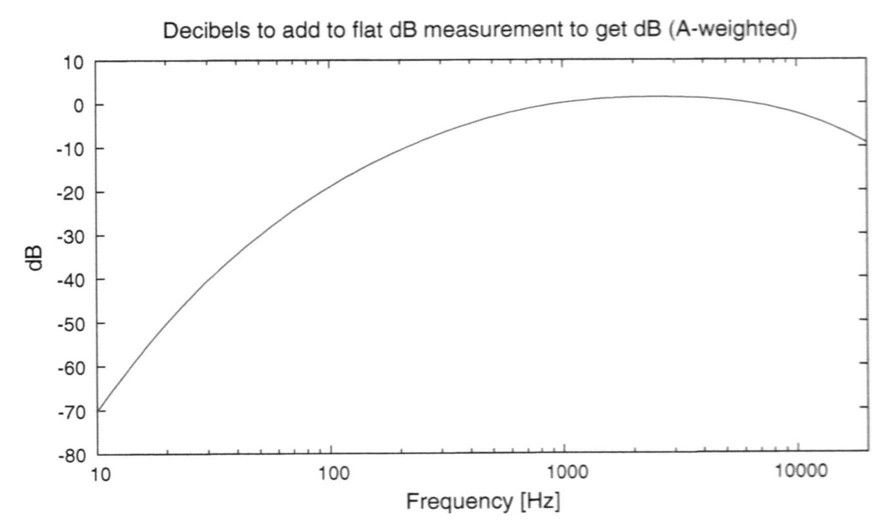

Figure 26.3: *For pure tones, one can get a better estimate of the loudness of a signal by adding A-weighting to the dB scale based on ratios of amplitudes.*

Formula from https://en.wikipedia.org/wiki/A-weighting: $1.9998 + 20\log_{10}\left(\frac{12200^2 f^4}{(f^2+20.6^2)\sqrt{(f^2+107.7^2)(f^2+737.9^2)}\,(f^2+12200^2)}\right)$.

The A-weighting scale is only defined up to 20 kHz. Many cheap A-weighted sound-pressure meters have an upper cutoff of about 8.5 kHz, to avoid misleading readings from frequencies included in the dBA scale that are not humanly audible.

Figure drawn with gnuplot [33].

Sound refers to pressure waves in air whose frequency is within human hearing range, though the term is often extended to *ultrasonic* waves that have a higher frequency than human hearing, to *infrasonic* waves that have a lower frequency than human hearing, and to pressure waves in other fluids (such as water) and even solids. (In solids, the wave is an oscillation in stress, rather than in pressure.)

Noise refers to any undesired signal that interferes with measuring or interpreting the desired signal. The signal can be of any type—pressure, temperature, voltage, current, strain, light intensity, ... The goal in most analog systems is to achieve a sufficient *signal-to-noise ratio* (SNR), so that the desired signal can be amplified, recorded, or analyzed without being too corrupted by the undesired noise.

Human hearing is not uniform across all frequencies, and so standard weighting curves have been created to measure sound pressure level so that 0 dB roughly corresponds to the threshold of hearing across the full range of human hearing. The most common weighting scheme for sound measurement in the United States of America is *A-weighting* (Figure 26.3) which is approximately -10.847 dB at 200 Hz, approximately 0 dB at 1 kHz, and peaks at about $+1.2713$ dB near 2.5125 kHz. That means that to get the same dB (A-weighted) value a signal must have $10^{10.847/20} = 3.486$ as large an amplitude at 200 Hz as at 1 kHz. The standard abbreviation for "dB (A-weighted)" is dBA.

The A-weighting scale approximates human hearing sensitivity fairly well up to about 10 kHz, but over-estimates human hearing sensitivity above that frequency—probably because of the difficulty in making analog filters with sharper cutoffs at the time that A-weighting was defined (around 1936).

A different weighting (ITU-R 468) is commonly used in Europe—it peaks around 6.3 kHz and drops off much more sharply at high frequencies than A-weighting. The ITU-R 468 seems to be more appropriate for wide-band and impulsive (click) noise measurement than A-weighting, which was designed around hearing response to low-volume, sine-wave tones [131].

An ordinary conversation in a quiet room is about 40 dBA–60 dBA—our lab tends to be a bit noisier, probably around 70 dBA. See the Wikipedia article on sound level [157] for a table of other common sound levels.

Worked Example:
Let's work out what the amplitude would be for a 200 Hz sine wave that is reported to have a loudness of 40 dBA(SPL).

If the frequency were 1 kHz, we could ignore the A-weighting, and just use the formula 40 dB = $20\log_{10}(A/20\,\mu\text{PaRMS})$, which simplifies to $A = 10^{40/20} 20\,\mu\text{PaRMS} = 2\,\text{mPaRMS}$. But at 200 Hz, our ears are less sensitive (−10.847 dB according to the A-weighting scheme), so we need more amplitude to get 40 dBA. We can solve $40\,\text{dBA} = 20\log_{10}(A/20\,\mu\text{PaRMS}) - 10.847\,\text{dB}$ to get

$$A = 10^{(40+10.85)/20} 20\,\mu\text{PaRMS} = 6.97\,\text{mPaRMS}$$

at 200 Hz. But that only gives us the RMS pressure—because the input is a sine wave, it has a crest factor of $\sqrt{2}$ (see Section 29.3), so the amplitude is ±9.86 mPa.

Exercise 26.2

What is the RMS pressure level for a 100 Hz sine-wave tone reported as having 70 dBA(SPL)? What is the amplitude?

26.4 Microphone sensitivity

A microphone converts sound pressure levels to voltage (or current) levels, and one of the most important specifications for a microphone is how much voltage one gets for a given pressure. Indeed, for any sensor the *sensitivity* of the sensor (its output divided by its input) is one of its most important attributes.

Since the input of a microphone is a pressure wave and the output is volts, we'd expect microphone sensitivity to be specified in V/Pa (volts per pascal). But if you look at microphone data sheets, you won't see exactly that information. For example, the SPU0410HR5H-PB MEMS microphone from Knowles Acoustics has sensitivity given as −42 dB @1 kHz (0 dB = 1 V/Pa) [58].

To understand this, we must understand the *decibel* notation for expressing ratios using logarithms. If we are looking at a ratio of voltages $r = V_1/V_2$, then we can express this in decibels as $(20\log_{10} r)$ dB. The specification −42 dB means a ratio of $10^{-42/20} \approx 0.00794$. But what is the ratio with respect to? The 0 dB (ratio of 1) specification says that it is with respect to a sensitivity of 1 V/Pa, so the specification for the MEMS microphone is that it provides 7.94 mV/Pa for a 1 kHz input.

Exercise 26.3

A microphone's sensitivity is often expressed in decibels, with a reference of 1 V/Pa. How many volts per pascal would you expect from a microphone that is specified as (-44 ± 2) dB?

Although most people encounter the term "decibel" primarily in the context of sound pressure levels, electrical engineers routinely use it for all sorts of ratios. If you are still not comfortable with decibels, you might want to watch the EEVblog video *Decibels (dB's) for Engineers—a Tutorial* at https://www.youtube.com/watch?v=mLMfUi2yVu8.

Electret microphones are slightly more complicated to determine the sensitivity for than MEMS microphones, because they are current-output devices not voltage-output devices, but the manufacturers insist on giving the sensitivity as a voltage output. The manufacturers get the voltage output through the simplest of current-to-voltage converters: a resistor connected to a voltage source (see Figure 26.4).

The resistor in the circuit can go by many names—you will see it referred to as a *bias resistor*, because it sets the DC level at which the microphone operates; a *sense resistor*, because it converts the changes in current from the microphone into changes in voltage; and a *load resistor*, because the *load line* for the resistor determines the DC voltage across the microphone.

Even in a very simple circuit like that of Figure 26.4, there are two design parameters: the supply voltage V_{pow} and the bias resistance R_{B}. This circuit is not technically a voltage divider, since the microphone is not a simple linear impedance, but the analysis by matching the current through the microphone with the current through the resistor is very similar to the analysis we do with voltage dividers.

We will often have one of the circuit parameters constrained by other parts of our system (for example, if we are powering the microphone from a Teensy board, we probably want V_{pow} to be 3.3 V).

Even if V_{pow} is fixed, there is no particular reason to use 2.2 kΩ for R_{B}—we may have good reasons for choosing smaller or larger values. So how do we decide?

The resistor R_{B} serves two purposes: to provide DC power to the microphone and to turn the fluctuating current from the microphone into a fluctuating voltage at V_{out}. This means that we need to do both a DC and an AC analysis. The DC analysis will tell us what voltage will be on V_{out} in the absence of any sound, and the AC analysis will tell us how much that voltage varies.

If we set R_{B} very large, the AC analysis would indicate large voltage swings for small current swings, but the DC analysis might show that there isn't enough voltage across the microphone for it to work at all. Similarly, if we set R_{B} very small, we can get a large enough DC voltage across the microphone, but the AC voltage swing (the signal we want from the sensor) may be very small and difficult to separate from noise.

This combination of DC and AC analysis with conflicting goals is typical for bias resistors—the details will vary in different situations, but we often have to trade off getting the best DC bias to get adequate AC performance or vice versa.

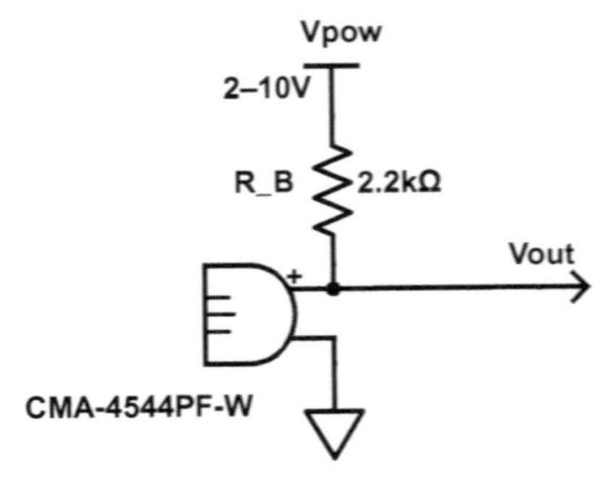

Figure 26.4: *An electret microphone requires a voltage source and a bias resistor to set the correct DC voltage across the microphone and to convert the current fluctuations to voltage fluctuations. The spec sheet for the CMA-4544PF-W assumes that the resistor* R_{B} *is 2.2 kΩ, but that is not necessarily the best choice.*

The bias voltage can be set more precisely using a transimpedance amplifier (see Chapter 23).

The small "+" next to the microphone indicates that the electret microphone is polarized—the "+" pin connects to the drain of the FET and must be at a higher voltage than the pin that connects to the source (here connected to ground). Scheme-it does not have polarized microphone symbols, so the "+" must be added as a label.

Figure adapted from the CMA-4544PF-W data sheet [15]. Figure drawn with Digi-Key's Scheme-it [18].

26.4.1 *Microphone DC analysis*

For the DC analysis, we need to characterize the microphone (which you will do in Lab 7), measuring the current through the microphone for any given voltage across the microphone. We can plot this as an I-vs.-V graph, as in Figure 26.5. The shape of this curve is one you'll see a lot, as it is characteristic of many transistors and other nonlinear devices.

If the microphone is connected as in Figure 26.4, with a 2.2 kΩ bias resistor to a 3.3 V supply, we can compute the current through the resistor as $(3.3\,\text{V} - V_{\text{out}})/2.2\,\text{k}\Omega$ (the *load line*, shown in Figure 26.5). If there is no current through the V_{out} output port, then we can determine V_{out} by where the currents match—that is, where the load line intersects the microphone's I-vs.-V curve.

Load lines are often given by the x- and y-intercepts of the line. When the current through the resistor is 0 A, then the voltages on both sides of the resistor are the same. This is the *open-circuit voltage* for the microphone, the voltage you would get if the microphone were replaced by an open circuit—in Figure 26.5 the open-circuit voltage is just $V_{\text{pow}} = 3.3\,\text{V}$. When the voltage $V_{\text{out}} = 0\,\text{V}$, then the current is $V_{\text{pow}}/R_{\text{B}}$, which is referred to as the *short-circuit current*—the current you would get through the resistor if the microphone were short-circuited by a wire. The short-circuit current is not shown in Figure 26.5, because the current would be 1.5 mA, which is way above the currents we are interested in for the microphone.

If we change R_{B} but leave V_{pow} unchanged, we rotate the load line around the point $(V_{\text{pow}}, 0\,\text{A})$ because the open-circuit voltage is determined by our power supply. Increasing R_{B} will make the current through the microphone and the DC output voltage V_{out} smaller, while decreasing the resistance will increase the current and the DC output voltage.

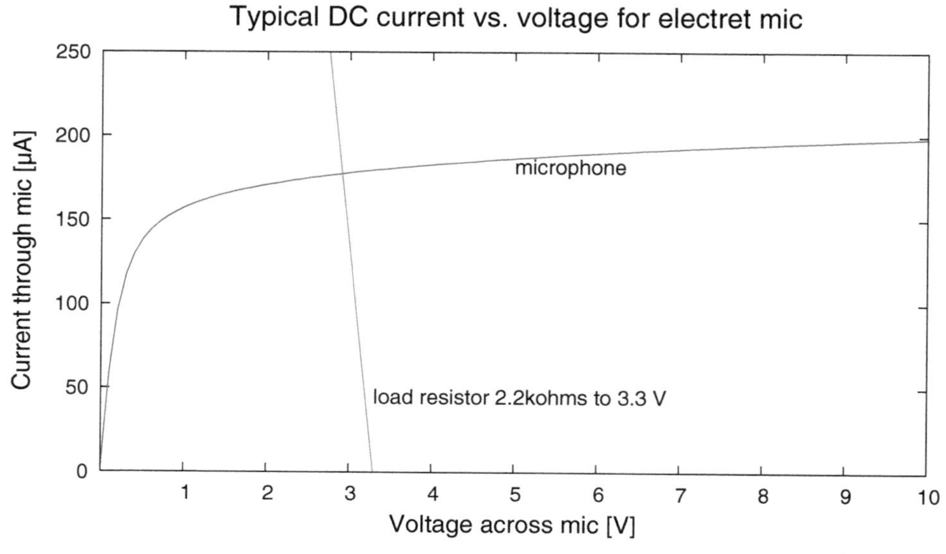

Figure 26.5: *The current for an electret microphone goes up with the voltage across the bias resistor. At low voltages, the current goes up roughly linearly with voltage (the* linear *or* triode *region of the FET), but at higher voltages the current increases only slowly with the voltage (the* saturation *region). Also shown is a load line for a 2.2 kΩ resistor connected to a 3.3 V power supply. The DC operating point for the microphone is where the curves intersect—here about 2.91 V at 177 μA, well into the saturation region, which is the region where electret microphones are used.* ***Warning: the curve here is from a model that has roughly the right shape, not from measurements—do not use it for your lab work.***

Figure drawn with gnuplot [33].

Electret microphones are usually used in the saturation region of the internal FET, where the current does not change much with the voltage, so that the current can be dependent only on the pressure variations in the sound.

In Figure 26.5, being in the saturation region requires around 1 V–10 V across the microphone with around 155 μA to 200 μA through the mic.

Exercise 26.4

Using the I-vs.-V curve of Figure 26.5, at what voltage does the FET in the microphone enter the saturation region? This is not a precise number, as the transition from linear to saturation is gradual—pick a number so that exceeding that value is definitely in the saturation region.
Using that value, what is the largest resistor you can use in the bias resistor circuit of Figure 26.4 with $V_{\text{pow}} = 3.3$ V?
What is the smallest resistor you could use in this circuit?
How would you choose a resistor in the legal range?

26.4.2 *Power-supply noise*

One major problem with using a simple bias resistor to set the microphone voltage is that any noise on the power supply will appear as noise on the output voltage. Because the current through the microphone is nearly constant, the voltage drop across the bias resistor is nearly constant, so the fluctuation in the output voltage due to noise will be essentially the same as the fluctuation in the power supply.

If there is a dedicated, linear power supply for the microphone and preamplifier, the noise on the power supply may be small enough to be ignored, but if a switching power supply is used or the power

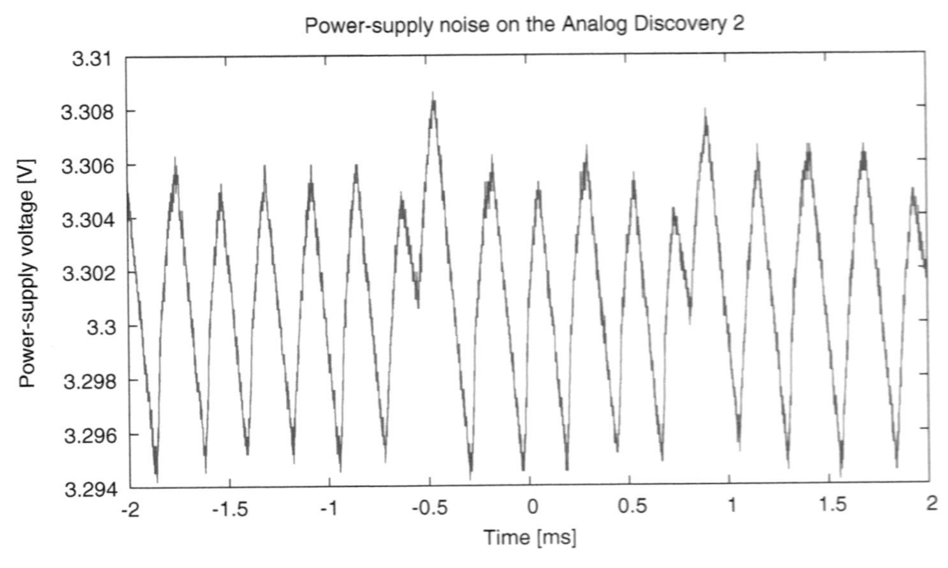

Figure 26.6: *The noise on the Analog Discovery 2 power supply set to 3.3 V with no load is less than ±10 mV, but that amount of noise could well be large compared to the signal from a microphone.*
Data collected with Analog Discovery 2 [19]. Figure drawn with gnuplot [33].

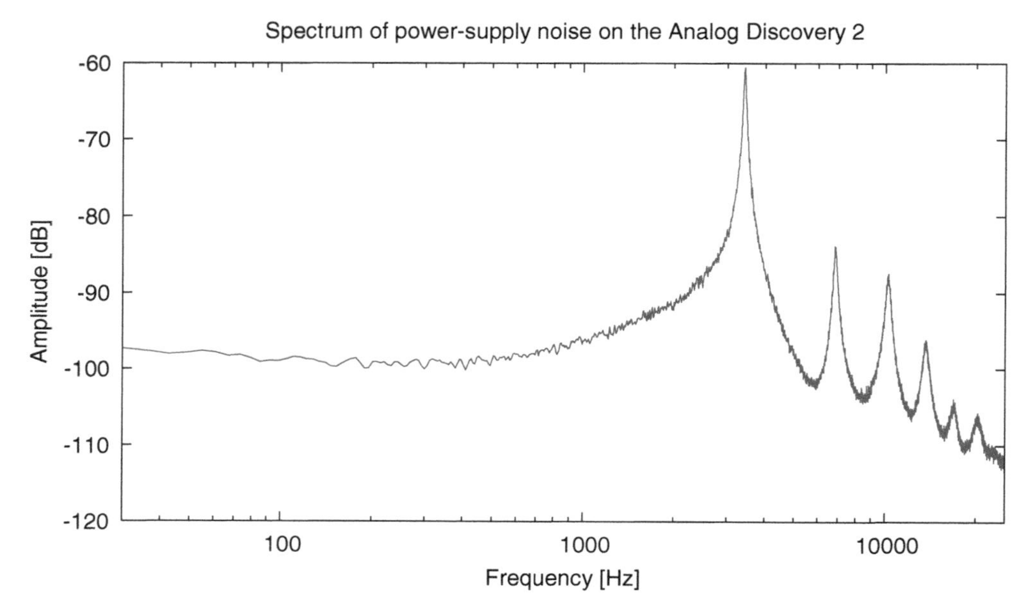

Figure 26.7: *A Fast Fourier Transform of the power-supply noise of the Analog Discovery 2, with the same setup as Figure 26.6 shows that the chaotic oscillation has a peak frequency around 3394 Hz. This frequency is in the range where it would be objectionable in an audio amplifier.*

Data collected with Analog Discovery 2 [19]. Figure drawn with gnuplot [33].

supply is shared with circuits that inject a lot of noise into the power supply (like the H-bridge of the power amplifier of Lab 11), then the noise may be a serious problem. Switching power supplies are far more efficient than linear ones, and so have become ubiquitous—it is a good idea not to rely on power supplies being low-noise linear supplies, unless you budget for the extra expense and power consumption.

Figure 26.6 shows the noise on the Analog Discovery 2 power supply set to 3.3 V with no load. The switching power supply keeps the voltage within 10 mV of the target voltage, but there is substantial chaotic oscillation within that range. Figure 26.7 shows a Fast Fourier Transform of the power-supply noise (averaged over many recordings), showing that the chaotic oscillation has a frequency around 3394 Hz—well within the range where it would be objectionable in an audio signal.

Putting a load on the switching power supply can greatly reduce the noise—even a few hundred microamps can make a big difference. We can reduce the power-supply noise further by filtering the power-supply voltage through a low-pass filter before using it to set the bias voltage of the microphone as shown in Figure 26.8.

26.4.3 *Microphone AC analysis*

If you connect the electret microphone and bias resistor to anything other than a high-impedance amplifier input, you need to figure out how the subcircuits interact. The easiest way to do this is to use Thévenin equivalence, as shown in Figure 26.9. The open-circuit voltage is $V_{\text{oc}} = V_{\text{power}} - RI_{\text{mic}}$ and the short-circuit current is $I_{\text{sc}} = V_{\text{power}}/R - I_{\text{mic}}$, so the Thévenin-equivalent resistance is just $V_{\text{oc}}/I_{\text{sc}} = R$.

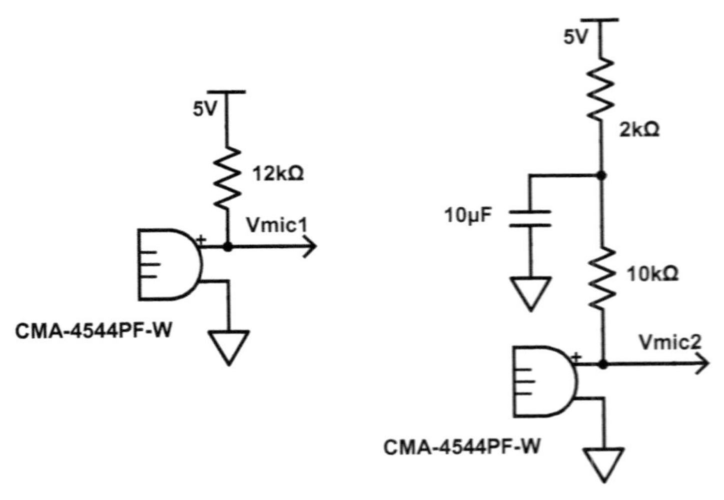

Figure 26.8: *This figure shows two schematics for biasing an electret microphone with a 12 kΩ resistance. The one on the left provides 12 kΩ for both the* DC *bias and for the* I-to-V *converter for* AC*, but does no filtering of power-supply noise. The one on the right has a 12 kΩ resistance for* DC *bias, but only 10 kΩ for the I-to-V converter, and filters the power-supply noise through a 7.958 Hz low-pass filter.*
Figure drawn with Digi-Key's Scheme-it [18].

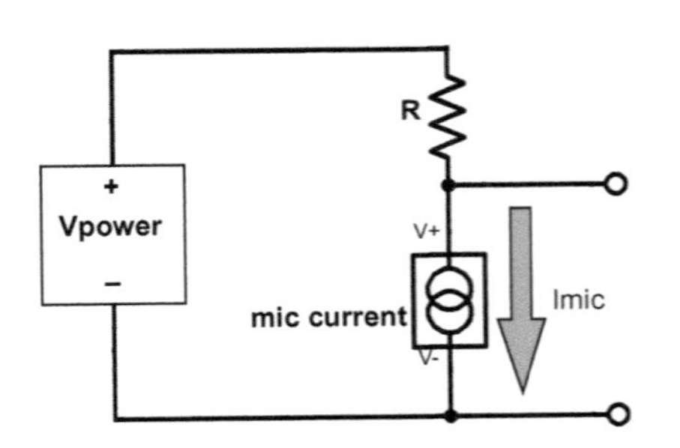

(a) *We can model an electret microphone as a current source, with current I_{mic}, to make a theoretical model of the microphone with a resistor to bias it. Figure drawn with Digi-Key's Scheme-it [18].*

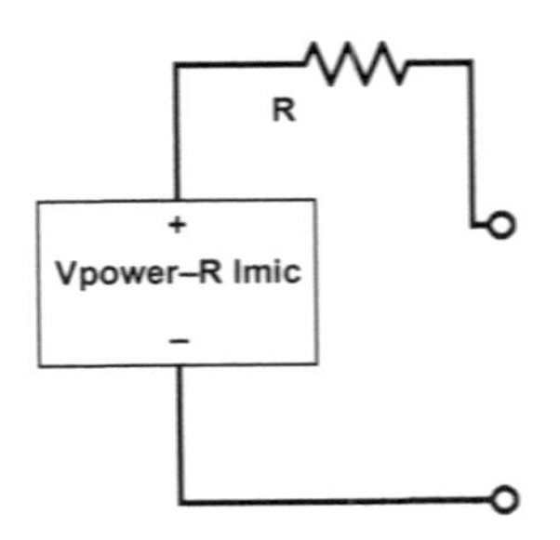

(b) *The Thévenin equivalent of the model in Figure 26.9a has a voltage source that varies as $-I_{\text{mic}}R$ with a series resistance R. Figure drawn with Digi-Key's Scheme-it [18].*

Figure 26.9: *Using Thévenin equivalence, the electret microphone with bias resistor can be modeled as a voltage source and resistor.*

Worked Example:
Design a high-pass active filter with a corner frequency of 16 Hz for an electret microphone used with a 2 kΩ bias resistor. We want the overall transimpedance gain from the AC microphone current to the output voltage to be about 20 kΩ.

The high-pass active filter needs to have a resistor R_{hi} and capacitor C_{hi} in series between the microphone output and the negative input of the amplifier. The resistance R_{hi} is essentially in series with the Thévenin-equivalent 2 kΩ of the microphone and bias circuit, so our RC time constant is $C_{\text{hi}}(R_{\text{hi}} + 2\,\text{k}\Omega) = 1/(2\pi 16\,\text{Hz}) = 9.947\,\text{ms}$.

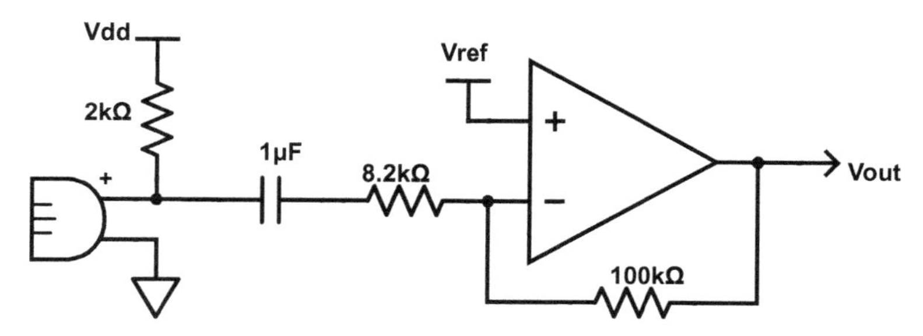

Figure 26.10: *This circuit provides a high-pass active filter with a corner frequency of 15.6 Hz and a passband gain of −9.8. The 2 kΩ bias resistor results in a 2 kΩ equivalent resistance for the microphone and bias circuit (see Figure 26.9), which must be included in the calculations for the corner frequency and gain of the active filter.*

If we choose $C_{\text{hi}} = 1\,\mu\text{F}$, we get $R_{\text{hi}} = 7.95\,\text{k}\Omega$, which is not a standard size. If we choose $8.2\,\text{k}\Omega$, then our corner frequency is $1/(2\pi(1\,\mu F)(10.2\,k\Omega)) = 15.6\,\text{Hz}$, and if we choose $7.5\,\text{k}\Omega$, we get $1/(2\pi(1\,\mu\text{F})(9.5\,\text{k}\Omega)) = 16.75\,\text{Hz}$. Because a $1\,\mu\text{F}$ capacitor is likely to be a cheap ceramic capacitor, and they tend to run smaller than nominal values more often than larger than nominal values, the larger resistor and lower corner frequency are probably the better design.

The current-to-voltage conversion is performed by two impedances: the $2\,\text{k}\Omega$ resistance to the power supply and $C_{\text{hi}}+R_{\text{hi}}$. Because they share the microphone output node, we have two equations:

$$2\,k\Omega\,I_{\text{2k}} = (R_{\text{hi}} + 1/(j\omega C_{\text{hi}}))\,I_{\text{hi}}\ ,$$
$$I_{\text{2k}} + I_{\text{hi}} = I_{\text{mic}}\ ,$$

where I_{mic} is the current through the microphone to ground, I_{2k} is the current through the bias resistor from the power supply, and I_{hi} is the current through the input impedance and feedback impedance of the filter.

At DC, the impedance of the capacitor is infinite, so $I_{\text{hi}} = 0\,\text{A}$ and the DC current through the microphone comes entirely from the $2\,\text{k}\Omega$ bias resistor.

In the passband, the impedance of C_{hi} is effectively zero, and the AC current is split between the bias resistor (I_{2k}) and R_{hi} (I_{hi}). With a little algebra we get that

$$I_{\text{hi}} = \frac{2\,\text{k}\Omega}{2\,\text{k}\Omega + R_{\text{hi}}} I_{\text{mic}}$$

in the passband. If we choose R_{hi} to be $8.2\,\text{k}\Omega$, then $I_{\text{mic}} = 5.1 I_{\text{hi}}$, which means that the feedback resistor for the filter needs to be 5.1 times the desired transimpedance gain of the circuit: $R_{\text{f}} = 20\,\text{k}\Omega\,5.1 = 102\,\text{k}\Omega$. Because $102\,\text{k}\Omega$ is not a standard value, we can use $100\,\text{k}\Omega$ and get a transimpedance gain of $19.6\,\text{k}\Omega$.

Figure 26.10 shows the schematic for the resulting design.

Exercise 26.5

We are not limited to using a bias resistor as in Figure 26.4 to convert current to voltage. We could also use a transimpedance amplifier, as in Figure 23.1. Design a transimpedance amplifier for an electret microphone with a current-vs.-voltage characteristic like that in Figure 26.5, draw a schematic, and do a DC analysis. The schematic should have no ports other than power, ground, and the output.
What is the voltage across the microphone? What is the output voltage for your circuit? What are the minimum and maximum gain that the transimpedance amplifier can have? The answer for this question is not unique—there are several equally correct answers, so explain your engineering judgment and what measurements you would need to make to answer the question for a different electret microphone.

Using a transimpedance amplifier instead of a bias resistor has a few advantages:

- The output of the transimpedance amplifier is very low impedance (essentially $0\,\Omega$), and so can connect to the input of an active or passive filter without affecting the filter characteristics. A bias resistor of size R, in contrast, provides an impedance of R, which needs to be added to whatever other impedance is in series with it to compute the effect of the filter.
- The bias voltage is fixed by a voltage divider that is independent of the resistor for the I-to-V gain, and the bias voltage remains fixed even as the current through the microphone fluctuates.
- Power-supply noise can be filtered out in generating the bias voltage for the transimpedance amplifier, using just a capacitor added to the voltage divider that sets the voltage. In contrast, the simple bias-resistor design provides no filtering of noise from the power supply, and adding a low-pass filter before the bias resistor puts some strong limitations on both the filter and the I-to-V resistor.
- Many microphones need to be connected with long wires to the rest of the amplification system. If the transimpedance amplifier is placed right at the microphone, then the long wires can be driven with a fairly large current for the desired signal, reducing the problem of capacitive coupling of noise into the wires.
- By using a potentiometer as the feedback resistor, the I-to-V gain of the transimpedance amplifier can be made variable, without affecting the bias voltage.

The downside of using a transimpedance amplifier is that it requires an op amp, which is more expensive and takes up more space than just a resistor.

Exercise 26.6

Redesign the example of Figure 26.10 using a transimpedance amplifier with $2\,\text{k}\Omega$ gain, instead of a $2\,\text{k}\Omega$ bias resistor. The target gain for the overall system is still $20\,\text{k}\Omega$ and target corner frequency is still 16 Hz. (Don't worry about whether the overall gain is positive or negative—we don't care about the polarity of the signal.)

In Section 26.1, we saw that the electret microphone varied the voltage across a capacitor by varying the capacitance. This voltage is connected as the gate-to-source voltage, V_{gs}, of a field-effect transistor (FET). The change in V_{gs} is proportional to the change in pressure on the microphone diaphragm. We want to see how the current through the microphone from the drain to the source (I_{d}) varies as the pressure (P) on the microphone varies. Since the gate voltage variation is proportional to the pressure variation, we know the $\partial I_{\text{d}}/\partial P$ is proportional to $\partial I_{\text{d}}/\partial V_{\text{gs}}$, which lets us apply standard FET models.

The output voltage variation for the microphone in Figure 26.4 can be determined by Ohm's law: $\Delta V_{\text{out}} = -R_{\text{B}}\Delta I_{\text{d}}$ (the IR voltage drop across the resistor increases, so the output voltage decreases). In that circuit $V_{\text{out}} = V_{\text{ds}}$, so in the linear region, we get $\Delta V_{\text{out}} \propto -R_{\text{B}} V_{\text{out}} \Delta P$, while in the saturation region, we get $\Delta V_{\text{out}} \propto -R_{\text{B}}\sqrt{I_{\text{d}}}\Delta P$, where the symbol $\propto$ means "is proportional to". Remember from the DC analysis that increasing R_{B} decreases I_{d} and V_{out}, so the dependence of the voltage on the resistance requires a little thought.

In the saturation region, the effect that dominates is that the sensitivity goes up as R_{B} increases, because the current I_{d} doesn't change much and $\sqrt{I_{\text{d}}}$ changes even less proportionately. In the linear region, the sensitivity is roughly constant, going up as you get closer to the saturation regions.

The result is that maximum sensitivity occurs near the transition between the linear and saturation regions. Unfortunately, that is also where the distortion due to nonlinear behavior of the FET will be highest. To minimize distortion, electret microphones are usually used well into the saturation region, so that $\Delta V_{\text{out}} \ll V_{\text{out}}$ and the microphone can reasonably be approximated as having current change proportional to pressure. This requires using a small enough R_{B} for V_{out} to stay in the saturation region.

26.4.4 *Sound pressure level*

Worked Example:
Let's figure out how sensitive an electret microphone is, given the information on its data sheet. We'll use the CME-1538-100LB microphone from CUI Inc. as an example.

> Warning: this is deliberately **not** the same microphone as we use in the labs—don't copy the "answer" from here, as it will be wrong for your microphone. (We usually use CMA-4544PF-W (CUI Inc) microphones, but check the parts list for the specific offering of the course, as we may buy whatever microphone happens to be cheapest.)

The data sheet for the CME-1538-100LB microphone gives the sensitivity of the microphone in a rather strange-looking format: "f = 1 KHz, 1 Pa 0 dB = 1 V/Pa min = −41 typ = −38 max = −35 dB" [16].

Let's unpack what that means. First, there is the notation "dB", which stands for decibels. A measurement in decibels is always based on a unitless ratio, in this case of sensitivities: $\frac{\text{V/Pa}}{\text{V/Pa}}$. The numerator is the AC voltage you get from the microphone with a pressure wave of amplitude 1 Pa at a frequency of 1 kHz. The denominator is a standard reference to compare to—in this case a sensitivity of 1 V/Pa.

The decibel scale itself is a logarithmic one,

$$\text{sensitivity dB} = 20\log_{10}\frac{V_{\text{mic}} \text{ at } 1\,\text{Pa}}{1\,\text{V/Pa}},$$

so 20 dB means a signal that is 10 times larger, and 40 dB means a signal that is 100 times larger. The specification −38 dB means a ratio of $10^{-38/20} = 12.59\text{E-3}$, so the typical sensitivity of the microphone is approximately 12.59 mV/Pa. The minimum and maximum ±3 dB translates to a range of $10^{-41/20} = 8.913$ mV/Pa to $10^{-35/20} = 17.78$ mV/Pa. Since the range is that wide, it might be best to express it without so many significant figures: 8.9–18 mV/Pa.
But the microphone is a current-output device, not a voltage-output device, so we need to know how the current is being converted to voltage. The schematic CUI uses (adapted as Figure 26.4) shows a bias resistor of 2.2 kΩ, so we can compute that the 12.59 mV/Pa is better expressed as 5.722 μA/Pa, 8.913 mV/Pa as 4.051 μA/Pa, and 17.78 mV/Pa as 8.083 μA/Pa. Again, we're carrying too many significant figures: we have 4–8 μA/Pa.

The sensitivity depends where on the I-vs.-V curve (similar to Figure 26.5) you operate the microphone—the sensitivity on the data sheet is presumably for a 2 V supply voltage, which is given as typical on the data sheet. Increasing the DC current through the microphone would increase the sensitivity proportionately, but increasing the current-to-voltage sense resistor would provide a larger increase in sensitivity, despite the drop in the DC current. The sensitivity is not very precisely specified: ±3 dB is about a factor of 2 variation in sensitivity from part to part.

To figure out the size of the signal from your microphone you need to combine several steps:

- How big is your input signal? (see example in Section 26.3)
- How sensitive is your microphone (in μA/Pa)?
- What is the gain of your I-to-V converter (in V/A = Ω)?

Exercise 26.7

Based on the specifications on the data sheet for a CMA-4544PF-W electret microphone, compute the microphone's output RMS voltage with a 2.2 kΩ sense resistor and an input of 64 dBA at 1 kHz. This computation only requires adjustment for the input loudness—the frequency of 1 kHz needs no correction for dBA weighting, and 2.2 kΩ is the bias resistor used on the data sheet.

Exercise 26.8

Based on the specifications on the data sheet for a CMA-4544PF-W electret microphone, compute the microphone's output RMS voltage with a 10 kΩ sense resistor and an input of 74 dBA at 60 Hz. This involves three correction factors: for the sense resistor, for the input loudness, and for dBA weighting for the frequency.

Exercise 26.9

Create a worksheet for computing the AC voltage expected from your microphone—that is, instructions for solving problems having the same format as Exercise 26.8. The worksheet should provide a general method for calculating the RMS AC voltage expected from a specified input sound pressure level expressed as dBA(SPL). You should be able to plug in the sensitivity of the microphone (from the data sheet) and the size of the pull-up resistor or transimpedance amplifier gain you use, as well as the loudness of the input to get the output voltage.
Your worksheet should be a fill-in-the-blank table that you use with a calculator. Please include the worksheet, filled out for Exercise 26.8. using the current-to-voltage conversion resistor you choose as part of your design report for Lab 7.

One major difference between this course and fill-in-the-worksheet labs that you have done in other courses is that in this course, *you* design the worksheet. Incidentally, providing the instructions for filling out such a worksheet is an excellent way to communicate to other engineers exactly how you did a computation. This allows them to modify the calculation for a change in design (such as needing to use a different microphone) and to check that your calculations make sense.

Chapter 27

Lab 7: Electret Microphone

Bench equipment: power supply, function generator, oscilloscope
Student parts: electret microphone, resistors, capacitors, breadboard, PteroDAQ, voltmeter

27.1 Design goal

For this lab, you will

- characterize the DC behavior of an electret microphone,
- design a simple circuit to bias the microphone appropriately to have an output that is centered at 1.4 V (or some other voltage that you choose in the saturation region), and
- observe the waveform of the microphone on the oscilloscope.

27.2 Characterizing the DC behavior

The expression "characterize the DC behavior of a component" means to produce a plot of the DC current through a component for different DC voltages across the component (an I-vs.-V plot). To do this characterization requires measuring the voltage across the component and the current through it simultaneously. The more pairs of measurements made, the more detailed the characterization can be.

What you learn about your microphone in this lab will be important for the design decisions you make in Lab 9. In particular, you will want to know how large a signal you can expect out of your microphone with different sense resistors and different loudnesses of input.

There are three ways to create data for I-vs.-V plots, depending on what tools are available:

- using the Analog Discovery 2 (the easiest),
- using PteroDAQ (fairly easy, but limited voltage range), and
- using a voltmeter (tedious).

Use whichever method works best for you.

Pre-lab 7.1

Prepare gnuplot scripts for plotting current versus voltage and equivalent resistance V/I versus voltage, including fitting different models (constant current, current linear with voltage, etc.). Think ahead to the format that PteroDAQ or the Analog Discovery 2 will use for formatting the data.
You'll probably have to tweak these scripts once you get real data, but you'll want to have a usable basis to start from, rather than spending lab time reading gnuplot documentation.

For Pre-lab 7.1, you are not expected to have any data yet, so you won't have values for the parameters that will let you get the curves correct, but you should be able to take the formulas given in Section 27.3 and type them as functions in gnuplot format. If you give the parameters reasonable initial values (eyeballed from Figure 26.5, for example), you will get plots that look at least sort of like the ones in Figure 27.5. Then, when you get the data, you can add fit commands to get the parameters right.

Remember to use mnemonic variable and parameter names in your gnuplot script. The variable `x` should only appear in plot commands, not in the definitions of the functions.

It can be very valuable to display data immediately after collecting it, so that you can plan the next stage of data collection: do you need more points at the low end? at the high end? Are there regions that need to be filled in? Are there anomalous points that need to be remeasured?

If we apply a voltage across the series chain of a sense resistor and microphone (as shown in Figure 27.1), the voltage across the resistor tells us what the current through the series chain is (which is the same for both the resistor and the microphone). Ideally, you would like the two voltages you measure (across the microphone and across the resistor) to be roughly equal, so that you can measure them with similar accuracy, but you may have to compromise on this to get sufficient voltage range across the microphone.

> Connecting clip leads to the microphone can be difficult, but the microphone plugs into the breadboard easily, allowing wires or double-ended header pins to be used as probe points. Be sure to get the polarity right, as shown in Figure 27.2.

Since we want to measure both the voltage across the microphone and the current through the microphone, we need two voltage measurements: the voltage across the microphone and the voltage across the resistor. The power supply or function generator has to deliver a voltage that is the sum of these two voltages, so we need a source capable of being set to a higher voltage than the largest microphone voltage we want on our I-vs.-V plot.

Each point on the I-vs.-V curve requires a different voltage from the voltage source. With a function generator, you can change the voltage continuously, and thus get a continuous I-vs.-V curve. With

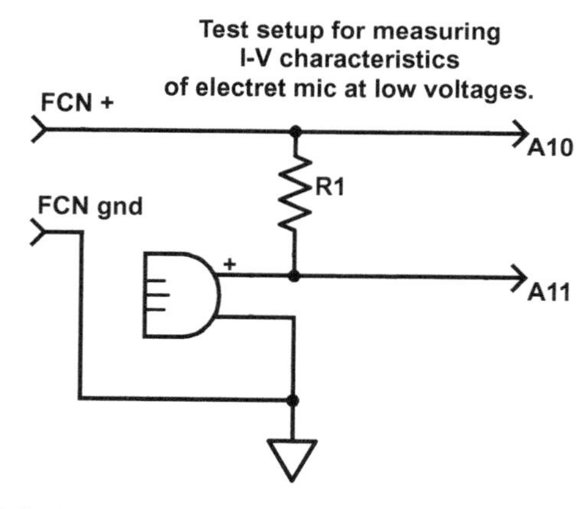

Figure 27.1: *Schematic for testing the DC characteristics of an electret microphone with a function generator or with a DC power supply. The function generator or power supply is connected to the inputs on the left, and DC voltages are measured across the resistor and across the microphone.*

The resistor value should be made as large as possible (for measuring small currents), while allowing the voltage across the microphone to get as high as 3 V.

Figure drawn with Digi-Key's Scheme-it [18].

Figure 27.2: *Back of the CMA-4544PF-W electret microphone. In this orientation, the lower terminal is* Terminal 2, *which should be connected to ground, and the upper terminal is* Terminal 1, *which should be connected via a series resistor to the positive voltage source (or Terminal 1 to the positive power rail and Terminal 2 via a series resistor to ground—either way, Terminal 1 should be the more positive voltage).*
The exact appearance of the back of the microphone seems to vary from year to year, but the dimensions and off-center positioning of the pins remains the same.

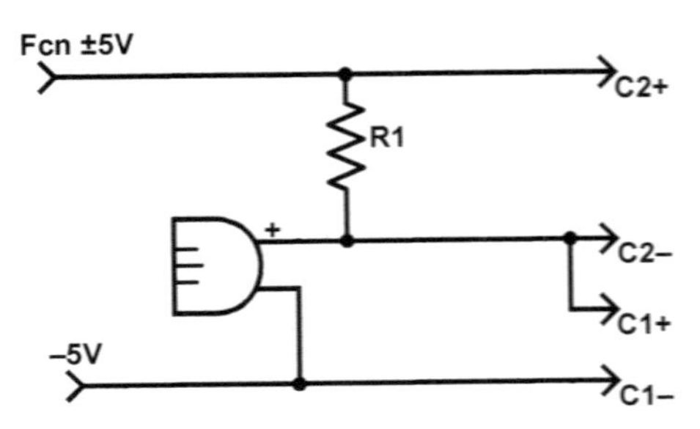

Figure 27.3: *With the function generator, power supply, and oscilloscope channels of the Analog Discovery 2, you can measure the microphone characteristics up to almost the entire 10 V range of the microphone (the 10 V range of the function generator minus the IR drop of the sense resistor).*
Figure drawn with Digi-Key's Scheme-it [18].

a fixed-voltage supply (such as a microcontroller power supply), you can use a potentiometer as an adjustable voltage divider, as shown in Figure 27.4, and use the output of the voltage divider to deliver different voltages to your microphone-plus-resistor series chain, and so get different points on the I-vs.-V curve.

27.2.1 *DC characterization with Analog Discovery 2*

If you have an Analog Discovery 2 USB oscilloscope, you can use it both as a function generator to provide a voltage source and to gather the data, using Channel 1 to measure the voltage across the microphone and Channel 2 to measure the voltage across a sense resistor to get the current.

Because the Analog Discovery 2 has true differential channels and a wide voltage range for the function generator, you can measure the I-vs.-V behavior for almost the whole legal range of the microphone. The function generator has a range from $-5\,\text{V}$ to $+5\,\text{V}$, so if the negative power supply is used to provide a $-5\,\text{V}$ connection to one end of the series chain, the total voltage across the chain can be varied from $0\,\text{V}$ to $+10\,\text{V}$, as shown in Figure 27.3.

You'll have to export the data (preferably as tab-separated text) to use it for model fitting—exporting is an option under the "File" menu.

You can even set up a "math channel" that computes C2/R and reports it in amps. Renaming the channels (an option hidden behind the gear icon on each channel) makes it easier to interpret the data, both on the screen as you collect it and later when you try to interpret the data and fit models to it. The "View" menu provides an "Add XY" option that lets you see the I-vs.-V curve as you collect it.

> If you try to collect the data fast by using a moderately high frequency for the function generator, you will get anomalous results—the current measurements will be different when the voltage is increasing than when the voltage is decreasing. (Try it and see!) You can minimize this hysteresis by using a very slow input wave—0.1 Hz or lower.

I believe that the hysteresis comes from a parasitic feedback circuit consisting of the drain-gate capacitance and a large resistance for the drain-channel leakage current (see Chapter 32 for more information about field-effect transistors). At very low frequencies (well below the corner frequency of the feedback circuit), we see the DC behavior of the FET in the microphone that we are interested in measuring. At frequencies above the corner frequency, the gate bias V_{gs} varies along with the drain-source voltage V_{ds}, giving a much larger fluctuation in current than we would expect from a DC analysis. Near the corner frequency, the V_{gs} is not in phase with V_{ds} resulting in different current measurements when V_{ds} is increasing than when V_{ds} is decreasing, because V_{gs} is different.

27.2.2 *DC characterization with PteroDAQ*

This section is only needed if you have to make measurements without an Analog Discovery 2.

Taking lots of repetitive measurements is not something people do well, but it is something that microcontrollers are very good at.

If you have a function generator, it can be set to a slow sine wave (several-second period) that sweeps from 0 to 3.3 V, and all the data can be collected with no further human input. You can use the circuit of Figure 27.1 to measure current and voltage (as long as the maximum voltage is no more than the 3.3 V limit of the microcontroller inputs).

If possible, the function generator output should be checked with an oscilloscope before wiring up PteroDAQ, to avoid applying voltages outside the acceptable range ($0\,\text{V} \leq V \leq 3.3\,\text{V}$) to the analog-to-digital inputs—it is easy to accidentally set up a function generator to a different range than you expect. The Analog Discovery 2 shows you the waveform when you set up the function generator, and so confirmation with an oscilloscope is not usually needed. The Agilent 33120A function generators that we used to use often confused students by providing twice the voltage that was entered on the controls, both for peak-to-peak and offset inputs.

PteroDAQ can be set up to measure both A11 and A10–A11, so both the voltage across the microphone and the voltage across the sense resistor can be simultaneously measured. The measurements can be made many times a second, getting different voltage-current pairs.

The sampling rate for the measurement and the period of the waveform should be set so that you get a few thousand measurements in one period of the input wave—if you sample too slowly, then your measurements will be at widely spaced voltages and you won't get a detailed curve. The period should be fairly large—perhaps 10 s or more, to avoid the erroneous measurements due to Miller feedback described in Section 27.2.1

If you don't have a function generator (for example, if you are doing the DC characterization at home before coming to lab), you can make your own slow sweep of voltage by using a potentiometer as a voltage divider as in Figure 27.4 and adjusting the potentiometer slowly from one end to the other.

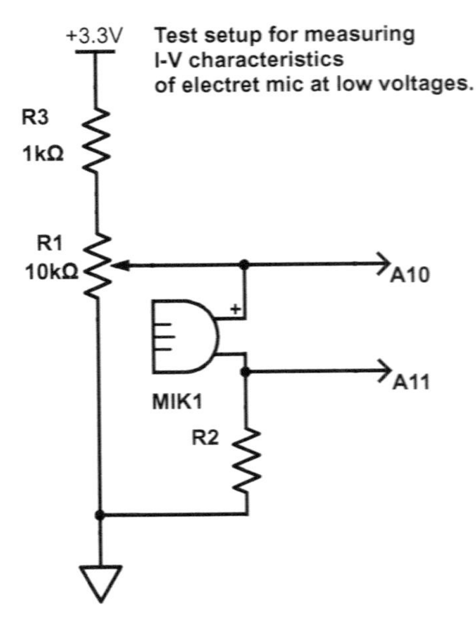

Figure 27.4: *Schematic for testing the DC characteristics of the microphone using a potentiometer. The voltage source is the microcontroller board of PteroDAQ. The current can be computed from V_{A11}/R_2 and the voltage across the microphone measured with a differential analog-to-digital channel. The resistor R_3 ensures that the voltages do not get too close to the upper power rail, where the differential ADC readings get inaccurate.*

I have arbitrarily swapped the microphone and resistor in this setup relative to Figure 27.1. You can use either order, but be sure to document what you do, and keep the columns of your data files in a consistent order.

Figure drawn with Digi-Key's Scheme-it [18].

If sweeping by hand, collect data at frequent time points (at 30 Hz, for example) as you adjust the potentiometer slowly from one end to the other. You should gather a few thousand data points—the order of the data points doesn't matter as we won't be using the time stamps. You can sweep back and forth until you've collected enough data.

One problem with using PteroDAQ for the DC characterization of the mic is that we are limited to the 0 V–3.3 V range of the microcontroller board, far less than the 10 V range of the microphone.

> You will need to choose a value for your sense resistor in Figure 27.1 or Figure 27.4 that will produce a voltage across the resistor of about 0.2 V with the largest current that you get through the microphone. That 0.2 V design point was chosen as a compromise between the precision with which the voltage can be measured by PteroDAQ and having a wide range of voltages across the microphone (0 V–3.1 V).

You may have to make a measurement with an arbitrary resistor, chosen based on the data sheet, measure the current with that resistor, then compute a resistance that will give you the desired voltage drop, and take another set of data with that resistor.

> Record metadata in the *notes* field that includes who collected the data, what the values of all the components were, and how to interpret the readings (voltage across microphone, voltage across series resistor of __kΩ, etc.).

27.2.3 *DC characterization with a voltmeter*

This section is only needed if you have to make measurements without an Analog Discovery 2.

You can measure current and voltage with just a voltmeter, a resistor, and a fixed power supply. Measuring with the multimeter and recording the data by hand in a lab notebook or by typing it into a file is tedious, and so relatively few data points will be taken. This makes it easy to miss important phenomena, so the automated methods using the Analog Discovery 2 or PteroDAQ are preferable.

Hand measurement is useful as a supplement to PteroDAQ measurements, if you have a higher voltage power supply, as you are then not limited to the 0 V–3.3 V range of the microcontroller A-to-D converter. It would be good to measure the current at several different voltages (say 1 V–9 V in steps of 1 V *across the microphone*). The data sheet specifies a maximum voltage for the microphone. Don't exceed it.

Record the voltage across the microphone (not just the power-supply voltage) and the voltage across the resistor in a table on your computer and plot using gnuplot. If your measurements are all for voltages over 1 V, you should see fairly constant current, increasing somewhat with higher voltage. Be sure to record the size of your sense resistor in the metadata! If you change the sense resistor for different measurements, then you might need to add an additional column for the resistance.

Getting hundreds of points this way is possible, but extremely tedious, and I don't recommend it, but measuring one or two points by hand helps in choosing the size of current-sensing resistor to use for automatic measurements.

Some students will be tempted to use the ammeter function of their multimeter, but I strongly recommend against it, for multiple reasons:

- If you accidentally connect your ammeter across a power supply, you will get enough current through the ammeter to blow the ammeter's fuse. Replacing the fuse on a bench multimeter is a nuisance, and replacing the fuse on cheap handheld multimeter may be extremely difficult (they are often soldered in place, because fuse holders cost too much for the very low markup on cheap multimeters).
- The voltmeter function of our multimeters is more precise and more accurate than the ammeter function.
- We are measuring small currents, so we can get better measurements by using a much larger sense resistor than the ones built into the ammeters. Large sense resistors can disrupt the circuit being measured, so general-purpose ammeters are designed to use as low a sense resistance as they can. We don't have that constraint for the measurements we're making.

> In this course, always measure current with your own sense resistor and a voltmeter (or DAQ or oscilloscope channel), rather than using an ammeter.

27.2.4 *Plotting results*

You have collected a lot of numbers from which you can derive either an I-vs.-V curve or an R-vs.-V curve (where the resistance here is the DC resistance $R = V/I$ not the dynamic resistance $\frac{dV}{dI}$). We're more interested in the I-vs.-V curve, as that is the traditional way to describe nonlinear devices.

Write a gnuplot script to plot the PteroDAQ or Analog Discovery 2 data as I-vs.-V using "with lines" to get interpolation between the closely spaced data points. Plot the multimeter-measured data (if any) on the same plot as the points. Plotting with log scales on both axes makes the data look simpler and easier to come up with models for. Using a linear scale for voltage may hide the low-voltage behavior.

If your results look like mine, you'll probably find that there are few data points below 0.05 V. Collect another data set with a larger sense resistor, to measure the smaller currents for microphone voltages up to 0.1 V. What size would be appropriate for that?

27.2.5 *Optional design challenge*

Try to devise a circuit that allows you to *safely* measure up to almost 10 V across the microphone using PteroDAQ. No voltage applied to the microcontroller pins should go outside the range 0 V–3.3 V.

Hints for the design challenge:

- You may find it easier to use single-ended measurements rather than differential ones, and do any necessary subtraction when you analyze the data.
- Almost everything we do in this course involves voltage dividers.
- *Check your circuit with a multimeter before hooking up the microcontroller board, to make sure that you won't accidentally deliver a large voltage to the board!*

27.3 Analysis

Between the DC characterization of Section 27.2 and the AC view of Section 27.4, we will try fitting models to the data collected.

Here are some models to try fitting:

Linear (resistance) model: $I = v/R_{\mathrm{M}}$. You may want to fit R_{M} for just the lower voltages across the microphone below about 0.1 V.

Saturation (current sink) model: $I = I_{\mathrm{sat}}$. You want to fit the saturation current I_{sat} for just the higher voltages across the microphone, above 2 V, perhaps.

Non-constant saturation model: The saturation region of the JFET in the microphone does not have a constant current, but one that increases with voltage. You can model this with a power law, $I_{\mathrm{sat}}(v) = (v/1\,V)^{\alpha} I_{\mathrm{sat1}}$, or with an affine function, $I_{\mathrm{sat}}(v) = I_{\mathrm{sat0}} + v/R_{\mathrm{slope}}$. Both of these models have two parameters: one to set the height of the curve and one to control the slope of the curve. The division by 1 V is needed to make the value being exponentiated unitless, so that the model has the right units (A), derived from the I_{sat1} parameter.

Again, only the saturation region at high voltage can be modeled this way.

Blended linear and saturation model:

$$I(v) = v/\sqrt{R_{\mathrm{M}}^2 + (v/I_{\mathrm{sat}}(v))^2}.$$

This makes a smooth transition from the linear model to any of the saturation models, with $\lim_{v\to 0} R(v) = R_M$ and $\lim_{v\to\infty} R(v) = v/I_{\mathrm{sat}}$. This model is similar in shape to a standard model for FET behavior, but uses a different mathematical formulation. The values used for the separate resistance and saturation current models can be reused here, or you can try to fit new parameters.

The current-vs.-voltage plots for four models are shown in Figure 27.5.

When I tried this lab, I got a very good fit blending a non-constant saturation-current model with the linear low-voltage model, except at very low voltages and currents, where the quantization errors of the analog-to-digital converters made measurement unreliable. You can do slightly better by having

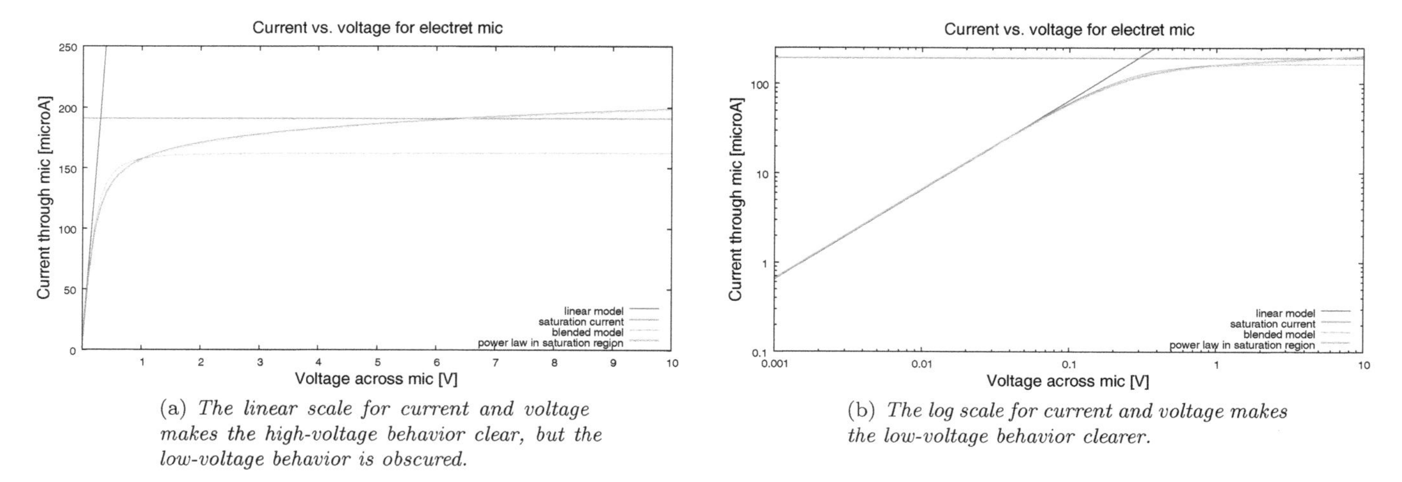

(a) *The linear scale for current and voltage makes the high-voltage behavior clear, but the low-voltage behavior is obscured.*

(b) *The log scale for current and voltage makes the low-voltage behavior clearer.*

Figure 27.5: *Current-vs-voltage plots for the electret mic. For this figure, I've deliberately suppressed the data and the parameters I got, to give the flavor of the models without giving you the "answers" for the lab. Your plots should, of course, include the data and you should use* sprintf *in the titles of the plotted curves to print the parameters of the models.*
Figure drawn with gnuplot [33].

power-law fits to the current for both the linear region and the saturation region (with different multiplicative constants and exponents).

27.4 Microphone to oscilloscope

The purpose of this section is to review how to use an oscilloscope and to get more familiar with AC and DC coupling of oscilloscopes. I also want you to be aware that attaching a measuring instrument to a circuit changes the circuit, so we will be using the input impedance of the oscilloscope as part of an RC high-pass filter.

You will be converting the current output of the electret microphone to a voltage output by putting the microphone in series with a bias resistor connected to a +3.3 V power supply (like a voltage divider, but with the microphone between the output and ground), or using a transimpedance amplifier (see Chapter 23).

You will look at the signal coming out of the microphone both with DC coupling, and after passing the signal through a DC-blocking capacitor.

Pre-lab 7.2

Choose an appropriate capacitor to get a high-pass filter with a corner frequency close to 2 Hz, using the input impedance of a 1× oscilloscope probe as the resistor (nominally 1 MΩ), as shown in Figure 27.6.

We are relying on the oscilloscope as part of the circuit—test equipment always modifies the circuit in some way, but we usually try to arrange things so that the modification does not affect what we are measuring. Here we are counting on that modification of circuit behavior to get the behavior we want.

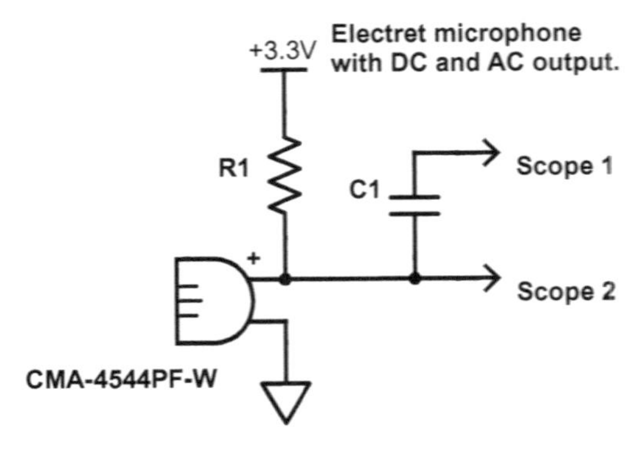

Figure 27.6: *Schematic for the microphone, the bias resistor, and the DC-blocking capacitor. You will need to fill in the values used. Figure drawn with Digi-Key's Scheme-it [18].*

Pre-lab 7.3

The input impedance of an Analog Discovery 2 scope is not purely resistive—instead of 1 MΩ it is approximately 1 MΩ in parallel with 24 pF [3]. How does the extra 24 pF affect the response of the high-pass filter designed in Pre-lab 7.2?

Pre-lab 7.4

Using your measurements or the formula you found for the I-vs.-V curve, pick a microphone voltage that is well into the saturation region (say 1.4 V). Look up how much DC current you'll have through the microphone at that voltage. Compute what resistance your bias resistor has to be to get that output voltage, remembering that for Ohm's law you need to be looking at the voltage across the resistor and the current through the resistor. Choose the closest standard resistance value from your collection of resistors, and measure the DC voltage you get with that value.

Pre-lab 7.5

With the resistance computed in Pre-lab 7.4, what AC RMS voltage do you expect for a 60 dBA(SPL) sound input at 1 kHz? (Use the worksheet developed in Exercise 26.9.)

Now hook up the oscilloscope probe to the output of the microphone and try to view the output of the microphone as a time-varying signal.

Using the Analog Discovery 2 without the optional BNC adapter board gives you only DC coupling. If you want to see the small variation on the signal, you need to adjust the offset of the channel to center the signal on the screen. The offset control is a little bit counterintuitive—if you want the center of the y-axis to be 1.4 V, you have to enter an offset of −1.4 V, as you are specifying how much change to make to the signal before amplifying it.

Once you have centered the signal, you can change the volts/division setting for the channel to zoom in to see the fluctuation. Using a fixed offset allows seeing both the DC and AC components of a signal, but with some signals with slowly drifting DC offsets, it can be hard to keep adjusting the offset to keep the signal on screen.

One interesting thing to do is to look at the Fast Fourier Transform (FFT) of the microphone voltage and of the power-supply voltage (one channel for each). If you do this in quiet conditions, you can see that the spectrum of the electrical noise at the microphone is the same as at the power supply (see Section 26.4.2). I like to set the FFT to do exponential averaging (of RMS or dB) with a weight of 25 to 100 (the higher the weight, the longer it takes for the average to settle, but the smoother the result). I also prefer the Blackman-Harris window to the default "flat-top" window, but the difference is small. The extra options for the FFT are exposed in WaveForms by clicking on the down-pointing green arrow in the FFT pane. For FFTs of the power-supply and microphone noise, you probably want to set the sampling rate (in the "Time" pane) to 50 kHz, so that you can get a range from 0 Hz to 25 kHz with a resolution of about 6 Hz.

If you put a low-pass filter in the bias circuit, as shown in Figure 26.8, but with appropriate resistor sizes for the bias voltage you want, then you should be able to see a noise reduction of about 30 dB from the power supply to the mic for medium-frequency noise (3 kHz–5 kHz), though 60 Hz interference may increase.

If you are interested only in the AC signal and not the DC bias, you can put the signal through a high-pass filter (which is referred to as *AC coupling* or *capacitive coupling*).

Use one channel of the Analog Discovery 2 to look at the signal directly (DC coupling) and the other channel to look at it after a capacitor whose size you computed for Pre-lab 7.2.

Try setting up (different) trigger conditions for each channel, to get a stable display on the scope.

> You can whistle or talk into the microphone or provide some other sound source. It may be easiest to use the scope if you have a known input signal. One way to get that is to hook up the function generator to an earbud or loudspeaker to put out a sine wave of known frequency, say between 100 Hz and 5 kHz, and put the loudspeaker near the microphone. Some students have found sound-generating apps on their phone to be useful also. You'll annoy the rest of the class less if you use frequencies around 200 Hz than if you use frequencies above 1 kHz.

What is the amplitude of the AC output of the microphone? The AC amplitude is half the peak-to-peak voltage—that is, it is the maximum voltage minus the minimum voltage, divided by 2. This amplitude will depend on how loud your sound input is. What is the RMS AC voltage?

The Analog Discovery 2 can provide several measurements derived from the signal trace using the "Measurements" option on the "View" menu of the oscilloscope.

The PteroDAQ system reports both the DC offset and the AC RMS values for a steady signal—the sparkline for the channel reports the mean and standard deviation (DC and RMS AC) for the entire recording. With a single channel you can use a sampling rate of 5 kHz–6 kHz with 32× averaging, which will provide adequate sampling rates for observing low-pitched sounds (up to about 1000 Hz—remember the effects of aliasing that you observed in Lab 3). Because the microcontroller board shares a single analog-to-digital converter for all channels, increasing the number of channels requires reducing either the hardware averaging or the sampling rate. The maximum sampling rate is often limited by how fast the Python program on the host computer can accept the data—see Table A.2 for some timing measurements I made.

On PteroDAQ, stretching the space for a channel vertically (by dragging the line below the channel down) may make the sound waves large enough to be visible on the sparkline. This isn't quite as useful as a real oscilloscope, but allows you to do the lab without access to an electronics lab with an oscilloscope.

You do **not** want to use a DC-blocking capacitor for the PteroDAQ input, as it needs a DC bias to keep the signal between 0 V and 3.3 V.

27.5 Demo and write-up

Demo the oscilloscope outputs to the instructor, showing that you know how to adjust the time base, the input scaling, and the triggering.

Turn in a write-up describing what you did. Remember that the instructor is *not* your audience—an engineering student who has not read the assignment is.

Provide at least

- a purpose of the report (the problem you are solving or the design information you are providing);
- the I-vs.-V or R-vs.-V curves for the electret microphone (which electret microphone?), including your data and your models—make sure that all parameter values for the models are reported;
- a *correct* schematic of the testing circuit, complete with component values (remember that the microphone you are using is a polarized microphone—you need to mark one of its ports with a + label);
- a *correct* schematic of the circuit used to connect to the oscilloscope, complete with component values (if electrolytic capacitors are used, show the correct orientation in the schematic);
- a plot of a short recording of sound (less than a second) using PteroDAQ or the Analog Discovery 2 and gnuplot.

Chapter 28

Impedance: Inductors

28.1 Inductors

Inductance results from combining two properties of conductors and magnetic fields:

- When a current passes through a conductor, it generates a magnetic field perpendicular to the conductor (as shown in Figure 28.1). When the current changes, the magnetic field changes proportionately.
- When a conductor is in a time-varying magnetic field perpendicular to the conductor, a voltage is produced that is proportional to the rate of change of the field.

That means that, for any conductor, the voltage from one end of the conductor to the other is proportional to the change in current:

$$v(t) = L\frac{\mathrm{d}i(t)}{\mathrm{d}t}.$$

The constant of proportionality, usually written as L, is called the *inductance* (or, more pedantically, *self-inductance*) of the conductor.

The sign here is positive, because I am using the same direction of current and voltage as we used for the capacitor (see Figure 28.2).

Coming up with a hydraulic analogy for an inductor is not quite as simple as for a capacitor, but there is a standard one: think of a large, massive paddle wheel which can be turned by the water flowing and which can in turn move the water.

The pressure difference of the water before and after the paddle wheel produces a force on the paddle wheel of pressure difference divided by the area of the paddles, which accelerates the paddle wheel. The pressure difference P times the area of the paddle A_p gives the force F, which can in turn be expressed as the mass of the paddle wheel times the acceleration: $F = ma$. Putting this all together gives us a formula for pressure difference: $P = F/A_p = ma/A_p$. So if voltage corresponds to pressure difference and current to flow velocity, then the derivative of current corresponds to acceleration, and the inductance L corresponds to m/A_p.

Just as a very massive paddle wheel is hard to start turning, but hard to stop once it is moving, tending to maintain a constant velocity, so an inductor tends to maintain a constant current.

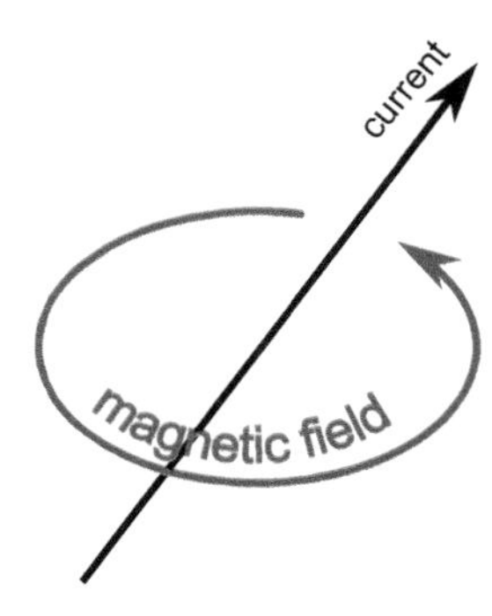

Figure 28.1: *This cartoon illustrates the application of the right-hand rule—if current is flowing in a straight wire, then the magnetic field wraps around the wire as shown. The name of the rule comes from a mnemonic for remember the direction of the field—if your right thumb points in the direction of the current, then when you curl the fingers of your right hand, they point in the direction of the field.*

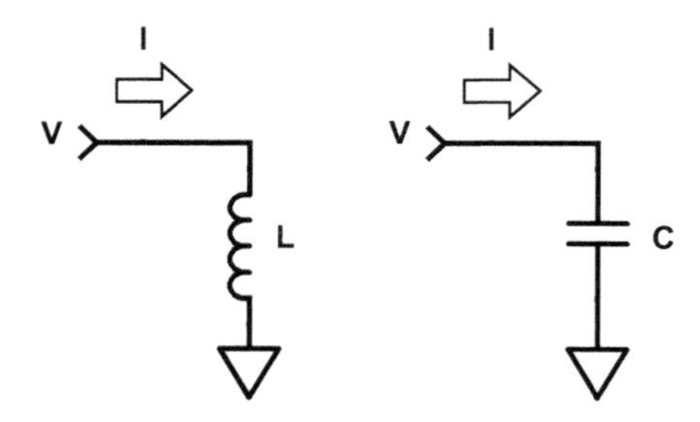

Figure 28.2: *If we use the same direction current flow for both inductors and capacitors as shown here, then the differential equations are* $v = L\frac{di}{dt}$ *and* $i = C\frac{dv}{dt}$. *A positive voltage causes the current through the inductor to increase, and a positive current causes the voltage across the capacitor to increase.*
Physicists often prefer to use the opposite convention for the direction of current flow in inductors, to emphasize that the voltage induced by a change in current tends to oppose the change in current—inductors tend to have a constant current, just as capacitors tend to have a constant voltage.

The unit of inductance is the *henry*, named after the 19th century physicist Joseph Henry, written as a capital H. The henry can be expressed in terms of various other electrical units:

$$H = \frac{\mathrm{V}}{\mathrm{A/s}} = \frac{\mathrm{Vs}}{\mathrm{A}} = \Omega s = \frac{\mathrm{s}^2}{\mathrm{F}} = \frac{\mathrm{Tm}^2}{\mathrm{A}},$$

where V is volts, A is amperes, s is seconds, Ω is ohms, F is farads, T is teslas, and m is meters.

In addition to the *self inductance*, described in this section, there is another notion of *mutual inductance* between two inductors whose magnetic fields are shared. Although mutual inductance is essential for understanding transformers, we will not be covering it (or transformers) in this course.

28.2 Computing inductance from shape

This section is not essential to the course, but provides some additional information about inductors that is sometimes useful for reasoning about inductors.

Shape	$G = \frac{2\pi}{\mu_0}\frac{L}{w}$
Circle	$\ln\left(\frac{8R}{r}\right) - 2 + Y/4$
Single-layer coil (solenoid approximation)	$\frac{w}{2l}$
Single-layer coil (Wheeler's approximation)	$\frac{w}{1.78R+1.97l}$
Parallel wires	$2\ln\left(\frac{S}{r}\right) + Y/2$
Parallel wires (high frequency)	$2\cosh^{-1}\left(\frac{S}{2r}\right)$
Coax cable (high-frequency approximation)	$\ln\left(\frac{R}{r}\right)$
Straight wire	$\ln\left(\frac{8w}{r}\right) - 1 + Y/4$

Table 28.1: *Inductance of various shapes can be estimated from the parameters of the shape:* $L = \frac{\mu_0}{2\pi}wG$*, where* μ_0 *is the magnetic constant,* w *is the length of the wire (just one of the wires for parallel wires or coax), and* G *is a geometric factor that depends on the shape of the wire.*
The magnetic constant used to be an exact value, but is now a measured value: $\frac{\mu_0}{2\pi} \approx 0.2\,\mu\text{H/m}$.
R *is the radius of a coil (at the center of the wires) or the inner radius of the outer conductor of a coax cable.*
r *is the radius of the wire.*
N *is the number of turns of a coil.*
D *is the center-to-center distance between two parallel wires.*
Y *is a skin factor, between 0 and 1 that is near 1 for low frequencies, but shrinks for higher frequencies. [130, 156] All the formulas here assume that the wires are in air or vacuum and that the relative permeability of the wires is approximately 1. Formulas rearranged from Wikipedia table [130].*

The inductance of a conductor depends mainly on the length of the wire and a multiplicative factor depending on the geometry of the conductor. We can write the inductance as

$$L = \frac{\mu_0}{2\pi}wG,$$

where μ_0 is the *magnetic constant*, w is the length of the wire, and G is a unitless geometric factor. The magnetic constant used to be defined as $\mu_0 = 4\pi 10^{-7}\frac{\text{Tm}}{\text{A}}$, but it is now a measured value which is very close to that value. We can use $\frac{\mu_0}{2\pi} \approx 0.2\,\mu\text{H/m}$.

Table 28.1 gives approximations for the geometric factor for several common wire configurations. I don't expect anyone to remember the geometric factors for inductance, and we won't use them directly—the main thing to note is that inductance increases linearly with wire length. When you need to keep inductance to a minimum (for example, to prevent ringing in the ground wire of an oscilloscope probe), keep wires short.

Inductance can be greatly increased if we coil up the wire to make the magnetic field more intense—short, tight coils increase the inductance more.

Several of the formulas have a correction for *skin effect*. At low frequency in thin wires (most of what we do in this class), the skin effect is negligible and $Y = 1$. At high frequencies (radio electronics) or in thick wires (power-grid electronics), the current mainly travels on the surface of the wire, and Y drops to 0. (Actually, at high frequencies things get even more complicated, but we'll only deal with low frequencies in this class.)

Many of the configurations in Table 28.1 are impractical to work with, and formulas have been devised for computing the inductance of many different conformations of inductors, including flat ones that can be built directly on circuit boards [73]—formulas and calculators for other common configurations are easily found on the web. We won't be computing inductance for layouts of wires in this class, but you should be aware that it is possible to design inductors to have specific inductances.

The air-coil inductors mentioned above often don't have enough inductance, either for their size or for the resistance of the wire, so another technique is used to increase inductance: adding a high-permeability *core* to the coil. The high-permeability material increases the magnetic field for a given current and greatly increases the inductance. The Wikipedia article on magnetic cores [134] has pictures of some of the standard shapes for magnetic cores.

28.3 Impedance of inductors

Most of the math in Section 28.2 was stuff to read once and forget—it was there just to give you three concepts: increasing wire length increases inductance, coiling up wires increases inductance, and adding ferromagnetic cores increases inductance. The math in this section is more important—you will use the impedance of inductors for design problems. The derivation of the impedance is something that you should understand, and the formula itself is important enough to memorize.

So, let's derive impedance for an inductor. Remember that impedance is the ratio of voltage to current and that we model both voltage and current with complex-valued sinusoids. For simplicity, let's say that the current through an inductor is $i(t) = e^{j\omega t}$. Then the voltage across the inductor is

$$\begin{aligned} v(t) &= L\frac{\mathrm{d}i(t)}{\mathrm{d}t} \\ &= L\frac{\mathrm{d}e^{j\omega t}}{\mathrm{d}t} \\ &= j\omega L\, e^{j\omega t} \\ &= j\omega L\, i(t)\ , \end{aligned}$$

and the impedance is

$$v(t)/i(t) = j\omega L\ .$$

> The impedance of an inductor is
>
> $$Z_{\mathrm{L}} = j\omega L\ .$$
>
> This formula is worth the trouble of memorizing.

For DC ($\omega \to 0$ radians/s), the impedance is $0\,\Omega$, and at very high frequencies the impedance is large.

Because long, thin wires are used to make inductors, and wires are not perfect conductors, we often end up having a resistance in the wire large enough to matter, so it is common to model real-world inductors as an inductance and a resistance in series, as shown in Figure 28.3.

The magnitude of the impedance is dominated by the resistance at low frequencies and by the inductance at high frequencies. We have a *corner frequency* where the two contribute equally: ($R = |j\omega L|$ or $f = \frac{R}{2\pi L}$). See Figure 28.4 for how the magnitude of impedance changes with frequency for a typical ferrite-core inductor (an air-core inductor would usually have a larger resistance for the same inductance). The plot was created with the gnuplot script in Figure 28.5.

Figure 28.3: *We often model actual inductors as an inductance (L_1) in series with the resistance of the wire used to make the inductor (R_1).*
Figure drawn with Digi-Key's Scheme-it [18].

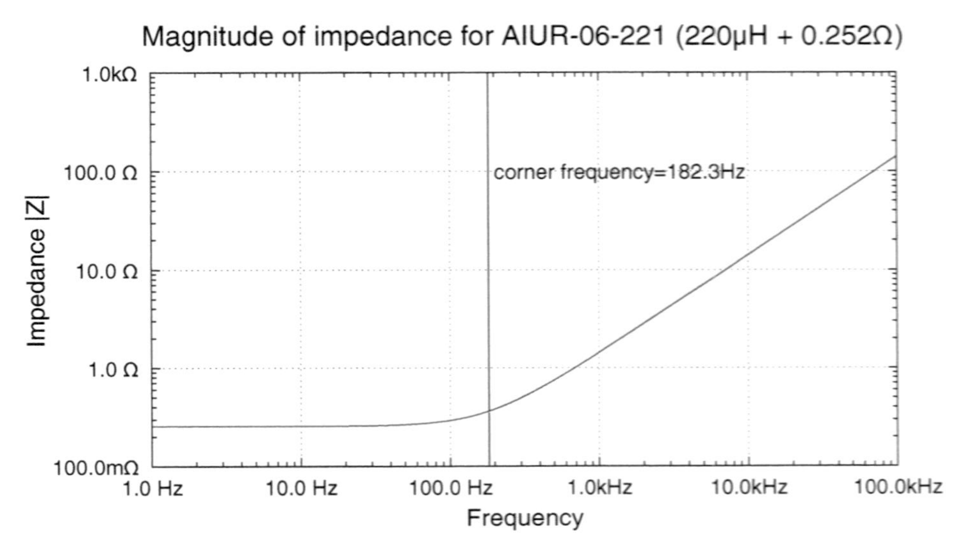

Figure 28.4: *Using the model of Figure 28.3 for the AIUR-06-221 inductor with $L_1 = 220\,\mu\text{H}$ and $R_1 = 0.252\,\Omega$, we can compute the impedance for any frequency. Below the corner frequency $R_1/(2\pi L_1)$, the impedance is mainly from the R_1 resistance, and above the corner frequency, the impedance is mainly from the L_1 inductance.*
Data collected with Analog Discovery 2 [19]. Figure drawn with gnuplot [33].

```
load '../definitions.gnuplot'

# values for inductor AIUR-06-221
L=220E-6
R=0.252

set title \
  sprintf("Magnitude of impedance for AIUR-06-221 (%.0f{/Symbol m}H + %.3f{/Symbol W})",\
          L*1e6,R) font 'Helvetica,16pt'
set margins 12,6,4,4

set ylabel "Impedance |Z|" font 'Helvetica,14pt'
set format y "%.1s%c{/Symbol W}"

set xlabel "Frequency" font  'Helvetica,14pt'
set format x '%.1s%cHz'
set xrange [1:1e5]

corner_freq = R/(2*pi*L)        # compute and show corner frequency
unset arrow
unset label
set arrow nohead from corner_freq,0.1 to corner_freq,1000
set label sprintf("corner frequency=%.1fHz",corner_freq) at corner_freq*1.1, 100

plot abs(R+Zl(x,L)) notitle
```

Figure 28.5: *This is the gnuplot script used to generate Figure 28.4, which uses the* `definitions.gnuplot` *script of Figure 7.4. The plot is of the* magnitude *of impedance: abs(R+Zl(x,L)).*

Capacitors usually have a maximum voltage, while inductors usually have a maximum current.

We'll be using a ferrite core inductor in Lab 11. The ferrite core puts a current limit on the inductor, called the *saturation current* for the inductor, because the magnetic field saturates in the core. Another current limit is a thermal one based on the diameter of the wire used: the *current rating* for the inductor. These two ratings are generally within a factor of about 5 of each other, but either may be the larger one. Designs should not exceed the tighter of the two limits.

If you are interested in reading more about inductance and inductors, start with the Wikipedia article on inductance [129].

28.4 LC resonators

Inductors are often paired with capacitors to make resonant circuits (also called *tanks* or *LC tanks*), as shown in Figure 28.6. The behavior of these circuits is easily derived from the formulas for their impedance.

Let's look at the series circuit first. At very low frequencies, the impedance of the inductor is negligible, and so the series circuit acts like the capacitor. At very high frequencies, the impedance of the capacitor is negligible, and so the series circuit acts like the inductor.

At intermediate frequencies, we can look at the formula for the series L_1C_1 impedance:

$$j\omega L_1 + \frac{1}{j\omega C_1} = \frac{1-\omega^2 L_1 C_1}{j\omega C_1},$$

which is zero when $\omega = 1/\sqrt{L_1C_1}$.

We can interpret the formula as saying that the series circuit switches from capacitor-like behavior at low frequencies (high impedance) to inductor-like behavior at high frequencies (again, high impedance) by passing through zero impedance at the resonant frequency.

The parallel circuit has the opposite behavior. At very low frequencies, the parallel circuit has very low impedance, because of the inductor, and at very high frequencies it has very low impedance because of the capacitor.

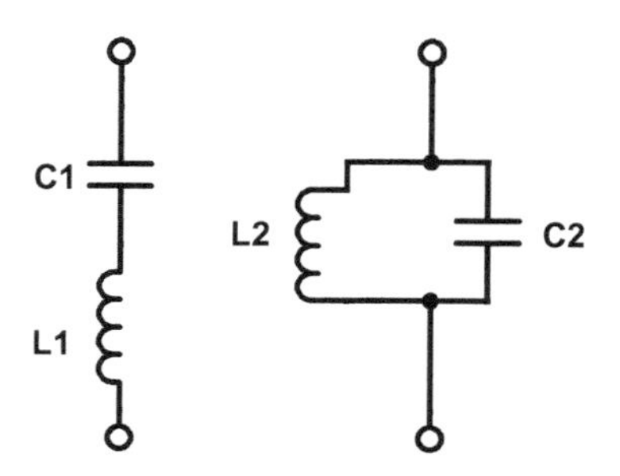

Figure 28.6: *Resonant circuits can be made by pairing an inductor and a capacitor. The series circuit has a minimum at the resonant frequency $\omega = 1/\sqrt{LC}$, while the parallel circuit has a maximum there.*

At intermediate frequencies, we can use the formula for the parallel L_2C_2 impedance:

$$\frac{1}{1/(j\omega L_2) + j\omega C_2} = \frac{j\omega L_2}{1 - \omega^2 L_2 C_2},$$

which becomes infinite at the resonant frequency, where $\omega = 1/\sqrt{L_2 C_2}$.

That means that the parallel circuit switches from inductor-like behavior at low frequencies (low impedance) to capacitor-like behavior at high frequencies (again, low impedance), passing through infinite impedance at the resonant frequency.

There is a nice online demo of voltage and current sloshing back and forth between an inductor and a capacitor in a series LRC resonant circuit at https://falstad.com/circuit/—it may be worth a few minutes to play with the simulator there to get a more intuitive view of how LC circuits behave.

Exercise 28.1

What is the resonant frequency of a 100 μH inductor in parallel with a 100 nF capacitor? Use gnuplot to plot the magnitude of the impedance from 10 Hz to 1 MHz on a log-log plot. [*Hint*: you need to use at least 10,000 samples to see the resonant spike clearly.]

Chapter 29

Loudspeakers

29.1 How loudspeakers work

A loudspeaker consists of a *voice coil*, a low-mass coil of wire, attached to a stiff, low-mass (often paper) *cone*. Movement of the coil moves the cone, which in turn moves the air, producing sound. The voice coil sits in a magnetic field that is radially oriented, generated by a permanent magnet. See the illustrations in Figures 29.1 and 29.2.

When current is passed through the coil, it produces a force that is perpendicular to both the wire and the magnetic field—that is, a vertical force in the orientation shown in Figure 29.1. The force produced is approximately proportional to the current (not exactly, because the magnetic field is not really uniform). This force moves the voice coil and the cone up or down. The supports for the cone (the "spider" and "surround") act as a spring to provide a restoring force, so that the displacement of the cone is roughly proportional to the current.

The schematic symbol is shown in Figure 29.3. There is no polarity symbol on the loudspeaker, as it will work no matter which way around the voice coil is connected, though the voice coil will move in opposite directions if the wires are swapped. Our ears are not particularly sensitive to the phase of sound waves, so reversing the loudspeaker is not usually perceptible. When connecting up several loudspeakers, however, it does matter which way the loudspeakers are connected, as the interference patterns from the interaction of sound waves from different speakers do depend on the phase of the sounds from the speakers.

See https://en.wikipedia.org/wiki/Voice_coil for more information about how loudspeakers work.

Each year for the parts kit, we get the cheapest mid-range loudspeaker we can find that can handle at least 5 W. Because these parts are often factory closeout specials, which parts we get vary from year to year, and good specifications are often unavailable. We often only get a nominal impedance for the loudspeaker (typically 8 Ω, though we may get 4 Ω or 16 Ω loudspeakers). Data sheets for loudspeakers often give two different values in ohms: the magnitude of the impedance and the DC resistance. In Lab 8 you will measure both. The impedance is not really a single number though, but a function of frequency, and you will model the impedance of the loudspeaker as a combination of different linear components (inductors, resistors, and capacitors).

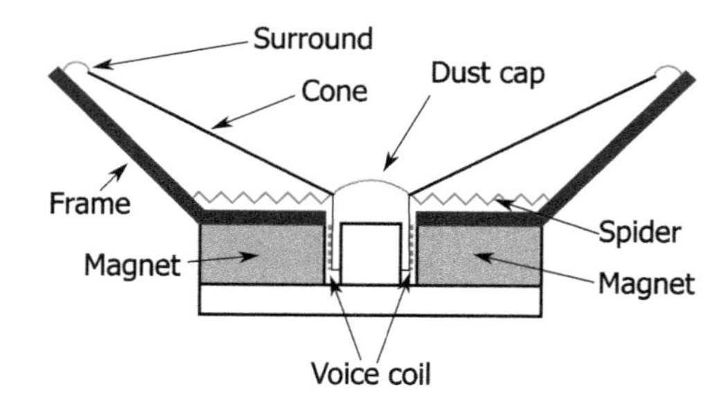

Figure 29.1: Cross-sectional view of a typical loudspeaker, showing the cone, the voice coil, and the permanent magnet. Figure adapted from https://en.wikipedia.org/wiki/File:Loud_Speaker_Schemata.svg (Creative Commons License 3.0 Attribution Share-Alike unported). Figure drawn with Inkscape [45].

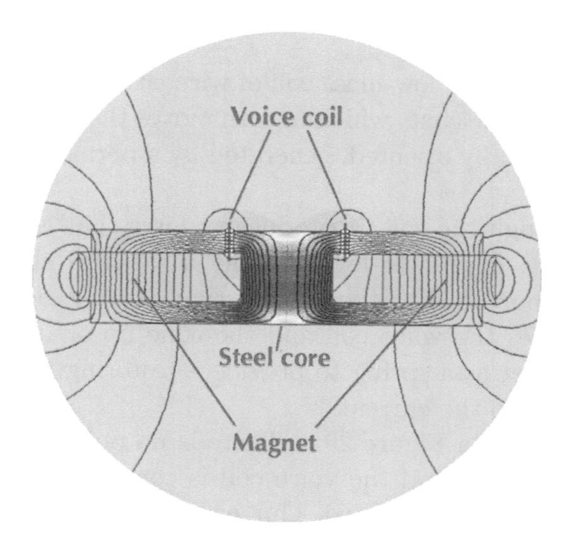

Figure 29.2: *Cross-sectional view of the magnetic field of a loudspeaker, as computed by Finite Element Method Magnetics (FEMM) [69]. The field is radial (always perpendicular to the wires of the voice coil) in the magnetic gap where the voice coil sits. Image adapted with permission from https://www.femm.info/examples/woofer/woofer-ans.gif.*

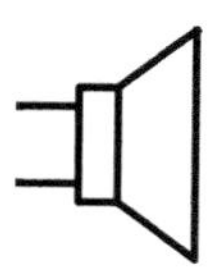

Figure 29.3: *The schematic symbol for a loudspeaker is intended to be physically iconic, with the trapezoid corresponding to the cone and the rectangle corresponding to the permanent magnet and voice coil. The two connections are to the two ends of the voice coil. Figure drawn with Digi-Key's Scheme-it [18].*

29.2 Models of loudspeakers

29.2.1 *Models as electronic circuits*

29.2.1.1 *R and RL models for loudspeaker*

The simplest model of a loudspeaker is that it is a constant impedance (the nominal $8\,\Omega$ or $4\,\Omega$ value). This model works fairly well, but only when the loudspeaker is used in the frequency range for which it is intended.

(a) *This model for a loudspeaker includes R_1 for the behavior at frequencies where the loudspeaker is usually used and L_1 for behavior at high frequencies.*

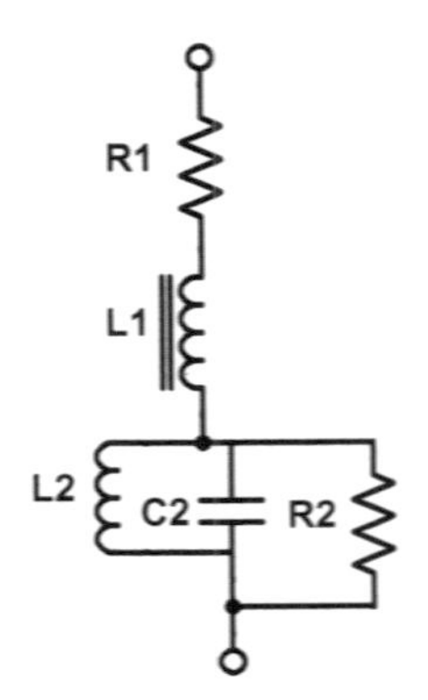

(b) *This model for a loudspeaker includes R_1 for the behavior at frequencies where the loudspeaker is usually used, L_1 for behavior at high frequencies, and L_2, C_2, R_2 for the mechanical resonance peak.*

Figure 29.4: *Two simple models that we can fit to the electrical measurements we make of our loudspeakers. Figure drawn with Digi-Key's Scheme-it [18].*

A slightly more sophisticated model has a voice-coil inductance in series with a DC resistance (as in Figure 29.4(a)). At low frequencies, this model is equivalent to the constant-impedance model, but above the corner frequency, the impedance goes up, behaving more like an inductor.

This model is just a simplified way of expressing the mathematical behavior of a complicated 2-port device—there is no physical counterpart to the node between the inductor and resistor that we could stick a voltmeter or oscilloscope on. Building mathematical models out of standard components allows easy construction of complicated functions, and sometimes allows reasoning about the real world by analyzing the model, but the model is always an analogy, not the reality.

Exercise 29.1

If you know the part number for the loudspeaker you will be using for Lab 8, look up a data sheet and try to determine power limit(s), the DC resistance, and the inductance that the speaker is supposed to have. If you don't have a data sheet, use typical small mid-range speaker values: $5\,\text{W}$, $8\,\Omega$, $0.5\,\text{mH}$.
Use gnuplot to plot the magnitude of the impedance of the loudspeaker from $1\,\text{Hz}$ to $1\,\text{MHz}$, modeling it as just a series inductor and resistor with the values on the data sheet (this should be a log-log plot). What is the lowest impedance over this range? the highest?
Use the definitions.gnuplot file given in Figure 7.4.

One of the key decisions an engineer has to make is what level of detail is needed in a model—how good does the analogy have to be for the design task? For some purposes, the simple resistor model of a loudspeaker is good enough, but for other purposes, such as determining the electrical behavior at high frequencies, adding the inductor is essential. But even the resistor-plus-inductor model is often not detailed enough.

29.2.1.2 *Loudspeaker model with RLC for mechanical resonance*

One of the other important properties of a loudspeaker has to do with its mechanical design: the mass of the voice coil and cone and the stiffness of the spring supporting them. As you may remember from your physics class, a mass and spring form a *harmonic oscillator* that vibrates at a resonant frequency. At that resonant frequency, even small currents produce large motions of the voice coil, in turn producing large voltages—that is, the impedance at that frequency is large. Loudspeakers are usually used only at frequencies above their mechanical resonant frequency, both to avoid distortion of the sound and to avoid damage to the loudspeaker.

The speakers we get for the lab are usually *mid-range* speakers, with a resonance around 200 Hz. *Tweeters* are designed for high frequencies and have a resonance above 1 kHz, while *woofers* are designed for low frequencies and have a resonance below 100 Hz.

> We can model the resonance peak of the loudspeaker with a parallel arrangement of a resistor, inductor, and capacitor. Since the resonance peak adds to the baseline impedance, we can put this RLC circuit in series with the DC resistance and inductance of our previous model, as shown in Figure 29.4(b).

In Section 28.4, we saw that an inductor and capacitor in parallel have an infinite impedance at their resonant frequency, and that frequency depends only on the product of the inductance and capacitance LC. If we then add a resistor in parallel, the impedance at the resonant frequency is just that resistance, with lower impedances at other frequencies. How rapidly the impedance drops on either side depends on the size of the capacitor—a large capacitance (hence small inductance) causes the impedance to drop more quickly near the peak.

Exercise 29.2

Plot the magnitude of impedance for $100\,\Omega \parallel 220\,\mu\text{H} \parallel 47\,\text{nF}$ and for $100\,\Omega \parallel 4.7\,\mu\text{H} \parallel 2.2\,\mu\text{F}$ from 100 Hz to 10 MHz on the same log-log plot. What is the resonant frequency for each? What is the impedance at the peak?

Exercise 29.3

Write out the equations for the complex impedance of each of the models in Figure 29.4.

29.2.1.3 *Loudspeaker model with nonstandard impedance*

The model in Figure 29.4(b) does a pretty good job of fitting the behavior of the loudspeaker, except at higher frequencies, because the high frequency impedance does not grow as fast as would be implied by an inductor.

We can get a much better fit if we use a nonstandard component in place of L_1—one that has an impedance of $(j\omega\ 1\,\mathrm{s})^\alpha R_\alpha$ instead of $j\omega L_1$.

The variables α and R_α are not standard named symbols like L for inductance—they are arbitrarily chosen here to let us specify a nonstandard model. The multiplication by 1 s makes the number being exponentiated unitless, so that R_α can be in units of ohms.

We can interpret R_α as the magnitude of impedance when $\omega = 1\,\mathrm{radian/s}$, that is, at about 0.1592 Hz. If we multiply α by 90°, we get the phase of the impedance.

The inductor-like component $(j\omega\ 1\,\mathrm{s})^\alpha R_\alpha$ for $0 < \alpha \leq 1$ used here exponentiates j as well as ω. That exponentiation means that the impedance on the complex plane lies along a line with angle $90°\alpha$, not on the real axis (like a resistor, with $\alpha = 0$) nor the imaginary axis (like a capacitor with $\alpha = -1$ or inductor with $\alpha = +1$). I have seen "semi-inductance" used as a name for R_α when α is fixed at 0.5.

We can even generalize the notion of an RC or L/R time constant to the nonstandard component, with time constant $(R_\alpha/R)^{1/\alpha}\ 1\,\mathrm{s}$, for $\alpha \neq 0$.

In Lab 8, we'll collect the data needed to fit each of the models described above, fit them with gnuplot, and see how good the model is. Figure 29.5 shows the three models for a 16 Ω bass shaker.

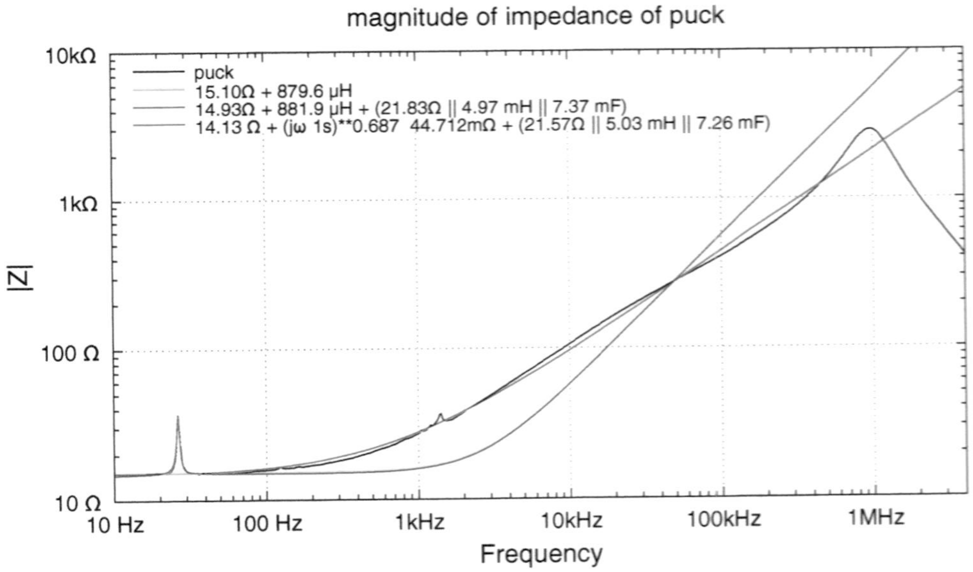

Figure 29.5: *The three proposed models are shown here for data collected from a "puck tactile transducer mini bass shaker", which has a very low resonant frequency. The secondary resonance (around 1.4 kHz) is not modeled.*

The simple R+L model (green) does not include the resonance peak. The R+L+RLC model (blue) fits the resonance peak, but is otherwise almost identical to the R+L model. The nonstandard inductance model (red) is almost identical to the R+L+RLC model at the resonance peak, but fits the data well over the entire range.

We can get a good fit to higher frequencies (up to ≈ 10 MHz) if we include a capacitor of about 100 pF in parallel with the final model.

Data collected with Analog Discovery 2 [19]. Figure drawn with gnuplot [33].

29.2.1.4 *Resonance with nonstandard impedances*

This section is optional, as we will not have to do any modeling of resonances with nonstandard impedances, though the methods can be applied to modeling high-frequency electrical behavior of loudspeakers (above 1 MHz) and to determining the resonant frequency of the LC-loudspeaker filter in Section 35.4.

In Section 28.4, we saw that we got interesting resonance behavior from combining an inductor and a capacitor, and in Section 29.2.1.3, we saw a way to capture inductance, capacitance, resistance as special cases of nonstandard impedance. Can we get resonance phenomena by combining nonstandard impedances?

Let's consider the case of two nonstandard impedances in series:

$$Z = (j\omega)^\alpha R_\alpha + (j\omega)^\beta R_\beta \ .$$

The first thing to do is to replace j by the polar form $e^{j\pi/2}$, getting

$$Z = e^{j\alpha\pi/2}(\omega\ 1\,\mathrm{s})^\alpha R_\alpha + e^{j\beta\pi/2}(\omega\ 1s)^\beta R_\beta \ .$$

Then we can apply Euler's formula to get the real and imaginary parts:

$$\Re(Z) = \cos(\alpha\pi/2)(\omega\ 1\,\mathrm{s})^\alpha R_\alpha + \cos(\beta\pi/2)(\omega\ 1\,\mathrm{s})^\beta R_\beta \ ,$$

$$\Im(Z) = \sin(\alpha\pi/2)(\omega\ 1\,\mathrm{s})^\alpha R_\alpha + \sin(\beta\pi/2)(\omega\ 1s)^\beta R_\beta \ .$$

We will get resonance whenever the imaginary part goes to zero:

$$0\,\Omega = (\omega\ 1\,\mathrm{s})^{\alpha-\beta}\left(R_\alpha \sin(\alpha\pi/2) + R_\beta \sin(\beta\pi/2)\right) \ ,$$

which can be solved for ω to get

$$\omega = \left(\frac{-R_\beta \sin(\beta\pi/2)}{R_\alpha \sin(\alpha\pi/2)}\right)^{1/(\alpha-\beta)} s^{-1} \ .$$

We get a resonance whenever α and β have opposite signs.

The special case of an inductor and a capacitor sets $\alpha = 1$, $R_\alpha = L/1\,\mathrm{s}$, $\beta = -1$, and $R_\beta = 1\,\mathrm{s}/C$, yielding

$$\omega = \left(\frac{R_\beta}{R_\alpha}\right)^{1/2} s^{-1} = \frac{1}{\sqrt{LC}} \ ,$$

which is the standard result.

We can also deal with parallel rather than series impedances by looking at the sum of admittances (admittance is one over impedance), instead of the sum of impedances. To get the admittances, the exponents α and β get negated and the coefficients R_α and R_β inverted, giving us

$$\omega = \left(\frac{-R_\alpha \sin(\beta\pi/2)}{R_\beta \sin(\alpha\pi/2)}\right)^{1/(\beta-\alpha)} \mathrm{s}^{-1} \ .$$

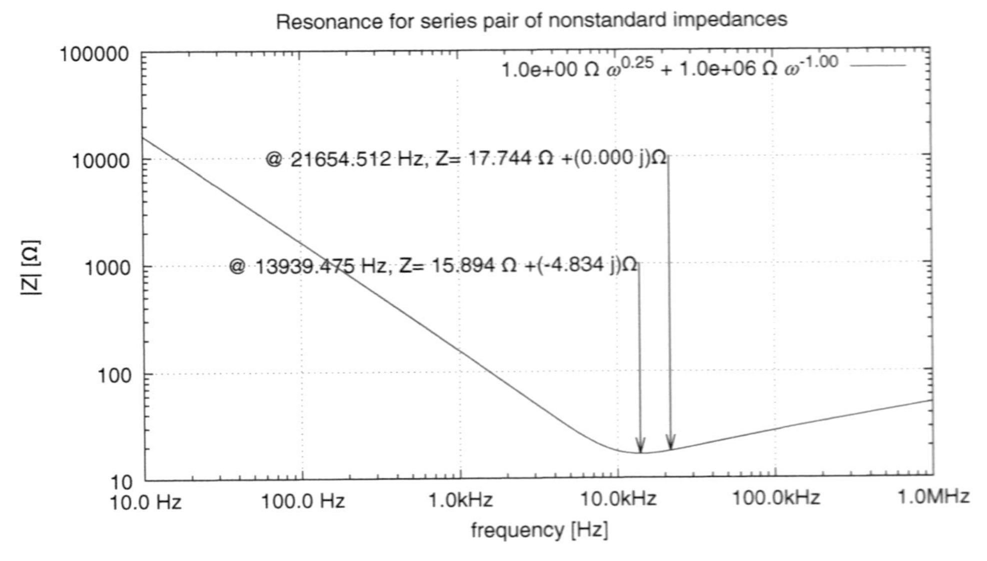

Figure 29.6: *This figure shows a resonance of two series impedances, where one is nonstandard, slightly inductor-like impedance, and the other is* $1\,\mu F$ *capacitor. The zero-phase definition of the resonant frequency is at a substantially higher frequency than the minimum-magnitude definition.*

The entire analysis so far is a simple one that just looks at where the imaginary part goes to $0\,\Omega$. The minimum magnitude does not necessarily occur at the same frequency as where the imaginary part disappears, unless $\alpha + \beta = 0$, though the frequencies are usually fairly close. A more complicated analysis that looks at where the magnitude of the impedance is minimized can be found on my blog [51]. The solution is more complicated—if we define $u = (\omega\ 1\,\mathrm{s})^{(\alpha-\beta)} R_\alpha / R_\beta$, then the series resonances are

$$u = \frac{-(\alpha+\beta)\cos\left(\frac{\pi(\alpha-\beta)}{2}\right) \pm \sqrt{(\alpha+\beta)^2\cos^2\left(\frac{\pi(\alpha-\beta)}{2}\right) - 4\alpha\beta}}{2\alpha},$$

when those solutions are real values for u. Figure 29.6 shows the difference between the two definitions of resonance.

29.2.2 *Fitting loudspeaker models*

Fitting a model to data is the process of choosing values for the parameter for the model to minimize the error of the model. The error is the difference between the data and the model, and the fitting algorithm is trying to minimize the sum of the square of the errors.

Think of a single-parameter constant model (trying to summarize the data in one number). If the errors are more positive than negative, then increasing the value for the model will reduce the squared errors. When the positive and negative errors are balanced (sum of squares of positive errors about the same as the sum of squares of negative errors), then you've got a fit. For a constant model, the least-square fit is the arithmetic mean of the data.

Visually, the error is a vertical distance on a linear plot, and the least-square fit provides a visually balanced fit. If the plot has a logarithmic scale on the vertical axis, then the visually balanced fit requires that the error whose square is being minimized be the difference between the *log* of the model and the *log* of the data.

Changing the definition of the error (from data – model to log(data) – log(model)) can change the fit substantially, particularly if the data values have a wide range.

The impedance spectrum that we are trying to fit (Figure 29.5) has a wide range of impedance values and a wide range of frequencies, so we plotted it on a log-log scale. To get a good fit on this plot, we want errors to be minimized on the log scale. Fitting the log of the model to the log of magnitude of the impedance—rather than directly fitting the model to the magnitude—avoids putting all the weight on the high-frequency, high-impedance data points, which is what would happen otherwise.

Taking logarithms makes the fitting be concerned about the *ratio* of the model and the data rather than *difference*. The differences are much larger when the impedance is larger, which would have given undue weight to the high impedances at high frequencies, if we had not taken logarithms. When fitting the logarithms of the model and the data, the errors are balanced when the data is plotted with a log scale on the y-axis.

To get the plot and fits of Figure 29.5, I used two data files: `puck-10Hz-10MHz.txt`, which has 601 samples logarithmically spaced over 6 decades (100 samples per decade), and puck-20Hz-40Hz.text, which has 101 samples linearly spaced (every 0.2 Hz) around the resonant peak. It would have been possible to fit the models using just the data from `puck-10Hz-10MHz.txt`, but the extra data around the resonant peak allows closer fitting of the parameters for the resonance.

Figure 29.7 has the script used to create Figure 29.5. Let's walk though the parts of the script. First, we include the definitions.gnuplot script from Figure 7.4—in this instance, that file was located one directory higher in the hierarchy, hence the "../" before the file name. Next we set the ranges for both x and y to include all the data, so that fitting is not limited to a narrow range of points (unless specifically narrowed in the fit command).

There are three blocks of text for defining and fitting the models. The models are defined using the formulas for complex impedance defined in definitions.gnuplot, and each is given with all the parameters of the model passed as parameters of the function. Using explicitly passed parameters makes it easy to ensure that there is no sharing of values between models—although both model2 and model3 use Rp, Lp, and Cp locally for the parallel RLC circuit, when we do the fitting we pass in different variables for the two models (R2, L2, and C2 or R3, L3, and C3).

> Never use the same variable names at the top level of the script for different models—the best fit for one model is unlikely to be the best fit for a different model!

Finding the right parameters for the models can be quite tricky, as automatic fitting procedures such as those in gnuplot may not do what you want them to. After defining the models, we need approximate starting values for the parameters, so that the fitting converges to reasonable values. We can also restrict the range of frequencies when fitting a model, so that the fitting is not confused by data for which the model is not applicable.

For example, when fitting the first R+L model, we want to ignore the resonant spike rather than raising the value of R_1 to average the spike with the frequencies around it. The easiest way to fix this problem is to specify a frequency range in the fitting commands, as shown in the first section of Figure 29.7. Fitting R_1 on just the low frequencies and L_1 on just the high frequencies avoids the problem of the resonance peak distorting the fit of the simple model. The exact ranges to use should be determined after looking at the data and seeing what range of frequencies the resonance peak covers.

I fit L_1 just on the middle of the data, avoiding both the resonant peak and the rapid drop after 1 MHz. To model the loudspeaker at higher frequencies, I would need to add a capacitor in parallel with

```
load '../definitions.gnuplot'
set xrange [*:*]
set yrange [*:*]

model1(f, R1, L1)= R1+Zl(f, L1)
R1=15; L1=900e-6    # initial guess for R+L loudspeaker parameters
fit [*:20] log(abs(model1(x, R1, L1))) 'puck-10Hz-10MHz.txt' \
      u 1:(log($3)) via R1
fit [100:1e6] log(abs(model1(x, R1, L1))) 'puck-10Hz-10MHz.txt' \
      u 1:(log($3)) via L1

f_r = 26.       # approximate resonant frequency
model2(f, Rs, Ls, Rp, Lp, Cp)=  Rs + Zl(f, Ls) + Zpar3(Rp, Zl(f, Lp), Zc(f, Cp))
Rs=R1; Ls=L1;   # copy from model 1
R2=36-R1; L2=1/(2*pi*f_r); C2=1/(2*pi*f_r)   # initial guesses
fit log(abs(model2(x, Rs, Ls, R2, L2, C2))) 'puck-20Hz-40Hz.txt' \
      u 1:(log($3)) via L2, C2, R2, Rs
fit [100:1e6] log(abs(model2(x, Rs, Ls, R2, L2, C2))) 'puck-10Hz-10MHz.txt' \
      u 1:(log($3)) via Ls

model3(f, Rs, Ls, Rp, Lp, Cp, power)= \
        Rs + Znonstandard(f, Ls, power) + Zpar3(Rp, Zl(f, Lp), Zc(f, Cp))
Rs3=Rs; R3=R2; L3=L2; C3=C2     # copy from model2
Ls3=45*Ls; power=0.7            # adjust from model2, based on slope
fit [100:700e3] log(abs(model3(x, Rs3, Ls3, R3, L3, C3, power))) \
        'puck-10Hz-10MHz.txt' u 1:(log($3)) via Ls3, power
fit log(abs(model3(x, Rs3, Ls3, R3, L3, C3, power))) \
        'puck-20Hz-40Hz.txt' u 1:(log($3)) via R3, L3, C3, Rs3

set title 'magnitude of impedance of puck' font 'Helvetica,14'
set key top left  reverse font 'Helvetica,10'
set samples 10000;     set style data lines
set grid;              set logscale xy
set xrange[10:4e6]
set format x '%.0s%cHz'
set xlabel 'Frequency' font 'Helvetica,14'
set format y '%.0s%c{/Symbol W}'
set yrange[10:10e3]
set ylabel '|Z|' font 'Helvetica,14'
OHM = "{/Symbol W}"

plot \
    'puck-10Hz-10MHz.txt' u 1:(($1<20 || $1>40)?$3:NaN) title 'puck' lc 'black',\
    'puck-20Hz-40Hz.txt' u 1:3 notitle lc 'black', \
    abs(model1(x, R1, L1)) \
        title sprintf('%.2f%s + %.1f {/Symbol m}H', R1,OHM, L1*1E6) lc 'green', \
    abs(model2(x, Rs, Ls, R2, L2, C2)) title sprintf(\
       '%.2f%s + %.1f {/Symbol m}H + (%.2f%s || %.2f mH || %.2f mF)', \
             Rs,OHM, Ls*1E6, R2, OHM, L2*1E3, C2*1E3) lc    'blue', \
   abs(model3(x, Rs3, Ls3, R3, L3, C3, power)) title sprintf(\
     '%.2f %s + (j{/Symbol w} 1s)**%.3f  %.3fm%s + (%.2f%s || %.2f mH || %.2f mF)', \
             Rs3,OHM, power, Ls3*1e3,OHM, R3,OHM, L3*1E3, C3*1E3) lc 'red'
```

Figure 29.7: *This gnuplot script fits three models to the data collected from a nominally 16Ω bass shaker, mounted on a small piece of plywood, then plots the models labeled with the model parameters.*

the whole model—because the model does not include such a capacitance, I am only using the data where the model can reasonably be fitted.

To get initial values for the parameters, we can estimate R_1 by looking at the measured impedance at low frequencies, and we can estimate L_1 by looking at the impedance at some high frequency f and using $|Z| = 2\pi f L$. These estimates do not have to be very precise, as the fitting will correct them.

For the second model, which added a resonance peak, we can copy the R_1 and L_1 values (as RS and LS), but we need to make estimates for R_2, L_2, and C_2 to fit the resonant peak, as the fitting procedure won't converge to reasonable values unless you have a resonant peak in your function near the right place.

You can get initial estimates by looking at the frequency of the peak, f, and using $2\pi f = 1/\sqrt{L_2 C_2}$. You can make the numerical values for your initial guess be the same for both L_2 and C_2: $L_2 = 1\,\text{HzH}/(2\pi f)$ and $C_2 = 1\,\text{HzF}/(2\pi f)$. That choice puts the peak in the right place, but doesn't give it the right shape—the parameter fitting in gnuplot can adjust the shape much more easily than it can find the peak. You can make an initial guess for the resistance by looking at the magnitude of the impedance at the peak: $R_2 = |Z_{\text{peak}}| - R_1$. These guesses are usually good enough to allow gnuplot's fitting procedures to converge in the area around the peak.

If you have collected closely spaced measurements around the resonance peak, use just those values to fit the peak. By including the series resistance in the fit, as well as the parallel R, L, and C, you can get a good fit for the shape of the peak.

Once you have fitted the peak, you probably want to refit the L_1 for the high frequencies, as the changed fit around the peak may have shifted the best fit a little for the series inductor.

Fitting the nonstandard inductor model starts by copying the parameters for the resonant peak from the second model, then fitting R_α and α simultaneously. A good initial estimate of α can be found from looking at the slope of the line on the Bode plot—for loudspeakers it is generally less than 1 decade/decade, often 0.5–0.7. To estimate R_α, I looked at the magnitude of impedance around 50 kHz, where the first two models were getting the impedance about right, and computed what R_α value would work with the initial estimate of α to fit the data there.

I then fit the inductor-like parameters over a slightly narrower range than before, getting a better fit in the audio range at the expense of not fitting the bump around 1 MHz. It took some trial and error to find a range that resulted in a good fit over as much of the data as the model could reasonably fit. Finally, I refit the resonant peak to compensate for the higher impedance of the inductor-like component rather than a simple inductor at those frequencies.

The next block controls how the plot is displayed. Most of the controls here should be familiar by now. We need a large number of samples to get a clean depiction of the very narrow resonant peak from the models, as the default number of samples misses the peak. Formatting the x-ticks with the fancy gnuplot commands for engineering notation produces a more readable labeling.

Although gnuplot supports Unicode now, importing Unicode characters into LaTeX with the `fancyvrb` package does not work, so I used the `{/Symbol W}` form to get Ω, `{/Symbol m}` to get μ and `{/Symbol w}` to get ω.

The plotting is straightforward except for one trick—using a conditional expression to suppress data (replacing the data with NaN, which is the special not-a-number value) from 20 Hz to 40 Hz from the `puck-10Hz-10MHz.txt` file, so that only the more detailed data from puck-20Hz-40Hz.txt is visible there.

If you collect phase information as well as magnitude information, you can fit impedance models more accurately. Fitting both phase and magnitude simultaneously is best done by doing all the fitting with complex number arithmetic, but the gnuplot fitting methods assume that the model and the data are real-valued, so that approach can't be used in gnuplot.

It is possible to fit models to magnitude and phase data in gnuplot, but the process is a bit tricky—you have to alternate fitting the magnitude data and the phase data, and you may have to limit the range of frequencies being fit or which parameters are being fit, to avoid over-fitting to either the phase or the magnitude. Other fitting programs may make fitting to complex-valued data easier.

With the models given in Section 29.2.1, I've generally done fairly well by fitting to just magnitude, with perhaps one iteration of adjusting the resonant peak L_2 and C_2 parameters using the phase information, which captures the frequency of the resonance more clearly.

Exercise 29.4

File https://users.soe.ucsc.edu/~karplus/bme51/book-supplements/CDM-10008-50Hz-5MHz.txt has impedance data for a CUI Inc. loudspeaker (model CDM-10008), collected using an Analog Discovery 2 with their impedance-analyzer board. Plot the magnitude of impedance versus frequency on a log-log plot, and fit the models of Figure 29.4 to the data. You only need to fit the primary resonance (around 1410 Hz), not the secondary resonances near 2720 Hz, 9610 Hz, and 1870 Hz. (Bonus points for fitting more of the resonances, as long as the fit is good.)

29.3 Loudspeaker power limitations

Even if we avoid the resonant frequency of a loudspeaker, providing too large a current can have two damaging effects: heating up the voice coil too much ("burning out" the speaker) or pushing the cone past the limits that the supports can tolerate, tearing the cone or the supports. The limits are usually expressed in terms of the amount of power the loudspeaker can accept. To avoid damaging the loudspeaker, you do not want to exceed the limit for *average power*for long and you should never exceed the limit for *maximum power*. Those limits won't be a problem for Lab 8, but they will be important constraints for Lab 11 in a few weeks.

In loudspeaker specifications, the limit for average power is sometimes referred to as the *RMS power*, but that is really a misnomer, as we are not taking the root of the mean of the square of the *power*, but only of the voltage or current.

So how is RMS involved in the computation of average power? In Section 4.6, we saw that for DC circuits, the power dissipated in a component is $P = VI$. We'd like a definition for AC power that allows a similarly simple computation to figure out how much energy is converted to heat. The total energy converted to heat in time τ for a DC circuit is just $P\tau$, but for an AC circuit we need to integrate power over time: $\int_0^\tau P(t)\,dt$.

For time-varying signals, we can compute *instantaneous power* (power as a function of time) as $p(t) = v(t)i(t)$. The instantaneous power should never exceed the peak-power limit for a loudspeaker. We can get the total energy in the signal by integrating, and the average power is the total energy divided by the duration integrated over.

For a resistive load, $p(t) = v^2(t)/R$, so the total energy is $\int_0^\tau v^2(t)\,dt/R$ and the average power is $\int_0^\tau v^2(t)\,dt/(R\tau)$. If we wanted a single number, V, that would give the average power as $P = V^2/R$ that number would have to be

$$V = \sqrt{\int_0^\tau v^2(t)\,dt/\tau}\ ,$$

which is precisely the definition of the *root-mean-square* (RMS) voltage.

We can use exactly the same reasoning for current, with $p(t) = i^2(t)R$, to get that average power is $\int_0^\tau i^2(t)\,dtR/\tau$, and the single number for a current such that average power $P = I^2R$ is

$$I = \sqrt{\int_0^\tau i^2(t)\,dt/\tau}\ ,$$

the RMS current.

Because $v(t) = i(t)R$ for a resistive load, we also have for RMS voltages that $V = IR$, and we can compute average power as RMS voltage times RMS current. This analysis in terms of RMS voltage or RMS current only works for resistive loads, where the voltage and current waveforms are always in

Waveform	RMS	Crest factor
DC: $V = A$	A	1
Sinusoid: $V = A\sin(\omega t)$	$A/\sqrt{2}$	$\sqrt{2}$
Square wave: $\pm A$	A	1
Pulse: 0 and A, duty cycle D	$A\sqrt{D}$	$1/\sqrt{D}$
Triangle wave: $\pm A$	$A/\sqrt{3}$	$\sqrt{3}$
Gaussian noise: mean 0, std. dev. σ	σ	∞

Table 29.1: *This table lists the* crest factors, *the ratio of peak values to RMS values, for some common waveforms.*

phase with each other. We'll revisit AC power in Section 35.1, to look at what happens in loads with complex impedances, where the voltage and current may be out of phase.

Let's look at what power is dissipated in a resistor for some common time-varying waveforms. If $v(t)$ is constant ($v(t) = k$), then the RMS voltage is the same constant k, so our definition works for DC signals. If $v(t) = A\sin(\omega t)$, then the instantaneous power is $A^2\sin^2(\omega t)/R$, and the average power depends on what interval we average over:

$$\begin{aligned} P &= A^2/R \int_0^{\tau} \sin^2(\omega t)\, dt/\tau \\ &= A^2/R \left(\frac{1}{2} - \frac{\sin(2\omega\tau)}{4\omega\tau}\right) . \end{aligned}$$

As we integrate over longer and longer times ($\tau \to \infty\,$s), the time-varying part of the average goes to zero, and we are left with $P = A^2/(2R)$. The RMS voltage for the sine wave $A\sin(\omega t)$ is thus $A/\sqrt{2}$.

What about if we look at a square wave that alternates between $+A$ and $-A$? The instantaneous power is either $(+A)^2/R$ or $(-A)^2/R$, which is always just A^2/R, so the RMS voltage is simply A.

The *crest factor* of a signal is the ratio of the AC amplitude to the AC RMS and depends on the shape of the waveform.

Crest factor is always at least 1, which is the crest factor for square waves. For sinusoids, the crest factor is $\sqrt{2}$. More crest factors can be found in Table 29.1.

The peak-to-peak voltage, being twice the amplitude for these symmetric waveforms ($V_{\text{pp}} = A - (-A)$), is twice the RMS voltage for square waves, but $2\sqrt{2}$ times the RMS voltage for sine waves.

Exercise 29.5

If you have a sine wave with 1.6 V DC ±1.5 V AC, what is the AC RMS voltage of the signal? AC+DC RMS?

Exercise 29.6

If you have an $8\,\Omega$ loudspeaker with a $5\,\mathrm{W}$ RMS power limit, what is the largest voltage sine wave you can give it? Give the voltage of the sine wave in three forms: RMS, amplitude, and peak-to-peak voltage. How much current will you have to supply at that voltage?

29.4 Zobel network

This section is optional material, as we will not be designing Zobel networks in this course.

Sometimes loudspeaker designers add circuitry to a loudspeaker to make the impedance seen by an amplifier more constant. This approach makes loudspeakers more interchangeable and allows the speaker design and the amplifier design to be done independently.

The most common such circuit is a *Zobel network*, which puts a series chain of a resistor and capacitor in parallel with the network.

At low frequencies, the capacitor's impedance is large and the loudspeaker's small, and so the overall impedance looks just like the loudspeaker. At high frequencies, the loudspeaker has high impedance and the capacitor low impedance, so the overall impedance looks just like the resistor.

One can get a fairly constant impedance by matching the resistor to the DC resistance of the loudspeaker and adjusting the capacitor to balance the increase in impedance of the loudspeaker with the decrease in impedance of the capacitor. For the model of the puck bass shaker in Figure 29.5, a resistor of $16\,\Omega$ in series with $10\,\mu\mathrm{F}$ does a good job of keeping the impedance fairly constant.

The biggest problem with Zobel networks is that they are very inefficient—at high frequencies all the power is dissipated in the resistor, heating it up and wasting power from the amplifier.

Chapter 30

Lab 8: Loudspeaker Modeling

Bench equipment: Analog Discovery 2 (optional: RC circuits with hidden component values)
Student parts: loudspeaker, resistors, breadboard, clip leads, ohmmeter, (optional: capacitors and inductors)

30.1 Design goal

This is a measurement and modeling lab. The goal is to provide an equivalent circuit for the loudspeaker in your parts kit, so that we can later design audio amplifiers for it.

To gather the data needed to build the model, we will do *impedance spectroscopy*, measuring the impedance of the loudspeaker across a range of frequencies.

30.2 Design hints

Since this is a measurement and modeling lab, rather than a design lab, the main things to think about are how to make and record the large number of measurements efficiently.

You also want to be sure that all the voltages you measure are in a reasonable range for the measuring equipment. If one of the voltages gets very small, you may have difficulty getting accurate measurements. You should be able to design your test setup ahead of time to ensure that your voltages stay within a reasonable range.

> When plotting the model fits, be sure to echo all the parameters of the fit to the plot, so that the reader of the design report can easily see the parameters. Don't forget to include the descriptions and formulas for the models in your report!

Because the resonance peak is often a fairly sharp spike on the graph, the default sampling used by gnuplot to plot a function may miss the peak of the spike. You can increase the number of samples with the gnuplot command `set samples 5000` to get sufficiently fine sampling to accurately plot the function.

30.3 Methods for measuring impedance

There are many ways of measuring resistance and impedance—in this lab we'll concentrate on using the impedance analyzer of the Analog Discovery 2, which is well designed for impedance spectroscopy.

30.3.1 *Using the impedance analyzer*

Starting with version 3.6.8 of WaveForms, the Analog Discovery 2 provides a mechanism for direct measurement of impedance. The user interface for the impedance analyzer is a bit complicated, because it allows compensation for the impedance of your measuring jig, potentially allowing more accurate measurement of the impedance of the loudspeaker itself. This compensation is mainly useful at higher frequencies, where the capacitance of the test setup (including the oscilloscope) interferes more with the measurement.

30.3.1.1 *Setting up the impedance analyzer*

The impedance analyzer requires putting the device being tested (often abbreviated DUT, for "device under test") in series with a known resistor, but by default expects the reference resistor to be a power of 10 from $10\,\Omega$ to $1\,\mathrm{M}\Omega$. You can change one of the preset values to a different resistance using the "Compensation" pop-up, to use resistances that are not powers of 10.

Entering a measured resistance for the reference resistor can improve the impedance measurement, if you can measure the resistance more accurately than relying on the nominal value—easy with a good ohmmeter, but cheap handheld multimeters are often less accurate than the 1% tolerance of metal-film resistors. You can also purchase 0.1% tolerance resistors, to get accurate references without needed a precision ohmmeter.

The idea of impedance spectroscopy is that a sine wave of known frequency is applied to the series chain, and the voltage across the device is measured, as is the voltage across the resistor (to get the current through the device). The ratio of the voltage and the current gives the impedance. Because the Analog Discovery 2 can also measure the phase difference between the waveforms, a full complex impedance can be computed.

The impedance analyzer was designed to be usable with shared-ground instruments (like using the BNC adapter board with probes), with both Channel 1 and the function generator across the series pair and Channel 2 across whichever device is connected to ground. If the resistor is closest to ground, select "W1-C1-DUT-C2-R-GND", but if the loudspeaker is closest to ground, select "W1-C1-R-C2-DUT-GND". Both channels' negative leads should be connected to ground. With either of these two settings, the voltage across whichever component is further from ground should be at least 10% of the total voltage, as that voltage will have to be calculated as the difference of a pair of measurements (Channel 1 minus Channel 2) increasing the noise on it.

Starting with version 3.9.1 of WaveForms, the impedance analyzer also supports differential voltage measurement, with the setting "W1-C1P-DUT-C1N-C2-R-GND", which has Channel 1 across the device under test and Channel 2 across the reference resistor. With this setup, there is a little less noise, because only one measurement is made for the device further from ground, and each channel can choose to be high-gain or low-gain independently.

Digilent also makes an impedance analyzer board that provides the reference resistors and relays to connect them to DUT—if you use this board, the "Adapter" option should be chosen. The setup for the adapter board has the DUT further from ground, so the reference resistor should not be larger than about 10 times the impedance of the DUT.

To avoid inaccurate measurement, both voltages (across the DUT and across the reference resistor) need to be reasonably large. That requires choosing the reference resistor wisely, so that it is close to the magnitude of impedance of the DUT. If the resistor is too large, then the voltage across the DUT will be small and hard to measure accurately. Similarly, if the resistor is too small, then the voltage across the resistor will be small and hard to measure accurately.

Generally, if the ratio of the magnitude of the impedance to the reference resistor is between 0.1 and 10, then the measurement will be good, and if the ratio is between 0.01 and 100, the measurement will still be usable. Ratios further from 1 suggest redoing the measurement with a different reference resistor.

> Be careful when using small reference resistors not to exceed the $\pm 30\,\mathrm{mA}$ current limits for the waveform generator—if you get clipping, then the impedance analyzer will not provide accurate results. Reduce the voltage used if you would otherwise have too much current.

If the impedance range of your DUT is large over your frequency range, you might not be able to keep the voltages close to each other over the full range. Pick a reference resistor that satisfies this constraint over the most important part of your frequency range. If you want to increase accuracy, you can redo part of the range with a different reference resistor.

The frequency sweeps for the impedance analyzer are specified as a start and stop frequency, and a number of samples or samples/decade. Under the gear icon, you can specify whether the steps are logarithmic (usual for sweeping a wide range of frequencies) or linear (useful for closely spaced frequencies around a resonant peak). For example, if we sweep from 10 Hz to 10 MHz at 100 steps per decade, we get 601 samples, logarithmically spaced. The extra sample is there because of *fencepost error*—dividing the six decades into 600 intervals results in 601 boundaries.

> Under the gear icon are other advanced options, such as averaging, settling time, and minimum number of periods. When measuring a resonant circuit near its resonant frequency, increasing the minimum number of periods may be necessary, as the circuit may take a while to reach its steady state.

To compensate for the impedance of the measurement setup, you need to select the frequency sweep you want, then click on the mysterious wrench icon to the right of the "Compensation" item in the toolbar. You need to short circuit the loudspeaker and run "Perform Short Compensation", remove the loudspeaker and run "Perform Open Compensation", then check the "Enable Open Compensation" and "Enable Short Compensation" boxes and run a single sweep of the impedance analyzer.

> Because the compensation is measuring the impedance of your test setup, including your wiring, the short circuit and open circuit should include as much of the wiring as possible—do the shorting and opening of the circuit right at the loudspeaker. Also, don't move your wires around after performing the compensation, as that will change their inductance and capacitance. Using a twisted pair of wires to the loudspeaker (rather than separate wires) helps keep the inductance and capacitance of the wiring small and consistent.

Any time that you change the frequency sweep, you need to redo the compensation, because the software records the measurements made for a particular sweep setting.

The impedance analyzer provides a lot of output options for impedance: magnitude of impedance, phase, real and imaginary parts, as well as modeled values for a simple inductor or capacitor in series or parallel with a resistor. For a complicated impedance like the loudspeaker, we are probably best off with the magnitude and phase of the impedance being output, as these are the easiest to use in fitting models with gnuplot.

> You can control which columns are output both by tool-bar buttons and the “Source” menu that appears in the “export” dialog box. Use the tool-bar buttons to show the $|Z|$ and phase displays and export using the “Impedance Analyzer” source. Make sure that you are saving the data you want in the file you export, as it is very easy to get confused and end up saving different data!

30.3.1.2 *How compensation works for the impedance analyzer*

> *This section is optional, as we will not be requiring you to do the compensation calculations yourself—you can rely on the compensation calculations built into WaveForms. An early version of this section appeared on my blog [47].*

We begin by modeling what we measure as a combination of the actual device under test Z_{DUT} and impedances associated with the test fixture, as shown in Figure 30.1. Because we have three measurements (open circuit, short circuit, and DUT), we are limited to models that have three impedances.

Let's look at the short-circuit compensation first. For the first model, if we replace the device under test with a short circuit, we measure an impedance of $Z_{\text{sc}} = Z_{\text{s1}}$, while for the second circuit we measure $Z_{\text{sc}} = Z_{\text{p2}}||Z_{\text{s2}}$.

In the first model, we can do short-circuit compensation as

$$Z_{\text{DUT}} = Z_{\text{m}} - Z_{\text{sc}},$$

where Z_{m} is the measured impedance with the DUT in place, and Z_{sc} is the measured impedance with a short circuit. For the second circuit, we would need to measure another value to determine the appropriate correction to get Z_{DUT}.

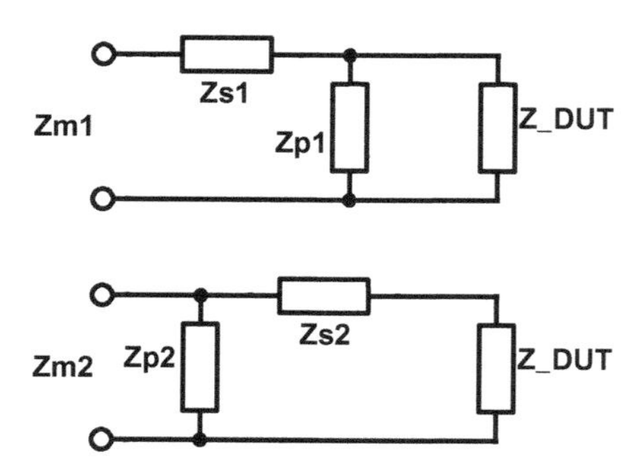

Figure 30.1: *Here are two ways of modeling the impedances of the test fixture when measuring the impedance of a device under test (DUT). We have a series impedance Z_{s} and a parallel impedance Z_{p}—the only difference is whether we put Z_{p} in parallel with the DUT or with the series chain of Z_{s} and the DUT.*

In either case, the measurements that we can make are of just the overall impedance Z_{m}, but we can estimate Z_{s} and Z_{p} by replacing the DUT with either a short circuit or an open circuit and measuring the resulting impedance.

Figure drawn with Digi-Key's Scheme-it [18].

For open-circuit compensation, in the first model we get $Z_{\rm oc} = Z_{\rm s1} + Z_{\rm p1}$ and in the second model we get $Z_{\rm oc} = Z_{\rm p2}$. So for the first model we would need another measurement to get $Z_{\rm DUT}$, but for the second model, $Z_{\rm m} = Z_{\rm oc} || Z_{\rm DUT}$, so

$$Z_{\rm DUT} = \frac{1}{1/Z_{\rm m} - 1/Z_{\rm oc}} = \frac{Z_{\rm m} Z_{\rm oc}}{Z_{\rm oc} - Z_{\rm m}} ,$$

where $Z_{\rm m}$ is the measured impedance with the DUT in place, and $Z_{\rm oc}$ is the measured impedance with an open circuit.

If we do both compensations, we can use either model, but the corrections we end up with are slightly different.

For the first model, we have

$$\begin{aligned} Z_{\rm m} &= Z_{\rm s1} + (Z_{\rm p1} || Z_{\rm DUT}) \\ &= Z_{\rm sc} + ((Z_{\rm oc} - Z_{\rm sc}) || Z_{\rm DUT}) . \end{aligned}$$

We can rearrange this equation to

$$Z_{\rm m} - Z_{\rm sc} = (Z_{\rm oc} - Z_{\rm sc}) || Z_{\rm DUT} ,$$

or

$$\frac{1}{Z_{\rm m} - Z_{\rm sc}} = \frac{1}{Z_{\rm oc} - Z_{\rm sc}} + \frac{1}{Z_{\rm DUT}} .$$

We can simplify that to

$$\begin{aligned} Z_{\rm DUT} &= \frac{1}{1/(Z_{\rm m} - Z_{\rm sc}) - 1/(Z_{\rm oc} - Z_{\rm sc})} \\ &= \frac{(Z_{\rm m} - Z_{\rm sc})(Z_{\rm oc} - Z_{\rm sc})}{Z_{\rm oc} - Z_{\rm m}} . \end{aligned}$$

If $Z_{\rm sc} = 0\,\Omega$, this formula simplifies to our open-compensation formula, and if $Z_{\rm oc} \to \infty\,\Omega$, the formula approaches our formula for short-circuit compensation.

For the second model, the algebra is a little messier. We have $Z_{\rm m} = Z_{\rm oc} || (Z_{\rm s2} + Z_{\rm DUT})$, which can be rewritten as

$$\frac{1}{Z_{\rm m}} - \frac{1}{Z_{\rm oc}} = \frac{1}{Z_{\rm s2} + Z_{\rm DUT}} ,$$

or

$$\begin{aligned} Z_{\rm s2} + Z_{\rm DUT} &= \frac{1}{1/Z_{\rm m} - 1/Z_{\rm oc}} \\ &= \frac{Z_{\rm m} Z_{\rm oc}}{Z_{\rm oc} - Z_{\rm m}} . \end{aligned}$$

We also have $1/Z_{\rm sc} = 1/Z_{\rm p2} + 1/Z_{\rm s2}$, so

$$\begin{aligned} Z_{\rm s2} &= \frac{1}{1/Z_{\rm sc} - 1/Z_{\rm oc}} \\ &= \frac{Z_{\rm sc} Z_{\rm oc}}{Z_{\rm oc} - Z_{\rm sc}} , \end{aligned}$$

and so

$$\begin{aligned}
Z_{\mathrm{DUT}} &= \frac{Z_{\mathrm{m}} Z_{\mathrm{oc}}}{Z_{\mathrm{oc}} - Z_{\mathrm{m}}} - \frac{Z_{\mathrm{sc}} Z_{\mathrm{oc}}}{Z_{\mathrm{oc}} - Z_{\mathrm{sc}}} \\
&= Z_{\mathrm{oc}} \left(\frac{Z_{\mathrm{m}}}{Z_{\mathrm{oc}} - Z_{\mathrm{m}}} - \frac{Z_{\mathrm{sc}}}{Z_{\mathrm{oc}} - Z_{\mathrm{sc}}} \right) \\
&= Z_{\mathrm{oc}} \left(\frac{Z_{\mathrm{m}} Z_{\mathrm{oc}} - Z_{\mathrm{sc}} Z_{\mathrm{oc}}}{(Z_{\mathrm{oc}} - Z_{\mathrm{m}})(Z_{\mathrm{oc}} - Z_{\mathrm{sc}})} \right) \\
&= \frac{Z_{\mathrm{oc}}^2 (Z_{\mathrm{m}} - Z_{\mathrm{sc}})}{(Z_{\mathrm{oc}} - Z_{\mathrm{m}})(Z_{\mathrm{oc}} - Z_{\mathrm{sc}})} \,.
\end{aligned}$$

Once again, when $Z_{\mathrm{sc}} = 0\,\Omega$, this formula simplifies to our formula for just open-circuit compensation, and when $Z_{\mathrm{oc}} \to \infty\,\Omega$, this approaches our formula for short-circuit compensation.

We can make the two formulas look more similar, by using the same denominator for both, making the formula for the first model

$$Z_{\mathrm{DUT}} = \frac{(Z_{\mathrm{oc}} - Z_{\mathrm{sc}})^2 (Z_{\mathrm{m}} - Z_{\mathrm{sc}})}{(Z_{\mathrm{oc}} - Z_{\mathrm{m}})(Z_{\mathrm{oc}} - Z_{\mathrm{sc}})} \,.$$

That is, the only difference is whether we scale by Z_{oc}^2 or correct the open-circuit measurement to use $(Z_{\mathrm{oc}} - Z_{\mathrm{sc}})^2$. At low frequencies (with any decent test jig), the open-circuit impedance is several orders of magnitude larger than the short-circuit impedance, so which correction is used hardly matters, but at 10 MHz, changing the compensation formula can make a bigger difference.

The WaveForms software allows exporting the measured values for the open-circuit or the short-circuit compensation right after measuring them. I exported them for a simple setup using short alligator-clip leads from a breadboard. The results of the measurements and fits to them are shown in Figure 30.2.

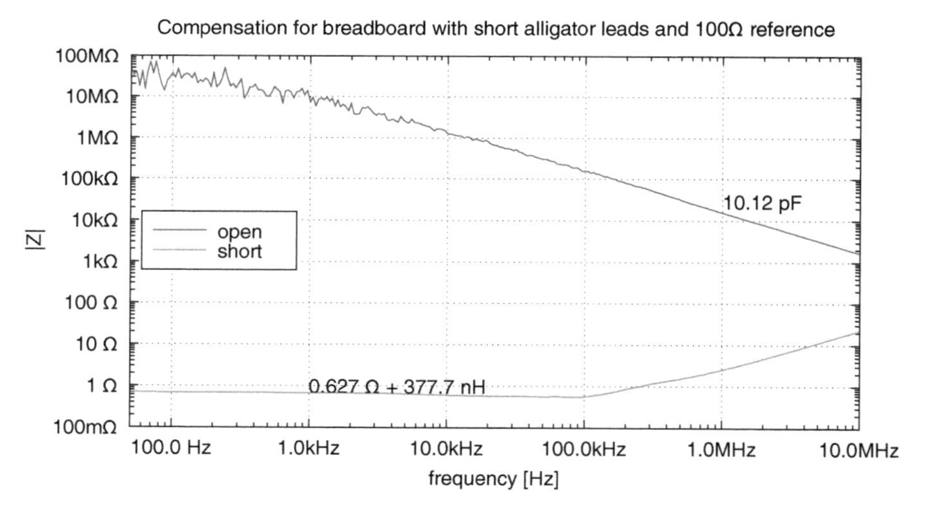

Figure 30.2: *Using a test fixture on a breadboard with short alligator-clip leads, I measured the open-circuit and short-circuit impedances. Because I was using a 100 Ω sense resistor, the large open-circuit impedance at low frequencies is noisy—the voltage across the resistor is too small to be measured well by the oscilloscope channel. I reduced the noise here by averaging 10 measurements when doing the compensation measurement (an option in WaveForms).*

The parallel impedance is approximated well by a single capacitor, and the series impedance is approximated well by a resistor and inductor in series.

Data collected with Analog Discovery 2 [19]. Figure drawn with gnuplot [33].

30.3.2 *Using voltmeters*

This section is optional—only needed if you don't have an Analog Discovery 2 available.

If you don't have an Analog Discovery 2, but do have a function generator, you can do $|Z|$ measurements with just a voltmeter, though most cheap voltmeters will not provide good measurements at high frequencies.

For measuring DC resistance, an ohmmeter passes a known DC current through the resistor and measures the voltage drop across the resistor—the resistance is determined by Ohm's law: $R = V/I$. To measure just the magnitude of impedance $|Z|$ rather than complex impedance Z, we can do something similar, using an AC current rather than a DC current, and measuring the RMS current and RMS voltage.

Because we don't have an AC current source available (our function generators are *voltage* sources), we will use a voltage divider arrangement, putting the impedance we are measuring in series with a known impedance (usually a resistor), and measuring the voltage across each subcircuit of the divider. The voltage across the resistor gives us the current, and we can divide the voltage across the loudspeaker by the current to get the magnitude of the loudspeaker's impedance at the frequency of the function generator. (To think about: Why do I say to use two voltmeters, rather than a voltmeter and an ammeter?)

Since the resistor and the loudspeaker are connected in series, the current through them is the same:

$$I = V_R/R = V_Z/Z \ .$$

Because we are not keeping track of the phase, but only measuring the magnitude of the waveform, we should more properly write

$$|I| = |V_R|/R = |V_Z|/|Z| \ .$$

That allows us to solve for the magnitude of the impedance:

$$|Z| = R\frac{|V_Z|}{|V_R|} \ .$$

In order to make the current and voltage measurements both be accurate, we want the voltages we measure across the loudspeaker and across the resistor to be comparable—if one is much smaller, it will likely be measured less accurately.

There is one small error in the analysis in the previous paragraph—we assumed that no current went through the meters. This approximation is good if the meter impedance is much larger than the resistor or the loudspeaker, but fails if the meter impedance is comparable or even smaller than the known resistor or the impedance being measured. For the loudspeaker measurement, we can ignore this problem, but we'll come back to it in Chapter 38 and Lab 12, where we want to measure higher impedances.

The impedance of most components or circuits will change with frequency, so we will usually want to measure the impedance at several frequencies, not just one, in order to make a more accurate model of the component.

It is possible to do this lab with just a function generator, a resistor, the loudspeaker to test, and one voltmeter. You can still do the measurements of V_Z and V_R—you'll just have to keep moving the leads of the voltmeter.

Measure the resistor you are using with the ohmmeter function of the multimeter, as the resistors in your kit are not exactly the resistance that they are labeled with, and the meters are probably more accurate than the labeling. If you are using a small resistor, make sure you subtract off the meter-lead resistance. Touch the probes to each other without the resistor to measure the meter-lead resistance. Some multimeters have a button for "nulling" the meter, to set the currently read value to be treated as zero.

With separate measurements for the two voltages, we only get $|Z|$, the magnitude of the impedance, and not the full complex impedance. It turns out that we can model the loudspeakers fairly well with just that information.

30.4 Characterizing an unknown RC circuit

This section of the lab may be optional—it depends on the instructor constructing a number of RC circuits with hidden R and C values, for students to characterize. If there is sufficient lab time, practicing characterization with simple RC circuits is valuable, but if lab time is short, jumping directly to characterizing loudspeakers in Section 30.5 is advised.

As an alternative to instructors constructing unknown RC circuits on PC boards, students can construct their own on breadboards, make measurements, and give them to their lab partners to fit and reconstruct the circuit.

For this section, the instructor needs to create a number of two-port series or parallel RC circuits, with time constants in the range 3 μs to 2 ms and an impedance at 10 kHz between 50 Ω and 20 kΩ. The component values are hidden by covering the circuit with heat-shrink tubing or enclosing them in some other opaque covering that allows marking each circuit with a unique ID. The circuits should each be different, with the component values recorded in a solution key.

Students are each issued an RC circuit, and they do impedance spectroscopy on the circuit from 40 Hz to 1 MHz. They then recreate the schematic using the shape of the impedance spectrum and fit the R and C values to the data. Because the components are close to ideal in this frequency range, the data should be very close to the models, making fitting easy (much easier than the loudspeakers of Section 30.5).

The students should first be issued an unknown RC circuit that consists of just two components (one resistor and one capacitor), which they will have to identify as a serial or parallel arrangement and figure out the R and C values.

If time permits, students can try to characterize more complicated circuits with three components in one of four different series-parallel arrangements: R_1 in series with a parallel arrangement of R_2 and C, R_1 in parallel with a series arrangement of R_2 and C, C_1 in series with a parallel arrangement of R and C_2, or C_1 in parallel with a series arrangement of R and C_2.

Bonus: if you have time on your hands, then try measuring the magnitude of impedance as a function of frequency for a known inductor. You may need to change your test setup slightly to measure this part, as its impedance is rather different from the RC circuits or from a loudspeaker. Be sure to stay well below the saturation current for the inductor, as its properties change as the ferrite core gets saturated. Does the data fit the model in Figure 28.4?

30.5 Characterizing a loudspeaker

Pre-lab 8.1

Write a gnuplot script that can plot the magnitude of impedance of your loudspeaker as a function of frequency, from the data you will collect in the lab. You will want to view this plot as you collect data, so have the script written (and tested on dummy data) *before* lab. If you are entering the collected data by hand, you'll want to figure out the format you'll be using. If your setup is going to change for different frequency ranges, you'll want an extra column for each parameter that you plan to change. If you will be using a single setup, then you can use gnuplot variables to name the parameters and set them once in the script (like the way that L and R are set in Figure 28.5).

- Mark your loudspeaker and your partner's and put metadata in the files so that you can later associate the data with the right speaker.
- Measure the DC resistance of your loudspeakers.
- Measure any other components you will use in your measurement circuit.
- Measure the magnitude of impedance of your loudspeaker (using the circuit you devised in pre-lab), at frequencies that are roughly equally spaced on a log scale (say 10, 15, 20, 30, 50, 75, 100, ..., or 10, 12, 15, 18, 22, 27, 33, 39, 47, 56, 68, 82, 100,... if you are measuring by hand) over the entire frequency range. You should have at least 30 evenly spaced measurements on the log frequency scale. (With the Analog Discovery 2, you can easily make 100 measurements per decade, and you can take the frequency down to 1 Hz.)

Make your measurements with the loudspeaker face up on the bench—putting the loudspeaker face down may change the mechanical characteristics (either from the cone being supported by the bench or from trapping air under the cone).

- Measure the magnitude of impedance of your loudspeaker at frequencies close to the resonant frequency of your loudspeaker. You want several measurements at very close frequencies, to get the shape of the peak—try to get the resonant peak itself to within 1 Hz. Even if you have a data sheet, the resonance may not be where the data sheet says it is, as the resonance is dependent on how the speaker is mounted and even minor variations in the stiffness of the paper cone. The loudspeaker may also vary in its properties as it ages (the paper springs may become less stiff as they are flexed a lot, for example).

30.6 Demo and write-up

We will look at models for loudspeakers that include the resonant peak (see Section 29.2.1) in class after collecting data in lab.

Your report should include not only the schematic diagram of your measuring circuit, but also schematics of any models of the loudspeaker that you fit to your data. Be sure to include the values of the components in your schematics!

Some things to discuss:

- Which models fit the data best? How did you do the fitting?
- When would you use the simpler models?
- What RMS voltage can you apply to your loudspeaker without exceeding the power limit? (It may be a function of frequency.)
- What is the largest peak-to-peak voltage for a square wave that you can apply to your loudspeaker without exceeding the RMS power limit?
- If you have phase measurements, how well do your models match the phase?

Chapter 31

Lab 9: Low-Power Audio Amplifier

Bench equipment: oscilloscope, function generator, power supply, soldering station, fume extractor

Student parts: electret microphone, loudspeaker, MCP6004 op amp, resistors, capacitors, potentiometer, breadboard, op-amp protoboard, screw terminals, clip leads, earbud headphone (optional)

31.1 Design goal

In this lab, you will design a low-power audio amplifier for an electret microphone, using a single quad op-amp chip and a single positive-voltage power supply. You will use this amplifier in the next lab as a preamplifier (often abbreviated to *preamp*) for a power amplifier, Lab 11.

You need to use a potentiometer to adjust the gain of your amplifier, so that at maximum gain you can amplify sounds from a quiet room or normal conversation (30 dBA–60 dBA) to get a rail-to-rail output signal with no load on the output, while at minimum gain you can drive your loudspeaker without clipping or otherwise distorting the signal (see Section 18.2.3).

Because one purpose of the gain control is to prevent clipping, you need to put the gain control early in the multi-stage amplifier, because having low gain after clipping has already occurred does not remove the clipping.

> The amplifier you design in this lab will later be reused as a preamplifier for the power amplifier in Lab 11, so you will solder the circuit onto a printed-circuit board, to make a reusable module. The notion of designing components that can be reused in future designs is an important one in all branches of engineering. Whenever you design something you should make a point of looking for ways that the design can be made reusable.

31.2 Power limits

Based on the measurements you did in Lab 7 (electret microphone), you should be able to make a simple circuit that produces voltage fluctuations from an electret microphone by adding a sense resistor connected to the positive power rail (or, on the other side of the microphone, to the negative voltage rail). The voltage fluctuations can be reasonably large, but the changes in current through the microphone remain quite small. In other words, this circuit has a high *source impedance*, and so is not capable of delivering much power. In particular, it is not capable of driving a loudspeaker.

To get enough power to drive a loudspeaker, you'll need an amplifier that can provide substantial current and voltage. For this lab, you'll use an op-amp chip (the MCP6004 quad op-amp chip) to make the amplifier. The output of any amplifier is going to be limited both by the supply voltage and by the maximum current output of the chips we build it from. These limitations are pretty strong for the MCP6004, and we won't be able to make very loud sounds with the designs in this lab. We'll look at one way around that problem in Chapter 35 and Lab 11.

> There are two different types of "maximum" reported for chips: the absolute maximum, above which you are likely to damage the chip, and the recommended maximum, above which the chip is likely to misbehave (though it won't be damaged until the absolute maximum is reached). Your designs need to *guarantee* that the absolute maximum is not exceeded, but should be designed around the recommended operating conditions.

If you can't find recommended operating conditions anywhere on a data sheet, it is generally safe to use about 85% of the absolute maximum (allowing for 10% variation in power-supplies and 5% variation in parts values).

We do not want to amplify any DC bias on the microphone, just the audio frequencies. We can choose among several different frequency ranges: human hearing (about 20 Hz to 20 kHz for young people), the response range for the microphone (from the microphone data sheet), the response range for the loudspeaker (from the loudspeaker data sheet), FM radio broadcast range (30 Hz–15 kHz), AM radio broadcast range ($\approx$ 3 kHz), traditional telephone (300 Hz–3.4 kHz), or wide-band telephone (50 Hz–7 kHz).

Preamplifiers are usually designed around the widest of the possible frequency ranges, so as not to limit their applicability. If you make a wide-band preamplifier, rather than one limited to the frequency range of your loudspeaker, you should still block DC signals, but allow through signals down to 10 Hz or 20 Hz.

You may want to test your amplifier with an earbud headphone, which may not have a clear resonance peak and may be usable down to 20 Hz, the lower limit of human hearing. Panasonic RP-HJE120-PPK earbud headphones are a 16 Ω load, fairly constant across the entire audio range. The data sheet misleadingly reports the "sensitivity" of the headphone as 97 dB/mW, when they mean 97 dB at 1 mW. If you increase the power, the volume goes up, but 100 times as much power is only 10 times the amplitude and so only an increase of 20 dB. There is a 10 dB increase in the sound for each 10-fold increase in input power.

Earbuds can be painfully loud even with the small power output of the amplifier in this lab (see Pre-lab 9.5).

You can use a pair of alligator clips (on the tip and the ground connection, see Figure 31.1) to connect to the earbuds.

Most phones use the current American Headphone Jack (AHJ) standard, which is tip=left, first-ring=right, second-ring=ground, sleeve=microphone, except in China, where the older OMTP standard is mandated (left, right, microphone, ground). The OMTP standard makes more sense if shielded cables are to be used, but the AHJ standard seems to have replaced it in most parts of the world.

31.3 DC bias

Amplifying the DC bias on the microphone can result in the op-amp output saturating (getting stuck at either the highest or lowest voltage). To get rid of any DC bias, you can use a simple high-pass filter

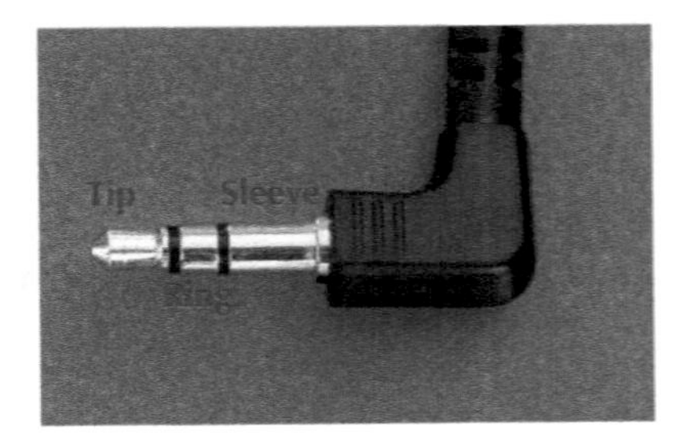

Figure 31.1: *A TRS (tip-ring-sleeve) plug for earbuds or other stereo audio applications. The tip is the left channel, the ring is the right channel, and the sleeve is ground.*
Phone headsets that add microphone signal use a TRRS (tip-ring-ring-sleeve) plug. The tip and ring next to it are still the left and right channels, but the microphone and ground may be connected in either of two ways: American Headphone Jack (AHJ) has the second ring as ground and the sleeve as the microphone. The OMTP standard in China has the second ring as the microphone and the sleeve as ground.

between the microphone circuit and the input to the amplifier. This can be as simple as a capacitor and resistor acting as a voltage divider, as described in Section 11.6. As a reminder, the impedances of the resistor and capacitor have the same magnitude when $\left|\frac{1}{j\omega C}\right| = R$, that is, when $\omega = \frac{1}{RC}$, where $\omega = 2\pi f$. If you want frequencies greater than f to be passed with little change in amplitude, then you want $RC > \frac{1}{2\pi f}$.

If your circuit consists of the microphone and sense resistor connected to an RC high-pass filter, then all four components need to be thought about together. For DC analysis, the capacitor separates the circuit into separate subcircuits, which can be analyzed independently, but for AC analysis, the capacitor does not separate them. If R, the resistor for the filter, is too low (less than the bias resistor for the mic), then at the frequencies passed by the filter, the AC I-to-V conversion is effectively done by R rather than by the bias resistor (more precisely, by R in parallel with the sense resistor), reducing the AC voltage from what is expected.

But the solution is not to make R arbitrarily large. If R is too high (larger than the input impedance of a 1× oscilloscope probe), then it will be difficult to measure the input voltage to the amplifier chip with an oscilloscope or with PteroDAQ, making debugging more difficult.

When there is no sound, the input to the amplifier should be in the middle of the voltage range. Think about ways you can accomplish this. Remember that a capacitor is an open circuit for DC.

It is not good to drive loudspeakers or headphones with a DC signal (that would push the cone away from the center-rest position), so you'll want the two wires to the loudspeaker to have the same voltage when there is no sound. When there is sound, the voltage across the loudspeaker should be sometimes positive and sometimes negative, averaging 0 V.

Because we are using a single, positive power supply, none of our signals are negative with respect to our ground node. That means that neither of the loudspeaker wires can be connected to ground. Whatever current goes through the loudspeaker needs to go through the components connected to the loudspeaker also. If these are op amps, then we'll probably be hitting the current limits of the op amps, which will change the voltage out of the op amp. Make sure that your amplifier design will not be thrown off by these voltage changes (in particular, make sure that these changes don't get further amplified).

31.4 Pre-lab assignment

Read the Wikipedia articles about sound pressure level [157], A-weighting [107], and op amps [142, 143]. Reread the data sheet for the microphone, and re-read your lab report from Lab 7.

Your first task is to design a block diagram for your amplifier. Remember that the blocks of a block diagram show the *function* of part of the design, not how it is implemented. I want to see things like "convert sound to current" and "block DC" rather than part numbers or components—those belong to a schematic diagram. The lines of the block diagram represent signals that carry information, and should be annotated with how that information is encoded (voltage and frequency ranges, for example).

Pre-lab 9.1

Make a block diagram of your circuit showing the major functions of the amplifier, going all the way from sound input to sound output.
Remember that you may need to provide V_{ref} for the high-pass filter and the negative-feedback amplifier (and that V_{ref} is not at 0 V).

Pre-lab 9.2

Come up with at least three different ways to adjust the gain of your amplifier using a 10 kΩ potentiometer. You may want to look at Section 19.4 again for ideas. Explain the advantages or disadvantages of each way.

Pre-lab 9.3

Look up the power-supply voltage limits for the MCP6004 op-amp chip. You want to make sure that your power-supply voltage stays well below the $V_{\mathrm{dd}} - V_{\mathrm{ss}}$ limits of the chip—it is best to stay at least 10%–15% below the *absolute* max ratings. You also need to stay above the minimum voltage requirement for the chip. What voltage will you use for the power supply? Why? Put it on your block diagram!

Pre-lab 9.4

Look up the maximum output current (the output short-circuit current) of the MCP6004 op amp. If your output is a 16 Ω earbud, which is going to limit the power more: the voltage limits from the power supply or the current limits of the op-amp output?

Pre-lab 9.5

Look up the specifications for the Panasonic RP-HJE120-PPK earbuds. With the largest 1 kHz sine wave that the MCP6004 op amp can provide to them without clipping, how loud would the earbuds be? Remember that for power, dB is $10 \log_{10}(P/P_0)$.

Pre-lab 9.6

Design your microphone current-to-voltage converter.

Once you have determined the power supply voltage, decide how you will hook up your microphone and how you will convert its current to voltage—possibly using a circuit similar to Figure 26.4, but with different power supply voltage and (probably) a different resistor. You could also use a design like those in Figure 23.1.

Use the current vs. voltage measurements from Lab 7 (electret microphone) to determine component values in your circuit needed to get a DC bias across the microphone well into the saturation region.

You would like to get large voltage swings for small changes in current, but you want to make sure that current fluctuations you expect do not produce such large voltage swings that the microphone bias moves out of the saturation region of the curve.

Pre-lab 9.7

Using the worksheet you designed for Lab 7 (with any modifications needed for the changes in design for this lab), compute the AC voltage swing at the output of the current-to-voltage converter for the loudest sound you want to amplify (maybe 70 dBA) and for the quietest (maybe 30 dBA). (Assume 1 kHz input, to avoid needing a dBA correction.)

Don't forget to include whatever scaling is necessary based on the current-to-voltage conversion of Pre-lab 9.6.

Convert any RMS voltage you get into a peak-to-peak voltage.

After the current-to-voltage converter, you will need to block the DC component of the signal, to avoid pinning the output to one of the voltage rails. A high-pass filter can block the DC and recenter the voltage at a different DC voltage, but you have to be careful that the filter does not interfere with the already-designed current-to-voltage converter.

To avoid interference, we want the input impedance of the high-pass filter to be at least 10 times higher than the output impedance of the current-to-voltage converter. If the converter is a transimpedance amplifier, its output impedance is near zero, so there is no difficulty (as long as we don't ask for too much current from the op amp), but if the simpler bias resistor was chosen, the output impedance is the value of the bias resistor.

A simple RC high-pass filter has an input impedance in the passband that is essentially just the value of the resistor (the impedance of the capacitor is close to zero in the passband), so the resistor of the high-pass filter needs to be much larger than the bias resistor.

Pre-lab 9.8

Design any DC-blocking (high-pass) filter you need between the microphone and the op amp.

Make your preamp have a flat response (dropping no more than 3 dB from the peak gain) down to 10 Hz–20 Hz. Remember that a *flat response* means a nearly constant gain with respect to frequency.

Make sure that this high-pass filter does not change the impedance of your current-to-voltage converter.

Once you have figured out how big the signal from your microphone will be, you need to figure out how much gain you need from the amplifier. Remember that at maximum gain you want a rail-to-rail output signal with no load on the output, and that at minimum gain you want to be able to drive the loudspeaker without hitting either current or voltage limits on the op amp.

Pre-lab 9.9

Given the AC voltages from your microphone computed in Pre-lab 9.7 and the current and voltage limits from Pre-lab 9.4 and Pre-lab 9.3, what are the minimum and maximum gain you want from your amplifier?
What would be the maximum gain for each stage if you used 1, 2, or 3 stages of amplification?

Pre-lab 9.10

Look up the input offset range of the MCP6004 op amps you are using and consider what would happen to the output of your amplifier if the input offsets were worst case.
If the offsets cause problems, what can you do about them?

Because we are amplifying much higher frequencies than we did for previous labs, we need to worry about the gain limitations of the individual op amps at high frequency: the gain-bandwidth product explained in Sections 18.2.2 and 19.5.

The bandwidth that we can get from a single op amp depends on the gain that we are asking—the lower the gain, the more bandwidth we can get. You are limited in the number of stages you can use, though, as the entire preamp has to fit on the op-amp prototyping board, which only has room for one 4-op-amp chip.

Pre-lab 9.11

Look up the gain-bandwidth product for the op-amp chips you'll be using (MCP6004) and determine the bandwidth possible for the gains you computed in Pre-lab 9.9.
Choose one design, and design the feedback circuits for the op amps to get the desired total gain.

Pre-lab 9.12

For each connection in the block diagram, figure out what values (pressure, voltage, current, etc.) you expect to have there, based on what you learned in Lab 7. Label the block diagram with these signal levels.
Label the voltages for the different signals in the block diagram with "DC ± AC amplitude" or "DC, AC RMS", like "1.5 V DC, 3.5 mV RMS" to indicate the DC bias and the AC voltage. (I'm not suggesting that "1.5 V DC, 3.5 mV RMS" is the correct label for any signal in your block diagram—it is merely an illustration of the format.)
You don't need to redraw the block diagram from Pre-lab 9.1 unless you've changed the design. Just go back and fill in the labels.

Pre-lab 9.13

Design your circuit using as many op amps as are needed. If you are using more than 4 op amps, you will need to redesign your block diagram.
Draw a schematic for the whole circuit. Put boxes around parts of the schematic to show the relationship to the block diagram. Put pin numbers on all the op-amp connections, to speed wiring and debugging.

You need to have a block diagram and circuit diagram before coming to lab, though it need not be the one that you end up with after debugging your design.

To test the design, you must have some expectations about what you should see on the oscilloscope or with the multimeter at each stage. It would help a lot if you wrote some of those expectations down before the lab time (like what DC voltage you expect to see at the microphone, and how large an AC signal you expect to see added to that). What about the DC voltage and AC signal after the input high-pass filter? At the output of the op amp?

31.5 Power supplies

You can power your audio amp either from an external power supply (such as the one in the Analog Discovery 2) or from the 3.3 V power supply on your microcontroller board. The Teensy LC is limited to about 100 mA (a limitation of its voltage regulator and heat dissipation on the board)—is that going to be a major limitation on the power to the loudspeaker, or are other limitations going to be reached first?

The Analog Discovery 2 is capable of delivering 700 mA or 2.1 W (whichever is less) from each of its two power supplies (one can be set from 0 V to 5 V, the other from 0 V to −5 V), when powered from an external 5 V power supply. It can supposedly supply up to 250 mW from each supply when USB-powered, but I have had trouble with intermittent problems with the Analog Discovery 2 when using only USB power.

If you use beefier bench supplies, such as the Agilent E3631 power supply, which is capable of delivering up to 50 V at 1 A, make sure that you have set the power-supply limits to your desired voltage (3.3 V? 5 V?—read the data sheet for the op amp!) and a reasonably low current (say 100 mA), before turning on the output of the power supply.

When using a bench power supply, you should always set the voltage and current limits before turning on the output of the supply—you do not want to damage your circuits or your test equipment by unintentionally providing a large voltage or a large current.

Many power supplies have an *output on/off* button to turn the power on and off to your circuit, without having to power down the whole power supply and lose the settings. In WaveForms 3, the “Supplies” tab has a green “run” arrow or a red “stop” sign to turn the power supply on and off.

31.6 Procedures

Build your amplifier from your schematics on a breadboard first. As with any multi-stage design, it helps to build the amplifier one stage at a time, testing each stage as you go. You can use a function generator with a low voltage setting to provide an input and an oscilloscope to look at the output—look at the voltage range to check the gain and offset, and the waveform to check for clipping.

If you use a larger input signal, so that the output clips, you can plot V_{out} vs. V_{in} for the amplifier. The gain is the slope of the linear portion of the curve, and you can probably see some nonlinear distortion even before the amplifier starts clipping.

When the whole amplifier is working, measure the gain of the amplifier from 10 Hz to 100 kHz using the *network analyzer* capability of the Analog Discovery 2.

The *network analyzer* function sweeps the frequency of the function generator, measuring two voltages and the phase relationship between them. Network analyzers are usually used for reporting filter characteristics, with the first channel connected across the input to the filter, and the second channel connected across the output from the filter. The function generator and Channel 1 are connected to the input of the amplifier, and Channel 2 is connected to the output.

The second channel is reported as "Relative to Channel 1", so it is a gain measurement V_{out}/V_{in}. If you want the gain to be reported without the conversion to decibels, you can use the "Units" menu to select "Gain" instead of the default "dB".

Because our multi-stage amplifier is a transimpedance amplifier, converting the current output of the microphone to a voltage, we need to measure the ratio of output voltage to the input current, but the network analyzer provides a known voltage to the input, not a current. By using a large resistor R_i in series with the function generator, we can convert the voltage to a current. Because we are expecting tiny currents from the microphone to produce fairly large voltage swings at the output of the amplifier, we need to make R_i be large so that reasonable voltages from the function generator produce the range of currents we expect to see. If we provide an input voltage from the function generator of $\pm V_f$ and the network analyzer reports a gain of $G = V_{out}/V_f$, then the transimpedance gain is $V_{out}/(V_f/R_i) = GR_i$.

The analysis above assumes that the current-to-voltage conversion for the microphone is being done by a transimpedance amplifier, which has essentially a $0\,\Omega$ input impedance (that is, the bias voltage does not change as the current changes). If you are using a bias resistor R_b instead, then the current through the input resistance also has to pass through the bias resistor before reaching a constant-voltage node, so the input current is $\pm V_F/(R_i + R_b)$ and the gain of the overall amplifier is $G(R_i + R_b)$.

Test the amplifier without the microphone or loudspeaker. Does the amplifier have the filter characteristics you expect from your design?

Repeat the measurement with a loudspeaker connected as a load—the load impedance of the loudspeaker will make a large difference in the behavior of the amplifier.

You can also test the amplifier including the microphone by using the function generator to drive a loudspeaker placed next to the microphone.

If you do not have an Analog Discovery 2, you can do some testing with PteroDAQ, as long as your power supply is limited to the 3.3 V range of the analog-to-digital converter. PteroDAQ is not capable of recording audio signals at a high sampling rate, but you should be able to see what happens to a 300 Hz signal with a 5 kHz sampling rate. A short recording of such a signal should allow you to see both the DC bias and the RMS AC signal level from the PteroDAQ data about the channel.

31.7 Soldering the amplifier

After you have a working amplifier on your breadboard, you need to lay out and solder the same amplifier on a printed-circuit board.

> Although the breadboard design and implementation is a joint project, each student should solder their own amplifier board—both for the soldering practice and so that every student has their own preamp to use in Lab 11, when you will have different lab partners.

For this lab, you will be soldering the circuit onto a *protoboard*—a printed-circuit board that is not designed for a single circuit but for prototyping various circuits (see Figures 31.2 and 31.3). Soldered wires and components are less likely to have hard-to-debug intermittent connections than breadboard connections, wires tend to be shorter (and so pick up less noise), and the soldered board provides a more permanent amplifier module.

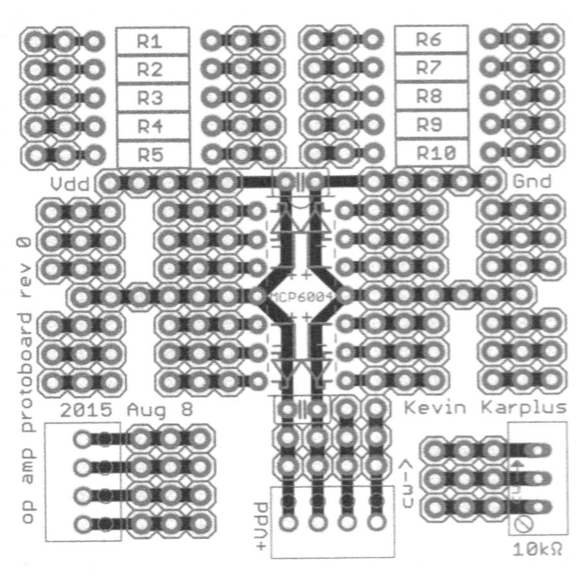

Figure 31.2: *A drawing of the protoboard, showing the printed-circuit-board layers. The blue lines are copper traces on the bottom (solder side) of the printed-circuit board, and the green circles are pads on both sides of the board surrounding drilled holes in the board, where components or wires can be soldered. The grey lines are silkscreen printing on the top surface of the board. (One can do silkscreen printing on the solder side also, but this board has none.) This board has no wiring on the top (component side) of the board. The rectangles at bottom left and bottom center are for screw terminals, and the rectangle at bottom right is for a "trimpot"—a screw-driver adjustable potentiometer.*

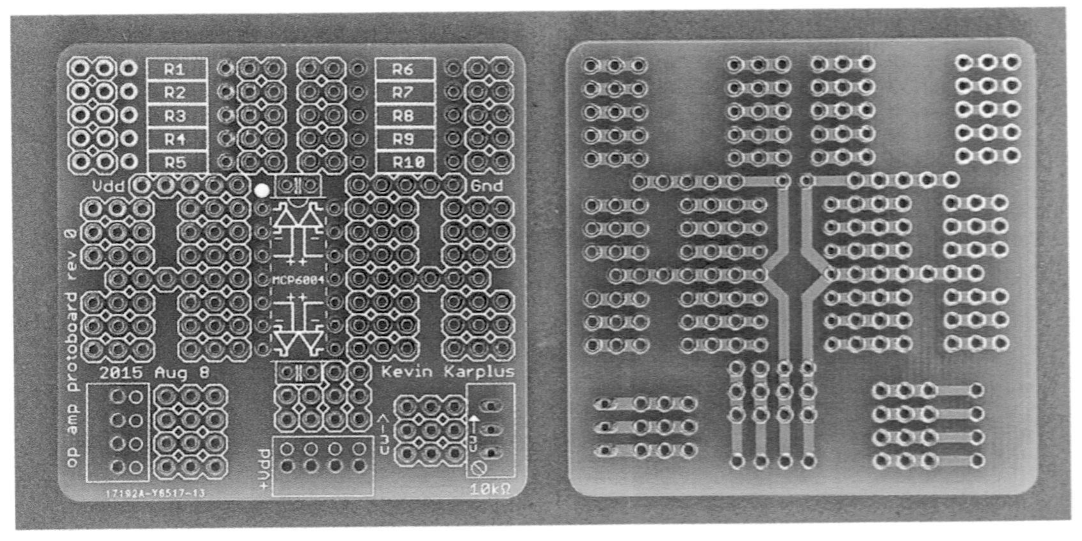

Figure 31.3: *The front and back of the op-amp protoboard shows the wiring clearly on the back (solder) side of the board, and the silk screening on the front.*

Because it is considerably more difficult to correct mistakes in design on a soldered board, you need to be much more careful than usual in designing your circuit, drawing up the schematics, and laying out where on the board all the components and wires will be.

Based on the "finished" schematics I've seen in lab reports, many students need to spend at least twice as much time checking schematics for errors as they usually do. Finding the errors while they are on paper is much less effort than finding them once you have soldered them in place.

> Start by putting pin numbers for each connection on your schematic. Double-check all your pin numbers against the data sheets for the parts. Be particularly careful about swapping the + and − input pins of an op amp!

When doing your layout, you need to make sure that every connection in your schematic is included. One way to do this is to make a *netlist* listing each node of the circuit with every connection that needs to be made: exactly which pins have to be connected to that node. When you have finished the netlist, check it by highlighting a copy of the schematic, one line of the netlist at a time, to make sure that everything on the schematic is included in the netlist and that the netlist does not connect things not on the schematic.

If you are using 4 op amps in a quad op-amp package, there are $4! = 24$ different ways of assigning the op amps in the schematic to the op amps in the quad package. It is very unlikely that an arbitrary assignment will be a particularly good one—look at where input will come from and where outputs will go to, then try to minimize the lengths of wires. If you change your mind about the assignment of op amps to simplify layout and routing, remember to go back and correct the pin numbers on the schematic and netlist!

To help with the layout, I've provided a PDF file [50] with a representation of the board scaled up to page size for experimenting with different layouts (where resistors and capacitors go, what wires need to be added, etc.). A small version of the PDF is shown in Figure 31.4.

Use the protoboard worksheet to figure out where to place each of the components you need for your circuit (this step is known as *layout*). Draw each wire you need to add on the worksheet (this step is known as *routing*), and verify that the connections you plan there match your netlist exactly.

Some people like to use colored pencils (or colored lines in PDF markup programs) to do the routing, to match the color coding of their schematics and physical wiring—this sort of redundancy helps in debugging.

> Please re-read Section 13.3, to make sure you understand the basic principles of color coding wires: all wires on the same node are the same color, and adjacent pins on different nodes should have different colors. Use red for your positive power supply only and black for ground only (you may not need either of these on the board—just for connecting power to the board). Wires between adjacent rows on the board can be short bare wires. Other wires should be color coded according to the node they belong to.

Color code your schematic first, then use those colors for your layout and your soldering.

Shorter wires are easier to trace and to debug than long wires, so it is generally a good idea to connect wires to the closest hole on the PC board that has the right node.

Some connections on your netlist may be handled by the PC board automatically (such as power to the chips) and not need explicit wires soldered to the board.

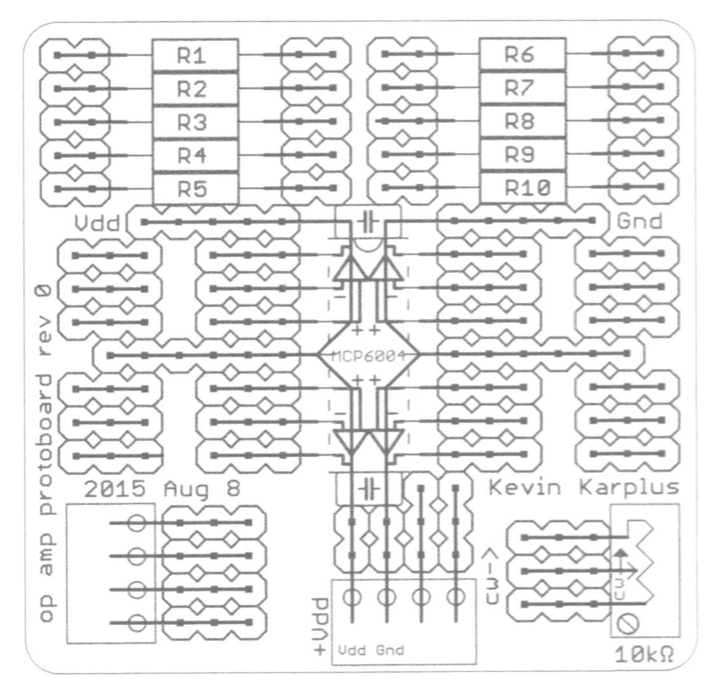

Figure 31.4: *Another representation of the board, intended to aid in laying out the components and wires of the protoboard. The red squares indicate where connections can be made, and the red lines indicate which squares are connected by the traces on the board. Each square should have exactly one wire or component lead ending on it—draw wires so that they do not pass over any other squares. If you want to connect two wires together, you need to use two squares that are connected together by the board.*
The Vdd and Gnd connections at the bottom of the board will often be coming from the board used for PteroDAQ, so the two other positions on that screw terminal should probably be the outputs of the board, which may be going back to the PteroDAQ data-acquisition system. The inputs can come from the screw terminal on the left.
This worksheet can be downloaded from https://users.soe.ucsc.edu/~karplus/bme51/pc-boards/op-amp-proto-rev0/op-amp-proto-rev0-worksheet-fullpage.pdf

The protoboard is designed for a single power supply, powering the MCP6004 chip. Since you are often going to be reading the output with the microcontroller board of PteroDAQ or the Analog Discovery 2, it makes sense to power the amplifier from the same instrument.

> Power and ground are provided through screw terminals or header pins at the bottom of the board—it would make sense for the other terminals in the same block of four to be used for any other communication needed to the microcontroller or oscilloscope. What information do you want to communicate? How will it be encoded?

The MCP6004 chip has power and ground already wired, and there are three connected holes available for each pin of the chip. There are also uncommitted groups of three holes each just outside the holes for the pins, which can be used for connecting wires and components when the holes next to the chip are not sufficient.

There are two places on the board for bypass capacitors, on either side of the MCP6004 chip. I like to use 4.7 μF ceramic capacitors as bypass capacitors, but somewhat smaller capacitors (0.1 μF) are also quite common. Putting bypass capacitors in minimizes problems you might have with noise propagating through the power-supply wiring, and are a standard part of almost all electronics design. It is almost certainly overkill to have bypass capacitors on both sides of this chip—if you decide to use only one, populate the one at the bottom of the board, near where power comes onto the board. That location puts the bypass capacitor before the noisy power gets to the op-amp chip.

The screw terminals at the bottom left of the chip have three connected holes for each terminal. This block of 4 screw terminals would be a good one to use for connecting the sensor input to the amplifier (microphone). The spacing of the wires on the microphone is 0.1″, the same as the spacing on the screw terminals, so the microphone can be placed directly into the screw terminals. Doing so will cover up the adjacent screw terminal hole, so you can't then use it for some other function. None of the screw terminals here have a dedicated function—they must be connected to the rest of the circuit with wires, capacitors, or resistors.

Warning: the microphone terminals are not symmetrically placed on the mic—figure out which way up you want the microphone before assigning the + and − polarity to the screw terminals for it.

There are 10 places on the board for soldering resistors. Each of these has two extra holes at each end, intended to be used for wires to connect it to somewhere else on the board. You may need more resistors than there are places pre-assigned for them—if so, you can solder resistors between any pair of holes on the board. Just make sure that you don't have long runs of bare wire from the resistor that can touch other wires and cause short circuits.

On the bottom right, there is a place for soldering in a trimpot (a screwdriver adjustable potentiometer), for use as the gain control.

There are no dedicated spots for adding capacitors, other than the two bypass capacitors. Because the small ceramic capacitors in your kit have 0.1″ lead spacing, they can be put between any two adjacent (but unconnected) holes on the board. The uncommitted blocks of 3 holes may prove to be very handy for connecting capacitors. Larger capacitors have a 0.2″ spacing between the leads and can either skip a row or be put across the gap between the uncommitted block and the blocks for the pins of the amplifier.

After you have finished doing the layout on paper and confirmed that all connections on your netlist have been drawn, you can make a *wiring list* that lists each wire that needs to be soldered in, giving the two end points for the wire and the color of the wire. Make sure that you have no more than one wire for each hole in the breadboard.

As you add each component or wire to the board, check it off on your wiring list, to make sure you don't omit any. Double check that you are hitting the right hole on the board. It is probably easiest to put in the chips first, then the capacitors and resistors, then the wires, and finally the screw terminals. It is a good idea to do short wires before long wires, especially if the wires need to cross.

If you organize your wiring list in the order you plan to make the connections, it should speed up the assembly of your circuit. If you re-order your wiring list, you need to check that it is consistent with the schematic again, to make sure you didn't accidentally lose a connection.

Some students think that they can save time by doing their soldering directly from their schematic, without making a netlist or wiring list. A few of them do save a little time that way, but most end up spending far more time debugging a non-working circuit, because of a missing or misplaced wire. Being slow and systematic in the layout and construction saves a lot of debugging time.

When you solder wires to the board, measure the wires carefully and make them very short—they should not stick up from the board more than about 6 mm (0.25"), except where they need to pass over some other component. Try to minimize the crossing over, in case you need to unsolder and resolder a component or wire.

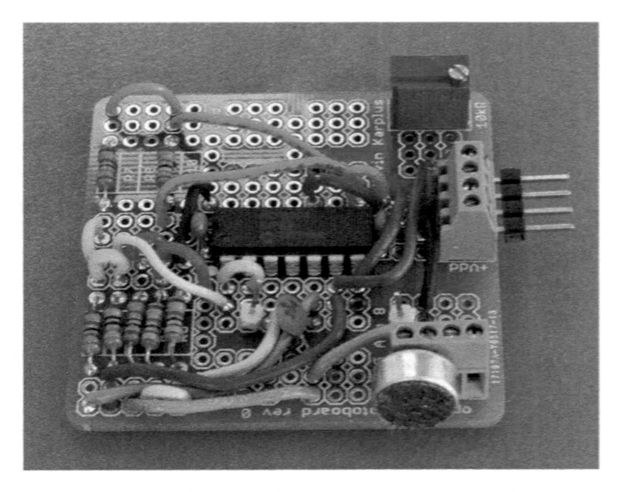

Figure 31.5: *A soldered microphone preamplifier, showing the microphone attached using screw terminals. The output and the connections to the power supply are done through the other screw terminal, shown here with double-ended header pins to connect to a breadboard. There are a couple of male header pins soldered into the middle of the board, to act as test points for attaching oscilloscope probes.*
*You should **not** try to reverse-engineer this circuit—the design here is for a slightly different specification than the goal of this lab. **Do** try to spot color-coding errors in this layout, as there is at least one.*

Figure 31.5 shows a reasonably well laid out and wired microphone preamplifier using the op-amp protoboard.

31.8 Bonus

If you finish the amplifier early, you can try adding the following features:

- Multiple op amps in parallel, to increase the output current capability, (this is easy on a breadboard, but probably not possible on the soldered prototyping board, as there won't be enough spare op amps).
- A tone control, to boost or cut the bass or treble. One of the most popular tone control circuits is the Baxandall tone control circuit, based on one published in 1952 [8]. The modern version of the circuit is well explained in an online tutorial [13].
- A "line" input that can accept the output signal from a phone or MP3 player. To design this, you'll need a cable connected to a TRS (tip-ring-sleeve) or TRRS (tip-ring-ring-sleeve) plug (see Figure 31.1) to connect to the phone and you'll need to measure the voltages output by your phone. You should expect something like 100 mV to 1 V peak-to-peak, if driving a high-impedance load.

  Some phones need a pull-down resistor on the microphone line (around 2 kΩ) to detect that a headset has been plugged in—lower resistances are used for signaling button presses on Android phones [4].

31.9 Demo and write-up

This audio-amplifier design report is cumulative with the microphone characterization of Lab 7. The characterization measurements of that lab are needed here to justify your choice of bias method and resistor sizes, and the sensitivity calculations are needed to justify choice of gain. Your design report must be a standalone report, as your reader has no access to prior lab reports—if there is information you need from a prior report, it must be included in this report.

Demonstrate to an instructor a working amplifier. Show the input and output of the amplifier as two traces on a dual-trace oscilloscope, or as two channels on PteroDAQ with a fairly high sampling rate.

Demonstrate the amplifier working by speaking into the microphone. (You may also be able to get some feedback squeal, but don't irritate people too much by doing it a lot.) Listening to the output of the amplifier with an earbud headphone instead of through a loudspeaker should give you a much clearer idea of any distortion that the amplifier is generating. You can also excite the microphone by playing sine waves from the frequency generator through another loudspeaker.

Your write-up should include the final design goal (which may have been modified somewhat from the goal in the pre-lab assignment), the design decisions and the calculations that supported them, a block diagram of the final design, a schematic diagram that matches the block diagram, photographs of the soldered boards, and results of any testing that was done.

Chapter 32

Field-effect Transistors

For controlling high-current devices (motors, loudspeakers, lights, etc.), most modern circuits use metal-oxide-semiconductor field-effect transistors (MOSFETs), which act as voltage-controlled switches.

A field-effect transistor (FET) is a three-terminal device, with a *source*, *drain*, and *gate*. The schematic symbols for MOSFETs are shown in Figure 32.1. The voltage difference between the gate and the source, V_{gs}, determines how much current can flow from the drain to the source. The gate itself has no DC connection to either the source or the drain—it is essentially one plate of a capacitor (the "metal" in MOS), with the other plate of the capacitor being the semiconductor that forms the channel between the source and drain. The "oxide" of MOS is the insulator separating the plates of the capacitor.

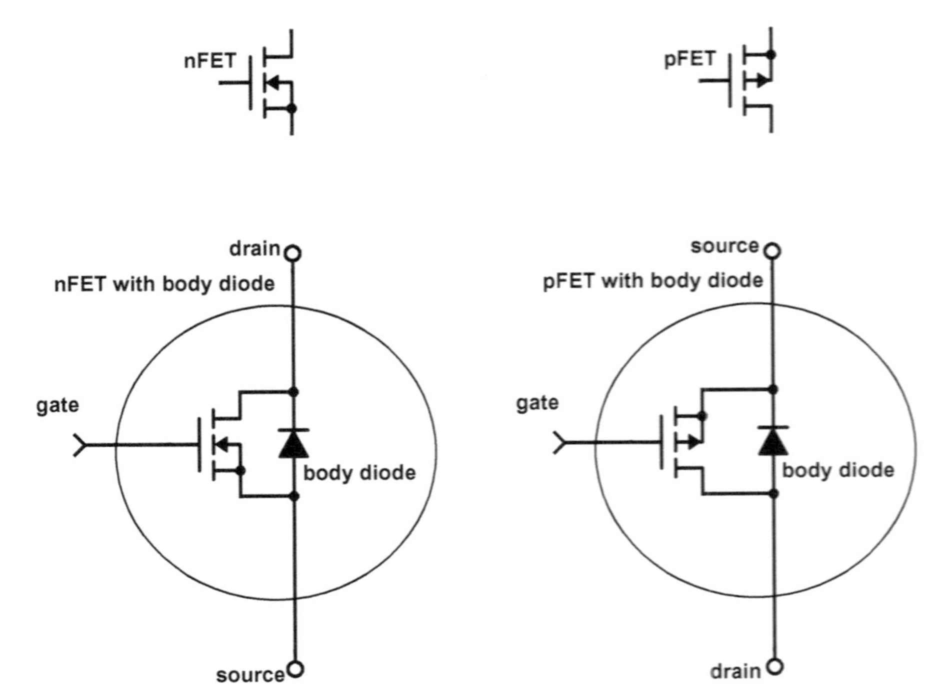

Figure 32.1: *Symbols for enhancement Metal-Oxide-Semiconductor Field-Effect Transistors (MOSFETs). The sources are indicated by connecting the central arrow to the source. For nFETs, the arrow points in (mnemonic: n is in). The gates are indicated by one plate of a capacitor, reminding us that the gate is only capacitively connected to the rest of the transistor—there is no DC connection. The broken line for the channel is used to indicate enhancement-mode FETs—the only type we'll discuss in this course.*
Figure drawn with Digi-Key's Scheme-it [18].

FETs are controlled by the voltage between the gate and the source, V_{gs}. When V_{gs} is sufficiently high, the transistor is *on*, and large currents can flow from the drain to the source. The *source* and *drain* have the names they do to correspond to the flow of the majority carriers in the semiconductor. In negatively-doped semiconductors, the charge carriers are electrons, and the source has a lower voltage than the drain (electrons move from the more negative to the more positive terminal). In a pFET, which is positively doped, the charge carriers are positively charged holes, and the source has a higher voltage than the drain.

Power FETs are often drawn with *body diodes* in parallel with the channel that are an inherent part of the process for making the transistors. FETs are usually used with the body diodes having reverse bias (not conducting), so an nFET source is more negative than the nFET drain, and a pFET source is more positive than the pFET drain. The arrow from the middle of the FET to the FET source corresponds to the direction of the body diode, so points in the opposite direction from current flow through the FETs when they are normally used. The body diodes are often important for use as fly-back diodes (see Figure 32.12).

Although we will be treating our FETs as switches (either "on" or "off", controlled by the gate-to-source voltage V_{gs}), more sophisticated models treat them as resistors (at low drain-source currents) or constant-current sources (at high drain-source currents), where the on-resistance R_{on} or saturation current depends on V_{gs}. Data sheets usually provide some values for the on-resistance and plots of I_{ds} vs. V_{gs}.

I considered having students measure these characteristics as part of Lab 10, but the measurements are hard to make with low-cost equipment, as R_{on} is quite small for power FETs. Test fixtures made with breadboards have parasitic resistances much larger than the R_{on} we want to measure. If we were trying to use the saturation-current model, it would be even worse, as the saturation currents are much larger than our cheap power supplies provide. Even if we had much beefier power supplies, the testing would require careful handling to avoid arc-welding our connection wires. The results of our tests would still not match the professional tests done for the data sheets, because those tests often require active cooling of the FETs to hold them at a constant temperature while very large test currents are used.

If anyone wants to try doing some measurements, it is not too difficult to measure *non-power* FETs, which have much larger on-resistances. For example, the circuit in Figure 32.2 can be used to measure R_{on} for a 2N7000 nFET. This circuit is not what manufacturers usually use, as they are more likely to test with a known V_{ds} or I_{ds} than with a constant load resistance. The measurements I got using the circuit are shown in Figure 32.3.

To get the "corrected" R_{on} values in Figure 32.3, I had to make some corrections for the breadboard test fixture I was using. I assumed that most of the error in measuring small voltages was an offset error in the input amplifiers of the oscilloscope.

As we did when measuring the loudspeakers, I wanted an open-circuit and a short-circuit correction. For the open-circuit correction, I removed the FET and measured the voltage across the load resistor. The current through the $150\,\Omega$ resistor should be approximately $150\,\mu$A, because of the $1\,\mathrm{M}\Omega$ resistance of Channel 1. To correct for offsets, I added $750\,\mu\mathrm{V} - V_{\text{r-open}}$ to all the measurements for Channel 2. For the short-circuit correction, I shorted the drain and source together with a short wire and measured the voltage across that wire. The voltage there should be 0 V, so I subtracted $V_{\text{ds-short}}$ from all measurements for Channel 1. These corrections mainly affect the R_{on} values at the extremes, where either V_{ds} or V_r is small and difficult for the Analog Discovery 2 to measure.

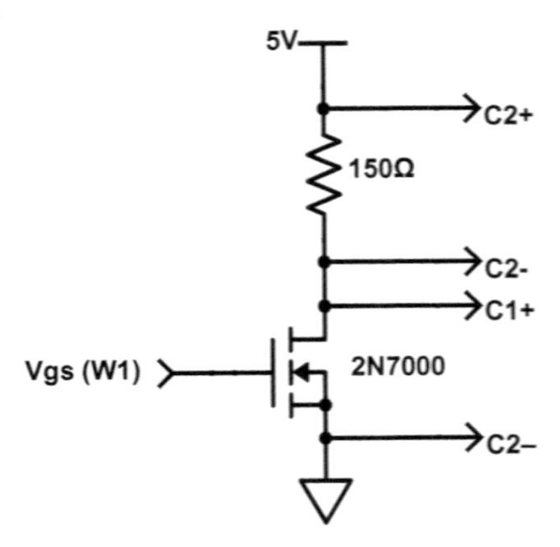

Figure 32.2: *This schematic shows the test setup I used to measure on-resistance for a 2N7000 nFET. The load resistor (150 Ω) is small enough to provide reasonably large test currents when the nFET is on, but large enough that the power dissipated in the resistor will not overheat a $\frac{1}{4}$-watt resistor. Professional tests of on-resistance are usually done with a constant current, rather than a constant load resistance.*

Figure drawn with Digi-Key's Scheme-it [18].

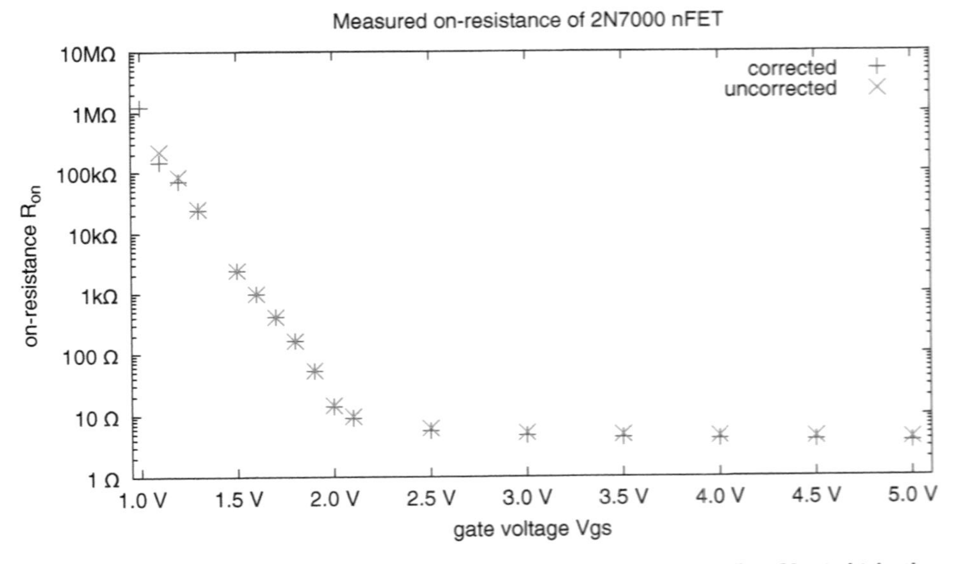

Figure 32.3: *The on-resistance R_{on} versus gate-source voltage V_{gs} for a 2N7000 nFET. When V_{gs} is high, the on-resistance is low and does not vary much as V_{gs} changes. When V_{gs} is low, the resistance goes up rapidly (the behavior is exponential).*

The voltage at which the transition between these two different modes of operation occurs is the threshold voltage *for the transistor.*

I used open-circuit and short-circuit corrections (see the main body text for explanation) to improve the measurements slightly.

Figure drawn with gnuplot [33].

In Figure 32.3, the on-state is pretty easy to see—for V_{gs} above the threshold voltage, R_{on} is small and almost constant. It is harder to define a clear off-state—you need to have V_{gs} well below the threshold to consider the nFET "off". When using an FET as a switch, we generally want to stay away from the threshold voltage, with V_{gs} much lower (for the FET to be off) or much higher (for the FET to be on).

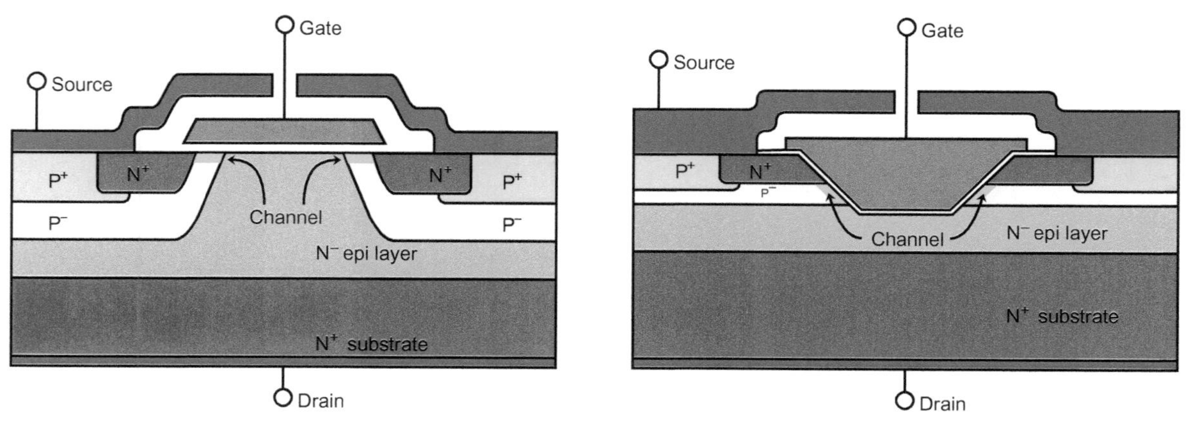

(a) *This cross section is of one cell of VD-MOS power nFET.* (b) *This cross section is of one cell of V-groove nFET.*

Figure 32.4: *These cross sections of a power nFET show the gate in orange and the source and drain connections in blue. The white area separating the gate from the source is a high-quality insulator (such as silicon dioxide).*
The channel is the thin green region bridging the P^- semiconductor near the gate. When the gate is sufficiently positive, holes are repelled from the region and electrons are attracted to it, making a conductive channel from the N+ region of the source to the N^- and N^+ regions of the drain. Larger gate voltages make the channel thicker, hence reducing its resistance.
The body diode junction is between the P^- body region and the N^- epi layer—it is kept reverse-biased by the P^+ and N^+ source and drain connections.
In both drawings, the N^+ substrate layer should be much, much thicker to be to the proper scale.
Figure 32.4a adapted from https://upload.wikimedia.org/wikipedia/commons/1/1e/Vdmos_cross_section_en.svg drawn by Cyril Buttay, licensed under Creative Commons Attribution-Share Alike 3.0 Unported.

32.1 Single nFET switch

Figure 32.4 shows cross-section views of a power nFET. The source and drain for the nFET are on opposite sides of the chip, with the source on top connected to the N^+ regions and the drain connected to the entire bottom of the chip. The body diode is the large junction between the P^- and N^- layers, connected through the P^+ and N^+ layers to the source and drain.

FETs come in many different packages, as shown in Figure 32.5.

In this class, you use FETs mainly as nonlinear, digital devices—that is, as switches that are either *on* or *off*. (We saw a different use of FETs in Lab 7, the electret-microphone lab, where the electret capacitor is attached to the gate of a JFET, and what we measured in that lab was the current from drain to source.)

One common way to control devices from a microcontroller (particularly for LEDs, heaters, and unidirectional motors) is to use a single *negatively-doped metal-oxide-semiconductor (nMOS) field-effect transistor (FET)*, or nFET, as a *low-side switch*. A low-side switch is a switch that controls the lower-voltage input to a load. The positive input to the load is connected directly to the positive power rail, but the negative input to the load is connected to the drain of an nFET, whose source is then connected to the negative power rail (see Figure 32.6).

Power FETs are used as switches rather than as linear transistors for two reasons: they are very energy-efficient if used that way (they dissipate little power as heat), and their current-vs.-voltage characteristics are very temperature-dependent, so designing linear circuits with power FETs is quite difficult (beyond the scope of this class).

Why are FETs power-efficient as switches? Let's look at them both when they are turned off and when they are turned on:

Figure 32.5: *Here are four of the most common packages for FETs. The SOT-23-3 surface-mount package is the most popular package for very cheap FETs (it is also referred to as TO-236-3 and SC-59). Transistors using this package are usually rated for 0.5 W to 3 W.*
The TO-92 (also referred to as TO-226-3) package is popular for breadboarding and is the classic package for non-power transistors. Transistors using this package have similar ratings to the one using SOT-23 packages, but generally cost more.
The TO-251-3 package (also know as IPak) is popular for moderately high-power transistors, as the metal drain connection can be connected to a heat sink. These are often difficult to use in breadboards, as the short, thick leads don't work with all breadboard contacts. Some transistors using this package are rated for as much as 150 W.
The TO-220 package is popular for higher power transistors, as it can be bolted directly to a heat sink. Some transistors using this package are rated for as much as 500 W.

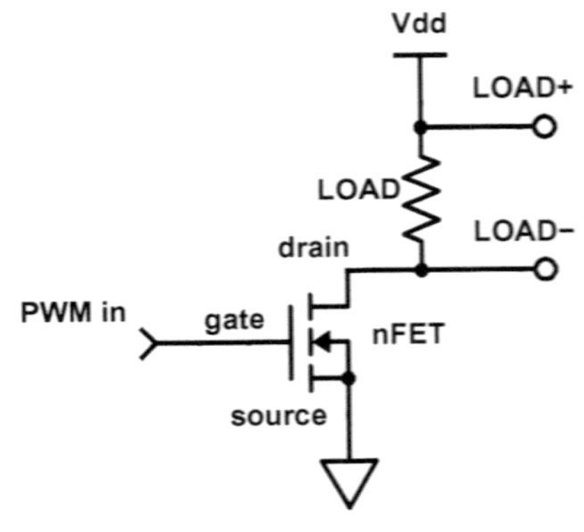

Figure 32.6: *A single nFET can be used to provide high-current switching from a logic-level signal out of a microcontroller. The on-resistance (R_{on}) of the nFET should be much lower than the load resistance R_{load}, so that power is delivered to the load rather than dissipated in the nFET.*
Figure drawn with Digi-Key's Scheme-it [18].

> When an FET is off (low V_{gs}), it has a very high resistance between the source and drain and passes little current.

I measured about 400 pA for an AOI514 nFET transistor with $V_{\text{gs}} = 0\,\text{V}$ and $V_{\text{ds}} = 9\,\text{V}$, though its data sheet allows up to $5\,\mu\text{A}$ for $V_{\text{gs}} = 0\,\text{V}$ and $V_{\text{ds}} = 30\,\text{V}$ if the transistor warms up to 55°C. Even with the data sheet's worst-case conditions, the transistor would only dissipate $30\,\text{V}\ 5\,\mu\text{A} = 0.15\,\text{mW}$ of heat when it is off. This much power dissipation is unimportant in a power amplifier—too small to be worth bothering with. For a micropower circuit that has to run for months on a small battery, it would be a lot, though.

When an FET is on, it has a very low resistance, and so the voltage drop across it is small.

The *on-resistance* of an FET is V_{ds}/I_d, measured at a low enough V_{ds} that the FET is in the linear region, not the saturation region (look back at Figure 26.5, the I-vs.-V curves for the FET in the electret microphone, for the linear vs. saturation distinction). If you need the on-resistance for an nFET at a gate voltage other than what is provided in a table on the data sheet, you can often get it from the plots provided on the data sheet that show either the resistance or the drain current (I_d) at a fixed voltage V_{ds} as a function of V_{gs}.

The data sheet for the AOI514 nFET specifies an on-resistance of at most $11.9\,\text{m}\Omega$ when $V_{gs} = 4.5\,\text{V}$ and $5.9\,\text{m}\Omega$ when $V_{gs} = 10\,\text{V}$. If you use a 9 V power supply and the FET to drive an $8\,\Omega$ loudspeaker, the current would be about 1.125 A, so the power dissipated in a turned-on FET would be $I^2R = 15\,\text{mW}$, while the power to the loudspeaker would be about 10 W. Again the power dissipation in the FET is negligible.

A pFET behaves the same way as an nFET, except that the voltages and currents are all negated—that is, the source is set to the most positive voltage on any node of the FET (rather than the most negative), because it is the source for the positively-charged holes that are the majority charge carrier in p-doped semiconductors. The pFET turns on when V_{gs} gets sufficiently negative, that is, when the gate voltage is enough lower than the source voltage.

In summary, FETs dissipate little power because they have extremely low current when off and low voltage drops when on. But when FETs are switching between the off and on states, they can have intermediate resistance values with large currents and large voltages and dissipate large amounts of power. When using FETs as switches, the goal is to have them spend as little time as possible in the intermediate region (subject to other constraints on the design), to minimize the power loss.

Exercise 32.1

Look up drain-source leakage current (with $V_{gs} = 0\,\text{V}$) for the nFET and pFET in your kits. Compute the power dissipation in each FET with a 9 V V_{ds} at the worst-case leakage current (−9 V for the pFET).

Which nFET or pFET you have will vary from year to year, as we choose the cheapest FETs with sufficiently high ratings for the power amp—most of the through-hole FETs we have used in the past are no longer available, but the surface-mount ones should be around for a while longer.

Exercise 32.2

Look up the on-resistance of the nFET and pFET in your kit when the transistor is turned on with small signals (say $V_{gs} = 4.5\,\text{V}$ for the nFET and $V_{gs} = -4.5\,\text{V}$ for the pFET).

If you put a single FET in series with a $6\,\Omega$ loudspeaker, what fraction of the voltage would be across the loudspeaker and what fraction across the FET? If you put 10 V DC across the loudspeaker and FET, what current would flow? How much power would be dissipated in the FET? How much in the loudspeaker?

(Your answer should be a two-by-four table: two choices of FET and four values: fraction of voltage, current, power in FET, power in loudspeaker.)

32.2 cMOS output stage

One standard configuration for an FET output stage is to have one nFET and one pFET, with their drains connected to the output and their sources connected to the power rails, as shown in Figure 32.7. This sort of pairing of an nFET and a pFET is called *cMOS*, which stands for *complementary Metal Oxide Semiconductor*. Using a cMOS output stage is very common in motor controllers, cMOS digital logic, and in class-D amplifiers like you'll be designing.

> In a cMOS output stage, the *sources* for the nFET and pFET are connected to the power rails. That ensures that the body diodes are reverse-biased and that no current flows when the transistors are off. The arrows on the transistors should point in the opposite direction to the current flow, as they correspond to the body diodes.

Some high-power drivers do not use cMOS output stages, but use nFETs for the high-side switching instead of pFETs. The nFETs have somewhat better electrical characteristics than pFETs, but the circuitry needed to control an nFET that is connected to the positive supply gets a little complicated. The problem is that the gate of the nFET has to be higher than the source to turn the transistor on, but we want the source and drain to be at almost the same voltage with the transistor on. That means that an nFET on the high side would need to have a gate that is driven several volts above the top power rail. There are *charge-pump* circuits that achieve that, but we don't need the improved performance of using nFETs on both the low side and the high side, so we'll stick with the simpler cMOS design.

The cMOS FET output stage works great if one of the two FETs is turned on and the other off, providing a low-resistance connection to the desired voltage rail and no connection to the other rail.

> Turning both FETs on at once, however, results in a very low resistance between the power rails and a very large current, because the transistors have only a few milliohms of on-resistance. The large current is called *shoot-through current.*
>
> Avoiding shoot-through current, which heats up the FETs to no useful effect, is one of the biggest challenges in designing high-power drivers.

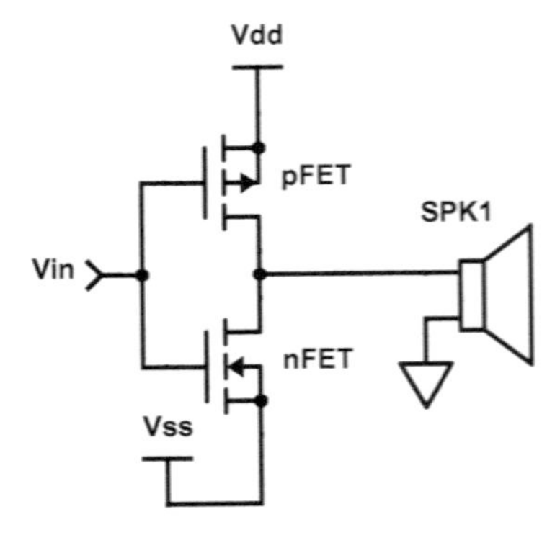

Figure 32.7: *A cMOS output stage, consisting of one pFET with its source connected to the positive voltage rail and one nFET with its source connected to the negative voltage rail, and the output from the two connected drains. The* ***sources*** *(which have the central arrows connected to them) are connected to the power rails.*
The other side of the loudspeaker is connected to a voltage halfway between the power rails, which is easy to do with a dual power supply. With a single power supply, it is more difficult to produce a halfway voltage source that is capable of producing the large currents needed for a loudspeaker. See Section 32.4 for a different solution.
Figure drawn with Digi-Key's Scheme-it [18].

If you don't prevent shoot-through current in your class-D amplifier, you can melt your breadboard, as I have done in erroneous designs (see Figure 32.8).

Various circuits can be used to ensure that you don't have both FETs on at once. The simplest of them is just connecting the two gates together and switching the gate voltage very rapidly from a voltage near the top power rail to a voltage near the bottom power rail. This design is used in *inverters*, whose name comes from their function of changing a high input to a low output and *vice versa*. The Schmitt-trigger inverters you used in the hysteresis oscillator probably have this sort of design in their output stage.

When the gate voltages are at the top power rail for a $\pm 6\,\mathrm{V}$ supply, the pFET has $V_{\mathrm{gs}} = 0\,\mathrm{V}$ and the nFET has $V_{\mathrm{gs}} = 12\,\mathrm{V}$, so the pFET is off and the nFET is on. When the gate voltages are at the bottom power rail, the pFET has $V_{\mathrm{gs}} = -12\,\mathrm{V}$ and the nFET has $V_{\mathrm{gs}} = 0\,\mathrm{V}$, so the pFET is on and the nFET is off. There are roughly three voltage ranges for the tied-together gates:

- below the nFET threshold (only the pFET is on),
- between the thresholds (both FETs on), and
- above the pFET threshold (only the nFET is on).

This simple inverter design has problems as the power supply voltages on the nFET and pFET sources increase, for two reasons:

- The nFET threshold is referenced to the nFET source (the lower power rail) and the pFET threshold is referenced to the pFET source (the upper power rail). Increasing the voltage difference between the power rails means that there is a wider range of gate voltages between the thresholds, where both FETs are on (and we get shoot-through current). Increasing the voltage change that has to occur on the gate increases the time it takes for the change to happen. The longer duration for the shoot-through current increases the power wasted as heat substantially.
- Increasing the drain-to-source voltage for the FETs during shoot-through current increases the current.

The circuits that drive the gates of the FETs have limited current capability. Typical current limits are $\pm 4\,\mathrm{mA}$ for a 74HC04 inverter or a TLC3702 comparator with a $5\,\mathrm{V}$ power supply, $\pm 5\,\mathrm{mA}$ for a Teensy LC output pin (normal drive), or $\pm 24\,\mathrm{mA}$ for a 74AC04 inverter.

Figure 32.8: *Shoot-through current through an FET can cause it to dissipate a lot of power as heat. Without a heat sink to dissipate the heat, the temperature can rise quite high. Although power FETs are usually designed to withstand a junction temperature of 175°C, breadboards melt at lower temperatures.*

The outputs of cMOS devices like these inverters and comparator outputs are often well modeled as either a resistor to power or a resistor to ground. Data sheets do not usually give the current-vs.-voltage curves or output resistances of the inverter output stages, so I measured them myself—see Figure 32.9.

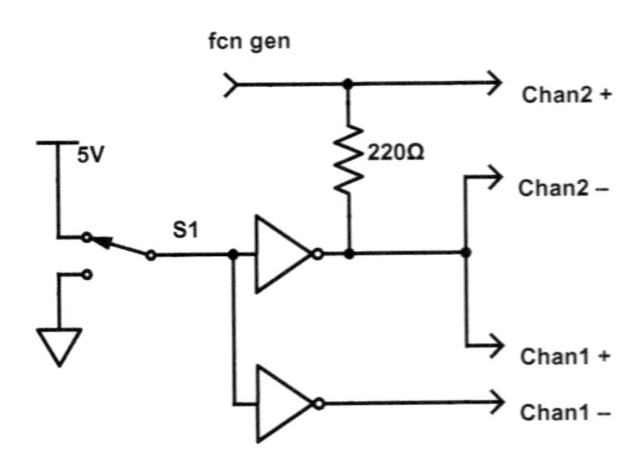

(a) *Test jig for measuring the current-vs.-voltage characteristics for the output of an inverter. Channel 1 measures the voltage difference between the loaded and the unloaded inverters. Channel 2 measures the current being supplied to the output of the inverter. The range of the test is limited by the current output of the function generator.*

Figure drawn with Digi-Key's Scheme-it [18].

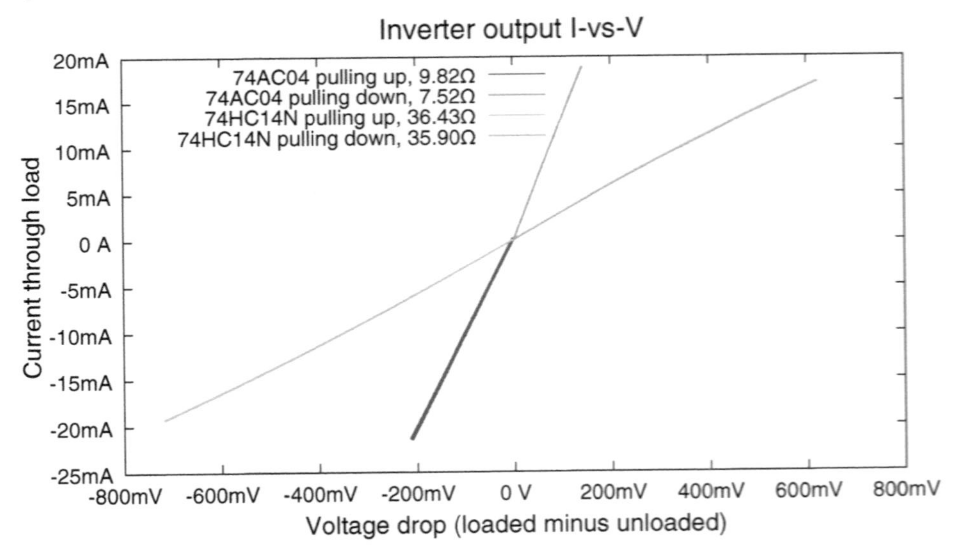

(b) *The current-vs.-voltage curves are given for the output stages of two different inverters, one from the 74AC logic family, one from the 74HC logic family. Positive voltages and currents are for a low output. Negative voltages represent the voltage below the high voltage that would have been obtained without a load.*

Data collected with Analog Discovery 2 [19]. Figure drawn with gnuplot [33].

Figure 32.9: *Test jig and measurements for current-vs.-voltage plots for two different cMOS inverter families. The 74HC family provides little more drive capability than a TLC3702 comparator, but the AC family provides substantially more current.*
The inverter output stages are reasonably modeled as either a resistor to ground or a resistor to the top power supply—about 36 Ω for the 74HC family and 7.5 Ω down and 9.8 Ω up for the 74AC family with a 5 V power supply. (Lower power-supply voltages will increase the effective resistance.)

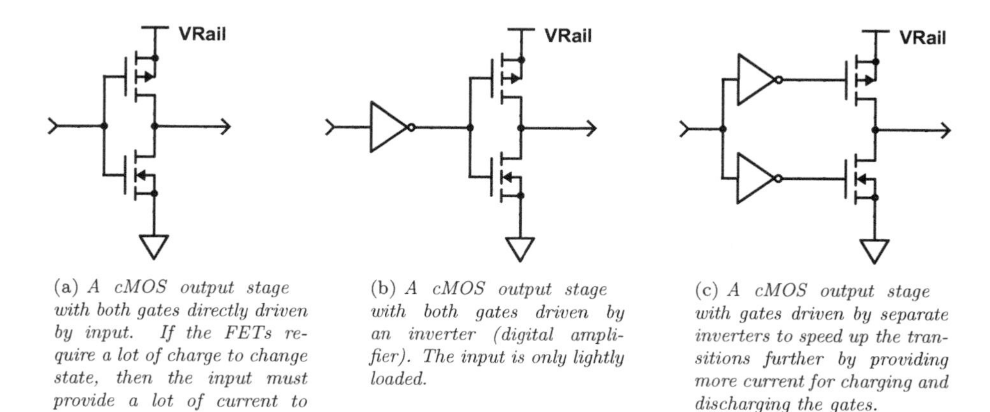

(a) *A cMOS output stage with both gates directly driven by input. If the FETs require a lot of charge to change state, then the input must provide a lot of current to switch quickly.*

(b) *A cMOS output stage with both gates driven by an inverter (digital amplifier). The input is only lightly loaded.*

(c) *A cMOS output stage with gates driven by separate inverters to speed up the transitions further by providing more current for charging and discharging the gates.*

Figure 32.10: *A cMOS output stage can be sped up by providing more current to drive the gates faster. The inverters are powered from the same power rails as the cMOS stage, so that their outputs swing from rail to rail. The inverters also change the meaning of the input, so that a high input turns on the pFET instead of the nFET. See Figure 35.8 for yet another circuit to drive the gates. Figure drawn with Digi-Key's Scheme-it [18].*

With a resistor model for the driving circuit, the RC time constant for charging the gates is independent of the power-supply voltage, so the time it takes to switch the cMOS stage is roughly independent of the power-supply voltage. (We'll see a better model for switching speed in Section 32.5.)

Unfortunately, while the input is switching, the power consumption is quite high, and the FETs get hot, so we don't usually use low-current drivers like comparators, but add extra circuitry to switch between on and off more rapidly.

Figure 32.10 shows a few of the many choices for driving the gates of the FETs. The inverters act as digital *current* amplifiers, increasing the current available for driving the gates, without changing the voltage range. By having a separate inverter for each gate, we can have more current charging or discharging the gate, switching the transistor on or off faster.

If you try to drive an nFET and pFET gate directly from the output of a low-current source such as a 74HC series logic device or a comparator, the rise time may be quite large, which is why one might want to use a design like Figure 32.10(b) or Figure 32.10(c), so that the rise and fall time of the gate voltage is kept quite low. Using a separate inverter for each gate reduces the capacitive load on each inverter, so one can get much faster transitions. Unfortunately, this method is limited to the voltage ranges acceptable to the inverter chips, which may be fairly small.

Another approach is to control the two gates separately with different control circuits. High-power drivers often introduce a short *dead time* between turning one FET off and turning the other FET on, to give the gates time to charge or discharge and avoid the shoot-through current that heats the FETs and wastes power. The relatively modest current levels we use in Lab 11 do not require this additional level of protection, as long as we provide enough drive current to switch the FETs off quickly.

32.3 Switching inductive loads

Many of the loads that we need large currents for (for example, motors and loudspeakers) are inductive, and switching inductive loads on and off requires some extra care. Let's look at what happens if we have a voltage source, an inductor, and a switch to connect or disconnect them, as shown in Figure 32.11. When we connect the voltage source and the inductor, the current through the inductor rises. If the inductor were an ideal inductor, the current would keep rising forever—because $V_+ = L_1 \frac{\mathrm{d}I}{\mathrm{d}t}$, we have $I(t) = 1/L \int_0^t V_+ \, dt = (V_t/L_1)t$. But because the inductor has a resistance R_1, the current initially rises

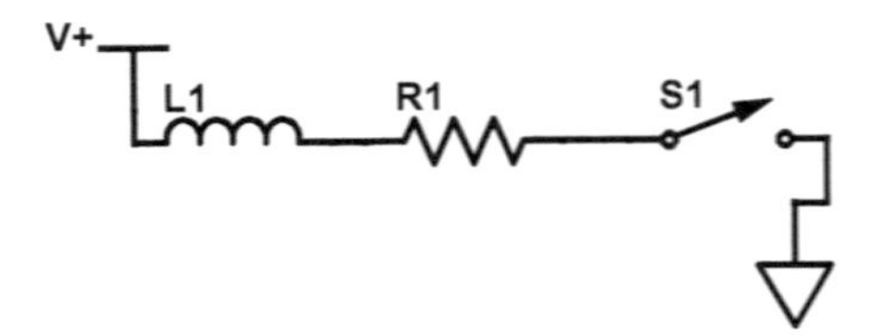

Figure 32.11: *The inductor L_1 has an internal resistance of R_1. When the switch is closed, the current through the inductor rises, asymptotically approaching $I = V_+/R_1$. When the switch is opened, the current continues to flow, causing the voltage across the switch to get larger than V_+.*

quickly, but then rises more slowly, asymptotically approaching $I = V_+/R_1$. (The differential equations are just like the ones for RC charging in Section 11.4, but swapping the roles of voltage and current.)

What happens when the switch is opened is a little more confusing—the current through the inductor does not instantly drop when the switch is opened, as that would require an infinite negative voltage drop across the inductor. Instead, the current continues to flow, but where does it go? If we think of the open switch as a capacitor (two conductors separated by an insulator), the current charges that capacitor, raising the voltage across the switch. Because the capacitance is small, the voltage increases very rapidly.

When the voltage across the switch gets higher than V_+, the voltage across the inductor becomes negative, and the current starts ramping down. If the inductor is large and the capacitance small, the voltage across the switch can get quite high before the current drops substantially. This spike in voltage when the switch is turned off is called an *inductive spike*.

Inductive spikes can be useful as a way to get voltages higher than the power supply, but they can also be quite damaging, as most electronic parts can be damaged by too-high voltages.

We can suppress inductive spikes by providing additional paths for the current when a switch is turned off. In motor-control circuits, these current paths are provided by *fly-back* or *free-wheeling* diodes, which divert the current back into the power supply (see Figure 32.12).

The diodes are normally reverse-biased and non-conducting, but when the FETs turn off and the inductor current causes the voltage to rise (or fall), the diodes clamp the voltage to about 0.7 V–1 V outside the power rails. In a cMOS output stage, the high-side diode on the pFET serves as a fly-back diode when the low-side nFET is turned off, and the low-side diode on the nFET serves as a fly-back diode with the high-side pFET is turned off.

The diodes are generally built into power FETs as *body diodes* (a natural consequence of the way FETs are constructed), so that no extra components are needed when using a cMOS output stage. But if you are using a single nFET as a switch, as in Figure 32.6, the built-in diode doesn't help. The problem is that the current through the inductor is toward the nFET drain, so that the spike on V_{ds} would be positive when the FET is turned off, but that still reverse-biases the body diode. If you have an inductive load driven by an nFET, then there should be a normally reversed-biased diode in parallel with the load, as shown in Figure 32.13. The voltage spike caused by turning off the transistor can then be clamped by the diode.

The injection of current back into the power supply can cause electrical noise to be distributed over the power wiring, so it is usual to put a large capacitor between the power rails near the FETs, to reduce the high-frequency components of the noise. Polymer electrolytic capacitors, with their low ESR (effective series resistance) and tolerance for high ripple currents are a popular choice, though they are

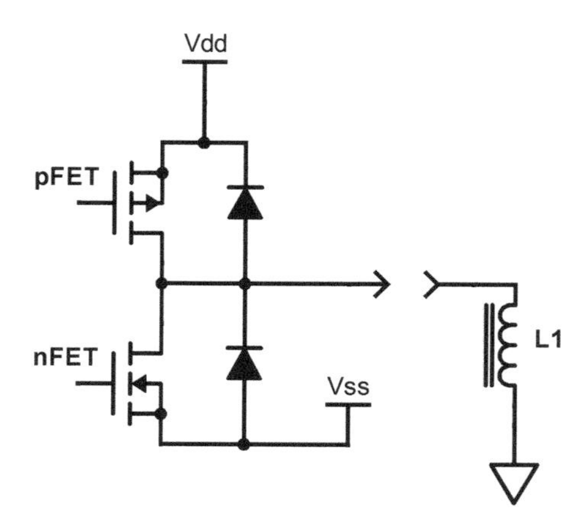

Figure 32.12: *A cMOS output stage for an inductive load includes fly-back diodes that keep the voltage across the inductor from going much beyond the power rails when the FETs are off, even though current continues to flow in the inductor. These fly-back diodes are provided automatically in power FETs as the* body diodes.

Figure drawn with Digi-Key's Scheme-it [18].

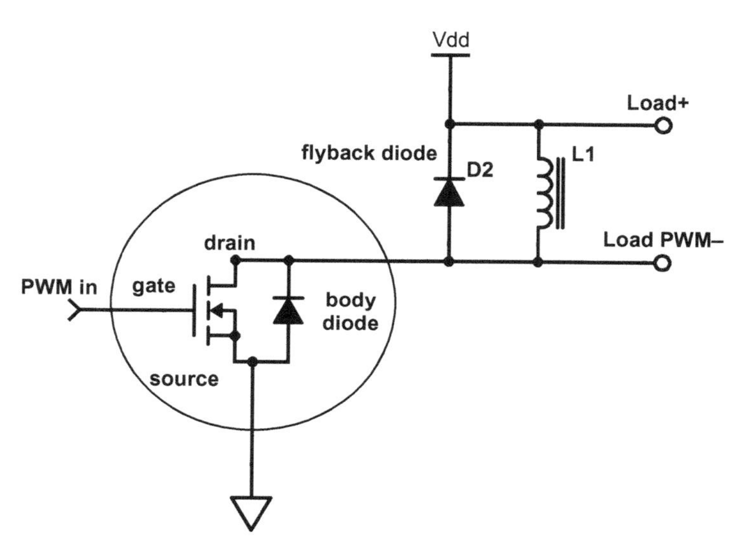

Figure 32.13: *In a single-nFET low-side switch, the body diode of the nFET keeps the output spikes from an inductive load to no more than about 0.7 V–1 V below ground, but an extra fly-back diode is needed to avoid positive voltage spikes more than 0.7 V–1 V above* V_{dd} *when the nFET is turned off.*

Both diodes are reverse-biased by the power supply, and so do not normally conduct.

Figure drawn with Digi-Key's Scheme-it [18].

usually paired with smaller ceramic capacitors to remove the higher frequency components (electrolytic capacitors have a rather high series inductance, and so a low self-resonant frequency).

32.4 H-bridges

An alternative to either the single nFET transistor of Section 32.1 or the complementary pair of Section 32.2 is to have four FETs arranged in an *H-bridge*, as shown in Figure 32.14.

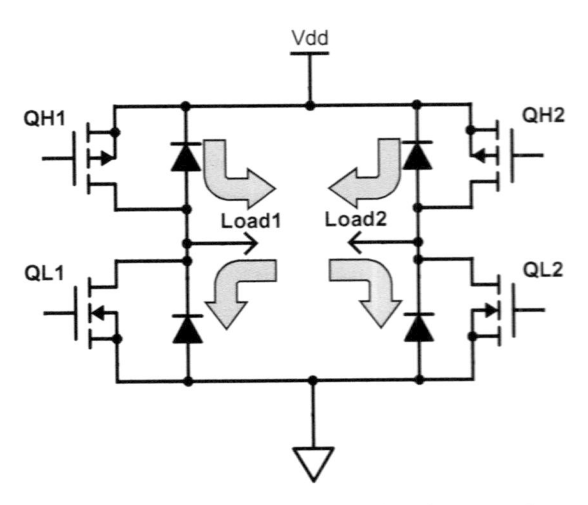

Figure 32.14: *H-bridge to provide bidirectional motor control from a single voltage supply.*
As with a cMOS driver, we never want QH1 and QL1 on at the same time, nor QH2 and QL2. The load (motor, loudspeaker, etc.) is connected between Load1 and Load2. If QH1 and QL2 are turned on, then current flows from Load1 to Load2 (blue arrows). If QH2 and QL1 are turned on, then current flows the other way, from Load2 to Load1 (orange arrows). If both QL1 and QL2 are turned on, then there is a loop through the ground line that short circuits the load, braking the motor.
The schematic shows the body diodes of the FETs, which serve as fly-back diodes—important for inductive loads like motors and loudspeakers (see Section 32.3).
Figure drawn with Digi-Key's Scheme-it [18].

> The H-bridge consists of two cMOS output stages, with the load connected between the outputs. An H-bridge allows current to be directed either way through the load, without needing a dual power supply (unlike the circuit in Figure 32.7, where a ground line is needed halfway between the power rails).

For example, if we wanted a ± 5 V square wave on the load, the cMOS stage of Figure 32.7 would need both a $+5$ V supply and a -5 V supply, while the circuit of Figure 32.14 only needs a single $+5$ V supply.

Using the same single power supply for all parts of a power amplifier, including the final output stage simplifies the design in Lab 11, so an H-bridge design is particularly attractive.

There are four common patterns for turning on the transistors of an H-bridge:

QH1 and QL2 on, others off: Load1 is connected to V_{dd} and Load2 is connected to ground, resulting in current flow through the load from Load1 to Load2.

QH2 and QL1 on, others off: Load2 is connected to V_{dd} and Load1 is connected to ground, resulting in current flow through the load from Load2 to Load1.

All FETs off: In this pattern, the load is not driven, and the only conduction is through the body diodes to suppress inductive spikes. This "floating" configuration allows fairly large inductive spikes ($V_{dd} + 2$ diode drops), which suppresses current through an inductor quickly. If the load is a motor, back-EMF generated by the motor at normal speeds will not be a large enough voltage to produce any current, and the motor will spin freely.

QL1 and QL2 on, others off: Both sides of the load are connected to ground, quickly dissipating any energy in the load. This configuration is often used for braking motors. (There is a symmetric configuration turning on just the pFETs instead of the nFETs, which would also brake a motor, but it is rarely used, because the nFETs do a better job of the braking, being able to handle more current and provide a lower resistance than the pFETs.)

H-bridges are particularly useful as motor controllers, as fairly simple control can be used to determine the direction and speed of the motor. Because motors often have very large currents, it is common to use four nFETs in an H-bridge, because low on-resistance nFETs are cheaper and faster to switch than pFETs with similar on-resistance.

The problem with using nFETs as high-side switches is that the gate voltage needs to be higher than top power rail. This is achieved with either a *bootstrap* or *charge pump* circuit which uses a capacitor and switches or diodes to raise the voltage. The charge pump alternately charges a capacitor then raises the voltage at the lower end of the capacitor to provide a higher voltage. The bootstrap design is a charge pump that relies on the PWM signal driving the H-bridge itself to do the pumping (and so limits the duty cycle to somewhat less than 100%). The design of charge pumps and bootstraps is beyond the scope of this book.

There are some designs that deliberately use pFETs on the high-side, to avoid the electromagnetic interference (EMI) that results from high-frequency charge pumps.

Because H-bridges are so common, one can buy chips that incorporate all the control circuitry, needing just the FETs to be added to complete the controller. These chips usually incorporate prevention of shoot-through current, thermal shutdown, over-voltage protection, under-voltage protection, and short circuit protection as well. One can even buy complete controller-plus-H-bridge chips that include both the control and the FETs, though cheaper, lower-current H-bridge chips often use bipolar transistors rather than FETs.

32.5 Switching speeds of FETs

This section is fairly advanced—a detailed understanding of how fast FETs switch may not be needed for the initial design of a class-D power amplifier, but this section will help understand the signals you will see on the oscilloscope. If you try to do a higher-power class-D amplifier, with larger voltages, then the switching speeds of the FETs might become very important.

Estimating how long it takes for an FET to switch can be tricky—although the gate looks a lot like a plate of a capacitor, simple RC charging models for the gate don't work very well. One problem is that the other plate of the capacitor is changing when the transistor turns on or off, and the voltage at the drain of the transistor is not constant. Also important is that the capacitance itself is not constant, but depends on both the gate and the drain voltages.

The simplest model that works fairly well is the *gate-charge* model, which has three parameters: the *gate source charge* Q_{gs}, the *gate drain charge* Q_{gd}, and the total gate charge Q_g, which is larger than the sum of the other two.

The gate-charge model is most easily explained by looking at the steps involved in turning on an FET. When turning on an nFET, the gate voltage ramps up until it reaches the threshold voltage, at which point the nFET starts to turn on and the drain current I_d ramps up while the drain voltage V_{ds} ramps down. When the transistor finishes turning on, both I_d and V_{ds} are constant, even if the gate voltage has not yet reached its final value. Figure 32.15 shows oscilloscope traces for V_{gs}, V_{ds}, and I_d for an nFET.

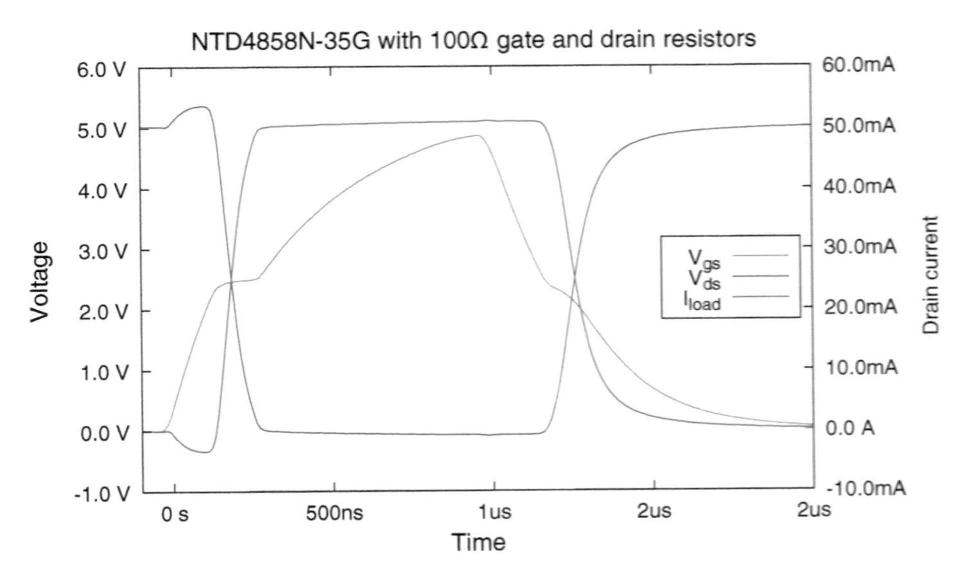

Figure 32.15: *Oscilloscope traces for the gate voltage (purple), drain voltage (blue), and load current (red) for an NTD4858N-35G nFET with a 100 Ω load resistor to 5 V on the drain.*
The current measured here is not just the current through the channel, but also some transient current through the gate-drain capacitance, which causes the small negative spike in current when the gate voltages rises before the FET turns on and a small positive current even after the transistor has turned off.
The gate is driven with a pulse from 0 V to 5 V through a resistance of 100 Ω (the resistance is to slow down the transitions, to make the Miller plateau visible).
The gate voltage initially rises fast, and capacitive coupling to the drain causes the drain to go above the top power rail momentarily, though not by very much.
Noise was reduced by averaging 1000 traces (the noise reduction was not really necessary for this signal, but cleans it up a little).
Data collected with Analog Discovery 2 [19]. Figure drawn with gnuplot [33].

The *Miller plateau* in Figure 32.15 is where the gate voltage remains nearly constant while the drain voltage makes its transition.

- The gate source charge Q_{gs} is how much charge must be transferred to the gate to begin turning on the transistor—raising the gate voltage to the Miller plateau.
- The gate drain charge Q_{gd} is how much more charge must be transferred to the gate to complete turning on the FET. (The area under the Miller plateau.)
- The total gate charge is an estimate of how much charge has to be added to the gate (in total) to get the gate all the way to its final voltage, and so is specified with a target voltage (often 4.5 V or 10 V).

We can estimate the time each part takes from the charge transferred divided by the current that is charging the gate (if the current is not constant, then you need to integrate it to get the charge). In Figure 32.15, waveforms are shown for an nFET, with the gate connected to a 0 V–5 V pulse through a 100 Ω resistor. The gate voltage rises along an RC charging curve until the nFET starts to turn on, then holds fairly steady as V_{ds} falls, then returns to an RC charging curve.

The capacitive coupling from the drain to the gate when V_{ds} drops slows the rise of the V_{gs} voltage until the drain has reached its new low voltage level. This flattening of the gate voltage is referred to as the *Miller plateau*, the feedback capacitance from the drain to the gate internal to the nFET as the *Miller capacitance*, and the gate voltage in the middle of the plateau as the *Miller voltage*.

The nearly constant gate voltage on the Miller plateau is a result of a negative-feedback loop through the drain-gate capacitance. As the drain voltage drops, it produces a current away from the gate. If V_{ds} drops too fast (taking more current than the signal driving the gate provides), it reduces the gate voltage and thus reduces the drain current, slowing down the drop. If V_{ds} drops too slowly, then the external gate current charges the gate and V_{gs} rises, making V_{ds} drop more quickly. The feedback loop keeps the gate voltage roughly constant.

The charge transferred to the gate before the output starts switching is the gate-source charge Q_{gs}, which I estimate from these curves to be about 6.3 nC (average of about 38 mA for 170 ns).

The charge transferred to the gate while the drain voltage is dropping is the gate-drain charge Q_{gd}, which I estimate from these curves to be about 3.3 nC (average of about 25.5 mA for 130 ns). My gate-charge estimates from these plots are somewhat different from the ones on the data sheet ($Q_{gs} = 4.7\,\text{nC}$ and $Q_{gd} = 5.2\,\text{nC}$), but their sum is almost the same. Differences in the conditions under which the parameters are measured and differences in where the plateau is defined to start probably account for the differences.

> The plateau in the gate voltage can be used to control how fast the drain voltage changes, $\frac{\mathrm{d}V_{ds}}{\mathrm{d}t}$, which is known as the *slew rate* of V_{ds}.

If the gate voltage is constant, the current into the gate goes entirely into charging the gate-drain capacitance, so

$$I_g = C_{gd}\frac{\mathrm{d}(V_{gs} - V_{ds})}{\mathrm{d}t}$$

which can be rewritten as

$$\frac{\mathrm{d}V_{ds}}{\mathrm{d}t} = -I_g/C_{gd} \ .$$

Here C_{gd} is the reverse transfer capacitance C_{rss}, which the NTD4858 data sheet gives as typically $C_{rss} = 200\,\text{pF}$.

In Figure 32.15, the current into the gate at the Miller plateau is approximately $(5\,\text{V} - 2.5\,\text{V})/100\,\Omega$, or 25 mA, so we would expect the slew rate to be about $\frac{-25\,\text{mA}}{200\,\text{pF}} \approx -125\,\text{V}/\mu\text{s}$, but the observed slew rate is only about $-50\,\text{V}/\mu\text{s}$, substantially slower than predicted.

The problem with our crude analysis of slew rate is that C_{rss} is far from being a constant—the data sheet reported it as measured at $V_{ds} = 12\,\text{V}$, but the test in Figure 32.15 has $V_{ds} \approx 2.5\,\text{V}$, where C_{rss} is about three times larger [80].

> We normally want as high a slew rate as we can get, since the energy dissipated in the nFET while it is turning on is proportional to the switching time (see Section 32.6). This means driving the gate of the FET with a low-impedance source.

There are times, however, when we want to keep the slew rate lower, to reduce radio-frequency interference by removing energy from the high-frequency part of the spectrum, for example. Too fast an edge can also cause ringing due to parasitic inductance in the wiring forming a high-frequency resonant LC circuit—a slower edge would have less energy at the resonant frequency.

When we want a slower output transition, we can either add a series resistor to the gate (to limit the gate current) as done for creating Figure 32.15 or add a gate-drain capacitor to provide more feedback capacitance.

I've neglected some important stuff in this section, in the interest of keeping the model simple enough for back-of-the-envelope calculations. There are more detailed explanations in application notes from FET manufacturers [101].

32.6 Heat dissipation in FETs

Power FET transistors are designed to handle large currents—for the NTD4858 FETs we used to use, up to 45 A, which are much larger currents than you'll be dealing with. The main limitation of the FETs when they are fully on, though, is with self-heating: if you try to put 45 A through an NTD4858, it will dissipate $(45\,\text{A})^2\ 6.2\,m\Omega = 12.6\,\text{W}$ and quickly heat up past its maximum temperature unless a really good heat sink is used to reduce the thermal resistance and dissipate the heat into the air.

Although the NTD4858N nFETs are supposedly capable of dissipating 2 W of power without damage, that is only true if it is mounted on a PC board with sufficient copper to act as a heat sink. We used the transistors in a breadboard with no heat sink, so they couldn't dissipate that much power. So how hot can the transistors be allowed to get, and how do we compute how hot they will get?

> Temperature calculations are done with *thermal resistance* calculations, with the temperature drop from one spot to another computed as the product of the power heating one spot and the thermal resistance (in °C/W) of the connection between the spots.

The most important temperatures for electronics are the *junction temperature*, which is the temperature internal to the part that is dissipating heat, and the *ambient temperature*, which is the temperature of the air that the heat is being dumped into.

For the NTD4858N nFET, the junction-to-ambient thermal resistance with a small heat sink (a square inch of 1 oz copper on a PC board) is 73.5°C/W, so keeping the junction below the 175°C maximum junction temperature limits us to about 2.4 W dissipated by the nFET (about 19.6 A with a resistance of 6.2 mΩ). Without a heat sink, as we used them, the junction-to-ambient thermal resistance is higher (probably around 150°C/W), and so you need to keep the power dissipation in the FETs even lower (under 1.2 W, or about 13.7 A). Since we only went up to about an amp (dissipating about 6 mW if continuously on), we didn't have much heating of the nFET when it was fully on.

The steady-state power dissipation for the FETs all the way on or all the way off is fairly small, but when we are putting out an intermediate voltage, the power dissipation can be high.

For example, if you were driving an 8 Ω load and used an FET to get the voltage only half way to the power rail, then the FET would have the same resistance and same current as the load, and each would be dissipating the same power (2.5 W in the example of a 9 V power supply). To maintain efficiency and prevent overheating the FETs, we try to switch between the on and off states as quickly as possible.

How much power is dissipated as heat in an FET used as a high-frequency switch?

> The power lost is a combination of two sources:
>
> - static power when current is flowing: $I_{\text{max}}^2 R_{\text{on}} D$, where D is the duty cycle—what fraction of time the current is flowing, and
> - switching power $E_{\text{sw}} f_{\text{PWM}}$, where E_{sw} is the energy lost in one cycle (turning the switch on and turning the switch off) and f_{PWM} is the frequency of the pulse driving the switch. The switching energy E_{sw} turns out to be proportional to the switching time T.

We can minimize the power loss (and heating of the FET) by choosing transistors with small R_{on}, but only if we can simultaneously keep the switching time small—the large transistors with small R_{on} also have large gate capacitance, which can limit how fast they switch.

We can make a crude estimate of the switching energy by assuming a resistive load on the drain of the FET and using the gate-charge model. For both rising and falling edges, the time when there is power being dissipated in the FET is roughly the time it takes to move the gate-drain charge Q_{gd} onto or off of the gate. If the gate current is I_g, then the switching time is $T = \frac{Q_{gd}}{I_g}$.

When turning on the FET, the voltage across the FET ramps down roughly linearly from $V_{\max}$ to 0 V, and the current ramps up roughly linearly from 0 A to $I_{\max}$, making the instantaneous power

$$P(t) = V_{\max} \frac{T-t}{T} I_{\max} \frac{t}{T} ,$$

where time $t = 0$ is when the transistor starts to turn on. Integrating over the switching time, $\int_0^T P(t)\,dt$, gets an energy of $V_{\max} I_{\max}(1/2 - 1/3)T$. That is, the total energy dissipated in each switching event is about $V_{\max} I_{\max} T/6$, and in each cycle (with two switching events, turning on and turning off) is $V_{\max} I_{\max} T/3$.

Summarizing, we get an estimate of switching energy for one cycle as

$$E_{\text{sw}} = \frac{V_{\max} I_{\max}}{3} \frac{Q_{gd}}{I_g} ,$$

where $V_{\max}$ is the drain-to-source voltage when the FET is off, $I_{\max}$ is the drain current when the FET is on, and I_g is the current that is charging (or discharging) the gate.

The analysis above makes a lot of simplifying assumptions, including assuming a resistive load. With a capacitive load or an inductive load, the current and voltage may make transitions at different times, resulting in times when both the current and the voltage are high, causing a higher switching energy loss. The energy loss still tends to be proportional to the switching time, but the constant of proportionality is different.

In the cMOS circuit described in Section 32.2, there is a possibility for a particularly nasty additional power loss: *shoot-through current* when both FETs are simultaneously on. Eliminating shoot-through

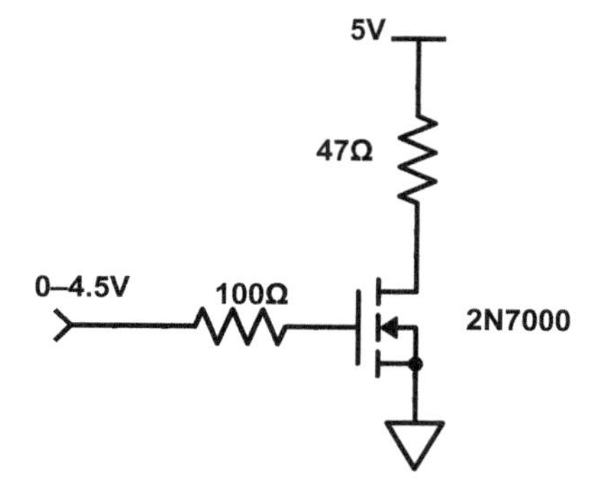

Figure 32.16: *This circuit is for Exercise 32.5. It shows an nFET switching a resistive load on and off, using a 4.5 V pulse. Figure drawn with Digi-Key's Scheme-it [18].*

current (or minimizing its duration, so that the average power lost to it remains small) is an important part of designing a class-D amplifier.

One can get nFETs with on-resistance less than 1 mΩ and current ratings of 300 A such as Infineon's IPLU300N04S4R8XTMA1, but their larger gates means it takes a lot more current to turn the transistors on and off quickly. (The gate charge at 10 V is around 290 nC, compared to 26 nC for the NTD4858N—see Section 32.5 for information about the gate-charge model.) Even with only 0.77 mΩ on resistance, a 300 A current would dissipate $(300\,\text{A})^2 0.77\,\text{m}\Omega = 69.3\,\text{W}$—one would need a really big heat sink with good thermal conductivity to keep such an FET cool enough to continue operating.

Exercise 32.3

You have a power FET with a junction-to-case thermal resistance of 3.6 K/W mounted on a heat sink with a 24.4 K/W thermal resistance to ambient. What is the total junction-to-ambient thermal resistance? What is the temperature of the junction if 1 W is dissipated, when the ambient temperature is 25°C? If the maximum temperature for the junction is 175°C and the ambient temperature is 35°C, how much power can be dissipated?

Exercise 32.4

Data sheets don't always give thermal resistance and maximum junction temperature as single numbers, but sometimes as a plot of maximum power as a function of ambient temperature. For the plot in Figure 32.17, what is the maximum junction temperature (in °C) and thermal resistance (in K/W)?

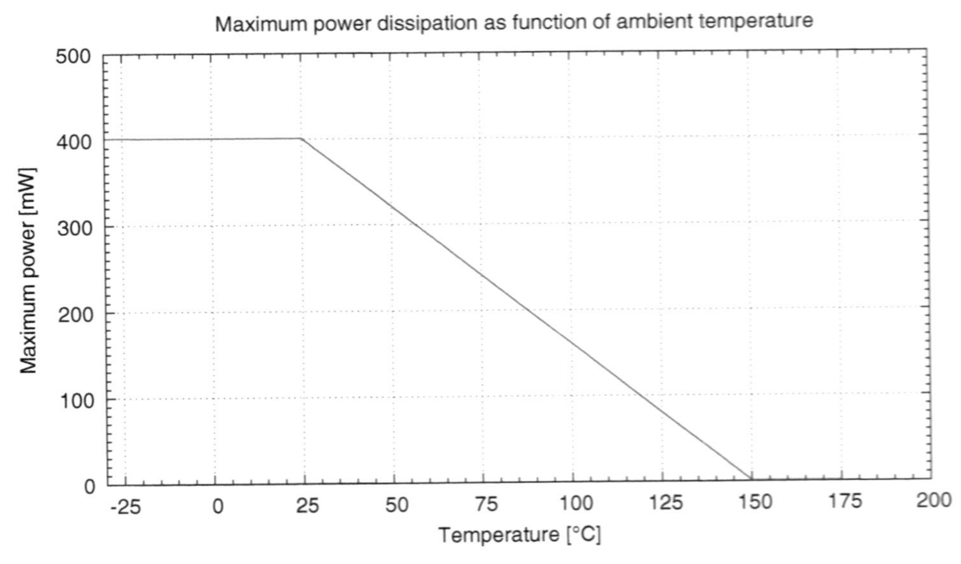

Figure 32.17: *Plot of maximum power dissipation for an unspecified power FET as a function of ambient temperature, for Exercise 32.4.*
Figure drawn with gnuplot [33].

Exercise 32.5

Look up the data sheet for the 2N7000 transistor. For the circuit of Figure 32.16 with a 100 kHz pulse input that is at 4.5 V for 40% of the period, then at 0 V for the rest of the period, determine the following:

- R_{on} for the nFET,
- the power dissipation in the 2N7000 nFET when the input is 4.5 V,
- the average power dissipation in the 2N7000 nFET due to on-resistance,
- the gate voltage when the transistor switches (the Miller plateau),
- the gate current at that gate voltage,
- the gate-drain charge Q_{gd} of the 2N7000,
- the approximate switching energy per cycle dissipated in the nFET,
- the average power dissipation due to switching.

Is the switching power a substantial contributor to the total power in this circuit?

Chapter 33

Comparators

A *comparator* is a very-high-gain, differential amplifier that is designed to be used without feedback. When the "+" input is at a higher voltage than the "−" input, then the output is high (at the top voltage), otherwise the output is low. It differs from an op amp in that it is designed to make transitions between low and high very quickly, while op amps have deliberately slowed transitions and are intended to be used in the middle voltages, away from the highest and lowest voltages.

33.1 Rail-to-rail comparators

For some applications (such as the power amplifier in Lab 11), you need to make large voltage changes rapidly, but our MCP6004 op amps are not capable of doing that. Not only are they limited to a maximum voltage swing of about 6 V, they have a designed-in speed limitation: a *slew rate* of 0.6 V/μs, so it would take them 10 μs to do their full 6 V swing. The 6 V limitation may not be a severe one for your power-amp design, but the 0.6 V/μs is.

Because of these limitations of the op amp, you have another chip in your parts kit: the TLC3702 dual comparator chip. It provides rail-to-rail output, like the MCP6004 op amp, but has a wider power-supply voltage range and a much higher slew rate—the rise and fall times are mainly limited by the current it can provide and the capacitance being driven, rather than by internal speed limitations of the chip.

The data sheet provides different ways of estimating how fast a capacitive load can be driven. Here are three approaches:

- With a 50 pF load, the typical rise and fall times are given as 125 ns and 50 ns, respectively. This is 80–200 times faster than the output edges of an MCP6004 op amp. These rise and fall times are good approximations when the capacitive load is small (50 pF or less), but do not work well for large loads like those we'll see in Chapter 35.
- We can look at the relationship between the DC voltage drop and the output current. According to the data sheet, with a 5 V power supply the TLC3702's output driver is roughly equivalent to a 58 Ω resistor to ground or a 65 Ω resistor to 5 V, and with a 16 V power supply the output is equivalent to a 24 Ω resistor to ground or a 30 Ω resistor to 16 V. Figure 33.1 shows measurements I made—the resistances I measured were somewhat lower than the typical ones on the data sheet—possibly because the ambient temperature was lower than 25°C—but probably within the typical range.

  These resistance values can be used with a known load capacitance to get RC time constants. This method is appropriate when driving large capacitive loads, where internal delays in the chip are a small part of the total time.

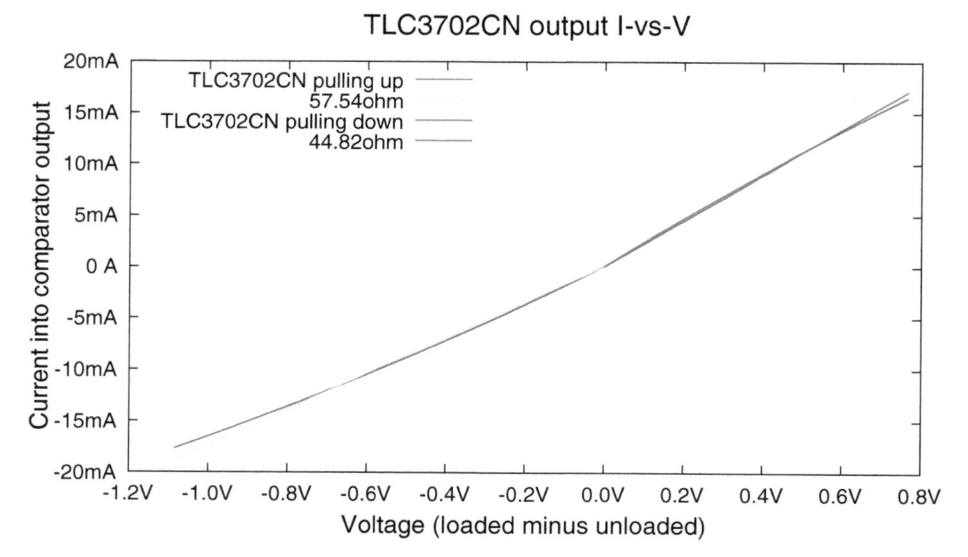

Figure 33.1: *To check the data sheet, I measured the output voltage of the TLC3702CN rail-to-rail comparator powered with 5 V using various load currents. The resistances, 58 Ω up and 45 Ω down, are somewhat lower than the typical values on the data sheet. The I-vs.-V curve is not quite straight, but shows some current saturation when the current exceeds ±13 mA.*

The test setup is similar to Figure 32.9a, but with the two inputs of the comparator tied to opposite power rails to get the appropriate output level.

Data collected with Analog Discovery 2 [19]. Figure drawn with gnuplot [33].

- We can look at the plot of typical rise and fall times vs. load capacitance. For a 5 V power supply, the rise time is about 115 ns + 130 ns/nF, and the fall time, about 45 ns + 110 ns/nF. For small loads, the constant part is the crucial one, but for large loads the part that is linear with capacitance is the crucial part.

  If you do unit conversions, you'll see that 130 ns/nF is 130 Ω, but that is *not* the output impedance since rise and fall time are measured from 10% to 90% of the voltage step, so corresponds to about 2.2 RC time constants:

$$e^{-t_1/(RC)} = 0.9$$
$$e^{-t_2/(RC)} = 0.1$$
$$t_2 - t_1 = \ln(0.9/0.1)RC \approx 2.2RC\ .$$

  The output impedance of the TLC3702 is thus about 59 Ω for rising edges and 50 Ω for falling edges, very close to what we computed from the DC I-vs.-V characteristics.

> One warning about the TLC3702 chip (and multi-amplifier chips in general)—don't leave inputs for unused comparators or amplifiers unconnected.

When I first set up a class-D amplifier using the TLC3702 chip, I used only one of the comparators and left the inputs to the other comparator floating. This circuit "worked", but produced a rather loud and annoying hum, even when the input was constant. At first I thought that the problem was in the preamplifier, but there was no sign of hum on the output of that stage. It turned out that the unconnected comparator was swinging rail-to-rail based on capacitive pickup of 60 Hz interference. This switching coupled enough noise into the internal power supply on the chip that the threshold for the comparator I was using shifted a little bit—that threshold shift had the same effect as adding 60 Hz interference to the input, producing the annoying hum. Tying the two inputs of the unused comparator to opposite power rails (so that the comparator inputs were far apart) eliminated the hum.

33.2 Open-collector comparators

This section is optional, as we will not be using open-collector comparators in Lab 11.

Another class of comparators is represented by the LM2903 dual comparator chip. The LM2903 has a different sort of output than the op amps and Schmitt triggers you've used so far—an *open-collector output.* When the positive input is higher than the negative input, there is essentially no current through the output, but when the positive input is lower than the negative input, the output can sink current. *Sinking* current means that current flows from whatever is connected to the open-collector output through the LM2903 to the low-voltage power rail for the LM2903. The voltage of the output depends on what is connected to the output pin, but is generally somewhat above the low-voltage power rail.

To get the output to ever move away from the lower power rail, you need a pull-up resistor to provide current when the LM2903 output is off and pull the output up. Sizing that pull-up is a design task—let's look at the constraints on it. The LM2903 and pull-up resistor need to drive the gates of the FETs high and low (enough to turn the FETs fully off) rapidly. The gates of FETs are essentially capacitors. You can look up the input capacitance on the data sheets for the FETs, though the gate-charge model of Section 32.5 provides a better calculation of the time it takes to turn on or off the transistors.

To choose values for the pull-up resistors, we need to know how the output of the LM2903 behaves. When the output is supposed to be high, there is no current through the LM2903 output, and the gate capacitance is charged through the pull-up resistor to whatever voltage the other end of the resistor is connected to. That is, the output-high voltage V_{OH} is just whatever the resistor is connected to. When the output is supposed to be low, the output-low voltage V_{OL} depends on the current through the LM2903 output transistor and the resulting IR drop across the pull-up resistor: $V_{\mathrm{OL}} = V_{\mathrm{OH}} - I_{\mathrm{sat}} R$.

The spec sheets for the LM2903 do not give us the output current vs. voltage characteristic, so I measured it with a 12 V power supply and the Analog Discovery 2 USB oscilloscope and function generator, using the setup in Figure 33.2 to get the results in Figure 33.3. The characteristic in Figure 33.3 is not well approximated by either a constant current or a constant resistance, but for V_{OL} higher than 1 V above the bottom power rail, we can model the LM2903 as a current source with a current of about $I_{\mathrm{sat}} = 16\,\mathrm{mA}$.

The 16 mA that I measured at 1.5 V is much more than the minimum spec of 6 mA. Minimum specs on a data sheet are often much more conservative than typical values, to account for variation due to temperature, aging of parts, or variation in the manufacturing. The saturation current curve seems to depend somewhat on the power-supply voltage for the LM2903, increasing with higher voltage, but I did not measure this effect.

The low-voltage region is not well modeled by a resistor, as the I-vs.-V line does not have the right slope, except in the region around 10 mV, where it behaves like a 120 Ω resistor. As a very crude approximation (within a factor of 2.2) one could use a resistance of 58 Ω for collector voltages below 1 V.

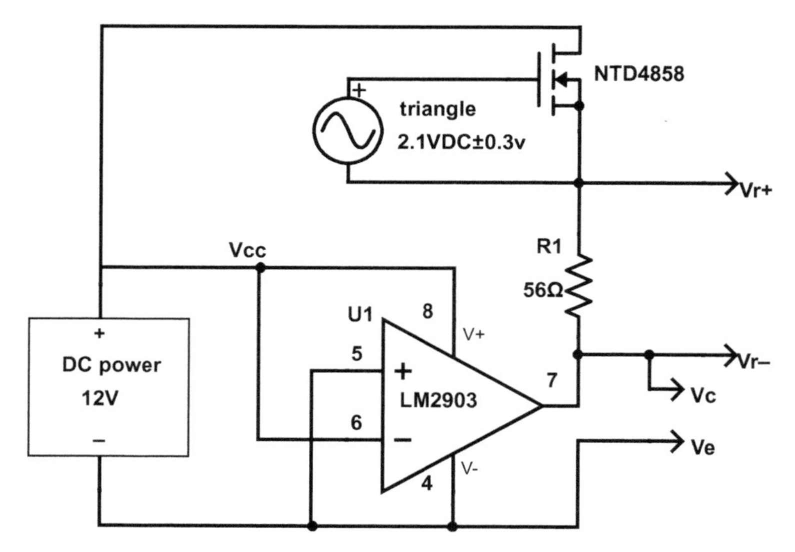

Figure 33.2: *Test setup for characterizing the open-collector output of an LM2903 comparator. Two values were used for resistor* R_1*: 56Ω for high collector voltage and 2.2kΩ for low collector voltage.*
The triangle-wave generator and nFET provide a fluctuating resistance in series with the 56Ω sense resistor, providing a load that varies from 56Ω to about 1 MΩ.
Figure drawn with Digi-Key's Scheme-it [18].

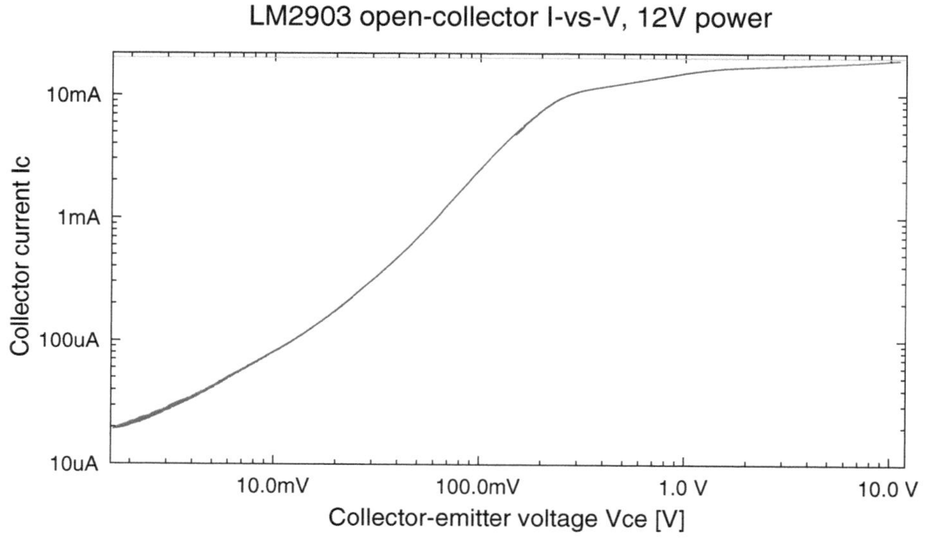

Figure 33.3: *The output of the LM2903 comparator has a current of around 15 mA with* $V_{OL} = 1$ *V and a maximum current around 19 mA, even for* $V_{OL} = 11$ *V, when the power supply is 12 V—the 20 mA line is the maximum current on the data sheet.*
Data collected with Analog Discovery 2 [19]. Figure drawn with gnuplot [33].

33.3 Making Schmitt triggers

This section is optional material—it will not be needed for Lab 11, though it is used in Chapter 36.

Comparators convert an analog signal to a single digital bit of information. That conversion is often more useful if hysteresis is included to reduce extra transitions due to noise, as we saw in Section 16.1.

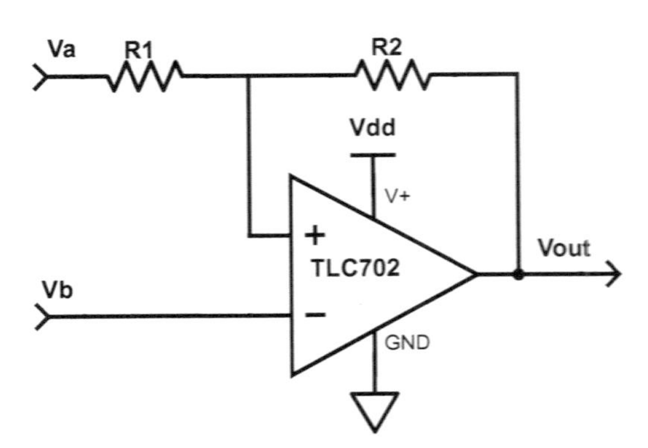

Figure 33.4: *This circuit uses* positive *feedback to provide the function of a Schmitt trigger. The output is always pinned at one of the two power rails,* V_{dd} *or ground, and the threshold at which the trigger switches depends on the output voltage. We have a high output when* $V_a R_2 + V_{out} R_1 > (R_1 + R_2) V_b$.
Figure drawn with Digi-Key's Scheme-it [18].

Making a Schmitt trigger out of a comparator is very straightforward for rail-to-rail comparators, and only slightly more complicated for open-collector or open-drain comparators. The trick is to use *positive* feedback, instead of negative feedback, as shown in Figure 33.4.

> The Schmitt trigger is the **only** circuit we use that has positive feedback. All other op-amp circuits in this book use only negative feedback.
>
> The positive feedback has the effect of converting the comparator to a one-bit memory, being stable with either the output high or the output low, until forced to change.

The positive feedback provides different thresholds for comparison of the inputs, depending on whether the voltage at the output is low or high. For a rail-to-rail comparator, the output high and low voltages are just V_{dd} and ground.

The positive input to the op amp can be determined by the voltage-divider formula:

$$V_p = \frac{V_a R_2 + V_{out} R_1}{R_1 + R_2} \, .$$

So, the output will go from low to high when $V_p > V_b$, that is, when

$$V_a R_2 > (R_1 + R_2) V_b \, ,$$

and the output will go from high to low when $V_p < V_b$, that is, when

$$V_a R_2 + V_{dd} R_1 < (R_1 + R_2) V_b \, .$$

> If V_b is held constant at V_{ref}, we have a non-inverting Schmitt trigger, with thresholds
>
> $$V_a > V_{ref} + V_{ref} R_1 / R_2 \qquad (33.1)$$
>
> for the low-to-high transition and
>
> $$V_a < V_{ref} - (V_{dd} - V_{ref}) R_1 / R_2 \qquad (33.2)$$
>
> for the high-to-low transition.

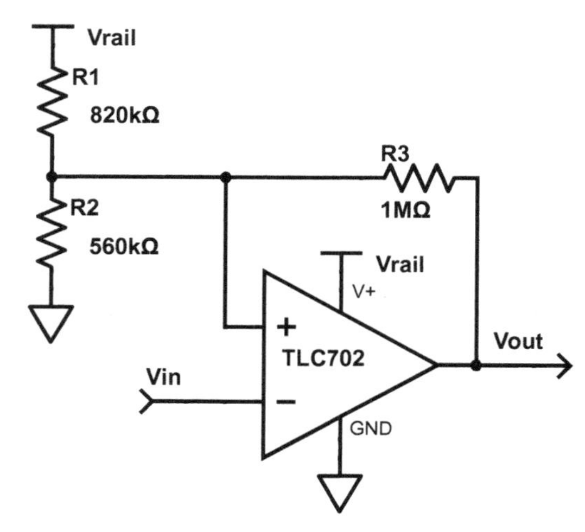

Figure 33.5: *A Schmitt-trigger inverter can be made from a rail-to-rail comparator by using positive feedback instead of negative feedback. For this circuit, the two thresholds are* $V_{IL} = 0.305V_{rail}$ *and* $V_{IH} = 0.554V_{rail}$.
Figure drawn with Digi-Key's Scheme-it [18].

If V_a is held constant at V_{ref}, we have an inverting Schmitt trigger, with thresholds

$$V_b < V_{ref}\frac{R_2}{R_1 + R_2} \tag{33.3}$$

for the low-to-high transition and

$$V_b > \frac{V_a R_2 + V_{dd} R_1}{R_1 + R_2} V_{ref} \tag{33.4}$$

for the high-to-low transition.

We can replace V_{ref} and R_1 with a voltage divider that is Thévenin-equivalent.

33.3.1 *Inverting Schmitt trigger with rail-to-rail comparator*

Let's analyze an example inverting Schmitt trigger using a rail-to-rail comparator, as shown in Figure 33.5.

Worked Example:
Compute the thresholds for the Schmitt trigger in Figure 33.5.

We can compute the Thévenin equivalent of the voltage divider formed by R_1 and R_2 in Figure 33.5: the voltage is $V_{ref} = V_{rail}\frac{560\,\text{k}\Omega}{560\,\text{k}\Omega + 820\,\text{k}\Omega} = 0.4058V_{rail}$, and the resistance is $560\,\text{k}\Omega \parallel 820\,\text{k}\Omega = 332.8\,\text{k}\Omega$. Since this is an inverting Schmitt-trigger arrangement, that gives us thresholds of $V_{in} < 0.7503V_{ref} = 0.3045V_{rail}$, and $V_{in} > (V_{ref}\ 1\,\text{M}\Omega + V_{rail}\ 332.8\,\text{k}\Omega)/(1.3328\,\text{M}\Omega) = 0.5542V_{rail}$.

Alternatively, we can compute the thresholds using just the voltage-divider formula, without recourse to Thévenin equivalence. If the output voltage is low, then R_3 is in parallel with R_2 and the threshold is

$$V_{IL} = \frac{R_2 \parallel R_3}{R_1 + (R_2 \parallel R_3)} V_{rail} \ .$$

For the values in Figure 33.5, we get $V_{IL} = 0.3045 V_{rail}$.

If the output voltage is high, R_3 is in parallel with R_1 and the threshold voltage is

$$V_{IH} = \frac{R_2}{R_2 + (R_1 \parallel R_3)} V_{rail} \ ,$$

which for the values in Figure 33.5 gives us $V_{IH} = 0.5542 V_{rail}$.

The same formulas can be turned around to design a Schmitt trigger with specific thresholds. The thresholds depend only on the power supply voltages and the ratios of sizes of the resistors, not their absolute sizes, so any solution we come up with can be scaled to another solution by multiplying all resistances by the same constant.

We have two equations and three unknowns:

$$\begin{aligned} V_{IL}/V_{rail} &= \frac{R_2 \parallel R_3}{R_1 + (R_2 \parallel R_3)} \\ &= \frac{R_2 R_3}{R_1 R_2 + R_1 R_3 + R_2 R_3} \\ V_{IH}/V_{rail} &= \frac{R_2}{R_2 + (R_1 \parallel R_3)} \\ &= \frac{R_1 R_2 + R_2 R_3}{R_1 R_2 + R_1 R_3 + R_2 R_3} \ . \end{aligned}$$

Taking the ratio of these formulas gives us $V_{IH}/V_{IL} = R_1/R_3 + 1$, which makes a good starting point for sizing resistors.

Worked Example:
What resistors could we use to get thresholds of 1 V and 2.8 V with a 5 V power supply?

The ratio of the thresholds is 2.8, so $R_1/R_3 = 1.8$. If we arbitrarily pick $R_3 = 100\,\text{k}\Omega$, we get $R_1 = 180\,\text{k}\Omega$, and $R_1 \parallel R_3 = 64.29\,\text{k}\Omega$. We can solve

$$2.8\,\text{V}/5\,\text{V} = \frac{R_2}{R_2 + 64.29\,\text{k}\Omega}$$

to get $R_2 = 81.82\,\text{k}\Omega$. If we round to the E12 or E24 series to get $R_2 = 82\,\text{k}\Omega$, we have

$$\begin{aligned} V_{IL} &= \frac{R_2 \parallel R_3}{R_1 + (R_2 \parallel R_3)} 5\,V = 1.001\,V \\ V_{IH} &= \frac{R_2}{R_2 + (R_1 \parallel R_3)} 5\,V = 2.803\,V \ , \end{aligned}$$

which is very close to our design goal (closer than we can expect to achieve with 1% resistors).

Exercise 33.1

Design a Schmitt trigger for a 3.3 V power supply that has thresholds at 1 V and 2 V. Use a rail-to-rail comparator and E24-series resistors.

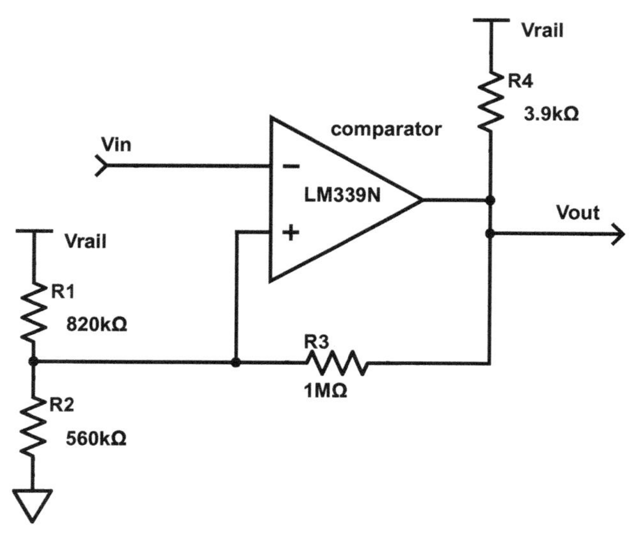

Figure 33.6: *A Schmitt trigger can be made from an open-collector comparator by using a pull-up resistor on the output and using much higher impedances for the positive feedback than for the pull-up resistor. The thresholds are* $V_{\mathrm{IL}} = 0.311 V_{\mathrm{rail}}$ *and* $V_{\mathrm{IH}} = 0.554 V_{\mathrm{rail}}$.
Figure drawn with Digi-Key's Scheme-it [18].

33.3.2 *Inverting Schmitt trigger with open-collector comparator*

We can make a Schmitt trigger with an open-collector comparator, as in Figure 33.6.

Worked Example:
Compute the thresholds for a Schmitt trigger made from an open-collector *comparator.*

For the high threshold, the circuit is essentially the same as before with $R_3 + R_4$ in parallel with R_1, giving us

$$V_{\mathrm{IH}} = \frac{R_2}{R_2 + (R_1 \parallel (R_3 + R_4))} V_{\mathrm{rail}} ,$$

which for the values in Figure 33.6 gives us $V_{\mathrm{IH}} = 0.554 V_{\mathrm{rail}}$. For the low threshold, the math gets a little messier, as the output does not go completely to ground.

To figure out how low the output voltage of the comparator gets, we need to know something about the characteristics of the open-collector output—how its voltage varies with the current it is sinking.

If the current sunk by the output transistor is less than 4 mA, then the output is guaranteed to stay below 400 mV. More usefully, based on the saturation-voltage-versus-output-current plot on the LM339N data sheet from TI [98], for currents up to about 10 mA, we can treat the output transistor as approximately a 100 Ω resistor.

We can use Thévenin-equivalence repeatedly to simplify the computation of the low threshold. The 3.9 kΩ pull-up resistor to V_{rail} and the 100 Ω internal resistance to ground can be combined and treated as a 97.5 Ω resistance to 0.025 V_{rail}. That can be combined with the 1 MΩ resistor and the 560 kΩ resistor to get a 359 kΩ resistor to 0.00897 V_{rail}. Finally, combining that with the 820 kΩ resistor to V_{rail} gives a threshold voltage of 0.311 V_{rail}.

The calculation for the low threshold for an open-collector Schmitt trigger can often be approximated by assuming that the pull-up resistance is much larger than the equivalent resistance of the transistor (in the example, 39 times larger), and that the feedback resistance is much larger than the pull-up resistance (in the example, 256 times larger). Then we can treat the output as going approximately rail-to-rail and use the easier calculations for the rail-to-rail comparator. In this example, the error is only 2%, which is less than the error from assuming that the output transistor of the comparator can be modeled as a resistor.

33.3.3 *Non-inverting Schmitt trigger with rail-to-rail comparator*

Worked Example:
Let's analyze the non-inverting Schmitt trigger shown in Figure 33.7.

The negative input of the op amp is connected to a voltage divider that provides a reference voltage of $V_{\text{dd}}/2$, with the capacitor C_1 helping to filter out noise from the reference voltage (forming a low-pass filter with corner frequency of 3.88 Hz).

We can apply to the formulas Equations 33.1 and 33.2, to get the thresholds $V_{\text{in}} > V_{\text{dd}}/2(1+0.33)$ and $V_{\text{in}} < V_{\text{dd}}/2(1-0.33)$, or $V_{\text{dd}}(0.5 \pm 0.165)$. In general, whenever $V_{\text{ref}} = V_{\text{dd}}/2$, the thresholds will be symmetric about V_{ref}.

We can check our work here by looking at the voltage at the positive input of the op amp at each threshold using the voltage-divider formula:

- When $V_{\text{out}} = 0\,\text{V}$ and $V_{\text{in}} = 0.665 V_{\text{dd}}$, the positive input to the op amp is $0.665 V_{\text{dd}} \frac{1\,\text{M}\Omega}{1\,\text{M}\Omega + 330\,\text{k}\Omega} = 0.5\ V_{\text{dd}}$.
- When $V_{\text{out}} = V_{\text{dd}}$ and $V_{\text{in}} = 0.335 V_{\text{dd}}$, the positive input to the op amp is $\frac{0.335 V_{\text{dd}}\ 1\,\text{M}\Omega + V_{\text{dd}}\ 330\,\text{k}\Omega}{1\,\text{M}\Omega + 330\,\text{k}\Omega} = 0.5\ V_{\text{dd}}$.

We have two reasons for using large resistors for R_1 and R_2:

- First, R_2 always has a current of $\pm V_{\text{dd}}/(2R_2)$, for a power dissipation of $V_{\text{dd}}^2/(2R_2)$, so keeping R_2 large avoids wasting power as heat.
- Second, the input impedance of this circuit is just $R_1 = 330\,\text{k}\Omega$, which means that anything connected to the input sees a 330 kΩ resistance to $V_{\text{dd}}/2$. We generally want fairly high impedances for voltage inputs to avoid changing the voltage, hence the use of a large resistor for R_1.

Exercise 33.2

Design a non-inverting Schmitt trigger with thresholds at $0.25 V_{\text{dd}}$ and $0.75 V_{\text{dd}}$.

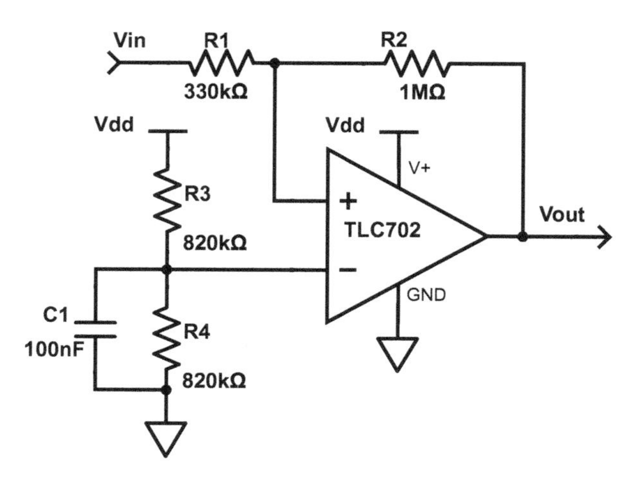

Figure 33.7: *This Schmitt trigger is non-inverting (as can be told by the input being fed through R_1 to the positive input of the op amp). Its thresholds are $0.335V_{\mathrm{dd}}$ and $0.665V_{\mathrm{dd}}$.*
Figure drawn with Digi-Key's Scheme-it [18].

Chapter 34

Lab 10: Measuring FETs

34.1 Goal: Determining drive for FETs as switches

In Chapter 32, we saw how FETs can be used as switches, saw how to pair an nFET with a pFET to make a cMOS output circuit, and how to use two cMOS output circuits to make an H-bridge.

In this lab, we'll look at driving FET gates with different drivers and examine the consequences on the drain voltages. There are two different sorts of tests we'll look at:

- driving a load with a single FET, and
- looking at shoot-through current for a pair of FETs.

34.2 Soldering SOT-23 FETs

In the first few years of the course, we used power FETs in large TO-251-3 and TO-220-3 packages (see Figure 32.5), but the ones we chose kept getting discontinued by the manufacturers, and substitutions got more and more expensive. Because we don't really need the high currents and high power dissipation of the large-package transistors, starting in 2018, we switched to using SOT-23-3 surface-mount FETs, as most of industry does for these power levels.

If you are using FETs in SOT-23 packages, you will have to solder them to a breakout board in order to connect them to a breadboard. Because the SOT-23 packages are so small, I made a breakout board that holds two of them, making half an H-bridge (one cMOS output). The layout is shown in Figure 34.1 and the schematic in Figure 34.2.

Soldering surface-mount devices (SMDs) by hand is more difficult than through-hole soldering, but the large lands used on the printed-circuit boards in Figure 34.1 make it reasonably easy. Here are the steps I followed to solder the boards:

- Put the board face up on the bench.
- Place one FET using sharp-pointed tweezers.
- Tape the FET and the drain side down with a tiny piece of blue painters' tape.
- Solder the source and gate.
- Remove the tape.
- Solder the drain.
- Repeat for the other FET.
- Put the headers through the holes (from the component side).
- Flip the board over and solder the header.
- Put the header pins into a breadboard at the edge of the board.
- Insert the capacitor from the component side.
- Prop the breadboard up so the solder side of the board is exposed.

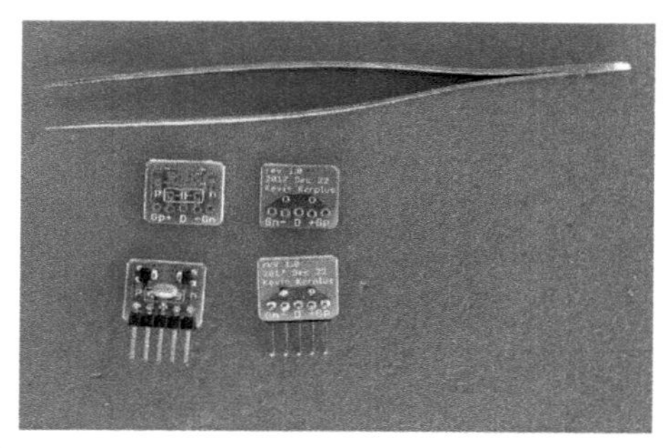

(a) *Breakout boards for surface-mount SOT-23 FETs.*

(b) *Close-up view of a half-H-bridge breakout board fully populated.*

Figure 34.1: *To use SMD components, like FETs in SOT-23 packages, a small printed-circuit board (a* breakout board*) is used. The one shown here implements half an H-bridge, as shown in Figure 34.2.*
The connected drains are the center pin, the ***gates are the outside pins****, and the sources are between the gates and the shared drain. The bypass capacitor connects the two sources, which should be connected to the two power supplies.*

- Solder the capacitor in place and trim its leads.

Figure 34.3 shows a slightly different sequence for soldering the board.

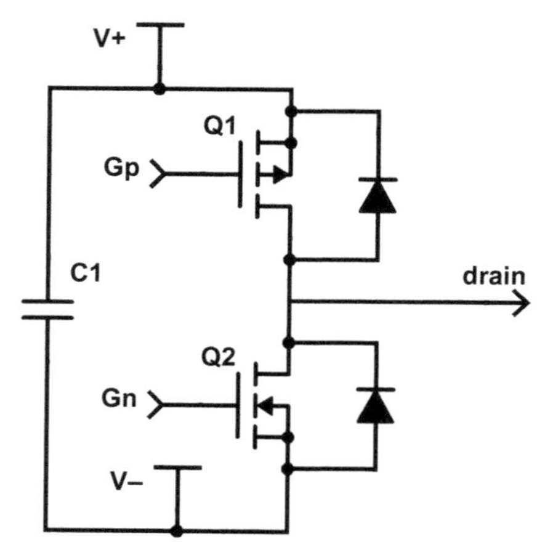

Figure 34.2: *Schematic for the half-H-bridge breakout board shown in Figure 34.1. The bypass capacitor helps keep inductive spikes from switching the H-bridge from propagating back through the power-supply wiring.*
Figure drawn with Digi-Key's Scheme-it [18].

> You will need to use these half-H-bridges again in Lab 11, so you should make sure that each partner has at least one working board to bring forward to that lab.

34.3 FETs without load (shoot-through current)

After soldering the half-H-bridges, the first tests are to ensure that the nFET and pFET are soldered on the correct sides and that you can turn on each FET separately.

- Put the half-H-bridge in the breadboard,
- connect the two gates together to make a cMOS inverter (see Section 32.2 and Figure 32.7),

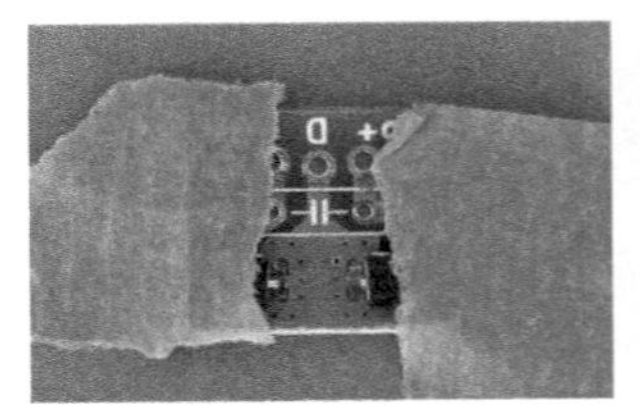

(a) *SOT-23 FETs taped in place with painter's blue tape, which leaves little glue residue when removed.*

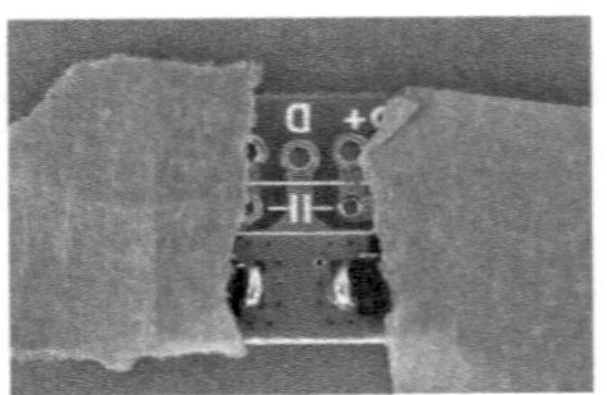

(b) *SOT-23 FETs taped in place after soldering the drain connections.*

(c) *After removing the blue tape, we see that the nFET moved during soldering (tape not pressed down enough).*

(d) *Reheating the drain connection with the soldering iron allowed nudging the nFET with tweezers into a better position.*

(e) *With one solder joint holding the transistor in place, the source and the gate can be soldered without needing tape.*

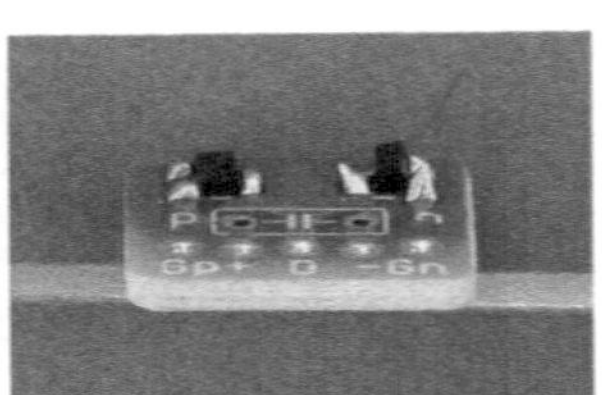

(f) *A side view shows that the nFET is not sitting flat against the printed-circuit board, a flaw known as* tombstoning.

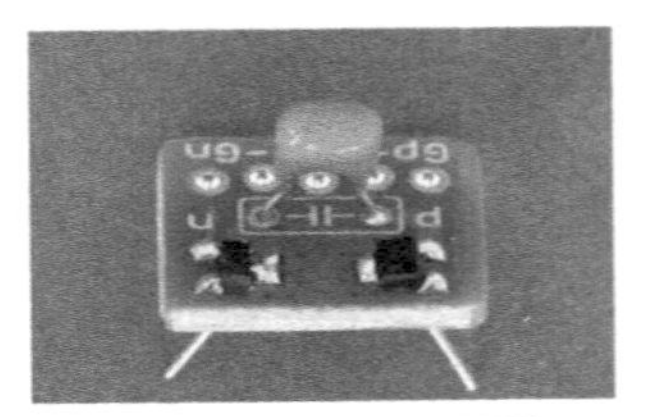

(g) *Squeezing the nFET and PC board together with tweezers and heating the solder joints reduced the tombstoning.*

Figure 34.3: *One strategy for soldering the SOT-23 FETs to the board is given here and a slightly different order of operations is suggested in the body text. In either case, the crucial thing to ensure is that the nFET and pFET are put on the correct sides of the board.*
A through-hole bypass capacitor of 10 μF is easily added.

- connect the gates to either the positive power rail or ground (make this easy to change),
- connect the pFET source to a positive power supply, and
- connect the nFET source to ground.
- Don't connect anything to the drain yet.

> Check to make sure that the FETs are not getting hot—if they are, turn off the power right away!

If the FETs are getting hot, it means one of two things: either both transistors are turned on at the same time, or you have connected power up backwards, so that both body diodes are conducting. The most common error is to have swapped the nFET and pFET when soldering, which reverses which source is which, having the same effect as wiring the power supply to the wrong source pins. Check your wiring carefully, and resolder the board if the FETs were swapped.

Once the board can be powered without overheating, connect two oscilloscope channels to measure the gate voltage and the drain voltage, each relative to ground.

Check to make sure that with the gate voltages at 0 V, the drain voltage goes to the positive power-supply voltage, and with the gate voltages at the positive power-supply voltage, the drain voltage goes to 0 V. This checks that both the nFET and the pFET are working as switches for the chosen power-supply voltage.

Now that the DC operation of the cMOS output stage has been checked, we want to check how fast the output switches and how serious a problem shoot-through current during switching is for different drivers of the gates.

We have four different drivers to test:

- gates driven directly by function generator;
- gates driven by rail-to-rail comparator, with the positive input from the function generator, and the negative input at half the power supply;
- gates driven by a 74AC04 inverter, whose input comes from the function generator; and
- each gate driven by a separate inverter, but with a common input on the two inverters.

For each driver, set up the function generator to provide a pulse at 1 kHz between 0 V and the power-supply voltage. Record traces for both the gate voltage and the drain voltage for each of the edges (rising and falling)—use a high sampling rate (100 MHz) for the oscilloscope, so that you can get good time resolution on the signals. Trigger on the drain voltage with the trigger level set to half the power-supply voltage and record twice: once triggering on a falling drain voltage and once on a rising drain voltage, so that rise and fall transitions can be synchronized and plotted on the same plot.

When you plot the transitions, use a short enough time interval for the xrange that you can see the transition—we want to measure how long it takes, not just see a vertical line on the plot. Put all the falling drain voltages on one plot and all the rising drain voltages on another.

How rapidly does the drain voltage change? How much time does the drain voltage spend not at either power rail? (That is approximately how long shoot-through current is present.) Does the gate voltage show a Miller plateau as the drain voltage switches?

For each driver, let the pulsed input run for a couple of minutes at a high frequency (say 100 kHz) and check frequently to see whether the FETs are getting warm. If the FETs are getting hot, turn the power off—you probably have too much shoot-through current when both FETs are on. Using a more powerful driver should make the gate voltage spend less time at intermediate voltages and reduce the power dissipation. Does it? Using a lower pulse frequency should result in fewer switching events per second, and again lower power dissipation.

For the comparator, look at the difference in output for the comparator between having the function generator provide a pulse signal at the input and having the function generator provide a triangle wave at the input. (Make sure you keep the function generator signal between the power rails.)

34.4 FET with load

Once you have examined the half-H-bridge without a load and found a driver circuit that can drive it at 100 kHz without shoot-through current overheating the FETs, let's try driving a load with a single FET, as shown in Section 32.1 and Figure 32.6.

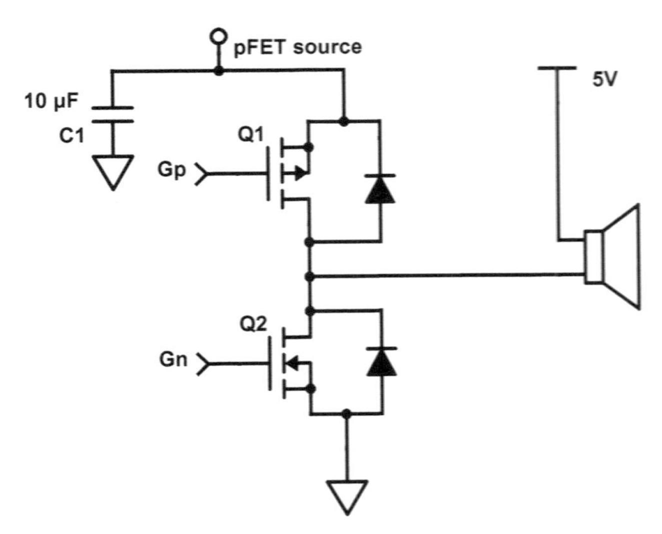

Figure 34.4: *Using a half-H-bridge board as a low-side switch driving a loudspeaker leaves us with two unused connections—the pFET gate and the pFET source. What we do with them makes a big difference in the behavior of the circuit.*

We'll use the loudspeaker as the load, since it is capable of dissipating a fair amount of power (generally 5 W–10 W, depending on what loudspeakers are cheap in a given year), unlike the $\frac{1}{4}$-watt resistors of the parts kit. It is also a somewhat inductive load, and so we'll be able to look at inductive spikes also (see Section 32.3).

Because our half-H-bridge breakout boards have both an nFET and a pFET on them, we have to decide what to do with the unused FET. Let's say we are driving the nFET as a low-side switch, as in Figure 34.4. What we connect the pFET gate and source to makes a big difference in the behavior of the circuit.

Try driving the nFET with a 100 kHz, 50% duty-cycle pulse (using the gate driver you chose in Section 34.3) using three different connections for the pFET:

floating gate, floating source: leave both the source and gate unconnected to anything (except the capacitor soldered on the board for the source),

gate and source together, but floating: connect the source and gate together to force the pFET to be turned off, but leave them unconnected to anything else (other than the capacitor soldered on the board), or

gate and source powered connect both the source and gate to the positive power supply.

For each wiring, use the oscilloscope to record the drain voltage and source voltage of the pFET, to figure out what is going on.

For the wiring with the gate and source powered, use the oscilloscope to record the drain voltage and the *gate* voltage of the nFET.

Do not connect your inverters or other driving circuit to the source or drain of the FETs!

It is also interesting to look at the effect of lower frequencies (the waveform changes shape), but if you use audio frequencies, be prepared for very loud sounds from the loudspeaker.

The case where the gate and source are connected to the power supply is easy to analyze, as the pFET is always off, and the body diode acts as a fly-back diode. The two cases where the source is not connected to the power supply are more interesting—try to figure out what causes the behavior you observe. At least one case should result in large inductive spikes. Be sure to explain the traces you observe in your report.

> Repeat the test, but driving the pFET with a pulse and having the nFET gate and source floating separately, connected together but floating, or both connected to ground. You will want to look at the drain and source of the nFET for all three cases, and the drain and gate of the pFET for the case where the nFET is grounded.

From the gate and drain traces, try to determine how long it takes for the FET to turn on or turn off. Can you see the Miller plateaus on the gate voltage traces? Do you get inductive spikes when the FETs turn on or turn off? How big are they?

Try to estimate the average current through the loudspeaker (and hence through the FET), based on the DC resistance of the loudspeaker, the power-supply voltage, and the 50% duty cycle of your pulse. Let the circuit run for a few minutes and see whether the FET is getting warm. (Check frequently, and turn the power off when the FET starts to get hot, to avoid damage to the FET from overheating.)

If you see a lot of ringing on the transitions, try adding large bypass capacitors to the power rails. Where you place the capacitors makes a difference, because of the resistance and inductance of the wiring. Ideally, you would like very low impedance between the bypass capacitor and the FET, with most of the impedance between the capacitor and the power supply.

34.5 Write-up

This assignment is mainly about understanding how FETs behave as switches, in preparation for using them in Lab 11. Your report should provide the measurements you made and interpretation of the results.

Draw schematics for your various test setups, and make sure you provide part numbers for all parts—especially the FETs, the comparators, and the inverters, because the properties of the parts are critical, and the generic symbols aren't enough information.

> Use the correct symbols for enhancement-mode nFETs and pFETs, and make doubly sure that you have the sources and drains correctly connected in your schematics.

Be careful in your terminology—talking about the voltage *across* or *of* a 3-terminal device like an FET makes no sense. It is important to use V_{gs} or V_{ds} or to talk about the gate voltage or the drain voltage.

34.6 Bonus lab parts

Bonus lab parts should only be attempted after the basic functions have been demonstrated. It may be better to move on to Lab 11 and finish it before attempting any bonus parts for this lab.

- Look at the power-supply voltage at the pFET source pin as the half-H-bridge switches without a load. How much voltage noise is being injected into the power-supply wiring?

- Try using other frequencies than 100 kHz. How high a frequency can you run at without overheating the FETs for each driver type without a loudspeaker as a load?
- Try adding a resistor between the function generator and the gates to limit the gate current and slow down the rise of the gate voltage. You will have to use a fairly low frequency to avoid overheating the FETs, but you should be able to get a good plot of the Miller plateau (as in Figure 32.15).
- Try driving both the nFET and the pFET, with the loudspeaker as load (with the other end at either ground or the power-supply voltage).
- Record the noise injected into the power-supply wiring by the switching of the loudspeaker.
- Try soldering a breakout board without the bypass capacitor and record the power-supply noise, to see how much the bypass capacitors on the breakout board reduce the noise on the power supply.
- Try measuring the current through the loudspeaker by putting a 0.5 Ω sense resistor in series with it. How does this compare with your estimate of the current from just the voltage and DC resistance of the speaker?
- Characterize the nFETs and pFETs in your kit by plotting I_{ds} vs. V_{ds} for various V_{gs}.
- Characterize the nFETs and pFETs in your kit by plotting I_{ds} vs. V_{gs} for various V_{ds}.
- Characterize the nFETs and pFETs in your kit by plotting R_{on} vs. I_{ds} for various V_{gs}. R_{on} is measured in the linear region of the curve, where V_{ds} is small.

Chapter 35

Class-D Power Amplifier

The goal for Lab 11 will be to design an amplifier for driving 10 W loudspeakers—this chapter provides the background necessary to do so efficiently.

35.1 Real power

Before designing a power amplifier for driving 10 W loudspeakers, we need to understand exactly what a 10-watt specification means in terms of how much current and voltage we can apply to the loudspeaker.

We looked at power for resistive loads in Section 29.3, and we'll repeat that analysis here for complex impedances. To compute the energy dissipated in an impedance you need to multiply the instantaneous voltage by the current, then integrate over time. To get the average power, you divide the energy by the time you integrated over.

For example, if the voltage across a device is $V(t) = A\cos(\omega t + \phi)$ and the current into the device is $I(t) = B\cos(\omega t + \psi)$ (the real parts of two sinusoids of the same frequency but possibly different phases), then their product is

$$P(t) = AB\cos(\omega t + \phi)\cos(\omega t + \psi)$$

or, using the trig identity $\cos(a)\cos(b) = 0.5(\cos(a+b) + \cos(a-b))$,

$$P(t) = 0.5AB\left(\cos(2\omega t + \phi + \psi) + \cos(\phi - \psi)\right) .$$

The power formula has a constant part that depends on the difference in phases, and a periodic part with frequency twice as high as the waveform we started with. If you average the power over full periods, the periodic part integrates to 0, and you have just the constant part: $0.5AB\cos(\phi - \psi)$.

Let's see how this formula compares with what we computed in Section 29.3 for resistive loads. For a resistive load, $V = IR$, so $A = RB$, and current and voltage are in phase ($\phi = \psi$). The cosine of the difference is 1, and the real power is $0.5AB$, which simplifies to $P = 0.5A^2/R = 0.5B^2R$, just as we computed in Section 29.3.

It turns out to be easier to analyze sinusoids using *RMS phasors*, representing

$$V(t) = A\cos(\omega t + \phi) \qquad \text{as} \qquad \bar{V} = (A/\sqrt{2})e^{j\phi} ,$$

and

$$I(t) = B\cos(\omega t + \psi) \qquad \text{as} \qquad \bar{I} = (B/\sqrt{2})e^{j\psi} .$$

That is, a voltage amplitude of A results in an RMS voltage of $A/\sqrt{2}$ and a current amplitude of B results in an RMS current of $B/\sqrt{2}$. The main point of using RMS phasors for voltage and current is that you can multiply them to compute power as $\bar{V}\bar{I}^*$, where the asterisk represents the complex conjugate (negating the imaginary part, which has the same effect as negating the phase).

Using the complex conjugate gives us the right result for voltage and current being in phase ($\phi = \psi$), no matter what phase that is.

You will sometimes see *amplitude phasors* (representing $A\cos(\omega t + \phi)$ as $\bar{V} = Ae^{j\phi}$ without the $\sqrt{2}$ correction) rather than *RMS phasors*. With amplitude phasors, the power computation real($\bar{V}\bar{I}^*$) gives us double the actual power.

If $\bar{V}$ and $\bar{I}$ are RMS phasors, then $\bar{V}\bar{I}^*$ is called the *complex power*, $|\bar{V}\bar{I}^*| = |\bar{V}|\,|\bar{I}|$ is called the *apparent power*, real($\bar{V}\bar{I}^*$) is called the *real power*, and imag($\bar{V}\bar{I}^*$) is called the *reactive power*.

Power engineers like to change the names of the units (watts for real power, volt-amps for apparent power, and vars for reactive power). They're all the same unit, but changing the name reminds you that they are not computed the same way and are not interchangeable. You will not be held responsible for those names in this course, and may use watts for all forms of power, but we will expect you to mean *real power* unless you explicitly modify the noun.

The real power is what you computed by doing the integral of the cosine waves—it represents the power dissipated in the device. A power source such as a sine wave generator dissipates negative power (the current into the device and the voltage have opposite signs)—negative power dissipation means that the device is putting power into the system, not taking it out.

The reactive power represents power that is being temporarily stored in the device during part of the cycle, and released in a different part of the cycle. If you have a purely reactive circuit (just capacitors and inductors with no resistors, so the impedance is purely imaginary), then the voltage and current are $\pm 90°$ out of phase, and the real power is $0.5AB\cos(\pi/2) = 0$. No energy is dissipated in a purely reactive circuit. Of course, real-world circuits always have some resistance, so a purely reactive circuit is an unobtainable mathematical idealization, as with so many of our electronic models.

If you know the RMS voltage phasor $\bar{V}$ for a sinusoid to a complex impedance Z, the RMS current phasor is just $\bar{V}/Z$, the complex power is $\bar{V}(\bar{V}/Z)^*$, and the real power is real($\bar{V}(\bar{V}/Z)^*$).

You'll want to compute the real power that your amplifier delivers to the loudspeaker, to make sure you don't exceed 10 W RMS (or whatever the RMS rating of your loudspeaker is). The RMS rating of a loudspeaker is based on how much power it can handle on a continuous basis, without overheating. For very short durations, it can handle a higher power, specified as the peak power, which is often about twice the RMS power.

Because the power rating of a loudspeaker is based on heating, you'll want to compute how much power is delivered at all frequencies, including those that do not produce useful sound, to make sure that you are not wasting power heating up the loudspeaker to no useful effect. Of particular interest is the PWM frequency discussed in Section 35.2.

Manually computing the complex impedance of a loudspeaker or a complicated network of resistors, capacitors, and inductors is tedious and unnecessary. Computers were invented for doing this sort of laborious mathematical calculation.

Exercise 35.1

Use gnuplot with the best-fitting model for the loudspeaker that you developed in Lab 8 to plot the real power dissipated in the loudspeaker as a function of frequency. Assume that the input is a sine wave with an amplitude of 1 V (which is an RMS voltage of about 0.707 V). This calculation requires the full complex impedance of the loudspeaker, not just its magnitude, since the current and voltage may not be in phase. Both the frequency on the x-axis and the power on the y-axis should be on log scales. Use the definitions in definitions.gnuplot (Figure 7.4).

35.2 Pulse-width modulation (PWM)

Linear amplifiers, such as the op amps we've been using so far, adjust the voltage or the current of a signal to be a multiple of the input. Doing this with a fixed-voltage power supply generally means dynamically changing the resistance between the power supply and the output, which in turn means dissipating power in the amplifier. The simplest (class A) amplifiers have very low power efficiency, delivering less than 30% of the power to the output. More sophisticated designs (class AB) are much more efficient, up to about 65% efficiency. Even then, the power lost as heat in a power amplifier is substantial, and large heat sinks are required to keep the circuit from overheating.

The class-D amplifiers we'll be designing in Lab 11 can have efficiencies well over 90%, so that little heat is dissipated in the amplifier—we'll be building without any heat sinks!

The idea behind class-D amplifiers is that you don't provide the output as a voltage that is a fixed multiple of your input voltage (the classic idea of a linear amplifier). Instead, you provide a pulse train that rapidly switches between the positive power rail and the negative power rail.

As we saw in Section 32.1, we can get voltage-controlled switches that dissipate very little power when they are fully on or fully off, so that switching between the power rails can be done with very little loss.

What you adjust in a class-D amplifier is how much time you spend at the positive rail and how much time you spend at the negative rail. This is referred to as *pulse-width modulation*, abbreviated PWM, because you are adjusting the duration of the high pulses [148]. A PWM signal is usually specified with two parameters: its *duty cycle* (what fraction of the period the output is high) and its *frequency*. Figure 35.1 shows a waveform with PWM frequency 100 kHz and an 80% duty cycle ($D = 0.8$).

If the high voltage is V_{hi}, the low voltage is V_{lo}, and the duty cycle is D, then the average voltage over one full period is

$$V_{\text{avg}} = DV_{\text{hi}} + (1 - D)V_{\text{lo}},$$

so the average voltage in Figure 35.1 is 4 V.

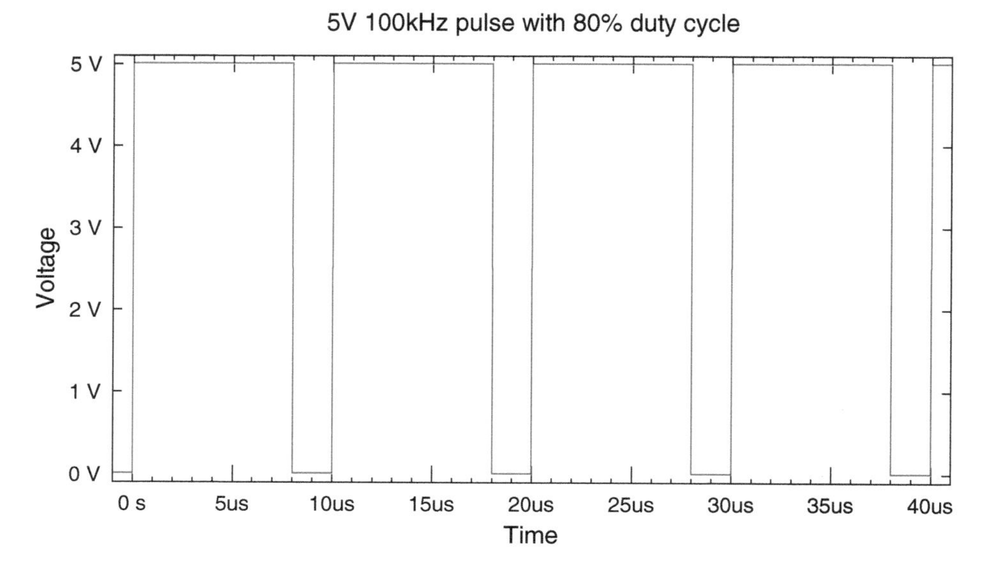

Figure 35.1: *This pulse waveform is high for 8 μs and low for 2 μs, which gives it a period of 10 μs (or a PWM frequency of 100 kHz) and a duty cycle of* $\frac{8\,\mu\text{s}}{10\,\mu\text{s}} = 0.8$. *The average voltage is* $0.8\ 5\,\text{V} + 0.2\ 0\,\text{V} = 4$ *V.*
Figure drawn with gnuplot [33].

It is fairly common in audio power amplifiers to have a symmetric power supply ($V_{\text{lo}} = -V_{\text{hi}}$), in which case $V_{\text{avg}} = (2D - 1)V_{\text{hi}}$. That is, at a duty cycle of 50%, the average voltage is halfway between the power rails (0 V), at 100% the average voltage is the top power rail (V_{hi}), and at 0% the average voltage is the lower power rail ($-V_{\text{hi}}$).

We want to set the PWM frequency high enough that our system doesn't respond much to the fluctuations in voltage but only to the average voltage, which is determined by the duty cycle. If you make the PWM frequency very high, it is difficult to make the switches that drive the output switch fast enough to follow the pulses (see Section 32.5). High-current switches are generally slower to operate than low-current ones, and power losses during switching can be minimized by switching less often.

If you make the PWM frequency too low, it is difficult to filter out the PWM pulses adequately without also filtering out some of your desired audio frequencies. Normally, a PWM frequency in the range 40 kHz to 2 MHz is chosen to balance these conflicting constraints for audio work.

Lower frequencies are usually chosen for motors or LED lighting, where there is no need for high frequency signals and efficiency is more important. Motors are often switched at 20 kHz, to avoid getting annoying sounds from the coils of the rotors—some older controllers use lower frequencies (around 2 kHz), resulting in an irritating whine when the motors are not run at full speed. LEDs are often switched at around 2 kHz–10 kHz, to avoid visible flicker.

Power FETs are available with very low on-resistances, and thus little self-heating with large currents, but they achieve this by increasing the width of the channel (the path from source to drain), which increases the area of the gate, in turn increasing the gate capacitance. The increased gate capacitance slows down how fast we can turn the transistor on and off, in turn limiting how high a PWM frequency we can use. This limitation for large power FETs is one of the main reasons why motor control is usually done with a fairly low PWM frequency.

Pulse-width modulation is often used for driving inductive loads like motors and loudspeakers. Both motors and loudspeakers are more dependent on the current through their coils than on the voltage across them, because the magnetic field strength of a coil is proportional to the current.

PWM works particularly well with inductive loads, because the rapid changes in voltage result in much slower changes in current. The current through an inductor is the integral of the voltage across the inductor (see Section 28.1), so the current ramps up when the voltage is positive and ramps down when the voltage is negative. Thus, one gets triangular fluctuations in the current around the average value, with the average value determined by the duty cycle of the PWM signal.

In addition to the inductance smoothing out the square-wave voltage to a triangle-wave current, the mass of the loudspeaker or motor keeps it from moving much in response to the high-frequency part of the waveform, so you only observe the sound waves or motor movement due to the lower-frequency components. With a high enough PWM frequency, the inductance and this mechanical filtering may be all that is needed to get smooth output.

In this class we won't rely entirely on mechanical filtering to remove the PWM signal but use an electronic filter as well (see Section 35.4).

Exercise 35.2

If you have a PWM signal with +3.3 V and 0 V as the high and low voltage levels, what is the average voltage for a duty cycle D? In particular, what is the average voltage for $D = 0.1$, $D = 0.5$, and $D = 1.0$?

Exercise 35.3

If you have a PWM signal with +2 and −2 V as the high and low voltage levels, what is the average voltage for a duty cycle D? In particular, what is the average voltage for $D = 0.1$, $D = 0.5$, and $D = 1.0$?

35.3 Generating PWM signals from audio input

Pulse-width modulation is used a lot for controlling motors and LEDs with computers, since it allows simple digital outputs (ON and OFF), and turns control problems into problems of controlling how long things are on. Time is generally easier to work with in computers than voltage or current are.

For class-D amplifiers, however, we don't generate the pulse-width modulated signal with a computer, but directly from a voltage input. You can get a PWM signal from a voltage input by using a comparator and a triangle wave generator—see Figure 35.2.

If the voltage of your input is higher than the triangle wave, then the comparator's output is high. If the voltage of your input is lower than the triangle wave, then the comparator's output is low. The fraction of time the output spends in the high state corresponds to how much of the time the triangle wave is lower than your input. Since the triangle wave spends an equal time at each voltage, the fraction of the time spent high (the duty cycle) is proportional to the input voltage, and the frequency of the PWM signal is the same as the frequency of the triangle wave.

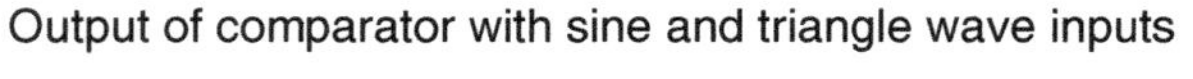

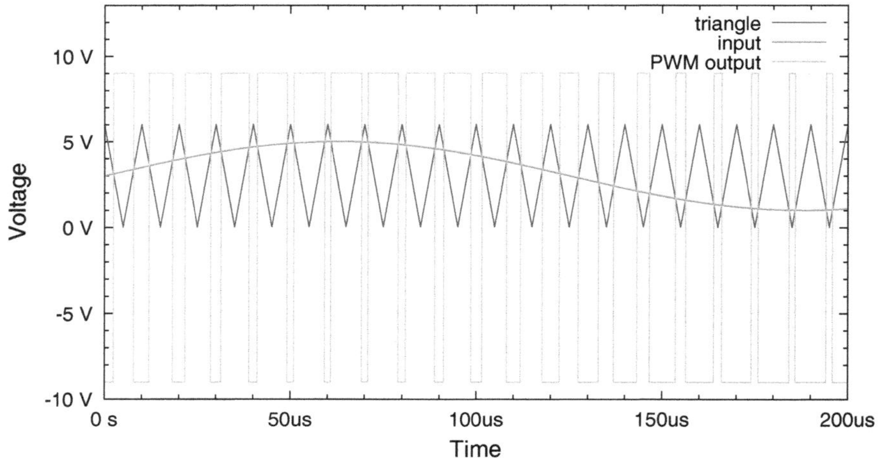

Figure 35.2: *The output of a comparator for different input voltages with a triangle wave on the negative input of the comparator. The comparator power inputs are set to the positive and negative rails of the power-amp stage (here, ±9 V).*
The DC offset and amplitude of the triangle wave affect the behavior of the comparator. Here I've used a ±3 V triangle wave centered at 3 V.
Figure drawn with gnuplot [33].

Figure 35.3 shows two ways to generate the signals needed for an H-bridge: Figure 35.3(a) using two comparators and Figure 35.3(b) using a single comparator and an inverter.

Worked Example:
We want to calculate how much power would be delivered to a $4\,\Omega$ loudspeaker by a PWM signal whose output voltage levels are −3 V and +3 V. Obviously, that depends on what the PWM signal is—let's assume it was generated using a comparator comparing a $1V_{\mathrm{pp}}$ 400 Hz sine wave and a $2V_{\mathrm{pp}}$ 200 kHz triangle wave, both centered at the same voltage.

Because both inputs to the comparator are centered at the same voltage, the center voltage does not matter (as long as the signals stay within the legal voltage range for the comparator). We can do our computations as if both were centered at 1 V, so the sine wave runs from 0.5 V to 1.5 V, while the triangle wave varies from 0 V to 2 V.

The duty cycle for the PWM signal will vary from 0.25 to 0.75. When the sine wave is at its lowest voltage, it will be higher than the triangle wave for only one quarter of the PWM period, but when it is at its highest voltage, it will be higher than the triangle wave for three-quarters of the PWM period.

We can make a crude approximation to how much power a loudspeaker gets from a PWM signal by approximating the loudspeaker as a simple resistor in the audio range and as a pure inductor at the PWM frequency and higher frequencies.

As a pure inductor, the loudspeaker would have only reactive power and no real power at the PWM frequency and its harmonics. That means that we can approximate the input voltage to the loudspeaker as being the average voltage during the period of the PWM signal.

The average voltage for a ±3 V PWM signal is $(2D-1)\,3\,\mathrm{V}$, so the loudspeaker is effectively seeing a sine wave that runs from $(2 \cdot 0.25 - 1)\,3\,V = -1.5\,\mathrm{V}$ to $(2 \cdot 0.75 - 1)\,3\,V = 1.5\,\mathrm{V}$. With an amplitude of 1.5 V, the RMS voltage is $(1.5/\sqrt{2})\,V \approx 1.061\,\mathrm{V}$, and the power is $V_{\mathrm{rms}}^2/R = 1.125\,\mathrm{V}^2/(4\,\Omega) = 281\,\mathrm{mW}$.

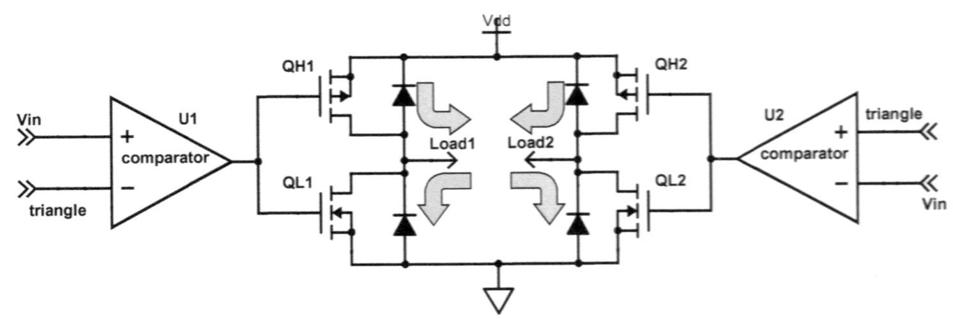

(a) *The two sides of an H-bridge can be driven by different comparators with swapped inputs.*

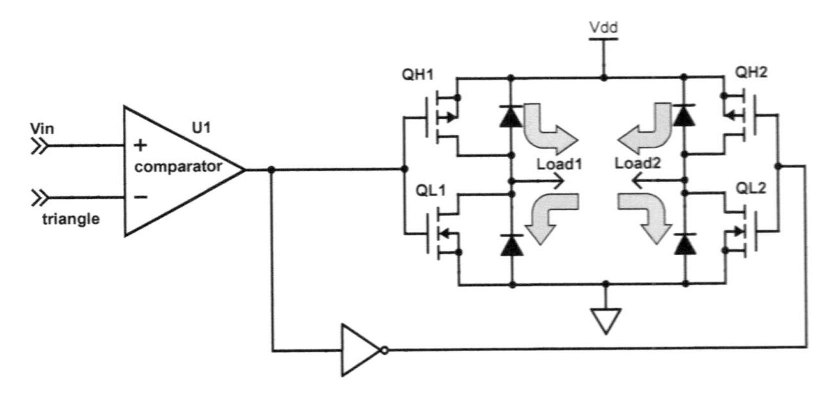

(b) *The two sides of an H-bridge can be driven from a comparator and an inverter. The design here is likely to have problems, because the output of the comparator is too heavily loaded and the input to the inverter is thus not a clean digital signal.*

Figure 35.3: *The two sides of an H-bridge need to have opposite signals—when one side is driven high, other is driven low. We can apply the schemes of Figure 32.10 for speeding up charging and discharging the gate using current amplifiers. Both sides of the H-bridge should be handled the same, to get similar transition times for both sides of the bridge.*
Figure drawn with Digi-Key's Scheme-it [18].

Exercise 35.4

How much power would be delivered to an 8 Ω loudspeaker by a PWM signal that switches between −5 and +5 V, if the PWM signal is generated from a $2V_{pp}$, 200 Hz square wave and a $3.3V_{pp}$, 300 kHz triangle wave, centered at the same voltage?
Assume that the loudspeaker is a pure inductor above 100 kHz and a pure resistor below 50 kHz.

35.4 Output filter overview

To avoid wasting power driving the loudspeaker at high frequency and to avoid radiating electromagnetic interference (EMI) from the wires to the speaker, we'll use an LC low-pass filter to deliver clean audio signals to our loudspeaker. This sort of filtering is fairly common in professionally designed class-D amplifiers [63]. The corner frequency of the filter needs to be between the highest input frequency to be amplified and the PWM frequency, which is a much higher frequency.

Motor drivers often don't bother with the LC filters, though such filters can reduce heating in the motor and make electromagnetic interference easier to control. One reason for omitting the filters is that motors require a lot of current, and power inductors capable of handling large currents are large and expensive. The AIUR-06-331 inductors I bought for the class have a saturation current of 1.9 A—plenty for our 10 W loudspeakers but not a lot for a motor. The RLB0912-221KL inductors have an even lower limit of 0.8 A.

You could skip the output filter in your design, connecting your loudspeaker directly to the amplifier outputs. The amplifier would work, but you would have high-frequency ripple in the current to the loudspeaker, some of which may cause radio-frequency interference with AM radios—long wires to the loudspeaker can act as antennas for broadcasting the radio-frequency signals. The filter designs in Figure 35.4 can remove the PWM signal and higher frequencies from the loudspeaker. The symmetric filter has a higher parts count, but ensures that both the long wires running to the speaker have little high-frequency signal on them, so will not radiate electrical noise.

To design the LC filter, we need to choose the inductor and the capacitor values. At one time, I was going to have students wind their own inductors around cardboard tubes, picking the number of turns of wire using an inductance calculator like the ones at http://www.pronine.ca/multind.htm and https://crystalradio.net/professorcoyle/professorcoylecyl.shtml. But after figuring out how much wire would be needed and making an attempt to wind one myself, I decided it would be better and cheaper if I just bought some inductors for the class. I ended up buying three inductors: 100 μH, 220 μH, and 330 μH (AIUR-06-101K, AIUR-06-221K, AIUR-06-331K made by Abracon).

The AIUR-06 inductors are high-power ferrite-core inductors, but they have one problem for our use: their leads are bit fat for most breadboards, so you might need to attach them to your circuit with clip leads. More recently, I bought RLB0914-101KL (100 μH) and RLB0912-221KL (220 μH) inductors made by Bourns, which have slightly smaller leads, but larger internal resistances.

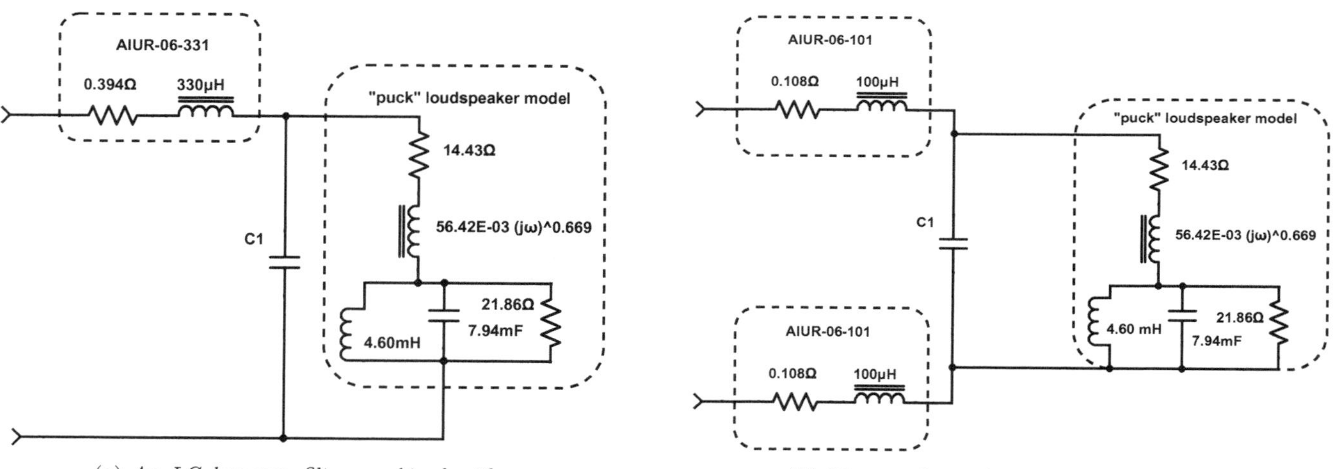

(a) *An LC low-pass filter combined with a loudspeaker model for a 16Ω "puck tactile transducer mini bass shaker."*

(b) *We can also make a symmetric filter, so that both wires running to speaker have passed through inductors. This design can be analyzed with the same math as for Figure 35.4a—just add the two inductors and their series resistors together.*

Figure 35.4: *The two input ports are connected to the two outputs of the H-bridge. C_1 must be an unpolarized (not electrolytic) capacitor in this design, because the voltage across it may be either negative or positive.*

The 0.394 Ω resistance is not a resistor that you add, but a model for the resistance of the wire in the inductor.

Figure drawn with Digi-Key's Scheme-it [18].

To make design easier for you, I've written a gnuplot script (Figures 7.4 and 35.5) that plots the real power into a loudspeaker using a specific inductor with various standard capacitor sizes. The script of Figure 35.5 is available at https://users.soe.ucsc.edu/~karplus/bme51/book-supplements/puck-low-pass-3m3H-choose-C.gnuplot Copying scripts from the PDF file for the book doesn't work, because of character substitutions made in the typesetting.

You must replace the loudspeaker model in the script (`Zloud`) with the model that you developed in Lab 8, as you will have a different loudspeaker than the one I modeled. The model included in the script is for a "puck" designed for shaking furniture or walls, which has a much lower resonant frequency and a different nominal impedance than the loudspeakers you will use. You'll also have to change the frequency where the peak power is observed, which you can determine by running the script with your model, and zooming in on the area just after the dip in power at the resonant frequency.

The resonances at the corner of the filter were far too sharp for this "puck" speaker with a 330 μH inductor (the largest I had on hand), so I modeled using a 3.3 mH inductor. Your loudspeaker will be a mid-range loudspeaker, and so you'll probably want a much smaller inductor for the filter than the one chosen for the puck—that means that the script should be modified to report the inductance in μH rather than mH.

The script also allows you to set the value for a small resistor in series with the loudspeaker for viewing current waveforms on the oscilloscope. Such a resistor does change the LC design somewhat.

You can also use the gnuplot script to help you choose a power-supply voltage that will deliver between 2 W and 10 W of power to the loudspeaker, using the detailed model of the loudspeaker and LC filter, rather than just an 8 Ω estimate. I show the results for one setting with the parameters from my puck loudspeaker model in Figure 35.6.

Remember that current will go up along with voltage for a fixed load, and so the real power will go up with the square of voltage. Also, the power computations in the gnuplot script are for sine waves, but the real power to the loudspeaker may be higher with other input signals (with square waves maximizing the power for a given power supply—twice the power of a sine wave).

35.5 Higher voltages for more power

To get more power to the loudspeaker, there are two options: increase the voltage to the loudspeaker or decrease the resistance of the loudspeaker. Although 6 Ω and 4 Ω loudspeakers are available, the scope for increasing power by lowering the resistance is limited. Lowering the resistance while increasing the power generally requires running thicker wires in the voice coil, increasing its mass, and making it more difficult to move at high frequencies. Also, as the resistance of the loudspeaker is lowered, the resistance of the transistors and of the wiring to the loudspeakers becomes more important, and the efficiency of the amplifier plus loudspeaker goes down. High-power loudspeakers are generally either 4 Ω or 8 Ω, and tweeters (high-frequency loudspeakers) are often 16 Ω.

Because of the limited scope for increasing power by lowering loudspeaker resistance, increasing the voltage of the power supply is generally a more satisfactory solution to delivering more power to a loudspeaker or motor. The H-bridge design of Figure 32.14 allows a peak-to-peak voltage of $2V_{dd}$, as we can make either side of the load V_{dd} higher than the other side.

```
load '../definitions.gnuplot'

#LOUDSPEAKER MODEL---use your model from Lab 8
# The model here is for a low-frequency "puck" for shaking furniture,
# and is NOT appropriate for other loudspeakers
Zloud(f) = Rnom+Znonstandard(f,Lhi,Lhipow)+Zpar(Rs,Zpar(Zl(f,Ls),Zc(f,Cs)))
Rnom=14.43      ;    Lhi= 56.42E-03    ;    Lhipow = 0.669
Ls=4.60E-3      ;    Cs=7.94E-3        ;  Rs=21.86

# CHOOSE A SERIES RESISTOR IF YOU PLAN TO MEASURE LOUDSPEAKER CURRENT
Rseries = 0.0 # extra resistor for displaying current on scope

# LC FILTER MODEL: voltage divider with Lf+Rf and (Cf || loudspeaker+Rseries)
    gain(f,Lf,Rf,Cf) = divider(Zl(f,Lf)+Rf, Zpar(Zloud(f)+Rseries,Zc(f,Cf)))
Lf=3.3e-3;   Rf=3.7     # parameters for  AIUR-06-332 inductor

# real power into loudspeaker with no filter (voltage from amplifier V at frequency f)
    nofilter_power(V,f)=real_power(V*divider(Rseries,Zloud(f)), Zloud(f))
# real power into loudspeaker with sine wave from amplifier V at frequency f
    loud_power(V,f,Lf,Rf,Cf) = \
        real_power(V*gain(f,Lf,Rf,Cf)*divider(Rseries,Zloud(f)), Zloud(f))
# real power into a resistor model (Rnom) for the loudspeaker
    R_power(V,f,Lf,Rf,Cf) = real_power(V \
             *divider(Zl(f,Lf)+Rf, Zpar(Rnom+Rseries,Zc(f,Cf))) \
             *divider(Rseries,Rnom), Rnom)

# total impedance, in case we want to plot that also
Ztotal(f,Lf,Rf,Cf) = Zl(f,Lf)+Rf + Zpar(Zloud(f)+Rseries,Zc(f,Cf))

 # CHOOSE YOUR SUPPLY VOLTAGE (note: 10V is too large for our components!)
V_supply=10    # +/- power supply voltage
V_RMS = V_supply/sqrt(2)        # RMS voltage for largest sine wave from H-bridge
peak_power_freq = 32.3  # approx frequency where power is maximized WITHOUT FILTER
# half the peak power (for Cf=3.3uF)
    half_power=0.5*loud_power(V_RMS,peak_power_freq,Lf,Rf,3.3e-6)

set title sprintf("Low-pass LC filter, 16-ohm puck+%.3gohm resistor, %.1fV to H-bridge",\
       Rseries,V_supply)
set yrange [1e-4:40]
set ylabel 'Real power to speaker'
set format y '%.1s%cW'
key_format = "L=%.3gmH+%.3gohm C=%.3guF"
plot half_power title sprintf("half-power=%.3gW",half_power), \
        nofilter_power(V_RMS,x) title "no filter", \
        loud_power(V_RMS,x,Lf,Rf,10e-6)   title sprintf(key_format, Lf*1e3, Rf, 10), \
        loud_power(V_RMS,x,Lf,Rf,4.7e-6)  title sprintf(key_format, Lf*1e3, Rf, 4.7), \
        loud_power(V_RMS,x,Lf,Rf,2.2e-6)  title sprintf(key_format, Lf*1e3, Rf, 2.2), \
        loud_power(V_RMS,x,Lf,Rf,1e-6)    title sprintf(key_format, Lf*1e3, Rf, 1), \
        loud_power(V_RMS,x,Lf,Rf,0.47e-6) title sprintf(key_format, Lf*1e3, Rf, 0.47), \
        loud_power(V_RMS,x,Lf,Rf,0.22e-6) title sprintf(key_format, Lf*1e3, Rf, 0.22), \
        loud_power(V_RMS,x,Lf,Rf,0.1e-6)  title sprintf(key_format, Lf*1e3, Rf, 0.1), \
        R_power(V_RMS,x,Lf,Rf,1e-6) title sprintf("1uF, %.3gohm model", Rnom) lt 6 dt "- "
```

Figure 35.5: *Gnuplot script to choose capacitor and inductor values for low-pass filter. Modify to incorporate your loudspeaker model, change the inductor, and adjust the power-supply voltage, peak frequency, and (maybe) the series resistor.*

Exercise 35.5

How much power could you deliver as a square wave to an $8\,\Omega$ loudspeaker with an H-bridge if your power rails were 0 V and 30 V? What average power could you deliver if the signal was a sine wave?

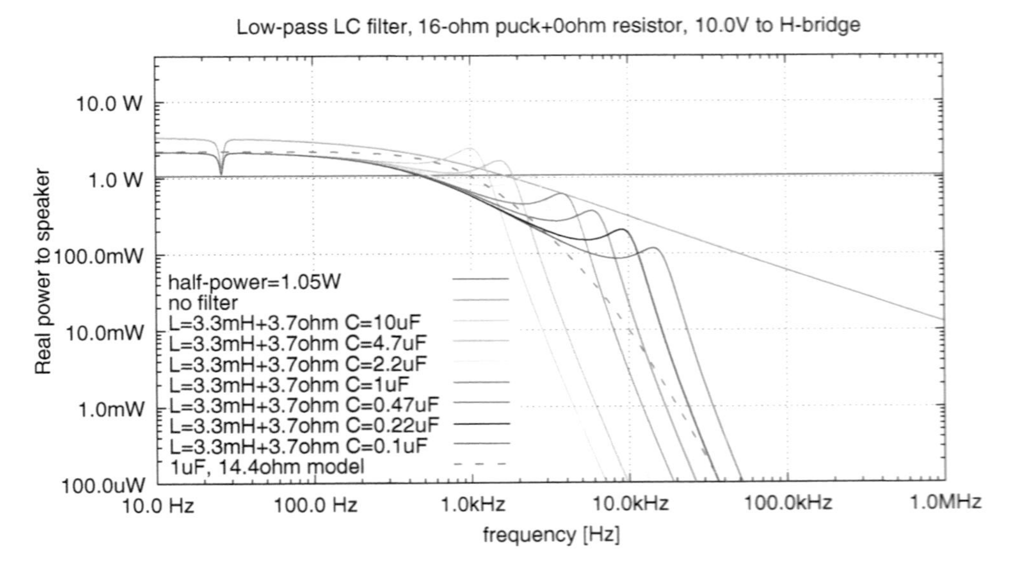

Figure 35.6: *The result of running the script in Figure 35.5 for modeling power to the loudspeaker. The power to the loudspeaker should be very low at the PWM frequency, fairly constant in the range of frequencies you want to amplify, and always below the power limit for the speaker. Here, the 4.7 μF capacitor gives the highest bandwidth, but the 10 μF capacitor could be a good choice if response at 500 Hz–1 kHz is more important than 1 kHz–2 kHz.*

The resonant peaks (ranging from 1 kHz to 20 kHz, depending on the capacitance) result from the parallel arrangement of the capacitor and the nonstandard inductor-like impedance of the loudspeaker. If the loudspeaker is replaced by a resistor with the DC resistance of the loudspeaker, there is no resonance.

The large inductor used here has a high series resistance—when combined with the 16 Ω puck loudspeaker, a large supply voltage is needed to get even 2 W of power out of the loudspeaker.

Figure drawn with gnuplot [33].

If you want to use higher power supply voltages, the TLC3702 comparators can handle reasonably high voltages. Unfortunately, the internal transition time limitations on the TLC3702 comparators appear to be due to slew rate limitations, causing the transitions to be slower at higher voltages, even though the output drive current goes up roughly linearly with power-supply voltages. This can result in unacceptably slow transitions if the TLC3702 is used to drive the FET gates directly.

We can't use an off-the-shelf inverter chip to do separate gate amplifiers at voltages above about 6 V, but we could build our own inverters out of smaller nFET and pFET transistors—ones that have smaller gate capacitance but larger on-resistances than the final-stage power nFETs and pFETs.

One of the big problems with using the simple designs discussed so far with a higher voltage is that the FET thresholds for turning on the transistors are referenced to the sources of the FETs, which are the power rails. As the power rails are moved further apart, the thresholds move with them, and we end up with a large range of gate voltages that would turn on both FETs. That large danger zone combined with slow transitions to get the gate voltage from one rail to the opposite rail results in a long duration for shoot-through current, and subsequent overheating of the FETs. Overheating is a problem even if we put in intermediate cMOS inverters as current amplifiers, and those inverters themselves can have problems with shoot-through current.

There are several solutions to the design problems for larger voltages, most of which involve driving the nFET and the pFET gates separately, as in Figure 35.7. With separate drivers for the gates, we can play with the voltage ranges or the timing for the gate signals, to avoid the shoot-through current.

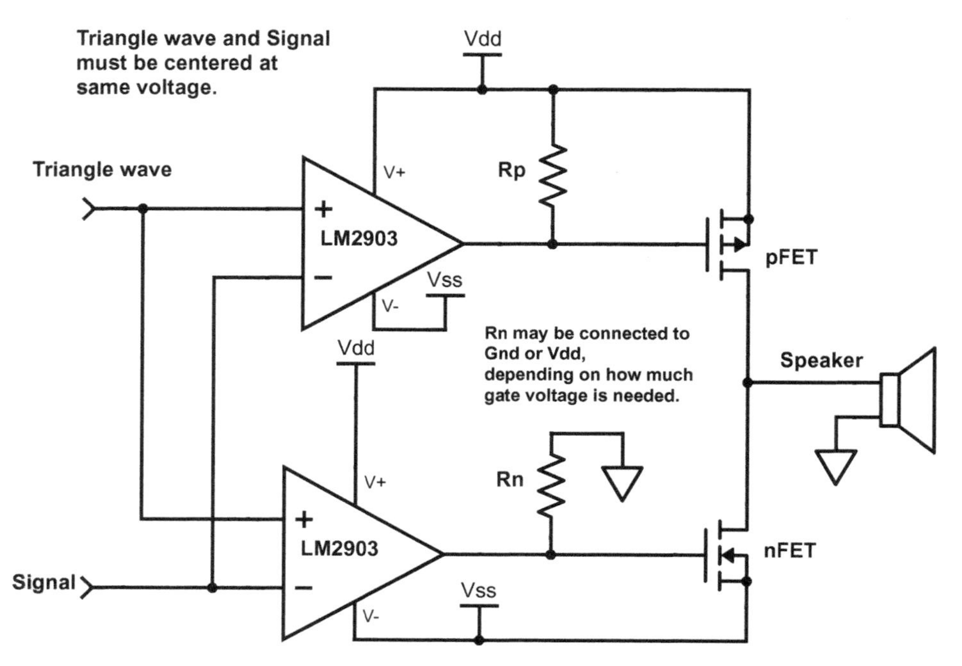

Figure 35.7: *By having separate control signals for the gates of the nFET and pFET, we can arrange the signals to the gates so that the FETs are not simultaneously on. The basic idea is to make sure that the FETs are turned off quickly, even at the cost of turning on more slowly.*
Figure drawn with Digi-Key's Scheme-it [18].

The design in Figure 35.7 has the pFET gate swinging from V_{ss} to V_{dd}, but the nFET gate swinging from V_{ss} to 0 V. Making R_p small turns off the pFET quickly, as does making R_n large, but both have the effect of slowing down the turn-on of the corresponding FETs.

We might do even better to limit the voltage range of the gates to stay near the power rails. For example, if the power rails are at 0 V and 30 V, we might want to have the nFET gate driven from 0 V to 5 V (turning on) while the pFET gate is driven from 25 V to 30 V (turning off). Having only a 5 V swing for each gate would allow much faster gate transitions than having a 30 V swing, while still turning on the FETs strongly.

If we use voltage regulators to produce 5 V and 25 V power supplies from the 30 V supply, we can simply duplicate the comparator circuitry for the nFET and the pFET drivers, one operating on the 0 V–5 V power rails, the other on the 25 V–30 V power rails. We have to be careful with level-shifting for the AC signals that are inputs to the current amplifiers, though, to make sure that we can't have both FETs being turned on at the same time.

A simpler solution is to have a single 5 V comparator and drivers for the nFETs, and take the output of the comparator through a different level-changing circuit for driving the pFETs. For example, we could use a small nFET with a voltage divider from the drain to the positive power supply, so that the pFET voltage swing is a controlled fraction of the total voltage swing, as shown in Figure 35.8.

We can use the open-collector output of an LM2903 comparator in place of Q_3 in Figure 35.8 to drive a pFET. It is possible to design a class-D amplifier using an LM2903 comparator (with appropriate resistors) for each nFET or pFET gate that needs to be driven, but getting the design right is trickier than using rail-to-rail comparators and inverters.

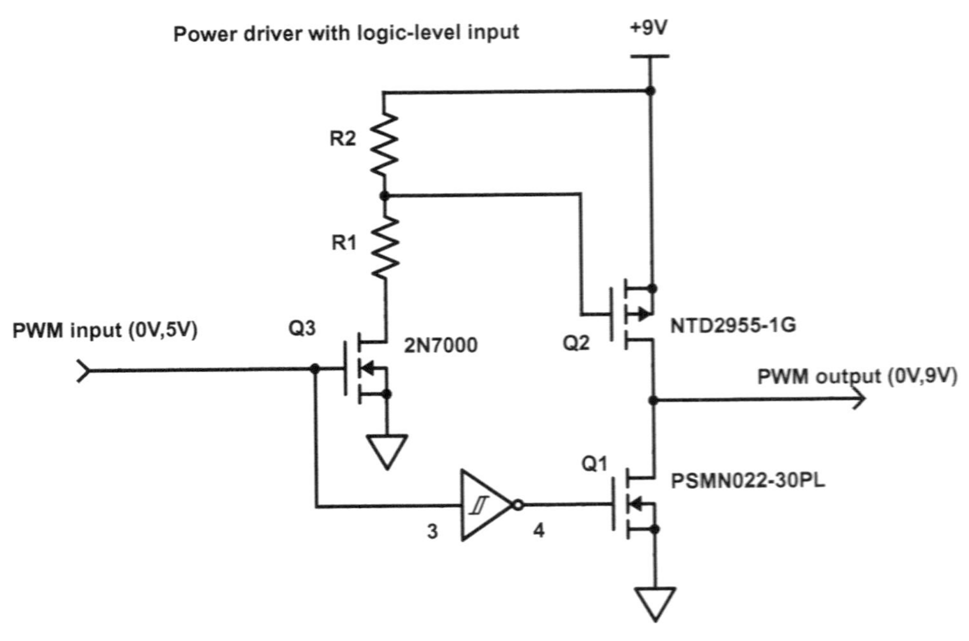

Figure 35.8: *Circuit for converting a logic-level PWM signal (0 V or 5 V) to a higher voltage PWM signal with more current capability. When the input is low, the output of the inverter is high, and the nFET Q_1 is turned on, but the nFET Q_3 is turned off, so R_2 pulls up the gate of Q_2, turning off the pFET Q_2. The result is a low-resistance path from the output to ground.*
When the input is high, the gate of Q_1 is driven low by the inverter, so Q_1 is turned off, but Q_3 is turned on, driving the gate of Q_2 to the voltage determined by the voltage divider R_1 and R_2. If that voltage is set reasonably, it turns on Q_2, and the output has a low-resistance path to the high voltage (+9 V in this schematic). The size of R_1 and R_2 depends on the power-supply voltage, how fast the gate of Q_2 needs to be driven, and the threshold voltage of Q_2.
This circuit can be made fast by using a low-input-capacitance nFET for Q_3, and sufficiently small resistors for R_1 and R_2 that Q_2 is turned off fast enough. If the resistors R_1 and R_2 are too small, then there will be too much current through Q_3 when it is on. Any current through Q_3 is wasted (not delivered to the load), so R_1 and R_2 should be as large as possible while still switching Q_2 fast enough.
Figure drawn with Digi-Key's Scheme-it [18].

Exercise 35.6

In Figure 35.8, what values should be used for R_1 and R_2 if you want to drive the gate of the pFET Q_2 on with $V_{gs} < -4\,\text{V}$, but you want to waste no more than 15 mA?

35.6 Feedback-driven class-D amplifier

This section is optional material—it is not essential for Lab 11.

There is another design for class-D amplifiers that does not use a triangle-wave generator. Instead it uses feedback from the unfiltered digital output and subtracts it from the input, making an "error signal", as shown in Figure 35.9. This error is integrated and compared to a constant voltage with a comparator that has some built-in hysteresis (a Schmitt trigger). The error integrator and Schmitt trigger provide the oscillation needed.

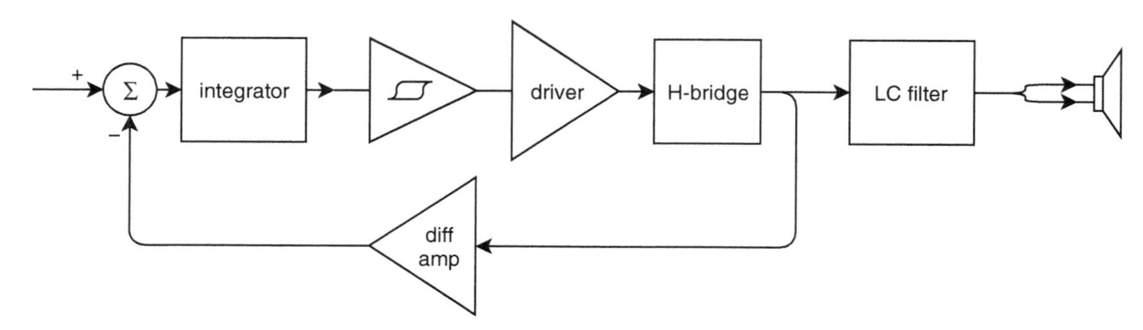

Figure 35.9: *Block diagram for a class-D amplifier that uses feedback from the digital output to produce an error signal that is integrated and thresholded with a Schmitt trigger to get a pulse signal.*
Figure drawn with draw.io [21].

Because the output of the H-bridge is a differential signal (not referenced to ground), the feedback needs to be processed through a differential amplifier—one that converts it to a signal with the same voltage range as the input signal from the preamplifier.

The feedback design can provide better power-supply rejection than a simple open-loop class-D amplifier, as noise from the power supply (such as 60 Hz hum) can be corrected by the feedback loop. The feedback loop can also correct for any distortion caused by dead time in the H-bridge driver.

Because the two outputs of the H-bridge are each either ground or the power rail, we can eliminate the differential amplifier in the feedback path by taking the feedback from just one side of the H-bridge. This connection still allows correction for the power-supply voltage, but not for dead time.

The computation of the error signal and its integration can be done with a *summing integrator*, which is a minor variation on the integrator of Section 36.1. From the basic formula relating current and voltage for a capacitor and Kirchhoff's current law, we can get the formula for the summing integrator in Figure 35.10:

$$C_1 \frac{\mathrm{d}V_{\mathrm{out}}}{\mathrm{d}t} = (2.5\,\mathrm{V} - V_1)/R_1 + (2.5\,\mathrm{V} - V_2)/R_2 \ .$$

We can write this as an integral:

$$V_{\mathrm{out}} = k + \int (2.5\,\mathrm{V} - V_1(t))/(R_1 C_1) + (2.5\,\mathrm{V} - V_2(t))/(R_2 C_1)\; dt \ ,$$

where k is an arbitrary constant that depends on the initial conditions of the circuit.

To turn the sum of the inputs into a difference, we need to negate $2.5\,\mathrm{V} - V_2$, which is easily done with an inverting amplifier.

By sending the feedback to V_2 in Figure 35.10, which has more weight than V_1, we can ensure that V_{out} always changes fast enough to provide a high enough PWM frequency, even when V_1 is at an extreme value. For the component values in Figure 35.10, the slowest slew rate is $57\,\mathrm{mV}/\mu\mathrm{s}$ and the fastest is $165.5\,\mathrm{mV}/\mu\mathrm{s}$. This limited difference in the slew rate limits the duty cycle to be in the range 25%–75%.

The response of an amplifier designed using the block diagram of Figure 35.9 is shown in Figure 35.11—it is close to what we expect from a model of the LC filter and loudspeaker, until we get close to the PWM frequency. The PWM is not at a single frequency, but is spread over a range of frequencies, whose width depends on the amplitude of the input. For low amplitude, the PWM is in a narrow band around 62 kHz, and the larger-than-expected output for the amplifier is similarly concentrated. For large amplitude inputs, the PWM is spread over a much wider range, and the larger-than-expected output is similarly spread out.

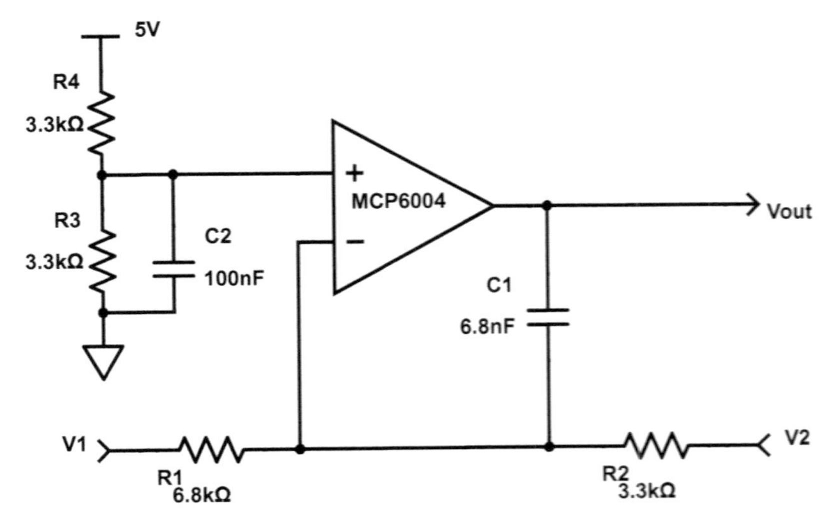

Figure 35.10: *Circuit that integrates the weighted sum of two inputs. The weight for V_i is $1/R_i$. Any number of inputs can be added at the summing junction (where R_1 and R_2 meet).*
Figure drawn with Digi-Key's Scheme-it [18].

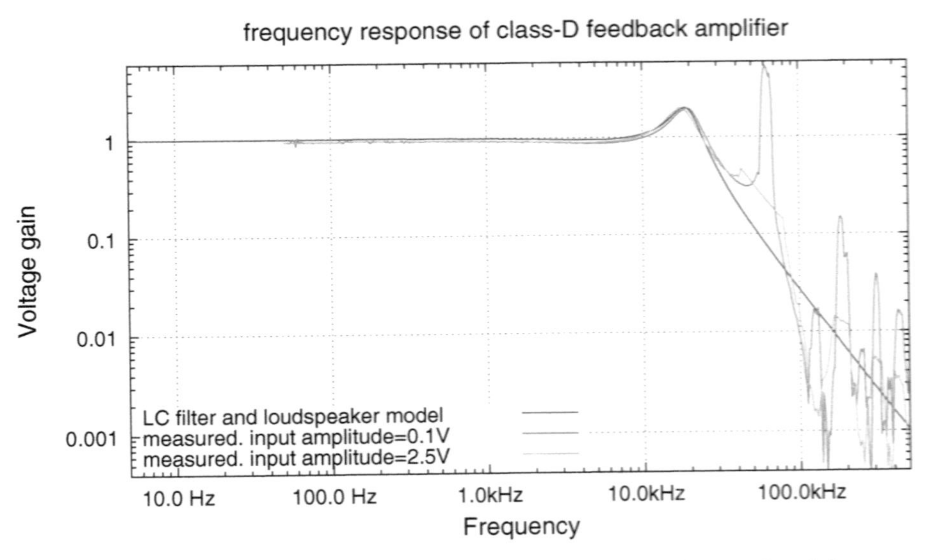

Figure 35.11: *The response of a class-D amplifier with feedback is close to what would be predicted from a model of the LC filter and loudspeaker, at least in the passband. At higher frequencies, the response is much larger than would be expected from the filter, because of the oscillation that is around 60 kHz for generating the PWM signal.*
The model here is similar to the one in Figure 35.6, but the voltage gain is being reported, rather than the real power. The power drops more quickly with frequency than the voltage does, both because of the V^2 term in computing power and because the impedance of the loudspeaker increases with frequency.
Figure drawn with gnuplot [33].

At frequencies above the PWM frequency, the input signal mixes with the oscillation to produce an alias in the audio range, so the amplifier must be used with a low-pass input filter.

The design of Figure 35.9 is similar to the design of sigma-delta analog-to-digital converter, which replaces the Schmitt trigger with a comparator and high-frequency sampling, and replaces the power output driver with a digital filter. A sigma-delta ADC produces a high-frequency bit stream which is then digitally low-pass filtered to produce the desired digital output. The feedback method results in

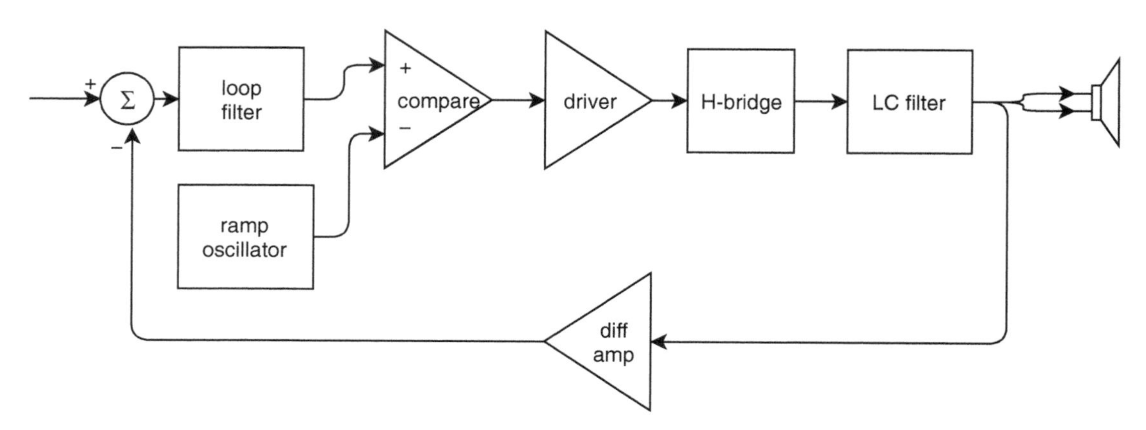

Figure 35.12: *This block diagram shows a class-D amplifier with feedback from after the LC filter, to get more uniform frequency response. The loop filter needs careful design, to ensure that overall feedback loop remains stable.*
Figure drawn with draw.io [21].

the digitization noise being moved to high frequencies, so the sigma-delta converter can have very high resolution for low-frequency signals with very little in the way of precise analog components, making it a popular device for bioinstrumentation.

Another design for a class-D amplifier with feedback takes the feedback from after the LC filter, as shown in Figure 35.12, so that any undesired characteristics of the LC filter can be corrected. Here I've also used a simple comparator with a separate ramp oscillator, rather than a Schmitt trigger. The loop filter needs to be designed carefully to get good frequency response and low harmonic distortion.

The differential amplifier in the feedback path is really needed in this design, because the output of the LC filter is a differential signal that cannot be easily represented by just one of the two wires.

Chapter 36

Triangle-Wave Oscillator

This chapter is optional material—it will not be needed for Lab 11.

In this chapter, we'll pull together apparently different threads in the book to design some useful circuits—triangle-wave oscillators for use in a class-D power amplifier and for analog audio synthesis.

- In Lab 4, we made an oscillator out of a Schmitt trigger, a resistor, and a capacitor.
- In Section 23.2, we looked at the exponential relationship between current and voltage for a semiconductor diode.
- In Section 33.3, we looked at how to make a Schmitt trigger out of a comparator.
- In Section 35.2, we looked at how to convert a voltage signal to a pulse-width-modulated signal using a comparator and a triangle-wave signal.

36.1 Integrator

The hysteresis oscillator that we made in Lab 4 was very simple and could easily be tuned to an approximate frequency, but its *shark's-fin* waveform is not very useful for converting voltage signals to pulse-width modulation signals, as it does not spend equal time at each voltage, so distortion would be introduced in the conversion.

The basic idea of the hysteresis oscillator is to charge a capacitor until we reach the higher threshold of a Schmitt trigger, then discharge it until we reach the lower threshold. The undesired shark's-fin waveform comes from the shape of the RC charging curve.

We can use the same basic idea to make either a saw-tooth or triangle-wave oscillator, if we replace the RC charging curve with a constant-current charging curve, which would give us a constant-slope voltage: $\frac{\mathrm{d}V}{\mathrm{d}t} = I/C$ (see Section 10.1). One simple way to achieve this is to make a negative-feedback amplifier whose feedback element is just a capacitor, as shown in Figure 36.1. This circuit is referred to as an inverting *integrator*, because the output voltage is the negative of the integral of the input voltage.

If $V_{\text{in}} > V_{\text{ref}}$, then current flows to the right through the resistor, and V_{out} must drop steadily for the capacitor to have the same current flow:

$$I = C\frac{\mathrm{d}(V_{\text{ref}} - V_{\text{out}})}{\mathrm{d}t} = -C_1\frac{\mathrm{d}V_{\text{out}}}{\mathrm{d}t} ,$$

assuming V_{ref} is constant. The current is set by Ohm's law, $I = (V_{\text{in}} - V_{\text{ref}})/R_1$, which gives us the formula for an inverting integrator:

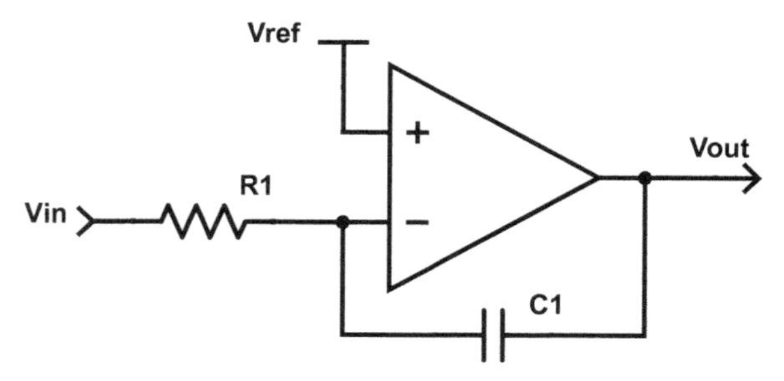

Figure 36.1: *An integrator is an op amp with a capacitor as a negative-feedback element. The current through the resistor is* $I = (V_{\text{in}} - V_{\text{ref}})/R_1$, *so the current through the capacitor must be the same, as no current flows through the input of the op amp. If* V_{ref} *is constant, then* $\frac{\mathrm{d}V_{\text{out}}}{\mathrm{d}t} = -I/C_1 = (V_{\text{ref}} - V_{\text{in}})/(R_1 C_1)$.
Figure drawn with Digi-Key's Scheme-it [18].

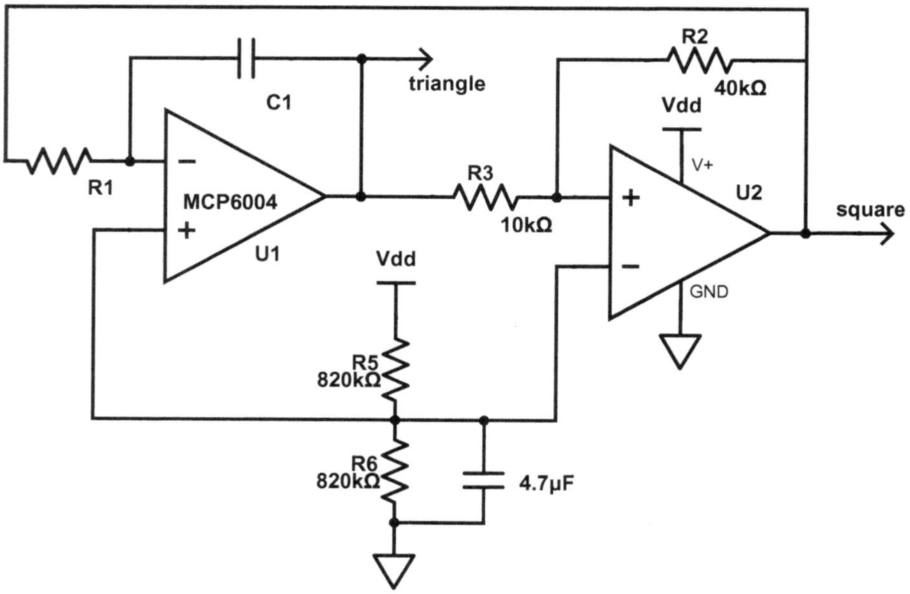

Figure 36.2: *This oscillator generates triangle waves and square waves at a fixed frequency, whose period is* $R_1 C_1$. *Op amp* U_1 *serves as an integrator, turning a constant current through* R_1 *into a voltage with constant slope. The Schmitt trigger implemented by* U_2 *decides when to reverse the current. Ideally,* U_2 *is a fast, rail-to-rail comparator, but an op amp can be used for lower frequency oscillators, where the slew-rate limitations of the op amp and the resulting slow transitions of the square wave may not matter much.*
Figure drawn with Digi-Key's Scheme-it [18].

The slope of the output of an inverting integrator is the voltage difference of the inputs divided by the time constant:

$$\frac{\mathrm{d}V_{\text{out}}}{\mathrm{d}t} = \frac{V_{\text{ref}} - V_{\text{in}}}{R_1 C_1} .$$

36.2 Fixed-frequency triangle-wave oscillator

We can make a simple fixed-frequency oscillator by combining an integrator with a Schmitt trigger, as shown in Figure 36.2.

The voltage reference in Figure 36.2 is a voltage divider with a capacitor that filters out noise from the power supply. The time constant for this filter is $0.5\ 820\,\text{k}\Omega\ 4.7\,\mu\text{F} = 1.927\,\text{s}$, so the reference takes about $10\,\text{s}$ to settle after any step change in the power-supply voltage. (Turning the power supply on is such a step change.) No unity-gain buffer is needed for the split-rail reference, as no current is taken from the voltage divider.

The Schmitt trigger implemented by U_2 is in one of two states: output high (V_{dd}) or output low (Gnd). The current through the resistor R_1 is $I = \pm V_{\text{dd}}/(2R_1)$, depending on the output of the Schmitt trigger. The integrator implemented by U_1 produces a voltage that rises or falls with slope $I/C = \pm V_{\text{dd}}/(2R_1C_1)$, with the voltage rising when the output of the Schmitt trigger is low. When the triangle output rises above the upper threshold of the Schmitt trigger, the square output switches to high, and the integrator starts a downward slope.

The time it takes the triangle wave to rise from the low threshold V_{lo} to the high threshold V_{hi} is $2R_1C_1(V_{\text{hi}} - V_{\text{lo}})/V_{\text{dd}}$, and the time to fall from the high threshold to the low one is the same, so that the total period is

$$\tau = 4R_1C_1\frac{V_{\text{hi}} - V_{\text{lo}}}{V_{\text{dd}}}\ .$$

Using Equations 33.1 and 33.2, we get that $V_{\text{hi}} = \frac{5}{8}V_{\text{dd}}$ and $V_{\text{lo}} = \frac{3}{8}V_{\text{dd}}$, for a period $\tau = R_1C_1$.

Of course, the 4:1 ratio of R_2 and R_3 in the Schmitt trigger is not easily realized with standard resistors: we have to either use two in series (say $R_2 = 20\,\text{k}\Omega + 20\,\text{k}\Omega$) or adjust to a standard value like $39\,\text{k}\Omega$ instead of $40\,\text{k}\Omega$, which would increase the period to $1.026R_1C_1$.

The Schmitt trigger should be implemented with a fast comparator, so that the square output is a clean square wave and the triangle wave has crisp corners.

Using an op amp instead of a fast comparator will make the transitions of the square wave slow, because of the slew-rate limitations. The intermediate values on the square wave transitions will result in rounding the tips of the triangle wave, which may be OK if the period of the triangle wave is large compared to the transition time of the square wave.

Figure 36.3 shows the triangle-wave waveforms for two different RC time constants, and Figure 36.4 shows how the slew-rate limitations of the op amp used for a Schmitt trigger round the peaks of the triangle wave.

The slew-rate limitation turns the square wave into a trapezoid, and rounds the corners of the triangle wave. This rounding of the corners in turn increases the time it takes for the capacitor of the integrator to charge, increasing the period. The period increases in a voltage-dependent manner, since the current is reduced from the ideal square wave for more time as the voltage swing of the Schmitt-trigger output increases.

Exercise 36.1

Design a triangle-wave oscillator that should have a frequency of approximately 67 kHz and a peak-to-peak voltage of 1.5 V using a single 3.3 V power-supply. Hint: use a rail-to-rail comparator to make the Schmitt trigger first.

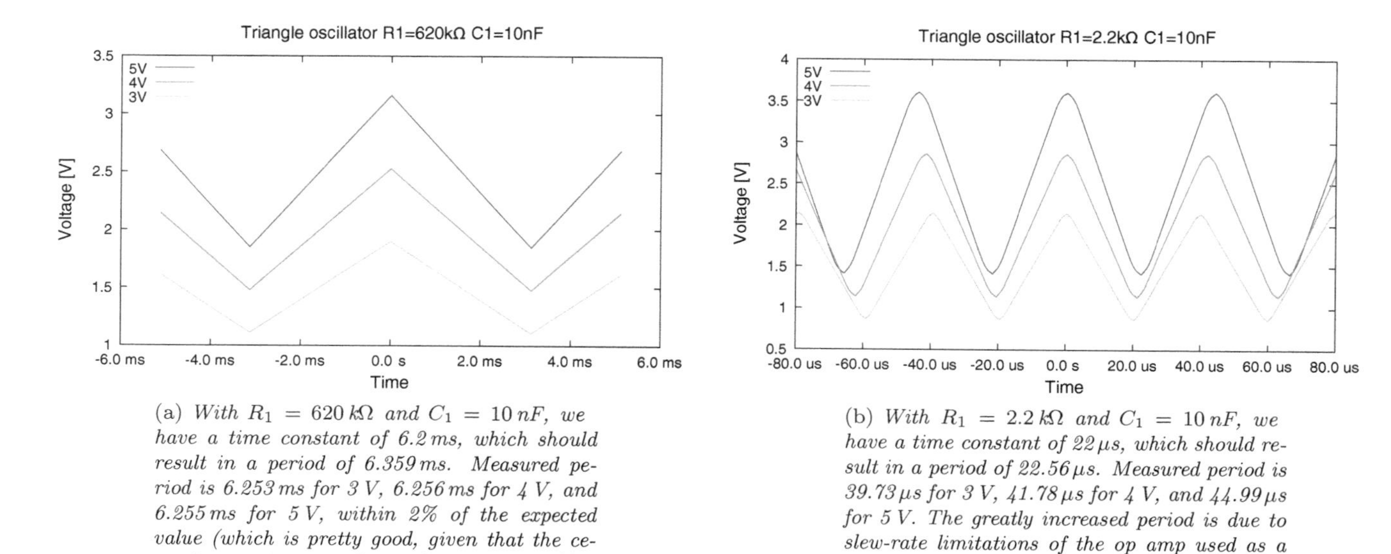

(a) *With* $R_1 = 620\,k\Omega$ *and* $C_1 = 10\,nF$, *we have a time constant of 6.2 ms, which should result in a period of 6.359 ms. Measured period is 6.253 ms for 3 V, 6.256 ms for 4 V, and 6.255 ms for 5 V, within 2% of the expected value (which is pretty good, given that the ceramic capacitor is only rated to about ±25%).*

(b) *With* $R_1 = 2.2\,k\Omega$ *and* $C_1 = 10\,nF$, *we have a time constant of 22* μs, *which should result in a period of 22.56* μs. *Measured period is 39.73* μs *for 3 V, 41.78* μs *for 4 V, and 44.99* μs *for 5 V. The greatly increased period is due to slew-rate limitations of the op amp used as a Schmitt trigger.*

Figure 36.3: *The waveforms here are from the circuit in Figure 36.2, except that* $R_2 = 39\,k\Omega$ *instead of 40* $k\Omega$, *so the period should be* $1.026 R_1 C_1$.

Data collected with Analog Discovery 2 [19]. Figure drawn with gnuplot [33].

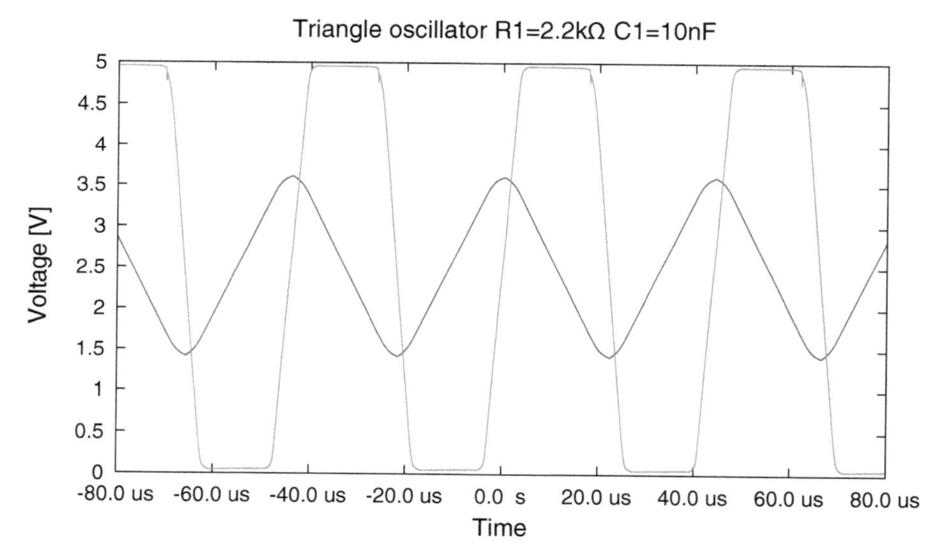

Figure 36.4: *The slew-rate limitations of the op amp used for the Schmitt trigger (nominally 0.6 V/* μs*) prevent a crisp edge on the square wave, which leads to rounded corners on the triangle wave and longer periods for higher voltage swings. The measured slew rates here are* $0.671\,V/\mu s$ *for the rising edge and* $-0.708\,V/\mu s$ *for the falling edge.*

Data collected with Analog Discovery 2 [19]. Figure drawn with gnuplot [33].

Exercise 36.2

Design a triangle-wave oscillator that should have a frequency of approximately 3 Hz with a 3 V peak-to-peak voltage using a single 5 V supply. Use an op amp to make the Schmitt trigger.

36.3 Voltage-controlled triangle-wave oscillator

We can make the triangle-wave oscillator into a voltage-controlled oscillator (VCO) by changing the magnitude of the square-wave current input to the integrator. As the current is increased, the capacitor in the integrator charges faster, and the frequency is increased. We can make the frequency be either linear with the voltage (Section 36.3.1) or exponential with the voltage (Section 36.3.3).

36.3.1 *VCO: Frequency linear with voltage*

Figure 36.5 shows one way to modify a triangle-wave oscillator to make a voltage-controlled oscillator. The op amp U_1 is the integrator, and U_2 is an inverting Schmitt trigger. The reference voltage to the integrator is $V_{\text{in}}/2$, so the current through R_6 and R_7 is

$$I_6 = \frac{V_{\text{in}} - V_{\text{in}}/2}{R_6 + R_7} .$$

When the output of the Schmitt trigger is low, the nFET is turned off, and all that current flows into the integrator.

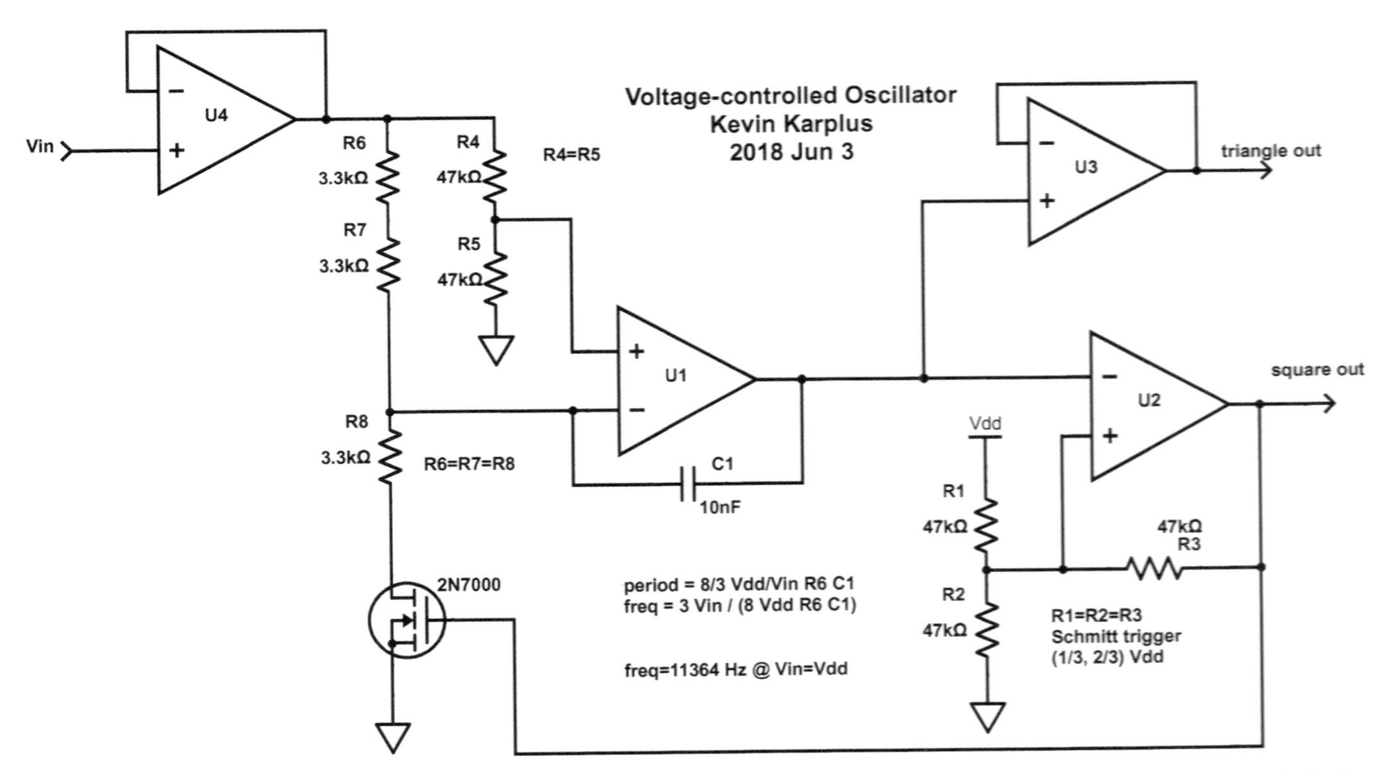

Figure 36.5: *Voltage-controlled oscillator whose frequency is approximately linear with the input voltage. Amplifier U_1 with feedback capacitor C_1 is an integrator, while U_2 with positive feedback R_1, R_2, and R_3 is an inverting Schmitt trigger whose output controls the 2N7000 nFET transistor.*

When the nFET is off, the current being integrated is $\frac{-V_{\text{in}}}{2(R_6+R_7)}$. When the nFET is on, the current being integrated is $\frac{V_{\text{in}}}{2(R_8+R_{\text{on}})} - \frac{V_{\text{in}}}{2(R_6+R_7)}$. If the on-resistance of the nFET is small, then the two currents have the same magnitude, but with opposite polarities. Figure drawn with Digi-Key's Scheme-it [18].

When the output of the Schmitt trigger is high, the nFET is turned on, and the current through R_8 is $V_{\text{in}}/(2R_8)$ (ignoring the on-resistance of the nFET), giving a current input to the integrator of

$$I_6 - I_8 = \frac{V_{\text{in}} - V_{\text{in}}/2}{R_6 + R_7} - \frac{V_{\text{in}}}{2R_8} .$$

If we set the resistors to be equal, $R_6 = R_7 = R_8 = R$, then the current into the integrator is $-V_{\text{in}}/(4R)$, which is the negative of the current when the nFET is off.

When the nFET is off, we have a falling voltage at the triangle-wave output

$$\frac{\mathrm{d}V}{\mathrm{d}t} = \frac{-V_{\text{ih}}}{4RC_1} ,$$

and when the nFET is on, we have a rising voltage of

$$\frac{\mathrm{d}V}{\mathrm{d}t} = \frac{V_{\text{in}}}{4RC_1} .$$

The thresholds of the inverting Schmitt trigger can be determined from Equations 33.3 and 33.4, after replacing the voltage divider by the Thévenin-equivalent: a $V_{\text{dd}}/2$ voltage source and $R_1 \parallel R_2 = 23.5\,\text{k}\Omega$:

$$V_{\text{tri}} < \frac{V_{\text{dd}}}{2} \frac{47\,k\Omega}{47\,k\Omega + 23.5\,k\Omega} = \frac{V_{\text{dd}}}{3} ,$$

$$V_{\text{tri}} > \frac{V_{\text{dd}}/2\;47\,k\Omega + V_{\text{dd}}\;23.5\,k\Omega}{47\,k\Omega + 23.5\,k\Omega} = \frac{2V_{\text{dd}}}{3} .$$

The value of the resistors R_1, R_2, and R_3 is not crucial—if they are all the same and in a reasonable range that does not demand too much current from the op amp, the thresholds will be 1/3 and 2/3 of V_{dd}.

From the voltage swing and the slope of the triangle wave, we can determine the period:

$$\tau = 2\frac{V_{\text{dd}}(2/3 - 1/3)}{V_{\text{in}}/(4RC_1)} = \frac{8}{3}\frac{V_{\text{dd}}}{V_{\text{in}}} RC_1 ,$$

which we can invert to get the frequency:

$$f = \frac{3}{8RC_1} \frac{V_{\text{in}}}{V_{\text{dd}}} .$$

Figure 36.6 shows the waveforms for the triangle-wave and square-wave outputs of the oscillator in Figure 36.5 with a 5 V power supply and 5 V input.

Figure 36.7 shows a plot of frequency of the oscillator of Figure 36.5 versus input voltage. With $R = 3.3\,\text{k}\Omega$, $C_1 = 10\,\text{nF}$, and $V_{\text{dd}} = 5\,\text{V}$, we expect 2273 Hz/V, but the measured frequencies are not quite linear with voltage. We can get a good fit to the measurements, though, if we add one more parameter: a fixed time per period overhead (partly due to the slew-rate limit of the op amps). Instead of fitting $f = aV_{\text{in}}$, we can fit $\tau = 1/(aV_{\text{in}}) + b$ or

$$f = \frac{aV_{\text{in}}}{1 + abV_{\text{in}}} .$$

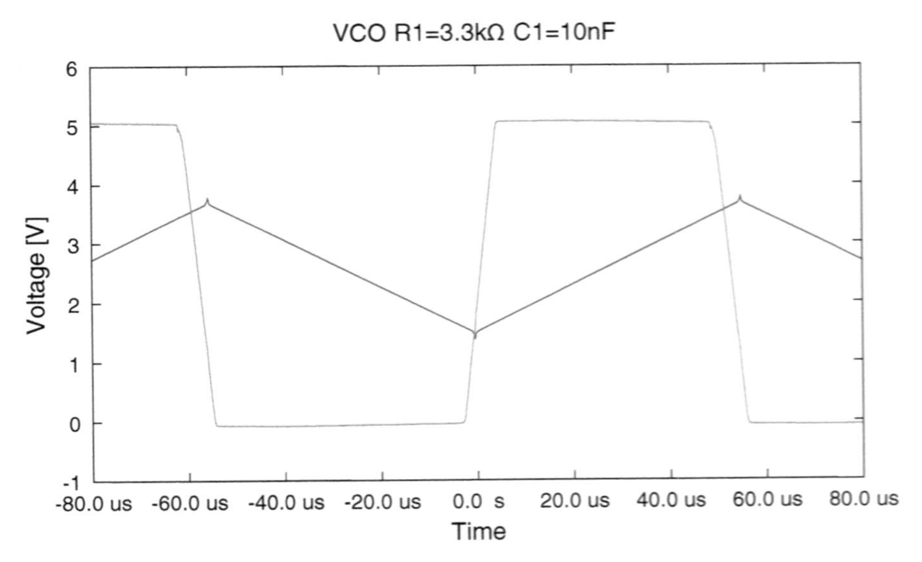

Figure 36.6: *The square-wave and triangle-wave outputs of the oscillator in Figure 36.5 (with $V_{dd} = 5$ V and $V_{in} = 5$ V) show the slew-rate limitations of the op amp in the square-wave output, but no rounding of the triangle wave—the nFET switches between on and off fairly sharply, despite the slow change of the square wave. There is even a little sharpening of the corner, possibly from capacitive coupling between the gate and drain of the nFET.*
Data collected with Analog Discovery 2 [19]. Figure drawn with gnuplot [33].

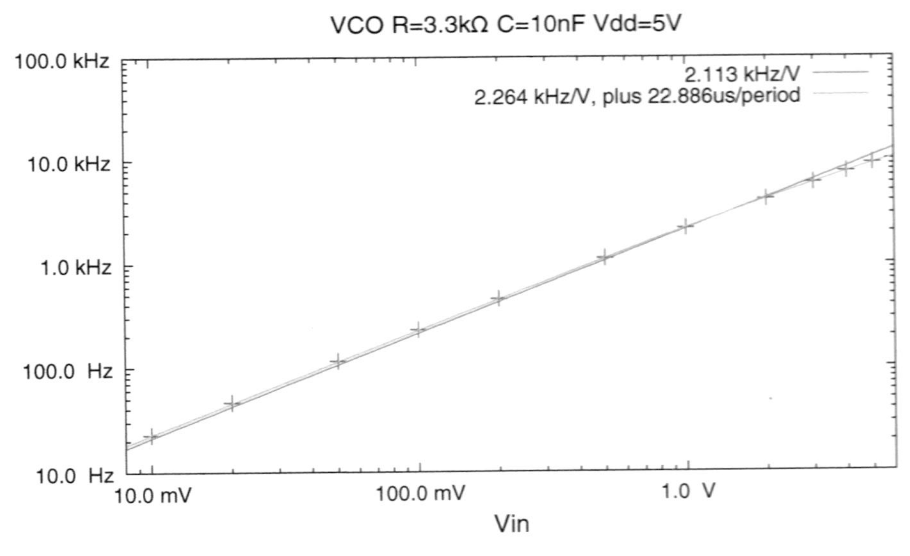

Figure 36.7: *The frequency of the VCO in Figure 36.5 is not quite linear with the input voltage, but we can get a very good fit to the measurements if we assume an extra time of about 23 μs per period.*
Data collected with Analog Discovery 2 [19]. Figure drawn with gnuplot [33].

One important consideration is that $R_8 = R$ cannot be very large, because there is a parasitic capacitance on the drain of the nFET, which needs to be charged from 0 V to $V_{dd}/2$ through R_8 when the nFET is turned off. This charging current delays the time for the reversal of current, changing the duty cycle and the period of the oscillator.

Figure 36.8 shows the waveforms for two different implementations, both with $RC_1 = 33\,\mu$s. One uses a small resistor (3.3 kΩ) and the other uses a large resistor (330 kΩ). The waveform for the large resistor

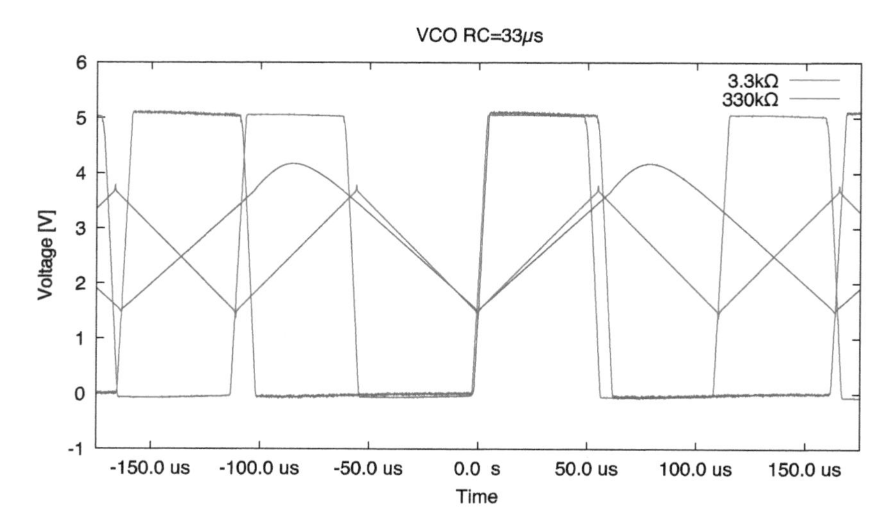

Figure 36.8: *Waveforms for two different VCO implementations with the same nominal RC_1 time constant of 33 μs. The red traces are from the implementation of Figure 36.5 with $R_1 = 3.3\,k\Omega$ and $C_1 = 10\,nF$. The blue traces are from $R_1 = 330\,k\Omega$ and $C_1 = 100\,pF$. The slopes of the triangle waves are slightly different (due to slightly different time constants), but the big difference is the overshoot of the blue triangle wave when the nFET turns off, due to charging the parasitic capacitance of the nFET drain through 330 kΩ.*

Data collected with Analog Discovery 2 [19]. Figure drawn with gnuplot [33].

shows a large overshoot for the integrator when the nFET turns off, corresponding to the charging of the parasitic capacitance of the nFET drain through the resistor.

36.3.2 *Sawtooth voltage-controlled oscillator*

I initially thought that the design could be changed to a saw-tooth (ramp-down) oscillator by making R_8 be $0\,\Omega$, causing the integrator output to rise at the slew rate of the op amp when the nFET is turned on, but this turns out not to work well. The problem is that turning on the nFET causes the voltage at the negative input of U_1 to drop to near 0 V and the op amp can't keep it at $V_{\mathrm{in}}/2$ as the simple DC feedback analysis assumes (see Figure 36.9). When the nFET turns off again, the node needs to recharge to $V_{\mathrm{in}}/2$. Initially, the voltage rises at the same rate as the output of the integrator, until either the input reaches $V_{\mathrm{in}}/2$ or the output reaches V_{dd}, whichever comes first. If the output reaches V_{dd}, then the negative input continues to charge as an RC circuit with $R = R_6 + R_7$ and $C = C_1 + C_{\mathrm{d}}$, where C_d is the drain capacitance of the nFET, with target voltage V_{in}, until the node reaches $V_{\mathrm{in}}/2$, at which point the output of the integrator can start dropping as intended. This behavior can be seen in Figure 36.9.

We can improve the sawtooth oscillator by putting in a small resistor for R_8. Using the same (slightly erroneous) analysis as before, when the nFET is on, the current input to the integrator is

$$I_6 - I_8 = \frac{V_{\mathrm{in}} - V_{\mathrm{in}}/2}{R_6 + R_7} - \frac{V_{\mathrm{in}}}{2R_8} ,$$

and the output of an ideal integrator rises at

$$\frac{\mathrm{d}V}{\mathrm{d}t} = \frac{V_{\mathrm{in}}(R_6 + R_7 - R_8)}{2(R_6 + R_7)R_8C_1} .$$

We can choose R_8 so that the ideal integrator slope is no more than the slew rate of the op amp for the highest input voltage we plan to use. For example, with $R_6 = R_7 = 3.3\,\mathrm{k}\Omega$, $C_1 = 10\,\mathrm{nF}$, and $R_8 = 220\,\Omega$, the slope is $\frac{\mathrm{d}V}{\mathrm{d}t} = 0.2197\,V_{\mathrm{in}}/\mu\mathrm{s}$. So for $V_{\mathrm{in}} < 2.73\,\mathrm{V}$, the op-amp can slew fast enough to integrate correctly, but at $V_{\mathrm{in}} = 3\,\mathrm{V}$ the slew-rate limitation means that the feedback can't match the input current, causing the small dip in the negative input seen in Figure 36.9.

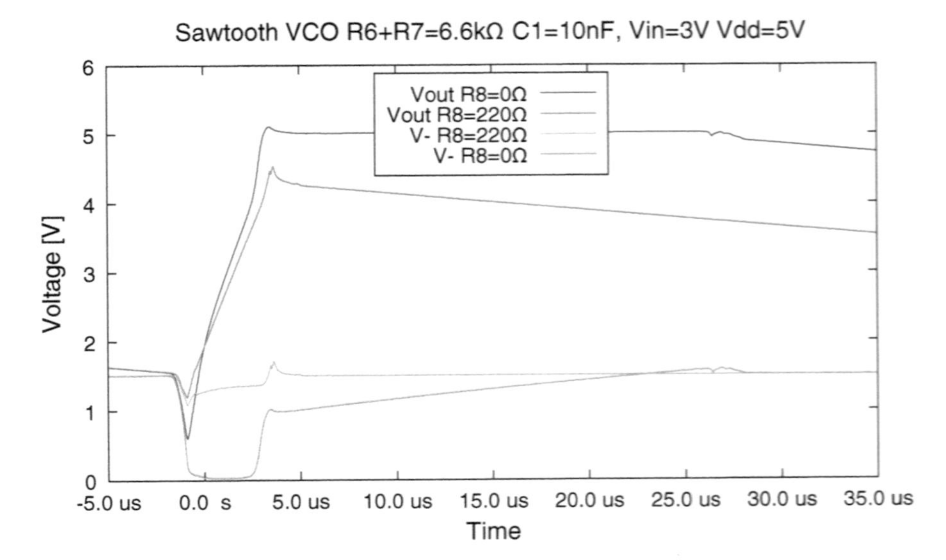

Figure 36.9: *The rising edge of a ramp-down sawtooth oscillator, using the schematic of Figure 36.5, with* $R_8 = 0\,\Omega$ *or* $R_8 = 220\,\Omega$. *The top two traces are for the output of the oscillator, and the bottom two for the negative input of* U_1, *which we had modeled as being a constant 1.5 V.*
With $R_8 = 0\,\Omega$, *the nFET pulls the negative input down to 0 V, and the output overshoots all the way to the top power rail, where it remains pinned until the negative input can charge back to 1.5 V. The capacitive coupling through* C_1 *can be seen causing a downward spike in the output just as the nFET turns on, and an upward spike as the nFET turns off.*
With $R_8 = 220\,\Omega$, *the nFET pulls the negative input down only a little bit, and the current is about as large as it can be without the integrator hitting the slew-rate limitations of* U_1. *The waveform is still not centered at 2.5 V, because the Schmitt trigger* U_2 *is still slew-rate limited, so takes a while to turn the nFET off after the upper threshold is crossed.*
Data collected with Analog Discovery 2 [19]. Figure drawn with gnuplot [33].

36.3.3 *VCO: Frequency exponential with voltage*

In many applications, especially musical ones, we want a wide-range VCO whose frequency is exponential with input voltage, not linear with voltage. The design in Section 36.3.1 relies on integrating a current, so if we could make a current that is exponentially related to the input voltage, we could use essentially the same design.

As you may remember from Section 23.2, the current through a semiconductor diode is very well approximated by

$$I \approx I_S e^{\frac{V_d}{nV_T}} ,$$

which is exponential with the voltage across the diode V_d. If we use diodes instead of resistors for converting the input voltage to current, we can get the exponential behavior we want.

> One subtle point is that the series connection of R_6 and R_7 in Figure 36.5 halves the current (due to Ohm's law), but putting two diodes in series does not halve the current. Instead of making R_8 have half the resistance of $R_6 + R_7$, Figure 36.10 shows two diodes in parallel to provide the necessary doubled current through the nFET when the nFET is turned on.

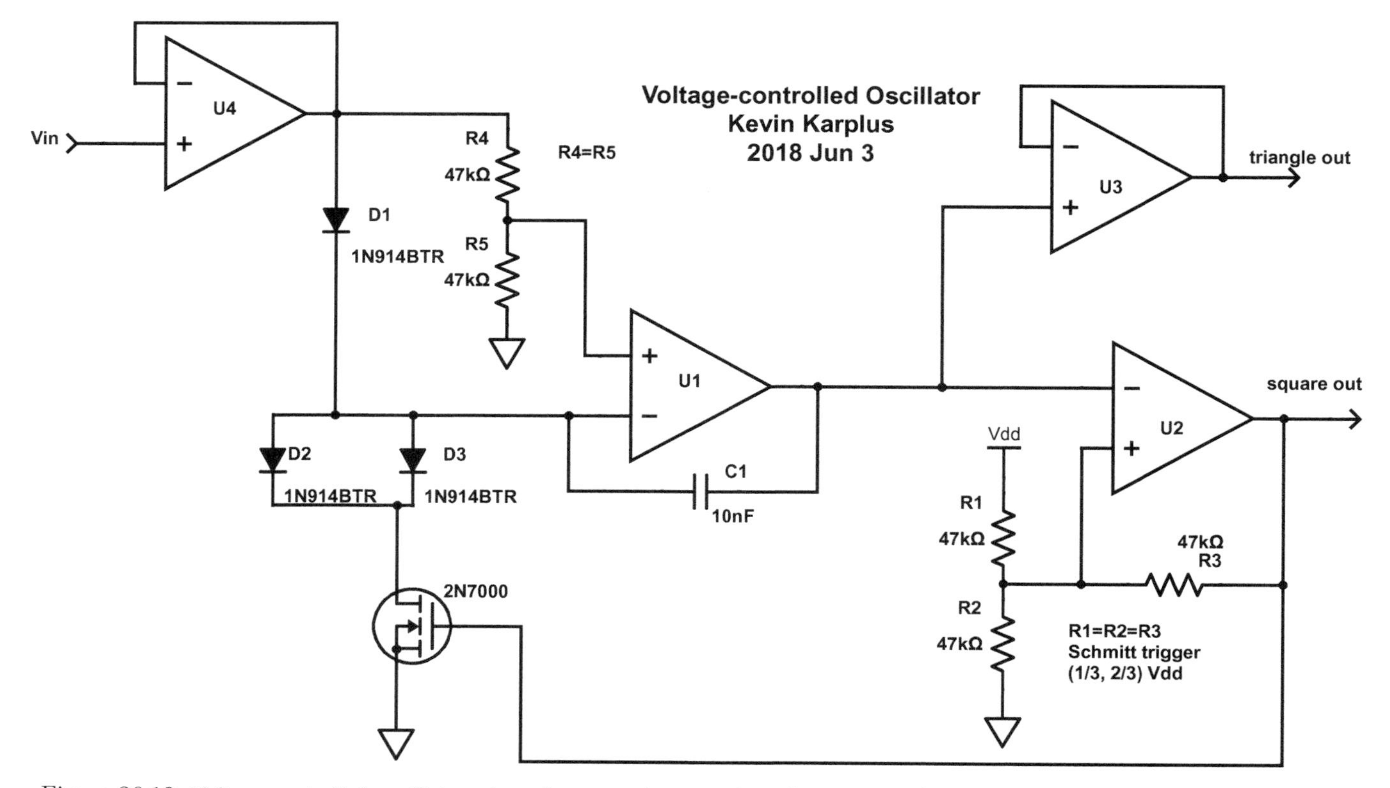

Figure 36.10: *Voltage-controlled oscillator whose frequency is approximately exponential with the input voltage. As in Figure 36.5, amplifier U_1 with feedback capacitor C_1 is an integrator, while U_2 with positive feedback R_1, R_2, and R_3 is an inverting Schmitt trigger whose output controls the 2N7000 nFET transistor.*

The current with the nFET turned off is determined by the I-vs.-V characteristics of diode D_1 with voltage drop $V_{\text{in}}/2$, until the current limit of the unity-gain buffer U_4 is reached. With the nFET turned on, twice that current is diverted to ground, so the integrator sees the negation of the current it sees with nFET off.

Figure drawn with Digi-Key's Scheme-it [18].

The unity-gain buffer implemented with op amp U_4 in Figure 36.10 is important for limiting the current through the diodes—if the input were directly connected to the diodes, then even fairly modest input voltages could result in too high a current that burns out the diodes.

Figure 36.11 shows how the VCO frequency varies with voltage for the circuit of Figure 36.10.

The design of Figure 36.10 has several flaws.

- For example, to get high frequency we need fairly high current, but then the $V_{\text{ds}} = IR_{\text{on}}$ voltage drop across the nFET results in reduced diode bias voltage, and even a small change in bias voltage causes a large change in the current through the diodes. That means that the symmetry of the triangle wave is lost at higher frequencies, and the period is larger than would be predicted from a simple exponential formula.
- Another flaw is that the scaling factor for the exponential nV_{T} is dependent on temperature, with $V_{\text{T}} = kT/q$ being linear with absolute temperature T. In a musical instrument, temperature may vary over 25°C–50°C, which results in a huge change in the tuning of the VCO. Modern electronic instruments rely on crystal oscillators to provide very precise and stable tuning, but old analog

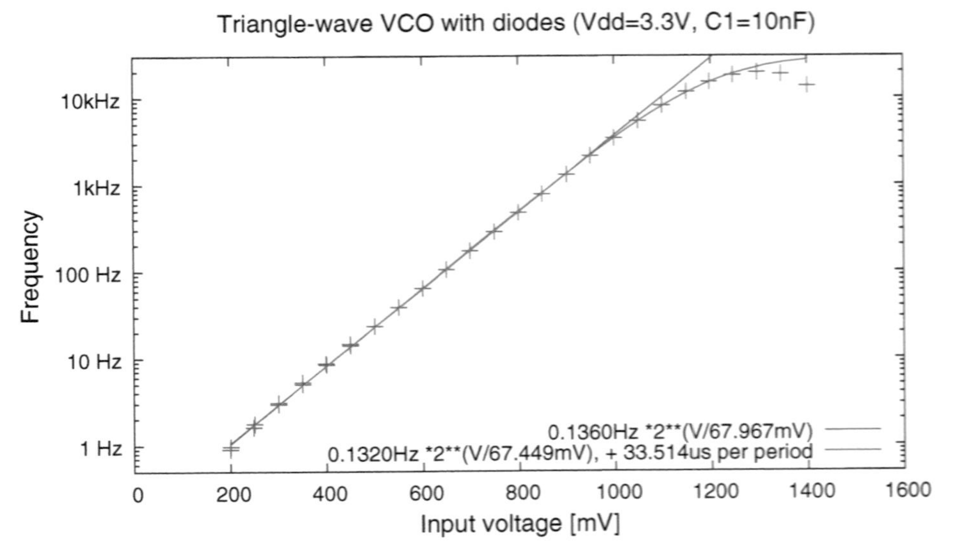

Figure 36.11: *Using diodes to convert the input voltage to a current, using the circuit of Figure 36.10 gives a frequency that is roughly exponential with voltage (at a rate of about an octave every 67.45 mV). As with Figure 36.5, the slew-rate limitations of the op amp used as a Schmitt trigger add an almost constant extra delay per period (here estimated as 33.5 µs). When the input voltage gets above about 1.25 V, the output current limit of the unity-gain buffer U_4 prevents further increase in frequency.*
Data collected with Analog Discovery 2 [19]. Figure drawn with gnuplot [33].

synthesizers used a variety of different methods to avoid having to retune constantly. Some of the methods included

- Heating the diode junctions to a fixed temperature (say 50°C) in a feedback loop, to keep the diode characteristics constant. Having multiple junctions on the same chip and using one of them as the temperature sensor in a servo loop helps ensure that all the diodes are at the desired temperature.
- Using resistors with well-characterized thermal coefficients in complicated circuits to try to compensate for the temperature dependence of the diodes. The compensation is difficult to make work over many octaves, and making sure that the resistor and the diode junction are at the same temperature is mechanically tricky.
- Using a spare diode in an array of diodes as a temperature sensor and using its current to adjust the current to the integrator.

- Even if temperature compensation of the diodes is arranged, other components of the system also have temperature coefficients that can change the frequency. For example, cheap ceramic capacitors and film capacitors often have a (400 ± 200) ppm/K (parts per million per degree) temperature coefficient, which over a 50° range results in up to a 0.3% change in capacitance (and, hence, in frequency). That may not sound like much, but it is half a semi-tone on a musical scale, which is a huge tuning error.
- Musicians all have different standards about how precise they want their instruments tuned, but a good general standard is tuning to within about ±0.1% (±1.7 cents, where one cent is a ratio of $2^{1/1200}$), though that is for the ratio of frequencies between different notes, rather than absolute pitch. Achieving this degree of tuning accuracy over several octaves and wide temperature range is difficult with analog components, but it is straightforward with a crystal oscillator and digital synthesis.

Chapter 37

Lab 11: Class-D Power Amp

Bench equipment: function generator, power supply, oscilloscope

Student parts: electret microphone, loudspeaker, preamplifier board from Lab 9, comparator, 2 nFETs, 2 pFETs, 2 half-H-bridge boards, 2 5-pin right-angle male headers, inverters, inductor, capacitors, breadboard, clip leads .

37.1 Design goal

In Lab 9, you made an audio amplifier for a microphone using an op amp, but it was not powerful enough to provide loud sounds from a loudspeaker. The op-amp output had limited current and the loudspeaker had a low impedance, so the power to the loudspeaker, I^2R, was small. With a 23 mA limit and 8 Ω loudspeaker, the maximum power was about 4 mW.

In this lab, you will design an amplifier that is capable of delivering at least 3 W peak (1.5 W RMS) to an 8 Ω loudspeaker, about 750 times more than the audio amplifier of Lab 9.

> The design goal given above is too vague for your report! Make it more explicit as you work out your design. Be sure to include what load you are driving, the power to the load, and the frequency range of the amplifier.

Because you will be using power field-effect transistors (FETs), a class-D amplifier design is probably the easiest to get working [115].

Design constraints: in addition to the parts in your kit, you should use the Analog Discovery 2 as a power supply and as a triangle-wave generator.

37.2 Pre-lab assignment

Start early on the pre-lab design for this lab, as there is a lot to do.

37.2.1 *Block diagram*

> **Pre-lab 11.1**
>
> Make some design specifications for your amplifier: What is the frequency range you wish to amplify? What is the power output of the amplifier? What AC voltage or current does your preamplifier provide?

Pre-lab 11.2

What frequency do you want your PWM signal to be? It should be at least three times the frequency of the highest frequency you want to amplify and above the range of human hearing (look up the range of human hearing, if you don't already know it). If the PWM frequency is too high, the FETs will be spending a greater portion of their time in the intermediate state where they are neither on nor off, and the efficiency of the amplifier will be reduced (and the FETs might get hot).

Pre-lab 11.3

As you did for the preamplifier lab, make a block diagram of the whole amplifier, from microphone to loudspeaker, giving the function of each block and any constraints (voltage, current, frequency, etc.) on the signals between the blocks. What are the voltage (or current) levels (DC and AC) for every signal that goes between blocks? Remember to include blocks for the power supply and for the function generator.
If you need to change DC voltage levels from one block to another, remember that you can use high-pass filters to do the conversion (as we did in the preamplifier).

Pre-lab 11.4

Draw a schematic for each block (including the preamplifier).
Draw a complete schematic of the entire amplifier, showing every component value and every pin number. You will be building and debugging from this schematic, so it must be as complete and accurate as you can make it. The preamplifier can be treated as a single component for this schematic, but make sure that all connections to the preamplifier (power and signal) are shown.

Part numbers are important—make sure every component (other than resistors and capacitors) has the correct part number on your schematic.

The part numbers must be included (and correct!) for the nFET and pFET—check the labeling of the components in your kit!

If you include an LC filter, as described in Section 35.4, then the part number, and not just the nominal inductance, is important for the inductor, as both the series resistance and the saturation current matter, and those can vary a lot between different inductors with the same inductance.

Be sure to document your loudspeaker model (with parameter values!) in your report, as the details of the loudspeaker impedance affect the design.

37.2.2 *Setting the power supply*

The Analog Discovery 2 has a dual power supply, capable of providing 700 mA or 2.1 W per channel (whichever is less). You can ignore the ground lead to make this a single power supply with a higher voltage and a total of 700 mA or 4.2 W. For example, if you wanted a 6 V supply, you could set the power supply to −3 V and +3 V, to get the full 4.2 W. Using +5 V and −1 V, however, would give you only 420 mA from the positive supply, limiting the total power to 2.52 W.

If you use both power supplies, make sure that you set the offset correctly for your triangle wave—the function generator is specified relative to the ground line of the Analog Discovery 2, not relative to the lower power rail.

Pre-lab 11.5

Look up the data sheets for the pFET and nFET in your parts kit (possibly the pFET SPP15P10PLHXKSA1 and the nFET PSMN022-30PL,127, if using TO-220 packages, possibly PMV20XNER and SSM3J332RLF if SOT-23 packages) and figure out the maximum power and current you could safely have them handle without a heat sink on a minimal PC-board footprint. Because power FETs are usually used with heat sinks, it may be difficult to figure out their maximum power *without* a heat sink. I found it useful to do Google searches with the package type and keywords like *thermal resistance without heat sink* to find appropriate thermal resistances from junction to ambient.

Also look at the gate voltages needed to turn the nFET or pFET fully on—this puts a minimum voltage constraint on your power supply.

Pre-lab 11.6

If you wanted to deliver 10 W (the maximum RMS power) to your loudspeaker, what is the maximum voltage and current you should apply? (You'll need to say what the loudspeaker resistance is.)

You should look at most powerful signals you can deliver to the speaker, which would be square waves that alternate between the positive and negative voltage rails.

If you want to deliver 3 W, what voltage and current would you need from the power supply?

Design the drivers for your H-bridge. Use the results of Lab 10 to decide what drivers to use—can you use just comparators or do you need to use inverters as current amplifiers?

Pre-lab 11.7

If you are using inverters as current amplifiers to drive the FET gates, check the voltage limitations on your inverter chips. Do they provide a stricter limitation on the power supply voltages than the desired output power requires? How much power could you deliver with the largest voltages acceptable to the inverter?

Pre-lab 11.8

Check the voltage limitations on your comparator chips. Do they provide a stricter limitation on the power supply voltages than the desired output power requires? How much power could you deliver with the largest voltages acceptable to the comparator?

The MCP6004 op-amp chips you used for the preamplifier have a 6 V power supply limit. If you choose larger voltages for the final stage, you need to figure out how to power the preamplifier safely.

When setting your power-supply voltages—one warning: be sure that the output of the preamplifier remains between the power rails for the next stage (the comparator). The preamplifier output has a peak-to-peak voltage that can be as big as the power supply to the preamplifier, and changing the DC bias may move this range to outside the power supply range to the comparator.

> Make sure that the triangle-wave voltage stays within the comparator's input voltage range and that the preamplifier output range stays within the range of the triangle wave, to avoid clipping.

37.3 Procedures

Build and test your circuit a block at a time, starting from the microphone. It is very hard to debug a large circuit when the problem could be anywhere.

Start by making sure your preamplifier works and provides the voltages you expect.

Check the function generator voltage with an oscilloscope before hooking it up to the comparator input. If you need a high-pass filter to recenter the voltage, test its output with the oscilloscope (DC-coupled) also.

> If you use only one comparator, make sure that the other comparator has its inputs tied to opposite power rails, so that it does not make random transitions that corrupt the power supply.

The LC filter and loudspeaker should be thought of as a single block, since the LC filter does something quite different without the loudspeaker as a load. Characterize the LC filter by using the network analyzer with Channel 1 across the input to the filter and Channel 2 across the loudspeaker (which is an important part of the filter!). Make sure that the voltage for the network analyzer is set low enough that you don't get clipping due to the ± 30 mA current limits of the waveform generator.

Plot the measured gain of the filter superimposed on the gain calculated from the component values and the loudspeaker model.

It is a good idea to start out the power amp section with a lower voltage (and current limit for power supplies that have current limits) than you intend to finally use, to avoid overheating components before the wiring has been debugged. Just make sure that the voltages are large enough that you can turn on the nFETs and the pFETs.

37.4 Demo and write-up

Demonstrate your amplifier working, producing undistorted amplified speech or music. Show the output voltage waveform (both before the LC filter and after it) on the oscilloscope.

Observe the waveforms at two different time scales: one where the PWM waveform is clearly visible, maybe at a sampling rate of 100 MHz, to see the rising and falling edges at maximum resolution, and one at a sampling rate of 100 kHz, where the audio output is clearly visible.

The Analog Discovery 2 has differential inputs, so you can measure voltages before and after the LC filter, or measure voltage and current simultaneously fairly easily, even though none of the nodes involved in these measurements is ground. Most other oscilloscopes have only one ground, so the ground leads from both channels have to be the same node in your circuit, and that may not be the ground node of your amplifier.

Document both the system-level design and the details of each block in your report. Remember that the reader of your design report has no access to other reports you may have written—this design relies heavily on Lab 7 through Lab 9, so this report should be a cumulative one, including the whole design, not just the most recently added piece. The results from earlier labs should be *included* in this report (not just referenced). Any errors that you made in previous reports should be fixed in this report.

Draw schematics for your various test setups, and make sure you provide part numbers for all parts—especially the FETs, the comparators, the inductors, and the inverters, because the properties of the parts are critical, and the generic symbols and nominal values aren't enough information.

37.5 Bonus lab parts

This section is optional—bonus activities to do only if you have spare time.

Bonus lab parts should only be attempted after the basic functions have been demonstrated.

- Try putting a $0.25\,\Omega$ resistor in series with your loudspeaker, so that you can use the oscilloscope to look at the current through the speaker as well as the voltage across it.
  Larger resistor values could cause problems with the amount of power dissipated in the resistor—if you have $0.7\,\text{A}$ of current, the power from a $1\,\Omega$ resistor would be $0.5\,\text{W}$, and our resistors can only handle $0.25\,\text{W}$. You will probably have to make the small resistor out of larger ones in parallel, because our parts kits don't contain such small resistors. Splitting the current over several resistors reduces the power dissipated in each one.
- Instead of a microphone input, use a headphone plug into an MP3 player, phone, or other sound source. Look up the specifications for the MP3 player and headphones to get voltage ranges and impedances, and figure out how this changes the preamplifier design! (or can the music source replace the preamplifier?) How would you combine L and R stereo channels into a single monaural signal?
- Use a dual power supply so that you can use just one nFET and one pFET, rather than an H-bridge. For example, if you use $\pm 3\,\text{V}$, most of your design can stay the same, but the power output will be only a quarter of an H-bridge with $6\,\text{V}$.
- Characterize the power stage (from the preamp output to the loudspeaker) using the network analyzer. You can use the second waveform generator of the Analog Discovery 2 to generate a triangle wave while the first waveform generator is used by the network analyzer. The differential channel for the output of the network is important here, as neither side of the loudspeaker is at ground.
- Characterize the entire amplifier (from the microphone input to the loudspeaker) using the network analyzer. You might need to use a voltage divider to scale down the first function generator's voltage to a small enough signal for the preamp.
- Characterize the acoustic behavior of the LC filter plus loudspeaker with the network analyzer with Channel 1 across the input of the filter and Channel 2 from a microphone listening to the loudspeaker (which may need to be amplified).
- Design a triangle-wave generator instead of using a function generator (see Chapter 36).
- Design a class-D amplifier using feedback from the H-bridge, rather than a triangle-wave generator and comparator (see Section 35.6).
- Use a higher voltage power supply for the final stage to get more power. This requires changing a lot of the circuitry, as many of the parts have $6\,\text{V}$ limits.

Chapter 38

Electrodes

An *electrode* in bioengineering is a transducer that converts between ionic current in an electrolyte and electron current in a metal wire. The transduction can go either way, using the electrodes to measure ionic current or to create it. In other fields, the term *electrode* may be used in slightly different contexts, such as converting electron flow in a wire to electron flow in a vacuum, but in this class, we'll only be using electrochemical electrodes. In all cases, an electrode converts between electron flow in metal and some other form of current.

All electrochemical electrodes work through redox reactions. For current to flow, there must be two electrodes, an *anode* and a *cathode*. A *reduction reaction* happens at the cathode, taking electrons from the wire, and an *oxidation reaction* happens at the anode, donating electrons to the wire.

I don't care whether you can remember which direction is reduction and which oxidation—this isn't a chemistry class. But it is useful to remember "anode" and "cathode", because a lot of parts are diodes (photodiodes, LEDs, fly-back diodes, etc.), and they are often labeled with those names. The cathode is where electrons come from the wire and enter the device (think of the "cathode-ray tubes" in old TV sets, which sprayed electrons from the cathode onto the phosphor screens).

38.1 Electrolytes and conductivity

An *electrolyte* is any substance that makes water conductive—generally a salt that dissociates into ions that can carry charges through the solution. In biological systems, these ions are generally sodium, potassium, calcium, magnesium, chloride, hydrogen phosphate, and hydrogen carbonate ions, but the H^+ and OH^- ions are also important charge carriers.

It is common in characterizing fresh water sources to measure the *ionic conductivity* of the water, to estimate how much electrolyte is dissolved in the water. At low concentrations of strong electrolytes, the conductivity is roughly linear with concentration, but the behavior is more complicated for weak electrolytes, which don't dissociate completely, and even strong electrolytes have nonlinear conductivity at high concentration [118, 138]. Table 38.1 and Figure 38.1 give approximate conductivity values for sodium chloride solutions.

Blood serum has a conductivity of about 13 mS/cm [42]. Whole blood has about half that conductivity, because the membranes of the blood cells are not conductive, and so the effective cross section of the conductor is reduced by the concentration of blood cells. The concentration of blood cells can be computed from the conductivity of the whole blood [42].

We want ionic conductivity to be a property of the solution, not of the measurement setup. If you think of two measuring electrodes as being the end plates of a cylinder of water, then the resistance of the water is going to be proportional to the distance between the electrodes (think of lots of smaller resistors in series), and inversely proportional to the area of the electrodes (think of lots of resistors in parallel).

Mass percent	mg/L	μS/cm(Foxboro 25°C)	μS/cm(CRC 20°C)
0.0001	1	2.2	
0.0003	3	6.5	
0.001	10	21.4	
0.003	30	64	
0.01	100	210	
0.03	300	617	
0.1	1000	1990	
0.3	3000	5690	
0.5	5000		8200
1.0	10000	17600	16000
2.0			30200
3.0		48600	
5.0		78300	70100
10.0		140000	126000
15.0			171000
20.0		226000	204000
25.0			222000
26.0		244000	

Table 38.1: *Conductivity of sodium chloride (NaCl) solutions as a function of concentration. This data is plotted in Figure 38.1.*
Maximum conductivity is at mass percentage 26%, which is the saturation concentration for NaCl. The Foxboro numbers are higher than the CRC ones, because they are taken at a higher temperature—even small differences in temperature affect ionic conductivity.
At very low concentrations, NaCl has a conductivity of about 126 μS/cm/*mM, but at 50 mM the conductivity has dropped to 111* μS/cm/*mM.*
Values from a table by Foxboro [26] and CRC [22].

When we measure resistance of a material, we get

$$R = \rho g,$$

where ρ is a constant (called *resistivity*) that depends on the properties of the material and g is a *geometric factor* that depends on the shape and size of the electrodes.

If we measure resistance between end plates of a cylindrical cell, the geometric factor is

$$g = s/A,$$

where s is the separation between the electrodes and A is the cross-sectional area of the cylinder.

The setup you will use in Lab 12 is *not* two parallel plates at the ends of a chamber filled with your solution, so the geometric factor is *not* just s/A.

The units of resistivity are Ωm or Ωcm. Resistance and resistivity are not the same thing—resistance depends on the placement and size of the electrodes, while resistivity is an inherent property of the material.

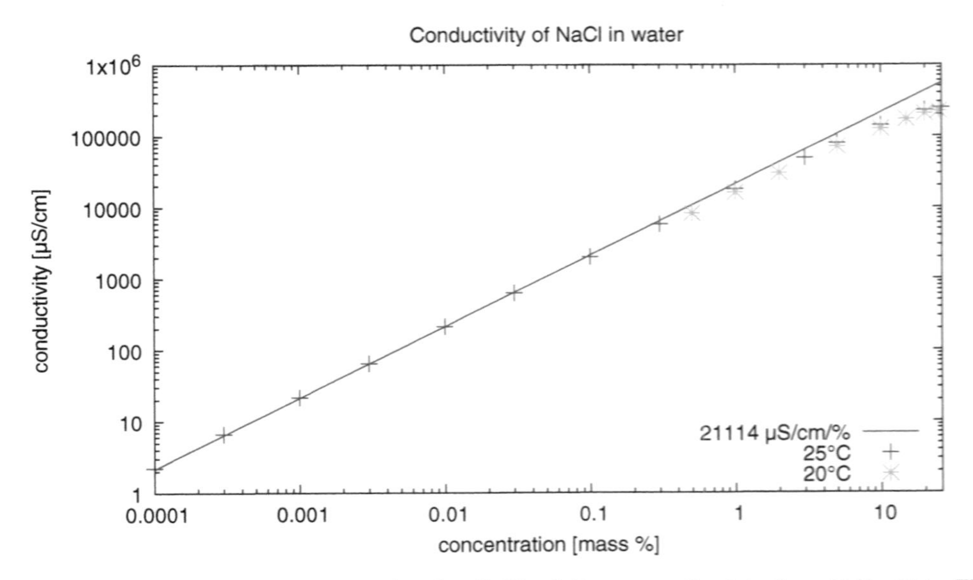

Figure 38.1: *This graph of conductivity vs. concentration for NaCl solutions uses the data from Table 38.1. The conductivity is essentially linear with concentration for low concentrations, but increases more slowly for high concentrations.*
Potassium chloride (KCl) is usually used for calibrating conductivity meters, because it has a much more linear conductivity vs. concentration plot.
Figure drawn with gnuplot [33].

For ultrapure distilled water, resistivity is often given in MΩcm at 25°C (the resistivity of pure water is very temperature-dependent [118]). A typical claim for a water system by Millipore is 18.2 MΩcm@25°C, though that can only be achieved with water that has dissolved monatomic gases—exposure to atmosphere dissolves CO_2 in the water forming some carbonic acid, which disassociates into H^+ and HCO_3^- ions [113], dropping the resistivity to about 13MΩcm [149].

Tap-water resistivity varies enormously depending on how "hard" the water is—that is, how much and what salts are dissolved in the water. For tap water and environmental water sources, the resistivity is not usually reported, but the *conductivity* (the multiplicative inverse of resistivity) is, because conductivity is roughly linear with the concentration of the electrolyte ions. The SI standard unit for conductivity is siemens per meter (S/m), though μS/cm is more commonly used for water measurements. Ultrapure water at 18.2 MΩcm has a conductivity of about 0.055 μS/cm, distilled water less than 11 μS/cm, tap water around 50–500 μS/cm, sea water around 50,000 μS/cm, and 1 M NaCl around 85,000 μS/cm.

Don't confuse conductivity with conductance—conductivity is the inverse of resistivity, and so is an inherent property of the material, while conductance is the inverse of resistance, and so depends on the size and placement of the electrodes.

For shapes other than a cylinder or prism with a constant cross section, it can be difficult to figure out the geometric factor that replaces s/A analytically, so for practical systems, rather than relying on precise geometric calculations, the measuring apparatus is calibrated with a solution of known conductivity every day.

If you want to try a theoretical calculation, for a geometry with two parallel cylindrical wires of diameter d, length L, and center-to-center separation s, the geometric factor that replaces s/A is approximately

$$\frac{1}{\pi L} \ln\left(s/d + \sqrt{(s/d)^2 - 1}\right) = \frac{1}{\pi L} \cosh^{-1}(s/d) \ ,$$

ignoring effects from the ends of the electrodes [78, p. 231]. Interestingly, this geometric factor does not depend on the diameter and separation separately, but on their ratio. The formula is based only on the sides of the cylinders—if the ends are in the electrolyte, the resistance is reduced, so this formula for the geometric factor is likely too large for our setup.

38.2 Polarizable and nonpolarizable electrodes

> A *polarizable electrode* ideally has no DC current flow between the electrode and the surrounding electrolyte. The electrode/electrolyte interface behaves like a capacitor.

The classic example of a polarizable electrode is a platinum electrode. The platinum electrode is also known as a hydrogen electrode, because the reaction is $2H_3O^+ + 2e^- \rightleftharpoons H_2 + 2H_2O$ [160].

Platinum electrodes are sometimes used as bioelectrodes because of their inertness, but platinum is too expensive for us—we'll look at stainless-steel electrodes, which appear to be moderately polarizable—that is, they have some conduction at DC, but not much.

The voltage on polarizable electrodes depends heavily on the distribution of charges near the electrode, and if the electrode is moved, the voltage can change substantially (similar to the changes one gets from the movement of the membrane in capacitor microphones). Because such *movement artifacts* are usually undesirable, nonpolarizable electrodes are often preferred for low-frequency measurements.

> A *nonpolarizable electrode* has no polarization, that is, DC current flows freely and the electrode-electrolyte interface behaves like a resistor.

A nonpolarizable electrode that can be used in salt solutions is the silver/silver-chloride electrode, which we'll look at in Lab 12. Ag/AgCl electrodes are popular for biomeasurements because they are close to ideal nonpolarizability, have a low half-cell voltage, and are generally non-toxic (or have low toxicity).

Real electrodes have characteristics that don't match either ideal (polarizable nor nonpolarizable), but fall somewhere in between, and we'll try to characterize that non-ideal behavior in Lab 12. In Lab 13, we'll use commercial disposable Ag/AgCl electrodes for the EKG.

Michael R. Neuman's chapter "Biopotential Electrodes" has further discussion of bio-electrodes [79]. A more comprehensive source is *Bioimpedance and bioelectricity basics*, particularly Chapters 2 (Electrolytics) and 3 (Dielectrics) [39]. These references are more detailed than we need for this course, but are good places to check for more information.

38.3 Stainless steel

There are many different types of stainless steel, each with somewhat different metallurgical characteristics. All of them are alloys of iron, chromium, and other metals. Generally, any steel alloy with 10.5% or more chromium by mass is considered a stainless steel [159].

The resistance to corrosion that gives it the *stainless* name comes mainly from a layer of chromium oxide on the surface of the steel that blocks oxygen from coming in contact with the metal. This oxide layer forms naturally in oxidizing conditions, but in reducing conditions it may not form, and the steel behaves differently, corroding much more readily. (The two states are referred to as *passive* and *active* in discussions of the corrosion of stainless steel, and the half-cell potentials are different.)

Surgical steel is often 316L stainless steel and is commonly used for implantable devices, because of its strength, its resistance to corrosion in the body, and very low toxicity [159, 163]. According to the material safety data sheet that came with the stainless-steel welding wire that I made stainless-steel electrodes from, 316L steel has 18%–20% chromium, 11%–14% nickel, 2%–3% molybdenum, 1%–2.5% manganese, 0.3%–0.65% silicon, $< 0.75\%$ copper, and $< 0.03\%$ carbon (the L stands for "low carbon"). Scalpels and cutting instruments are often made from 440 and 420 stainless steels which are harder than 316L and so take a better edge, but are not as corrosion-resistant. Stainless steel is not commonly used for electrodes, though, and we'll try to see why in Lab 12.

Because stainless steel is a complicated alloy, it is not clear what the electrochemical reactions are when stainless steel is put in salt water. The most important reactions for stainless steel as an electrode are probably $Fe + 2Cl^- \rightleftharpoons FeCl_2 + 2e^-$ and $Fe \rightleftharpoons Fe^{2+} + 2e^-$ for electrons entering the wire and $2H_3O^+ + 2e^- \rightleftharpoons H_2 + 2H_2O$ for electrons leaving the wire. I'm not certain of those reactions—finding good information about how stainless steel reacts in NaCl solutions is surprisingly difficult, in part because there are so many different stainless steels, and the usual interest is in how fast the steel corrodes, rather than in what electrochemistry is involved.

> For polarizable electrodes, there is very little current due to the redox reactions with the electrode—most of the current we see at high frequencies is due to changes in the distribution of charges in the solution without electrons entering or leaving the metal.

38.4 Silver/silver chloride

The electrodes most commonly used in the nanopipette and nanopore labs at UCSC (and generally for measuring potentials in biological systems) are Ag/AgCl electrodes. These are relatively cheap to make and are nonpolarizing. The reaction is $Ag \rightleftharpoons Ag^+ + e^-$; $Ag^+ + Cl^- \rightleftharpoons AgCl$. Because the Ag^+ is not released into solution but reacts on the surface, we should merge these equations to get $Ag + Cl^- \rightleftharpoons AgCl + e^-$.

Both the silver and the silver chloride have low solubility and so remain on the electrode—the flow of current (electrons in the wire and chloride ions in the solution) affects the ratio of silver to silver chloride on the surface of the electrode.

Applying a DC current for a long time can strip all the AgCl off of one electrode, changing the characteristics of the electrode. When there is no silver chloride left on the surface, the reaction for electrons leaving the wire changes to $2H_3O^+ + 2e^- \rightleftharpoons H_2 + 2H_2O$. Even applying a DC current for a fairly short time can change the Ag/AgCl ratio on the surface, changing the electrode characteristics, so the DC component of the current is kept very small when Ag/AgCl electrodes are used for measurement.

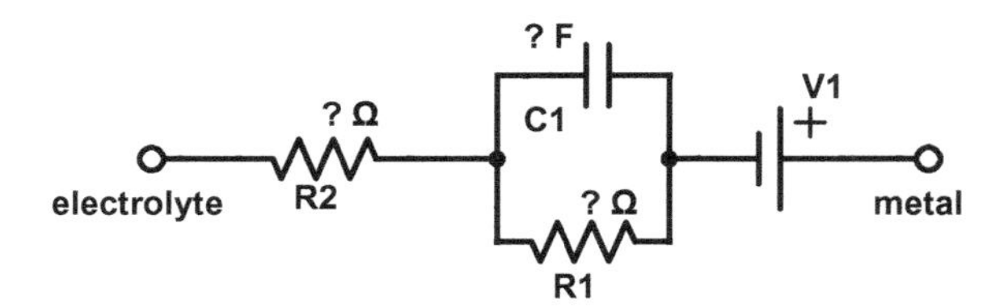

Figure 38.2: *Simplified equivalent circuit for a polarizable electrode. I don't know the half-cell potential—different sources give different redox potentials for stainless steel in salt water. Luckily, for our application the half-cell potentials cancel, because we are using the same materials for both electrodes in the same concentration of electrolyte.* R_1 *and* C_1 *should correspond to the electrode/electrolyte interface—*R_1 *would be infinite for a perfectly polarizing electrode and 0 for a perfectly nonpolarizing electrode.* R_2 *models the bulk resistance of the electrolyte—it should decrease roughly inversely to the concentration of the ions in solution. Figure drawn with Digi-Key's Scheme-it [18].*

See the Wikipedia article on silver chloride electrodes [154] for more information on these electrodes, including their half-cell potentials.

You will be making your own silver/silver-chloride electrodes by electroplating AgCl onto fine silver wires. *Fine silver* refers to silver that is at least 99.9% pure silver—there are many other grades of silver, such as *sterling silver*, which is 92.5% silver.

38.5 Modeling electrodes

Just as we modeled loudspeakers as series and parallel combinations of standard components in Lab 8, we can model electrodes with standard components. The most commonly used model for an electrode/electrolyte interface is the circuit shown in Figure 38.2. The model has three parts:

A battery for the half-cell potential: This is the part that electrochemistry classes teach about with the Nernst equation [140]. For our purposes, however, it is the least interesting part of the model. When measuring salinity by conductivity (as in Lab 12), we'll have two identical batteries back-to-back canceling each other out. When measuring EKG or muscle signals (as in Lab 13), the batteries may not cancel exactly as the electrolyte concentrations may be somewhat different on different electrodes, resulting in a nuisance DC bias that our amplifiers will have to eliminate.

Bulk resistance: R_2 in the model corresponds to the bulk resistance of the electrolyte—what we are interested in measuring in Lab 12.

Surface properties: R_1 and C_1 correspond to the surface properties of the electrode/electrolyte interface. A polarizable electrode will have large resistance, so that the interface behaves mainly like a capacitor, while a nonpolarizable electrode will have a small resistance, behaving mainly like a resistor. The resistance and capacitance come from insulating films (often oxides) that grow on the surfaces of the electrode. Steady DC current can cause the oxide film to get thicker or thinner, changing both the resistance and the capacitance.

For a pair of electrodes measuring conductivity of a solution, the two electrode-electrolyte interfaces are in series, back-to-back, so the half-cell potentials cancel (assuming the same electrolyte concentration at both electrodes). Because we are canceling out the half-cell potentials, we can lump the two electrode-electrolyte interfaces together, and just model the whole electrode-electrolyte-electrode system as a resistor in series with a parallel arrangement of a capacitor and resistor. This is the standard model used for EKG measurements, for example. In real EKG applications, the salt concentrations on the skin may vary from electrode to electrode, so the half-cell potentials may not be identical—EKGs have to be able to handle a DC component that is large compared to the signal of interest.

Exercise 38.1

Make a gnuplot function for computing the complex impedance for a pair of electrodes as a function of frequency and the three parameters R_1, R_2, C_1. (Use `definitions.gnuplot` from Figure 7.4.)
Try plotting the magnitude (absolute value) vs. frequency on a log-log scale with various values of the parameters. Try, for example, $R_1 = 20\,\Omega$, $C_1 = 10\,\mu\text{F}$, $R_2 = 10\,\Omega$ and $R_1 = 150\,\Omega$, $C_1 = 20\,\mu\text{F}$, and $R_2 = 30\,\Omega$. How can you quickly estimate the three parameters from visual inspection of the plot? These estimates may be important for fitting the model to data, as gnuplot's curve-fitting techniques can be easily confused if the initial estimates are too far off.

In some cases, the parallel resistor R_1 will be very large—stainless steel can grow a fairly effective insulating layer (mainly chromium oxide). When R_1 is very large, then the R_1C_1 time constant will be large—possibly too large for the corner frequency to be visible on a plot at the frequencies we measure. In that case, it is simplest to pretend that R_1 is infinite, and just omit it from the model (making an *ideally polarizable* electrode model).

Sometimes the parallel resistor R_1 will be quite small, and the RC time constant so small that the capacitor is irrelevant at the frequencies we are interested in. In that case, we can omit the capacitor from the model, but then we have a constant impedance $R_1 + R_2$ over all frequencies. Because we can't distinguish the surface resistance from the bulk electrolyte resistance, we can model the whole electrode-electrolyte system as just a single resistor (making an *ideally nonpolarizable* electrode model).

In the loudspeaker lab (Lab 8), we found that the loudspeaker did not behave like a standard inductor, and we had to introduce a nonstandard component that was inductor-like, but with an exponent on ω. We can use the same nonstandard component, $(j\omega\ 1\,\text{s})^{\alpha} R_{\alpha}$, but with a negative exponent to get a capacitor-like component for modeling electrodes. Remember that the 1 s multiplied by the angular frequency gives a unitless quantity, so that R_{α} is in ohms—in fact, it is the magnitude of impedance for the component at about 0.1592 Hz, when $\omega = 1$ rad/s. This nonstandard component is a generalization of inductors ($\alpha = 1$ and $R_{\alpha} = L/1\,\text{s}$), resistors ($\alpha = 0$ and $R_{\alpha} = R$), and capacitors ($\alpha = -1$ and $R_{\alpha} = 1\,\text{s}/C$).

Exercise 38.2

Make a gnuplot function that models the magnitude of impedance of a capacitor-like component, $(j\omega\ 1\,\text{s})^{\alpha} R_{\alpha}$ with $-1 < \alpha < 0$ (just the component, without series or parallel resistors).
What is the shape of the curve on a log-log plot? Under what circumstances would you use this model rather than the model of Figure 38.2? How can you get a quick estimate of the parameters α and R_{α} from a set of data points, to provide as initial values to gnuplot for fitting data?

To measure the impedance of electrodes, we'll use the same setup as in Lab 8 using two channels or two voltmeters along with a current-sense resistor to measure the voltage and current simultaneously (Figure 38.3). Polarizable electrodes may have a high impedance, which means that the input impedance of the measuring device may have an effect on the measurements.

The Analog Discovery 2 impedance analyzer can compensate (to some extent) for the meter and wiring impedances by doing two preliminary runs: one with the electrodes shorted together (to compensate for wiring inductance) and one with the electrodes not in the solution (to compensate for wiring and meter capacitance). These compensation runs need to be redone every time the sense resistor or frequency range is changed.

Exercise 38.3

Use gnuplot to plot the ratio of the two voltmeter channels (electrode voltage/resistor voltage) for the measurement circuit on the left side of Figure 38.3, assuming that the electrodes are $4\,\text{k}\Omega + (5\,\text{M}\Omega \parallel 10\,\text{nF})$ and the sense resistor is $10\,\text{k}\Omega$.
Plot the ratio from 10 Hz to 1 MHz under two assumptions: that the voltmeters are ideal, infinite-impedance devices and that the voltmeters are $1\,\text{M}\Omega \parallel 30\,\text{pF}$.

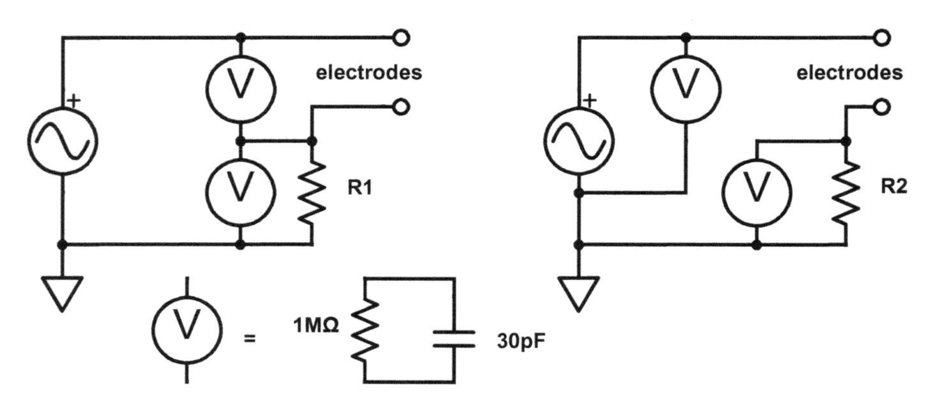

Figure 38.3: *Two circuits that we can use for measuring the impedance of electrodes in Lab 12. The circuit on the left uses one channel to measure voltage across the electrodes and the other to measure current. The circuit on the right is suitable for use with instruments that assume that the channels have a shared reference.*
The setups are essentially the same as we used for measuring loudspeaker impedance in Lab 8, but the impedances of the electrodes may be much higher, requiring a larger sense resistor R_1 or R_2. That means that the impedance of the voltmeters (or oscilloscope channels) may matter—particularly at high frequencies, where the capacitance may result in surprisingly small impedances for the meter. The Analog Discovery 2 has about 30 pF of capacitance on the input channels when using the signal cable assembly.
Figure drawn with Digi-Key's Scheme-it [18].

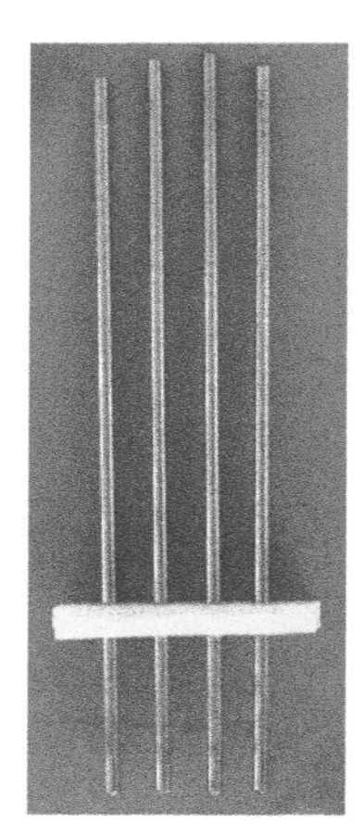

(a) *Four electrodes in a row for resistivity measurement. Outer two provide current, inner two measure voltage.*

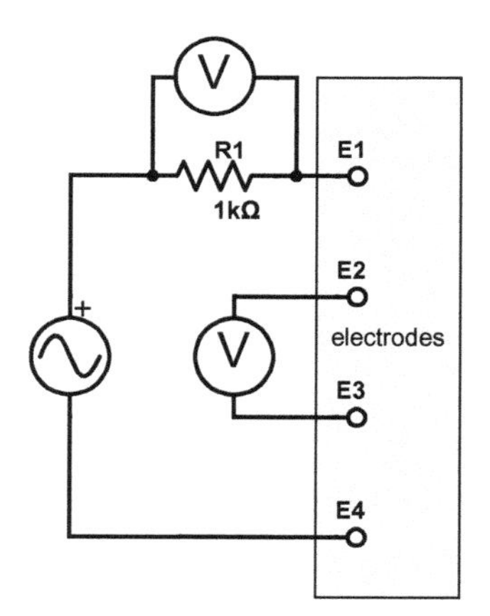

(b) *The circuit for measuring resistivity with four electrodes is similar to the circuit of Figure 38.3 for two electrodes. Figure drawn with Digi-Key's Scheme-it [18].*

Figure 38.4: *To measure resistivity of a solution with less effect from electrode impedance, one can use four electrodes in a line. AC current is passed through the outer pair. The voltage is measured across the inner pair. Because there is essentially no current through the middle two electrodes, the surface chemistry there does not affect the readings.*
Figure drawn with Digi-Key's Scheme-it [18].

38.6 Four-electrode resistivity measurements

Although this chapter is mainly about the impedance of electrodes, it is worth mentioning a slightly different approach for measuring conductivity, using four electrodes instead of two, for which the impedance of the electrodes is less important.

The four electrodes are arranged in a straight line, and a measured current is passed through the solution being tested using the outer pair of electrodes. The voltage is then measured using the inner pair of electrodes, as shown in Figure 38.4. Because the current needed for voltage measurement is very small, the impedance of the middle electrodes does not affect the measurement much. As a result, the ratio of the voltage across the middle electrodes and the current through the outer electrodes is nearly constant across a wide range of frequencies, even for polarizable electrodes.

The voltage across the middle electrodes is smaller than across the outer electrodes (which would be the voltage used for two-electrode measurements), by a factor of about four. As with the two-electrode measurements, calibration of the probe is done with a solution of known conductivity, rather than by theoretical calculations from the geometry.

Chapter 39

Lab 12: Electrodes

Bench equipment: Analog Discovery 2, stainless-steel electrodes, wire holder, cups, secondary containment, stock NaCl solutions ($\approx$150 mL/station), concentrated NaCl for plating ($\approx$250 mL/station), micrometer, calipers, thermometer (optional)

Student parts: silver wire, EKG electrodes, resistors, breadboard, clip leads

39.1 Design goal

This lab is not a design lab but a measurement and modeling lab. The goal is to provide equivalent circuits for two different types of electrodes in salt water: stainless steel and silver/silver-chloride. In addition, we'll look at the effect of electrolyte concentration on the impedance of the electrodes. As with Lab 8, we'll be looking at impedance over a range of frequencies. This experimental technique is sometimes referred to as *impedance spectroscopy* and is a common way to characterize the properties of an electrochemical system.

You will also learn how to make your own silver/silver-chloride electrodes out of fine silver wire.

39.2 Design hint

Since this is a measurement lab, rather than a design lab, the main things to think about ahead of time are how to make and record the large number of measurements efficiently. You should also set up your plotting and fitting scripts ahead of time, so that you can see how good your data is as you collect it.

> Because the electrodes vary so much, we do not have solid theory to guide us on the choice of sense resistor to use as a known impedance. It may be best to set up your breadboard so that a choice of resistors can be made by moving just one wire, and run the short/open compensation for each resistor before making any measurements.
>
> Be sure to draw a schematic of your setup!

39.3 Stock salt solutions

When doing this lab on-campus, we have wet-lab staff make up stock NaCl solutions using lab-grade NaCl and lab-grade pure water. We generally have four different concentrations: 1 M, 0.1 M, 20 mM, and 5 mM. In addition to these stock solutions, students also measured tap water, which in Santa Cruz usually has slightly higher conductivity than the 5 mM NaCl solution.

For at-home labs, you will have to make up your own stock solutions, generally without access to lab-grade materials or tools. For the purposes of this lab, it is not essential to have well-calibrated stock solutions, so you can make do with table salt and a bottle of distilled water from the grocery store.

Different sources of salt are used for culinary purposes and none of them are pure NaCl. Sea salt has a wide mix of different salts, and commercial table salt usually adds NaI to prevent thyroid problems from iodine deficiency and adds silica (very fine sand) to prevent clumping. The result is that you can't have a really good idea of the concentration of salts that you get by adding even carefully measured culinary salt into water, whether you measure by weight or volume.

American kitchens generally do not have accurate scales, unlike many European kitchens, so I'll provide advice here for making stock solutions measuring only volume. I measured out sea salt with US tablespoon measures and found that I was getting about 16.45 g/Tbsp (with a standard deviation about 0.25 g/Tbsp). I found an online converter between volume and mass for various culinary salts [92], and my measurements were roughly consistent with their estimate of 16.8 g/Tbsp for sea-salt crystals, though not with their estimate of 14.75 g/Tbsp for fine-ground sea salt.

A collection of five 1 L clean soda bottles (anything in the range 350 mL to 2 L) is useful for making and storing your stock salt solutions.

Because sea water has about 35 g/L of salt to get a conductivity of about 5 S/m at 25°C [100], we can approximately reconstruct sea-water concentration with about $4\frac{1}{4}$ Tbsp ($12\frac{3}{4}$ tsp) in a 2 L bottle of distilled water. To get well into the nonlinear region of conductivity vs. concentration, we'd like our highest concentration to be two or three times as high. If you are using sea salt, 5 Tbsp dissolved in 1 L distilled water makes a good concentrated stock solution. Measure and record what you use (including what type of salt you use) to make the concentrated solution, but don't worry about exactly matching what others do.

Use that stock solution and dilute a portion of it 5:1 (say $\frac{1}{4}$ cup solution with 1 cup distilled water), to get a 0.2 strength solution. Do this dilution repeatedly to get a series of 1, 0.2, 0.04, 0.008, and 0.0016 times your initial concentration. Using distilled water, rather than tap water, is important to get low concentrations of dissolved salts. In most parts of the country, your lowest concentration solution should have lower conductivity than tap water.

Serial dilution is easy, but errors are cumulative, so your lowest concentration solution might be rather inaccurate. If you have other measuring instruments (such as syringes for measuring pet medication), you can do direct dilution to make your stock solutions (1.6 mL of stock solution and enough distilled water to make up 1 L for the lowest concentration, for example).

39.4 Pre-lab assignment

Pre-lab 12.1

Review your electrochemistry and read up on Ag/AgCl electrodes. Try to find some useful online sources other than those mentioned in Chapter 38—list them, and be sure to cite them properly in your lab write-up.

Pre-lab 12.2

Compute the surface area of the stainless-steel electrodes you will use. The stainless electrodes are about 2.7 cm long to the plastic spacer and 1/8″ (3.18 mm) in diameter. The spacer should hold them 2 cm apart.

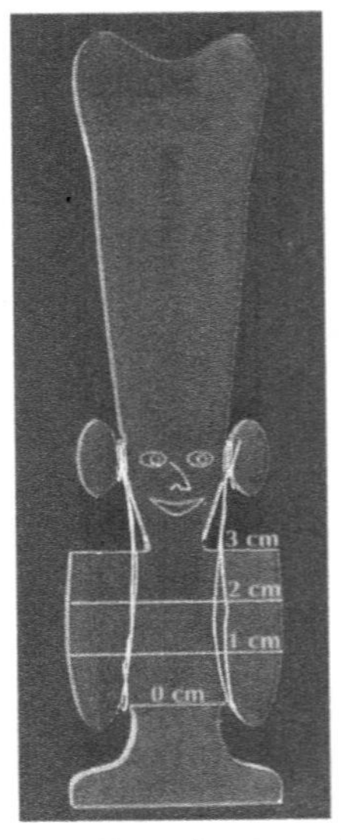

(a) *Electrode holder showing silver wire wrapped around it. Cut the silver wire into two pieces 6″ (15 cm) long. Wipe any finger grease off of it with a paper towel. Wrap one wire around the left ear, through the left armpit, and back up around the ear again. Do the same with the other wire on the right ear.*

(b) *This side view of the electrode holder in tinted water shows the path of the wire around the holder. The yellow highlight shows how having the water depth to the 2 cm line gives a length of 4 cm plus the thickness of the holder.*

Figure 39.1: *The electrode holder makes it relatively easy to get the same area of the Ag/AgCl electrodes immersed in different concentrations of salt water.*

For labs on campus, calipers and micrometers should be available to measure the electrodes more accurately. The depth-gauge portion of the calipers (the stiff metal part that sticks out the end of the calipers) is particularly useful for measuring the length of the electrodes to the plastic spacer. For at-home labs, you may have to make do with less precise measurements using rulers.

The wire holders used at UCSC for the silver/silver-chloride wires are laser-cut from acrylic sheet that is either 4.4 mm or 5.3 mm thick (depending on when they were made). Fine silver wire is wrapped around them in two places as shown in Figure 39.1(a), and the holder immersed in salt water up to one of the depth lines, which are 1 cm, 2 cm, or 3 cm above the bottom of the wire, so a U-shaped piece of the wire is in contact with the electrolyte, as shown in Figure 39.1(b). The U-shaped wire can be modeled as if it were a straight wire of the same diameter and total length, ignoring the bends, which don't change the area enough to make a difference.

Pre-lab 12.3

Compute the length of wire that is immersed and the surface area of the Ag/AgCl electrodes you will use. Assume that the fine silver wire in your kits is 24-gauge AWG (look up the diameter), and you will be folding it around a wire holder that could be either 4.4 mm or 5.3 mm thick and you will be immersing the wires to a depth of 1 cm, 2 cm, or 3 cm (depending on which calibration line you use on the electrode holder).

Do sanity checks on your computation—make sure you are computing area, not volume. Make sure that the units are consistent—the area of a small length of wire is not going to be the size of your laptop!

39.5 Procedures

Warning: salt water and expensive electronics equipment do not play nicely together. All liquids must be kept contained in secondary containment tubs throughout the lab. Do not remove the cups from the tubs!

39.5.1 *Characterizing stainless-steel electrodes*

The impedance of the electrodes will depend on the condition of the oxide coat on the stainless steel, which varies a lot even among the very similar electrodes used in lab, so you will want to do your entire set of measurements with one set of electrodes in a single lab session.

The electrolyte conductivity is temperature-dependent, so measure and record the temperature of the solutions, if you have a thermometer available.

We'll want to make the measurements starting with the lowest concentration of salt water and working our way up, because we'll be reusing the same electrodes for each different concentration.

Small amounts of solution left on the electrodes will change the concentration of the next solution slightly, and adding small amounts of water to a high concentration will change the results less than adding small amounts of salt to a low concentration.

The measurements will be made in much the same way as for the loudspeaker measurements of Lab 8, but we'll want to go to lower frequencies (down to 1 Hz) and we don't expect to see a resonance peak like we did with the loudspeakers. Lower frequencies take longer per measurement and we don't need fine resolution of frequencies, so doing 10–20 measurements per decade is probably sufficient for this lab.

Because the oscilloscope impedance may matter, we'll want to use both open and short compensation with the Analog Discovery 2 impedance analyzer.

The default 1 kHz–1 MHz range for the impedance analyzer is not appropriate for this lab. Many of the phenomena we want to model occur at very low frequencies—below 50 Hz and possibly below 10 Hz.

You can use the default range to do a quick sweep to get an estimate of the impedance for choosing a sense resistor, but reset to a range that includes low frequencies before doing the open-circuit and short-circuit compensation.

Remember that you can use sense resistors that are not powers of 10 (see Section 30.3.1.1).

Start with tap water (which will have a low electrolyte concentration, though possibly not the lowest). Immerse the electrodes up to the plastic spacer, as shown in Figure 39.2, so that you have a known area of electrode submerged. You can rest the alligator clips on the rim of the cup to help stabilize the electrodes.

Pick an arbitrary sense resistor, and do a quick sweep with the impedance analyzer from 100 Hz to 1 MHz—don't bother doing open and short compensations for this sweep, as it is just to get a rough idea of the range of impedances, not to make a careful measurement. Once you have done a quick

Figure 39.2: *Stainless-steel electrodes made from 1/8″ 316L stainless-steel TIG welding rod, showing proper depth for immersing electrodes to get repeatable surface area and use of alligator clips to stabilize the electrodes in the cup.*

sweep, choose a sense resistor that is near the midrange of the measurements from the quick sweep. For example, if your measurements range from $10\,\Omega$ to $2\,\mathrm{k}\Omega$, you might want to choose $100\,\Omega$ as your sense resistor.

The reason for picking a sense resistor near the middle of the possible range is that you want voltages across the sense resistor and across the electrodes to be similar in magnitude. If the resistors are close to the same size, then the voltage ranges on the two channels are about the same. A 10:1 ratio of impedances is fine, but a 1000:1 ratio results in very small voltages on one of the channels and serious quantization error.

Your sense resistance does not need to be a power of 10, though WaveForms 3 has those values as defaults. You can change the resistance values in the compensation pull-down dialog box.

If one of the two voltages is not measured directly, but computed by subtraction in the impedance analyzer, then it will have a larger measurement error—if you can arrange for that voltage to be the larger of the two voltages, you will probably have slightly more accurate measurements. The W1-C1-DUT-C2-R-GND setting measures the voltage across the sense resistor directly with Channel 2, but the voltage across the electrodes (the device under test) is $V_{c1} - V_{c2}$ and so has somewhat larger error. Similarly the W1-C1-R-C2-DUT-GND setting measures the voltage across the electrodes directly, but the voltage across the sense resistor is $V_{c1} - V_{c2}$. The W1-C1P-DUT-C1N-C2-R-GND setting measures both voltages directly.

Your open-circuit and short-circuit compensation are intended to correct for the capacitance and inductance of your wiring, so you should make the wiring consistent by twisting together the wires to the electrodes. You should also make the open circuit or short circuit right at the electrodes (disconnecting one alligator clip or connecting both alligator clips to the same electrode), so that the parasitic capacitance and inductance for the compensation are as close as possible to the capacitance and inductance of the actual test.

Twist your clip leads together. After choosing your sense resistor, do open and short compensation with that sense resistor, and make the open circuit or short circuit right at the electrodes. Measure the impedance of the electrodes across a wide range of frequencies.

You can record either magnitude and phase of impedance or real and imaginary parts. Either allows fitting models, but it may be slightly easier to work with magnitude and phase information, as we can do a lot of our fitting with just magnitude information.

If you don't have an Analog Discovery 2, you can make measurements with just a voltmeter, but your frequency range will be much more limited—cheap hand-held meters are often designed for AC measurements around 50 Hz–60 Hz and may not work outside 10 Hz–1000 Hz. You also will get only magnitude information, not phase information. If measuring by hand, you should measure at several frequencies per decade, roughly uniformly spaced in log space (say, 3, 5, 7.5, 10, 15, 20, 30, ... , 300E3, or using the same E6 series used for resistors and capacitors in Table 4.2). Sorry, you'll have to record this by hand—I've not figured out a reasonable way to automate AC measurements using PteroDAQ.

Repeat the measurements for the other salt concentrations, starting from the lowest and going to higher concentrations (so as not to transfer salt left on the electrodes into the lower concentrations).

You should do a quick sweep with the sense resistor for your previous concentration to see if it is still suitable for the higher concentration. Change sense resistors if necessary.

39.5.2 *Interpreting results for stainless-steel electrodes*

Plot the magnitude of the impedance $|Z_{\text{electrode}}|$ as a function of frequency on a log-log plot. Fit the models developed in Chapter 38 to the data. Remember to plot the data and the models on the same plot, so that the reader can see how good (or bad) the fit is. Showing just the data or just the model is not very useful, and showing them in separate plots defeats much of the point of a visual representation, which is to take advantage of your visual system to make sense out of the data. You can aid your reader's visual interpretation of the plots by using points for the data and lines for the models, color-coding both according to the concentration of the NaCl.

The shape of the curve tells you which models are worth fitting—choosing appropriate models is part of the assignment! For example, if the impedance is constant, independent of frequency, then a single resistor, R, is all you need for the model, but if there is a steep drop at low frequencies before a constant value, then you need a resistor in series with a capacitor, $R + C$. If there is a flat region, and then a drop with increasing frequency, you need a resistor in parallel with a capacitor, $R \parallel C$. If there is a flat region, a dropping region, then another flat region, you want both series and parallel resistors, $(R_1 \parallel C_1) + R_2$.

Remember that what you are fitting is a model for a *pair* of electrodes, not a single electrode. You have no data that corresponds to only one electrode.

For each type of model, make one plot with all the concentrations. Color-code the model fits for a given concentration the same as the data for that concentration and report the values of all the parameters of the model in the legend for the plot.

Which parameters change a lot with concentration? which only a little? If you were to use these electrodes to measure the conductivity of the solution, what frequency or frequencies would you want to measure at, and why?

Interpret the model(s) you fitted to say what the bulk electrolyte resistance is. How does that resistance depend on concentration? Can you come up with a formula to convert your measurements to NaCl concentration?

Use the known concentrations and the temperature to determine what the conductivity of the electrolyte should be, and make a plot of the resistance from your model fits vs. the "correct" conductivity. How can you convert your measurements into conductivity units (μS/cm)?

What is the conductivity of the tap water? What is the ionic concentration of the tap water, pretending that it is also an NaCl solution (even though it isn't)?

39.5.3 *Electroplating silver wire with AgCl*

Wipe the silver wire to make sure all finger oil has been removed. Wrap the silver wire around the electrode holder as shown in Figure 39.1(a). Wipe the wire again after winding it around the electrode holder.

You will be plating AgCl onto the anode, with one molecule of AgCl created for each electron transferred to the anode, so you want to measure the total charge transferred. The easiest way to measure the charge is to use a constant current and time how long it runs.

In past years, I had students measure the current with a series resistor and adjust the voltage until they got the current they desired, but many did a poor job of this control—making mistakes like assuming the voltage across the electrodes was fixed or needed to be a certain value and failing to keep the current even approximately constant.

Instead of relying on human beings to adjust the voltage to keep the current constant, we will instead use a simple analog circuit, as shown in Figure 39.3. The circuit is a simple negative-feedback amplifier, with the pair of electrodes as the feedback element. Whatever current goes through the electrode also goes through the sense resistor, and the negative-feedback loop tries to keep the voltage across the sense resistor equal to the input voltage. The input voltage controls the current, with $I = V_{\mathrm{ir}}/R_{\mathrm{sense}}$, so the amplifier is called a *transconductance amplifier*.

The design in Figure 39.3 requires that the op amp provide all the current, and the voltage across the electrodes is limited by the power supply to the op amp. If we use a 3.3 V supply (say from the Teensy LC), then an MCP6004 op amp can only provide about 10 mA–11 mA with this circuit, even with a short-circuit load. A 1 mA output requires a 100 mV input. That input can be provided either by an external source (such as a function generator set to DC) or by a voltage divider. Luckily, for this lab the voltages and currents available from a 3.3 V power supply are plenty.

You want to electroplate your wire at about 1 mA/(cm)^2, and you computed the area of each electrode in Pre-lab 12.3, so you should be able to figure out what current you need and what input voltage to provide your circuit.

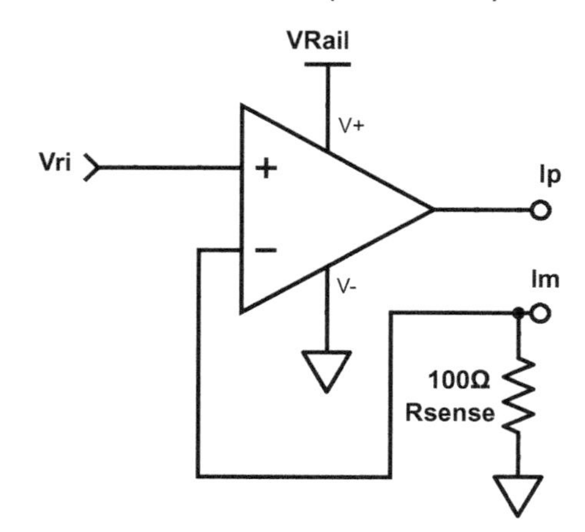

Figure 39.3: *This circuit provides a controlled current from Ip to Im. The current is* $I = V_{\mathrm{Im}}/R_{\mathrm{sense}} = V_{\mathrm{ir}}/R_{\mathrm{sense}}$. *For* $R_{\mathrm{sense}} = 100\,\Omega$, *a* $100\,mV$ *input specifies a* $1\,mA$ *current. The current is limited by the current capabilities and by the voltage limitations of the op-amp output.*

> Before connecting up the electrodes, connect the Im and Ip nodes to analog inputs on the Teensy, so that you can record the voltage across the electrodes and the current through them as you do the electroplating.
>
> You can test your constant-current circuit by putting a resistor, R_{test} in place of the electrodes. The voltage at Im should match your input voltage and the difference between the two nodes should be IR_{test}. The circuit should work as long as $R_{\mathrm{test}} < (3.3\,\mathrm{V} - V_{\mathrm{ir}})/I$. (A $1\,\mathrm{k}\Omega$ resistor makes checking the current easy, as the voltage across the resistor is just the current in milliamps.)

Put the electrode holder upright in a cup within the secondary containment tub, and carefully add a concentrated NaCl solution until it is past the shoulders but not deep enough for the alligator clips to touch the solution. This NaCl solution does not have to be a precise strength or even particularly pure, as long as there is a large Cl^- concentration. I usually use tap water with ordinary table salt to make this plating solution for the class, though using a purer salt solution would be more normal procedure if the electrodes were to be used in an experiment.

> Start recording the voltage and current before making the final electrical connection to the electrodes.
>
> Time the plating! Remember to record in your notes how long and at what current you did the plating, so that someone else could replicate it.
>
> After a couple of minutes you should have a fairly uniform dark grey coating of AgCl on the positive wire. What matters is the charge transferred, as each electron transferred from the cathode to the anode results in one AgCl molecule being formed at the anode.

If the plating is spotty or mottled, it probably means that your silver wire was not clean—oil from your fingers can block the salt water from properly contacting the wire. Rewipe the wire and plate some

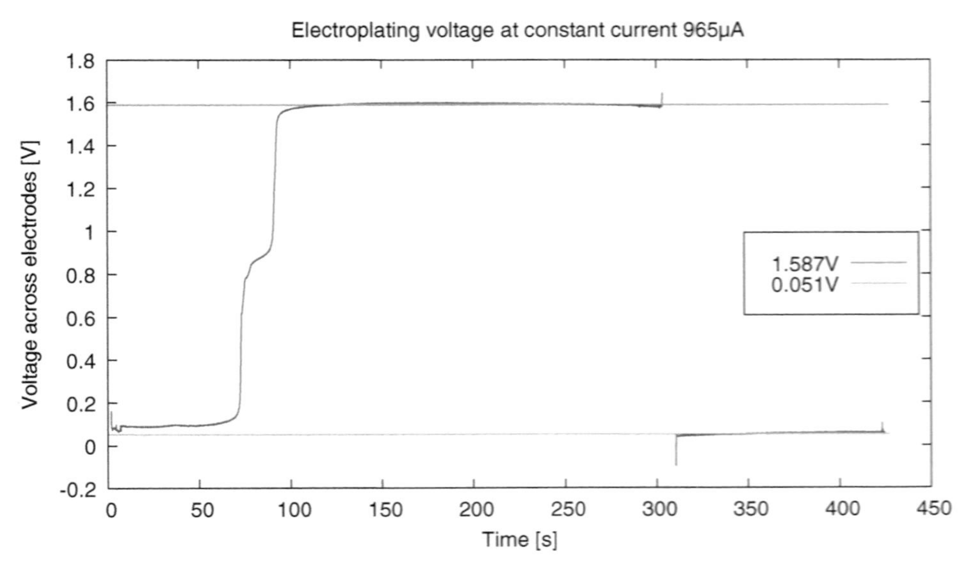

Figure 39.4: *This plot shows a record of the voltage across the electrodes during an electroplating run. Initially there was a little AgCl on the cathode, which was stripped off in the first 70 s–90 s. After about 5 min, the electrode connections were swapped, and AgCl was transferred from what was previously the anode to the new anode.*

more. If you were doing this lab with real biological samples, the wires would probably be cleaned with acetone or other organic solvent, and rinsed with deionized water, both before and after plating.

> Swap the wires and plate the other wire at the same current, which will **not** be at the same voltage! As you are plating the second wire, you are unplating the first wire. You want to run the second plating for half the total charge of the first, so that half the initial plating has been removed from the first electrode and the two electrodes are about equally plated.

Figure 39.4 gives an example of a voltage record for plating in the two directions. The reason that the voltages are different for the first and second platings is that different reactions are happening at the cathode. When the cathode has AgCl on it already, the reaction is just to turn that into Ag and release a Cl^- ion, but if the cathode has no silver-chloride on the electrode to remove, a hydrogen reaction ($2H^+ + 2e^- \to H_2$) has to happen at the cathode, and that reaction has a different half-cell voltage.

There is another method for producing the AgCl layer, by immersing the fine silver wires in chlorine bleach. The household bleach that I have is an 8.25% solution of sodium hypochlorite (with probably a little NaOH to keep the pH high and slow the decomposition of the hypochlorite ions). The reaction of silver with sodium hypochlorite is probably

$$OCl^- + 2Ag(s) + H_2O \rightleftharpoons AgCl(s) + Ag^+ + 2OH^-$$

or

$$2OCl^- + 4Ag(s) + H_2O \rightleftharpoons 2AgCl(s) + Ag_2O(s) + 2OH^- ,$$

resulting not only in insoluble silver chloride, but also either dissolved silver or slightly soluble silver oxide. Soaking the silver wires for 15 to 20 minutes in the bleach results in a uniform grey coating on the wires.

Using bleach to create Ag/AgCl electrodes is common practice in wet labs, but bleach is a bit too hazardous for a lab that has no sink and no eyewash station nearby, which is why we use the safer electroplating technique to create the AgCl layer. If you are doing the lab at home and choose to use bleach to make your AgCl layer, wear safety goggles, wear old clothes that you don't mind bleach damage to, be very careful with the bleach, and rinse the electrodes thoroughly before using them.

39.5.4 *Characterizing Ag/AgCl electrodes*

Silver chloride is sensitive to light (silver halides are the photo-sensitive substance in black-and-white film), so you want to do all your measurements of your Ag/AgCl electrodes in one session—the electrodes will change their characteristics if you do the measurements on different days.

Immerse the plated electrodes in known concentrations of NaCl (not the same solution as used for electroplating, which may be somewhat depleted of Cl^- ions), up to one of the calibrated lines on the electrode holder (1 cm, 2 cm, 3 cm at the shoulders). It may be easiest to hold the holder upright in the cup first, and gradually pour in liquid. If reagents are in short supply, using the shallowest immersion (the 1 cm line) is best, to reduce the amount of reagent needed.

Do the same characterization experiments with the Ag/AgCl electrodes that you did with the stainless-steel electrodes. If the impedance of the electrodes is low, you may be tempted to use very small sense resistors to read the current, but be careful not to exceed the current limitations of the Analog Discovery 2 function generator ($|I| < 30\,\text{mA}$).

Pre-lab 12.4

If the impedance analyzer is set to use a $\pm 1\,\text{V}$ signal and the impedance of the electrodes is small, $|Z| < 1\,\Omega$, what is a reasonable range of values for the sense resistor?

Some students try measuring the DC resistance of the electrodes with an ohmmeter. Warning: when measuring DC resistance, don't run an ohmmeter until the value settles!

Ohmmeters work by passing a known current through the device being tested and measuring the voltage. That current will strip AgCl off of one of the electrodes, changing the characteristics of the electrodes—the value won't stop changing until all the AgCl is stripped from the electrode!

39.5.5 *Characterizing EKG electrodes*

Take two of the disposable EKG electrodes and stick them together face-to-face, trying to seal them together all the way around. Clip alligator leads onto them and characterize the EKG electrodes the same way you did the home-made Ag/AgCl electrodes. You don't have any concentration here to vary, as the electrodes come with their own gel electrolyte in little sponges over the Ag/AgCl electrodes.

39.6 Demo and write-up

Because this is a measurement and fitting lab, you should report the full measurement conditions, including the test voltage, the size of the sense resistor, the sizes of the electrodes, and the depth of immersion.

Your report should include the schematic diagram of the equivalent circuit for each type of electrode, the plots and fits of the electrode impedance, and your interpretation of the results.

Plots should be very clear about what parts are data, what parts are models, and which models go with which data. Plotting the data with points and the models with lines of the same color helps, and also makes clear how many measurements you made. Keep your color coding of the concentrations consistent across all plots—changing the meaning in each figure is very confusing to the reader! Don't forget to include the values of any parameters that you fit with sufficient precision to be usable (one significant figure is not enough).

The write-up should be understandable by another engineer who has not read this assignment—in particular, you must include the formulas for the models, especially if you use the nonstandard component with impedance $R_\alpha(j\omega\ 1\,\mathrm{s})^\alpha$. Be careful when you refer to solutions with expressions like "0.1 M"—that term only makes sense if the solute is named: "0.1 M NaCl". Also, if you made your solutions using kitchen salt and imprecise measurements, it may be better to name your concentrations without claiming a precise molarity.

Some things to discuss:

- Which models fit the data best?
- Which parts of the data may be measurement artifacts, rather than real properties of the electrodes?
- What do the parameters of the models tell you about the salt water and about the electrodes?
- What is the ionic concentration of the tap water (pretending that the electrolytes in it are NaCl)?
- How do your results for stainless-steel and Ag/AgCl electrodes compare?
- What electrodes would you use and what frequency would you use if you were trying to measure the concentration of salt water electronically? (Conductivity is a standard way to measure the salinity of drinking water.)

Chapter 40

Instrumentation Amps

This chapter describes the internals of the instrumentation amplifiers that were introduced in Section 18.5, describing how they can be built out of op amps. Two popular designs are included: one using three op amps (Section 40.1) and one using only two (Section 40.2).

40.1 Three-op-amp instrumentation amp

The classic instrumentation amplifier design uses three op amps, as shown in Figure 40.1. Because both inputs connect only to op-amp inputs, the input bias currents can be made very low, particularly if FET-input op amps are used.

The design looks complicated, but we can simplify the analysis by breaking it down into two stages as shown in the block diagram of Figure 40.2.

The first stage of the three-op-amp amplifier, made with op amps U1 and U2 and the resistors R_1, R_2, and R_{gain}, takes a differential input and provides a differential output. Whatever gain is done by the amplifier is done in the first stage.

The second stage, made with op amp U3 and resistors R_3 through R_6, converts the differential signal to a single voltage referenced to V_{ref}.

The first stage is a little tricky to analyze, as we have two coupled negative-feedback loops. If R_{gain} were infinite, the two op amps would not be coupled—each would be a unity-gain buffer, and we would have $V_{\text{b}} = V_{\text{p}}$ and $V_{\text{a}} = V_{\text{m}}$. With a finite R_{gain}, we have to look at voltage dividers in the feedback loops. Because the negative-feedback loops keep the two inputs of U1 at V_{m} and the two inputs of U2 at V_{p}, we have

$$V_{\text{m}} = \frac{R_{\text{gain}} V_{\text{a}} + R_1 V_{\text{p}}}{R_{\text{gain}} + R_1} \,,$$

$$V_{\text{p}} = \frac{R_{\text{gain}} V_{\text{b}} + R_2 V_{\text{m}}}{R_{\text{gain}} + R_2} \,.$$

If we match the internal resistors, $R_1 = R_2 = R$, we can simplify the equations to

$$V_{\text{a}} = \frac{V_{\text{m}}(R_{\text{gain}} + R) - V_{\text{p}} R}{R_{\text{gain}}} \,,$$

$$V_{\text{b}} = \frac{V_{\text{p}}(R_{\text{gain}} + R) - V_{\text{m}} R}{R_{\text{gain}}} \,.$$

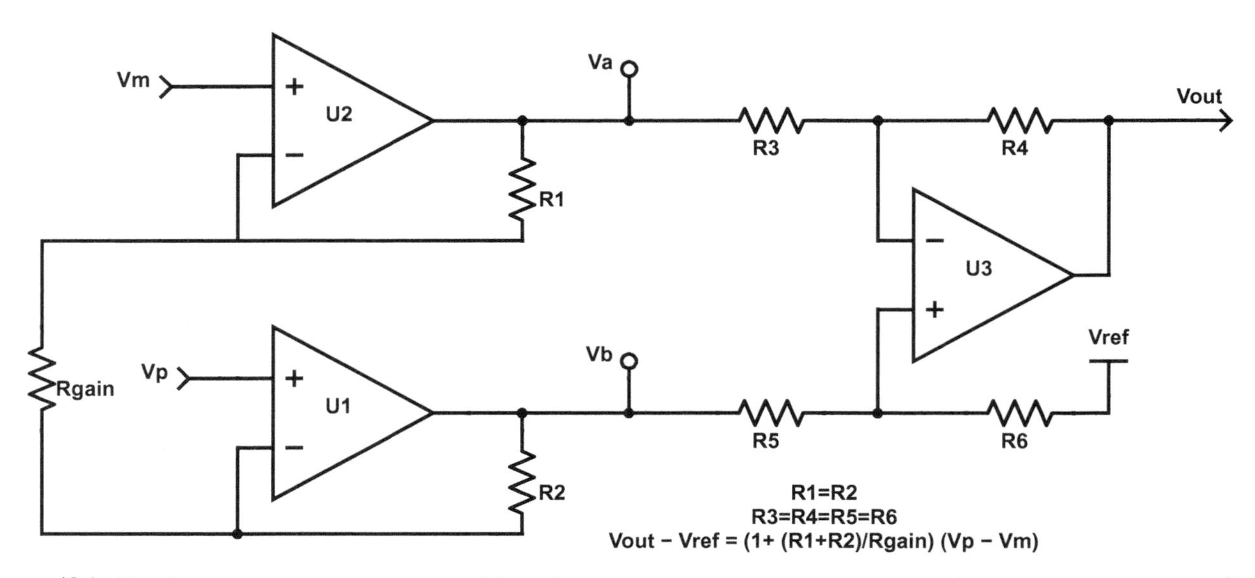

Figure 40.1: *The 3-op-amp instrumentation amplifier relies on a single external resistor to set the gain, with gain* $= 1 + (R_1 + R_2)/R_{\text{gain}}$.
Figure drawn with Digi-Key's Scheme-it [18].

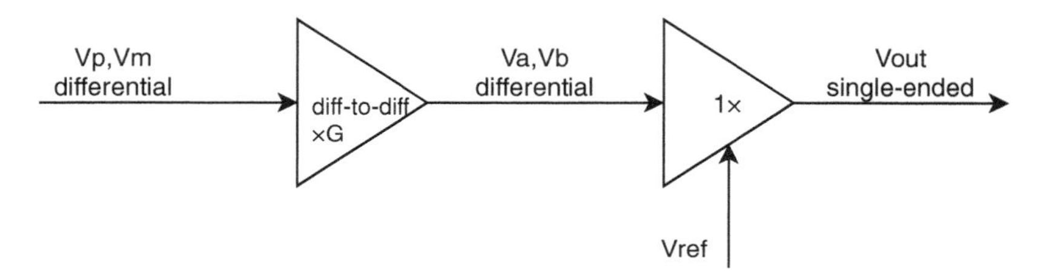

Figure 40.2: *The 3-op-amp instrumentation amplifier can be analyzed as two stages: the first stage takes a differential input and provides a differential output. Whatever gain is done by the amplifier is done in the first stage. The second stage converts the differential signal to a single voltage referenced to* V_{ref}.
Figure drawn with draw.io [21].

Let's look at the sum and difference of these intermediate signals:

$$V_{\text{b}} + V_{\text{a}} = V_{\text{p}} + V_{\text{m}} \ ,$$

$$V_{\text{b}} - V_{\text{a}} = (V_{\text{p}} - V_{\text{m}}) \frac{R_{\text{gain}} + 2R}{R_{\text{gain}}} \ .$$

so the *common-mode signal* (the average of the two inputs) is unchanged, but the *differential signal* is amplified by $1 + 2R/R_{\text{gain}}$.

The second stage is easy to analyze as a simple negative-feedback amplifier. The two inputs to the op amp U3 are voltage-divider outputs, with the negative input set to $\frac{R_4 V_{\text{a}} + R_3 V_{\text{out}}}{R_3 + R_4}$ and the positive input to $\frac{R_6 V_{\text{b}} + R_5 V_{\text{ref}}}{R_5 + R_6}$. Since the negative feedback keeps the inputs to the op amp essentially equal, we have

$$\frac{R_4 V_{\text{a}} + R_3 V_{\text{out}}}{R_3 + R_4} = \frac{R_6 V_{\text{b}} + R_5 V_{\text{ref}}}{R_5 + R_6} \ .$$

In the design of three-op-amp instrumentation amplifiers, it is usual to match all four of the resistors in the second stage $R_3 = R_4 = R_5 = R_6$, which allows a lot of simplification, down to $V_a + V_{out} = V_b + V_{ref}$ or

$$V_{out} - V_{ref} = V_b - V_a = (V_p - V_m)\frac{R_{gain} + 2R}{R_{gain}} .$$

Unlike the negative-feedback op-amp circuits we've used previously, there are three inputs: V_p, V_m, and V_{ref}, with the output voltage V_{out} centered at V_{ref} and varying with the difference $V_p - V_m$.

One could also get a gain of g in the second stage by setting $R_4 = R_6 = gR_3 = gR_5$, so that $V_{out} - V_{ref} = g(V_b - V_a)$, but this seems to be rarely done, possibly because it is easier to match equal resistors than to provide matching ratios of resistance.

Because the common-mode signal is not eliminated until the second stage of the amplifier, setting the gain high on the first stage limits how much common-mode signal can be tolerated—both V_a and V_b have to stay within the limits of the op amps that drive them.

Exercise 40.1

If you have a signal with $|V_p - V_m| < 4\,\text{mV}$ and want to amplify the signal to fill the range $1\,\text{V} < V_{out} < 2\,\text{V}$ for the analog-to-digital converter of the PteroDAQ microcontroller board that you are using, what gain would you need?

If the internal resistors of the instrumentation amp (R_1 through R_6 in Figure 40.1) are $30\,\text{k}\Omega$, what external gain resistor do you need? What does V_{ref} need to be?

Draw a schematic using an instrumentation-amp symbol (not showing the internal op amps of the instrumentation amp) to achieve the desired amplification. Assume that you are powering the amplifier from the 3.3 V signal from the microcontroller board.

Exercise 40.2

For the amplifier of Exercise 40.1, if the common-mode voltage $(V_p+V_m)/2$ is V_c, what is the range of voltages for the intermediate nodes V_a and V_b, assuming that the amplifier is implemented as a 3-op-amp instrumentation amp, with all equal resistors in the final op amp? What is the legal range for V_c, to keep both V_a and V_b within the legal range for the op amps U1 and U2?

40.2 Two-op-amp instrumentation amp

The instrumentation amplifier we used for Lab 5, the INA126PA from Texas Instruments, consists of 2 op amps internally, one for each of the differential inputs, as shown in Figure 40.3. At about \$2.70 per amplifier, these instrumentation amplifiers are much more expensive than op amps (the MCP6004 chips have 4 op amps for only \$0.36), but are still on the low end of prices for instrumentation amps. (Surface-mount parts are now much cheaper than through-hole parts—generally about half the price, but the ratio of instrumentation-amp cost to op-amp cost remains about the same.)

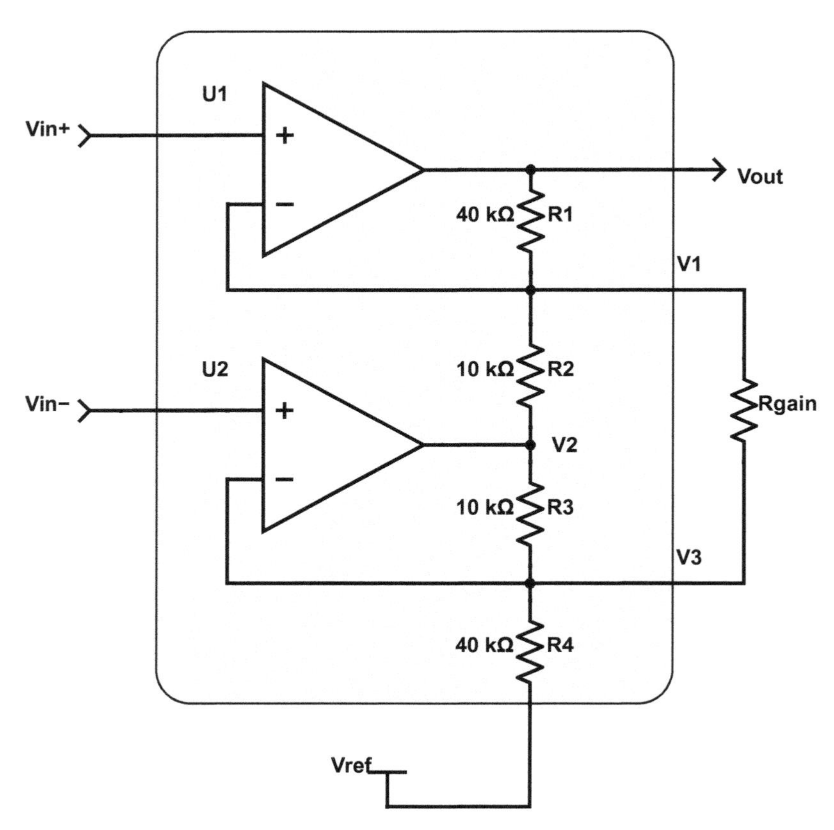

Figure 40.3: *The 2-op-amp instrumentation amplifier used in the INA126P and INA126PA chips relies on a single external resistor to set the gain, with gain* $= 5 + 80\,\mathrm{k}\Omega/R_{\mathrm{gain}}$.
Figure drawn with Digi-Key's Scheme-it [18].

Because both inputs go into op-amp inputs, with no feedback connections to the inputs, the instrumentation amp has a very high input impedance (typically $10^9\,\Omega \parallel 4\,\mathrm{pF}$) and very low input bias currents (typically $-10\,\mathrm{nA}$). Because the on-chip components have been very carefully matched, the common-mode rejection is quite good (typically a gain of $-94\,\mathrm{dB} = 20\mathrm{E}{-6}$ for the average of $V_{\mathrm{in}+}$ and $V_{\mathrm{in}-}$).

It is worthwhile to analyze the circuit for the instrumentation amp in terms of the op amps that comprise it. The negative-feedback loops will try to ensure that $V_1 = V_{\mathrm{in}+}$ and $V_3 = V_{\mathrm{in}-}$. We then have the following equations from Kirchhoff's current law for the two inverting inputs to the op amps:

$$(V_{\mathrm{out}} - V_{\mathrm{in}+})/R_1 + (V_2 - V_{\mathrm{in}+})/R_2 + (V_{\mathrm{in}-} - V_{\mathrm{in}+})/R_{\mathrm{gain}} = 0\,\mathrm{A}\ , \tag{40.1}$$

$$(V_2 - V_{\mathrm{in}-})/R_3 + (V_{\mathrm{in}+} - V_{\mathrm{in}-})/R_{\mathrm{gain}} + (V_{\mathrm{ref}} - V_{\mathrm{in}-})/R_4 = 0\,\mathrm{A}\ . \tag{40.2}$$

If we add design constraints that $R_1 = R_4$ and $R_2 = R_3$, we can subtract Equation 40.2 from Equation 40.1, getting

$$(V_{\mathrm{out}} - V_{\mathrm{ref}} + V_{\mathrm{in}-} - V_{\mathrm{in}+})/R_1 + (V_{\mathrm{in}-} - V_{\mathrm{in}+})/R_2 + 2(V_{\mathrm{in}-} - V_{\mathrm{in}+})/R_{\mathrm{gain}} = 0\,\mathrm{A}$$

or

$$(V_{\mathrm{out}} - V_{\mathrm{ref}})/R_1 = (V_{\mathrm{in}+} - V_{\mathrm{in}-})(1/R_1 + 1/R_2 + 2/R_{\mathrm{gain}})\ ,$$

which simplifies to

$$\frac{V_{out} - V_{ref}}{V_{in+} - V_{in-}} = 1 + R_1/R_2 + 2R_1/R_{gain} \ .$$

If we plug in $R_1 = 40\,\mathrm{k\Omega}$ and $R_2 = 10\,\mathrm{k\Omega}$, we get the gain equation

$$G = \frac{V_{out} - V_{ref}}{V_{in+} - V_{in-}} = 5 + 80\,\mathrm{k\Omega}/R_{gain} \ ,$$

which is what the INA126 data sheet claims for the gain [97].

As with the 3-op-amp instrumentation amp, the output voltage V_{out} is centered at V_{ref}, which is independent of V_{in+} and V_{in-}. The reference voltage is an input to the amplifier, not an output, and the amplifier will take current from V_{ref}, which should come from a low-impedance source. The current is only $I = (V_{in-} - V_{ref})/40\,\mathrm{k\Omega}$, which is less than $\pm 150\,\mu\mathrm{A}$, if $|V_{in-} - V_{ref}| < 6\,\mathrm{V}$, so we don't need a powerful voltage source—the output of a unity-gain buffer usually provides plenty of current.

Exercise 40.3

Design a 2-op-amp instrumentation amplifier that has a gain of $6 + 150\,\mathrm{k\Omega}/R_{gain}$.

The analysis above shows that the 2-op-amp instrumentation amp provides an output voltage that is independent of the common-mode voltage, but that analysis depends on perfect op amps and perfectly matched resistors. In reality, the legal common-mode voltage range is limited because of limits on the output voltage of the second amplifier—the one whose output is not visible (V_2 in Figure 40.3).

We can derive an equation for V_2 by adding the two equations we got from Kirchhoff's current law, Equations 40.2 and 40.1, instead of subtracting them:

$$(V_{out} - V_{in+})/R_1 + (V_2 - V_{in+})/R_2 + (V_2 - V_{in-})/R_3 + (V_{ref} - V_{in-})/R_4 = 0\,\mathrm{A} \ .$$

If we use the same design constraints as before, $R_1 = R_4$ and $R_2 = R_3$, and define the common-mode voltage $V_{cm} = (V_{in+} + V_{in-})/2$, we get

$$\frac{V_{out} + V_{ref} - (V_{in+} + V_{in-})}{R_1} + \frac{2V_2 - (V_{in+} + V_{in-})}{R_2} = 0\,\mathrm{A} \ ,$$

$$\frac{V_{out} + V_{ref} - 2V_{cm}}{R_1} + 2\frac{V_2 - V_{cm}}{R_2} = 0\,\mathrm{A} \ ,$$

$$V_2 - V_{cm} = -\frac{R_2}{R_1}\left(\frac{V_{out} + V_{ref}}{2} - V_{cm}\right) \ ,$$

$$V_2 = V_{cm}\left(1 + \frac{R_2}{R_1}\right) - \frac{R_2}{R_1}\left(\frac{V_{out} + V_{ref}}{2}\right) \ .$$

This formula is not really very clean, as it involves both V_2 and V_{out}, but we have already determined that $V_{out} = GV_\Delta$, where $V_\Delta = V_{in+} - V_{in-}$ and $G = 1 + R_1/R_2 + 2R_1/R_{gain}$. Substituting for V_{out}, we can then get a formula for $V_2 - V_{ref}$:

$$\begin{aligned} V_2 - V_{ref} &= V_{cm}\left(1 + \frac{R_2}{R_1}\right) - V_{ref} - \frac{R_2}{R_1}V_{ref} - \frac{R_2}{R_1}\frac{GV_\Delta}{2} \\ &= \left(1 + \frac{R_2}{R_1}\right)(V_{cm} - V_{ref}) - \frac{R_2}{R_1}\frac{GV_\Delta}{2} \ . \end{aligned}$$

The voltage V_2 depends on both the input signal V_Δ, which gets amplified by $-\frac{R_2}{R_1}\frac{G}{2}$, and the common-mode voltage V_{cm}, which gets amplified by $\left(1+\frac{R_2}{R_1}\right)$.

One interesting special case is when $R_{\text{gain}} = \infty\,\Omega$. In that case, $G = 1 + R_1/R_2$, and the formula simplifies to

$$\begin{aligned} V_2 - V_{\text{ref}} &= \left(1+\frac{R_2}{R_1}\right)(V_{\text{cm}} - V_{\text{ref}}) - \frac{R_2}{R_1}\frac{V_\Delta(1+R_1/R_2)}{2} \\ &= \left(1+\frac{R_2}{R_1}\right)(V_{\text{cm}} - V_{\text{ref}} - V_\Delta/2) \\ &= \left(1+\frac{R_2}{R_1}\right)(V_{\text{in}-} - V_{\text{ref}})\ . \end{aligned}$$

which translates the limits on V_2 into somewhat tighter limits on $V_{\text{in}-}$.

The general case is messier to analyze, but for $R_1 = 40\,\text{k}\Omega$ and $R_2 = 10\,\text{k}\Omega$, the signal gain to V_2 is $G/8$ and the common-mode gain is 1.25.

> This common-mode gain is somewhat higher than the common-mode gain of one for the intermediate nodes of the 3-op-amp instrumentation amplifier, but the signal gain at V_2 is only one eighth of the final gain, while the 3-op-amp design has the full signal gain at the intermediate nodes. That means that the 2-op-amp design generally has less restriction on its input range than the 3-op-amp design.

Chapter 41

Electrocardiograms (EKGs)

41.1 EKG basics

An electrocardiogram (abbreviated ECG or EKG—both are acceptable in English, with ECG being more common[a]) measures small differential voltages using electrodes on the patient's skin to view the electrical behavior of the heart muscles. For EKG purposes, the heart is modeled as an electrical dipole that changes its orientation and magnitude as waves of depolarization pass through the muscles of the heart.

The single-channel EKG usually uses three electrodes: the pair of electrodes that carry the differential signal and an extra body electrode that provides a reference voltage to keep the differential pair between the power rails of the instrumentation amp. The body electrode is an *output* from our circuitry to set the common-mode voltage for the differential *input* electrodes. For our purposes, it suffices to generate a V_{ref} voltage with a unity-gain buffer, as we've done for other amplifiers, for the body electrode.

The medically standard 12-lead EKG has ten electrodes: one on each ankle and wrist and six horizontally across the chest [90]. The designation "12-lead" means that there are 12 differential channels being recorded, not that there are 12 wires—six of the channels are in the vertical plane, based on the four limb electrodes, the other six are from the six chest electrodes. The standard placement requires that the patient be lying down, with arms and legs relaxed to avoid picking up electromyogram (EMG) signals from skeletal muscles in the arms and legs.

For monitoring people moving around (as in a stress EKG), the limb leads are moved in—the wrist electrodes are moved to the chest near the shoulders and the ankle electrodes are moved either to the abdomen just above the pelvic bone or to the lower back (to reduce EMG interference from the hip and thigh muscles).

In Lab 13, you won't be connecting the electrodes to your wrists and ankles, as keeping your arms relaxed is very difficult to do when you are adjusting the circuit, turning recording on or off, or typing up results. Instead, we'll use the electrode locations commonly used for emergency EKGs: in the hollows just below the collar bones avoiding the pectoral muscles and shoulder muscles, and on the abdomen just above and inside the left iliac crest (the bone at the top of the pelvis). You can then choose two of the electrodes as the signal electrodes (for leads I, II, or III) and use the third one as a body electrode, as shown in Figure 41.1. With this placement of the electrodes, only the pectoral muscles are likely to provide interference with the EKG signal, and they are fairly easy to keep relaxed when you are sitting.

[a]I use EKG, the older acronym that comes from the German name, in part because my first electrocardiogram was ordered by a Prussian pediatrician.

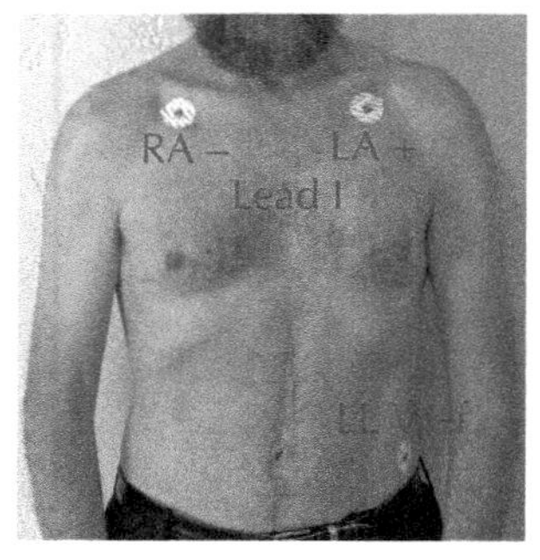

(a) *Lead I is LA–RA.*

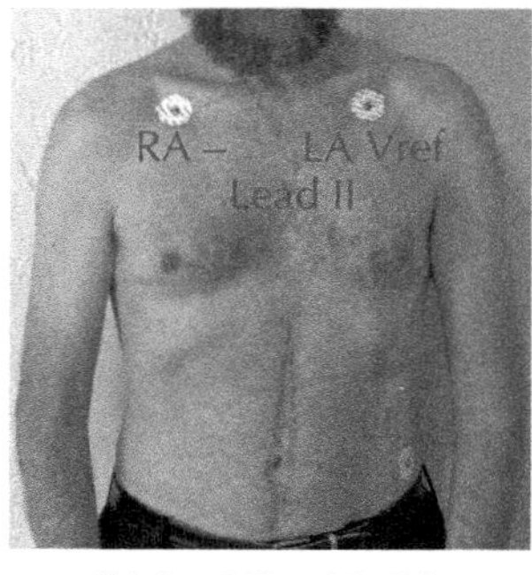

(b) *Lead II is LL–RA.*

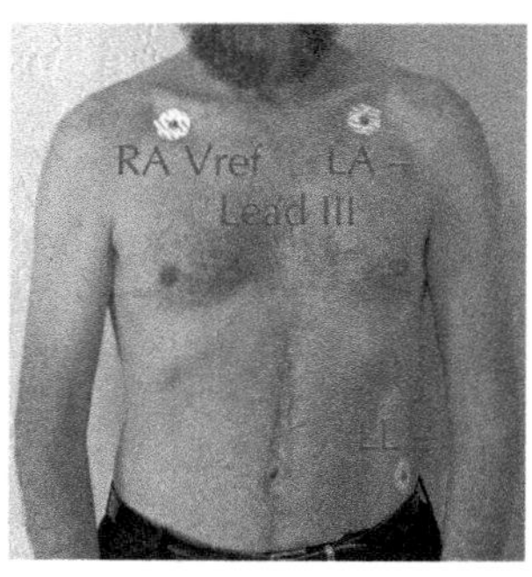

(c) *Lead III is LL–LA.*

Figure 41.1: *This figure shows the approximate electrode placement for the three electrodes. The two "arm" electrodes go just below the collarbone, trying to avoid any muscle that might add extraneous signals. The "left-leg" electrode goes just above the iliac crest (the top of the pelvic bone). If you are just recording Lead I (LA − RA), then the placement of the LL electrode is not very important—almost anywhere on the body will do. The LL electrode is important if you want to record Lead II (LL − RA) or Lead III (LL − LA).*

We will be looking at lead I (left arm minus right arm, often abbreviated LA−RA) or lead II (left leg minus right arm, LL−RA) which have the classic EKG signal shown in Figure 41.2. The 200 ms by 0.5 mV grid of Figure 41.2 is standard for EKG traces and is usually printed as a 5 mm by 5 mm square. In fact, many of the guides for interpreting EKGs refer to the voltage and time both in *cm*, assuming that doctors and nurses can't deal with millivolts and milliseconds. The standard scaling is 1 mV/cm and 400 ms/cm, which is probably not the scaling here.

The vertical spike of the large R part of the wave is typically around 0.5 mV–1 mV, depending mainly on the quality of the electrode-skin contact. The initial P wave corresponds to depolarization of the atrium, which starts the push of blood from the atria to the ventricles. The R spike corresponds to the depolarization of the ventricles, which begins the squeeze that pushes blood out of the heart. The T wave corresponds to repolarization of the ventricles. See Ashley and Niebauer's explanation for more details of the physiological events the various parts of the ECG correspond to [6].

We don't want to DC-couple the whole amplifier, as the ± 1 mV signal we're interested in is often accompanied by a ± 300 mV DC offset, due to different half-cell potentials at the skin-electrode contact. The DC blocking is usually done after the first stage of amplification by the instrumentation amp, and the initial gain is limited to avoid saturating the instrumentation amplifier with the DC offset.

Cardiac monitors usually have a passband of 0.5 Hz to 40 Hz or even narrower (1 Hz–30 Hz), and diagnostic 12-lead EKGs usually have a wider range (0.05 Hz–100 Hz or 0.05 Hz–150 Hz) [14, 86].

The wider frequency response provides a detailed look at the heart signals and is necessary for some sorts of diagnosis (such as elevated ST response), but the narrower frequency response of cardiac monitors filters out high-frequency noise from skeletal muscles and low-frequency artifacts from breathing and other motions, and so is more suitable when looking just at cardiac rhythm or cardiac rate.

Capacitive coupling of ambient electrical interference to the signal wires can be reduced by twisting the wires together and surrounding the twisted pair with a conductive shield, biased to the common-mode voltage of the wires. This common-mode bias can be generated from the instrumentation amplifier itself, by splitting the R_{gain} resistor in two resistors, each half of R_{gain}, and using a unity-gain buffer

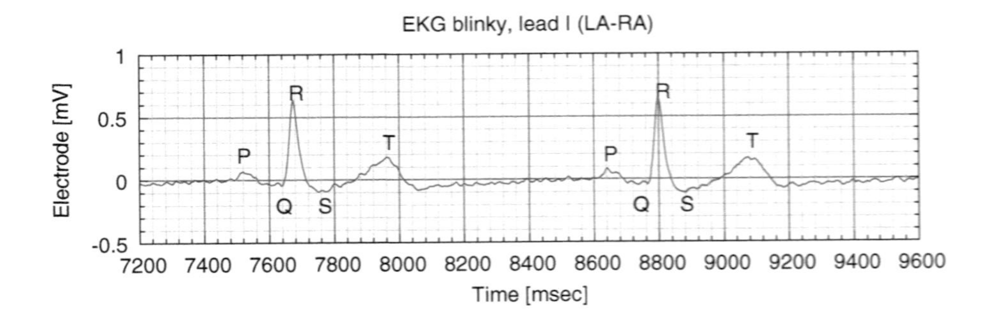

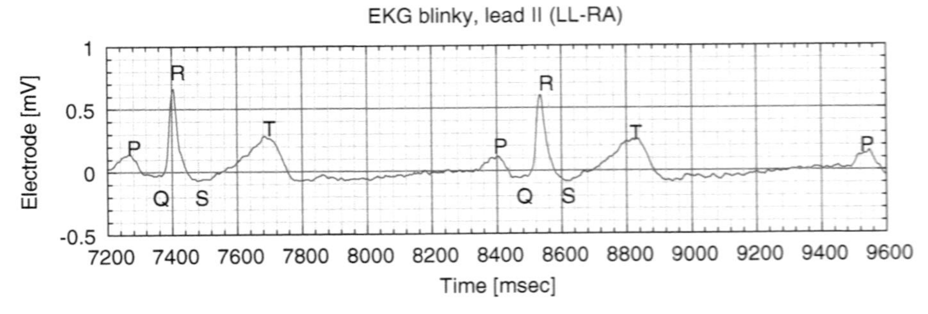

Figure 41.2: *An EKG trace of Kevin Karplus's heartbeat. Despite the same time scales, these traces were recorded at different times and correspond to different heartbeats. The period in both the traces is about 1125 ms for a pulse rate of 0.89 Hz or 53.3 beats/minute. The gnuplot commands to create a grid like that seen here are given in Figure 41.3.*
Data collected with PteroDAQ using a Teensy LC board [46]. Figure drawn with gnuplot [33].

```
set margins 9,6,1,1

set xtics 200
set mxtics 5
set ytics 0.5
set mytics 5
set style line 1 linetype 1 linecolor rgb 'grey40' lw 1
set style line 2 linetype 1 linecolor rgb 'grey70' lw 0.5
set grid  xtics ytics mxtics mytics ls 1, ls 2

# depends on xrange & yrange: number of boxes high/boxes wide
set size ratio 3./12.
```

Figure 41.3: *These are the gnuplot commands used to set up the grid and change the shape of the plots for Figure 41.2. The "size" command does not guarantee that the grid will come out square—you'll have to tweak the size or the xrange and yrange to make the grid square.*

whose input is connected to the middle of the R_{gain} resistance. This shielding is not absolutely needed for an EKG—we do unshielded leads in Lab 13, but if you are trying to amplify much smaller signals, such as from an electroencephalogram (EEG), then shielding becomes more important.

Common-mode noise can also be reduced by using an active bias circuit, instead of just biasing the body electrode to V_{ref}. Supposedly, feeding back the common-mode voltage with a gain of about -40 as the body bias will reduce the interference substantially, but I have not had much luck with this technique. Although the active bias circuit did indeed reduce the common-mode noise substantially, the signal on the bias wire was capacitively coupled back into the EKG signal leads, creating an interference signal that appeared as a differential voltage, not just a common-mode voltage. The result was a much, much noisier output signal than when I just used V_{ref} on the bias electrode. For the technique to work well, it may be necessary to shield the EKG signal pair (with the shield set to the common-mode voltage) and run the bias wire outside the shield, to minimize coupling to the signal pair.

41.2 Safety

Medical equipment, such as an EKG machine, is required to be isolated from the line voltage, to avoid the dangers of electric shock. Blood and muscle are good conductors of electricity, but skin is not (think of your body as a bag of salt water—the bag is not conductive, but everything inside is). Cuts in the skin eliminate the resistance of the skin, making the body much more conductive.

AC voltages are more dangerous than DC ones, in part because of capacitive coupling through the skin, and in part because muscles and nerves respond more to varying voltages than to constant ones. It only takes a small current through the heart to cause problems: about 30 mA at 60 Hz or 300 mA–500 mA DC can cause fibrillation [124]. About 10 mA at 60 Hz through the arm can cause violent muscular contractions that make it impossible for the person to release a wire.

To prevent such problems, it is common to put a large resistor (like 1 MΩ) between EKG electrodes and the EKG amplifier.

All electrode leads should be connected through resistors to make sure that any current from the EKG to the body stays under 50 μA (the limit according to the ANSI/AAMI ES1-1993 standard, as reported by Company-Bosch and Hartmann [14]), even if the wires are accidentally connected to the highest voltages in the EKG circuitry. The American Heart association calls for a stricter limit, of a maximum of 10 μA "between any patient electrode and either power line ground or the accessible part of the electrocardiograph ... even in the presence of a single fault" [61].

Because the instrumentation amplifier for the EKG has a very high input impedance, this extra resistance in the signal leads does not result in noticeable changes in the voltage observed.

Adding 1 MΩ to the lead that biases the body, however, could cause problems with not providing enough current to keep the electrode voltages within the range needed by the instrumentation amplifier. The body and signal leads can have fairly large capacitive coupling to the 60 Hz ambient electrical fields, and the resulting voltage may be large enough to make the common-mode signal fall outside the range your instrumentation amplifier can handle.

Think of your body as the output node of a voltage divider, with a capacitor to a 60 Hz source and the bias-lead resistor to virtual ground—if the capacitance and resistance are both large, then the 60 Hz voltage on the output will be large. Reducing the resistance reduces the signal on your body.

Reducing the resistance of the bias lead can result in safety concerns, particularly if there is a path through your power supply to the AC power lines. The pre-lab exercise in Lab 13 has you compute the lowest resistance you can get away with. You can reduce the hazard by powering your EKG circuit from a battery (such as a laptop battery) with no connection to AC power. Battery operation also reduces power-supply noise from being injected into your electronics.

To protect circuitry or patients from higher voltages, it is common to add protection diodes to a circuit, as shown in Figure 41.4.

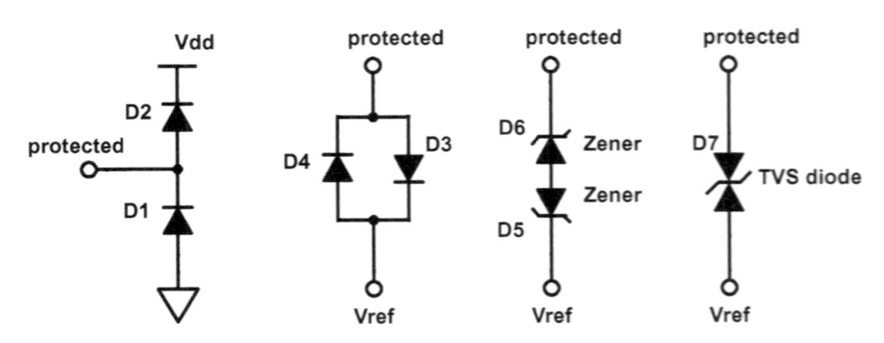

Figure 41.4: *Four diode-clamp circuits for protecting a node from excessive voltages. D_1 and D_2 limit the voltage swing to be from a little below ground to a little above V_{dd}, with the amount outside the voltage range determined by the current and the diode characteristics. This sort of protection is very common for inputs to integrated circuits.*

D_3 and D_4 limit the voltage swing to a narrow window around V_{ref}, with the amount of swing determined by the current and diode characteristics. For the tiny currents of EKGs, this clamping method may eliminate the desired signal.

D_5 and D_6 limit the voltage to a wide window around V_{ref} ($V_{ref} \pm V_{Zener}$) by using a pair of Zener diodes.

D_7 is roughly equivalent to the circuit using D_5 and D_6, but using a single transient-voltage-suppression *diode designed for clamping voltages in a specific range.*

Figure drawn with Digi-Key's Scheme-it [18].

Two of the circuits use ordinary diodes to limit the voltage range, either to the power-supply range or to a narrow range around a reference voltage. The other two use special *Zener* diodes that conduct in the reverse of the normal direction when the back bias exceeds a well-defined *Zener* voltage.

It is common for medical equipment to use optoisolators to separate the parts connected to the patient from the parts connected to the computers and power lines. An *optoisolator* is a paired LED and phototransistor (or photodiode) that communicates information via photons, rather than electrons. They provide excellent protection against stray current paths. Some optoisolators can be used for transferring analog signals, but many have a nonlinear response optimized for digital signals.

Unfortunately, optoisolators are not good for providing power to the electronics on the patient side of the isolation barrier. For that, there are two main solutions: batteries and isolated power supplies. Battery power is a cheap solution, but has reliability issues—batteries always seem to fail when they are needed most, and replacing batteries is both expensive and a nuisance. Isolated power supplies use small transformers to transfer power through magnetic fields, rather than through direct conduction of electrons. Isolated power supplies are slightly more expensive than other power supply designs (used for electronics that people do not come in contact with). Power supplies designed and tested to meet medical standards usually cost about twice as much as other power supplies with similar ratings, but even cheap wall warts usually meet minimal isolation resistance standards.

41.3 Action potentials

Where does the signal that we're seeing on an EKG come from? There are two ways to answer this: one based on what happens in individual cells and one based on the waves of activity in the heart.

> We can look at a cell membrane as a capacitor: the lipid bilayer is a good insulator, allowing no ions to pass it, and both the cytosol inside the cell and the extracellular fluid are good conductors, containing a lot of salt. Ion pumps can move charge from one side of the membrane to the other, charging the capacitor, and ion channels can let ions flow back, discharging the capacitor.

In resting states, cardiac muscle cells pump sodium and calcium ions out of the cell and potassium ions into the cell, resulting in a positive charge on the outside relative to the inside of about 90 mV. When sodium channels are opened, the sodium ions rush in, depolarizing the membrane (discharging the capacitor). This individual-cell activity is referred to as an *action potential.*

The sodium channels then close and the cell enters a plateau phase, where potassium ions flow out and calcium ions flow in. It is during this phase that the cardiac muscles contract, triggered by the calcium. The sodium action potential lasts less than a millisecond, but the calcium action potential can last for hundreds of milliseconds [108].

Then the calcium channels close and the cell rapidly repolarizes as the potassium continues to flow out. The ion pumps then gradually restore the initial balance of ions, with an excess of sodium and calcium outside the cell and an excess of potassium inside the cell. You can find a more detailed explanation of the entire process in the Wikipedia article on cardiac action potentials [114].

The action potentials refer to voltages across a cell membrane, but our EKG electrodes have no access to the inside of cells. When a cell depolarizes, the sudden influx of positively charged sodium ions increases the charge inside the cell, but decreases the charge outside the cell. As the depolarization wave travels across the heart, the tissue in front of the wave is positively charged relative to the tissue behind the wave.

> The dipole moment between the tissue in front of and behind the wave is what the EKG is detecting, so a depolarization wave traveling toward the + electrode and away from the − electrode will result in an upward (positive) spike in the EKG waveform.

The large ventricular wave traveling down the septum between the ventricles produces the R spike—the septum is usually oriented somewhat to the left and downwards, resulting in a positive R spike on leads I and II. The repolarization wave has the opposite polarity, but travels in the opposite direction, so it also results in a positive spike (the T spike). One of the best explanations of heart dipole and repolarization waves I've seen (including an animation of the depolarization wave) is on Natalie's Casebook [72].

The online *Textbook of Cardiology* has a very detailed explanation of the relationship between action potentials and EKG waveforms on the "Cardiac Arrhythmias" page [60]. A more basic explanation can be found on the Wikipedia EKG page [126], but the explanation by Ashley and Niebauer may be clearer and more informative [6].

Chapter 42

Lab 13: EKG

Bench equipment: oscilloscope, soldering station, fume extractor

Student parts: MCP6004 op amp, resistors, capacitors, breadboard, op-amp protoboard, screw terminals, EKG electrodes, clip leads, PteroDAQ

42.1 Design goal

For this lab, you will design and solder a one-channel electrocardiogram circuit, not using a prebuilt instrumentation amplifier, but making your own instrumentation amplifier out of op amps.

> This is not a medical-grade EKG, as it does not include electrical isolation, electrostatic protection for the electronics, or calibration signals. It is also only a single channel, not the 12 channels of a standard EKG.

You don't get to use the INA126P instrumentation-amp chip for this lab. You have to design and construct your own instrumentation amplifier out of MCP6004 op amps. Because you aren't using the INA126P, you don't have its output range limitations, but the much more relaxed ones of the MCP6004 chips.

42.2 Pre-lab assignment

All electrode leads should be connected through resistors to make sure that any current from the EKG to the body stays under $50\,\mu$A (the limit according to the ANSI/AAMI ES1-1993 standard, as reported by Company-Bosch and Hartmann [14]), even if the wires are accidentally connected to the highest voltages in the EKG circuitry. The American Heart association calls for a stricter limit, of a maximum of $10\,\mu$A "between any patient electrode and either power line ground or the accessible part of the electrocardiograph ... even in the presence of a single fault" [61].

> **Pre-lab 13.1**
>
> Choose the size for your current-limiting resistors on the EKG electrode wires. What is the smallest resistance you can use on the reference electrode to ensure that the ANSI/AAMI limits are met? (The signal wires can use much higher resistances if you wish.)

In Lab 12, we did not have to worry about the half-cell potential of our electrodes, since the two electrodes were very similar and shared a common electrolyte, so the half-cell potentials canceled. With

EKG electrodes, each electrode has its own electrolyte gel, but the concentration of electrolytes in the gel will vary depending on such things as how sweaty the skin is there and how salty the sweat is. Because of this difference in electrolytes between electrodes, EKG electrodes can have as much as a 200 mV–300 mV difference in their half-cell potentials.

> That means that the DC offset of your input signal could be anywhere from −300 mV to +300 mV, and this offset is in the differential voltage (not just the common-mode voltage). Your instrumentation amplifier will amplify that DC offset, and so you can't have a large gain in your first stage.

Some designs have high-pass filtering before the first stage, but those designs are difficult to get right, because of the high impedance of the voltage source—filtering after first-stage amplification is usually easier.

If you use a high-pass filter to block the DC offset, you need to consider what cutoff frequency it should use. If the cutoff frequency is too high, you can lose part of your signal, but if the cutoff frequency is too low, then the EKG will take a long time to recover from any sudden changes in the DC offset. Such sudden changes in DC offset are common in EKG signals as a result of movement (sometimes called *motion artifacts*).

As a rough rule of thumb, it takes about 5 RC time constants for a simple RC filter to recover from a sudden change, so the lower bound for your corner frequency is set by how long you are willing to tolerate a loss of signal due to a motion artifact.

Because your first-stage instrumentation amplifier can't have a large gain, you will need a second-stage amplifier, after eliminating the amplified DC offset.

Pre-lab 13.2

Figure out what gain you need to achieve to record your EKG signal with PteroDAQ. Your input signals will probably be up to 2 mV peak-to-peak, but may have a 300 mV DC offset (in either direction). The DC offset is an offset to the differential signal, not just to the common mode. Split the gain into different stages, making sure that the DC offset won't saturate any stage. You will be using your own amplifier design made from MCP6004 op amps, not the INA126P, so the limitations of the INA126P amplifier are irrelevant to this design problem.

You will most likely also need a low-pass filter in the second stage, to prevent clipping due to amplifying 60 Hz interference. A low-pass filter as an anti-aliasing filter is a good idea in any case—choose your corner frequency based on your sampling rate.

Pre-lab 13.3

Draw a block diagram for your EKG, complete with signal levels.

Because you need a second-stage amplifier, you probably can't use a 3-op-amp instrumentation amp, as you'll run out of op amps (you only get 4 on the protoboard). Your instrumentation amplifier should use resistors you have—the 40 kΩ used internally by the INA126P is not a standard value in any of the common resistor series (see Table 4.2), so don't use that value!

Pre-lab 13.4

Design your own instrumentation amplifier using op amps. Consider the design in Section 40.2, which uses only two op amps, allowing you to have two left for creating V_{ref} or second-stage amplification.

Pre-lab 13.5

Have a complete, clear schematic (preferably drawn with a tool like Scheme-it [18]) before coming to the lab.

You need to have three electrodes connected in your complete schematic, as described in Section 41.1, but many students in the past only showed two in their initial design.

Pre-lab 13.6

Plot the gain of your amplifier as a function of frequency, taking into account any low-pass or high-pass filtering in the amplifier. What is the gain at 1 Hz? at 60 Hz?

Think about the layout of your printed-circuit board—don't just throw wires and components around randomly. For example, all the electrode wires should be together, so that they can be twisted into a bundle to reduce 60 Hz pickup. Similarly, the power wires and the output wires that communicate with the microcontroller board should also all be together. There are two places on the printed-circuit board for screw terminals (or right-angle male header pins)—how will you use them effectively?

How will you assign the op amps in your schematic to the op amps in the layout? Although the four op-amp positions on the MCP6004 chip are electrically interchangeable, your wiring will be much simpler and shorter with some assignments than with others (if you use all four op amps, there are $4! = 24$ different possible assignments). Similarly, choosing the placement of your resistors can lead to either short, clean wiring or messy tangles.

Pre-lab 13.7

Design a layout for the circuit on the op-amp protoboard worksheet [50].

42.3 Procedures

You can get an idea of the signal you want to amplify by recording an EKG signal without any amplification—connect the two sense electrodes to a differential oscilloscope channel of the Analog Discovery 2 and the bias electrode to ground. (Or connect the sense electrodes to the differential inputs of PteroDAQ and use a couple of resistors (100 kΩ to 220 kΩ) to make a voltage divider for the bias electrode.)

The small signal should be just barely visible, though it may be buried in 60 Hz interference and quantization noise. After recording the signal with 1200 Hz sampling, I was able to recover the EKG signal with digital filtering (low-pass at 100 Hz and notch filter rejecting 58 Hz to 62 Hz) [49]. I don't recommend trying to use just the least-significant bits of an analog-to-digital converter like this, but it is amazing how much information can be recovered by digital filtering.

Make a breadboard version of your EKG circuit, to make sure that everything works as designed.

You can test the low-gain first stage of your EKG with the Analog Discovery 2, but to test the high-gain second stage or both stages together, you may want to make a small test fixture. The function generators are not good at producing very small signals, but you can use a voltage divider to reduce the output of the function generator. If you want to be able to control both the differential DC offset and the common-mode, you can use two function generator channels to drive the voltage divider, as shown in Figure 42.1. Make sure that the two channels are synchronized and 180° out of phase.

Disconnect your voltage divider from the instrumentation amplifier when using the EKG electrodes—the 1 kΩ resistor is such a low impedance that it would effectively short-circuit the electrodes!

If you want to use the network analyzer with the test fixture of Figure 42.1, you run into a small problem—the network analyzer of the Analog Discovery 2 controls only Wavegen 1 not Wavegen 2. You can set Wavegen 2 to be a constant voltage for the network analyzer, but then the common-mode voltage varies as well as the differential voltage (the common-mode voltage only varies by half as much as the differential voltage, but you can't tell whether you are amplifying the differential signal or the common-mode signal).

If you connect both inputs of your instrumentation amplifier to Wavegen 1, you can measure the common-mode rejection of your amplifier design. There is no need to produce very small input signals for

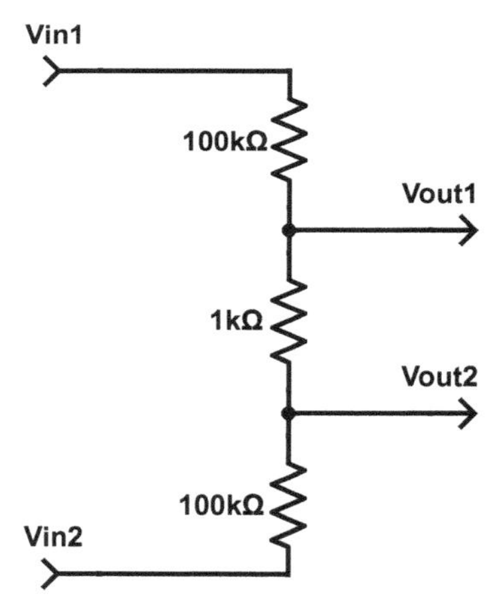

Figure 42.1: *You can make a dummy EKG signal that has only a few millivolts of differential signal with an arbitrary common mode.*

With the Analog Discovery 2, you can use two synchronized function generators, with $V_{\text{in}1}$ connected to one and $V_{\text{in}2}$ connected to the other. If the amplitudes and offsets of both are identical, with the second channel having a phase of 180°, then the output of this divider will be a differential signal with common mode equal to the average of the offsets and differential amplitude 2/201 of the input amplitudes. An amplitude of $\pm 100.5\,mV$ on each function-generator channel will result in an amplitude of $\pm 1\,mV$ on the differential output, and an offset of 1.65 V on each channel will result in a 1.65 V common mode for the differential signal.

If you have only a single function generator, then you can connect the other input to a constant voltage. The common mode will not be constant, but will be a signal with half the input amplitude. Such a test signal is still usable, but it does not allow you to see whether you are amplifying the differential signal or the common-mode signal.

Figure drawn with Digi-Key's Scheme-it [18].

this test, as your amplifier should have gain much less than 1 for common-mode signals (the INA126PA has a common-mode rejection of at least 74 dB, and your home-brew amplifier should be able to get a rejection of at least 33 dB).

> Make a wiring harness by twisting three color-coded wires together and attaching an alligator clip to each. You need to twist the wires together to reduce AC pickup by the leads. Wires should be about 100 cm (3′) long with the last 20 cm (8″) before the alligator clips not twisted together, so that the leads can spread out to the electrodes.

Prep your skin by cleaning the spots where you want the electrodes. Using some Scotch tape to peel off a layer of dead skin (part of the *stratum corneum*) is supposedly effective in reducing the resistance of the skin. Other methods include gentle abrasion with fine sandpaper [81], shaving, and non-alcohol wipes [85]. I've had good luck just washing the skin and drying with vigorous rubbing with a towel. Do dry the skin thoroughly before attempting to attach the electrode—not only does this remove any loose dead skin cells, but the electrodes don't stick well to wet skin.

> Remember that it is safer to have the EKG circuit completely isolated from the power lines. If you are using PteroDAQ or a USB oscilloscope, running off a laptop battery is safer than using a power supply.

Not only is running off of batteries safer, but some laptop power supplies also create a lot of electrical noise. Running off of batteries can reduce noise problems substantially.

Once your breadboard version is working, populate and solder your PC board. Don't take apart your breadboard—you may want to use it to confirm that your electrodes are working correctly.

Here are some hints for laying out your PC board:

- Don't solder in electrode wires or output connections—that's what the screw terminals are for. If you don't want to use screw terminals for the power and output connections, you can solder male headers sticking out the bottom of the board to plug into a breadboard (like the male headers we put on the microcontroller board).
- Pay attention to the wiring that is already present on the back of the printed-circuit board—don't short out your safety resistors!
- When deciding where to place resistors, try to put the shared nodes of series connections next to each other to keep wiring short. A long series chain can be folded up so that shared nodes are adjacent, as shown in Figure 42.2.
- If you run out of the spaces dedicated for resistors, you can use other holes on the board, but don't run resistors for long distances—the bare wires can short. Use holes either 0.1″ apart (flying resistors) or 0.4″ apart (flush to board). The feedback connections for op amps are conveniently 0.1″ apart, which is attractive for flying resistors.
- Color code your schematic and your wiring, as described in Section 13.3.
- Check the wiring carefully—a lot of errors can be trivially debugged just by counting how many things are connected together for each node. If a resistor has no connections at one end or has too many connections, something is obviously wrong.

I highly recommend that everyone solder their own EKG board (not just one per pair of partners), so that they have something that they can demonstrate to friends and relatives after the end of the course.

42.4 Demo and write-up

Using the soldered EKG board, record several seconds of heartbeat using at least a 240 Hz sampling rate (lower sampling rates may miss the narrow R spike, unless you have done some low-pass filtering to broaden the spike.)

Plot the data with properly labeled x and y axes. Be sure that your y-axis shows the voltage at the electrodes, *not* the voltage after amplification. I expect to see EKG signals that are around 1 mV, not a volt! Remember to put what lead was used for the recording in the title of each plot.

Use a digital filter (with a modification of the script from Figure 20.7) to eliminate DC drift and 60 Hz interference. If you are trying to get cardiac-monitor bandwidth, you can use a band-pass filter with cutoffs of 0.5 Hz and 50 Hz, but if you want diagnostic EKG bandwidth, you will need a wider band-pass filter and an additional band-stop filter that cuts out a narrow band around 60 Hz (say 58 Hz to 62 Hz) and possibly yet another filter that cuts out a band around 120 Hz, if your band-pass filter includes frequencies that high.

To add an additional band-stop filter after the band-pass filter in Figure 20.7, make new variables for the new cutoff frequencies, and copy the lines using `signal.iirfilter` and `signal.sosfiltfilt` for the band-pass filter. Change the variables in the `iirfilter` to the new cutoff frequencies and change the `btype` from "bandpass" to "bandstop". Change the input of the `sosfiltfilt` call from the `values` to `bandpass`, the already computed output of the band-pass filter, and save the result in a new variable `bandstop`. In the for-loop that outputs values, use the `bandstop` array rather than the `bandpass` array. You may want to play with the first parameter of `iirfilter`, which determines the order of the filter—a higher order makes for sharper drop-off at the cutoff frequencies, but too high an order can result in ringing artifacts and numerical instability.

Remember that both cutoff frequencies of a digital band-pass or band-stop filter must be below the Nyquist frequency, so if you want a filter with a cutoff frequency of 62 Hz, your sampling frequency needs to be above 124 Hz (and should be even higher, because cutoff frequencies very near 0 Hz or the

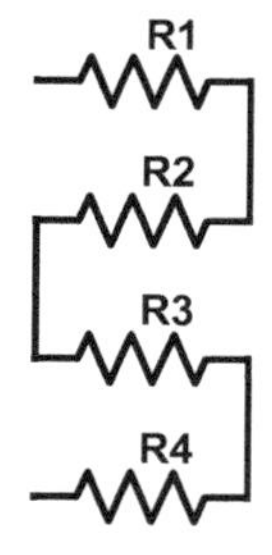

Figure 42.2: *A series chain of resistors can be folded up to make the ends of resistors that are connected adjacent in the layout. Figure drawn with Digi-Key's Scheme-it [18].*

Nyquist frequency make for numerically unstable filters). I recommend a sampling rate of 300 Hz to 1200 Hz for EKG signals if you want to use the diagnostic EKG bandwidth, and 120 Hz to 600 Hz if you want to use cardiac-monitor bandwidth.

Plot the EKG signals both before and after applying the digital filtering.

Compute the pulse rate from the recording. This should be a precise computation, based on the time between peaks, not an approximate one based on peaks in a fixed time interval. Watch out for fencepost errors!

The report should, of course, include the block diagram and schematics for the one-channel EKG, along with any design notes about why the design ended up the way that it did.

Appendix A: PteroDAQ Documentation

PteroDAQ is a data-acquisition system that works with a microcontroller board connected via a USB cable to a host computer. The microcontroller board can be any of a number of different boards. The choice of board provides some differences in functionality besides the channel numbers listed in Table A.1 (e.g., voltage ranges on the inputs, resolution of analog-to-digital converters, and sampling frequency). The pinout for each board is available from the manufacturer of the board—from PJRC for the Teensy boards [87], from NXP for the FRDM-KL25Z board [31], and from the various manufacturers for Arduino and Arduino-clone boards.

	Channels				
Board names	Analog	Differential	Digital	Frequency	Trigger
Arduino Uno Arduino Duemilanove Arduino Diecimila Arduino Ethernet Arduino Pro Arduino LilyPad	8	0	20	0	2
Arduino Mini Arduino Nano Arduino Pro Mini Arduino Fio	10	0	20	0	2
Arduino Mega	17	0	14	0	6
Arduino Leonardo Arduino Yun Arduino Micro Arduino Robot Arduino Esplora Arduino LilyPad USB	14	0	23	0	5
FRDM-KL25Z	18	2	54	2	18
Teensy 3.1 Teensy 3.2	24	2	18	5	18
Teensy LC	15	1	14	3	12

Table A.1: *This table lists the boards supported by the PteroDAQ data-acquisition system, with some indications of the capabilities of each board. The table is derived from the PteroDAQ* `boards.py` *file, using the names that PteroDAQ uses, and so the names used may not be the official names of the boards. The "trigger" channels are the digital inputs that can be used to start a sample, rather than using the timer. Warning: the FRDM-KL25Z code has not been maintained recently and may no longer be compatible with the newer versions of the mbed compiler and library.*

The microcontrollers used in this course contain an *analog-to-digital converter (ADC)*, that converts voltages into numbers. Depending on which microcontroller board is used, the converters may produce 10-bit numbers (0 through 1023) or 16-bit numbers (0 through 65535). The full-scale voltage range is either 0 V to 3.3 V or 0 V to 5 V, again depending on the microcontroller board.

The FRDM-KL25Z and Teensy boards have 16-bit ADC converters (that is, the range of the numbers is 0 to $2^{16}-1 = 65535$), while the Arduino boards have only a 10-bit resolution (that is, values from 0 to 1023). The FRDM-KL25Z can measure voltages from any of several different pins (PTB0, PTB1, PTB2, PTB3, PTC0, PTC1, PTC2, PTD1, PTD5, PTD6, PTE20, PTE21, PTE22, PTE23, PTE29, PTE30) as *single-ended* inputs—that is the value reported is the difference between the voltage at the pin and the ground pin on the microcontroller. The Teensy boards can measure voltage using A0 through A12 (Teensy LC) or A0 through A14 (Teensy 3.1 or 3.2). The Arduino boards can measure voltages on any of the 6 analog input pins A0 through A6 (the Arduino Mega has A0 through A15).

The FRDM-KL25Z also has two *differential* channels: E20 – E21 and E22 – E23. The outputs of these channels are from $-2^{15} = -32768$ to $2^{15} - 1 = 32767$. The Teensy boards also have differential channels: A10 – A11 on all and A12 – A13 on Teensy 3.1 and 3.2. The Arduino boards do not have differential channels.

The voltage at the analog pins must be limited to be between the power rails of the microcontroller. That is, on the FRDM-KL25Z and Teensy boards $0\,\text{V} \leq V_{\text{A}} \leq 3.3\,\text{V}$ and on most Arduino boards $0\,\text{V} \leq V_{\text{A}} \leq 5\,\text{V}$ (some Arduino boards are 3.3 V). Putting negative voltages or larger positive voltages on the board may damage the chips on the board. A differential channel may have a negative number (for example, if PTE20 $<$ PTE21), but the voltages on each pin must stay between 0 V and +3.3 V with respect to the ground pin of the microcontroller.

With the currently supported microcontroller boards, it is not possible for PteroDAQ to measure voltages simultaneously on different inputs, but the microcontroller can switch the ADC rapidly from one input to another, so that measurements can be made at *almost* the same time. Some microcontrollers (including the one on the Teensy 3.1/3.2) have multiple analog-to-digital converters, and PteroDAQ is likely to be extended to handle multiple ADCs, to get nearly simultaneous measurement and a faster sampling rate (see Section 9.1 for an explanation of sampling rate and why it is important).

The ADC is *ratiometric*, which means that the output value for the 16-bit analog-to-digital converters is $65536 V_{\text{A}}/V_{\text{REFH}}$, where V_{A} is the voltage at the analog input pin, and V_{REFH} is a reference voltage being compared to. (For differential inputs on the FRDM-KL25Z and Teensy boards, the scaling is $32768 V_{\text{A}}/V_{\text{REFH}}$, as the 65536 values map to twice as wide a range: $-V_{\text{REFH}}$ to $+V_{\text{REFH}}$.) The default reference voltage in the PteroDAQ software we are using is the power-supply voltage to the board, which the system measures to convert the raw readings into volts.

If the power-supply voltage is likely to fluctuate, then it is worth using one channel to record the 1 V bandgap reference built into the Teensy LC board—you can then correct for the fluctuation (if it is not too fast) by dividing the analog channel reading by the 1 V reading.

The ARM-based boards (FRDM-KL25Z and Teensy) can be used to measure the frequency of digital signals, by counting the number of rising edges in each sample time, but that capability has not been implemented for the Arduino boards. The FRDM-KL25Z and Teensy LC boards can measure up to 3.9 MHz, but use a 20-bit counter to count pulses, so the frequency being measured must be no more than a million times the sampling rate. The Teensy 3.1/3.2 can measure up to 4 MHz, but uses only a 15-bit counter, so the measured frequency can only be up to 30,000 times the sampling rate.

For each channel, PteroDAQ provides a *sparkline* that plots the last few samples recorded (updated 10 times a second). PteroDAQ also displays the last sample measured, the average of all the samples (labeled "DC"), and the standard deviation of all the samples (labeled "RMS").

The *RMS* stands for "Root Mean Square", which is a common way for measuring the magnitude of time-varying signals. There are two common ways of defining RMS voltage used in voltmeters: AC and AC + DC.

- The *AC+DC* definition squares the observed voltage, averages the squared signal over some time period, then takes the square root. As shown in Section 35.1, this definition is useful for computing how much heat is dissipated in a component.
- The *AC* definition first computes the average voltage, then squares the difference from that average, and again takes the square root of the average of the square—this is the same concept as standard deviation in statistics, and serves the same purpose of measuring how much a signal is changing.

Because many of our sensors and circuits will have a small changing signal with a large DC offset, the AC definition is usually the more useful one for us, and that is what PteroDAQ reports. You can compute the AC + DC RMS voltage, if needed, as $V_{\text{AC+DC}} = \sqrt{V_{\text{DC}}^2 + V_{\text{RMS}}^2}$.

The DC and RMS values can be used to measure steady-state signals (either DC or AC) with higher precision than single samples. In some of the places where an AC voltmeter is called for in the labs, PteroDAQ can be substituted—the crucial questions are whether the voltages are in a range suitable for the analog-to-digital conversion and whether the aliasing due to PteroDAQ's sampling causes high frequency signals to be misinterpreted.

There are some limitations of the FRDM-KL25Z and Teensy boards that are important for this course:

- The microcontroller can only record voltages, frequencies, digital values, and time stamps, so we need to convert other signals (like current or capacitance) into signals that we can measure.
- The highest voltage allowed on any pin is 3.3 V and the lowest is 0 V.
- There are only 15–24 channels that can be used for single-ended voltage input (and only 1–2 that can be used for differential input).
- The resolution is only 16 bits (65536 different values). With V_{REFH} at the default value of 3.3 V, the step size is about 50 μV for single-ended measurements and 101 μV for differential measurements. The accuracy is much worse than the resolution, but is difficult to estimate ahead of time—repeatability is probably only ±1 mV, and the absolute accuracy probably only ±3%. On the Arduino boards, the resolution is only 10 bits (1024 different values) and the step size about 5 mV.
- The ADC gets inaccurate when near the power rails or reporting small numbers—you usually want to have all inputs stay at least 2.5 mV from either power rail (50 counts on 16-bit ADC), and measure differential signals that are at least 1 mV (20 counts).
- Readings are not simultaneous, but the analog-to-digital converter is quickly switched from one input to another. It takes between 2.2 μs and 128 μs for each conversion on the FRDM-KL25Z and Teensy boards, depending on how much hardware averaging is done.

If you pick too high a sampling rate for the number of channels and degree of hardware averaging selected, so that the microcontroller can't do the conversions in time, PteroDAQ reports "Warning: triggering too fast, next trigger before finished". On the FRDM-KL25Z and Teensy LC boards at 32× hardware averaging, this happens at about 6.9 kHz, and on the Teensy 3.1/3.2 at about 10.3 kHz.

At 4× averaging, much higher sampling rates are possible, and limitations on the host computer occur before the sampling-rate limit is reached (the Teensy 3.1/3.2 can do 35 kHz sampling if the host computer can read the data fast enough).

- The FRDM-KL25Z has limited memory size for queuing recordings and will drop data points if its queue is full. Each sample requires 8 bytes for a time stamp and 2 bytes per analog channel, so the 8 kB queue (half the RAM) on the board can only hold about 820 samples of a single channel or 683 samples of 2 channels. That means it can hold about 2.7 s worth of data at a 300 Hz sampling rate, or over a minute of data at 10 Hz.

  The Teensy 3.1 and 3.2 have a 32 kB queue, and the Teensy LC has only a 4 kB queue. This buffer can be filled as fast as the processor can collect data, and the PteroDAQ software transfers the data as quickly as it can to the host computer, which results in different speeds with different microcontroller boards and different host computers.

- When not doing large amounts of hardware averaging, the main limiting factor on the sampling rate for the PteroDAQ system is how fast the host computer can read the samples over the USB connection, to keep the queue from filling up. This is dependent on the microcontroller board, the operating system on the host, and the Python program on the host.

  The FRDM-KL25Z at 32× averaging can generally record a single channel at 6.3 kHz indefinitely, even with slow host computers—the limitation is mainly from the time taken to do the conversions.

  At 4× averaging, with my old, slow MacBook Pro laptop, I can record for about 20 s at 17 kHz or 35 s at 15 kHz, before the FRDM-KL25Z queue fills up and the PteroDAQ system warns that samples are being dropped. On faster machines, the sampling rate can go to 20 kHz, but there is always a chance that the host operating system will temporarily take enough processing power away from the Python job that a few samples will be lost, even at only 10 kHz.

  The larger queue on the Teensy 3.1/3.2 does not seem to help much—I still see samples dropped after the first 500,000 or so at 15 kHz–17 kHz. On a faster (2.7 GHz Core i5) iMac, I can get up to 5,000,000 samples at 20 kHz or 2,000,000 at 25 kHz, but eventually the operating system makes the Python program take too long a pause and the queue on the Teensy fills up.

  Table A.2 shows how small a sample period I could use on several boards to run for several minutes with my iMac as a host computer. I used 1× and 32× sampling on the boards that supported hardware averaging, The Teensy boards were compiled for fast code rather than small code, at the default speed settings (48 MHz for the Teensy LC and 96 MHz for the Teensy 3.1).

  For some applications, missing a few samples and having an irregular sampling rate is not a major problem, and the system adapts to the communication rate, producing samples as fast as they are consumed by the host computer.

- The FRDM-KL25Z and Teensy analog-to-digital converters were designed assuming that the input comes from a low-impedance source—one that can provide a reasonably large current to charge or discharge capacitors.

  With a high-impedance source (over about 1 kΩ), hardware averaging does not improve the signal-to-noise ratio, as noise injected onto the analog pin by the sampling hardware itself is not removed by the source before the next sample is taken. For high-impedance sources, using no

Board	1 analog channel	2 analog channels	3 analog channels
Arduino (ATMega386)	267 μs	415 μs	563 μs
FRDM KL25Z (1×)	47 μs	50 μs	56 μs
Teensy LC (1×)	44 μs	48 μs	56 μs
Teensy 3.1 (1×)	44 μs	48 μs	51 μs
FRDM KL25Z (32×)	146 μs	276 μs	405 μs
Teensy LC (32×)	148 μs	276 μs	403 μs
Teensy 3.1 (32×)	134 μs	258 μs	381 μs

Table A.2: *Measured fastest sample periods for various microcontroller boards running PteroDAQ. The three NXP processors all have similar performance, because they have the same underlying ADC hardware. The Teensy 3.1 was overclocked to 96 MHz, to get similar performance on the ADC as the Teensy LC running at 48 MHz. The extra processor power of the Teensy 3.1 makes a small difference with* 32× *averaging, but at* 1× *averaging, the speed is limited by how fast the Python program on the host can run to keep the data queue from overflowing, so all three boards take about 50 µs a sample, with only slight increases as the number of channels increases.*

hardware averaging and a slow sampling rate may result in more accurate measurement than doing the hardware averaging.

Section 19.2 discusses a way to avoid having to connect high-impedance inputs directly to the analog-to-digital converter, to avoid this problem.

- Many of the labs in this book call for using the microcontroller board as a power supply for the circuit being tested, to ensure that voltages being measured stay within the allowed input voltage range.

  The FRDM-KL25Z 3.3 V power supply uses a low-dropout voltage regulator (LDO) that can handle up to 1 A, and so using the board as a power supply is limited primarily by the current limit from the USB power source (500 mA).

  The Arduino boards use the same series of LDOs, so can supply up to 1 A when powered by a separate power supply, but only 500 mA when powered over the USB line. The 5 V supply from the USB line is often quite noisy, making a rather poor voltage reference.

  The Teensy LC and 3.1 boards use a much more limited voltage regulator built into the microcontroller chip itself. They are limited to providing about 100 mA as a 3.3 V power supply. The Teensy 3.2 is almost identical to the Teensy 3.1, but can supply 250 mA from its 3.3 V supply.

Appendix B: Study Sheet

There is very little to memorize in this class. Here are the few concepts that are worth having instantly available in your memory. Your studying should not be memorizing these few formulas, but using them repeatedly to solve design and analysis problems.

B.1 Physics

$$Q = CV \ ,$$
$$I(t) = \frac{\mathrm{d}Q(t)}{\mathrm{d}t} \ ,$$
$$V = IR, \text{Ohm's law} \ .$$

B.2 Math

$$j = \sqrt{-1} \ ,$$
$$|x + jy| = \sqrt{x^2 + y^2} \ ,$$
$$e^{j\theta} = \cos(\theta) + j\sin(\theta) \ ,$$
$$\frac{\mathrm{d}e^{j\omega t}}{\mathrm{d}t} = j\omega e^{j\omega t} \ ,$$
$$\omega = 2\pi f \ ,$$

where ω is angular frequency in radians/sec, and f is frequency in Hz

B.3 Op amps

For the generic op-amp amplifier in Figure B.1, only the approximations when open-loop gain $A \to \infty$ are worth memorizing:

Inverting: $V_{\text{out}} - V_p \approx \frac{Z_\text{f}}{Z_\text{i}}(V_p - V_m)$.

Non-inverting: $V_{\text{out}} - V_m \approx \frac{Z_\text{f}+Z_\text{i}}{Z_\text{i}}(V_p - V_m)$.

Transimpedance (set $Z_\text{i} = 0$ and look at I_m from amplifier into input node V_m): $V_{\text{out}} - V_p \approx Z_\text{f} I_m$

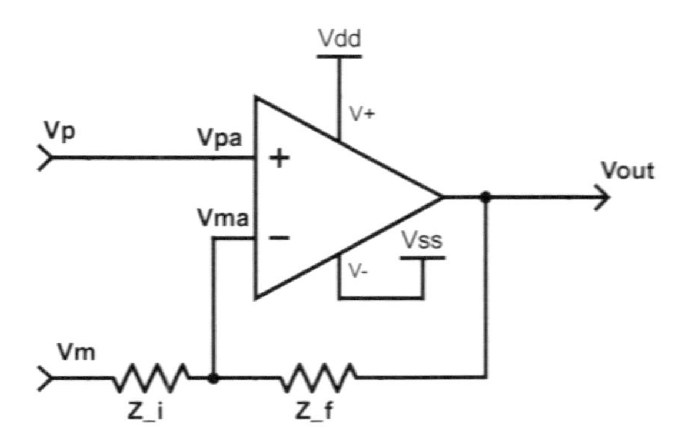

Figure B.1: *Generic negative-feedback op-amp circuit, used for inverting and non-inverting amplifiers. The negative feedback tries to keep the plus and minus inputs of the op amp at the same voltage:* $V_p = \frac{V_m Z_\text{f} + V_{\text{out}} Z_\text{i}}{Z_\text{f}+Z_\text{i}}$.

B.4 Impedance

$$v(t) = i(t)Z$$
$$Z = R, \text{resistor}$$
$$Z = \frac{1}{j\omega C}, \text{capacitor, angular frequency} = \omega$$
$$Z = j\omega L, \text{inductor, angular frequency} = \omega$$
$$Z_{\text{series}} = Z_1 + Z_2$$
$$Z_{\text{parallel}} = Z_1 \parallel Z_2 = \frac{1}{\frac{1}{Z_1} + \frac{1}{Z_2}} = \frac{Z_1 Z_2}{Z_1 + Z_2}$$
$$\text{gain} = \frac{V_{\text{out}}}{V_{\text{in}}} = \frac{Z_{\text{down}}}{Z_{\text{up}} + Z_{\text{down}}}, \text{voltage divider}$$
$$2\pi f = \omega = \frac{1}{RC}, \text{corner frequency for RC}$$
$$2\pi f = \omega = \frac{1}{\sqrt{LC}}, \text{resonant frequency for LC}$$

Gain of a simple RC circuit (one R, one C) is $\sqrt{2}/2$ at the corner frequency.

References

[1] Agilent Technologies.
User's Guide: Agilent 33120A 15MHz Function/Arbitrary Waveform Generator. 6th ed. Publication number 33120-90006. Agilent Technologies, Inc., Mar. 2002.
URL: https://www.keysight.com/us/en/assets/9018-04436/user-manuals/9018-04436.pdf (visited on 27 Sept. 2021) (cit. on pp. 202, 240).

[2] American Medical Association and American Heart Association.
The One Graphic You Need for Accurate Blood Pressure Reading. 8 Feb. 2019.
URL: https://www.ama-assn.org/delivering-care/hypertension/one-graphic-you-need-accurate-blood-pressure-reading (visited on 2 Nov. 2019) (cit. on p. 297).

[3] *Analog Discovery 2 Reference Manual.* Digilent. URL:
https://digilent.com/reference/test-and-measurement/analog-discovery-2/reference-manual (visited on 27 Sept. 2021) (cit. on pp. 210, 383).

[4] Android. *3.5mm Headset: Accessory Specification.* 2 Aug. 2022.
URL: https://source.android.com/docs/core/accessories/headset/plug-headset-spec (visited on 27 Aug. 2022) (cit. on p. 431).

[5] Art of Problem Solving. *LaTeX:About.*
URL: https://artofproblemsolving.com/wiki/index.php/LaTeX:About (visited on 15 June 2014) (cit. on p. 78).

[6] Euan A. Ashley and Josef Niebauer. In Ashley and Niebauer. *Cardiology Explained.* London: Remedica, 2004. Chap. 3, Conquering the ECG.
URL: https://www.ncbi.nlm.nih.gov/books/NBK2214/ (visited on 29 Aug. 2022) (cit. on pp. 534, 538).

[7] Charles F. Babbs. "Oscillometric Measurement of Systolic and Diastolic Blood Pressures Validated in a Physiologic Mathematical Model."
BioMedical Engineering OnLine 11(56) (2012). DOI: 10.1186/1475-925X-11-56. URL:
https://biomedical-engineering-online.biomedcentral.com/articles/10.1186/1475-925X-11-56 (visited on 27 Sept. 2022) (cit. on p. 283).

[8] Peter J. Baxandall. "Negative-Feedback Tone Control."
Wireless World (Oct. 1952), pp. 402–405 (cit. on p. 431).

[9] Akshay Bhat. *Stabilize Transimpedance Amplifier Circuit Design.* Application Note 5129. Maxim Integrated. 3 Feb. 2012. URL:
https://www.maximintegrated.com/en/design/technical-documents/app-notes/5/5129.html (visited on 5 Sept. 2020) (cit. on p. 326).

[10] BitScope. *BitScope Model 10.*
URL: https://bitscope.com/product/BS10/ (visited on 7 Apr. 2015) (cit. on p. 210).

[11] Russ Brit. “Skin Tone Not a New Issue for Heart Monitors.” *MarketWatch* (30 Apr. 2015).
URL: https://www.marketwatch.com/story/skin-tone-not-a-new-issue-for-heart-monitors-2015-04-30 (cit. on p. 311).

[12] Lovely Chhabra *et al.*
“Mouse Heart Rate in a Human: Diagnostic Mystery of an Extreme Tachyarrhythmia.” *Indian Pacing and Electrophysiology Journal* 12(1) (2012), pp. 32–35. ISSN: 0972-6292 (cit. on p. 284).

[13] Eric Coates. *Learn about Electronics: Module 4.2 Amplifier Controls.* 9 Dec. 2020. URL: https://www.learnabout-electronics.org/Amplifiers/amplifiers42.php (visited on 29 Aug. 2022) (cit. on p. 431).

[14] Enrique Company-Bosch and Eckart Hartmann.
“ECG Front-End Design is Simplified with MicroConverter.” *Analog Dialogue* 37 (Nov. 2003).
URL:
https://www.analog.com/en/analog-dialogue/articles/ecg-front-end-design-simplified.html (visited on 13 June 2019) (cit. on pp. 534, 536, 539).

[15] CUI Inc. *CMA-4544PF-W Electret Condenser Microphone.* CUI Inc. 17 Jan. 2020. URL: https://www.cuidevices.com/product/resource/cma-4544pf-w.pdf (visited on 5 Sept. 2020) (cit. on p. 365).

[16] CUI Inc. *CME-1538-100LB Electret Condenser Microphone.* CUI Inc. 24 Apr. 2020. URL: https://www.cuidevices.com/product/resource/cme-1538-100lb.pdf (visited on 5 Sept. 2020) (cit. on p. 372).

[17] Deranged Physiology. *Normal Arterial Waveforms.* URL:
https://derangedphysiology.com/main/cicm-primary-exam/required-reading/cardiovascular-system/Chapter%20760/normal-arterial-line-waveforms (visited on 20 Mar. 2021) (cit. on p. 355).

[18] Digi-Key. *Scheme-it: Free Online Schematic Drawing Tool.* 2014.
URL: https://www.digikey.com/schemeit/ (visited on 6 Sept. 2020)
(cit. on pp. 34, 40, 46, 47, 49, 53–55, 64, 84, 88, 89, 154, 159, 167, 171, 173, 179, 181, 188, 195, 197, 201, 207, 214, 220–222, 226, 227, 240, 248, 254, 260, 265, 271, 273, 274, 290, 302, 304, 306, 309, 314, 318, 319, 323, 328, 331, 335, 343, 365, 369, 376, 377, 379, 383, 391, 396, 397, 412, 433, 435, 437, 439, 442, 444, 445, 450, 456–458, 460, 462, 464, 477, 478, 482, 483, 485, 488, 491, 496, 510, 512, 528, 530, 537, 541, 542, 544, 575).

[19] Digilent.
Analog Discovery 2 100MSPS USB Oscilloscope, Logic Analyzer and Variable Power Supply.
URL: https://digilent.com/shop/analog-discovery-2-100ms-s-usb-oscilloscope-logic-analyzer-and-variable-power-supply/ (visited on 27 Sept. 2021) (cit. on pp. 151, 229–231, 255, 275, 276, 324, 325, 345, 367, 368, 391, 399, 414, 447, 454, 456, 490, 493–495, 497).

[20] Digilent. *WaveForms Reference Manual.*
URL: https://digilent.com/reference/software/waveforms/waveforms-3/reference-manual (visited on 27 Sept. 2021) (cit. on p. 211).

[21] *Draw.IO.* Moved to https://app.diagrams.net/, but the draw.io link redirects.
URL: https://www.draw.io (visited on 16 July 2017)
(cit. on pp. 61, 62, 69, 84, 87, 200, 256, 258, 261, 484, 486, 528, 575).

[22] *Electrical Conductivity of Aqueous Solutions.*
URL: https://diverdi.colostate.edu/all_courses/CRC%20reference%20data/electrical%20conductivity%20of%20aqueous%20solutions.pdf (visited on 27 Sept. 2021) (cit. on p. 506).

[23] John A. Evans and William A. Whitelaw.
"The Assessment of Maximal Respiratory Mouth Pressures in Adults."
Respiratory Care 54(10) (Oct. 2009), pp. 1348–1359.
URL: https://rc.rcjournal.com/content/54/10/1348 (visited on 18 Nov. 2022) (cit. on p. 279).

[24] Everlight. *PD204-6C Technical Data Sheet: 3mm Silicon PIN Photodiode T-1.*
No longer available from Everlight directly. 2005.
URL: https://media.digikey.com/pdf/Data%20Sheets/Everlight%20PDFs/PD204-6C.pdf
(visited on 30 Jan. 2023) (cit. on p. 305).

[25] Nick Faulkner. *The Blame Game: Things were Done, Mistakes were Made.* 2 May 2013. URL:
https://nickfalkner.com/2013/05/02/the-blame-game-things-were-done-mistakes-were-made/
(visited on 5 July 2020) (cit. on p. 76).

[26] Foxboro. *Conductivity Ordering Guide.* 3 Oct. 1999. URL:
https://pdf4pro.com/cdn/conductivity-ordering-guide-supersedes-august-1-1986-43e9eb.pdf
(visited on 5 Sept. 2020) (cit. on p. 506).

[27] Freescale Semiconductor.
200 kPa On-Chip Temperature Compensated Silicon Pressure Sensors MPX2202 Series.
Oct. 2012.
URL: https://www.nxp.com/docs/en/data-sheet/MPX2202.pdf (visited on 27 Aug. 2022)
(cit. on p. 289).

[28] Freescale Semiconductor. *K20 Sub-Family Data Sheet.* K20P81M100SF2.
Freescale Semiconductor (NXP). Feb. 2013. URL:
https://www.nxp.com/docs/en/data-sheet/K20P81M100SF2.pdf (visited on 5 Sept. 2020)
(cit. on p. 59).

[29] Freescale Semiconductor.
Kinetis KL26 Sub-Family 48 MHz Cortex-M0+ Based Microcontroller.
KL26P121M48SF4 Rev. 5. Freescale Semiconductor (NXP). Aug. 2014. URL:
https://www.nxp.com/docs/en/data-sheet/KL26P121M48SF4.pdf (visited on 14 Nov. 2018)
(cit. on pp. 59, 200).

[30] Freescale Semiconductor. *Kinetis KL26 Sub-Family Reference Manual.*
KL26P121M48SF4RM Rev. 3.2. Freescale Semiconductor (NXP). Oct. 2013 (cit. on p. 231).

[31] Freescale Semiconductor (NXP). *FRDM-KL25Z Quick Reference Card.*
URL: https://www.nxp.com/docs/en/supporting-information/FRDM-KL25Z-Quick-Reference-Card.pdf (visited on 15 Nov. 2018) (cit. on p. 547).

[32] L.A. Geddes *et al.*
"Characterization of the Oscillometric Method for Measuring Indirect Blood Pressure."
Annals of Biomedical Engineering 10 (1982), pp. 271–280 (cit. on p. 283).

[33] *gnuplot homepage.* Use http, as gnuplot refuses https connection.
URL: http://www.gnuplot.info/ (visited on 19 Jan. 2021)
(cit. on pp. 13, 38, 46, 56, 67, 84, 129, 140, 142–144, 151, 156, 161, 162, 175–177, 180, 182, 183, 206, 218–220, 223, 224, 229–233, 251, 252, 255, 264, 275, 276, 281, 284–286, 305, 307, 308, 310, 316, 318, 322, 324, 325, 333, 338, 339, 345, 354, 357, 363, 366–368, 382, 391, 399, 414, 435, 447, 451, 454, 456, 474, 476, 481, 485, 490, 493–495, 497, 507, 535, 575).

[34] Google. *Google Books Ngram Viewer.* URL: https://books.google.com/ngrams/ (visited on 18 Nov. 2022) (cit. on pp. 102, 112, 114, 115).

[35] Google. *Google Books Ngram Viewer (bandpass).* URL: https://books.google.com/ngrams/graph?content=bandpass%2Cband-pass%2Cband+pass (visited on 22 Mar. 2018) (cit. on p. 102).

[36] Google. *Google Books Ngram Viewer (textbook).* URL: https://books.google.com/ngrams/graph?content=text+book%2Ctextbook%2Ctext-book&year_start=1750 (visited on 14 Mar. 2021) (cit. on p. 102).

[37] Google. *Google Books Ngram Viewer (try and).* URL: https://books.google.com/ngrams/graph?content=try+to%2Ctry+and (visited on 26 Mar. 2018) (cit. on p. 109).

[38] Google. *Google Books Ngram Viewer (x axis).* URL: https://books.google.com/ngrams/graph?content=x+axis%2C+x-axis%2C+y+axis%2C+y-axis&year_start=1800&year_end=2020 (visited on 12 June 2019) (cit. on p. 102).

[39] Sverre Grimnes and Ørjan Grøttem Martinsen. *Bioimpedance and Bioelectricity Basics.* 3rd ed. Academic Press, 2014. ISBN: 978-0124114708 (cit. on p. 508).

[40] Hanwei Electronics Co. Ltd. *Technical Data MQ-3 Gas Sensor.* 2017. URL: https://www.sparkfun.com/datasheets/Sensors/MQ-3.pdf (cit. on p. 60).

[41] Tom Harris and Wesley Fenlon. *How Light Emitting Diodes Work.* 31 Jan. 2002. URL: https://electronics.howstuffworks.com/led.htm (visited on 11 Dec. 2018) (cit. on p. 302).

[42] Frederic G. Hirsch *et al.* "The Electrical Conductivity of Blood I. Relationship to Erythrocyte Concentration." *Blood* 5(11) (1950), pp. 1017–1035. DOI: 10.1182/blood.V5.11.1017.1017. URL: https://ashpublications.org/blood/article/5/11/1017/7012/THE-ELECTRICAL-CONDUCTIVITY-OF-BLOOD-I (visited on 5 Sept. 2020) (cit. on p. 505).

[43] Hobby Projects. *About Oscilloscope: Oscilloscope Trigger Controls.* 2011. URL: https://hobbyprojects.com/oscilloscope_tutorial/oscilloscope_trigger_controls.html (visited on 5 Dec. 2019) (cit. on p. 213).

[44] Adam E. Hyatt, Paul D. Scanlon, and Masao Nakamura. *Interpretation of Pulmonary Function Tests: A Practical Guide.* 3rd ed. Lippincott Williams & Wilkins, 2011 (cit. on p. 279).

[45] *Draw Freely—Inkscape.* URL: https://inkscape.org/ (visited on 25 Nov. 2018) (cit. on pp. 72, 84, 91, 396, 575).

[46] Abraham Karplus and Kevin Karplus. *PteroDAQ.* URL: https://github.com/karplus/PteroDAQ (visited on 17 Apr. 2021) (cit. on pp. 144, 206, 223, 232, 233, 281, 284–286, 307, 308, 318, 354, 535).

[47] Kevin Karplus. *Compensation in Impedance Analyzer.* 18 July 2020. URL: https://gasstationwithoutpumps.wordpress.com/2020/06/05/compensation-in-impedance-analyzer/ (visited on 5 June 2020) (cit. on p. 412).

[48] Kevin Karplus. *Descaffolding.* 12 Feb. 2013. URL: https://gasstationwithoutpumps.wordpress.com/2013/02/12/descaffolding/ (cit. on p. x).

[49] Kevin Karplus. *EKG without amplifier.* 18 Dec. 2017. URL: https://gasstationwithoutpumps.wordpress.com/2017/12/18/ekg-without-amplifier/ (visited on 5 Dec. 2019) (cit. on p. 541).

[50] Kevin Karplus. *Op amp protoboard rev 0 worksheet.* 8 Aug. 2015. URL: https://users.soe.ucsc.edu/~karplus/bme51/pc-boards/op-amp-proto-rev0/op-amp-proto-rev0-worksheet-fullpage.pdf (cit. on pp. 428, 541).

[51] Kevin Karplus. *Resonance for Non-linear Impedance.* 9 June 2020. URL: https://gasstationwithoutpumps.wordpress.com/2020/06/09/resonance-for-non-linear-impedance/ (visited on 22 June 2021) (cit. on p. 401).

[52] Kevin Karplus. *Showing is better than telling, but not by much.* 14 Apr. 2013. URL: https://gasstationwithoutpumps.wordpress.com/2013/04/14/showing-is-better-than-telling-but-not-by-much/ (cit. on p. ix).

[53] KEMET Corporation. *Introduction to Capacitor Technologies: What is a Capacitor?* 2013. URL: https://www.kemet.com/content/dam/kemet/lightning/documents/ec-content/technical-archive/What-is-a-Capacitor.pdf (visited on 10 Mar. 2023) (cit. on p. 146).

[54] KEMET Corporation. *Single-Ended Aluminum Electrolytic Capacitors: ESK +85°C.* 17 Nov. 2020. URL: https://content.kemet.com/datasheets/KEM_A4004_ESK.pdf (visited on 29 May 2021) (cit. on pp. 149, 151).

[55] Kingbright. *Phototransistor WP3DP3BT Datasheet.* Version 5. 7 Oct. 2016. URL: https://www.kingbrightusa.com/images/catalog/SPEC/wp3dp3bt.pdf (visited on 24 Nov. 2018) (cit. on p. 119).

[56] Henry Knipe and Rachael Nightingale. *Ultrasound Transducer.* Sept. 2017. URL: https://radiopaedia.org/articles/ultrasound-transducer (cit. on p. 259).

[57] Henry Knipe and Patricia O'Gorman. *Propagation Speed.* URL: https://radiopaedia.org/articles/propagation-speed (visited on 5 Sept. 2020) (cit. on p. 258).

[58] Knowles Acoustics. *Ultra-Mini SiSonic^{TM} Microphone Specification With MaxRF Protection.* 27 Mar. 2013. URL: https://media.digikey.com/pdf/Data%20Sheets/Knowles%20Acoustics%20PDFs/SPU0410-HR-5H-PB.pdf (visited on 27 Aug. 2022) (cit. on p. 364).

[59] Donald E. Knuth. *The TEXbook.* Addison-Wesley Professional, 1984. ISBN: 978-0201134483 (cit. on p. 77).

[60] Sébastien Krul. *Cardiac Arrhythmias.* URL: https://www.textbookofcardiology.org/wiki/Cardiac_Arrhythmias (visited on 5 Sept. 2020) (cit. on p. 538).

[61] Michael M. Laks *et al.* "Recommendations for Safe Current Limits for Electrocardiographs: A Statement for Healthcare Professionals From the Committee on Electrocardiography, American Heart Association." *Circulation* 93 (1996), pp. 837–839. DOI: 10.1161/01.CIR.93.4.837. URL: https://www.ahajournals.org/doi/full/10.1161/01.cir.93.4.837 (visited on 5 Sept. 2020) (cit. on pp. 536, 539).

[62] Leslie Lamport. *LATEX: A Document Preparation System.* 2nd ed. Addison-Wesley Professional, 1994. ISBN: 978-0201529838 (cit. on pp. 77, 78, 118).

[63] W. Marshall Leach. In W. Marshall Leach, Jr. *Introduction to Electroacoustics and Audio Amplifier Design.* Second Edition—Revised Printing. Kendall/Hunt, 2001. Chap. The Class-D Amplifier. URL: https://leachlegacy.ece.gatech.edu/ece4435/f01/ClassD2.pdf (visited on 27 Aug. 2022) (cit. on p. 477).

[64] LEGO Group. *Fair Play.* URL: https://www.lego.com/en-us/legal/notices-and-policies/fair-play/ (visited on 20 Mar. 2021) (cit. on p. 356).

[65] Philipp Lehman. *The biblatex Package.* Version 3.12. 30 Oct. 2018. URL: http://mirrors.ibiblio.org/CTAN/macros/latex/contrib/biblatex/doc/biblatex.pdf (visited on 5 Sept. 2020) (cit. on p. 119).

[66] *matplotlib.* URL: https://matplotlib.org/ (visited on 25 Nov. 2018) (cit. on pp. 32, 575).

[67] Maxim Integrated. *Penetration Depth Guide for Biosensor Applications.* 30 Mar. 2018. URL: https://www.maximintegrated.com/en/design/technical-documents/app-notes/6/6433.html (visited on 27 Nov. 2019) (cit. on p. 311).

[68] Mayo Clinic. *Ankle-brachial Index.* URL: https://www.mayoclinic.org/tests-procedures/ankle-brachial-index/about/pac-20392934 (visited on 9 Dec. 2018) (cit. on p. 282).

[69] David Meeker. *Finite Element Method Magnetics: Woofer—Analysis of a Woofer Motor.* URL: https://www.femm.info/wiki/Woofer (visited on 5 Sept. 2020) (cit. on p. 396).

[70] Dan Meyer. *Asilomar #4: Be Less Helpful.* 2009. URL: https://blog.mrmeyer.com/2009/asilomar-4-be-less-helpful/ (cit. on p. x).

[71] Microchip. *MCP6001/R/U/2/4, 1 MHz, Low-Power Op Amp.* 14 May 2019. URL: https://ww1.microchip.com/downloads/en/DeviceDoc/MCP6001-1R-1U-2-4-1-MHz-Low-Power-Op-Amp-DS20001733L.pdf (visited on 5 Sept. 2020) (cit. on pp. 263, 322).

[72] James Moffatt. *E C G Basics.* URL: https://www.nataliescasebook.com/tag/e-c-g-basics (visited on 19 Oct. 2018) (cit. on p. 538).

[73] S. Mohan *et al.* "Simple Accurate Expressions for Planar Spiral Inductances." *IEEE Journal of Solid-State Circuits* 34(10) (Oct. 1999), pp. 1419–1424. URL: https://web.stanford.edu/~boyd/papers/inductance_expressions.html (visited on 5 July 2020) (cit. on p. 390).

[74] Matt A. Morgan and Mirjan M. Nadrljanski. *Ultrasound Frequencies.* Sept. 2017. URL: https://radiopaedia.org/articles/ultrasound-frequencies (cit. on p. 259).

[75] *MQ303A Alcohol Sensor.* 2017. URL: https://raw.githubusercontent.com/SeeedDocument/Grove-Alcohol_Sensor/master/res/MQ303A.pdf (cit. on p. 60).

[76] Randall Munroe. *.NORM Normal File Format.* URL: https://xkcd.com/2116/ (visited on 6 July 2019) (cit. on p. 91).

[77] Randall Munroe. *Curve Fitting.* URL: https://xkcd.com/2048/ (visited on 6 July 2019) (cit. on p. 96).

[78] Munir H. Nayfeh and Morton K. Brussel. *Electricity and Magnetism.* Copyright 1985. Dover Publications, 2015. ISBN: 978-0-486-78971-2 (cit. on p. 508).

[79] Michael R. Neuman. "Biopotential Electrodes." In *The Biomedical Engineering Handbook: Second Edition.* Ed. by Joseph D. Bronzino. Boca Raton: CRC Press LLC, 2000. Chap. 48. URL: https://web.archive.org/web/20220330061138if_/http://www.fis.uc.pt/data/20062007/apontamentos/apnt_134_5.pdf (visited on 29 Aug. 2022) (cit. on p. 508).

[80] Nexperia. *Understanding Power MOSFET Data Sheet Parameters.* 6 July 2020. URL: https://assets.nexperia.com/documents/application-note/AN11158.pdf (visited on 27 Aug. 2022) (cit. on p. 448).

[81] Craig D. Oster. *Proper Skin Prep Helps Ensure ECG Trace Quality.* 2005. URL: https://multimedia.3m.com/mws/media/358372O/proper-skin-prep-ecg-trace-quality-white-paper.pdf (visited on 5 Dec. 2018) (cit. on p. 543).

[82] Overleaf. *Documentation.* URL: https://www.overleaf.com/learn (visited on 14 Nov. 2018) (cit. on p. 78).

[83] Oxford University Press. *Oxford Learner's Dictionaries.* 2019. URL: https://www.oxfordlearnersdictionaries.com/us (visited on 6 July 2019) (cit. on p. 102).

[84] Richard Palmer. *DC Parameters: Input Offset Voltage (V_{IO}).* SLOA059. Texas Instruments. Mar. 2001. URL: https://www.ti.com/lit/an/sloa059/sloa059.pdf (visited on 27 Aug. 2022) (cit. on p. 252).

[85] Philips Healthcare. *Improving ECG Quality.* Sept. 2008. URL: https://www.documents.philips.com/doclib/enc/fetch/577817/577869/Improving_ECG_Quality_Application_Note_(ENG).pdf (visited on 24 Apr. 2023) (cit. on p. 543).

[86] Physio Control. *DIAG Frequency Response Necessary to Reproduce ST Segment.* 2009. URL: https://emtlife.com/attachments/diag-frequency-response-necessary-to-reproduce-st-segment-3304790-a-3-pdf.1603 (visited on 18 Nov. 2022) (cit. on p. 534).

[87] PJRC. *Teensy Pin Assignments.* URL: https://www.pjrc.com/teensy/pinout.html (visited on 15 Nov. 2018) (cit. on p. 547).

[88] John Prymak *et al.* "Why That 47 uF Capacitor Drops to 37 uF, 30 uF, or Lower." In *Proceedings CARTS USA 2008, 28th Symposium for Passive Electronics.* Newport Beach, CA: Electronics Components, Assemblies & Materials Association, Mar. 2008. URL: https://www.kemet.com/content/dam/kemet/lightning/documents/ec-content/technical-archive/Why-47-uF-capacitor-drops-to-37-uF-30-uF-or-lower.pdf (visited on 18 Nov. 2022) (cit. on pp. 148–150).

[89] Purdue Online Writing Lab. *Extended Rules for Using Commas.* 2020. URL: https://owl.purdue.edu/owl/general_writing/punctuation/commas/extended_rules_for_commas.html (visited on 27 Feb. 2021) (cit. on p. 110).

[90] Andrew Randazzo. *The Ultimate 12-Lead ECG Placement Guide (with Illustrations).* 1 June 2016. URL: https://www.primemedicaltraining.com/12-lead-ecg-placement/ (visited on 5 Dec. 2018) (cit. on p. 533).

[91] Arie Van Rhijn. *LMV1012: Integrated Circuits for High Performance Electret Microphones.* SNAA114. Texas Instruments. 2011. URL: https://www.ti.com/lit/wp/snaa114/snaa114.pdf (cit. on p. 361).

[92] *Salts Weight vs. Volume Conversion.* URL: http://convert-to.com/types-of-salts-weight-volume-converters (visited on 24 Apr. 2023) (cit. on p. 516).

[93] Warren Schultz. *Interfacing Semiconductor Pressure Sensors to Microcomputers.* URL: https://www.nxp.com/webapp/Download?colCode=AN1318&location=null (visited on 18 Nov. 2022) (cit. on pp. 289, 290).

[94] Stanford Medicine. *Measuring and Understanding the Ankle Brachial Index (ABI).* URL: https://stanfordmedicine25.stanford.edu/the25/ankle-brachial-index.html (visited on 9 Dec. 2018) (cit. on p. 282).

[95] Milan Stork and Jiri Jilek. "Cuff Pressure Waveforms: Their Current and Prospective Application in Biomedical Instrumentation." In *Biomedical Engineering, Trends in Electronics, Communications and Software.* Ed. by Anthony N. Laskovski. InTech, 8 Jan. 2011. ISBN: 978-953-307-475-7. DOI: 10.5772/13475. URL: https://www.intechopen.com/chapters/12908 (visited on 27 Aug. 2022) (cit. on p. 283).

[96] Robert Talbert. *Examples and the Light Bulb.* 25 Mar. 2013. URL: https://www.chronicle.com/blognetwork/castingoutnines/2013/03/25/examples-and-the-light-bulb/ (cit. on p. ix).

[97] Texas Instruments. *INAx126 MicroPower Instrumentation Amplifier, Single and Dual Versions.* Dec. 2015. URL: https://www.ti.com/lit/ds/symlink/ina126.pdf (visited on 8 Dec. 2018) (cit. on pp. 252, 531).

[98] Texas Instruments. *LM339, LM239, LM139, LM2901 Quad Differential Comparators.* Version Rev. U. Nov. 2018. URL: https://www.ti.com/lit/ds/symlink/lm139.pdf (visited on 8 Dec. 2019) (cit. on p. 460).

[99] *Theory and Practice of pH Measurement.* PN 44-6033/rev. D. Emerson Process Management. Dec. 2010. URL: https://www.emerson.com/documents/automation/manual-theory-practice-of-ph-measurement-rosemount-en-70736.pdf (visited on 5 Sept. 2020) (cit. on p. 257).

[100] Robert H. Tyler *et al.* "Electrical Conductivity of the Global Ocean." *Earth, Planets, and Space* 69(156) (14 Nov. 2017). DOI: 10.1186/s40623-017-0739-7. URL: https://earth-planets-space.springeropen.com/articles/10.1186/s40623-017-0739-7 (visited on 27 Aug. 2022) (cit. on p. 516).

[101] Vishay Siliconix. *Power MOSFET Basics: Understanding Gate Charge and Using it to Assess Switching Performance.* Application Note 608A. Vishay Siliconix. 16 Feb. 2016. URL: https://www.vishay.com/docs/73217/an608a.pdf (visited on 24 Dec. 2018) (cit. on p. 449).

[102] Vyaire Medical. *MicroGard®.* URL: https://www.vyaire.com/products/microgard-ii-pft-filter (visited on 5 Sept. 2020) (cit. on p. 280).

[103] J. Watson. "The Tin Oxide Gas Sensor and its Applications." *Sensors and Actuators* 5 (1984), pp. 29–42 (cit. on p. 60).

[104] Grant Wiggins. *Autonomy and the need to back off by design as teachers.* 12 Feb. 2013. URL: https://grantwiggins.wordpress.com/2013/02/12/autonomy-and-the-need-to-back-off-by-design-as-teachers/ (cit. on p. x).

[105] Wikibooks. *LaTeX.* URL: https://en.wikibooks.org/wiki/LaTeX (visited on 15 June 2014) (cit. on p. 78).

[106] wikiHowStaff. *How to Solder Aluminum.* 18 June 2019. URL: https://www.wikihow.com/Solder-Aluminum (visited on 24 Aug. 2019) (cit. on p. 242).

[107] Wikipedia. *A-weighting.* URL: https://en.wikipedia.org/wiki/A-weighting (visited on 19 Dec. 2016) (cit. on p. 421).

[108] Wikipedia. *Action potential.*
URL: https://en.wikipedia.org/wiki/Action_potential (visited on 26 May 2015) (cit. on p. 538).

[109] Wikipedia. *Blood pressure.*
URL: https://en.wikipedia.org/wiki/Blood_pressure (visited on 7 Sept. 2014) (cit. on p. 281).

[110] Wikipedia. *Callendar-Van Dusen equation.*
URL: https://en.wikipedia.org/wiki/Callendar%E2%80%93Van_Dusen_equation (visited on 29 Aug. 2022) (cit. on p. 57).

[111] Wikipedia. *Capacitance.*
URL: https://en.wikipedia.org/wiki/Capacitance (visited on 4 July 2014) (cit. on p. 146).

[112] Wikipedia. *Capacitive coupling.*
URL: https://en.wikipedia.org/wiki/Capacitive_coupling (visited on 7 Apr. 2015) (cit. on p. 211).

[113] Wikipedia. *Carbonic Acid.*
URL: https://en.wikipedia.org/wiki/Carbonic_acid (visited on 28 Dec. 2018) (cit. on p. 507).

[114] Wikipedia. *Cardiac Action Potential.*
URL: https://en.wikipedia.org/wiki/Cardiac_action_potential (visited on 16 Dec. 2017) (cit. on p. 538).

[115] Wikipedia. *Class-D amplifier.*
URL: https://en.wikipedia.org/wiki/Class_d_amplifier (visited on 17 Feb. 2013) (cit. on p. 499).

[116] Wikipedia. *Color blindness.*
URL: https://en.wikipedia.org/wiki/Color_blindness (visited on 6 Sept. 2015) (cit. on p. 206).

[117] Wikipedia. *Concert Pitch.*
URL: https://en.wikipedia.org/wiki/Concert_pitch (visited on 2 July 2017) (cit. on p. 11).

[118] Wikipedia. *Conductivity (electrolytic).*
URL: https://en.wikipedia.org/wiki/Conductivity_(electrolytic) (visited on 18 Dec. 2015) (cit. on pp. 505, 507).

[119] Wikipedia. *Decoupling capacitor.*
URL: https://en.wikipedia.org/wiki/Bypass_capacitor (visited on 4 July 2014) (cit. on p. 185).

[120] Wikipedia. *Diode modelling.*
URL: https://en.wikipedia.org/wiki/Diode_modelling (visited on 28 Sept. 2016) (cit. on p. 315).

[121] Wikipedia. *Direct coupling.*
URL: https://en.wikipedia.org/wiki/Direct_coupling (visited on 7 Apr. 2015) (cit. on p. 211).

[122] Wikipedia. *Electret.* URL: https://en.wikipedia.org/wiki/Electret (visited on 16 June 2014) (cit. on p. 359).

[123] Wikipedia. *Electret microphone.*
URL: https://en.wikipedia.org/wiki/Electret_microphone (visited on 16 June 2014) (cit. on p. 359).

[124] Wikipedia. *Electric shock.*
URL: https://en.wikipedia.org/wiki/Electric_shock (visited on 26 May 2015) (cit. on p. 536).

[125] Wikipedia. *Electrical resistance.*
URL: https://en.wikipedia.org/wiki/Electrical_resistance (visited on 15 June 2014) (cit. on p. 36).

[126] Wikipedia. *Electrocardiography.*
URL: https://en.wikipedia.org/wiki/EKG (visited on 7 Mar. 2013) (cit. on p. 538).

[127] Wikipedia. *Electrolytic capacitor.*
URL: https://en.wikipedia.org/wiki/Electrolytic_capacitor (visited on 25 June 2014) (cit. on p. 149).
[128] Wikipedia. *Engineering notation.*
URL: https://en.wikipedia.org/wiki/Engineering_notation (visited on 29 Mar. 2018) (cit. on p. 78).
[129] Wikipedia. *Inductance.*
URL: https://en.wikipedia.org/wiki/Inductance (visited on 6 Aug. 2014) (cit. on p. 392).
[130] Wikipedia. *Inductance: Self-inductance of thin wire shapes.*
URL: https://en.wikipedia.org/wiki/Inductance#Self-inductance_of_thin_wire_shapes (visited on 1 Aug. 2020) (cit. on p. 389).
[131] Wikipedia. *ITU-R 468 Noise Weighting.*
URL: https://en.wikipedia.org/wiki/ITU-R_468_noise_weighting (visited on 13 Oct. 2016) (cit. on p. 363).
[132] Wikipedia. *JFET.* URL: https://en.wikipedia.org/wiki/JFET (visited on 29 Oct. 2017) (cit. on p. 362).
[133] Wikipedia. *Lanczos Resampling.*
URL: https://en.wikipedia.org/wiki/Lanczos_resampling (visited on 22 Oct. 2018) (cit. on p. 141).
[134] Wikipedia. *Magnetic core.*
URL: https://en.wikipedia.org/wiki/Magnetic_core (visited on 6 Aug. 2014) (cit. on p. 390).
[135] Wikipedia. *Microphone.*
URL: https://en.wikipedia.org/wiki/Microphone (visited on 16 June 2014) (cit. on p. 359).
[136] Wikipedia. *Microphone: Condenser microphone.* URL:
https://en.wikipedia.org/wiki/Microphone#Capacitor_microphone (visited on 27 Aug. 2022) (cit. on p. 359).
[137] Wikipedia. *MIDI Tuning Standard.*
URL: https://en.wikipedia.org/wiki/MIDI_tuning_standard (visited on 2 July 2017) (cit. on p. 14).
[138] Wikipedia. *Molar conductivity.*
URL: https://en.wikipedia.org/wiki/Molar_conductivity (visited on 19 Dec. 2015) (cit. on p. 505).
[139] Wikipedia. *Near-infrared Window in Biological Tissue.*
URL: https://en.wikipedia.org/wiki/Near-infrared_window_in_biological_tissue (visited on 29 Sept. 2016) (cit. on p. 311).
[140] Wikipedia. *Nernst Equation.*
URL: https://en.wikipedia.org/wiki/Nernst_equation (visited on 19 Dec. 2015) (cit. on p. 510).
[141] Wikipedia. *Nyquist-Shannon sampling theorem.* URL:
https://en.wikipedia.org/wiki/Nyquist-Shannon_sampling_theorem (visited on 19 June 2014) (cit. on p. 139).
[142] Wikipedia. *Operational amplifier.*
URL: https://en.wikipedia.org/wiki/Operational_amplifier (visited on 19 Dec. 2016) (cit. on p. 421).
[143] Wikipedia. *Operational amplifier applications.* URL:
https://en.wikipedia.org/wiki/Operational_amplifier_applications (visited on 19 Dec. 2016) (cit. on p. 421).

[144] Wikipedia. *Oscilloscope.*
URL: https://en.wikipedia.org/wiki/Oscilloscope (visited on 20 July 2014) (cit. on p. 209).

[145] Wikipedia. *Phasor.* URL: https://en.wikipedia.org/wiki/Phasor (visited on 7 Aug. 2014) (cit. on p. 153).

[146] Wikipedia. *Photodiode.*
URL: https://en.wikipedia.org/wiki/Photodiode (visited on 4 Feb. 2013) (cit. on p. 304).

[147] Wikipedia. *Photometry (optics).*
URL: https://en.wikipedia.org/wiki/Photometry_(optics) (visited on 23 Aug. 2019) (cit. on p. 303).

[148] Wikipedia. *Pulse-width Modulation.*
URL: https://en.wikipedia.org/wiki/Pulse-width_modulation (visited on 27 Aug. 2022) (cit. on p. 473).

[149] Wikipedia. *Purified water.*
URL: https://en.wikipedia.org/wiki/Purified_water (visited on 3 Jan. 2017) (cit. on p. 507).

[150] Wikipedia. *Relaxation Oscillator.*
URL: https://en.wikipedia.org/wiki/Relaxation_oscillator (visited on 27 Aug. 2022) (cit. on pp. 221, 236).

[151] Wikipedia. *Response of Silicon Photodiode.* URL:
https://en.wikipedia.org/wiki/File:Response_silicon_photodiode.svg (visited on 27 Aug. 2022) (cit. on p. 310).

[152] Wikipedia. *Sallen-Key topology.*
URL: https://en.wikipedia.org/wiki/Sallen%E2%80%93Key_topology (visited on 7 Nov. 2018) (cit. on p. 343).

[153] Wikipedia. *Shockley diode equation.*
URL: https://en.wikipedia.org/wiki/Shockley_diode_equation (visited on 28 Sept. 2016) (cit. on p. 315).

[154] Wikipedia. *Silver chloride electrode.*
URL: https://en.wikipedia.org/wiki/Silver_chloride_electrode (visited on 19 Apr. 2015) (cit. on p. 510).

[155] Wikipedia. *Sinc Filter.*
URL: https://en.wikipedia.org/wiki/Sinc_filter (visited on 22 Oct. 2018) (cit. on p. 141).

[156] Wikipedia. *Skin Effect: Impedance of a Round Wire.* URL:
https://en.wikipedia.org/wiki/Skin_effect#Impedance_of_round_wire (visited on 1 Aug. 2020) (cit. on p. 389).

[157] Wikipedia. *Sound Level: Sound Pressure Level.* URL:
https://en.wikipedia.org/wiki/Sound_pressure#Sound_pressure_level (visited on 27 Aug. 2022) (cit. on pp. 362, 363, 421).

[158] Wikipedia. *Spectral sensitivity.*
URL: https://en.wikipedia.org/wiki/Spectral_sensitivity (visited on 3 May 2014) (cit. on p. 302).

[159] Wikipedia. *Stainless steel.*
URL: https://en.wikipedia.org/wiki/Stainless_steel (visited on 19 Apr. 2015) (cit. on p. 509).

[160] Wikipedia. *Standard hydrogen electrode.*
URL: https://en.wikipedia.org/wiki/Standard_hydrogen_electrode (visited on 19 Dec. 2015) (cit. on p. 508).

[161] Wikipedia. *Steinhart-Hart equation.*
URL: https://en.wikipedia.org/wiki/Steinhart-Hart_equation (visited on 18 June 2014) (cit. on p. 56).

[162] Wikipedia. *Sunlight.* URL: https://en.wikipedia.org/wiki/Sunlight (visited on 26 Sept. 2016) (cit. on p. 315).

[163] Wikipedia. *Surgical steel.*
URL: https://en.wikipedia.org/wiki/Surgical_steel (visited on 19 Apr. 2015) (cit. on p. 509).

[164] Wikipedia. *Thermistor.*
URL: https://en.wikipedia.org/wiki/Thermistor (visited on 4 July 2014) (cit. on p. 56).

[165] Wikipedia. *Thévenin's Theorem.*
URL: https://en.wikipedia.org/wiki/Th%C3%A9venin%27s_theorem (visited on 27 Aug. 2022) (cit. on p. 49).

[166] S.H. Wilson *et al.* "Predicted Normal Values for Maximal Respiratory Pressures in Caucasian Adults and Children." *Thorax* 39(7) (July 1984), pp. 535–538 (cit. on p. 279).

[167] Wolfram Alpha LLC. *Wolfram|Alpha: Computational Intelligence.*
URL: https://www.wolframalpha.com/ (visited on 1 Feb. 2021) (cit. on p. 174).

[168] Al Wright. *Printed Circuit Board Surface Finishes—Advantages and Disadvantages.*
URL: https://www.epectec.com/articles/pcb-surface-finish-advantages-and-disadvantages.html (visited on 18 July 2017) (cit. on p. 25).

[169] Würth Elektronik. *151034BS03000 WL-TMRW THT Mono-color Round Waterclear.* 2 Nov. 2018.
URL: https://www.we-online.com/components/products/datasheet/151034BS03000.pdf (visited on 29 Jan. 2023) (cit. on p. 303).

Index

74AC04: *see* inverter
74HC14N: *see* Schmitt trigger

A-weighting: 363
AC coupling: *see* capacitive coupling
acknowledgment: v, 118–119
action potential: 537–538
admittance: 400
affine: *see* functions, affine
aliasing: 141–144, 319–320
alligator clips: 125–126
ambient temperature: *see* temperature, ambient
amplifier: 247–278
  class-D: 471–486
  common-mode rejection ratio: 253, 530
  current: 247
  differential: 247–248
  input bias current: 252–253, 527, 530
  input offset voltage: 252, 255–256, 264, 267–270
  instability: 321–326, 332–334
  instrumentation: 259–262, 527–532
  inverting: 267
  multi-stage: 253–256
  negative feedback: 265–272
  non-inverting: 267
  operational: 263–264, 527–532
  power-supply rejection ratio: 253, 484
  rail-to-rail: 251
  saturation: 255
  transconductance: 247, 521–522
  transimpedance: 247, 313–321
amplitude: *see* voltage, amplitude
amplitude envelope: 282–283
Analog Discovery 2: 151, 197, 229–231, 242, 255, 275–276, 324–325, 342, 345, 367–368, 391, 399, 414, 425–426, 429, 447, 454–456, 490, 493–495, 497, 541–542
  current limit (function generator): 188, 524
  current limit (power supply): 425, 500–501
  function generator: 188–190, 202–204, 206, 236, 240, 243–244
  impedance analyzer: 405, 409–414, 511, 518–520
analog-to-digital converter: 69–70, 547–551
  differential: 379
  sigma-delta: 485
angular frequency ω: 67
anode: 302, 505
Arduino: 30–31, 548–549
attenuation: 11

B-equation: *see* thermistor, B-equation
beat frequency: 139–141
bias resistor: *see* resistor, bias
biblatex: 117–120
block diagram: 60–62, 85–88, 200
Bode plot: 160, 166, 174–177, 179–180, 182–183, 201, 325, 330, 333, 335, 338–339, 341, 345
body diode: *see* diode, body
Boltzmann's constant: 58, 315
breadboard: 129–133
breakout board: xii, 289, 360, 463–464, 467–469
bus: 129
bypass capacitor: *see* capacitor, bypass

capacitive coupling: 200–202, 211–212, 384, 534
capacitive sensor: 230–234

capacitor: 145–152
  bypass: 184–185, 240–241, 429–430, 463–465, 468
  ceramic: 147–148
  charging and discharging: 166–169, 221–225
  DC-blocking: 171, 200–202, 211–212, 421
  dissipation factor: 150–151
  electrolytic: 149–152, 201, 443
    bipolar: 150
  feedback: 226, 228
  marking: 147
capacitor microphone: *see* microphone, capacitor
Cartesian coordinates: 15
cathode: 302, 505
centering a signal: *see* recentering a signal
ceramic capacitor: *see* capacitor, ceramic
charge rate: *see* capacitor, charging and discharging
chemiresistor: 39, 59
citation: 117–120
clipping: 199, 250–252, 272–273, 319, 355, 425–426, 502, 540
cMOS: 101, 439–442
CMRR: *see* amplifier, common-mode rejection ratio
color coding (wire): 189, 194–196, 204, 293–294, 428
common-mode voltage: 248, 253, 528–529, 531–532, 534–536
comparator: 218, 453–461, 475–476, 481–483
compensation
  impedance analyzer: 412–414
complex impedance: *see* impedance
complex numbers: 14–16, 152–157
component: 34, 63
conductance: 35
conductivity: 505–508, 513
corner frequency: 21, 160–161, 166, 171, 320, 390, 391, 477, 553
countable and uncountable nouns: 100–103, 105
crest factor: 406
current: 34–35
  RMS: 405
  saturation, FET: 361, 366
  saturation, inductor: 392
  shoot-through: 439, 450
  short-circuit: 50–51, 170, 366
current amplifier: *see* amplifier, current
current-limiting resistor: *see* resistor, current-limiting
cutoff frequency: 160, 278, 284–287, 338

DAQ (data-acquisition system): 68–70
dark current: *see* photodiode, dark current
data sheet: 122–123
DC bias: 65, 87, 293, 365–367
DC coupling: 211–212, 384
debugging: x–xi, 23, 37, 88, 132–133, 191–198, 293–296, 350, 421, 424, 428, 430, 454
decade: 11, 14, 37, 308, 316, 417, 518, 520
decibel (dB): 10, 12, 362–364, 372–374
decoupling capacitor: *see* capacitor, bypass
definitions.gnuplot: 93–94, 162–165
depletion region: *see* diode, depletion region
diastolic pressure: *see* pressure, blood (systolic and diastolic)
dielectric constant: 146–149, 231, 233, 237
differential
  amplifier: *see* amplifier, differential
  measurement: *see* voltage, differential
  pair: *see* voltage, differential
  signal: *see* voltage, differential
  voltage: *see* voltage, differential
dimensional analysis: 8–10, 381, 399
diode: 301–305
  body: 434, 443–444, 446
  depletion region: 301, 304–305
  fly-back: 434, 443–446
  reverse bias: 301–302, 304–307, 314, 434, 436, 439, 443–444
  Shockley model: 315
  temperature sensor: 58–59
  voltage: 301
  Zener: 537
discharge rate: *see* capacitor, charging and discharging
dissipation factor $RC\omega$: *see* capacitor, dissipation factor

distortion: 187, 210, 250–252, 372, 398, 426, 432, 484, 487
 harmonic: 251, 486
dominant wavelength: *see* LED, dominant wavelength
downsampling: 203–205
drain: *see* FET, drain
draw.io: 61–62, 69, 84, 86–87, 90, 200, 256, 258, 261, 484, 486, 528
duty cycle: 473–474, 484
dynamic microphone: *see* microphone, dynamic

edge: 168
EEG (electroencephalogram): 535
EKG (electrocardiogram): 533–537, 539–545
 electrode placement: 533–534
electret microphone: *see* microphone, electret
electrode: 505–513
 non-polarizable: 508–511
 polarizable: 508–511, 513
electrolyte: 505–508
electrolytic capacitor: *see* capacitor, electrolytic
EMG (electromyogram): 533
EMI (electromagnetic interference): 446, 477
engineering design: 2
engineering notation: 78
ϵ_0: *see* permittivity of free space ϵ_0
ESR (effective series resistance): 150–152, 156, 443
Euler's formula, viii: 15–16, 152–153

fall time: 168–169
fancyvrb LaTeXpackage: 99, 404
farad (unit, F): 9, 145
fencepost error: 285, 357, 411, 545
FET (field-effect transistor): 361–362, 366–369, 433–452
 drain: 433
 gate: 433
 gate-charge model: 446–448
 linear region: 361–362, 366
 saturation region: 361–362, 366–367
 source: 433
FFT (Fast Fourier Transform): 368, 384
filter
 band-pass: 173–179, 282–283, 288, 320, 334–341
 band-stop: 181–183
 high-pass: 159, 166, 265, 282, 330, 334
 high-pass, cascaded: 334
 infinite impulse response: 287
 LC: 477–479
 low-pass: 139–140, 143, 159–166, 185, 265, 282–283, 288, 314, 328–330, 477–479, 540, 544
 low-pass, cascaded: 329–330
 notch: 181
 order: 327
 Python program: 286–288
fitting models: 96, 401–405
flat response: 216, 423
floating insertions: 82–84
floating-point notation: 78
flyback diode: *see* diode, flyback
flying resistor: *see* resistor, flying
four-wire resistance measurements: 127, 513
FRDM-KL25Z: 243, 320, 547–548, 551
free-wheeling diode: *see* diode, flyback
frequency
 negative: 68, 141–143
frequency channel (PteroDAQ): 70, 547
frequency modulation (FM): 227, 232–233, 242
function generator: 68, 187–190
functions
 affine: 12, 105, 381
 constant: 12, 105
 exponential: 12
 linear: 12, 14, 105
 logarithmic: 14
 power-law: 14, 381–382
fundamental frequency: 251, 320, 356

gain: 17, 21, 52, 160–166, 173, 179–181, 183, 191–192, 201, 215–220, 230, 248, 250–251, 253, 256–264, 270, 272, 278, 280, 289–290, 292–296, 299, 306, 308, 313–317, 319, 321, 326, 328, 345, 351, 354, 356, 371, 373–374, 419, 423–426, 430, 485, 527–532, 534–535, 540–541, 553
 open-loop: 264, 322
 voltage divider: 46, 53–54

gain-bandwidth product: 249–250, 256, 259, 274–278, 322
gate-charge model: *see* FET, gate-charge model
gate: *see* FET, gate
gauge pressure: *see* pressure, gauge
gnuplot: 91–99, 128, 162–166, 205, 380–381, 402–404
    installing: 31–32
ground: 33–34, 46, 48–49, 61, 63–64, 89–90, 189, 195, 216, 222, 226–230, 247, 421, 539

H-bridge: 112, 444–446, 475–476, 478–480, 484, 501, 503
    half: 439–442, 463–467
harmonic distortion: *see* distortion, harmonic
harmonic oscillator: 398
harmonics: 251, 284, 320
henry (unit, H): 9, 388
hertz (unit, Hz): 9, 67
high-pass filter: *see* filter, high-pass
hum (60 Hz): 143–144, 227, 231–232, 283, 296, 319–320, 331, 352–353, 454, 484, 536
hydraulic analogy: 42–43, 145, 387
hysteresis: 217–220, 238–240
    oscillator: *see* oscillator, hysteresis
    voltage: 218–219

I-vs.-V plot: 35, 307–309, 315, 317–318, 349, 365–366, 369, 373, 375–385, 436, 438, 441, 455–456, 496
I^2C digital interface: 2, 287
impedance: 152–157, 390–393
    input: 162, 166, 170, 197, 214, 216, 257, 259, 270, 273, 328–329, 331, 334, 339–341, 360, 382–383, 421, 423, 461, 530, 536
    nonstandard: 398–399, 404, 511
    output: 187–188, 226, 257, 267, 360
    parallel: 153–154
    series: 153–154
impedance analyzer: xiv, 409–414, 518–520
impedance spectroscopy: 409, 515
inductance: 387–393
    mutual: 388
    self: 387
inductive load: 442–444
inductive spike: 443–444, 466–468

inkscape: 72, 396
input bias current: *see* amplifier, input bias current
input impedance: *see* impedance, input
input offset voltage: *see* amplifier, input offset voltage
input resistance: *see* resistance, input
integrator: 487–488
    summing: 484–485
interference
    60 Hz: *see* hum (60 Hz)
    electromagnetic: *see* EMI (electromagnetic interference)
inverter: 219, 221, 223, 226, 232, 235, 240, 246, 440–442, 464–466, 468, 475–477, 481–483, 501, 503
inverting amplifier: *see* amplifier, inverting

junction temperature: *see* temperature, junction

Kelvin clip: *see* four-wire resistance measurements
Kirchhoff's current law: 35, 45, 49, 343, 530–531

Laplace transform: 343–345
LaTeX, 4, 77, 82–83, 85, 100, 111, 113–114, 117, 120, 567
LC tank: 392–393
LED (light-emitting-diode): 302–304, 347–354, 474, 537
    dominant wavelength: 302–304
    peak wavelength: 302–304, 310–311
LEGO®, 112, 347, 350–351
level shifting: *see* recentering a signal
load: 169
load line: 365–367
load resistor: *see* resistor, load
logarithms: 10–14, 82, 94, 315–320, 362–364, 372–373, 402, 411
loudspeaker: 395–407
    cone: 395–398, 405
    dust cap: 396
    measurement: 409–418
    models: 396–405, 409–418
    schematic: 396

spider: 395–396
surround: 395–396
voice coil: 395–396, 398, 405
low-pass filter: *see* filter, low-pass
low-side switch: 436, 443–444
lumen: 303

magnetic constant μ: 389
matplotlib: 140
MCP6004: *see* amplifier, operational
metric prefixes: 7–8, 11, 78–79, 94, 112
metric units: 7–9
mho, ℧: 35
microphone: 359–374
capacitor: 359
dynamic: 359
electret: 359–361
MEMS: 360
models for electret: 365–372, 381
piezoelectric: 360
sensitivity: 364–365
types: 359–360
Miller
capacitance: 447
effect: 308
plateau: 446–448, 452, 466, 468–469
voltage: *see* Miller, plateau
model: 2, 33, *see also* functions
fitting: 128–129, 381–382, 417–418
Moiré pattern: 141
MPX2050DP pressure sensor: 287, 289
mutual inductance: *see* inductance, mutual

n-doped semiconductor: 301–302, 304–306, 434
naked formulas: 79
negative thermal coefficient (NTC): 54–55, 123
network analyzer: xiv, 299, 356, 426, 502–503
node: 3, 33–35, 46, 48, 63–64, 194–196
noise: 70, 105–106, 165, 185, 187, 230, 236, 243, 246, 321–322, 325–326, 363, 454, 461, 468–469, 489, 535–536, 541, 543
injection: 197–198
nominal value: 106, 184
non-inverting amplifier: *see* amplifier, non-inverting
nonstandard impedance: *see* impedance, nonstandard
noun cluster: 115–116
NPN transistor: 306–307, 314, 357
Nyquist frequency: 139–143, 287, 545

octave: 11, 14, 316, 497
offset voltage: *see* amplifier, input offset voltage
ohm (unit, Ω): 9, 35
Ohm's law: viii, 20, 32, 35, 39–42, 45–46, 49, 112, 153, 304, 313, 372, 383, 487, 495
on-resistance: 437–439, 446, 451–452, 474, 481, 491–492
op amp: *see* amplifier, operational
open-circuit voltage: *see* voltage, open-circuit
open-collector output: 455–456, 460–461, 482–483
open-loop gain: *see* gain, open-loop
optimization: 18–19, 124–125
optoisolator: 537
oscillator
hysteresis: 221–228, 231, 234–235, 238–246
sawtooth: 494–495
triangle-wave: 487–497
voltage-controlled: 490–497
oscilloscope: xv, 209–216
analog: 209
Analog Discovery 2: xiv, xv, 130, 151, 210–215, 229–231, 255, 275–276, 299, 324–325, 345, 367–368, 377–378, 383–384, 391, 399, 410, 414, 447, 454, 456, 490, 493–495, 497
BitScope BS10: xv, 210
digital: 209–211
input impedance: 214–216
Kikusui COS5041: xv
Tektronix TDS3052: xv
triggering: 212–213
USB: xv, 210–211
oscilloscope probe: 212, 214–216

p-doped semiconductor: 301–302, 304–306, 434, 438
parallel connection: 34–35, 39, 42
pascal (unit, Pa): 9, 18, 290, 364

patch-clamp amplifier: 313
peak wavelength: *see* LED, peak wavelength
permittivity of free space ϵ_0: 146, 233
pH meter: 257–258
phase: 67
phase margin: 322, 325–326
phase oscillator: 68
phasor: 153
  amplitude: 472
  RMS: 471–472
photodarlington: 306
photodiode: 304–305, 313
  dark current: 305
photoresistor: 39, 59, 306, 308
phototransistor: 144, 305–309, 314–315, 317, 319, 347–350, 352–353, 357
piezoelectric microphone: *see* microphone, piezoelectric
piezoresistor: 39, 59
PNP transistor: 306
polar coordinates: 15
port: 46, 63–64, 85–86, 89–90
positive thermal coefficient (PTC): 57
potentiometer: 51–52, 54, 239, 272–274, 376, 378–379, 427
power: 42, 405–407
  apparent: 472
  average: 405–407
  complex: 472
  instantaneous: 405
  reactive: 472
  real: 471–473
  RMS: 405
power rails: 171
power supplies: 425
power-law model: *see* functions, power-law
preamp: 419–432, 499–502
prepositions: 34, 64, 106–107, 111
pressure
  absolute: 287
  blood (mean arterial): 282–283, 286
  blood (systolic and diastolic): 281–288, 354
  breath: 279–280
  differential: 287
  gauge: 287
propagation delay: 229–230
PSRR: *see* amplifier, power-supply rejection ratio
PteroDAQ: 30–31, 59, 61, 68–70, 134–136, 144, 199, 203–204, 206, 210, 223, 232–233, 235, 238–239, 242, 246, 281, 284–286, 307–308, 318, 354, 378–379, 381, 384–385, 535, 547–551
  installation: 30–31
  sampling rate: 550
pull-down resistor: *see* resistor, pull-down
pull-up resistor: *see* resistor, pull-up
pulse rate: 282–286, 299, 357, 535, 545
pulse-width modulation (PWM): 473–486, 500, 502

RC time constant: 146, 160–169, 171, 176–178, 200–202, 222–223, 227, 235–236, 242, 323, 332, 338, 340, 453–454, 511
recentering a signal: 171–173, 200–202, 421, 482–483
relaxation oscillator: *see* oscillator, hysteresis
resistance: 35–39
  dynamic: 35
  input: 45
  thermal: *see* thermal resistance
resistance-temperature detector (RTD): 39, 57
resistivity: 506–507, 513
resistor: 36–39
  bias: 39, 365, 373
  color code: 36
  current-limiting: 39
  flying: 131–132
  load: 39, 365, 437
  pull-down: 39
  pull-up: 39
  sense: 38–39, 365, 511, 518, 520
  shunt: 38
  standard values: 37
resistors
  parallel: 39–42
  series: 39–42, 45
resonant frequency: 392–393, 398–402, 411, 417, 479
responsivity: 17
reverse bias: *see* diode, reverse bias
rise time: 168–169
RMS phasor: *see* phasor, RMS
RMS voltage: *see* voltage, RMS

roll-off: 327
root-mean-square voltage: *see* voltage, RMS

sampling: 139–144
  frequency: 139–143, 199–207, 242, 298, 319–320
  period: 139
  rate: 139
saturation current: *see* current, saturation
saturation voltage: *see* voltage, saturation
sawtooth oscillator: *see* oscillator, sawtooth
Scheme-it: 53, 84, 88, 90–91, 207, 365, 541, 567
Schmitt trigger: 218–219, 221–225, 227, 231, 235–246, 456–461, 483
scientific notation: 78–79
Seebeck effect: 58
self inductance: *see* inductance, self
semi-inductance: 399
sense resistor: *see* resistor, sense
sensitivity: 17, 56, 364
sensor: 1–2
  alcohol: 59–60
  chemical: 59–60
  ionic current: 505–513
  optical: 59, 304–309
  pH: 257–258
  pressure: 279–290
  resistance-based: 54–60
  sound: 359–374
  strain: 59
  temperature: 54, 59
  touch: 222, 227, 231, 234–235, 238, 241, 243, 246
series connection: 34–35, 39, 42
settling time: 168–169, 411
Shockley diode model: *see* diode, Shockley model
short-circuit current: *see* current, short-circuit
shunt resistor: *see* resistor, shunt
siemens (unit, S): 8–9, 35, 112, 248, 507
signal: 63
signal-to-noise ratio (SNR): 213, 363
silver
  fine: 510
  sterling: 510
silver/silver chloride electrode: 509–510, 515–517, 521–524
sinusoid: 152–153, 156, 161, 390, 471–472
slew rate: 448, 453–454, 481, 484, 488–490, 492–495, 497
soldering: xiii, 24–29, 235, 244, 246, 349–351, 426–431, 463–464
sound pressure level (SPL): 362, 372–374
source: *see* FET, source
sparkline: 70, 298, 353, 384–385, 549
speed of sound: 258
SPI digital interface: 2, 287
stainless steel: 509
Steinhart-Hart equation: *see* thermistor, Steinhart-Hart equation
step function: 168
steradian: 303
summing integrator: *see* integrator, summing
summing junction: 485
systolic pressure: *see* pressure, blood (systolic and diastolic)

teaching design: 2–4
Teensy microcontroller: xii, 30–31, 59, 61, 69–70, 135, 199, 204, 210, 223, 231, 240, 365, 425, 548–551
temperature
  ambient: 449
  junction: 449
  measurement: 54–59, 121–137
thermal resistance: 449
thermistor: 39, 54–57, 121–137
  B-equation: 55
  Steinhart-Hart equation: 56
thermocouple: 57–58
Thévenin equivalence: 49–52, 169–171, 173, 269
thin space: 78, 81, 113
threshold voltage: 435
transconductance amplifier: *see* amplifier, trans-conductance
transfer function: 159, 217, 249, 251
transimpedance amplifier: *see* amplifier, transimpedance
triangle-wave oscillator: *see* oscillator, triangle-wave

trimpot: *see* potentiometer
tweeter: 398

ultrasound imaging: 258–259
unity-gain buffer: 192, 270–272, 319, 489, 496–497, 531, 533, 535

VCO: *see* oscillator, voltage-controlled
virtual ground: 171, 266, 271–272, 296
voice coil: *see* loudspeaker, voice coil
voltage: 33–34, 63–68
- AC: 65–68
- amplitude: 66, 384
- DC: 65
- differential: 65, 69, 211–212, 247, 293
- open-circuit: 50–52, 366
- peak-to-peak: 66, 252, 384
- RMS: 65–66, 384, 405, 549
- saturation: 307–308
- single-ended: 244, 247–249, 293, 381, 548

voltage amplifier: *see* amplifier, voltage
voltage divider: 45–54, 121–137, 159–162, 192, 197, 200–201, 214, 216, 228–230, 265, 271–274, 294, 332, 376–377, 457–458, 461, 489, 503, 527–529, 542

Wheatstone bridge: 289–290
wire stripping: 126, 133
woofer: 398

Zener diode: *see* diode, Zener
Zobel network: 407

Colophon

This book was typeset by the author using LaTeX and the Computer Modern family of fonts.

Schematics diagrams were created using Scheme-it, a free schematic editor provided by Digi-Key Corporation [18]. Block diagrams were created using https://www.draw.io [21]. Graphs were created primarily using gnuplot [33], with one graph (Figure 9.2(b)) using Matplotlib [66]. Other line art was created with hand-written SVG code, with a few uses of Inkscape [45]. Photographs were taken by the author, using a Canon G10 digital camera.